Electromagnetism for Chemistry

James K. Baird

Electromagnetism for Chemistry

Springer

James K. Baird
Department of Chemistry
University of Alabama in Huntsville
Huntsville, AL, USA

ISBN 978-3-031-88507-5 ISBN 978-3-031-88508-2 (eBook)
https://doi.org/10.1007/978-3-031-88508-2

This Springer imprint is published by the registered company Springer Nature Switzerland AG
The registered company address is: Gewerbestrasse 11, 6330 Cham, Switzerland

In memory of
my wife,
Peggy Flanagan Baird
and my parents,
Paul Erwin Baird
and
Helen Kern Baird
and
my parents-in-law,
Zenas Ozro Flanagan
and
Cleo Eastep Flanagan

Preface

The standard graduate physical chemistry curriculum involves a number of topics borrowed from physics. Among these are: (1) classical thermodynamics, (2) statistical mechanics, and (3) quantum mechanics. These subjects are covered in the corresponding physical chemistry courses often titled chemical thermodynamics, statistical thermodynamics, and quantum chemistry, respectively. Conspicuously missing from this list is classical electromagnetism. Although electromagnetism is a core subject in both the undergraduate and graduate physics curricula, it is largely missing from the chemistry curriculum.

Chemists make use of the concepts of electromagnetism when they consider polar molecules, molecular polarizability, electrolytic solutions, colloid stability, spectroscopy, spectrophotometry, light scattering, X-ray diffraction, optical rotation, and intermolecular forces. In these chemical applications, the principles of electromagnetism seldom occur in isolation from the other branches of physics. For example, temperature plays a crucial role, so thermal effects often come into play. For this reason, the book contains a chapter on thermodynamics and statistical mechanics. When the properties of isolated molecules are under consideration, electromagnetism intersects with both classical mechanics and quantum mechanics. Hence, chapters covering the relevant portions of these two branches of physics have also been included.

The book is designed to be readable by a chemist, engineer, or physicist who has some familiarity with mathematics through undergraduate courses in vector calculus and the calculus of several variables. Every mathematical result is derived stepwise from these principles. It is the author's belief that, "If you can derive an equation, you can use it in any context." In support of the derivations in the chapters, there is an extensive Mathematical Methods Appendix covering trigonometry, hyperbolic functions, derivatives, integrals, coordinate systems, complex numbers, series expansions, matrix algebra, Kronecker delta, Dirac delta function, Fourier transforms, and the Kramers–Kronig relations. Nowhere is it asserted that this or that result is obvious or follows by inspection.

In particular, the author calls the reader's attention to Chap. 14, which introduces the London theory of dispersion forces between molecules, the Forster theory of resonance energy transfer, and the Hamaker theory of interactions between colloidal particles. When searching the literature for proofs of these famous results, the author found many of the mathematical details to be missing. It is his hope that Chap. 14 corrects these omissions.

The text has been extensively proofread for errors. As no work of mankind is perfect, the author would be most grateful to hear from readers who have located the pesky errors that remain.

Huntsville, USA James K. Baird

Acknowledgements The author would like to express his gratitude to the following individuals who made this book possible. The figures are the work of my former University of Alabama in Huntsville colleague, Dr. Rob McFeeters, and three UAH graduate students, Megan Gober, Sijay Huang, and Ellis Lee. The latter three, plus my UAH faculty colleague, Dr. Carmen Scholz, provided extensive proofreading, as did UAH graduate student Chris Lyons, and colleagues, Clayton Craft of Yale University, and Jim Butler of MIT. I would like to thank Jeffrey Champoux for editorial help and the University of Alabama in Huntsville for a sabbatical leave during the Fall Term 2017, when writing on the book began.

Competing Interests The author has no competing interests to declare that are relevant to the content of this manuscript.

Contents

Chapter 1
Electrostatics

This chapter begins with a recount of the histories of electricity and magnetism as they apply to chemistry. The Coulomb Law, the electrostatic field, and electric potential are introduced. Next comes the concept of superposition of charge. The centimeter, gram, second (cgs) system of units is introduced and used to calculate the classical radius of the electron. The electric current is identified as a source of magnetism. This observation leads to the consideration of a cgs unit of current not based on the cgs unit of charge. This inconvenience is resolved by introducing the modern system of meter, kilogram, second, Coulomb (MKSQ) units. Faraday's law of electrolysis is then invoked in order to determine the charge of the electron. Linear, surface, and volume charge distributions are introduced. As an example of a linear charge distribution, the electric field in the vicinity of a charged double strand DNA molecule is calculated. The electrostatic potential and electric field of a point electric dipole are introduced, and the energy of interaction with another point dipole is calculated. The chapter closes with the derivation of the formula representing the electric field of the point electric quadrupole. This is supplemented by a discussion of Gauss' law, perfect conductors, and Legendre polynomials, which are used to represent the electrostatic potential in spherical polar coordinates. The chapter ends with a proof of Earnshaw's theorem concerning the stability of matter.

1.1 Historical Connection Between Chemistry and Electricity

As early as 600 BCE, the ancient Greeks were aware of the phenomenon of static electricity. Both the Greeks and Romans left records of their observations of the behavior of amber when rubbed with a cloth. In modern times, the first mention of the existence of static electricity is probably that due to William Gilbert (English, 1544–1603), who served as court physician to Queen Elizabeth I. In the next century,

J. K. Baird, *Electromagnetism for Chemistry*,
https://doi.org/10.1007/978-3-031-88508-2_1

C. F. C. Du Fay (French, 1698–1739) noted that when a glass rod was rubbed with silk, the glass gained a vitreous charge (now called positive) while the silk gained a resinous charge (now called negative). In addition, when an ebonite rod was rubbed with fur, the ebonite obtained a negative charge while the fur gained a positive charge. The distinction between the two types of charges could be inferred by observing that two ebonite rods repelled one another as did two glass rods, but an ebonite rod and a glass rod attracted.

Although not properly understood at the time, a hint of the connection between chemistry and electricity emerged when Luigi Galvani (Italian, 1737–1798) observed the effect of charge on the action of a frog's leg. Subsequently, electrochemical cells capable of converting a chemical reaction into electric current were invented by Alessandro Volta (Italian, 1745–1827). When a number of cells were connected in series, a substantial voltage could be produced. This arrangement of cells functioning in tandem reminded Benjamin Franklin (American, 1706–1790) of a battery of cannons firing together. He is credited with associating the term, "battery," with the Voltaic pile. The earliest battery was a stack of alternating zinc and silver disks separated by brine-soaked cloth. In 1800, William Nicholson (English, 1753–1815) used the current from a Voltaic pile to decompose water into hydrogen and oxygen. In 1807, Sir Humphrey Davy (English, 1778–1829) passed a current through molten potash, commonly known as potassium hydroxide, and liberated potassium, a new element. Davy's protégé, Michael Faraday (English, 1791–1867), passed a current through an aqueous electrolyte solution and showed that the mass of the material deposited at the cathode was proportional to the total charge that had been sent through the circuit. With the help of his friend, William Whewell (English, 1794–1866), who knew Greek, he coined the now familiar terms, *electrolysis*, *electrolyte*, *electrode*, *anode*, *cathode*, and *ion*. On the basis of these experiments, the Irish physicist, George Johnstone Stoney (Irish, 1826–1911) gave the name *electron* to the smallest possible charge.

While studying the effects of external electric and magnetic fields on the electric currents that emanate from a heated cathode suspended in a vacuum chamber, J. J. Thomson (English, 1856–1940) discovered in 1898 that the current traveling from the cathode to the anode consisted of negatively charged particles. By applying a magnetic field perpendicular to the direction of the current and measuring the deflection, he was able to determine the charge to mass ratio of the particle that was carrying the current. On the basis of Thomson's experiments, this particle was regarded as being the same as Stoney's electron. In 1913, the American physicist, R. A. Millikan (1868–1953) carried out experiments in which he injected charged oil droplets into a vacuum chamber and stabilized them against free fall by applying an electric field. He found that the droplets carried a multiple of a unit charge. Millikan equated the unit charge with the charge of the electron. With the electron charge thus established, the mass of the electron could be calculated from Thomson's ratio.

Ernest Rutherford, (1871–1937), a physicist with ties to both England and New Zealand carried out an experiment in 1907, which proved that the atom consisted of a positively charged, heavy (compared to the electron) particle surrounded by electrons. The heavy charged particle, being at the center of the atom, was termed the nucleus.

Aside from the nucleus and the electrons, the remainder of the atom was empty space. This atomic model formed the basis for the theory of the hydrogen atom that was proposed in 1913 by Niels Bohr (Danish, 1885–1962). Although impressively successful in predicting the electromagnetic spectrum of the hydrogen atom, Bohr's model was only partially successful in understanding the electromagnetic spectra in the case of multi-electron atoms. The theoretical treatment of multi-electron atoms had to wait on the introduction of the wave mechanics.

It is safe to say that one encounters numerous electrostatic and electromagnetic effects in chemistry and physics. In addition to the efforts of those named above, one should mention the accomplishment of Biot, Coulomb, Gauss, Green, Heaviside, von Helmholtz, Henry, Joule, Kelvin, Kirchhoff, Laplace, Lenz, Neumann, Ohm, Poisson, Riemann, Savart, and Tesla. By 1926, when the quantum theory of W. Heisenberg (German, 1901–1976) and the wave mechanics of E. Schrodinger (Austrian, 1887–1961) became known, it was apparent that the basis for an electromagnetic theory of chemistry was in place. It would require a combination of these theories with the law of C. A. de Coulomb (French, 1736–1806), the electromagnetic field equations of J. C. Maxwell (Scottish, 1831–1879), and the nuclear atomic model to achieve the desired result.

This chapter concentrates on the effects of static electric charges. Nonetheless, it will be necessary to mention the effects of moving charges, and that will lead us to the consideration of magnetism and its role in motivating the creation of the meter, kilogram, second, Coulomb (MKSQ) system of the electromagnetic units.

1.2 Units

Using a torsion balance, Charles Coulomb showed that the force of attraction or repulsion between two charged bodies was inversely proportional to the square of the distance between them. On the basis of his experiments, the force, F_{ab}, acting between of any two objects bearing charges, q_a, and q_b, respectively, which are separated by a distance, r_{ab}, could be represented by the equation,

$$F_{ab} = k_e \frac{q_a q_b}{r_{ab}^2} \tag{1.2.1}$$

Equation (1.2.1) is known as Coulomb's Law. In agreement with the historical observations of static electricity, Eq. (1.2.1) predicts that the electrostatic force is attractive when the charges are of opposite sign but repulsive when they have the same sign. Acting along the radial direction from point a to point b, F_{ab} is said to be a *central* force. The coefficient, k_e, which is related to assumptions concerning the electric permittivity of vacuum, was chosen to be equal to unity.

The historical choice of $k_e = 1$ permits the units of charge, q_a, and q_b, to be expressed in terms of the basic (cgs) units of mass, length, and time.

In the cgs system, the unit of distance is the centimeter, the unit of mass is the gram, and the unit of time is the second. The unit of force is the dyne $= \text{g} \cdot \text{cm} \cdot \text{s}^{-2}$, and the unit of energy is the erg $=$ dyne $\cdot$ cm. In 1900, Max Planck (German, 1858–1947) discovered the connection between the quantum of energy, and the frequency of oscillation of the electric and magnetic fields in electromagnetic radiation. Since then, it has been the habit of theoretical physicists and chemists studying atomic and molecular phenomena to use the cgs units when making calculations. In order to read this vast historical literature, some familiarity with the cgs system is needed.

With the assumption that the electric permittivity of free space is unity, one sets $k_e = 1$. The mechanical units of charge are then determined by letting $q_a = q_b = q$ and setting $F_{ab} = 1$ dyne $= 1\,\text{g} \cdot \text{cm} \cdot \text{s}^{-2}$. and $r_{ab} = 1$ cm. in Eq. (1.2.1). When the resulting equation is solved for q, one finds

$$q = (F_{ab} r_{ab}^2)^{1/2} = 1\,\text{g}^{1/2} \cdot \text{cm}^{3/2} \cdot \text{s}^{-1} \tag{1.2.2}$$

The combination of mechanical units on the right hand side of Eq. (1.2.2) defines the "electrostatic unit" of charge, abbreviated, "esu", where esu $= \text{g}^{1/2} \cdot \text{cm}^{3/2} \cdot \text{s}^{-1}$. In the cgs units, the magnitude of the charge of the electron is $e = 4.8 \times 10^{-10}$ esu.

Example Express the esu in terms of the erg.

Equations involving the calculation of the energy abound in chemistry and physics. For this reason, it is important to be able to express the esu in terms of the cgs unit of energy, which is the erg $=$ dyne $\cdot$ cm $= \text{g} \cdot \text{cm}^2 \cdot \text{s}^{-2}$. The square of the esu yields

$$(\text{esu})^2 = \text{g} \cdot \text{cm}^3 \cdot \text{s}^{-2} = \text{erg} \cdot \text{cm}$$

or

$$\text{esu} = \text{erg}^{1/2} \cdot \text{cm}^{1/2}$$

The calculation of atomic quantities using cgs units is illustrated by the following example.

Example Classical Radius of the Electron [1].

Consider a full charge, q_a, interacting with a differential charge, dq_b, at a distance, r. According to Coulomb's Law, the differential force dF acting on dq_b is

$$dF = \frac{q_a dq_b}{r^2} \tag{1.2.3}$$

The work, d^2W, required to change the distance between the two charges by an amount, dr, is

$$d^2W = -dFdr = -\frac{q_a dq_b}{r^2} dr \tag{1.2.4}$$

Symbol, d^2W, indicates that the work is a differential in both distance, dr, and charge, dq_b. If we imagine bringing q_a and dq_b together from the relative separation, $r = \infty$, the work, dW, required is

$$dW = \int_{\infty}^{r} d^2W = -\int_{\infty}^{r} \frac{q_a dq_b}{r^2} dr = \frac{q_a dq_b}{r} \tag{1.2.5}$$

where dW is the first order differential with respect to dq_b. Suppose now that q_a is the instantaneous charge spread over the surface of a small sphere of radius, r. To consider a differential increase in this charge, we can let $dq_b = dq_a$. The energy required to charge the sphere to its full capacity, Q_a is then given by the integral,

$$W = \int_{0}^{Q_a} \frac{q_a dq_a}{r} = \frac{Q_a^2}{2r} \tag{1.2.6}$$

Equation (1.2.6) can be interpreted as the electrostatic energy of a charge, Q_a, acting on itself when confined to a sphere of radius, r. Although self energy may be difficult to imagine, it gives rise to the concept of the classical radius of the electron. According to Eq. (1.2.6) with $Q_a = e = 4.8 \times 10^{-10}$ esu, we can picture the energy stored by an electron interacting with itself as being given by $e^2/2r$, where r is the "classical" radius of the electron. To evaluate r using Eq. (1.2.6), we assume that the self energy, W, is equivalent to energy associated with the rest mass of the electron. If m is the electron rest mass and c is the vacuum speed of light, the energy is mc^2 (A. Einstein, German–American, 1879–1955). Hence,

$$\frac{e^2}{2r} = mc^2 \tag{1.2.7}$$

After solving Eq. (1.2.7) for r and using $c = 3 \times 10^8$ m/s and $m = 9.1 \times 10^{-28}$ g, the value of r is calculated to be

$$r = \frac{e^2}{2mc^2} = \frac{\left(4.8 \times 10^{-10}\,\text{esu}\right)^2\left(\text{erg} \cdot \text{cm}/(\text{esu})^2\right)}{2\left(9.1 \times 10^{-28}\,\text{g}\right)\left(3 \times 10^{10}\text{cm/s}\right)^2} = 1.4 \times 10^{-13}\,\text{cm} \tag{1.2.8}$$

where the unit conversion factors, $(\text{esu})^2 = \text{erg} \cdot \text{cm}$ and $\text{erg} = \text{g} \cdot \text{cm}^2 \cdot \text{s}^{-2}$, are used. The interpretation of r as the classical radius of the electron is not entirely fictional. The ratio, e^2/mc^2, appears as the unit of length in Thomson's theory of the scattering of X-rays by electrons.

The cgs system of units needs no modification so long as the phenomena of interest are solely electrostatic in origin. Modification is indicated, however, when the charges are moving and magnetic effects are encountered. Until the nineteenth century, the sciences of electricity and magnetism grew up more or less separately.

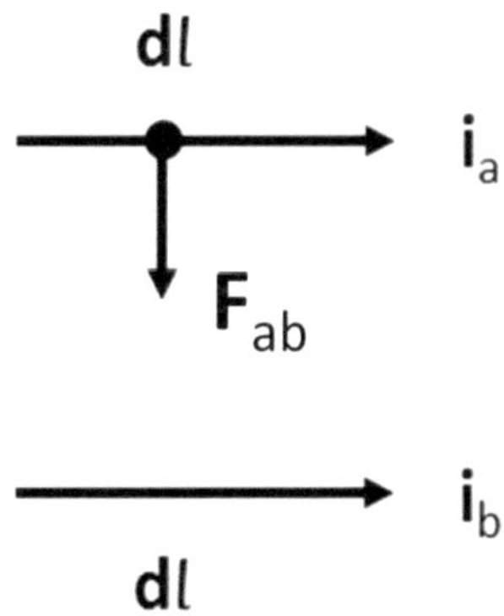

Fig. 1.1 Force, $\mathbf{F}_{ab}$, operating between two parallel wires, each of differential length, $d\ell$, with the one carrying current, $\mathbf{i}_a$, and the other carrying current, $\mathbf{i}_b$

In the early 1800's, however, Hans Christian Oersted (Danish, 1777–1851), noticed that a magnetized needle could be deflected by placing it near an electric current. Andre Marie Ampere (French, 1775–1836) then observed that a wire carrying a current exerted a force on a parallel wire carrying a current. Ampere's result could be summarized by the equation,

$$\frac{dF_{ab}}{d\ell} = 2k_m \frac{i_a i_b}{d_{ab}} \tag{1.2.9}$$

Equation (1.2.9) gives the rate of change, $dF_{ab}/d\ell$, of the force, F_{ab}, with respect to a differential length, $d\ell$, of a wire when the two wires carrying the currents i_a and i_b, respectively, are a distance d_{ab} apart. See Fig. 1.1. The multiplicative factor of two in Eq. (1.2.9) is a constant of integration.

Jean Baptiste Biot (French, 1774–1862) and Felix Savart (French, 1791–1841) found experimentally that if the currents were measured in the units of esu/s, the magnetic force coefficient, k_m, could not be set equal to unity. To maintain k_m at unity, the currents, i_a and i_b, had to be reckoned in terms of a new "electromagnetic unit" of current, abbreviated "emu". With both k_e and k_m set equal to unity, there was thus effectively one unit for static charge and another unit to account for the magnetic effect of a moving charge [2].

At mid century, Wilhelm Weber (German, 1804–1891) and Friedrich Kohlrausch (German, 1840–1910) investigated the connection between these two units by performing an experiment in which a charge measured in esu was allowed to pass through a galvanometer that was calibrated in emu. They found that the current in emu could be related to the current in esu/s by setting the $emu = (esu)/c$, where $c = 3 \times 10^{10}\,\mathrm{cm \cdot s^{-1}}$ is the vacuum speed of light. With the esu of charge taken as the fundamental unit of charge, Eq. (1.2.9) could then be rewritten in the form,

$$\frac{dF_{ab}}{d\ell} = 2\frac{i_{a,e} i_{b,e}}{c^2 d_{ab}} \tag{1.2.10}$$

where $i_{a,e}$ and $i_{b,e}$, are expressed in the units of esu/s. With this choice, both the charge and the current could then be expressed in cgs mechanical units. The constant, k_m, in Eq. (1.2.10) assumed the value $k_m = 1/c^2$. Magnetism was thus based on Eq. (1.2.10) while electrostatics had its basis in the Coulomb law,

$$F_{ab} = \frac{q_{a,e}q_{b,e}}{r_{ab}^2} \tag{1.2.11}$$

which is Eq. (1.2.1) written with $k_e = 1$. The subscript, e, on the currents in Eqs. (1.2.10) and the charges in Eq. (1.2.11), serves as a reminder that the charge is expressed in terms of esu.

The appearance of c in Eq. (1.2.10) strongly hints at the connection between light and electricity and magnetism. This connection will be made clearer when we discuss Maxwell's equations governing the electric and magnetic fields.

In 1901, another set of units was introduced by Giovanni Giorgi (Italian, 1871–1950). He replaced the cgs system of mechanical units by the MKSQ system, where Q is the common symbol for electric charge. In the Giorgi system, the esu is replaced by an independent unit, called the Coulomb, abbreviated as "C". In the new system, the dependence of electromagnetic phenomena on the vacuum speed of light is partitioned between Eqs. (1.2.10) and (1.2.11) by writing them in the forms,

$$\frac{dF_{ab}}{d\ell} = \frac{\mu_o i_{a,M} i_{b,M}}{2\pi d_{ab}^2} \tag{1.2.12}$$

and

$$F_{ab} = \frac{q_{a,M} q_{b,M}}{4\pi \varepsilon_o r_{ab}^2} \tag{1.2.13}$$

respectively. The subscript, M, denotes MKSQ units. The factor, μ_o, is identified with the magnetic permeability of free space, while the factor, ε_o, is referred to as the electric permittivity of free space. Expressed in terms of the MKSQ unit of energy, which is the Joule, where $\mathrm{J} = \mathrm{kg} \cdot \mathrm{m}^2 \cdot \mathrm{s}^{-2}$, the two fundamental constants are

$$\mu_o = 4\pi \times 10^{-7}\,\mathrm{J} \cdot \mathrm{s}^2 \cdot \mathrm{C}^{-2} \cdot \mathrm{m}^{-1} \tag{1.2.14a}$$

and

$$\varepsilon_0 = 8.85419 \times 10^{-12}\,\mathrm{J}^{-1} \cdot \mathrm{C}^2 \cdot \mathrm{m}^{-1} \tag{1.2.14b}$$

respectively. The vacuum light speed is then given by

$$c^2 = \frac{1}{\mu_o \varepsilon_o} \tag{1.2.15}$$

In Eqs. (1.2.12) and (1.12.13) the distances, $d\ell$, d_{ab}, and r_{ab}, are expressed in meters, and the force, F_{ab}, is expressed in the unit, Newton $= \mathrm{kg} \cdot \mathrm{m} \cdot \mathrm{s}^{-2}$. The currents, $i_{a,M}$ and $i_{b,M}$, have the units of Amperes, while the charges, $q_{a,M}$ and $q_{b,M}$ have the units of Coulomb = Ampere · second. The Ampere · second is taken to be the fundamental unit of charge, because in general it is easier to measure current and time than it is to measure charge.

In the Giorgi proposal, any attempt to define charge in terms of mechanical units is abandoned. There are now four independent units, which are mass, length, time, and charge. The Giorgi system thus defines an MKSQ system. The electron charge in MKSQ units is $e = 1.6 \times 10^{-19}$ C. Since its introduction, the Coulomb as the unit of charge has received wide acceptance and is now part of what has become known as SI (Systeme Internationale) units.

Despite the widespread adoption of SI units, the competition between cgs and MKSQ units lingers. In the case of an atom or a molecule, the electrostatic interactions of the electrons with one another and with the nucleus make a larger contribution to the energy than any of the magnetic interactions. Indeed, for many purposes, the magnetic energy can be ignored. In such cases, the repetitious inclusion of $4\pi\varepsilon_o$ in writing Coulomb's law can be avoided if Eq. (1.2.11) is used in preference to Eq. (1.2.13). For this reason, Eq. (1.2.11) dominates in atomic and molecular theory. Nonetheless, the trend in science and engineering as a whole has been to abandon the use of cgs units in electromagnetism in favor of the MKSQ system.

Example Use electrolysis data and the MKSQ system of units to calculate the charge of the electron.

When a voltage is applied to molten silver chloride, the reduction reaction at the cathode is $Ag^+ + e \rightarrow Ag$. See Fig. 1.2.

If a current of 0.50 A is passed for 32.16 min through the melt, it is found that 1.0787 g of metallic silver is deposited at the cathode. The molar mass of silver is 107.87 g/mol. The Avogadro number is $6.022 \times 10^{23}\ \mathrm{mol}^{-1}$. Faraday's Law of electrolysis states that the mass of silver deposited at the cathode is given by

$$\text{Mass} = (\text{current} \times \text{time} \times \text{molar mass})/(\text{electron charge} \times \text{Avogadro number})$$

This equation can be solved for the electron charge, e:

$$e = (\text{molar mass} \times \text{current} \times \text{time})/(\text{mass} \times \text{Avogadro number})$$

Inserting the experimental information, the magnitude of the charge of the electron in Coulombs is found to be

$$\begin{aligned} e &= \left(\frac{107.87\ \mathrm{g/mol}}{1.0787\ \mathrm{g}}\right)\left(\frac{0.5\ \mathrm{C}}{\mathrm{s}}\right)\left(\frac{32.16\ \mathrm{min} \times 60\ \mathrm{s}}{\mathrm{min}}\right)\left(\frac{\mathrm{mol}}{6.022 \times 10^{23}}\right) \\ &= 1.6 \times 10^{-19}\ \mathrm{C} \end{aligned}$$

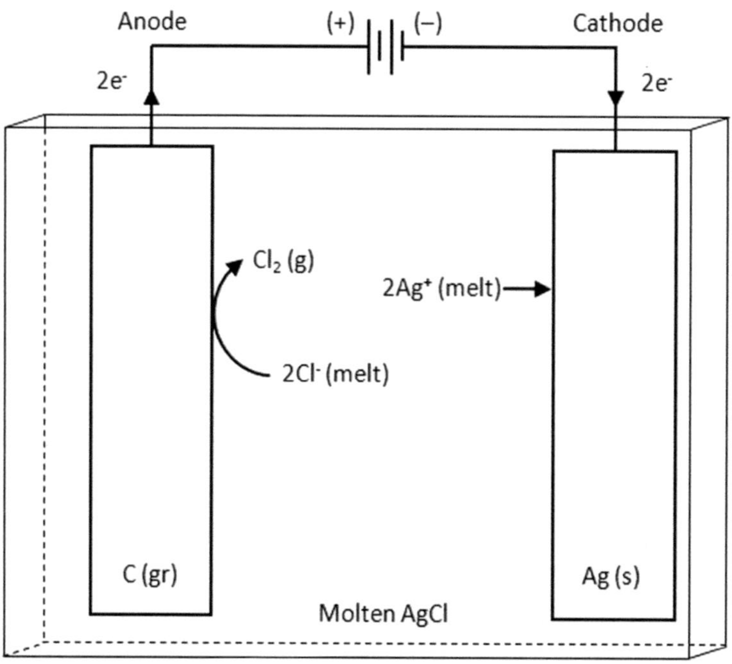

Fig. 1.2 Electrolysis of molten silver chloride. The anode is graphite carbon, while the cathode is solid silver

1.3 The Electrostatic Force and the Electric Field

According to Eq. (1.2.13), Coulomb's Law expressed in MKSQ units is

$$F_{ab} = \frac{1}{4\pi\varepsilon_o}\frac{q_a q_b}{r_{ab}^2} \tag{1.3.1}$$

where for convenience we have suppressed the subscript, M, on the charges. Like all forces, the Coulomb force has both magnitude and direction, making it a vector. The vector nature of F_{ab} can be made explicit by writing Coulomb's law, Eq. (1.3.1), in the form,

$$\mathbf{F}_{ab} = \frac{1}{4\pi\varepsilon_o}\frac{q_a q_b}{r_{ab}^2}\hat{\mathbf{r}}_{ab} \tag{1.3.2}$$

In Eq. (1.3.2), $\mathbf{F}_{ab}$ is the force vector, and $\hat{\mathbf{r}}_{ab}$ is a unit vector in the direction pointing from a, where q_a is located, to b where q_b is located.

Frequently, the origin of coordinates does not coincide with either of the points, a or b. In such case, the Cartesian (Rene Descartes, French 1596–1650) coordinates (x_a, y_a, z_a) of the point a are located by the vector $\mathbf{r}_a$, while the Cartesian coordinates

(x_b, y_b, z_b) of the point b are located by the vector, $\mathbf{r}_b$. If $\hat{\mathbf{e}}_x$, $\hat{\mathbf{e}}_y$ and $\hat{\mathbf{e}}_z$ are unit vectors pointing along the x, y, and z axes, respectively, then a generic vector locating either of these points can be written for $i = a, b$ as

$$\mathbf{r}_i = x_i\hat{\mathbf{e}}_x + y_i\hat{\mathbf{e}}_y + z_i\hat{\mathbf{e}}_z \tag{1.3.3}$$

The vector pointing from point a to point b is then

$$\mathbf{r}_{ab} = \mathbf{r}_b - \mathbf{r}_a = (x_b - x_a)\hat{\mathbf{e}}_x + (y_b - y_a)\hat{\mathbf{e}}_y + (z_b - z_a)\hat{\mathbf{e}}_z \tag{1.3.4}$$

By the Pythagorean theorem, the length of this vector is

$$r_{ab} = |\mathbf{r}_b - \mathbf{r}_a| = \left[(x_b - x_a)^2 + (y_b - y_a)^2 + (z_b - z_a)^2\right]^{1/2} \tag{1.3.5}$$

We can define a vector of unit length, $\hat{\mathbf{r}}_{ab}$, by dividing $\mathbf{r}_{ab}$ by its length, r_{ab}. The result is written as

$$\hat{\mathbf{r}}_{ab} = \frac{\mathbf{r}_{ab}}{r_{ab}} = \frac{(x_b - x_a)\hat{\mathbf{e}}_x + (y_b - y_a)\hat{\mathbf{e}}_y + (z_b - z_a)\hat{\mathbf{e}}_z}{\left[(x_b - x_a)^2 + (y_b - y_2)^2 + (z_a - z_b)^2\right]^{1/2}} \tag{1.3.6}$$

If the force exerted at b is divided by q_b, the result is termed the electric field strength, $\mathbf{E} = \mathbf{E}_{ab} = \mathbf{F}_{ab}/q_b$. The latter can be represented on the basis of Eq. (1.3.2) as

$$\mathbf{E} = \frac{1}{4\pi\varepsilon_o}\frac{q_a}{r_{ab}^2}\hat{\mathbf{r}}_{ab} \tag{1.3.7}$$

The force acting on a charge q_b is thus

$$\mathbf{F}_{ab} = q_b\mathbf{E} \tag{1.3.8}$$

If $q_a > 0$ is fixed in space, and the charge, $q_b > 0$, referred to as a test charge, is free to move, then the trajectory of q_b will be a straight line directed away from a; if $q_a < 0$, the trajectory of q_b will be directed toward a. The trajectories for $q_b > 0$ are called lines of electric force and are diagrammed in Fig. 1.3.

The following are properties of lines of electric force:

(1) The lines of electric force originate on positive charges and terminate on negative charges. If no charge of the opposite sign is present, the lines go off to infinity where they terminate.
(2) The number of lines of force generated by a positive charge or absorbed by a negative charge is proportional to the magnitude of the charge.
(3) The lines of force do not cross.

In the case of a point charge located at a, the magnitude $E_{ab} = |\mathbf{E}|$ of the electric field (force per unit charge) as measured at $\mathbf{r}_b$ is

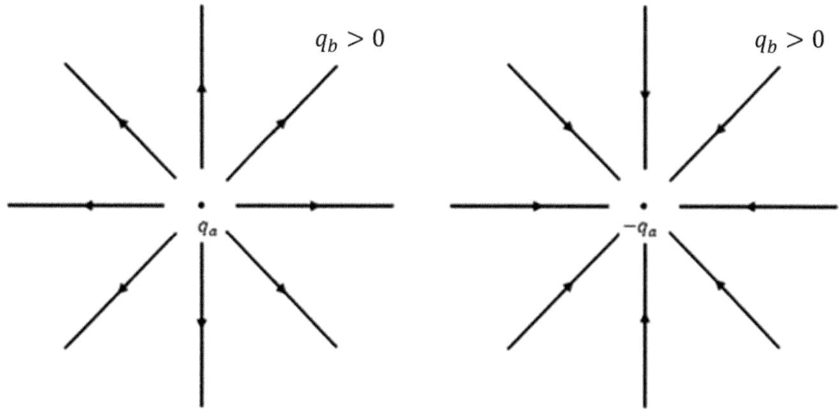

Fig. 1.3 Electric field lines emanating from an isolated positive charge (left) and from an isolated negative charge q_a (right). The test charge is $q_b > 0$

$$E_{ab} = \frac{1}{4\pi\varepsilon_o}\frac{q_a}{r_{ab}^2} \tag{1.3.9a}$$

The product of the electric field strength and the area of the sphere of radius, r_{ab}, surrounding the charge is

$$4\pi r_{ab}^2 E_{ab} = q_a/\varepsilon_o \tag{1.3.9b}$$

This product is independent of r_{ab}. If we regard E_{ab} as the area density of field lines passing through the surface of the sphere of radius, r_{ab}, then according to Eq. (1.3.9b), the total number of field lines, $4\pi r_{ab}^2 E_{ab}$, crossing the surface is proportional to q_a. Because the factor of 4π in the denominator of Eq. (1.3.9a) cancels in Eq. (1.3.9b), the MKSQ units are said to be "rationalized".

1.4 Law of Superposition of Charge

So far, we have considered the electric field due to an isolated point charge, q_a. If additional charges, q_b and q_c, are present, the total electric field, $\mathbf{E}(\mathbf{r}_d)$, at some point, d, due to the three charges is

$$\mathbf{E}(\mathbf{r}_d) = \frac{q_a}{4\pi\varepsilon_o r_{ad}^2}\hat{\mathbf{r}}_{ad} + \frac{q_b}{4\pi\varepsilon_o r_{bd}^2}\hat{\mathbf{r}}_{bd} + \frac{q_c}{4\pi\varepsilon_o r_{cd}^2}\hat{\mathbf{r}}_{cd} \tag{1.4.1}$$

Equation (1.4.1) expresses the law of superposition of charge, in that the total electric field at any given point is equal to the sum of the electric fields generated by each charge.

Although any arrangement of charges is necessarily made up of discrete multiples of the electron charge, it is sometimes convenient to consider continuous distributions of charge. This is achieved by replacing the discrete arrangement of charges by a charge density which is a continuous function of position. The charge can vary within a volume, on a surface, or along a line. The respective continuous charge densities are defined by the equations

$$\rho = \frac{dq}{dV} \quad \text{(volume)} \tag{1.4.2}$$

$$\sigma = \frac{dq}{dA} \quad \text{(surface)} \tag{1.4.3}$$

$$\lambda = \frac{dq}{d\ell} \quad \text{(line)} \tag{1.4.4}$$

where ρ is a volume charge density, σ is a surface charge density, and λ is a linear charge density. In Eqs. (1.4.2)–(1.4.4), the differential of charge is dq. The corresponding differentials of volume, area, and length are dV, dA, and $d\ell$, respectively. We can illustrate the applicability of Eq. (1.4.4), for example, by considering the following calculation:

Example The DNA Molecule as the Source of an Electric Field.

The monomer of the DNA polymer is termed a nucleotide and consists of a phosphate group bonded to a pentose ring, which in turn is bonded to a nitrogen containing base. See Fig. 1.4.

One of the four oxygen atoms of the phosphate group forms a bond to the 3' carbon of the pentose ring, while another of the oxygen atoms makes a bond to the 5′ carbon atom of the pentose ring of the next nucleotide unit. At physiological pH, each phosphate group carries one negative electron charge. We can estimate the electric field due to the aggregate charge of the repeating phosphate groups by modeling the DNA molecule as a thin, infinitely long rod having a uniform charge density, $\lambda < 0$, per unit length.

Consider a point a on the rod and a point b in space where the electric field is to be calculated. As depicted in Fig. 1.5, the origin of coordinates is located at "0".

The charge on a differential length of the rod bounded by any two points x and $x + dx$ is λdx. The incremental contribution, $d\mathbf{E}(\mathbf{r}_b)$, to the electric field, $\mathbf{E}(\mathbf{r}_b)$, due to this differential of charge at x is

$$d\mathbf{E}(\mathbf{r}_b) = \frac{1}{4\pi\varepsilon_o}\frac{\lambda dx}{r_{ab}^2}\hat{\mathbf{r}}_{ab} \tag{1.4.5}$$

We can resolve $d\mathbf{E}(\mathbf{r}_b)$ into a component, $dE_\perp$, perpendicular to the x-axis and a component, $dE_\parallel$, parallel to the x-axis. The component perpendicular to the x-axis is given by

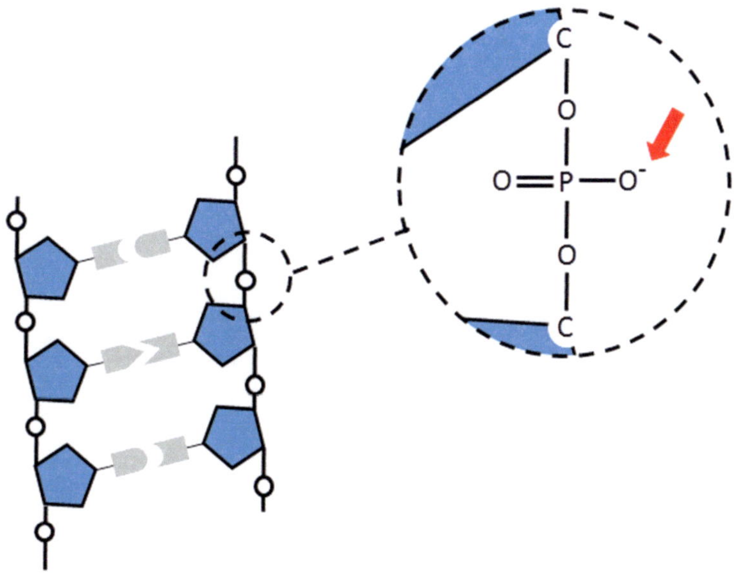

Fig. 1.4 Schematic structure of a DNA molecule. The open circles represent singly charged phosphate groups, $-PO_4^-{-}$. The pentagons represent pentose rings, while the other shapes represent the "lock-in-key" hydrogen bonded coordinations formed by pairs of the nitrogen containing bases

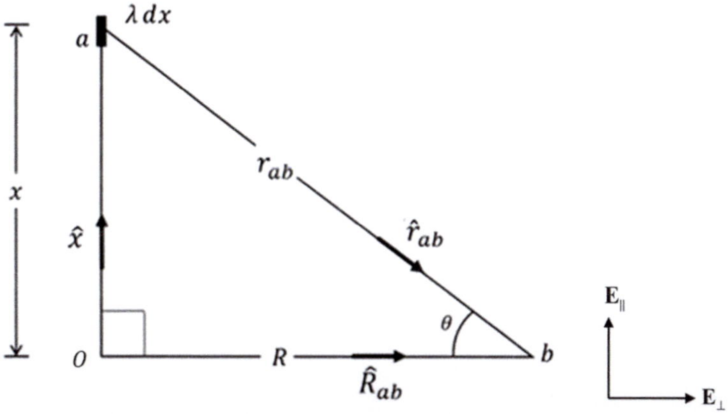

Fig. 1.5 A DNA molecule represented as a charged rod lying along the x-axis. The differential length at position, a, is dx. The linear charge density along the rod is λ. The unit vector parallel to the rod is $\hat{\mathbf{x}}$. The distance from a to the point of observation at b is r_{ab}. The unit vector parallel to this direction is $\hat{\mathbf{r}}_{ab}$. The normal distance from the origin at 0 to b is R, while the unit vector in this direction is $\hat{\mathbf{R}}_{ab}$. The angle between r_{ab} and R is θ. The component of the electric field parallel to the rod is $\mathbf{E}_{\parallel}$, and that perpendicular to the rod is $\mathbf{E}_{\perp}$

$$dE_{\perp} = \hat{\mathbf{R}}_{ab} \cdot d\mathbf{E}(\mathbf{r}_b) = \frac{1}{4\pi\varepsilon_o} \frac{\lambda dx}{r_{ab}^2} \hat{\mathbf{R}}_{ab} \cdot \hat{\mathbf{r}}_{ab} \tag{1.4.6}$$

where $\hat{\mathbf{R}}_{ab}$ is the unit vector in the direction of R. Because $\lambda < 0$, a positive test charge at b is attracted to the rod, so $dE_{\perp}$ points toward the rod. According to the figure, the dot product,

$$\hat{\mathbf{R}}_{ab} \cdot \hat{\mathbf{r}}_{ab} = \cos\theta \tag{1.4.7}$$

Upon substitution of Eq. (1.4.7) into Eq. (1.4.6), we obtain

$$dE_{\perp} = \frac{1}{4\pi\varepsilon_o} \frac{\lambda dx}{r_{ab}^2} \cos\theta \tag{1.4.8}$$

Because of the right triangle relation, the distance r_{ab} satisfies

$$r_{ab}^2 = R^2 + x^2 \tag{1.4.9}$$

with

$$x = R\tan\theta \tag{1.4.10}$$

We can substitute Eq. (1.4.10) into Eq. (1.4.9) to obtain

$$r_{ab}^2 = R^2\left(1 + \tan^2\theta\right) = R^2\sec^2\theta \tag{1.4.11}$$

The differential of Eq. (1.4.10) for fixed R is

$$dx = Rd(\tan\theta) = R\sec^2\theta d\theta \tag{1.4.12}$$

Upon substitution of Eqs. (1.4.11) and (1.4.12) into Eq. (1.4.8), we arrive at

$$dE_{\perp} = \frac{\lambda}{4\pi\varepsilon_o} \cdot \frac{R\sec^2\theta d\theta}{R^2\sec^2\theta} \cdot \cos\theta = \frac{1}{4\pi\varepsilon_o} \cdot \frac{\lambda\cos\theta d\theta}{R} \tag{1.4.13}$$

According to the law of superposition of charge, the total electric field perpendicular to the rod at b is given by the integral of Eq. (1.4.13). Assuming that the rod is infinitely long, the range of θ is $-\pi/2 \leq \theta \leq \pi/2$. Hence,

$$E_{\perp} = \frac{\lambda}{4\pi\varepsilon_o R} \int_{-\pi/2}^{+\pi/2} \cos\theta d\theta = \frac{\lambda}{4\pi\varepsilon_o R} [\sin\theta]_{-\pi/2}^{+\pi/2} = \frac{2\lambda}{4\pi\varepsilon_o R} \tag{1.4.14}$$

or

$$E_\perp = \frac{\lambda}{2\pi\varepsilon_o R} \tag{1.4.15}$$

We next calculate the electric field component, $dE_\parallel$, parallel to the rod as

$$dE_\parallel = \hat{\mathbf{x}} \cdot d\mathbf{E}(\mathbf{r}_b) = \frac{1}{4\pi\varepsilon_o}\frac{\lambda dx}{r_{ab}^2}\hat{\mathbf{x}} \cdot \hat{\mathbf{r}}_{ab} \tag{1.4.16}$$

where $\hat{\mathbf{x}}$ is the unit vector in the x-direction. The indicated dot product is

$$\hat{\mathbf{x}} \cdot \hat{\mathbf{r}}_{ab} = -\cos\left(\frac{\pi}{2} - \theta\right) = -\sin\theta \tag{1.4.17}$$

If we substitute Eqs. (1.4.11), (1.4.12), and (1.4.17) into Eq. (1.4.16), we obtain

$$dE_\parallel = \hat{\mathbf{x}} \cdot d\mathbf{E}(\mathbf{r}_b) = -\frac{\lambda}{4\pi\varepsilon_o} \cdot \frac{R\sec^2\theta d\theta}{R^2\sec^2\theta} \cdot \sin\theta d\theta \tag{1.4.18}$$

Again assuming that the rod is infinitely long, the integral of Eq. (1.4.18) is

$$E_\parallel = -\frac{\lambda}{4\pi\varepsilon_o R}\int_{-\pi/2}^{+\pi/2}\sin\theta d\theta = \frac{\lambda}{4\pi\varepsilon_o R}[\cos\theta]_{-\pi/2}^{+\pi/2} = 0 \tag{1.4.19}$$

According to Eq. (1.4.19), the contributions to $E_\parallel$ from sections of the rod at dx and $-dx$ are equal in magnitude but opposite in direction and thus cancel one another.

Although the concept of a charged rod in vacuum makes for an interesting electrostatic calculation, there is little likelihood that a single DNA molecule could be suspended in vacuum, with its net charge distribution intact. Nevertheless, a negatively charged DNA strand is stable in aqueous solution in the presence of neutralizing positively charged counter ions. In aqueous solution, two new effects come into play, both of which serve to alter Eq. (1.14.15). First, when a charge distribution is transferred from vacuum to a dielectric, the electric field is reduced by a multiplicative factor of $1/\varepsilon_r$, where ε_r is the relative electric permittivity of the dielectric. In the case of water at room temperature, the value of ε_r is about 80. The second effect arises by virtue of the principle of charge neutrality. To maintain neutrality, the negative charge on the DNA strand in aqueous solution must be balanced by the positive charges on the counter ions in the solution. The DNA is completely ionized at physiological pH, which indicates that most counter ions do not attach. Nonetheless, as the counter ions drift toward the strand under the action of the electric field of the strand, their local concentration builds up. This local increase in concentration exceeds the counter ion concentration in the bulk and serves as the driver of a counter ion diffusion current in the direction away from the strand. At equilibrium, the inward drift due to the electric field is balanced by the outward drift due to diffusion. The result is a stationary distribution of charge known as a *plasma*. We will take plasma effects into account when we consider ionic solutions in Chap. 5.

1.5 Electrostatic Energy of Point Charges

Early in the science of mechanics, the term *energy* was associated solely with the energy of motion. That is to say, only kinetic energy was recognized. According to this restricted definition, an object at rest had no energy. Nonetheless, it was realized that two objects at rest might have different energies by virtue of their relative positions in a gravitational field. For example, a ball at rest at the top of an inclined plane was presumably in a different state of energy than a ball at rest at the bottom. At rest, neither had any kinetic energy, but if the one at the top were released, it would develop kinetic energy as it rolled down the plane. Thus, a second form of energy was indicated. Since kinetic energy was still king, in order to distinguish the energy of an object at rest in a gravitational field from the energy of an object in motion, the gravitational field was regarded as providing the "potential" for energy. Ever since, the energy associated with a gravitational field has been known as "potential" energy.

An electrostatic field of force confers potential energy on a charged particle. If not otherwise restrained, charged particles exerting electrostatic forces on one another will move. As we have seen, the work, dW, associated with the displacement, $d\mathbf{r}_b$, of a charge q_b at b due to the force, $\mathbf{F}_{ab}$, produced by charge, q_a, at a is

$$dW = -\mathbf{F}_{ab} \cdot d\mathbf{r}_b \tag{1.5.1}$$

If there are no other forces acting, then $\mathbf{F}_{ab}$ and $d\mathbf{r}_b$ will be parallel, and the dot product is given by $\mathbf{F}_{ab} \cdot d\mathbf{r}_b = F_{ab} dr_{ab}$ where F_{ab} is the length of $\mathbf{F}_{ab}$. The change in work, dW, depends upon position and is equivalent to a change in the electrostatic potential energy, dU_e, so Eq. (1.5.1) can be rewritten as

$$dU_e = -F_{ab} dr_b \tag{1.5.2}$$

According to Eq. (1.5.2)

$$F_{ab} = -\frac{dU_e(r_{ab})}{dr_{ab}} \tag{1.5.3}$$

We can substitute Eq. (1.3.1) into Eq. (1.5.3) to obtain

$$\frac{dU_e}{dr_{ab}} = -\frac{q_a q_b}{4\pi \varepsilon_o r_{ab}^2} \tag{1.5.4}$$

By virtue of Eq. (1.5.3), the addition of a constant to $U_e(r)$, does not change F_{ab}. This is equivalent to saying that there is no absolute zero of energy. We are thus free to set $U_e = 0$ at $r_{ab} = \infty$. According to Eq. (1.5.4), the total change in potential energy associated with bringing the charges, q_a and q_b, from an infinite separation to a finite separation, r_{ab} can then be calculated as

$$U_e(r_{ab}) = -\frac{q_a q_b}{4\pi\varepsilon_o}\int_{\infty}^{r_{ab}}\frac{dr}{r^2} = \frac{q_a q_b}{4\pi\varepsilon_o r_{ab}} \tag{1.5.5}$$

For a system of N charges, the total energy of their mutual interactions is given by

$$U_e = \frac{1}{2}\sum_{i\neq j}^{N}\sum_{j=1}^{N}\frac{q_i q_j}{4\pi\varepsilon_o r_{ij}}, \quad i \neq j \tag{1.5.6}$$

The meaning of the factor of ½ is made clear, if we let $N = 3$ and write out the terms in the double sum explicitly. First, we sum over the index, j. The result is

$$\begin{aligned} U_e &= \frac{1}{2}\sum_{i\neq j}^{3}\left(\frac{q_i q_1}{4\pi\varepsilon_o r_{i1}} + \frac{q_i q_2}{4\pi\varepsilon_o r_{i2}} + \frac{q_i q_3}{4\pi\varepsilon_o r_{i3}}\right) \\ &= \frac{1}{2}\sum_{i\neq 1}^{3}\frac{q_i q_1}{4\pi\varepsilon_o r_{i1}} + \frac{1}{2}\sum_{i\neq 2}^{3}\frac{q_i q_2}{4\pi\varepsilon_o r_{i2}} + \frac{1}{2}\sum_{i\neq 3}^{3}\frac{q_i q_3}{4\pi\varepsilon_o r_{i3}} \end{aligned} \tag{1.5.7}$$

After evaluating the remaining three sums over the index, i, the result is

$$U_e = \frac{1}{2}\left(\frac{q_2 q_1}{4\pi\varepsilon_o r_{21}} + \frac{q_3 q_1}{4\pi\varepsilon_o r_{31}} + \frac{q_1 q_2}{4\pi\varepsilon_o r_{12}} + \frac{q_3 q_2}{4\pi\varepsilon_o r_{32}} + \frac{q_1 q_3}{4\pi\varepsilon_o r_{13}} + \frac{q_2 q_3}{4\pi\varepsilon_o r_{23}}\right) \tag{1.5.8}$$

Because of Eq. (1.3.5), $r_{12} = r_{21}$, $r_{13} = r_{31}$, and $r_{32} = r_{23}$. Thus, Eq. (1.5.8) simplifies to

$$U_e = \frac{q_1 q_2}{4\pi\varepsilon_o r_{12}} + \frac{q_1 q_3}{4\pi\varepsilon_o r_{13}} + \frac{q_2 q_3}{4\pi\varepsilon_o r_{23}} \tag{1.5.9}$$

Because two charges can interact only once, the factor of 1/2 in Eq. (1.5.6) serves to remove the effect of double counting.

The next example illustrates how Coulomb's law can be used to calculate the lattice energy of a one dimensional ionic crystal.

Example Electrostatic Contribution to the Lattice Energy of a Linear Ionic Crystal.

In a three dimensional crystal, the symmetrical arrangement of a few atoms defines the unit cell, which serves as the building block for the crystal. If the atoms in a crystal

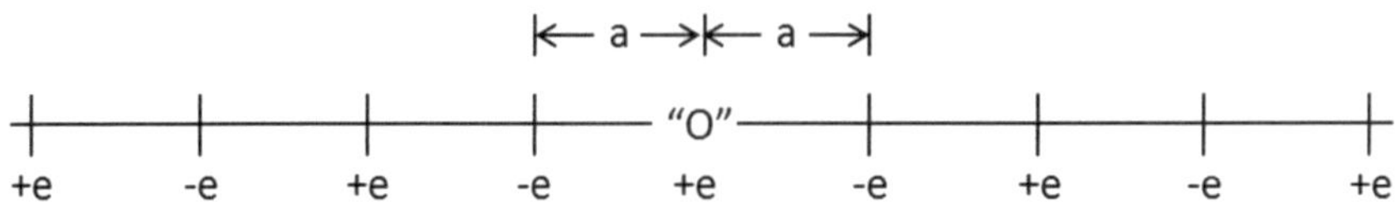

Fig. 1.6 A one dimensional crystal composed of univalent ions of opposite charge spaced a distance a apart. The positive ion located at "0" serves as the origin of coordinates

are in the form of ions, the material is held together principally by electrostatic forces. The electrostatic cohesive energy that binds the crystal can be calculated by summing all of the pair wise electrostatic interactions among the ions. To simplify the calculation of this energy, we will restrict our attention to a one-dimensional arrangement consisting of equally spaced univalent ions of opposite charge. In this one dimensional array, a pair of oppositely charged ions defines the unit cell. See Fig. 1.6.

On either side of a positive ion in this arrangement, there is a negative ion, and on either side of a negative ion, there is a positive ion. We select the positive ion at "0" as the origin of coordinates for the calculation. The electrostatic energy of the linear array of ions is then

$$\begin{aligned} U_e = &-\frac{e^2}{4\pi\varepsilon_o a} - \frac{e^2}{4\pi\varepsilon_o a} + \frac{e^2}{4\pi\varepsilon_o(2a)} + \frac{e^2}{4\pi\varepsilon_o(2a)} \\ &- \frac{e^2}{4\pi\varepsilon_o(3a)} - \frac{e^2}{4\pi\varepsilon_o(3a)} + \frac{e^2}{4\pi\varepsilon_o(4a)} + \frac{e^2}{4\pi\varepsilon_o(4a)} - \cdots \end{aligned} \tag{1.5.10}$$

Equation (1.5.10) can be factored to read

$$U_e = -\frac{2e^2}{4\pi\varepsilon_o a}\left(1 - \frac{1}{2} + \frac{1}{3} - \frac{1}{4} + \cdots\right) \tag{1.5.11}$$

The sum of fractions on the right hand side of Eq. (1.5.11) can be evaluated by considering the series representation of $\ln(1+x)$, which is

$$\ln(1+x) = x - \frac{x^2}{2} + \frac{x^3}{3} - \frac{x^4}{4} + \cdots \tag{1.5.12}$$

Upon substitution of $x = 1$ into Eq. (1.5.12), one obtains

$$\ln 2 = 1 - \frac{1}{2} + \frac{1}{3} - \frac{1}{4} + \cdots \tag{1.5.13}$$

Which permits Eq. (1.5.11) to be rewritten as

$$U_e = -\frac{Me^2}{4\pi\varepsilon_o a} \tag{1.5.14}$$

where

$$M = 2\ln 2 = 1.39 \tag{1.5.15}$$

The symbol M is termed the Madelung constant (Erwin Madelung, German, 1881–1972). The value $M = 2\ln 2 = 1.39$, which we have calculated, is characteristic of a linear array of univalent ions. According to Eqs. (1.5.14) and (1.5.15), the electrostatic energy of the ionic linear array is only a 39% greater than the potential energy of two oppositely charge univalent ions positioned a distance a apart.

1.6 Electrostatic Potential of a Point Charge

Using Eq. (1.5.5), we define the electrostatic potential, $\Phi(\mathbf{r}_b)$, at point b due to a charge q_a located at a separation, r_{ab}, as

$$\Phi(\mathbf{r}_b) = \Phi(x_b, y_b, z_b) = \frac{U_e}{q_b} = \frac{q_a}{4\pi\varepsilon_o r_{ab}} \tag{1.6.1}$$

Although the subscripts a and b are helpful in visualizing the local nature of Φ, they are non-essential from a mathematical point of view, so we shall consider hereafter a point charge, q, located at the origin of coordinates. The distance from q to the point where the potential is to be determined is r. According to Eq. (1.6.1), the potential is a constant on the surface of any spherical shell of radius r surrounding q. These spherical surfaces are called "equipotentials". See Fig. 1.7.

Because we are assuming that the position of the point charge coincides with the origin of coordinates, we can rewrite Eq. (1.6.1) as

$$\Phi(r) = \frac{q}{4\pi\varepsilon_o r} \tag{1.6.2}$$

As $\Phi(r)$ is a function of r alone, the electric field, $E(r)$, at r has only a radial component, which is given by

$$E(r) = -\frac{d\Phi(r)}{dr} = -\frac{d}{dr}\left(\frac{q}{4\pi\varepsilon_o r}\right) = \frac{q}{4\pi\varepsilon_o r^2} \tag{1.6.3}$$

Because $\Phi(r)$ is depends solely on r, it is spherically symmetric, and is termed a *central* potential. As illustrated in Fig. 1.7 the electric field lines, $E(r)$, associated with q are normal to the equipotentials, which are circles.

It is also useful to calculate the electric field in Cartesian coordinates. The calculation serves to confirm Eq. (1.6.3). In Cartesian coordinates, the gradient operator, ∇, is

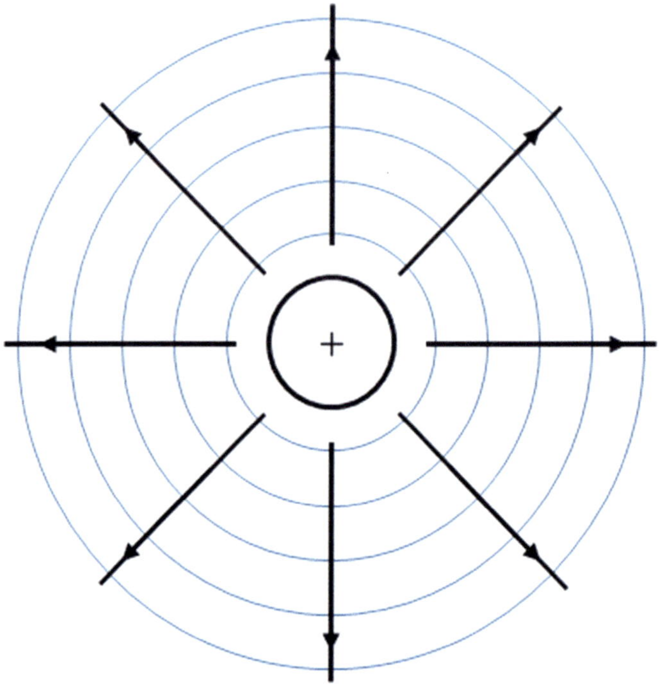

Fig. 1.7 Equipotentials (circles) and field lines (arrows) associated with a point positive charge

$$\nabla = \hat{\mathbf{e}}_x \frac{\partial}{\partial x} + \hat{\mathbf{e}}_y \frac{\partial}{\partial y} + \hat{\mathbf{e}}_z \frac{\partial}{\partial z} \tag{1.6.4}$$

Taking into account the possibility of components of the electric field in all three of the Cartesian directions,

$$\mathbf{E}(x, y, z) = -\nabla\Phi = -\hat{\mathbf{e}}_x \frac{\partial \Phi(x, y, z)}{\partial x} - \hat{\mathbf{e}}_y \frac{\partial \Phi(x, y, z)}{\partial y} - \hat{\mathbf{e}}_z \frac{\partial \Phi(x, y, z)}{\partial z} \tag{1.6.5}$$

According to Eqs. (1.6.2) and (1.6.4), the electric field at position **r** can be written as

$$\mathbf{E}(r) = -\nabla \frac{q}{4\pi\varepsilon_o r} = -\frac{q}{4\pi\varepsilon_o} \nabla\left(\frac{1}{r}\right) \tag{1.6.6}$$

The right hand side of Eq. (1.6.6) can be calculated in Cartesian coordinates as

$$\nabla\left(\frac{1}{r}\right) = \left(\hat{\mathbf{e}}_x \frac{\partial}{\partial x} + \hat{\mathbf{e}}_y \frac{\partial}{\partial y} + \hat{\mathbf{e}}_z \frac{\partial}{\partial z}\right)\left(\frac{1}{r}\right)$$

$$
\begin{aligned}
&= \left[\frac{\partial}{\partial r}\left(\frac{1}{r}\right)\right]\left(\hat{\mathbf{e}}_x \frac{\partial r}{\partial x} + \hat{\mathbf{e}}_y \frac{\partial r}{\partial y} + \hat{\mathbf{e}}_z \frac{\partial r}{\partial z}\right) \\
&= -\left(\frac{1}{r^2}\right)\left(\hat{\mathbf{e}}_x \frac{\partial r}{\partial x} + \hat{\mathbf{e}}_y \frac{\partial r}{\partial y} + \hat{\mathbf{e}}_z \frac{\partial r}{\partial z}\right) = -\frac{\nabla r}{r^2}
\end{aligned} \tag{1.6.7}
$$

where

$$
\nabla r = \hat{\mathbf{e}}_x \frac{\partial r}{\partial x} + \hat{\mathbf{e}}_y \frac{\partial r}{\partial y} + \hat{\mathbf{e}}_z \frac{\partial r}{\partial z} \tag{1.6.8}
$$

Now because,

$$
r = \left(x^2 + y^2 + z^2\right)^{1/2} \tag{1.6.9}
$$

we can compute,

$$
\nabla r = \left(\hat{\mathbf{e}}_x \frac{\partial}{\partial x} + \hat{\mathbf{e}}_y \frac{\partial}{\partial y} + \hat{\mathbf{e}}_z \frac{\partial}{\partial z}\right)\left(x^2 + y^2 + z^2\right)^{1/2} = \frac{1}{2}\left[\frac{2x\hat{\mathbf{e}}_x + 2y\hat{\mathbf{e}}_y + 2z\hat{\mathbf{e}}_z}{(x^2 + y^2 + z^2)^{1/2}}\right] = \frac{\mathbf{r}}{r} \tag{1.6.10}
$$

where in constructing Eq. (1.6.10), we have used the Cartesian coordinate representation of $\mathbf{r}$,

$$
\mathbf{r} = x\hat{\mathbf{e}}_x + y\hat{\mathbf{e}}_y + z\hat{\mathbf{e}}_z \tag{1.6.11}
$$

If we use $\hat{\mathbf{r}} = \mathbf{r}/r$, then the combination of Eqs. (1.6.7), and (1.6.10) yields,

$$
\nabla\left(\frac{1}{r}\right) = -\frac{\mathbf{r}}{r^3} = -\frac{\hat{\mathbf{r}}}{r^2} \tag{1.6.12}
$$

By substitution of Eq. (1.6.12) into Eq. (1.6.6), we arrive at the vector representation of the electric field in the form,

$$
\mathbf{E}(\mathbf{r}) = \frac{q\mathbf{r}}{4\pi\varepsilon_o r^3} = \frac{q\hat{\mathbf{r}}}{4\pi\varepsilon_o r^2} \tag{1.6.13}
$$

Equation (1.6.13) confirms the assertion that the electric field lines are everywhere in the radial direction.

1.7 Electrostatic Potential and Electric Field of a Point Electric Dipole

The electron pair forming the covalent bond between the two atoms in HCl is known to be positioned closer to the Cl atom than the H atom. When a charge distribution is such that the center of negative charge does not coincide with the center of positive charge, as in HCl, the charge distribution is said to possess an *electric dipole moment*. In standard chemical notation, the magnitude, δ, and the sign of the displaced electron charge is depicted by writing the formula as $H^{\delta+}Cl^{\delta-}$. As a supplement, or sometimes in place of the δ notation, a vector is drawn pointing from $+\delta$ to $-\delta$, as in $\overrightarrow{HCl}$, in order to indicate the direction of the electron displacement. This notation, however, does not coincide with the direction of the electric dipole as is standard in electrostatics. In electrostatics, the electric dipole is regarded as pointing from the negative charge to the positive charge, and the arrow used to represent it is drawn in the $-\delta$ to $+\delta$ direction. This convention is deeply ensconced in electrostatics, and to avoid introducing unfamiliar minus signs into our formulae, we shall adopt it.

Two point charges, equal in magnitude, but opposite in sign, which are separated by a distance, constitute an *electric dipole.* For ease in calculation with this model, we can assume that the charges are confined to the z-axis. The charge $+q$ is located at $z = +d$, a point denoted as α, while the charge $-q$ is located at $z = -d$, at a point denoted as β. See Fig. 1.8.

The magnitude of the charge times the distance of separation equals the magnitude of the *electric dipole moment*, p, which in this case is $p = 2qd$. The electric dipole is said to be a *point dipole* if we can let the value of d become infinitesimally small while maintaining the product, qd, constant. Because the dipole is electrically neutral overall, it is said to have *zero monopole moment.*

The origin of coordinates located at "0" corresponds to $z = 0$. We shall calculate the electrostatic potential and electric field at a point located by the Cartesian coordinates (x, y, z) The charge, $+q$, located at α has the Cartesian coordinates, $(x_\alpha, y_\alpha, z_\alpha) = (0, 0, +d)$. The vector distance from this point to (x, y, z) is

$$\begin{aligned}\mathbf{r}_\alpha &= (x - x_\alpha)\hat{\mathbf{e}}_x + (y - y_\alpha)\hat{\mathbf{e}}_y + (z - z_\alpha)\hat{\mathbf{e}}_z \\ &= x\hat{\mathbf{e}}_x + x\hat{\mathbf{e}}_y + (z - d)\hat{\mathbf{e}}_z\end{aligned} \tag{1.7.1}$$

The length of the vector $\mathbf{r}_\alpha$ is

$$\begin{aligned}r_\alpha &= \left[x^2 + y^2 + (z - d)^2\right]^{1/2} \\ &= \left(x^2 + y^2 + z^2 - 2zd + d^2\right)^{1/2} \\ &= \left(r^2 - 2zd + d^2\right)^{1/2}\end{aligned} \tag{1.7.2}$$

where we have used

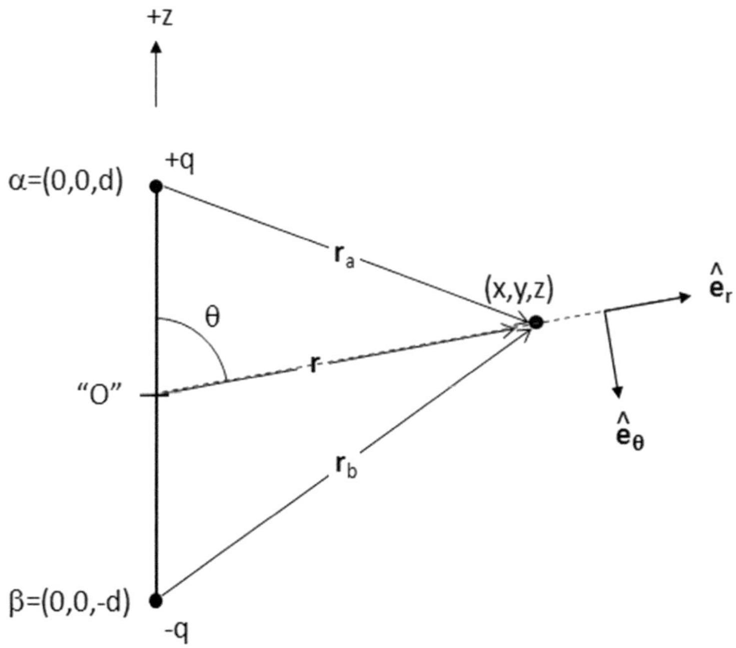

Fig. 1.8 Electric dipole formed by a charge of $+q$ located at the point α having coordinates $(0, 0, d)$ and a charge of $-q$ located at the point β having coordinates $(0, 0, -d)$. The origin of coordinates is located at "0". The vector distance from "0" to the point of observation with coordinates (x, y, z) is $\mathbf{r}$. The vector distance from α to the point of observation is $\mathbf{r}_a$, while the vector distance from β is $\mathbf{r}_b$. The angle between $\mathbf{r}$ and the $+z$-axis is θ. The unit vector in the $\mathbf{r}$ direction is $\hat{\mathbf{e}}_r$. The unit vector perpendicular to $\hat{\mathbf{e}}_r$ and pointing in the direction of increasing θ is $\hat{\mathbf{e}}_\theta$

$$r = \left(x^2 + y^2 + z^2\right)^{1/2} \tag{1.7.3}$$

The point at (x, y, z) can also be reckoned with respect to the charge, $-q$, at β which is positioned at $(x_\beta, y_\beta, z_\beta) = (0, 0, -d)$. This vector distance is

$$\begin{aligned} \mathbf{r}_\beta &= (x - x_\beta)\hat{\mathbf{e}}_x + (y - y_\beta)\hat{\mathbf{e}}_y + (z - z_\beta)\hat{\mathbf{e}}_z \\ &= x\hat{\mathbf{e}}_x + x\hat{\mathbf{e}}_y + (z + d)\hat{\mathbf{e}}_z \end{aligned} \tag{1.7.4}$$

According to Eq. (1.7.4), the length of $\mathbf{r}_\beta$ is

$$\begin{aligned} r_\beta &= \left[x^2 + y^2 + (z + d)^2\right]^{1/2} \\ &= \left(x^2 + y^2 + z^2 + 2zd + d^2\right)^{1/2} \\ &= \left(r^2 + 2zd + d^2\right)^{1/2} \end{aligned} \tag{1.7.5}$$

By the law of superposition of charge, the electrostatic potential at (x, y, z) due to the two charges is

$$\Phi(x, y, z) = \frac{q}{4\pi\varepsilon_o r_\alpha} - \frac{q}{4\pi\varepsilon_o r_\beta} \tag{1.7.6}$$

By virtue of Eqs. (1.7.2) and (1.7.5), Eq. (1.7.6) can be written in Cartesian coordinates as

$$\Phi(x, y, z) = \frac{q}{4\pi\varepsilon_o} \frac{1}{\left(r^2 - 2zd + d^2\right)^{1/2}} - \frac{q}{4\pi\varepsilon_o} \frac{1}{\left(r^2 + 2zd + d^2\right)^{1/2}} \tag{1.7.7}$$

When r^2 is factored out of each denominator in Eq. (1.7.7), one has

$$\begin{aligned}\Phi(x, y, z) = &\frac{q}{4\pi\varepsilon_o r} \frac{1}{\left[1 - 2z(d/r^2) + (d/r)^2\right]^{1/2}} \\ &- \frac{q}{4\pi\varepsilon_o r} \frac{1}{\left[1 + 2z(d/r^2) + (d/r)^2\right]^{1/2}}\end{aligned} \tag{1.7.8}$$

Substitute

$$z = r\cos\theta \tag{1.7.9}$$

into Eq. (1.7.8). The result is

$$\begin{aligned}\Phi(r, \theta) = &\frac{q}{4\pi\varepsilon_o r}\left[1 - 2\cos\theta(d/r) + (d/r)^2\right]^{-1/2} \\ &- \frac{q}{4\pi\varepsilon_o r}\left[1 + 2\cos\theta(d/r) + (d/r)^2\right]^{-1/2}\end{aligned} \tag{1.7.10}$$

We next recall the binomial theorem, which can be written in the form

$$(1 + s)^n = 1 + ns + \frac{n(n-1)}{2!}s^2 + \frac{n(n-1)(n-2)}{3!}s^3 + \cdots \tag{1.7.11}$$

Let

$$s = -2\cos\theta(d/r) + (d/r)^2 \tag{1.7.12}$$

With $n = -1/2$, we can use Eqs. (1.7.11) and (1.7.12) to expand the first trinomial in Eq. (1.7.10) in powers of (d/r). The result is

$$\left[1 - 2\cos\theta\left(\frac{d}{r}\right) + \left(\frac{d}{r}\right)^2\right]^{-1/2}$$

$$\begin{aligned}
&\approx 1 - \frac{1}{2}\left[-2\cos\theta\left(\frac{d}{r}\right) + \left(\frac{d}{r}\right)^2\right] \\
&\quad + \frac{(-1/2)(-3/2)}{2!}\left[-2\cos\theta\left(\frac{d}{r}\right) + \left(\frac{d}{r}\right)^2\right]^2 + \cdots \\
&= 1 + \cos\theta\left(\frac{d}{r}\right) - \frac{1}{2}\left(\frac{d}{r}\right)^2 \\
&\quad + \frac{3}{8}\left[4\cos^2\theta\left(\frac{d}{r}\right)^2 - 4\cos\theta\left(\frac{d}{r}\right)^3 + \left(\frac{d}{r}\right)^4\right] + \cdots \\
&\approx 1 + \cos\theta\left(\frac{d}{r}\right) - \frac{1}{2}\left(\frac{d}{r}\right)^2 + \frac{3}{2}\cos^2\theta\left(\frac{d}{r}\right)^2 + \cdots \\
&= 1 + \left(\frac{d}{r}\right)\cos\theta + \frac{3\cos^2\theta - 1}{2}\left(\frac{d}{r}\right)^2 + \cdots
\end{aligned} \tag{1.7.13}$$

In proceeding from the second line of Eq. (1.7.13) to the third, we have dropped all terms of order $(d/r)^3$ and higher. This approximation is consistent with the concept of a *point electric dipole*, where $(d/r) \ll 1$ for all values of r. If we substitute

$$s = 2\cos\theta(d/r) + (d/r)^2 \tag{1.7.14}$$

into Eq. (1.7.11) with $n = -1/2$ and expand the second trinomial in Eq. (1.7.10) in the binomial series, we obtain

$$\left[1 + 2\cos\theta\left(\frac{d}{r}\right) + \left(\frac{d}{r}\right)^2\right]^{-1/2} \approx 1 - \left(\frac{d}{r}\right)\cos\theta + \frac{3\cos^2\theta - 1}{2}\left(\frac{d}{r}\right)^2 + \cdots \tag{1.7.15}$$

Substitution of Eqs. (1.7.13) and (1.7.15) into Eq. (1.7.10) followed by dropping terms in $(d/r)^2$, leads to the result,

$$\Phi(r, \theta) = \frac{2qd\cos\theta}{4\pi\varepsilon_o r^2} \tag{1.7.16}$$

Since the electric dipole moment, p, is defined as

$$p = 2qd \tag{1.7.17}$$

Equation (1.7.16) can be rewritten as

$$\Phi(r, \theta) = \frac{p\cos\theta}{4\pi\varepsilon_o r^2} \tag{1.7.18}$$

Equation (1.7.18) represents the electrostatic potential of a point electric dipole.

Note that Eq. (1.7.18) involves r and polar angle θ, which are the natural variables in spherical polar coordinates. To calculate the electric field in the vicinity of a point electric dipole, we need the gradient operator in spherical polar coordinates, which is

$$\nabla = \hat{\mathbf{e}}_r \frac{\partial}{\partial r} + \hat{\mathbf{e}}_\theta \frac{\partial}{r\partial\theta} + \hat{\mathbf{e}}_\phi \frac{1}{r\sin\theta}\frac{\partial}{\partial\phi} \tag{1.7.19}$$

as shown in the Mathematical Methods Appendix. If Eq. (1.7.19) is applied to Eq. (1.7.18), we notice that because $\Phi(r, \theta)$ is independent of the azimuthal angle, ϕ, only the r and θ derivatives in Eq. (1.7.19) are involved in calculating the electric field. Hence, the electric field is simply

$$\mathbf{E} = -\nabla\Phi = -\frac{p}{4\pi\varepsilon_o}\left(\hat{\mathbf{e}}_r \frac{\partial}{\partial r} + \hat{\mathbf{e}}_\theta \frac{\partial}{r\partial\theta}\right)\left(\frac{\cos\theta}{r^2}\right) \tag{1.7.20}$$

After completion of the indicated differentiations, Eq. (1.7.20) reads

$$\mathbf{E} = \frac{p}{4\pi\varepsilon_o r^3}(2\cos\theta\hat{\mathbf{e}}_r + \sin\theta\hat{\mathbf{e}}_\theta) \tag{1.7.21}$$

where the unit vectors $\hat{\mathbf{e}}_r$ and $\hat{\mathbf{e}}_\theta$, are as shown in Fig. 1.8. According to Eq. (1.7.21), the radial component, E_r, of $\mathbf{E}$ is

$$E_r = \frac{p\cos\theta}{2\pi\varepsilon_o r^3} \tag{1.7.22a}$$

The polar component, E_θ, is

$$E_\theta = \frac{p\sin\theta}{4\pi\varepsilon_o r^3} \tag{1.7.22b}$$

The electric field components represented by Eqs. (1.7.22), can be spatially mapped by considering their effects on a positive test charge. See Fig. 1.9 where the equipotentials and electric field vectors are illustrated.

Consider the polar angle as shown in Fig. 1.8. According to Eqs. (1.7.22a) and (1.7.22b), a path along the positive z-axis, where $\theta = 0$, coincides with $E_\theta = 0$. Consequently, if a positive test charge is located on the positive z-axis at a distance $z > d$ from the origin, it will experience a net force that is entirely in the + z-direction, which is away from point, α. At an angle of $\theta = \pi/4$, the positive test charge will experience a force component in the radial direction repelling it away from the dipole and a force component in the positive θ-direction causing it to orbit the origin at "0". On the equator at $\theta = \pi/2$, the angular component has a maximum, while the radial component is zero. At an angle of $\theta = 3\pi/4$, the angular component is still in the positive θ- direction, but the radial component is now attractive. Finally,

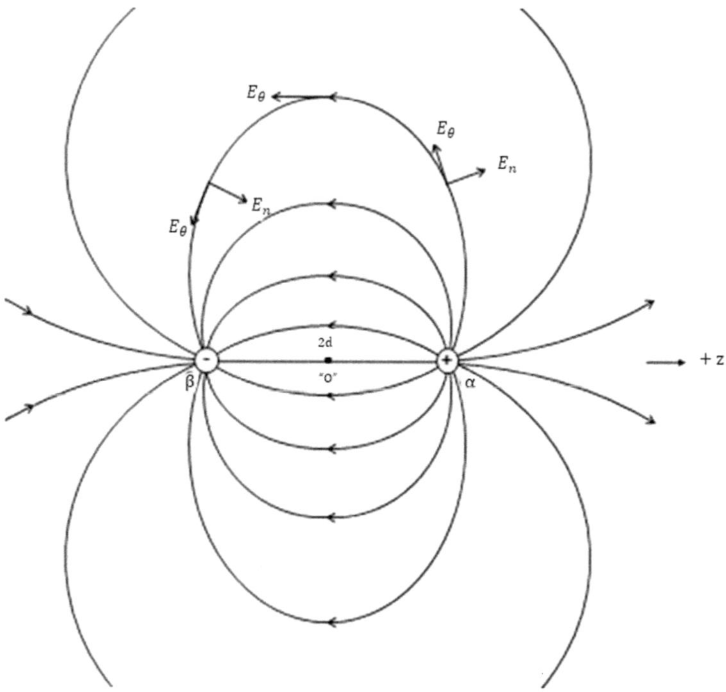

Fig. 1.9 Equipotentials of a dipole with point charges arrayed along the z-axis at positions α and β, which are separated by the distance $2d$. The curves represent the equipotentials. The tangential component of the electric field is labeled E_θ. The normal component is labeled E_n. If we take the limit $d \to 0$, while holding the product, $p = 2dq$ constant, the extended dipole in Fig. 1.9 shrinks to a point dipole located at "0" and the tangential and normal components of the electric field will point accurately in the θ- and r-directions, respectively

when the positive test charge is on the negative z-axis at $\theta = \pi$ with $z < -d$, the force is entirely attractive and toward the point at β. Because **E** in Eq. (1.7.21) has components in both the radial and angular directions, $\Phi(r, \theta)$ in Eq. (1.7.18) is said to be a *non-central* potential.

1.8 Potential Energy of a Dipole in an External Electric Field

It is of interest to calculate the potential energy of an electric dipole, which is acted on by an external electric field. We assume that the external electric field is derivable from an electrostatic potential, $\Phi_{ext}(\mathbf{r})$, where the distance $\mathbf{r}$ has its origin at the center of the dipole. With respect to this origin, the $+q$ charge is located by the vector, $\mathbf{d}$ while the $-q$ charge is located by the vector, $-\mathbf{d}$. According to the law of superposition of charge, the electrostatic energy, U_e, of the dipole in the external electric field is

$$U_e = q\Phi_{ext}(\mathbf{d}) - q\Phi_{ext}(-\mathbf{d}) \tag{1.8.1}$$

If the length of $\mathbf{d} = (d_x, d_y, d_z)$ is small, we can write $\Phi_{ext}(\mathbf{d})$ as a Taylor series (Brook Taylor, English, 1685–1731),

$$\Phi_{ext}(\mathbf{d}) = \Phi_{ext}(0) + \left(\frac{\partial \Phi_{ext}}{\partial x}\right)_0 d_x + \left(\frac{\partial \Phi_{ext}}{\partial y}\right)_0 d_y + \left(\frac{\partial \Phi_{ext}}{\partial y}\right)_0 d_z \tag{1.8.2a}$$

By replacing (d_x, d_y, d_z) by $(-d_x, -d_y, -d_z)$, Eq. (1.8.2a) becomes,

$$\Phi_{ext}(-\mathbf{d}) = \Phi_{ext}(0) - \left(\frac{\partial \Phi_{ext}}{\partial x}\right)_0 d_x - \left(\frac{\partial \Phi_{ext}}{\partial y}\right)_0 d_y - \left(\frac{\partial \Phi_{ext}}{\partial y}\right)_0 d_z \tag{1.8.2b}$$

Equations (1.8.2a) and (1.8.2b) can be consolidated to read,

$$\Phi_{ext}(\pm\mathbf{d}) = \Phi_{ext}(0) \pm \nabla\Phi_{ext}(0) \cdot (d_x\hat{\mathbf{e}}_x + d_y\hat{\mathbf{e}}_y + d_z\hat{\mathbf{e}}_z) \tag{1.8.3}$$

With the use of

$$\mathbf{d} = d_x\hat{\mathbf{e}}_x + d_y\hat{\mathbf{e}}_y + d_z\hat{\mathbf{e}}_z \tag{1.8.4}$$

Equation (1.8.3) can be rewritten in vector form as

$$\Phi_{ext}(\pm\mathbf{d}) = \Phi_{ext}(0) \pm \nabla\Phi_{ext}(0) \cdot \mathbf{d} \tag{1.8.5}$$

Upon substitution of Eq. (1.8.5) into Eq. (1.8.1), the result is

$$U_e = q\Phi_{ext}(0) + q\nabla\Phi_{ext}(0) \cdot \mathbf{d} - q\Phi_{ext}(0) + q\nabla\Phi_{ext}(0) \cdot \mathbf{d} \tag{1.8.6}$$

The monopole terms $q\Phi_{ext}(0)$ and $-q\Phi_{ext}(0)$ cancel, so that Eq. (1.8.6) simplifies to

$$U_e = 2q\mathbf{d} \cdot \nabla\Phi_{ext}(0) \tag{1.8.7}$$

The electric field at the origin of coordinates is

$$\mathbf{E} = -\nabla\Phi_{ext}(0) \tag{1.8.8}$$

By virtue of Eq. (1.7.17), which states that $p = 2qd$, the electrostatic potential energy of the electric dipole is

$$U_e = -\mathbf{p} \cdot \mathbf{E} \tag{1.8.9}$$

where $\mathbf{p} = 2q\mathbf{d}$ is the electric dipole moment vector.

1.9 Interaction Energy of Two Point Electric Dipoles

The interaction between dipoles on neighboring molecules makes a contribution to the cohesive energy of polar liquids such as water. It is thus of interest to obtain a formula for the potential energy associated with this interaction.

Consider a unit vector, $\hat{\mathbf{r}}$, which makes an angle θ with an electric dipole vector, $\mathbf{p}$. The projection of $\mathbf{p}$ onto $\hat{\mathbf{r}}$ is

$$\mathbf{p} \cdot \hat{\mathbf{r}} = \mathbf{p} \cdot \mathbf{r}/r = p \cos\theta \tag{1.9.1}$$

where the unit vector, $\hat{\mathbf{r}} = \mathbf{r}/r$. Consider the arrangement of electric dipoles shown in Fig. 1.10.

According to Eqs. (1.7.18) and (1.9.1), the electrostatic potential, $\Phi(\mathbf{r}_b)$, sensed by the dipole at b due to a point electric dipole located at a is

$$\Phi(\mathbf{r}_b) = \frac{\mathbf{p}_a \cdot \mathbf{r}_{ab}}{4\pi\varepsilon_o r_{ab}^3} \tag{1.9.2}$$

According to Eq. (1.6.5), the electric field at point b is

$$\mathbf{E}(\mathbf{r}_b) = -\nabla_b \Phi(r_{ab}) = -\nabla_b \left(\frac{\mathbf{p}_a \cdot \mathbf{r}_{ab}}{4\pi r_{ab}^3} \right) \tag{1.9.3}$$

where in Cartesian coordinates,

$$\nabla_b = \hat{\mathbf{e}}_x \frac{\partial}{\partial x_b} + \hat{\mathbf{e}}_y \frac{\partial}{\partial y_b} + \hat{\mathbf{e}}_z \frac{\partial}{\partial z_b} \tag{1.9.4}$$

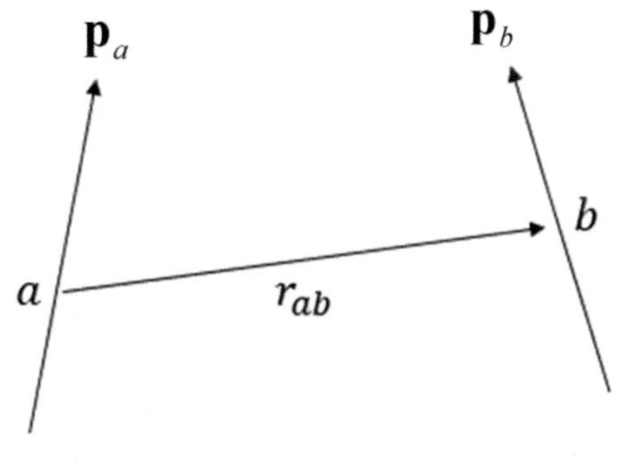

Fig. 1.10 Electric dipoles $\mathbf{p}_a$ and $\mathbf{p}_b$ located a vector distance r_{ab} apart

The ratio, $(\mathbf{p}_a \cdot \mathbf{r}_{ab})/r_{ab}^3$, can be differentiated as a product. The evaluation begins with

$$\mathbf{E}(\mathbf{r}_b) = -\nabla_b\left(\frac{\mathbf{p}_a \cdot \mathbf{r}_{ab}}{4\pi\varepsilon_o r_{ab}^3}\right) = -\frac{1}{4\pi\varepsilon_o}\left[\frac{\nabla_b(\mathbf{p}_a \cdot \mathbf{r}_{ab})}{r_{ab}^3} + (\mathbf{p}_a \cdot \mathbf{r}_{ab})\nabla_b\left(\frac{1}{r_{ab}^3}\right)\right] \tag{1.9.5}$$

The first term within the brackets on the right hand side of Eq. (1.9.5) can be calculated as follows: The dot product of the electric dipole vector, $\mathbf{p}_a = (p_{ax}, p_{ay}, p_{az})$, with the radius vector, $\mathbf{r}_{ab}$ is

$$\mathbf{p}_a \cdot \mathbf{r}_{ab} = p_{a,x}(x_b - x_a) + p_{a,y}(y_b - y_a) + p_{a,z}(z_b - z_a) \tag{1.9.6}$$

Expanded in terms of components, the gradient of this dot product is

$$\begin{aligned}\nabla_b(\mathbf{p}_a \cdot \mathbf{r}_{ab}) &= \left(\hat{\mathbf{e}}_x\frac{\partial}{\partial x_b} + \hat{\mathbf{e}}_y\frac{\partial}{\partial y_b} + \hat{\mathbf{e}}_z\frac{\partial}{\partial z_b}\right)\\ &\quad \times \left[p_{a,x}(x_b - x_a) + p_{a,y}(y_b - y_a) + p_{a,z}(z_b - z_a)\right]\\ &= p_{a,x}\hat{\mathbf{e}}_x + p_{a,y}\hat{\mathbf{e}}_y + p_{a,z}\hat{\mathbf{e}}_z\\ &= \mathbf{p}_a\end{aligned} \tag{1.9.7}$$

The second term within the brackets on the right hand side of Eq. (1.9.5) can be calculated with the result,

$$\begin{aligned}\nabla_b\frac{1}{r_{ab}^3} &= \left(\hat{\mathbf{e}}_x\frac{\partial}{\partial x_b} + \hat{\mathbf{e}}_y\frac{\partial}{\partial y_b} + \hat{\mathbf{e}}_z\frac{\partial}{\partial z_b}\right)\left(\frac{1}{r_{ab}^3}\right)\\ &= -\left(\frac{3}{r_{ab}^4}\right)\left(\hat{\mathbf{e}}_x\frac{\partial r_{ab}}{\partial x_b} + \hat{\mathbf{e}}_y\frac{\partial r_{ab}}{\partial y_b} + \hat{\mathbf{e}}_z\frac{\partial r_{ab}}{\partial z_b}\right) = -3\frac{\nabla_b r_{ab}}{r_{ab}^4}\end{aligned} \tag{1.9.8}$$

The gradient of the distance r_{ab} is

$$\begin{aligned}\nabla_b r_{ab} &= \left(\hat{\mathbf{e}}_x\frac{\partial}{\partial x_b} + \hat{\mathbf{e}}_y\frac{\partial}{\partial y_b} + \hat{\mathbf{e}}_z\frac{\partial}{\partial z_b}\right)\left[(x_b - x_a)^2 + (y_b - y_a)^2 + (z_b - z_a)^2\right]^{1/2}\\ &= \frac{1}{2}\left[\frac{2(x_b - x_a)\hat{\mathbf{e}}_x + 2(y_b - y_a)\hat{\mathbf{e}}_y + 2(z_b - z_a)\hat{\mathbf{e}}_z}{\left((x_b - x_a)^2 + (y_b - y_a)^2 + (z_b - z_a)^2\right)^{1/2}}\right]\\ &= \frac{\mathbf{r}_{ab}}{r_{ab}}\end{aligned} \tag{1.9.9}$$

Upon substitution of Eq. (1.9.9) into Eq. (1.9.8), the result is

$$\nabla_b\frac{1}{r_{ab}^3} = -3\frac{\mathbf{r}_{ab}}{r_{ab}^5} \tag{1.9.10}$$

Next substitute Eqs. (1.9.7) and (1.9.10) into Eq. (1.9.5). The result is

$$\mathbf{E}(\mathbf{r}_b) = -\frac{1}{4\pi\varepsilon_o}\left[\frac{\mathbf{p}_a}{r_{ab}^3} - \frac{3(\mathbf{p}_a \cdot \mathbf{r}_{ab})}{r_{ab}^5}\mathbf{r}_{ab}\right] \tag{1.9.11}$$

According to Eq. (1.8.9), the electrostatic energy of interaction of the two dipoles equals $-\mathbf{p}_b \cdot \mathbf{E}(\mathbf{r}_b)$. Upon substitution of Eq. (1.9.11) into Eq. (1.8.9), one obtains the final result,

$$U_e(r_{ab}) = \frac{1}{4\pi\varepsilon_o}\left[\frac{\mathbf{p}_a \cdot \mathbf{p}_b}{r_{ab}^3} - 3\frac{(\mathbf{p}_a \cdot \mathbf{r}_{ab})(\mathbf{p}_b \cdot \mathbf{r}_{ab})}{r_{ab}^5}\right] \tag{1.9.12}$$

1.10 The Electrostatic Potential of a Point Electric Quadrupole

As illustrated in Fig. 1.11, a linear electric quadrupole can be formed by two opposed electric dipoles lying along the z-axis.

In the linear quadrupole model in Fig. 1.11, the negative end of each dipole is located at the center. The coordinates of this point are $(0, 0, 0)$, not shown. The net charge at $(0, 0, 0)$ is $-2q$. The remainder of the quadrupole consists of a charge of $+q$ located at point a having coordinates $(x_a, y_a, z_a) = (0, 0, +d)$ and a second charge of $+q$ located a point b having coordinates, $(x_b, y_b, z_b) = (0, 0, -d)$. The overall charge is zero, and because the electric dipoles are pointing in opposite directions, the net electric dipole moment is also zero.

The electrostatic potential produced at (x, y, z) by the quadrupole has a contribution from each of the three charge centers and can be written as

$$\Phi(x, y, z) = \frac{q}{4\pi\varepsilon_o}\frac{1}{\left[x^2 + y^2 + (z-d)^2\right]^{1/2}} + \frac{q}{4\pi\varepsilon_o}\frac{1}{\left[x^2 + y^2 + (z+d)^2\right]^{1/2}} - \frac{2q}{4\pi\varepsilon_o}\frac{1}{\left(x^2 + y^2 + z^2\right)^{1/2}} \tag{1.10.1}$$

Equation (1.10.1) can also be expressed in the form,

$$\Phi(x, y, z) = \frac{q}{4\pi\varepsilon_o}\frac{1}{\left(r^2 - 2zd + d^2\right)^{1/2}} + \frac{q}{4\pi\varepsilon_o}\frac{1}{\left(r^2 + 2zd + d^2\right)^{1/2}} - \frac{2q}{4\pi\varepsilon_o r} \tag{1.10.2}$$

where $r^2 = \left(x^2 + y^2 + z^2\right)^{1/2}$. Upon factoring out r^2 from the polynomials in parentheses in the denominators of the first two terms in Eq. (1.10.2), one obtains

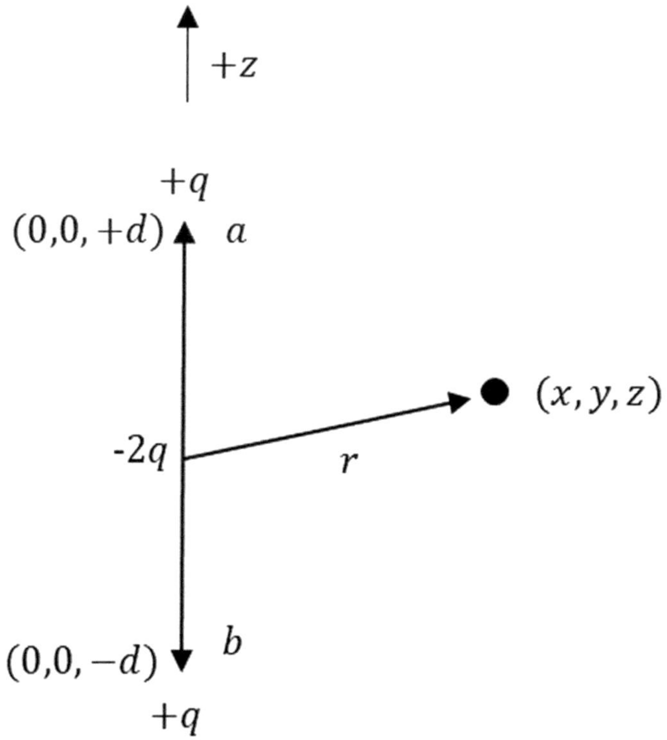

Fig. 1.11 A linear electric quadrupole formed by two opposed electric dipoles. A charge of $+q$ is located on z-axis at a with coordinates $(0, 0, +d)$, and a charge of $+q$ is located at b with coordinates $(0, 0, -d)$, respectively. At the center of the linear quadrupole, the negative ends of the dipoles overlap to make a net charge of $-2q$. The point of observation at (x, y, z) is a distance, **r**, measured from the location of the charge of $-2q$

$$\Phi(x, y, z) = \frac{q}{4\pi\varepsilon_o r}\frac{1}{\left[1 - 2z(d/r^2) + (d/r)^2\right]^{1/2}} + \frac{q}{4\pi\varepsilon_o r}\frac{1}{\left[1 + 2z(d/r^2) + (d/r)^2\right]^{1/2}} - \frac{2q}{4\pi\varepsilon_o r} \tag{1.10.3}$$

Equation (1.10.3) is the quadrupole analog of Eq. (1.7.8). When Eqs. (1.7.13) and (1.7.15) are substituted into Eq. (1.10.3), the result is

$$\Phi(r, \theta) = \frac{q}{4\pi\varepsilon_o r}\left[1 + \cos\theta\left(\frac{d}{r}\right) + \left(\frac{3\cos^2\theta - 1}{2}\right)\left(\frac{d}{r}\right)^2\right] + \frac{q}{4\pi\varepsilon_o r}\left[1 - \cos\theta\left(\frac{d}{r}\right) + \left(\frac{3\cos^2\theta - 1}{2}\right)\left(\frac{d}{r}\right)^2\right]$$

$$- \frac{2q}{4\pi\varepsilon_o r} \tag{1.10.4}$$

In Eq. (1.10.4), the monopole terms (those proportional to $(1/r)$ cancel, and the dipole terms (proportional to (d/r) also cancel. All that remains are the quadrupole terms, which sum to

$$\Phi(r, \theta) = \frac{2qd^2}{4\pi\varepsilon_o r^3} \frac{3\cos^2\theta - 1}{2} \tag{1.10.5}$$

If we define the electric quadrupole moment, Q, as

$$Q = 2qd^2 \tag{1.10.6}$$

Equation (1.10.5) can be transformed to read

$$\Phi(r, \theta) = \frac{Q}{4\pi\varepsilon_o r^3} \frac{3\cos^2\theta - 1}{2} \tag{1.10.7}$$

Equation (1.10.7) is a formula for the electrostatic potential of a point electric quadrupole.

The charge distribution in the carbon dioxide molecule provides a good example of an electric quadrupole. In the case of each carbon to oxygen bond, the bonding electrons lie closer to the oxygen atom than to the carbon atom, creating a bond electric dipole. Carbon dioxide is a linear molecule and the two bond dipoles, which point toward the carbon atom, are oppositely directed. The overall charge of the molecule is zero, and since the two bond dipoles are oppositely disposed, the net molecular electric dipole moment is also zero. The first non-zero multipole moment in the carbon dioxide molecule is the electric quadrupole moment.

1.11 Potential, Electric Field, and Interaction Energy of Volume Charge Distributions

One commonly encounters the problem of calculating the electrostatic potential, $\Phi(\mathbf{r})$, at a point, $\mathbf{r}$, located outside an arbitrary volume charge distribution. Let the charge density be $\rho(\mathbf{r}')$, at a point $\mathbf{r}'$, located within a differential volume, d^3r', occupied by the distribution. The charge contained in d^3r' is $\rho(\mathbf{r}')d^3r'$. This charge is located a distance $|\mathbf{r} - \mathbf{r}'|$ from a point, $\mathbf{r}$, where the electrostatic potential is being evaluated. According to the superposition principle, the potential at $\mathbf{r}$ is given by the integral

$$\Phi(\mathbf{r}) = \frac{1}{4\pi\varepsilon_o} \int \frac{\rho(\mathbf{r}')d^3r'}{|\mathbf{r} - \mathbf{r}'|} \tag{1.11.1}$$

In rectangular coordinates, $d^3r' = dx'dy'dz'$. The integral notation in Eq. (1.11.1) has the meaning $\int d^3r' \equiv \int dx' \int dy' \int dz'$, where the indicated integration is carried out over the volume occupied by the charge density. The corresponding electric field is

$$\mathbf{E}(\mathbf{r}) = -\nabla\Phi(\mathbf{r}) = \frac{-1}{4\pi\varepsilon_o}\nabla\int\frac{\rho(\mathbf{r}')d^3r'}{|\mathbf{r}-\mathbf{r}'|} = \frac{-1}{4\pi\varepsilon_o}\int d^3r'\rho(\mathbf{r}')\nabla\left(\frac{1}{|\mathbf{r}-\mathbf{r}'|}\right) \tag{1.11.2}$$

By virtue of Eq. (1.6.12), we have

$$\nabla\frac{1}{|\mathbf{r}-\mathbf{r}'|} = -\frac{(\mathbf{r}-\mathbf{r}')}{|\mathbf{r}-\mathbf{r}'|^3} \tag{1.11.3}$$

Substitute Eq. (1.11.3) into Eq. (1.11.2) to obtain,

$$\mathbf{E}(\mathbf{r}) = \frac{1}{4\pi\varepsilon_o}\int\rho(\mathbf{r}')\frac{(\mathbf{r}-\mathbf{r}')}{|\mathbf{r}-\mathbf{r}'|^3}d^3r' \tag{1.11.4}$$

which is the electric field at a position, **r**, outside the charge distribution.

By analogy with Eq. (1.5.6), the electrostatic potential energy of interaction of two charge distributions is

$$U_e = \frac{1}{2}\int\int\frac{\rho(\mathbf{r}')\rho(\mathbf{r}'')d^3r'd^3r''}{4\pi\varepsilon_o|\mathbf{r}'-\mathbf{r}''|} \tag{1.11.5}$$

1.12 Gauss' Law

Figure 1.12 illustrates an arbitrary volume charge distribution.

The value of **E** expresses the density of electric field lines passing through the surface at $d\mathbf{S}$. As such, the number of electric field lines passing normal to $d\mathbf{S}$ is $\mathbf{E}\cdot d\mathbf{S}$. If $\rho(\mathbf{r})$ is the local value of the charge density, the charge contained in d^3r is $\rho(\mathbf{r})d^3r$. By analogy with our calculation of the flux associated with a point charge (Eq. 1.3.9b), the conservation of electric field lines is expressed by

$$\int\mathbf{E}\cdot d\mathbf{S} = (1/\varepsilon_o)\int\rho(\mathbf{r})d^3r \tag{1.12.1}$$

In the Mathematical Methods Appendix, you will find a proof of what is called variously the divergence theorem, or the Gauss/Green theorem (George Green, English, 1793–1841) and (Carl Friedrich Gauss, German, 1777–1855). The general statement of this theorem is

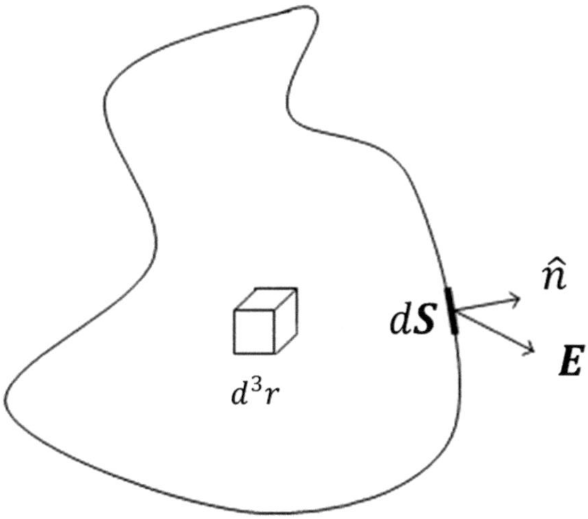

Fig. 1.12 Arbitrary volume charge distribution. A differential element of the volume of the charge distribution is indicated by d^3r. A differential element of surface area surrounding the volume is indicated by $d\mathbf{S} = dS\hat{\mathbf{n}}$ where $\hat{\mathbf{n}}$ is a unit vector which functions as the outward drawn normal to the surface element. The local value of the electric field on the surface element is $\mathbf{E}$

$$\int \mathbf{E} \cdot d\mathbf{S} = \int \nabla \cdot \mathbf{E} d^3 r \tag{1.12.2}$$

If Eqs. (1.12.1) and (1.12.2) are to mean the same thing, we must have

$$\nabla \cdot \mathbf{E} = (1/\varepsilon_o)\rho(\mathbf{r}) \tag{1.12.3}$$

Equation (1.12.3) is known as Gauss' Law.

If in Eq. (1.12.3), we replace $\mathbf{E}$ with the left hand side of Eq. (1.6.5), we obtain

$$\nabla^2 \Phi(\mathbf{r}) = -\rho(\mathbf{r})/\varepsilon_o \tag{1.12.4}$$

which is known as Poisson's equation (Simeon Denis Poisson, French, 1781–1840). In the special case that $\rho(\mathbf{r}) = 0$,

$$\nabla^2 \Phi(\mathbf{r}) = 0 \tag{1.12.5}$$

which is known as Laplace's equation (Pierre Simon Laplace, French 1749–1827).

1.13 Perfect Conductors

The charge carriers in metals are free electrons. When an electric field is applied to a perfect conductor, the electrons move to the surface so as to cancel the electric field inside the conductor. If there is no electric field inside a perfect conductor, then

$$\mathbf{E} = -\nabla \Phi = 0 \tag{1.13.1}$$

Equation (1.13.1) implies that Φ cannot change from point to point within the conductor. As such Φ must have the same value everywhere within the conductor including the surface. Because the value of Φ is independent of location, the surface of the conductor serves as an equipotential.

1.14 Legendre Polynomials

In the process of calculating the electrostatic potential of the point dipole and the point quadrupole, we utilized Eq. (1.7.13). With $x = (d/r)$, Eq. (1.7.13) permits us to define a function $f(x)$ that can be written in the form,

$$f(x) = \frac{1}{\left(1 - 2x\cos\theta + x^2\right)^{1/2}} = 1 + x\cos\theta + x^2\frac{3\cos^2\theta - 1}{2} + \cdots \tag{1.14.1}$$

The coefficients of the powers of x in Eq. (1.14.1) are polynomials in $\cos\theta$. These polynomial coefficients can be given the identifications,

$$P_0(\cos\theta) = 1 \tag{1.14.2}$$

$$P_1(\cos\theta) = \cos\theta \tag{1.14.3}$$

$$P_2(\cos\theta) = \frac{3\cos^2\theta - 1}{2} \tag{1.14.4}$$

The coefficient set $\{P_n(\cos\theta)\}$ are called the Legendre polynomials (Adrien Marie Legendre, French 1752–1833), and $f(x)$ in Eq. (1.14.1) is said to be their generating function. In a subsequent chapter, we shall make use of these Legendre polynomials in order to solve Eq. (1.12.5).

1.15 Earnshaw's Theorem and the Stability of Matter

Samuel Earnshaw (English, 1805–1888) showed that a collection of point charges, which are free to move, cannot maintain a fixed configuration sustained *solely* by their mutual electrostatic interactions. This statement is known as Earnshaw's theorem. The proof is based upon Eq. (1.3.8).

The potential energy of a point charge, q_b, at position, $\mathbf{r}_b$ is $q_b\Phi(\mathbf{r}_b)$. The charge will be at rest at any point where the net force acting on it is zero. This point locates a minimum of the potential energy. By virtue of Eq. (1.3.8), the condition for zero net force and a potential minimum is met if $-q_b(\nabla\Phi(\mathbf{r}))_{\mathbf{r}=\mathbf{r}_b} = q_b\mathbf{E}(\mathbf{r}_b) = 0$. However, according to Eq. (1.3.7), if $\mathbf{E}(\mathbf{r}_b) = 0$, there cannot be a net charge, q_a, nearby.

Earnshaw's theorem is the main reason why electrostatic forces by themselves are insufficient to produce stable atoms.

Although a charge distribution cannot be in equilibrium when all charges are at rest, it is the case in quantum mechanics that a charge distribution can exist in a steady state when the charges are moving. This is the basis for Bohr's model of the hydrogen atom where the electron and the proton orbit their common center of mass. In the Bohr model (Niels Bohr, Danish, 1885–1962), the Coulomb force acting between the electron and the proton is balanced by the centrifugal force associated with the electron's motion. Although the proton and the electron orbit their common center of mass, the proton is much heavier than the electron, which makes the electron orbit appear to be a circle with the proton at its center.

Bohr was aware of Planck's (Max Planck, German, 1858–1947) hypothesis that the energy, ε, contained in a light wave of frequency, ν, could be reckoned as $\varepsilon = nh\nu$, where n is an integer. At its first appearance (1900), Planck's constant h was an adjustable parameter. We now recognize it as a fundamental constant of nature with the value $h = 6.626 \times 10^{-27}$ erg · s. In the tradition of atomic physics, we shall develop the Bohr atomic model using cgs units. According to Louis de Broglie (French, 1892–1987), the wavelength, λ, associated with the motion of the electron is related to its linear momentum, p, by

$$\lambda = \frac{h}{p} \tag{1.15.1}$$

where the use of p in this context is not to be confused with its use to represent an electric dipole moment. For the length of a standing wave to be commensurate with a circular orbit of circumference $2\pi r$, the radius, r, and the wavelength must satisfy,

$$n\lambda = 2\pi r \tag{1.15.2}$$

where n is an integer. Combine Eqs. (1.15.1) and (1.15.2) to obtain

$$p = \frac{n\hbar}{r} \tag{1.15.3}$$

where $\hbar = h/2\pi$. According to Eq. (1.2.11), the attractive electrostatic force acting between the proton and the electron of charge, e, has the magnitude, e^2/r^2. Because the electron is orbiting the proton, there is a centrifugal force, mv^2/r. If the centrifugal force and the electrostatic force are to balance one another, it must be the case that

$$\frac{mv^2}{r} = \frac{e^2}{r^2} \tag{1.15.4}$$

If we use the relation,

$$p = mv \tag{1.15.5}$$

to relate the momentum, p to the mass, m, and speed, v, of the electron, we can write Eq. (1.15.3) in the form,

$$mv = \frac{n\hbar}{r} \tag{1.15.6}$$

By combining Eqs. (1.15.4) and (1.15.6), we obtain

$$r = \frac{n^2\hbar^2}{me^2} \tag{1.15.7}$$

The total energy, E, of the electron in its orbit is the sum of the kinetic, $(1/2)mv^2$, and potential energy, $-e^2/r$, that is,

$$E = \frac{1}{2}mv^2 - \frac{e^2}{r} \tag{1.15.8}$$

Substitute Eq. (1.15.4) into Eq. (1.15.8) to obtain,

$$E = -\frac{e^2}{2r} \tag{1.15.9}$$

Our final result for the quantized energy, E_n, is obtained by substituting Eq. (1.15.7) into Eq. (1.15.9). The result is

$$E_n = -\frac{1}{2}\frac{e^4 m}{n^2\hbar^2} \tag{1.15.10}$$

In the context of Eq. (1.15.10), n is said to be the principle quantum number. The first Bohr orbit is described by $n = 1$. Upon substitution of $n = 1$, $e = 4.8 \times 10^{-10}\,\text{esu}$, $m = 9.109 \times 10^{-28}\,\text{g}$, and $\hbar = 1.055 \times 10^{-27}\,\text{erg} \cdot \text{s}$. into Eq. (1.15.10), we obtain

$$E_1 = -\frac{1}{2}\frac{\left(4.8 \times 10^{-10}\,\text{esu}\right)^4\left(9.109 \times 10^{-28}\,\text{g}\right)}{\left(1.055 \times 10^{-27}\,\text{erg} \cdot \text{s}\right)^2} = -2.17 \times 10^{-11}\,\text{erg} \tag{1.15.11}$$

As 1 eV = 1.6×10^{-12} erg, the result in Eq. (1.15.11) converts to $E_1 = -13.6\,\text{eV}$, which is the binding energy of an electron in the first Bohr orbit. The Bohr theory shows how the consequences of Earnshaw's theorem can be obviated if the charges are moving in quantized orbits. By virtue of Eq. (1.15.11), one can conclude that Coulomb's law, and the values of Planck's constant and the mass and charge of the electron conspire to place chemical phenomena on the electron volt scale of energies. This contrasts with nuclear phenomena, which involve energies measured in MeV, and elementary particle phenomena, which involve energies measured in GeV.

References

1. J.C. Slater, N.H. Frank, *Electromagnetism* (McGraw Hill, New York, 1947), p. 101
2. J.S. Jayson, The Daniell cell, Ohm's law, and the emergence of the international system of units. Am. J. Phys. **82**, 60–65 (2014)

Recommended Additional Reading

1. A. Hinchliffe, R.W. Munn, Molecular Electromagnetism (Wiley, Chichester, West Sussex, UK, 1985)
2. D.J. Griffiths, *Introduction to Electrodynamics*, 4th edn. (Pearson Education Ltd., India, 2013)
3. S. Kielich, General molecular theory and electric field effects in isotropic dielectrics. in *Specialist Periodical Reports, Dielectrics and Related Molecular Processes*, ed. by M. Davies, vol. 1 (The Chemical Society, London, UK, 1972)
4. H.W. Meyer, *A History of Electricity and Magnetism* (MIT Press, Cambridge, MA, 1971)

Chapter 2
Thermodynamics and Statistical Mechanics

The subjects of thermodynamics and statistical mechanics are ordinarily considered separately from those of electricity and magnetism. Nonetheless, in many chemical phenomena, it is the case that the thermal properties and the electromagnetic properties are intimately related. This connection occurs when the electromagnetic measurements involve macroscopic samples of matter held at constant temperature and pressure. To account for the thermal properties of matter, this chapter includes discussions of the first and second laws of thermodynamics and introduces the thermodynamic functions used to express these laws. The functions include the internal energy, the entropy, the enthalpy, the free energies, and the chemical potential. The laws of thermodynamics are applied to describe the properties of ideal gases, which are then used as the basis for the description of the properties of ideal and real solutions, including the concepts of activity and the activity coefficient. The consideration of solutions will lead us to the Gibbs–Duhem equation which relates the chemical potentials of the solvent and solute. Upon completion of the survey of thermodynamics, statistical mechanics will be introduced and shown to serve as the molecular basis for thermodynamics. Statistical mechanical topics that will be introduced include the canonical ensemble, the grand canonical ensemble, and the theory of fluctuations. With the connection between thermodynamics and statistical mechanics established, the chapter closes with a discussion of the osmotic compressibility, which is the thermodynamic source of the concentration fluctuations that produce light scattering. The latter is observed in the case of solutions of macromolecules.

J. K. Baird, *Electromagnetism for Chemistry*,
https://doi.org/10.1007/978-3-031-88508-2_2

2.1 Introduction to Thermodynamics and Statistical Mechanics

The temperature concept, which is central to thermodynamics and statistical mechanics, does not appear explicitly in any of the laws of electromagnetism. A chapter on thermodynamics and statistical mechanics would thus seem to be out of place in a book devoted to electromagnetism. Nonetheless, in the case of real gases, liquids, solids, and plasmas, where the atomic collisions permit an exchange of energy, the electromagnetic properties become implicit functions of the temperature. This temperature dependence is of interest in chemistry. Below, we develop the theoretical framework that can be used to describe it.

Thermodynamics deals with the behavior of macroscopic samples of matter, largely without reference to variations on the molecular scale. Einstein asserted that the laws of thermodynamics are so fundamental that they would likely never require revision. Our foray into thermodynamics will require only a few concepts, namely the internal energy, the entropy, the free energies, the chemical potential, and the concept of activity as applied to solutes in solution.

In contrast to thermodynamics, statistical mechanics is based on the assumption that matter is not entirely homogeneous on the molecular scale. The goal of statistical mechanics is to calculate the properties of macroscopic samples of matter by exploiting statistical averages over the atomic and molecular properties.

We can imagine the subdivision of a macroscopic quantity of material into smaller samples, each having the same volume, the same temperature, and the same number of molecules. These samples can exchange heat and will consequently exhibit a distribution of energies from which an average energy value characteristic of the entire material can be calculated. In statistical mechanics, this particular collection of samples is known as the *canonical ensemble*. In subsequent chapters, we will exploit the canonical ensemble when we calculate the electrical properties of electrolytes and dielectrics. The ensemble will also play a key role when we consider the absorption and emission of electromagnetic radiation.

It is also possible to imagine the subdivision of a material into smaller samples having the same volume, the same temperature, the same molecular escaping tendency, but different numbers of molecules. If these samples are allowed to exchange heat and molecules, they will exhibit a distribution in the number of molecules per unit volume from which the average number density associated with the macroscopic material can be calculated. Because the number density is associated with an average, it can also be described by a variance. As such, under special conditions, we can expect to observe the effects of macroscopic scale fluctuations in the molecular density in a pure substances and fluctuations in the concentrations of the components in a mixture. In statistical mechanics, such a collection of samples is known as the *grand canonical ensemble*. In a subsequent chapter, we will show that variations in number density and concentration are capable of scattering light.

2.2 First Law of Thermodynamics

In Chap. 1, we derived a formula for the electrostatic potential energy, U_e, of interaction of two distributions of charge. We now introduce the concept of the internal energy, U, which differs from U_e in that it contains both potential energy and kinetic energy. The internal energy is reckoned with respect to the object's center of mass. The energy of the center of mass motion is of no consequence in thermodynamics. The first law of thermodynamics states that the internal energy is conserved.

On a microscopic scale, each atom or ion in a sample of matter can move from side to side, front to back, or top to bottom. That is to say, each atom has three degrees of freedom. In Fig. 2.1, which illustrates the structure of a solid on the atomic scale, the atoms are regarded as masses connected to one another by springs representing the chemical bonds. The oscillations of the atoms about their equilibrium positions can be assumed to be governed Hooke's law (Robert Hooke, English, 1635–1703).

One mole of any solid contains at least Avogadro's number, N_A, of atoms, which can support a total of $3N_A$ modes of vibration. If we knew the energies in all of these modes, we could sum them and make an accurate calculation of the internal energy. Except in special cases, this task is impossibly complex. Yet, without knowing all $3N_A$ modes of motion in detail, we can still make progress. The idea is to divide the modes into two classes.

In the *first* class are those vibrational modes which fail to preserve the volume, and are responsible for the change in work. This type of collective motion is associated with a net change in the product of the length × width × height of the material, which is equivalent to a change in volume. The resulting volume dilation or contraction is

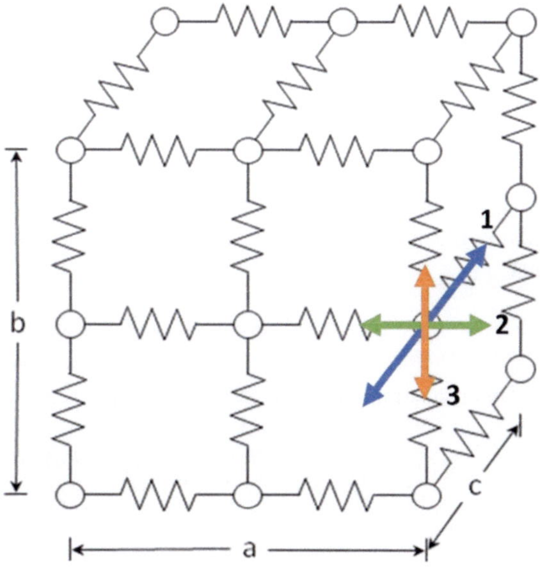

Fig. 2.1 A simple crystalline unit cell with orthorhombic symmetry consisting of atoms connected by bonds modeled as springs. The linear dimensions of the unit cell are *a*, *b*, *c*. The typical atom can vibrate in the mutually perpendicular directions denoted by 1, 2, 3, respectively

ordinarily resisted/abetted by the external atmospheric pressure, which acts normal to the surfaces bounding the sample. The differential change in *work*, dW, due to an external force, f_x, acting to produce a differential change in dimension, dx, is given by

$$dW = -f_x dx \tag{2.2.1}$$

Upon multiplying and dividing the right hand side of Eq. (2.2.1) by the cross sectional area, A_{yz}, of the surface with normal parallel to dx, we have

$$dW = -\left(\frac{f_x}{A_{yz}}\right)(A_{yz}dx) \tag{2.2.2}$$

The external pressure, P, is defined by

$$P = \frac{f_x}{A_{yz}} \tag{2.2.3}$$

while the change in volume, V, is

$$dV = A_{yz}dx \tag{2.2.4}$$

Hence, we can rewrite Eq. (2.2.2) in the form

$$dW = -PdV \tag{2.2.5}$$

In the *second* class are the vibrational modes, which preserve the volume, that is modes for which $dV = 0$. To achieve $dV = 0$, any increase in the length of the object in one dimension must be balanced by a decrease in the lengths in one or both of the other dimensions. Although these modes of vibration do no work, they do, nonetheless, store energy. We call this energy, *heat*.

The power of the first law of thermodynamics lies in the statement that any change in internal energy must involve either a change in heat or a change in work or both. That is, if dW is the change in work, and dQ is the change in heat, the first law states that

$$dU = dQ + dW \tag{2.2.6}$$

Upon substitution of Eq. (2.2.5) into Eq. (2.2.6), we obtain

$$dU = dQ - PdV \tag{2.2.7}$$

Although a change in internal energy, dU, may be hard to calculate theoretically using atomic models, it can easily be determined experimentally by measuring the change in heat, dQ_V, when a process is carried out in a fixed volume calorimeter. See Fig. 2.2.

Because the walls of the calorimeter are rigid, the heat which is measured corresponds to $dV = 0$. By virtue of Eq. (2.2.7), we thus have

$$dU = dQ_V \tag{2.2.8}$$

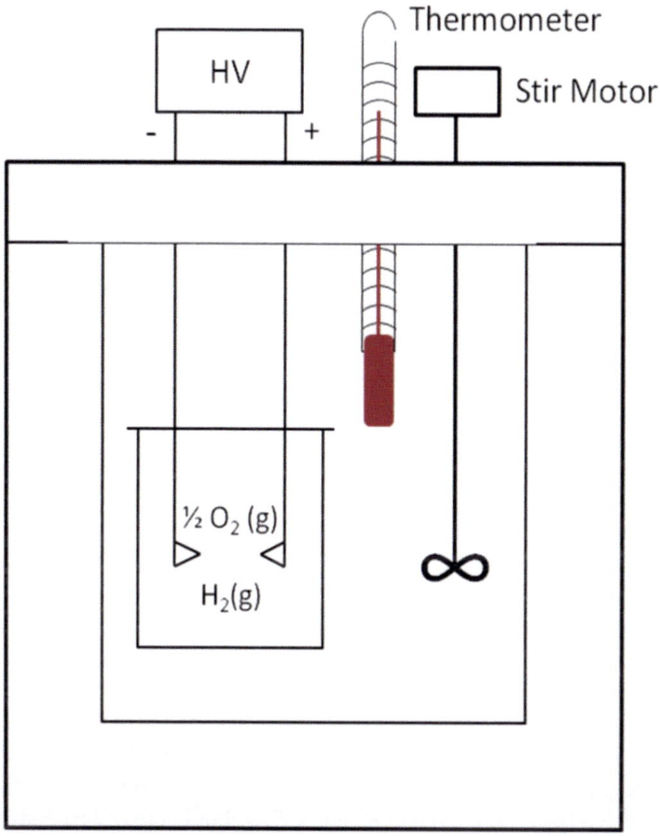

Fig. 2.2 Bomb calorimeter. The calorimeter has rigid walls, which maintain its volume fixed. The inner volume of the calorimeter, which contains the closed sample vessel, the thermometer, and a stirring motor, is ordinarily filled with water. A stoichiometric mixture of gaseous hydrogen and oxygen is shown in the sample vessel. The reaction to form water is initiated when the external high voltage power supply produces a spark between the electrodes in the sample vessel. If the electrical energy, the mass of the starting gases, and the heat capacity of the calorimeter, including the water, and all of the internal structures are known, then the rise in temperature upon explosion of the mixture of hydrogen and oxygen is proportional to the internal energy of formation of water

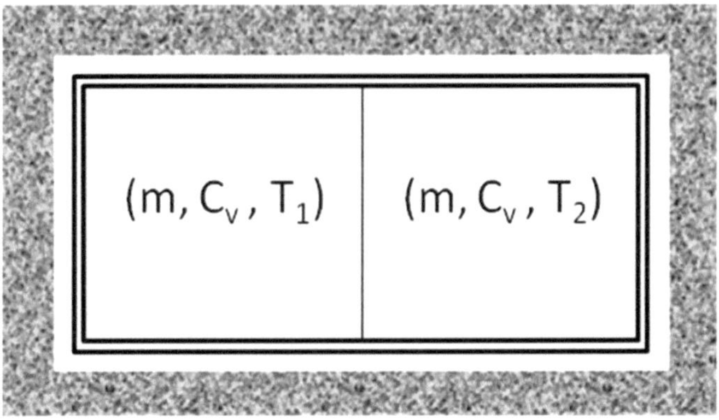

Fig. 2.3 Identical blocks of matter having mass, m, and heat capacity, C_V. The block on the left has temperature, T_1. The block on the right has temperature, T_2. The blocks are separated by a thermally insulating but removable partition. The container has rigid walls and is thermally isolated from the surroundings

where dQ_V is the heat change at constant volume. The change in heat, dQ_V, is determined by the change in temperature, dT, and the heat capacity, C_V, of the calorimeter, where

$$dQ_V = C_V dT \tag{2.2.9}$$

Being based upon the experimental procedures implied by Fig. 2.2, Eq. (2.2.8) is said to be the "operational" definition of dU.

The first law of thermodynamics is equivalent to the law of conservation of energy. We can see this by considering two equal mass blocks of the same material in a thermally isolated container having a fixed volume. See Fig. 2.3.

Because the volume of the container cannot change, the system consisting of the two blocks cannot do any work on the surroundings. As the blocks are thermally isolated from the surroundings, the only possible interaction between them is the transfer of heat from one block to the other. The heat capacity of each block is C_V. If one block is initially at temperature, T_1, and the other is initially at a higher temperature, T_2, then upon removal of the partition separating them, they will exchange heat until they reach a final common temperature, T_{com}. For energy to be conserved, we have according to Eqs. (2.2.8) and (2.2.9),

$$\Delta U = \Delta Q = C_V(T_{com} - T_1) + C_V(T_{com} - T_2) = 0 \tag{2.2.10}$$

The final temperature implied by Eq. (2.2.10) is

$$T_{com} = (1/2)(T_1 + T_2) \tag{2.2.11}$$

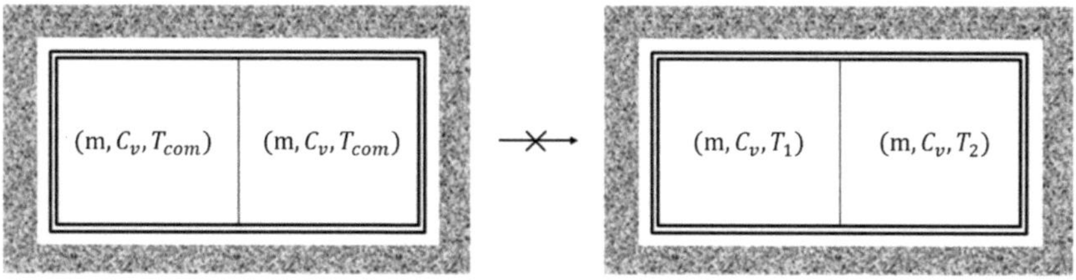

Fig. 2.4 A forbidden spontaneous heat transfer. On the left, the two blocks are shown at a common temperature, T_{com}. In the imagined process, one block spontaneously falls to a temperature, $T_1 < T_{com}$, transferring heat to the other, which rises to a temperature, $T_2 > T_{com}$. This type of heat transfer is never observed

Energy is also conserved if we consider the process where both blocks start at temperature, $T_{com} = (1/2)(T_1 + T_2)$, and then exchange heat. This is the reverse of the process shown in Fig. 2.3. As illustrated in Fig. 2.4, one block spontaneously warms to temperature T_2, while the other cools to temperature, T_1. The partition is then reinserted.

The energy conservation equation for the process illustrated in Fig. 2.4 is

$$\Delta U = \Delta Q = C_V(T_1 - T_{com}) + C_V(T_2 - T_{com}) = 0 \tag{2.2.12}$$

So far as the first law is concerned, Eqs. (2.2.10) and (2.2.12) describe feasible events. Nonetheless, the process by which one block spontaneously warms while the other block spontaneously cools has never been observed. Something is missing. The answer to that question is found in the second law.

2.3 The Second Law of Thermodynamics

In deriving the second law, we search for the differential of a function whose value is capable of distinguishing between the processes described by Eqs. (2.2.10) and (2.2.12). Since the phenomena shown in Figs. 2.3 and 2.4 involve heat transfer, the desired function presumably involves dQ in some way. Experiment and theory confirm that the appropriate differential is

$$dS = \frac{dQ}{T} \tag{2.3.1}$$

where S, is called the entropy. In cases where the volume is held constant, we can use Eq. (2.2.9) to write Eq. (2.3.1) in the form

$$dS = \frac{C_V dT}{T} \tag{2.3.2}$$

We now compute the entropy change associated with the permitted process illustrated in Fig. 2.3. The total entropy change, ΔS_{total}, can be written as the sum of two parts,

$$\Delta S_{total} = \Delta S_1 + \Delta S_2 \tag{2.3.3}$$

where ΔS_1 is the change in entropy of the block initially at temperature, T_1, while ΔS_2 is the change in entropy of the block initially at temperature, T_2. If we apply Eq. (2.3.2), to each block and integrate from the appropriate initial temperature to the final temperature, T_{com}, the results are

$$\Delta S_1 = C_V \int_{T_1}^{T_{com}} \frac{dT}{T} = C_V \ln\left(\frac{T_{com}}{T_1}\right) \tag{2.3.4a}$$

and

$$\Delta S_2 = C_V \int_{T_2}^{T_{com}} \frac{dT}{T} = C_V \ln\left(\frac{T_{com}}{T_2}\right) \tag{2.3.4b}$$

respectively. Substitution of Eq. (2.3.4a) and Eq. (2.3.4b) into Eq. (2.3.3) gives an expression for the total entropy change,

$$\Delta S_{total} = \Delta S_1 + \Delta S_2 = C_V \ln\left(\frac{T_{com}^2}{T_1 T_2}\right) \tag{2.3.5}$$

If we can prove that $T_{com}^2 > T_1 T_2$, then Eq. (2.3.5) predicts that the entropy change is positive in the case of the spontaneous process based on Fig. 2.3. The postulated inequality is demonstrated in the following theorem.

Theorem *If* $T_{com} = (1/2)(T_1 + T_2)$, *then* $T_{com}^2 > T_1 T_2$.

Proof The proof is based on the method of contradiction. In this method, one demonstrates that the reverse of the statement of the theorem is false. The reverse being false, the statement itself must be true. To begin the proof, assume that

$$T_{com}^2 < T_1 T_2 \tag{2.3.6a}$$

Upon substitution of Eq. (2.2.11), the inequality in (2.3.6a) reads

$$(1/2)^2 (T_1 + T_2)^2 < T_1 T_2 \tag{2.3.6b}$$

Upon clearing the factor of $(1/2)^2$ and expanding the remaining binomial, we get

$$T_1^2 + 2T_1 T_2 + T_2^2 < 4T_1 T_2 \tag{2.3.6c}$$

or

$$T_1^2 - 2T_1T_2 + T_2^2 < 0 \tag{2.3.6d}$$

from which it follows that

$$(T_1 - T_2)^2 < 0 \tag{2.3.6e}$$

Since T_1 and T_2 are both real numbers, the square of their difference can never be negative. Hence, the assumption in Eq. (2.3.6a) must be false. We thus conclude that

$$T_{com}^2 > T_1T_2. \quad \blacksquare \tag{2.3.6f}$$

One may note that $T_{com} = (1/2)(T_1 + T_2)$ is the arithmetic mean of T_1 and T_2, while $\sqrt{T_1T_2}$ is the geometric mean. We have thus proved that the arithmetic mean exceeds the geometric mean. Moreover, by virtue of Eq. (2.3.5), we have also shown that for a permitted process, $\Delta S_{total} > 0$.

In the case of the process defined by the energy conservation condition expressed by Eq. (2.2.12) and illustrated in Fig. 2.4, the two contributions to the total entropy are

$$\Delta S_1 = C_V \int_{T_{com}}^{T_1} \frac{dT}{T} = -C_V \ln\left(\frac{T_{com}}{T_1}\right) \tag{2.3.7a}$$

and

$$\Delta S_2 = C_V \int_{T_{com}}^{T2} \frac{dT}{T} = -C_V \ln\left(\frac{T_{com}}{T_2}\right) \tag{2.3.7b}$$

respectively. The sum is

$$\Delta S_{total} = \Delta S_1 + \Delta S_2 = -C_V \ln\left(\frac{T_{com}^2}{T_1T_2}\right) < 0 \tag{2.3.8}$$

Hence, for the forbidden process shown in Fig. 2.4, we find $\Delta S_{total} < 0$.

It is a general result that the total entropy change is positive in a permitted process, whereas it is negative in a forbidden process. Accordingly, a process at thermal equilibrium corresponds to $T_1 = T_2 = T_{com}$ for which $\Delta S_{total} = 0$.

2.4 Entropy of Mixing of Ideal Gases

Consider a container which has a fixed volume and is thermally isolated from its surroundings. See Fig. 2.5.

Since the molecules in an ideal gas do not exert intermolecular forces on one another, the internal energy in Fig. 2.5 is the same after mixing of the two gases as it was before. Expressed in terms of the first law of thermodynamics, this statement implies

$$dU = dQ - PdV = 0 \tag{2.4.1}$$

The entropy change associated with mixing can be calculated by using Eq. (2.4.1) to evaluate Eq. (2.3.1). The result is

$$dS = \frac{dQ}{T} = \frac{PdV}{T} \tag{2.4.2}$$

The ideal gas law states that

$$P = \frac{nRT}{V} \tag{2.4.3}$$

where n is the number of moles of gas, and R is the gas law constant. This equation can be used to replace P on the right hand side of Eq. (2.4.2) with the result,

$$dS = nR\frac{dV}{V} \tag{2.4.4}$$

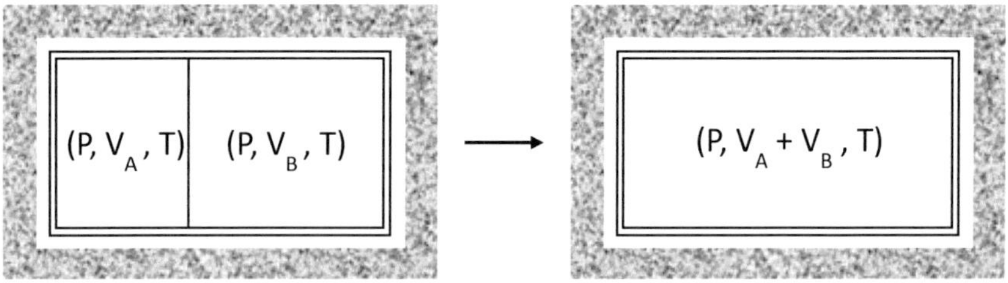

Fig. 2.5 A container with a thin partition dividing it into two compartments having volumes V_A and V_B, respectively. In the compartment of volume, V_A, there are n_A moles of ideal gas, A, while in the compartment of volume, V_B, there are n_B moles of ideal gas, B. The pressure, P, is the same on both sides of the partition as is the temperature, T. When the partition is removed, the two gases intermix by diffusion so that at equilibrium, each gas occupies the total volume, $V_A + V_B$. The container has rigid walls and is thermally isolated from its surroundings

According to Eq. (2.4.4), the change in entropy associated with the isothermal expansion of the n_A moles of ideal gas A from volume V_A to volume $V_A + V_B$ is

$$\Delta S_A = n_A R \int_{V_A}^{V_A+V_B} \frac{dV}{V} = n_A R \ln\left(\frac{V_A + V_B}{V_A}\right) \tag{2.4.5a}$$

while the change in entropy in the isothermal expansion of the n_B moles ideal gas B from volume V_B to volume $V_A + V_B$ is

$$\Delta S_B = n_B R \int_{V_B}^{V_A+V_B} \frac{dV}{V} = n_B R \ln\left(\frac{V_A + V_B}{V_B}\right) \tag{2.4.5b}$$

The total entropy change is

$$\Delta S_{total} = \Delta S_A + \Delta S_B \tag{2.4.6}$$

Upon substitution of Eqs. (2.4.5a) and (2.4.5b), Eq. (2.4.6) can be written in the form

$$\Delta S_{total} = n_A R \ln\left(\frac{V_A + V_B}{V_A}\right) + n_B R \ln\left(\frac{V_A + V_B}{V_B}\right) \tag{2.4.7}$$

The initial volume of each gas can be calculated using Eq. (2.4.3) with the results,

$$V_A = \frac{n_A RT}{P} \tag{2.4.8a}$$

and

$$V_B = \frac{n_B RT}{P} \tag{2.4.8b}$$

respectively. When Eqs. (2.4.8a) and (2.4.8b) are substituted into Eq. (2.4.7), the result is

$$\Delta S_{total} = n_A R \ln\left(\frac{n_A + n_B}{n_A}\right) + n_B R \ln\left(\frac{n_A + n_B}{n_A}\right) \tag{2.4.9}$$

The mole fractions, X_A and X_B, are defined by

$$X_A = \frac{n_A}{n_A + n_B} \tag{2.4.10a}$$

and

$$X_B = \frac{n_B}{n_A + n_B} \tag{2.4.10b}$$

respectively, where

$$X_A + X_B = 1 \tag{2.4.11}$$

Upon substitution of Eqs. (2.4.10a) and (2.4.10b) into Eq. (2.4.9), we obtain for the total entropy change,

$$\Delta S_{total} = -n_A R \ln X_A - n_B R \ln X_B \tag{2.4.12}$$

To obtain a result that does not depend upon the size of the system, we divide ΔS_{total} by the total number of moles, $n_A + n_B$, and define the molar entropy of mixing, $\Delta_{mix} S$, by

$$\Delta_{mix} S = \frac{\Delta S_{total}}{n_A + n_B} \tag{2.4.13}$$

Substitution of Eq. (2.4.12) into Eq. (2.4.13) followed by use of Eqs. (2.4.10a) and (2.4.10b) leads to the result,

$$\Delta_{mix} S = -R(X_A \ln X_A + X_B \ln X_B) \tag{2.4.14}$$

for the molar entropy of mixing of two ideal gases.

2.5 The Free Energies and the Enthalpy

An equation which combines the first and second laws can be obtained by substituting Eq. (2.3.1) into Eq. (2.2.7). The result is

$$dU = TdS - PdV \tag{2.5.1}$$

This equation is associated with the name of Rudolf Clausius (German, 1822–1888). By virtue of its dependence on the differentials, dS, and dV, Eq. (2.5.1) implies that the internal energy is a function of the entropy and the volume; namely,

$$U = U(S, V). \tag{2.5.2}$$

Whereas the volume of a system is readily measured, the direct determination of the entropy is not so easy. It thus makes sense to recast the Clausius equation in terms of a pair of more compatible variables, say the volume and the temperature. To make this transformation, we subtract the total differential of TS from both sides of Eq. (2.5.1). The result is

$$dU - d(TS) = TdS - PdV - TdS - SdT = -SdT - PdV \tag{2.5.3}$$

The left hand side is the differential of

$$A = U - TS \tag{2.5.4}$$

where A is the Helmholtz free energy introduced by Hermann Helmholtz (German, 1821–1894). According to Eqs. (2.5.3) and (2.5.4),

$$dA = -SdT - PdV \tag{2.5.5}$$

Equation (2.5.5) indicates that

$$A = A(T, V) \tag{2.5.6}$$

where the independent variables are the temperature and the volume.

Since it is useful to have a second free energy function, which depends upon the pressure instead of the volume, we can transform from V to P by adding the exact differential of PV to both sides of Eq. (2.5.5). The result is

$$dA + d(PV) = -SdT - PdV + PdV + VdP = -SdT + VdP \tag{2.5.7}$$

The new free energy function,

$$G = A + PV \tag{2.5.8}$$

is due to J. Willard Gibbs (American, 1839–1903). With this identification, Eq. (2.5.7) becomes

$$dG = -SdT + VdP \tag{2.5.9}$$

which implies that

$$G = G(T, P) \tag{2.5.10}$$

Still another useful function, called the enthalpy, can be generated by adding the total differential of PV to both sides of Eq. (2.5.1). The result is

$$dU + d(PV) = TdS - PdV + PdV + VdP = TdS + VdP \tag{2.5.11}$$

The left hand side of Eq. (2.5.11) defines the differential of the enthalpy, H, where

$$H = U + PV \tag{2.5.12}$$

We can use Eq. (2.5.12) to rewrite Eq. (2.5.11) as

$$dH = TdS + VdP \tag{2.5.13}$$

which implies that

$$H = H(S, P) \tag{2.5.14}$$

Next use Eq. (2.5.4) to write Eq. (2.5.8) in the form,

$$G = A + PV = U - TS + PV \tag{2.5.15}$$

Another representation of G results upon substitution of Eq. (2.5.12) into Equation (2.5.15). The result is

$$G = H - TS \tag{2.5.16}$$

According to Eq. (2.5.10), the total differential of G can be written as

$$dG = \left(\frac{\partial G}{\partial T}\right)_P dT + \left(\frac{\partial G}{\partial P}\right)_T dP \tag{2.5.17}$$

Term by term comparison of Eq. (2.5.17) with Eq. (2.5.9) yields the following results:

$$\left(\frac{\partial G}{\partial T}\right)_P = -S \tag{2.5.18a}$$

$$\left(\frac{\partial G}{\partial P}\right)_T = V \tag{2.5.18b}$$

We now consider two states of a sample of matter at the same temperature and pressure. The differences in their Gibbs free energies, entropies, and enthalpies are ΔG, ΔS, and ΔH, respectively. According to Eq. (2.5.16), these differences satisfy

$$\Delta G = \Delta H - T\Delta S \tag{2.5.19a}$$

By virtue of Eq. (2.5.18a),

$$\left(\frac{\partial \Delta G}{\partial T}\right)_P = -\Delta S \tag{2.5.19b}$$

Now use the product rule of differentiation to compute the derivative,

$$\left(\frac{\partial(\Delta G/T)}{\partial T}\right)_P = \frac{1}{T}\left(\frac{\partial \Delta G}{\partial T}\right) - \frac{\Delta G}{T^2} \tag{2.5.19c}$$

Upon substitution of Eqs. (2.5.19a) and (2.5.19b), the evaluation of Eq. (2.5.19c) leads to

$$\left(\frac{\partial(\Delta G/T)}{\partial T}\right)_P = \frac{-\Delta S}{T} - \frac{\Delta H - T\Delta S}{T^2} \tag{2.5.19d}$$

which when simplified reads,

$$\left(\frac{\partial(\Delta G/T)}{\partial T}\right)_P = \frac{-\Delta H}{T^2} \tag{2.5.20}$$

Equation (2.5.20) is known as the Gibbs–Helmholtz equation. Equation (2.5.20) will prove useful in Chap. 3 when computing the electrostatic contribution to the solvation energy of an ion.

2.6 Gibbs Free Energy for the Pure Ideal Gas and the Pure Solid

At constant temperature, the differential of the Gibbs energy in Eq. (2.5.9) assumes the form,

$$dG = VdP \tag{2.6.1}$$

If we assume an ideal gas equation of state,

$$V = \frac{nRT}{P} \tag{2.6.2}$$

then substitution of Eq. (2.6.2) into Eq. (2.6.1) has the result,

$$dG = nRT\frac{dP}{P} \tag{2.6.3}$$

The differential equation represented by Eq. (2.6.3) gives the change, dG, in the Gibbs energy, when at constant temperature the gas pressure changes by the amount, dP. It does not tell us the value of G at any one pressure, however. To integrate Eq. (2.6.3) and determine G at some temperature and pressure, we will need an initial condition. This condition is supplied by an experimental measurement of the free energy of the gas under arbitrary, but agreed upon conditions of pressure and temperature. These conditions are said to define "the standard state". The usual standard state pressure is $P^o = 1$ bar. We will denote the value of G at $P^o = 1$ bar as $G^o(T)$ and for the moment let the temperature be arbitrary. With this initial condition, the integral of Eq. (2.6.3) takes the form,

$$\int_{G^o(T)}^{G(T,P)} dG = nRT \int_{P^o}^{P} \frac{dP}{P} \tag{2.6.4}$$

Evaluation of the integrals in Eq. (2.6.4) leads to the result,

$$G(T,P) = G^o(T) + nRT \ln\left(\frac{P}{P^o}\right) \tag{2.6.5}$$

The "constant" of integration, $G^o(T)$, is still a function of temperature. When evaluated at 298.15 K, it is proportional to the standard Gibbs energies of formation, $\Delta_f G^o$, for the gas. Values of standard Gibbs energies of formation are listed in the thermodynamic function tables in physical chemistry textbooks. These tables are restricted to one mole of material, however, which implies that $G^o(298.15) = n\Delta_f G^o$, where n is the number of moles. The molar Gibbs free energy $\mu = G/n$ is known as the chemical potential. According to Eq. (2.6.5), the chemical potential of an ideal gas is given by

$$\mu(T,P) = \mu^o(T) + RT \ln\left(\frac{P}{P^o}\right) \tag{2.6.6}$$

where $\mu^o = \Delta_f G^o$. Since the data in the standard thermodynamic function tables are specified at a temperature of 298.15 K, the numerical applications of Eqs. (2.6.5) and (2.6.6) are likewise restricted to this temperature.

It is advantageous to retain the logarithmic form in Eqs. (2.6.5) and (2.6.6) even for substances that do not obey the ideal gas law equation of state. This is done by defining an otherwise arbitrary function, a, known as the *activity*, which satisfies,

$$\mu(T,P) = \mu^o(T) + RT \ln a \tag{2.6.7}$$

In the case of an ideal gas, $a = P/P^o$, but in the case of liquids, solids, and solutions, the temperature, pressure, and composition dependence of a must be determined by experiment or by using an empirical equation of state. In order to determine a as a function of changes in pressure, the differential of Eq. (2.6.7) reads

$$d\mu(T,P) = RTd \ln a \tag{2.6.8}$$

If we consider $\mu = G/n$ with n fixed, Eq. (2.6.1) reads

$$d\mu = \overline{V} dP \tag{2.6.9}$$

where $\overline{V} = V/n$ is the molar volume. Equating Eqs. (2.6.8) and (2.6.9), we get

$$RTd\ln a = \overline{V}dP \qquad (2.6.10)$$

In the case of most pure solids and liquids, experiments at pressures near atmospheric show that the molar volume is nearly independent of the pressure. This observation permits us to integrate Eq. (2.6.10) from the standard state pressure, P^o, where by definition, $a = 1$, to the pressure, P, where the chemical potential is $a = a(T, P)$. At constant temperature, the integral of Eq. (2.6.10) is

$$RT\int_1^a d\ln a = \overline{V}\int_{P^o}^P dP \qquad (2.6.11)$$

After integration and division by RT, the result is

$$\ln a = \frac{P^o\overline{V}}{RT}\left[\frac{P}{P^o} - 1\right] \qquad (2.6.12)$$

The form of Eq. (2.6.12) includes explicitly the condition, $a = 1$ when $P = P^o$.

2.7 Chemical Potential of a Solute

According to Eq. (2.5.2), the internal energy of a pure material is a function of the entropy and the volume. Anyone who has poured concentrated sulfuric acid into water to make a solution is well aware of the heat that is released as the acid is added. This observation indicates that in the case of solutions, the internal energy is a function of the concentrations of the components. For the sake of simplicity, we will consider a mixture of just two components, denoted as substances (1) and (2) with mole numbers, n_1 and n_2, respectively. To take into account the effect of changes in composition, Eq. (2.5.1) is extended so as to include the changes, dn_1 and dn_2, in the numbers of moles of the components. The resulting change, dU, in the internal energy, U, is achieved by multiplying dn_1 and dn_2 by the chemical potentials, μ_1 and μ_2, respectively, and adding the resulting products to Eq. (2.5.1). The result is

$$dU = TdS - PdV + \mu_1 dn_1 + \mu_2 dn_2 \qquad (2.7.1)$$

It should be noted that when Eq. (2.7.1) is applied to a solution, the pure component relations, $\mu_1 = G_1/n_1$ and $\mu_2 = G_2/n_2$, no longer apply. Rather μ_1 and μ_2 must be calculated so as to reflect the specific behaviors exhibited by components in solution.

The chemical potential of a solute in the solution can be represented as a function of its activity by integrating Eq. (2.6.8). Thus, the chemical potential of solute component, i, is

$$\mu_i = \mu_i^o(T) + RT \ln a_i \quad i = 1, 2 \tag{2.7.2}$$

where in Eq. (2.7.2), a_i is the activity, and μ_i^o is the chemical potential in the standard state of a solute. In the case of a solute, the activity has the definition,

$$a_i = \gamma_i (m_i/m^o) \tag{2.7.3a}$$

where m_i is the molality, $m^o = 1$ molal is the concentration in the standard state, and γ_i, is the *activity coefficient*, which is a function of concentration. Upon substitution of Eq. (2.7.3a) into Eq. (2.7.2), one obtains

$$\mu_i = \mu_i^o(T) + RT \ln(m_i/m^o) + RT \ln \gamma_i \tag{2.7.3b}$$

The values of $\mu_i^o(T)$ and γ_i are in general determined by experiment. Figure 2.6 summarizes measurements of μ_i for a solute as a function of $\ln(m_i/m^o)$.

In the limit that m_i approaches zero, γ_i is unity, and according to Eq. (2.7.3b), μ_i varies linearly with $\ln(m_i/m^o)$. When this occurs, the mixture is said to form a dilute ideal solution. According to Eq. (2.7.3b) with $\gamma_i = 1$, the intercept on the μ_i axis

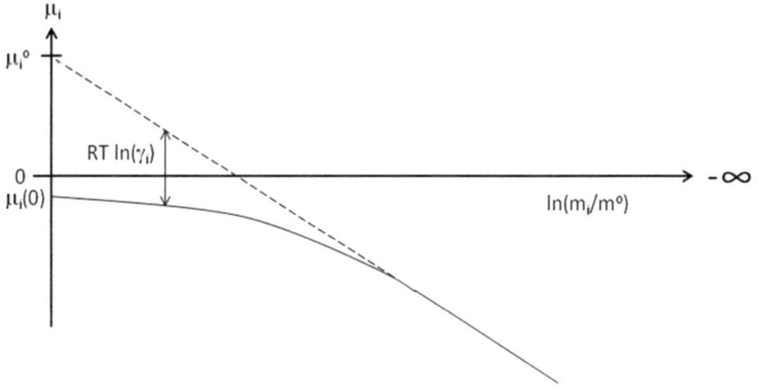

Fig. 2.6 Plot of the chemical potential, μ_i of solute, i, as a function of $\ln(m_i/m^o)$ for the concentration range where $m_i \leq m^o$. The temperature is T and the gas law constant is R. The value of m_i is decreasing from left to right. The solid line represents the experimental data. The dashed straight line represents the chemical potential of an ideal solution corresponding to $\gamma_i = 1$. In the dilute solution regime where $m_i \ll m^o$, the experimental data conform to the dashed straight line. The intercept of the dashed straight line on the ordinate at the concentration, $m_i = m^o$, is μ_i^o, the chemical potential of i in the standard state. The difference between the dashed and solid lines at any given concentration equals $RT \ln \gamma_i$. The intercept made by the solid line on the ordinate at concentration, $m_i = m^o$, is $\mu_i(0)$, which corresponds to the actual chemical potential of i at $m_i = m^o$

of a plot of μ_i vs. $\ln(m_i/m^o)$ is μ_i^o. To determine the value of μ_i^o experimentally, the measurements of μ_i obtained in the dilute solution range, where $\gamma_i = 1$, are extrapolated to $\ln(m_i/m^o) = 0$ (i.e. $m_i = m^o$). It should be noted that the value of μ_i^o depends upon both the solute and the solvent. The values of the standard state Gibbs free energy of formation, $\Delta_f G^o(solute, solvent)$, for various solutions are tabulated in the thermodynamic function tables of physical chemistry textbooks. In the notation of Eq. (2.7.3b), $\mu_i^o(298.15) = \Delta_f G^o(solute, solvent)$. In Chap. 6, we will use the laws of electrostatics to evaluate γ_i in the case of strong electrolytes.

We can generalize Eq. (2.5.9) to a mixture of two components by writing it as

$$dG = -SdT + VdP + \mu_1 dn_1 + \mu_2 dn_2 \tag{2.7.4}$$

Equation (2.7.4) implies that $G = G(T, P, n_1, n_2)$. Being an analytic function of its variables, the total differential of $G(T, P, n_1, n_2)$ can be written in the form

$$dG = \left(\frac{\partial G}{\partial T}\right)_{P,n_1,n_2} dT + \left(\frac{\partial G}{\partial P}\right)_{T,n_1,n_2} dP + \left(\frac{\partial G}{\partial n_1}\right)_{T,P,n_2} dn_1 + \left(\frac{\partial G}{\partial n_2}\right)_{T,P,n_1} dn_2 \tag{2.7.5}$$

A term by term comparison of Eqs. (2.7.4) and (2.7.5), allows us to make the following identifications,

$$S = -\left(\frac{\partial G}{\partial T}\right)_{P,n_1,n_2} \tag{2.7.6a}$$

$$V = \left(\frac{\partial G}{\partial P}\right)_{T,n_1,n_2} \tag{2.7.6b}$$

$$\mu_1 = \left(\frac{\partial G}{\partial n_1}\right)_{T,P,n_2} \tag{2.7.6c}$$

$$\mu_2 = \left(\frac{\partial G}{\partial n_2}\right)_{T,P,n_1} \tag{2.7.6d}$$

Equations (2.7.6c) and (2.7.6d) are particularly important, because they serve to define the chemical potentials of the two components forming the mixture.

2.8 Gibbs–Duhem Equation

We next derive the famous equation describing mixtures that was proposed by J. W. Gibbs and Pierre Duhem (French, 1861–1916). The Gibbs–Duhem equation will prove useful in developing the theory of the concentration fluctuations, which are responsible for the light scattering behavior of macromolecular solutions.

Consider a mixture of two components. The total differential of the internal energy is

$$dU = \left(\frac{\partial U}{\partial S}\right)_{V,n_1,n_2} dS + \left(\frac{\partial U}{\partial V}\right)_{S,n_1,n_2} dV + \left(\frac{\partial U}{\partial n_1}\right)_{S,V,n_2} dn_1 + \left(\frac{\partial U}{\partial n_2}\right)_{S,V,n_1} dn_2 \tag{2.8.1}$$

A term by term comparison of Eqs. (2.7.1) and (2.8.1) leads to

$$T = \left(\frac{\partial U}{\partial S}\right)_{V,n_1,n_2} \tag{2.8.2a}$$

$$-P = \left(\frac{\partial U}{\partial V}\right)_{S,n_1,n_2} \tag{2.8.2b}$$

$$\mu_1 = \left(\frac{\partial U}{\partial n_1}\right)_{S,V,n_2} \tag{2.8.2c}$$

$$\mu_2 = \left(\frac{\partial U}{\partial n_2}\right)_{S,V,n_1} \tag{2.8.2d}$$

The internal energy depends upon the size of the system. If we double the number of moles of each component, the volume will double, as will the entropy, and the internal energy. In general, if we replace n_1 and n_2 by λn_1 and λn_2, respectively, where λ is a real positive number, we will need to replace the volume, V, by λV, the entropy, S by λS, and the energy U by λU. In mathematical terms, the process can be written as

$$\lambda U(S, V, n_1, n_2) = U(\lambda S, \lambda V, \lambda n_1, \lambda n_2) \tag{2.8.3}$$

A function which satisfies Eq. (2.8.3) is said to be homogeneous. The differential properties of this type of function were investigated by the mathematician, Leonhard Euler (Swiss, 1707–1783). Use of Eq. (2.8.3) and Euler's theorem leads to a special representation of U in terms of its partial derivatives. This relation is found by writing the derivative of both sides of Eq. (2.8.3) with respect to λ as,

$$\frac{\partial}{\partial \lambda}[\lambda U(S, V, n_1, n_2)] = \frac{\partial}{\partial \lambda} U(\lambda S, \lambda V, \lambda n_1, \lambda n_2) \tag{2.8.4a}$$

The derivative on the left hand side of Eq. (2.8.4a) is elementary. The result is

$$\frac{\partial}{\partial \lambda}[\lambda U(S, V, n_1, n_2)] = U(S, V, n_1, n_2) \tag{2.8.4b}$$

The derivative on the right hand side Eq. (2.8.4a) can be evaluated by using the chain rule. This result is

$$\frac{\partial}{\partial\lambda}U(\lambda S, \lambda V, \lambda n_1, \lambda n_2) = \left(\frac{\partial U}{\partial(\lambda S)}\right)\left(\frac{\partial(\lambda S)}{\partial\lambda}\right) + \left(\frac{\partial U}{\partial(\lambda V)}\right)\left(\frac{\partial(\lambda V)}{\partial\lambda}\right) + \left(\frac{\partial U}{\partial(\lambda n_1)}\right)\left(\frac{\partial(\lambda n_1)}{\partial\lambda}\right) + \left(\frac{\partial U}{\partial(\lambda n_2)}\right)\left(\frac{\partial(\lambda n_2)}{\partial\lambda}\right) \tag{2.8.4c}$$

On the right hand side of Eq. (2.8.4c), we have suppressed mention of all variables other than the one being differentiated. Substitute Eqs. (2.8.4b) and (2.8.4c) into Eq. (2.8.4a) to obtain

$$U(S, V, n_1, n_2) = \left(\frac{\partial U(\lambda S)}{\partial(\lambda S)}\right)S + \left(\frac{\partial U(\lambda V)}{\partial(\lambda V)}\right)V + \left(\frac{\partial U(\lambda n_1)}{\partial(\lambda n_1)}\right)n_1 + \left(\frac{\partial U(\lambda n_2)}{\partial(\lambda n_2)}\right)n_2 \tag{2.8.4d}$$

Equation (2.8.4d) must hold for all values of λ, including $\lambda = 1$. Upon setting $\lambda = 1$, the result with all variables written out in full is

$$U(S, V, n_1, n_2) = \left(\frac{\partial U(S, V, n_1, n_2)}{\partial S}\right)S + \left(\frac{\partial U(S, V, n_1, n_2)}{\partial V}\right)V + \left(\frac{\partial U(S, V, n_1, n_2)}{\partial n_1}\right)n_1 + \left(\frac{\partial U(S, V, n_1, n_2)}{\partial n_2}\right)n_2 \tag{2.8.5}$$

Equation (2.8.5) is a statement of Euler's theorem for homogeneous functions and constitutes the integral of Eq. (2.7.1).

Next substitute Eqs. (2.8.2a) through (2.8.2d) into Eq. (2.8.5). The result is

$$U = TS - PV + \mu_1 n_1 + \mu_2 n_2 \tag{2.8.6}$$

We can obtain an equivalent formula for the Helmholtz free energy by rearranging Eq. (2.8.6) to read

$$A = U - TS = -PV + \mu_1 n_1 + \mu_2 n_2 \tag{2.8.7}$$

A formula for the Gibbs free energy can be obtained by writing Eq. (2.8.7) in the form,

$$G = A + PV = \mu_1 n_1 + \mu_2 n_2 \tag{2.8.8}$$

Equation (2.8.8) shows that the Gibbs energy of a mixture is a simple linear combination of the chemical potentials of its components.

The total differential of Eq. (2.8.6) is

$$dU = TdS - PdV + \mu_1 dn_1 + \mu_2 dn_2 + SdT - VdP + n_1 d\mu_1 + n_2 d\mu_2 \quad (2.8.9)$$

If Eqs. (2.7.1) and (2.8.9) are to be consistent, we must have

$$SdT - VdP + n_1 d\mu_1 + n_2 d\mu_2 = 0 \quad (2.8.10)$$

Equation (2.8.10) is known as the Gibbs–Duhem equation.

2.9 Statistical Mechanics in the Canonical Ensemble

As mentioned in the introduction to this chapter, statistical mechanics is a mathematical scheme for calculating the macroscopic thermodynamic properties of a material from the properties of the individual molecules which it contains. The experimentally observable quantities of a macroscopic sample of matter are represented in terms of statistical mean values calculated over microscopic subdivisions of the sample. A large number of molecules are involved in the averaging process. Due to the large numbers involved, the statistical variance about the mean is ordinarily very small.

This ambitious program begins with the division of a homogeneous sample of matter into smaller pieces called *systems*. The collection of systems is termed an *ensemble*. For mathematical simplicity, we shall consider an ensemble consisting of just three systems. Each system has the same fixed volume, contains the same number of molecules, and is at the same temperature. An ensemble with these restrictions is said to be *canonical* and is illustrated in Fig. 2.7.

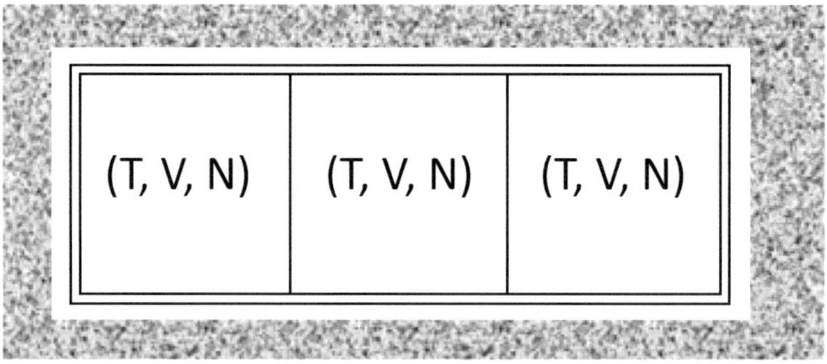

Fig. 2.7 Subdivision of a sample of matter into three systems, each having the same temperature, T, the same volume, V, and the same number of molecules, N. Subdivided in this fashion, the three systems are said to form a canonical ensemble. The sample of matter is in a container with rigid walls, which are thermally isolated from its surroundings. The solid line partitions separating the systems are assumed to be permeable to the flow of heat

Let P_1 be the probability of choosing system number *one* for observation, P_2 be the probability for choosing system number *two*, and P_3 be the probability of choosing system number *three*. According to Gibbs, the entropy content of an ensemble consisting of these three systems is

$$S = -k_B P_1 \ln P_1 - k_B P_2 \ln P_2 - k_B P_3 \ln P_3 \tag{2.9.1}$$

where k_B is Boltzmann's constant (Ludwig Boltzmann, Austrian, 1844–1906). Equation (2.9.1) is reminiscent of Eq. (2.4.14) for the entropy of mixing of ideal gases. Whereas the terms in Eq. (2.4.14) are on the molar scale with each proportional to gas law constant R, the terms in Eq. (2.9.1) are on the molecular scale with each proportional to $k_B = R/N_A$, where N_A is the Avogadro number (Amadeo Avogadro, Italian, 1776–1856). The analogy between the mole fractions, $\{X_i\}$, in Eq. (2.4.14) and the probabilities, $\{P_i\}$ in Eq. (2.9.1) is made complete by noting the sum of the probabilities is

$$P_1 + P_2 + P_3 = 1 \tag{2.9.2}$$

The internal energies of the systems are u_1, u_2, and u_3, respectively. The internal energy, U, characteristic of the ensemble is the weighted sum,

$$U = P_1 u_1 + P_2 u_2 + P_3 u_3 \tag{2.9.3}$$

If the ensemble is isolated from its surroundings, then according to the second law of thermodynamics, the probabilities, P_1, P_2, and P_3, must assume values which maximize S. In general, the location of the corresponding extremum is determined by setting $(\partial S/\partial P_1) = (\partial S/\partial P_2) = (\partial S/\partial P_3) = 0$ and solving the resulting equations for P_1, P_2, and P_3. In the case of the canonical ensemble, however, the values of P_1, P_2, and P_3, that maximize S must also satisfy Eqs. (2.9.2) and (2.9.3). These two equations are said to serve as "constraint" equations. In cases where constraints are involved, the method of Lagrange multipliers (Joseph–Louis Comte de Lagrange, French 1736–1813) can be used to find the location of the extremum. In applying the Lagrange method, one begins by combining the entropy and the constraints in the form of the auxiliary function, $F(P_1, P_2, P_3)$, where

$$\begin{aligned} F(P_1, P_2, P_3) = {} & S(P_1, P_2, P_3) + \alpha(U - P_1 u_1 - P_2 u_2 - P_3 u_3) \\ & + \gamma(1 - P_1 - P_2 - P_3) \end{aligned} \tag{2.9.4}$$

In Eq. (2.9.4), the coefficients, α and γ, are arbitrary parameters, known as "Lagrange multipliers", whose values can be determined as required. By virtue of Eqs. (2.9.2) and (2.9.3), the second and third terms on the right hand side of Eq. (2.9.4) are identically zero. Thus, the extremum of F is located at the same point as the extremum of S. Substitute Eq. (2.9.1) into Eq. (2.9.4) to obtain

$$F(P_1, P_2, P_3) = -k_B P_1 \ln P_1 - k_B P_2 \ln P_2 - k_B \ln P_3 \\ + \alpha(U - P_1 u_1 - P_2 u_2 - P_3 u_3) + \gamma(1 - P_1 - P_2 - P_3) \quad (2.9.5)$$

According to Eq. (2.9.5), the three partial derivatives of F with respect to the probabilities are

$$\frac{\partial F}{\partial P_i} = -k_B\left(\ln P_i + \frac{P_i}{P_i}\right) - \gamma - \alpha u_i = 0 \quad (i = 1, 2, 3) \quad (2.9.6)$$

The general solution to Eq. (2.9.6) is of the form,

$$k_B \ln P_i = -k_B - \gamma - \alpha u_i \quad (i = 1, 2, 3) \quad (2.9.7a)$$

or

$$P_i = \exp\left(-\frac{k_B + \gamma}{k_B}\right)\exp\left(-\frac{\alpha u_i}{k_B}\right) \quad (i = 1, 2, 3) \quad (2.9.7b)$$

Equations (2.9.2), (2.9.3), and (2.9.7a) constitute five equations expressed in terms of the five unknowns, P_1, P_2, P_3, α, and γ. In order to assign interpretations to the multipliers, α, and γ, we require Eq. (2.9.7b) to be consistent with the predictions of thermodynamics. To begin this analysis, compute the total differential of Eq. (2.9.1), which is

$$dS = -k_B[\ln P_1 + 1]dP_1 - k_B[\ln P_2 + 1]dP_2 - k_B[\ln P_3 + 1]dP_3 \\ = -k_B \ln P_1 dP_1 - k_B \ln P_2 dP_2 - k_B \ln P_3 dP_3 - k_B[dP_1 + dP_2 + dP_3] \quad (2.9.8)$$

By virtue of Eq. (2.9.2) the total differential,

$$dP_1 + dP_2 + dP_3 = 0 \quad (2.9.9)$$

Equation (2.9.8) thus simplifies to read

$$dS = -k_B \ln P_1 dP_1 - k_B \ln P_2 dP_2 - k_B \ln P_3 dP_3 \quad (2.9.10)$$

Upon substitution of Eq. (2.9.7a), Eq. (2.9.10) becomes

$$dS = (k_B + \gamma + \alpha u_1)dP_1 + (k_B + \gamma + \alpha u_2)dP_2 + (k_B + \gamma + \alpha u_3)dP_3 \\ = (k_B + \gamma)(dP_1 + dP_2 + dP_3) + \alpha(u_1 dP_1 + u_2 dP_2 + u_3 dP_3) \quad (2.9.11)$$

By invoking Eq. (2.9.9) again, Eq. (2.9.11) simplifies to read

$$dS = \alpha(u_1 dP_1 + u_2 dP_2 + u_3 dP_3) \quad (2.9.12)$$

The total differential of Eq. (2.9.3) is

$$dU = u_1 dP_1 + u_2 dP_2 + u_3 dP_3 \tag{2.9.13}$$

We are now ready to compare the predictions of the canonical ensemble with those of thermodynamics. A representation of α in terms of a thermodynamic derivative is obtained by dividing Eq. (2.9.12) by (2.9.13). The result is

$$\left(\frac{\partial S}{\partial U}\right) = \alpha \tag{2.9.14}$$

This same derivative can be calculated in thermodynamics. Rearrange Eq. (2.5.1) to read

$$dS = \frac{dU}{T} + \frac{P}{T}dV \tag{2.9.15}$$

Since Eq. (2.9.15) implies that $S = S(U, V)$, we can write the total differential of S in the form

$$dS = \left(\frac{\partial S}{\partial U}\right)dU + \left(\frac{\partial S}{\partial V}\right)dV \tag{2.9.16}$$

By comparing the right hand sides of Eqs. (2.9.15) and (2.9.16), we conclude

$$\left(\frac{\partial S}{\partial U}\right) = \frac{1}{T} \tag{2.9.17}$$

Hence, by virtue of (2.9.14) and (2.9.17),

$$\alpha = \frac{1}{T} \tag{2.9.18}$$

which serves to give a thermal interpretation to the Lagrange multiplier, α.

To evaluate γ, substitute Eq. (2.9.7b) into Eq. (2.9.2). The result is

$$1 = P_1 + P_2 + P_3 = \exp\left(-\frac{k_B + \gamma}{k_B}\right)\left[\exp\left(\frac{-u_1}{k_B T}\right) + \exp\left(\frac{-u_2}{k_B T}\right) + \exp\left(\frac{-u_3}{k_B T}\right)\right] \tag{2.9.19}$$

where we have used Eq. (2.9.18). Next introduce the sum,

$$Z = \sum_{j=1}^{3} \exp\left(-\frac{u_j}{k_B T}\right) \tag{2.9.20}$$

The sum in Eq. (2.9.20) serves to define the canonical partition function Z. On the basis of Eq. (2.9.19),

$$\exp\left(-\frac{k_B+\gamma}{k_B}\right)=\frac{1}{Z} \tag{2.9.21}$$

Equation (2.9.21) determines the Lagrange multiplier, γ. Equations (2.9.18) and (2.9.21) can be substituted into Eq. (2.9.7b) to obtain

$$P_i=\frac{e^{-u_i/k_BT}}{Z} \tag{2.9.22}$$

Equation (2.9.22) is an expression for the probabilities, P_i, $(i=1,2,3)$, in terms of the energies, u_i, $(i=1,2,3)$ and the temperature, T. The exponential in the numerator of the fraction in Eq. (2.9.22) is known as the Boltzmann factor.

Although the probabilities, themselves, can be used to calculate the thermal properties of macroscopic samples of matter, the Helmholtz free energy proves more useful when the predictions of statistical mechanics are to be compared with thermodynamics. To obtain a statistical mechanical formula for A, we begin with the entropy as expressed by Eq. (2.9.1) and substitute Eq. (2.9.7a). The result is

$$\begin{aligned}S&=(k_B+\gamma)P_1+\alpha P_1u_1+(k_B+\gamma)P_2+\alpha P_2u_2+(k_B+\gamma)P_3+\alpha P_3u_3\\&=(k_B+\gamma)(P_1+P_2+P_3)+\alpha(P_1u_1+P_2u_2+P_3u_3)\end{aligned} \tag{2.9.23}$$

We can use Eqs. (2.9.2) and (2.9.3) to write Eq. (2.9.23) to the form,

$$S=(k_B+\gamma)+\alpha U \tag{2.9.24a}$$

Use Eq. (2.9.18) to replace α in Eq. (2.9.24a). The latter then reads,

$$S=(k_B+\gamma)+\frac{U}{T} \tag{2.9.24b}$$

which can be rearranged in the form,

$$U-TS=-(k_B+\gamma)T \tag{2.9.24c}$$

Upon use of Eq. (2.5.4), Eq. (2.9.24c) becomes

$$A=-(k_B+\gamma)T \tag{2.9.24d}$$

Solve Eq. (2.9.21) for $(k_B+\gamma)$ to obtain

$$(k_B+\gamma)=k_B\ln Z \tag{2.9.24e}$$

Upon combining Eqs. (2.9.24d) and (2.9.24e), our final result is

$$A = -k_B T \ln Z \tag{2.9.25}$$

Equation (2.9.25) relates the Helmholtz free energy, A, to the canonical partition function, Z.

At this point, we note that the definition of Z in Eq. (2.9.20) applies when the systems making up the ensemble are distinguishable. This is the case in a crystalline solid where the atoms, serving as systems, are located at fixed lattice points. The assumption is inappropriate in the case of a gas or a liquid where the molecules can readily interchange places. Multiplication of the right hand side of Eq. (2.9.20) by $1/N!$ accounts for the indistinguishability of the molecules in the case of gaseous or liquid systems. Since our application of the canonical ensemble will be restricted to gases and liquids, we shall replace Eq. (2.9.20) with

$$Z = \frac{1}{N!}\sum_{j=1}^{3} \exp(-\beta u_j) \tag{2.9.26}$$

where we have also introduced the standard symbol, $\beta = 1/k_B T$. Calculations of thermal properties of a material using the canonical ensemble rest upon the ability of evaluate Z as defined in Eq. (2.9.26). We provide an example below.

Example Canonical Partition Function and Helmholtz Free Energy for Two Identical Particles Interacting by Way of a Potential Energy Function.

The kinetic energy of a particle of mass, m, is $(1/2)mv^2$ where v is the speed of the particle. The linear momentum is $p = mv$. With these identifications, the kinetic energy is $p^2/2m$. For a pair of particles, the total kinetic energy, KE, is

$$KE = \frac{p_1^2}{2m} + \frac{p_2^2}{2m} \tag{2.9.27a}$$

where p_1 and p_2 are the magnitudes of the linear momenta of the two particles. As there is motion in three dimensions, there are three components $p_{1,x}$, $p_{1,y}$, $p_{1,z}$ associated with momentum, $\mathbf{p}_1$, and three components associated with the momentum $\mathbf{p}_2$. In terms of components, the square of the total momentum of particle "1" is

$$p_1^2 = p_{1,x}^2 + p_{1,y}^2 + p_{1,z}^2 \tag{2.9.27b}$$

As similar result holds for p_2. Expanded in terms of components, the kinetic energy in Eq. (2.9.27a) is

$$KE = \frac{p_{1,x}^2 + p_{1,y}^2 + p_{1,z}^2}{2m} + \frac{p_{2,x}^2 + p_{2,y}^2 + p_{2,z}^2}{2m} \tag{2.9.27c}$$

If the locations in space are described by the vectors, $\mathbf{r}_1$ and $\mathbf{r}_2$, respectively, the potential energy of interaction of the pair of particles is given by $PE = U_2(\mathbf{r}_1, \mathbf{r}_2)$. Their total energy, u, is then given by

$$u = KE + PE = \frac{p_{1,x}^2 + p_{1,y}^2 + p_{1,z}^2}{2m} + \frac{p_{2,x}^2 + p_{2,y}^2 + p_{2,z}^2}{2m} + U_2(\mathbf{r}_1, \mathbf{r}_2) \quad (2.9.27\text{d})$$

As there are an infinite number of possible values for the variables, $\mathbf{p}_1$ and $\mathbf{p}_2$, and $\mathbf{r}_1$ and $\mathbf{r}_2$, that are consistent with any given value of u, we are justified in replacing the sum over states in Eq. (2.9.26) by an integral. The components of the position vectors are $\mathbf{r}_1 = (x_1, y_1, z_1)$ and $\mathbf{r}_2 = (x_2, y_2, z_2)$. The corresponding volume elements are $d^3r_1 = dx_1dy_1dz_1$, and $d^3r_2 = dx_2dy_2dz_2$.

To replace the sum in Eq. (2.9.26) by an integral, we need to consider how the Heisenberg uncertainty principle forces a discreteness on the otherwise continuous nature of the space and momentum coordinates. The product of coordinate differentials, $dp_{1,x}dx_1$, is said to be a differential element in a plane, known as the phase plane. According to the Heisenberg uncertainty principle, the lower limit on the product, $dp_{1,x}dx_1$, is provided by the Planck constant, h. Given that the smallest phase plane element has area, h, the number of quantum states in an arbitrary element of area, $dp_{1,x}dx_1$, is necessarily $dp_{1,x}dx_1/h$. To convert the sum in Eq. (2.9.26) to an integral, we must include for each particle all of the states in three such phase planes. For the first particle, the number of states is $(dp_{1,x}dx_1/h)(dp_{1,y}dy_1/h)(dp_{1,z}dz_1/h)$, and for the second particle it is $(dp_{2,x}dx_2/h)(dp_{2,y}dy_2/h)(dp_{2,z}dz_2/h)$. Taking into account all of these states, the resulting integral is

$$\begin{aligned} Z = \frac{1}{h^6 2!} &\int_{-\infty}^{+\infty} dp_{1,x} \int_{-\infty}^{+\infty} dp_{1,y} \int_{-\infty}^{+\infty} dp_{1,z} \\ &\times \int_{-\infty}^{+\infty} dp_{2,x} \int_{-\infty}^{+\infty} dp_{2,y} \int_{-\infty}^{+\infty} dp_{2,z} \\ &\times \exp\left[-\beta\left(\frac{p_{1,x}^2 + p_{1,y}^2 + p_{1,z}^2}{2m} + \frac{p_{2,x}^2 + p_{2,y}^2 + p_{2,z}^2}{2m}\right)\right] \\ &\times \int dx_1 \int dy_1 \int dz_1 \int dx_2 \int dy_2 \int dz_2 e^{-\beta U_2(\mathbf{r}_1, \mathbf{r}_2)} \end{aligned} \quad (2.9.28\text{a})$$

where we have included a factor of $N! = 2!$ in the denominator as prescribed by Eq. (2.9.26). After factoring, Eq. (2.9.28a) reads,

$$Z = \frac{1}{h^6 2!} \int_{-\infty}^{+\infty} dp_{1,x} e^{-\beta p_{1,x}^2/2m} \int_{-\infty}^{+\infty} dp_{1,y} e^{-\beta p_{1,y}^2/2m} \int_{-\infty}^{+\infty} dp_{1,z} e^{-\beta p_{1,z}^2/2m}$$

$$\times \int_{-\infty}^{+\infty} dp_{2,x} e^{-\beta p_{2,x}^2/2m} \int_{-\infty}^{+\infty} dp_{2,y} e^{-\beta p_{2,y}^2/2m} \int_{-\infty}^{+\infty} dp_{2,z} e^{-\beta p_{2,z}^2/2m}$$

$$\times \int d^3 r_1 \int d^3 r_2 e^{-\beta U_2(\mathbf{r}_1, \mathbf{r}_2)} \tag{2.9.28b}$$

To evaluate the integrals over the momenta, we use a standard integral in the Mathematical Methods Appendix. This integral is

$$\frac{1}{h} \int_{-\infty}^{+\infty} dp_{1,x} e^{-\beta p_{1,x}^2/2m} = \frac{1}{h} \sqrt{\frac{\pi}{\beta/2m}} = \sqrt{\frac{2\pi m k_B T}{h^2}} \tag{2.9.28c}$$

Equation (2.9.28b) involves the product of six integrals of the type shown in Eq. (2.9.28c). With the repeated use of Eq. (2.9.28c), Eq. (2.9.28b) reads,

$$Z = \frac{1}{2!} \left(2\pi m k_B T/h^2\right)^{3/2} \left(2\pi m k_B T/h^2\right)^{3/2} \int d^3 r_1 \int d^3 r_2 e^{-\beta U_2(\mathbf{r}_1, \mathbf{r}_2)} \tag{2.9.29a}$$

The remaining multiple integral over the coordinates is termed the "configuration integral", and for the case of two particles it is given the designation,

$$Z_2 = \int d^3 r_1 \int d^3 r_2 e^{-\beta U_2(\mathbf{r}_1, \mathbf{r}_2)} \tag{2.9.29b}$$

With the identification of Z_2 in Eq. (2.9.29b), Eq. (2.9.29a) can be rewritten as

$$Z = \frac{1}{2!} \left(2\pi m k_B T/h^2\right)^{3/2} \left(2\pi m k_B T/h^2\right)^{3/2} Z_2 \tag{2.9.29c}$$

Upon substitution of Eq. (2.9.29c) into Eq. (2.9.25), the result is

$$A = -k_B T \ln\left\{\frac{1}{2!}\left[\left(2\pi m k_B T/h^2\right)^{3/2}\right]^2\right\} - k_B T \ln(Z_2) \tag{2.9.30}$$

We now turn to the case of N particles. Equation (2.9.30) can be generalized to read

$$A = -k_B T \ln\left[\frac{1}{N!}\left(2\pi m k_B T/h^2\right)^{3N/2}\right] - k_B T \ln Z_N \tag{2.9.31a}$$

Expand the first logarithm so that Eq. (2.9.31a) can be written in the form,

$$A = k_B T \ln N! - k_B T \ln\left[\left(2\pi m k_B T/h^2\right)^{3N/2}\right] - k_B T \ln Z_N \tag{2.9.31b}$$

Now use Stirling's asymptotic approximation for large N (see the Mathematical Appendix), which reads,

$$\ln N! \approx N \ln N - N \tag{2.9.31c}$$

With the use of Eq. (2.9.31c), Eq. (2.9.31b) becomes

$$A = k_B TN \ln N - k_B TN - k_B T \ln\left[\left(2\pi m k_B T/h^2\right)^{3N/2}\right] - k_B T \ln Z_N \tag{2.9.31d}$$

Equation (2.9.31d) can then be rewritten as

$$A = k_B TN \ln N - k_B T \ln e^N - k_B T \ln\left[\left(2\pi m k_B T/h^2\right)^{3N/2}\right] - k_B T \ln Z_N \tag{2.9.31e}$$

Combine the second and third terms into a single logarithm, and then add and subtract $k_B T \ln V^N$. The result is

$$\begin{aligned} A = &- Nk_B T \ln\left[\left(2\pi m k_B T/h^2\right)^{3/2} e\right] + Nk_B T \ln N \\ &- k_B T \ln V^N - k_B T \ln Z_N + k_B T \ln V^N \end{aligned} \tag{2.9.32a}$$

Next, absorb each of the terms involving $k_B T \ln V^N$ into the log terms to their left, so that the result reads

$$\begin{aligned} A = &- Nk_B T \ln\left[\left(2\pi m k_B T/h^2\right)^{3/2} e\right] + Nk_B T \ln\left(\frac{N}{V}\right) \\ &- k_B T \ln\left(\frac{Z_N}{V^N}\right) \end{aligned} \tag{2.9.32b}$$

In Eq. (2.9.32b), the "configuration" integral, Z_N, has the definition,

$$Z_N = \int d^3 r_1 \int d^3 r_2 \cdots \int d^3 r_n e^{-\beta U_N(\mathbf{r}_1, \mathbf{r}_2, \ldots, \mathbf{r}_N)} \tag{2.9.33a}$$

The system of N particles will behave ideally if $U_N(\mathbf{r}_1, \mathbf{r}_2, \ldots, \mathbf{r}_N) \equiv 0$. In this case, Eq. (2.9.33a) reduces to

$$Z_N = \int d^3 r_1 \int d^3 r_2 \cdots \int d^3 r_n = V^N \tag{2.9.33b}$$

When Eq. (2.9.33b) is the case, the Helmholtz energy, A, assumes the value, A^{id}, where the latter is the free energy of an ideal gas. Upon substitution of Eq. (2.9.33b) into Eq. (2.9.32b), A^{id} is found to be

$$A^{id} = -Nk_BT\ln\left[\left(2\pi mk_BT/h^2\right)^{3/2}e\right] + Nk_BT\ln\left(\frac{N}{V}\right) \tag{2.9.34}$$

Equation (2.9.34) permits Eq. (2.9.32b) to be written in the form,

$$A = A^{id} - k_BT\ln\left(\frac{Z_N}{V^N}\right) \tag{2.9.35}$$

Equation (2.9.35) will prove useful in Chap. 6 where the theory of strong electrolytes is considered.

2.10 The Planck Frequency Distribution

As we shall see later, an electromagnetic wave consists of a plane wave front spanned by mutually perpendicular electric and magnetic field vectors. The direction of propagation of the wave is perpendicular to the plane defined by the two fields. The amplitudes of the electric and magnetic vectors vary sinusoidally in both space and time. Electromagnetic radiation interacts with matter in various ways. Below, we describe the thermal equilibrium which exists between electromagnetic radiation and matter when the latter is heated to a temperature where it glows. The radiation which is emitted under these conditions spans the infrared and visible bands of the spectrum.

If one looks through the peephole in the door of a heated pottery kiln, light appears to be emitted from everywhere within the kiln. Being enclosed, the light is necessarily emitted and absorbed by the objects in the kiln and by the walls of the kiln. The radiation and matter in the kiln are in thermodynamic equilibrium with one another at the temperature of the kiln. The spectrum of frequencies being emitted and absorbed is independent of the chemical nature of the materials involved. In order to understand the frequency dependence of the spectrum, which is a function of the temperature alone, Planck made the assumption, which was radical for its time (1900), that the energy transferred among the objects in the kiln occurred in units proportional to the frequency, ν, of the light. He let the unit of energy be $\varepsilon = h\nu$, where h was originally an adjustable parameter but is now realized to be a constant of nature, known as Planck's constant. On the basis of this assumption, Planck was able to fit the frequency dependence of the spectrum by assuming that the energy was absorbed and emitted in amounts given by $\varepsilon_n = n\varepsilon$, where $n = 0, 1, 2, 3$, etc. According to Eq. (2.9.20), the evaluation of the canonical partition function for these quanta reads,

$$Z = \sum_{n=0}^{\infty} \exp(-\beta\varepsilon_n) \tag{2.10.1}$$

where $\beta = 1/k_B T$. Upon substitution of Planck's relation, $\varepsilon_n = n\varepsilon$ into Eq. (2.10.1), the result can be represented in the form of a geometric series, i.e.

$$Z = \sum_{n=0}^{\infty} e^{-\beta n\varepsilon} = \sum_{n=0}^{\infty} \left[e^{-\beta\varepsilon}\right]^n = 1 + \left(e^{-\beta\varepsilon}\right) + \left(e^{-\beta\varepsilon}\right)^2 + \left(e^{-\beta\varepsilon}\right)^3 + \cdots \tag{2.10.2}$$

The sum of a general geometric series is

$$\frac{1}{1-x} = 1 + x + x^2 + x^3 + \cdots \tag{2.10.3}$$

By setting $x = e^{-\beta\varepsilon}$ in Eq. (2.10.3), Eq. (2.10.2) can be written as

$$Z = \sum_{n=1}^{\infty} e^{-\beta n\varepsilon} = \frac{1}{1 - e^{-\beta\varepsilon}} \tag{2.10.4}$$

By virtue of Eq. (2.9.22), the probability of observing an electromagnetic wave with energy $\varepsilon_n = n\varepsilon$ is then,

$$P_n = \frac{e^{-\beta\varepsilon_n}}{Z} = \frac{e^{-n\beta\varepsilon}}{Z} \tag{2.10.5}$$

If the radiant energy is transferred in quanta of size, $\varepsilon_n = n\varepsilon$, then on the basis of Eq. (2.10.5), the average energy, E, of the radiation is given by

$$E = \sum_{n=0}^{\infty} P_n \varepsilon_n = \frac{1}{Z} \sum_{n=0}^{\infty} n\varepsilon e^{-n\beta\varepsilon} \tag{2.10.6}$$

The sum on the right hand side of Eq. (2.10.6) can be evaluated by noting that $n\varepsilon e^{-n\beta\varepsilon} = -\left(\partial\left(e^{-n\beta\varepsilon}\right)/\partial\beta\right)$. Upon substitution of this identity, Eq. (2.10.6) reads

$$\sum_{n=0}^{\infty} n\varepsilon e^{-n\beta\varepsilon} = \sum_{n=0}^{\infty} \left(-\frac{\partial}{\partial\beta}\right) e^{-n\beta\varepsilon} \tag{2.10.7}$$

We can factor the derivative operator, $(-\partial/\partial\beta)$, out of the sum on the right, so that Eq. (2.10.7) reads

$$\sum_{n=0}^{\infty} n\varepsilon e^{-n\beta\varepsilon} = -\frac{\partial}{\partial\beta}\left(\sum_{n=0}^{\infty} e^{-n\beta\varepsilon}\right) = -\frac{\partial Z}{\partial\beta} \tag{2.10.8}$$

where we have recognized that the sum in parentheses is just the partition function, Z. Substitute Eq. (2.10.8) into Eq. (2.10.6) to obtain

$$E = -\frac{1}{Z}\frac{\partial Z}{\partial \beta} = -\frac{\partial \ln Z}{\partial \beta} \tag{2.10.9}$$

When the far right hand side of Eq. (2.10.4) is substituted into the far right hand side of Eq. (2.10.9), one obtains

$$E = -\frac{\partial}{\partial \beta}\ln\left(\frac{1}{1 - e^{-\beta\varepsilon}}\right) = \frac{\partial}{\partial \beta}\ln\left(1 - e^{-\beta\varepsilon}\right) = \frac{\varepsilon e^{-\beta\varepsilon}}{(1 - e^{-\beta\varepsilon})} \tag{2.10.10}$$

Upon introduction of Planck's hypothesis,

$$\varepsilon = h\nu \tag{2.10.11}$$

we can rewrite Eq. (2.10.10) in the form,

$$E(\nu) = \frac{h\nu e^{-\beta h\nu}}{(1 - e^{-\beta h\nu})} \tag{2.10.12}$$

If the numerator and denominator of the fraction in Eq. (2.10.12) are divided by the Boltzmann factor, $e^{-\beta h\nu}$, the result is

$$E(\nu) = \frac{h\nu}{(e^{\beta h\nu} - 1)} \tag{2.10.13}$$

Before Planck derived Eq. (2.10.13), Lord Rayleigh (John William Strutt, English, 1842–1919) and James Jeans (English, 1877–1946) had invoked wave theory in an attempt to understand the frequency dependence of the thermal radiation. They regarded the pottery kiln to be a cubical enclosure containing a number of light waves of various wavelengths. The sinusoidal spatial oscillations of a light wave traveling parallel to one of the edges of the box could be represented as $\sin(k_1x_1)$ where x_1 is a spatial coordinate along the edge, and k_1 is a parameter, which is one component of the three dimensional wave vector defining the direction of propagation of the wave. Since the enclosure had three space dimensions, the dependence of the wave on the other two space dimensions, x_2 and x_3, could be represented by the functions, $\sin(k_2x_2)$ and $\sin(k_3x_3)$, respectively. The complete wave vector, $\mathbf{k}$, consisted of the three components, $\mathbf{k} = (k_1, k_2, k_3)$. If the light wave was to be contained between identical walls separated by the distance, L, then the sine functions would have to exhibit the same value on opposite walls. If one wall was located at $x_1 = 0$, where $\sin(k_10) = 0$, then the light wave impinging on the opposite wall, located at $x_1 = L$, would need to satisfy $\sin(k_1L) = 0$ also. This latter condition requires $k_1L = 2\pi n_1$, where $n_1 = 1, 2, 3 \ldots$. Since n_1 changes by integral amounts, the smallest change in k_1 is $\Delta k_1 = 2\pi/L$. This argument also applies to the other components of the wave vector, so we have $\Delta k_2 = 2\pi/L$ and $\Delta k_3 = 2\pi/L$, as well. As illustrated in Fig. 2.8, we can imagine a $\mathbf{k}$-space consisting of three dimensions, each measured in units of $2\pi/L$.

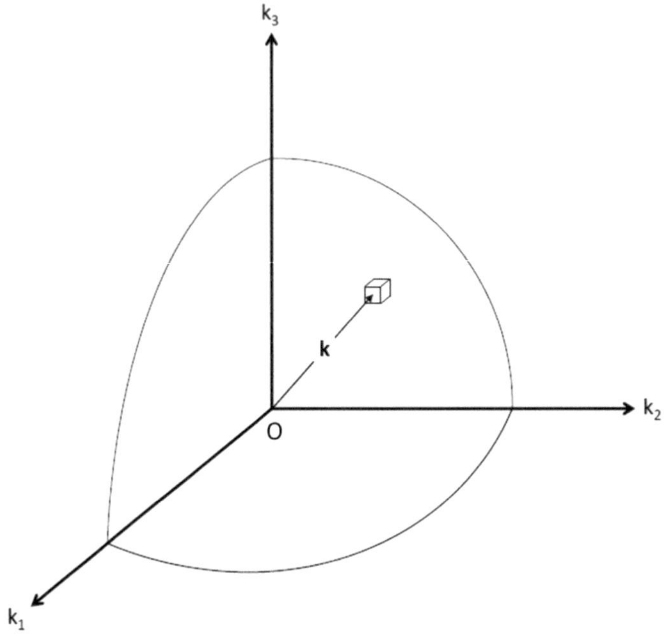

Fig. 2.8 A spherical surface in **k**-space. A differential cubical volume located on the surface of the sphere is identified by its coordinates (k_1, k_2, k_3). The vector from "O" to the differential volume is designated by **k**

Consider a vector, **k**, with its tail resting on the origin at $\mathbf{k} = 0$ and its point located on the surface of the sphere having radius, $k = |\mathbf{k}|$. Because of the discrete nature of the length of **k**, the point commands a cubical box having sides of length, $2\pi/L$, and volume $(2\pi/L)^3$. In a sphere of radius, k, with volume $4\pi k^3/3$, there are thus $(4\pi k^3/3)/(2\pi/L)^3$ such boxes. Because light has two polarizations, there are two light waves with the same value of **k** ending in each box. Hence, the total number, N, of light waves in a sphere of volume $4\pi k^3/3$ is

$$N = 2\left(\frac{4\pi k^3}{3}\right)\left(\frac{L^3}{8\pi^3}\right) \tag{2.10.14}$$

The local density of wave vectors within the range k to $k + dk$ is the derivative, dN/dk, which can be computed using Eq. (2.10.14),

$$\frac{dN}{dk} = 2\frac{d}{dk}\left(\frac{4\pi k^3}{3}\right)\left(\frac{L^3}{8\pi^3}\right) \tag{2.10.15}$$

If the light has wavelength, λ, then a sine wave representing the wave is periodic in space with period, λ. That is to say, at $x = \lambda$, the argument of the sine wave is 2π. A wave with this periodicity can be represented by $\sin(2\pi x/\lambda)$; however, if the wave is propagating in the k-direction, it is also represented by $\sin(kx)$. These two representations can be made identical by setting

$$k = \frac{2\pi}{\lambda} \tag{2.10.16}$$

If ν is the frequency and c is the speed of light in vacuum, then the wavelength satisfies

$$c = \lambda\nu \tag{2.10.17}$$

Equations (2.10.16) and (2.10.17) can be combined to give

$$k = \frac{2\pi\nu}{c} \tag{2.10.18}$$

The number of waves per unit frequency interval is then obtained by combining Eqs. (2.10.15) and (2.10.18) so as to compute, $dN/d\nu$. Using the chain rule, the result is

$$\frac{dN}{d\nu} = \frac{dN}{dk}\frac{dk}{d\nu} = \left[2\frac{d}{dk}\left(\frac{4\pi k^3}{3}\right)\left(\frac{L^3}{8\pi^3}\right)\right]\left[\frac{d}{d\nu}\left(\frac{2\pi\nu}{c}\right)\right] = \left(\frac{k^2L^3}{\pi^2}\right)\left(\frac{2\pi}{c}\right) \tag{2.10.19}$$

Now use Eq. (2.10.18) to replace the remaining dependence on k which appears on the right hand side of Eq. (2.10.19). Upon introducing the volume, $V = L^3$, the result is

$$\frac{dN}{d\nu} = \left[\frac{(2\pi\nu/c)^2V}{\pi^2}\right]\left(\frac{2\pi}{c}\right) = \frac{8\pi\nu^2V}{c^3} \tag{2.10.20}$$

The energy density, ρ, per unit volume and per unit frequency is defined by

$$\rho = \frac{E(\nu)}{V}\frac{dN}{d\nu} \tag{2.10.21}$$

Rayleigh and Jeans used k_BT to represent $E(\nu)$. Planck improved on the Rayleigh–Jeans theory by equating $E(\nu)$ to the expression in Eq. (2.10.13). Upon using Eq. (2.10.20) to evaluate $dN/d\nu$, Planck derived the form,

$$\rho(\nu) = \frac{8\pi h\nu^3/c^3}{(e^{\beta h\nu} - 1)} \tag{2.10.22}$$

Equation (2.10.22) is known as the Planck energy distribution. It will be exploited in Chap. 11 when we consider the probabilities governing the absorption and emission of radiation by an atom.

2.11 Statistical Mechanics in the Grand Canonical Ensemble

In the grand canonical ensemble, each system has the same fixed volume, the same temperature, and the same chemical potential. See Fig. 2.9.

As shown in Fig. 2.9, the individual systems are assumed to be able to exchange molecules, making the number of molecules in each system variable. The flexibility permitted by the exchange of molecules allows the effects of fluctuations in density to be calculated. It is these density fluctuations which are responsible for light scattering.

To prevent our equations from becoming too long, only two systems are considered in Fig. 2.9. The entropy of an ensemble of two systems is

$$S(P_1, P_2) = -k_B P_1 \ln P_1 - k_B P_2 \ln P_2 \tag{2.11.1}$$

The probabilities, P_1 and P_2, add to unity, so

$$P_1 + P_2 = 1 \tag{2.11.2}$$

If the internal energies of the two systems are u_1 and u_2, respectively, the ensemble average internal energy, U, is

$$U = P_1 u_1 + P_2 u_2 \tag{2.11.3}$$

In any given configuration, the number of molecules in the first system is N_1 while the number of molecules in the second system is N_2. The ensemble average number of molecules, N, in a system is

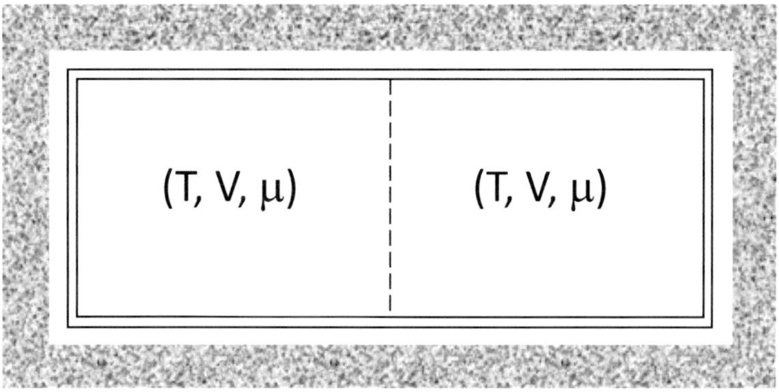

Fig. 2.9 Subdivision of a sample of matter into two systems, each having the same temperature, T, the same volume, V, and the same chemical potential, μ. Subdivided in this fashion, the two systems are said to form a grand canonical ensemble. The two systems are in a container with rigid walls, which is thermally isolated from its surroundings. The dashed line partition between the two systems is assumed to be permeable to heat and molecules

$$\overline{N} = P_1N_1 + P_2N_2 \tag{2.11.4}$$

Given that the ensemble is isolated from its surroundings, we can use the method of Lagrange multipliers to calculate the values of the probabilities, P_1 and P_2, which maximize S in Eq. (2.11.1) subject to the constraints in Eqs. (2.11.2), (2.11.3), and (2.11.4).

We begin by introducing the auxiliary function $F(P_1, P_2)$ defined by

$$\begin{aligned} F(P_1, P_2) &= S(P_1, P_2) + \alpha(U - P_1u_1 - P_2u_2) \\ &\quad + \gamma\left(\overline{N} - P_1N_1 - P_2N_2\right) + \delta(1 - P_1 - P_2) \\ &= -k_BP_1 \ln P_1 - k_BP_2 \ln P_2 + \alpha(U - P_1u_1 - P_2u_2) \\ &\quad + \gamma\left(\overline{N} - P_1N_1 - P_2N_2\right) + \delta(1 - P_1 - P_2) \end{aligned} \tag{2.11.5}$$

where the coefficients, α, γ, δ are undetermined multipliers. As pointed out in Sect. 2.9, the extremum of F is the same as the extremum of S. In the case of two systems, the conditions that determine this extremum are $(\partial F/\partial P_1) = (\partial F/\partial P_2) = 0$. On the basis of Eq. (2.11.5), the required partial derivatives of F are

$$\frac{\partial F}{\partial P_i} = -k_B\left[\ln P_i + \frac{P_i}{P_i}\right] - \alpha u_i - \gamma N_i - \delta = 0 \quad \text{for} \quad i = 1, 2 \tag{2.11.6}$$

An expression for $\ln P_i$ is obtained by solving Eq. (2.11.6). The result is

$$k_B \ln P_i = -k_B - \delta - \alpha u_i - \gamma N_i \quad \text{for} \quad i = 1, 2 \tag{2.11.7a}$$

Now form the antilog of Eq. (2.11.7a). The result is

$$P_i = \exp\left(-\frac{k_B + \delta}{k_B}\right) \exp\left(\frac{-\alpha u_i}{k_B}\right) \exp\left(\frac{-\gamma N_i}{k_B}\right) \tag{2.11.7b}$$

Equation (2.11.7b) must satisfy Eq. (2.11.2), which implies that

$$\begin{aligned} P_1 + P_2 = \exp\left(-\frac{k_B + \delta}{k_B}\right)&\left[\exp\left(\frac{-\alpha u_1}{k_B}\right) \exp\left(\frac{-\gamma N_1}{k_B}\right)\right. \\ &\left. + \exp\left(\frac{-\alpha u_2}{k_B}\right) \exp\left(\frac{-\gamma N_2}{k_BT}\right)\right] = 1 \end{aligned} \tag{2.11.8}$$

We can solve Eq. (2.11.8) for

$$\exp\left(-\frac{k_B + \gamma}{k_B}\right) = \frac{1}{\exp\left(\frac{-\alpha u_1}{k_B}\right) \exp\left(\frac{-\gamma N_1}{k_B}\right) + \exp\left(\frac{-\alpha u_2}{k_B}\right) \exp\left(\frac{-\gamma N_2}{k_B}\right)} \tag{2.11.9}$$

Substitute Eq. (2.11.9) into Eq. (2.11.7b) to obtain,

$$P_i = \frac{\exp\left(\frac{-\alpha u_i}{k_B}\right)\exp\left(\frac{-\gamma N_i}{k_B}\right)}{\exp\left(\frac{-\alpha u_1}{k_B}\right)\exp\left(\frac{-\gamma N_1}{k_B}\right)+\exp\left(\frac{-\alpha u_2}{k_B}\right)\exp\left(\frac{-\gamma N_2}{k_B}\right)} \tag{2.11.10}$$

where $i = 1, 2$. Although Eq. (2.11.10) is an explicit formula for the probability of observing the i-th system, it still depends upon the Lagrange multipliers, α, and γ, which need to be evaluated.

Upon consulting Eq. (2.9.10), the total differential of Eq. (2.11.1) is seen to be

$$dS = -k_B \ln P_1 dP_1 - k_B \ln P_2 dP_2 \tag{2.11.11}$$

Substitute Eq. (2.11.7a) into Eq. (2.11.11) to obtain

$$dS = (k_B + \delta)\sum_{i=1}^{2} dP_i + \alpha\sum_{i=1}^{2} u_i dP_{ii} + \gamma\sum_{=1}^{2} N_i dP_i \tag{2.11.12}$$

The total differentials of Eqs. (2.11.2), (2.11.3), and (2.11.4) are respectively

$$\sum_{i=1}^{2} dP_i = 0 \tag{2.11.13a}$$

$$\sum_{i=1}^{2} u_i dP_i = dU \tag{2.11.13b}$$

$$\sum_{i=1}^{2} N_i dP_i = d\overline{N} \tag{2.11.13c}$$

By virtue of Eqs. (2.11.13a) through (2.11.13c), Eq. (2.11.12) simplifies to

$$dS = \alpha dU + \gamma d\overline{N} \tag{2.11.14}$$

The generalization of Eq. (2.5.1) to an ensemble of systems exchanging molecules is

$$dU = TdS - PdV + \overline{\mu} d\overline{N} \tag{2.11.15}$$

where $\overline{\mu}$ is the chemical potential per molecule. Solve Eq. (2.11.15) for dS under conditions where the volume is constant and $dV = 0$. The result is

$$dS = \frac{1}{T}dU - \frac{\overline{\mu}}{T}d\overline{N} \tag{2.11.16}$$

Now, compare Eqs. (2.11.14) and (2.11.16) term by term. One obtains

$$\alpha = \frac{1}{T} \tag{2.11.17a}$$

and

$$\gamma = -\frac{\overline{\mu}}{T} \tag{2.11.17b}$$

Equations (2.11.17a) and (2.11.17b) determine the Lagrange multipliers, α and γ. When these two equations are substituted into Eq. (2.11.10), the probability of the i-th system takes the form,

$$P_i = \frac{\exp\left(\frac{-u_i}{k_B T}\right)\exp\left(\frac{\overline{\mu}N_i}{k_B T}\right)}{\sum_{j=1}^{2}\exp\left(\frac{-u_j}{k_B T}\right)\exp\left(\frac{\overline{\mu}N_j}{k_B T}\right)} \quad \text{for} \quad i = 1, 2 \tag{2.11.18}$$

At this point, it will prove useful to introduce the activity parameter, $\overline{a}$, which satisfies

$$\overline{\mu} = k_B T \ln \overline{a} \tag{2.11.19}$$

The assignment of the term "activity" to $\overline{a}$ follows from the similarity between Eqs. (2.11.19) and (2.7.2). Unlike Eq. (2.7.2), however, the chemical potential defined by Eq. (2.11.19) does not involve an additive experimentally determined standard state chemical potential, μ^o. It is in this sense that $\overline{a}$ is sometimes referred to as the "absolute" activity. Solving Eq. (2.11.19) for $\overline{a}$ and substituting the result into Eq. (2.11.18) gives

$$P_i = \frac{\overline{a}^{N_i}\exp\left(\frac{-u_i}{k_B T}\right)}{\sum_{j=1}^{2}\overline{a}^{N_j}\exp\left(\frac{-u_j}{k_B T}\right)} \tag{2.11.20}$$

The denominator in Eq. (2.11.20) is known as the grand canonical partition function. It is given the symbol, Ξ:

$$\Xi = \sum_{j=1}^{2}\overline{a}^{N_j}\exp\left(\frac{-u_j}{k_B T}\right) \tag{2.11.21}$$

Equation (2.11.21) permits the calculation of all of the thermal properties of a material subject to density fluctuations.

2.12 Theory of Fluctuations

As we shall see in detail in Chap. 8, light passing through matter is scattered by fluctuations in the molecular number density. The fluctuation, ΔN, in N is defined by

$$\Delta N = N - \langle N \rangle \tag{2.12.1a}$$

where $\langle N \rangle$, has the same meaning as $\overline{N}$, above. $\langle N \rangle$ serves as the ensemble mean value of the number of molecules in the typical system. Since the number of molecules, N, in a given system can be either greater than the mean or less than the mean, the fluctuation, ΔN, can have either sign. It is thus possible for the ensemble average, $\langle \Delta N \rangle$, to be zero. As we shall see, the light scattering properties of the ensemble do not depend upon the average value of ΔN but rather upon the average of its square, $\langle (\Delta N)^2 \rangle$, known as the variance, which is defined by

$$\langle (\Delta N)^2 \rangle = \langle (N - \langle N \rangle)^2 \rangle \tag{2.12.1b}$$

The notation in Eq. (2.12.1b) permits expansion of the binomial on right hand to read,

$$\begin{aligned} \langle (\Delta N)^2 \rangle &= \langle N^2 - 2N\langle N \rangle + \langle N \rangle^2 \rangle \\ &= \langle N^2 \rangle - 2\langle N \rangle \langle N \rangle + \langle N \rangle^2 \\ &= \langle N^2 \rangle - 2\langle N \rangle^2 + \langle N \rangle^2 \end{aligned} \tag{2.12.1c}$$

Combining terms involving $\langle N \rangle^2$ leads to the result,

$$\langle (\Delta N)^2 \rangle = \langle N^2 \rangle - \langle N \rangle^2 \tag{2.12.2}$$

Both averages on the right hand side of Eq. (2.12.2) can be calculated in the grand canonical ensemble.

The probability that a system contains N_j molecules is P_j. The ensemble average number of molecules, $\langle N \rangle$, is

$$\langle N \rangle = \sum_{j=1}^{2} N_j P_j = \sum_{j=1}^{2} \frac{N_j \overline{a}^{N_j} \exp(-u_j/k_B T)}{\Xi} = \frac{1}{\Xi} \sum_{j=1}^{2} N_j \overline{a}^{N_j} \exp(-u_j/k_B T) \tag{2.12.3}$$

where we have used Eqs. (2.11.20) and (2.11.21) to evaluate P_j. The product, $N_j \overline{a}^{N_j}$, can be replaced by the derivative,

$$N_j \overline{a}^{N_j} = \overline{a} \frac{\partial \overline{a}^{N_j}}{\partial \overline{a}} \tag{2.12.4}$$

Use of Eq. (2.12.4) permits Eq. (2.12.3) to be rewritten as

$$\langle N \rangle = \frac{1}{\Xi} \overline{a} \frac{\partial}{\partial \overline{a}} \left[\sum_{j=1}^{2} \overline{a}^{N_j} \exp(-u_j / k_b T) \right] \tag{2.12.5}$$

By reference to Eq. (2.11.21), we can write Eq. (2.12.5) as

$$\langle N \rangle = \frac{\overline{a}}{\Xi} \left(\frac{\partial \Xi}{\partial \overline{a}} \right) \tag{2.12.6}$$

Equation (2.11.19) can be inverted to obtain

$$\overline{a} = e^{\beta \overline{\mu}} \tag{2.12.7}$$

where

$$\beta = 1/k_B T \tag{2.12.8}$$

Upon substitution of Eq. (2.12.7) into Eq. (2.12.6), we can use the chain rule to write

$$\langle N \rangle = \frac{e^{\beta \overline{\mu}}}{\Xi} \left(\frac{\partial \Xi}{\partial \overline{\mu}} \right) \left(\frac{\partial \overline{\mu}}{\partial \overline{a}} \right) \tag{2.12.9a}$$

The derivative of Eq. (2.12.7) is

$$\left(\frac{\partial \overline{a}}{\partial \overline{\mu}} \right) = \beta e^{\beta \overline{\mu}} \tag{2.12.9b}$$

Substitute Eq. (2.12.9b) into Eq. (2.12.9a), to obtain

$$\langle N \rangle = \frac{e^{\beta \overline{\mu}}}{\Xi} \left(\frac{\partial \Xi}{\partial \overline{\mu}} \right) \left(\frac{1}{\beta e^{\beta \overline{\mu}}} \right) = \frac{1}{\Xi} \left(\frac{\partial \Xi}{\partial \overline{\mu}} \right) \left(\frac{1}{\beta} \right) \tag{2.12.9c}$$

Finally, upon use of Eq. (2.12.8), Eq. (2.12.9c) becomes

$$\langle N \rangle = \frac{k_B T}{\Xi} \left(\frac{\partial \Xi}{\partial \overline{\mu}} \right) \tag{2.12.10}$$

Equation (2.12.10) gives us an expression for $\langle N \rangle$.

We now turn to the calculation of $\langle N^2 \rangle$. In analogy with Eq. (2.12.3), we have the result

$$\langle N^2 \rangle = \sum_{j=1}^{2} N_j^2 P_j = \sum_{j=1}^{2} \frac{N_j^2 \overline{a}^{N_j} \exp(-u_j/k_B T)}{\Xi} = \frac{1}{\Xi} \sum_{j=1}^{2} N_j^2 \overline{a}^{N_j} \exp(-u_j/k_B T) \tag{2.12.11a}$$

Factor the argument of the sum in Eq. (2.12.11a) so it reads,

$$\langle N^2 \rangle = \frac{1}{\Xi} \sum_{j=1}^{2} (N_j \overline{a}^{N_j}) N_j \exp(-u_j/k_B T) \tag{2.12.11b}$$

Upon use of Eq. (2.12.4), Eq. (2.12.11b) becomes

$$\langle N^2 \rangle = \frac{1}{\Xi} \sum_{j=1}^{2} \overline{a} \frac{\partial (\overline{a}^{N_j})}{\partial \overline{a}} N_j \exp(-u_j/k_B T) \tag{2.12.11c}$$

Next factor the operator $\overline{a}(\partial/\partial\overline{a})$ out of the sum in Eq. (2.12.11c) to obtain,

$$\langle N^2 \rangle = \frac{1}{\Xi} \overline{a} \frac{\partial}{\partial \overline{a}} \left[\sum_{j=1}^{2} N_j \overline{a}^{N_j} \exp(-u_j/k_b T) \right] \tag{2.12.11d}$$

Equation (2.12.3) can be rewritten in the form,

$$\Xi \langle N \rangle = \sum_{j=1}^{2} N_j \overline{a}^{N_j} \exp(-u_j/k_B T) \tag{2.12.12}$$

Substitution of Eq. (2.12.12) into Eq. (2.12.11d) then leads to

$$\langle N^2 \rangle = \frac{1}{\Xi} \overline{a} \frac{\partial}{\partial \overline{a}} [\Xi \langle N \rangle] \tag{2.12.13}$$

which gives us an expression for $\langle N^2 \rangle$.

To obtain a compact formula for $\langle (\Delta N)^2 \rangle$, we will need to expand the derivative on the right hand side of Eq. (2.12.13). Calculation of the indicated derivative using the chain rule permits Eq. (2.12.13) to be written in the form,

$$\langle N^2 \rangle = \frac{1}{\Xi} \overline{a} \frac{\partial}{\partial \overline{\mu}} [\Xi \langle N \rangle] \left(\frac{\partial \overline{\mu}}{\partial \overline{a}} \right) \tag{2.12.14a}$$

Upon substitution of Eqs. (2.12.7) and (2.12.9b), Eq. (2.12.14a) assumes the form,

$$\langle N^2\rangle = \frac{1}{\Xi}e^{\beta\overline{\mu}}\frac{\partial}{\partial\overline{\mu}}[\Xi\langle N\rangle]\left(\frac{1}{\beta e^{\beta\overline{\mu}}}\right) = \frac{1}{\Xi}\frac{\partial}{\partial\overline{\mu}}[\Xi\langle N\rangle]\left(\frac{1}{\beta}\right) \tag{2.12.14b}$$

Next, expand the derivative of the product on the right hand side of Eq. (2.12.14b). Upon use of Eq. (2.12.8), the result is

$$\langle N^2\rangle = \langle N\rangle\frac{k_BT}{\Xi}\left(\frac{\partial\Xi}{\partial\overline{\mu}}\right) + \frac{k_BT}{\Xi}\Xi\left(\frac{\partial\langle N\rangle}{\partial\overline{\mu}}\right) \tag{2.12.14c}$$

By comparison with Eq. (2.12.10), the first term on the right hand side of Eq. (2.12.14c) is $\langle N\rangle^2$; hence,

$$\langle N^2\rangle = \langle N\rangle^2 + k_BT\left(\frac{\partial\langle N\rangle}{\partial\overline{\mu}}\right) \tag{2.12.14d}$$

According to the definition of the variance in Eq. (2.12.2),

$$\langle(\Delta N)^2\rangle = \langle N^2\rangle - \langle N\rangle^2 = k_BT\left(\frac{\partial\langle N\rangle}{\partial\overline{\mu}}\right) \tag{2.12.15}$$

Equation (2.12.15) forms the basis for the calculation of fluctuations in number density.

2.13 The Fluctuation $\langle(\Delta N)^2\rangle$ in a Compressible Fluid

In the case of a gas or a compressible liquid, $\langle(\Delta N)^2\rangle$, manifests itself as a density fluctuation. As such, one might expect $\langle(\Delta N)^2\rangle$ to be related to the compressibility of the fluid. For an open system capable of exchanging mass with its surroundings, Eq. (2.5.5) can be generalized to read,

$$dA = -SdT - PdV + \overline{\mu}d\overline{N} \tag{2.13.1}$$

Because the Helmholtz energy, $A(T, V, \overline{N})$, is an analytic function of its variables, the total differential of A can be expanded in the form,

$$dA = \left(\frac{\partial A}{\partial T}\right)dT + \left(\frac{\partial A}{\partial V}\right)dV + \left(\frac{\partial A}{\partial \overline{N}}\right)d\overline{N} \tag{2.13.2}$$

Comparing Eqs. (2.13.1) and (2.13.2) term by term, we can conclude that

$$S = -\left(\frac{\partial A}{\partial T}\right) \tag{2.13.3a}$$

$$P = -\left(\frac{\partial A}{\partial V}\right) \tag{2.13.3b}$$

$$\overline{\mu} = \left(\frac{\partial A}{\partial \overline{N}}\right) \tag{2.13.3c}$$

Consider the case of fixed temperature. On the basis of Eqs. (2.13.3b) and (2.13.3c), and the analytical property of the second derivatives of A, we then can calculate the result,

$$-\left(\frac{\partial P}{\partial \overline{N}}\right)_T = \left(\frac{\partial^2 A}{\partial \overline{N} \partial V}\right)_T = \left(\frac{\partial^2 A}{\partial V \partial \overline{N}}\right)_T = \left(\frac{\partial \overline{\mu}}{\partial V}\right)_T \tag{2.13.4}$$

Equation (2.13.4) serves as a proof of the identity,

$$-\left(\frac{\partial P}{\partial \overline{N}}\right)_T = \left(\frac{\partial \overline{\mu}}{\partial V}\right)_T \tag{2.13.5}$$

For our system of a single chemical component, the Gibbs–Duhem equation, as expressed by Eq. (2.8.10), assumes the form,

$$SdT - VdP + \overline{N} d\overline{\mu} = 0 \tag{2.13.6}$$

Upon holding the temperature fixed, and dividing both sides of Eq. (2.13.6) by $d\overline{N}$, one obtains

$$V\left(\frac{\partial P}{\partial \overline{N}}\right)_T = \overline{N}\left(\frac{\partial \overline{\mu}}{\partial \overline{N}}\right)_T \tag{2.13.7}$$

By eliminating $(\partial P/\partial \overline{N})_T$ between Eqs. (2.13.5) and (2.13.7), we get the result,

$$\left(\frac{\partial \overline{\mu}}{\partial \overline{N}}\right)_T = -\frac{V}{\overline{N}}\left(\frac{\partial \overline{\mu}}{\partial V}\right)_T \tag{2.13.8}$$

Again with the temperature fixed, divide both sides of Eq. (2.13.6) by dV. The result is

$$V\left(\frac{\partial P}{\partial V}\right)_T = \overline{N}\left(\frac{\partial \overline{\mu}}{\partial V}\right)_T \tag{2.13.9}$$

If we eliminate $(\partial \overline{\mu}/\partial V)_T$ between Eqs. (2.13.8) and (2.13.9), we find

$$\left(\frac{\partial \overline{\mu}}{\partial \overline{N}}\right)_T = -\frac{V^2}{\overline{N}^2}\left(\frac{\partial P}{\partial V}\right)_T \tag{2.13.10}$$

The isothermal compressibility, κ_T, is defined by

$$\kappa_T = -\frac{1}{V}\left(\frac{\partial V}{\partial P}\right)_T \tag{2.13.11}$$

Substitute Eq. (2.13.11) into the reciprocal of Eq. (2.13.10) to obtain

$$\left(\frac{\partial \overline{N}}{\partial \overline{\mu}}\right)_T = \frac{\overline{N}^2 \kappa_T}{V} \tag{2.13.12}$$

A final formula for the fluctuation in the number of molecules in a system can be calculated by equating $\overline{N}$ with the experimentally measurable N and substituting Eq. (2.13.12) into Eq. (2.12.15). This leads to the result,

$$\langle(\Delta N)^2\rangle = \frac{N^2 k_B T}{V}\kappa_T \tag{2.13.13}$$

Equation (2.13.13) shows that $\langle(\Delta N)^2\rangle$ depends upon N, V, T, and κ_T, all of which are measurable.

2.14 Solute Concentration Fluctuations

Consider a liquid solution consisting of components (1) and (2). Component (1) is in excess and serves as the solvent, while component (2) is the solute. For simplicity, we shall assume that the fluctuations in the concentration of the solvent are minimal in comparison with fluctuations in the concentration of the solute. This is the situation in the case of many aqueous solutions. We can then rewrite Eq. (2.12.15) in the form

$$\langle(\Delta N_2)^2\rangle = \langle N_2^2\rangle - \langle N_2\rangle^2 = k_B T\left(\frac{\partial\langle N_2\rangle}{\partial \overline{\mu}_2}\right) \tag{2.14.1}$$

where the subscript refers to the second component. To convert the chemical potential per molecule, $\overline{\mu}_2$, to chemical potential per mole, μ_2, we write

$$\mu_2 = N_A \overline{\mu}_2 \tag{2.14.2}$$

where N_A is Avogadro's number. The average number of molecules, $\langle N_2\rangle$, can be converted to molar concentration, c_2, by use of the equation,

$$c_2 = \frac{n_2}{V} = \frac{\langle N_2\rangle}{N_A V} \tag{2.14.3}$$

where n_2 is the number of moles of the solute, V is the volume of the solution, and c_2 is the volume concentration of the solute in the units of mol/dm^3. Using Eqs. (2.14.2) and (2.14.3), Eq. (2.14.1) can be transformed to read

$$\left\langle (\Delta N_2)^2 \right\rangle = N_A^2 V k_B T \left(\frac{\partial c_2}{\partial \mu_2} \right) \tag{2.14.4}$$

2.15 Osmotic Compressibility

According to Eq. (2.14.4), the fluctuation, $\left\langle (\Delta N_2)^2 \right\rangle$, can be expressed in terms of the rate of change of the chemical potential with respect to the concentration, $(\partial \mu_2 / \partial c_2)$. The concentration dependence of the osmotic pressure is also sensitive to this derivative. It thus makes sense to express $\left\langle (\Delta N_2)^2 \right\rangle$ in terms of the osmotic pressure, which is readily measured. As before, we let component (1) be the solvent and component (2) be the solute.

To determine the osmotic pressure, a compartment containing solute (2) dissolved in component (1) is separated from a reservoir of pure (1) by a membrane which is permeable solely to (1). See Fig. 2.10.

The liquids on opposite sides of the membrane would be at chemical equilibrium if they had the same composition. In an attempt to approach this state of equilibrium, which on the solution side would correspond to infinite dilution, component (1) diffuses through the membrane from the solvent compartment into the solution compartment. The transfer of volume across the membrane increases the pressure in the solution compartment to the value, $P + \Pi$, where the extra pressure, $\Pi = \rho g h$, as illustrated in Fig. 2.10, is termed the osmotic pressure. The increase in pressure causes the chemical potential of component (1) in the solution compartment to come to equilibrium with the chemical potential of component (1) in the solvent compartment.

The pure component (1) at pressure, P, on the solvent side has chemical potential $\mu_1^o(P)$ and activity unity. The activity, a_1, of component (1) on the solution side, where the pressure is $P + \pi$, is given by

$$\mu_1(P + \pi) = \mu_1^o(P + \pi) + RT \ln a_1 \tag{2.15.1}$$

Because component (1) is serving as the solvent on the solution side, the pure form serves as its standard state in solution. Solvent equilibrium across the membrane is achieved when

$$\mu_1(P + \Pi) = \mu_1^o(P) \tag{2.15.2}$$

If we substitute Eq. (2.15.1) into the left hand side of Eq. (2.15.2), the result is

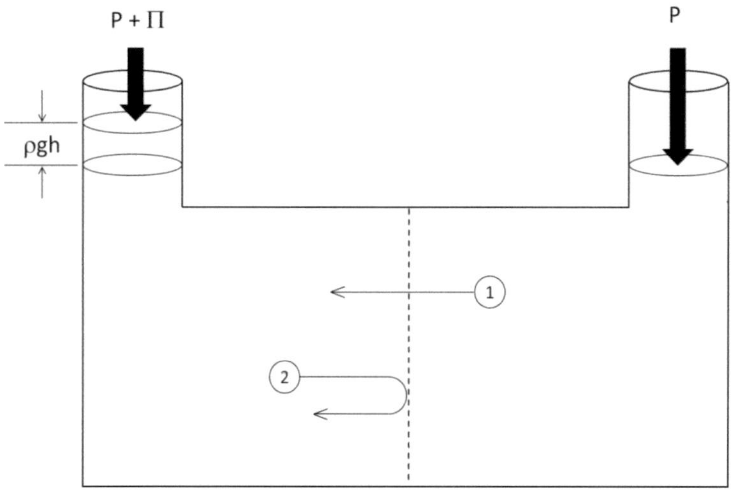

Fig. 2.10 Osmotic pressure experiment. The semipermeable membrane represented by the dashed line separates the solvent compartment on the right containing pure (1) from the solution compartment on the left containing a mixture of (1) and (2). The external atmospheric pressure is P. There is a net flow of (1) from right to left, which serves to increase the height of the liquid in the side arm on the left. The height of the liquid on the left exceeds that on the right by the amount, h. If g is the acceleration due to gravity, and ρ is the density of the liquid on the left, then the extra height corresponds to a hydrostatic pressure, ρgh, which balances the osmotic pressure driving component (1) from right to left across the membrane

$$\mu_1^o(P+\Pi)+RT\ln a_1=\mu_1^o(P) \tag{2.15.3}$$

Next, expand $\mu_1^o(P+\Pi)$ in a Taylor series about the ambient pressure, P. Retaining only the first two terms, this Taylor series reads,

$$\mu_1^o(P+\Pi)=\mu_1^o(P)+\left(\frac{\partial\mu_1^o}{\partial P}\right)(P+\Pi-P)=\mu_1^o(P)+\left(\frac{\partial\mu_1^o}{\partial P}\right)\Pi \tag{2.15.4a}$$

If G_1^o is the Gibbs energy of component (1) in its standard state, then according to Eqs. (2.7.6b) and (2.7.6c),

$$\left(\frac{\partial\mu_1^o}{\partial P}\right)=\left(\frac{\partial^2 G_1^o}{\partial P\partial n_1}\right)=\left(\frac{\partial^2 G_1^o}{\partial n_1\partial P}\right)=\left(\frac{\partial V_1^o}{\partial n_1}\right)=\overline{V}_1^o \tag{2.15.4b}$$

where $\overline{V}_1^o$ is the molar volume of the pure solvent. Using Eq. (2.15.4b), we can rewrite Eq. (2.15.4a) as

$$\mu_1^o(P+\Pi)=\mu_1^o(P)+\overline{V}_1^o\Pi \tag{2.15.4c}$$

Substitution of Eq. (2.15.4c) into Eq. (2.15.3), yields

$$-RT\ln a_1 = \overline{V}_1^o\Pi \tag{2.15.5}$$

At constant temperature and pressure, Eq. (2.8.10) reads,

$$n_1 d\mu_1 + n_2 d\mu_2 = 0 \tag{2.15.6a}$$

Rewrite Eq. (2.15.6a) as

$$d\mu_2 = -\frac{n_1}{n_2} d\mu_1 \tag{2.15.6b}$$

With the variable, $P + \pi$, suppressed, the differential of Eq. (2.15.1) is

$$d\mu_1 = RTd\ln a_1 \tag{2.15.7}$$

Upon substitution of Eq. (2.15.7) into Eq. (2.15.6b), one obtains

$$d\mu_2 = -\frac{n_1}{n_2} RTd\ln a_1 \tag{2.15.8}$$

The differential of Eq. (2.15.5) is

$$-RTd\ln a_1 = \overline{V}_1^o d\Pi \tag{2.15.9}$$

Upon substitution of Eq. (2.15.9) into Eq. (2.15.8), we have

$$d\mu_2 = \frac{n_1\overline{V}_1^o}{n_2} d\Pi \tag{2.15.10}$$

If the solution is sufficiently dilute, then the volume, V, of the solution is approximately equal to $n_1\overline{V}_1^o$. After making this substitution into Eq. (2.15.10), we obtain

$$d\mu_2 = \frac{V}{n_2} d\Pi \tag{2.15.11}$$

Now introduce the molar concentration, $c_2 = n_2/V$, and divide both sides of Eq. (2.15.11) by dc_2. The result is

$$\left(\frac{\partial\mu_2}{\partial c_2}\right) = \frac{1}{c_2}\left(\frac{\partial\Pi}{\partial c_2}\right) \tag{2.15.12}$$

Equation (2.15.12) can be substituted into Eq. (2.14.4) with the result,

$$\left\langle(\Delta N_2)^2\right\rangle = N_A^2 V k_B T c_2 \left(\frac{\partial c_2}{\partial\Pi}\right) \tag{2.15.13}$$

The derivative, $(\partial c_2/\partial \Pi)$, is sometimes termed the "osmotic compressibility". It is the force that drives mixing. This mixing force is the origin of the strong light scattering observed in a binary liquid mixtures that exhibit a critical mixing temperature. To the naked eye, the mixture exhibits a hazy orange color in transmitted light and a hazy blue color in scattered light. This light scattering phenomenon is known as *opalescence.*

When the solute is a polymer, the solution is non-ideal even at fairly low concentrations. To take into account the non-ideality, the osmotic pressure is represented as a power series in the concentration,

$$\Pi = RT\left(c_2 + bc_2^2 + \cdots\right) \tag{2.15.14}$$

Equation (2.15.14) is said to be a virial expansion. The coefficient of c_2 is unity, which is sometimes said to be the first virial coefficient. The coefficient, b, which multiplies c_2^2, is said to be the second virial coefficient. The adjective, "virial", refers to "force". The concentration derivative of Eq. (2.15.14) is given by

$$\left(\frac{\partial \Pi}{\partial c_2}\right) = RT(1 + 2bc_2) \tag{2.15.15}$$

Upon substitution of Eq. (2.15.15) into Eq. (2.15.13), the solute number fluctuation is found to be

$$\left\langle (\Delta N_2)^2 \right\rangle = N_A^2 V k_B T \frac{c_2}{RT[1 + 2bc_2]} \tag{2.15.16}$$

As we shall see in Chap. 8, when combined with light scattering experiments, Eq. (2.15.16) can be used to determine the number average molecular weight of a polymer.

Recommended Additional Reading

1. S.E. Wood, R. Battino, *Thermodynamics of Chemical Systems* (Cambridge University Press, Cambridge, UK, 1990)
2. D. Kondepudi, I. Prigogine, *Modern Thermodynamics*, 2nd edn (Wiley, Chichester, West Sussex, UK, 2015)
3. H. Eyring, D. Henderson, B.J. Stover, E. Eyring, *Statistical Mechanics and Dynamics* (Wiley, New York, 1964)
4. D.A. McQuarrie, *Statistical Mechanics* (Harper and Rowe, New York, 1976)
5. R. Kubo, *Statistical Mechanics.* Advanced Course with Problems and Solutions (North Holland Publishing Co., Amsterdam, Netherlands, 1965)
6. M.C. Gupta, *Statistical Thermodynamics* (Wiley, New York, 1990)

Chapter 3
Continuum Theory of Dielectrics

A dielectric is a material, which consists entirely of chemically bound charges that lack the freedom to move between molecules. As a consequence, the material is incapable of sustaining a steady current, and functions instead as an insulator. Dielectrics can be classified as polar and non-polar. The molecules in a non-polar dielectric do not have permanent electric dipole moments. In a uniform applied electric field, the electrons and nuclei in the atoms in a non-polar dielectric suffer only displacements from their equilibrium positions. By contrast, in a polar dielectric, the molecules possess permanent electric dipoles. In gaseous and liquid polar dielectrics and some solids, the molecules can rotate under action of the applied field. In both polar and non-polar dielectrics, the applied electric field produces charge distribution distortions, which are known as "polarization." The combination of the external electric field and the electric field associated with this internal polarization serves to define a new vector called the displacement. Lorenz, Debye, and Onsager, as well as many others, have attempted to link the charge distortion in a single molecule to the displacement observed in the macroscopic dielectric. In these models, a given molecule is regarded as carving out a cavity for itself in an otherwise continuous dielectric. The molecule in the cavity interacts with the applied field as well as the polarization raised in the surrounding continuous dielectric, thus sensing a local electric field which exceeds the applied field. These so-called semi-continuum theories provide a basis for deriving equations for the density dependence of the electric permittivity, also known as the dielectric constant. When combined with experimental data, the local field concept can be used to determine molecular permanent electric dipole moments, molecular polarizabilities, dipole solvation energies, and the effects of the solvent on the absorption spectrum of a solute.

J. K. Baird, *Electromagnetism for Chemistry*,
https://doi.org/10.1007/978-3-031-88508-2_3

3.1 Introduction

Molecular charges which are not free to translate through a material are said to be *bound* charges. When an electric field having a strength less than the threshold for electric breakdown is applied to such a material, the atomic and molecular orbitals become distorted, but the electrons and nuclei do not leave their parent molecules. That is to say there is no ionization. Rather, the external electric field displaces the charges to create temporary electric dipoles known as induced polarization. Depending upon the polarity of the chemical bonds, the molecules in the material may also carry permanent electric dipoles. Molecular permanent dipoles are free to rotate in gases, liquids, and in a few solids, such as ice. Within the limits set by their rotational Brownian motion, the dipoles will align themselves parallel to the applied electric field. In dielectrics, the combination of the displacement of positive and negative charges and the re-orientation of molecular permanent electric dipoles is termed simply *polarization.*

In ionic solids, the applied electric field can displace the ions slightly off their lattice positions in the unit cell. In certain cases, the result is permanent electric polarization similar to the permanent magnetization exhibited in ferromagnetism. Barium titanate ($BaTiO_3$) is an example. In an applied electric field, the oxide ions are essentially fixed, while the Ba^{2+} and Ti^{4+} ions shift their positions in the direction of the applied field. Unlike gases, liquids, and most other ionic solids, these displacements of charge persist after the external electric field is removed. This phenomenon of permanent electric polarization is known as *ferroelectricity* in analogy with ferromagnetism.

A conductor is capable of carrying a steady electric current. By contrast, a dielectric functions as an insulator under conditions where there are no ionization processes acting. This is the case in the absence of the ionization produced by an electric discharge or the ionization present in the case of an aqueous solution containing a dissolved acid, base, or salt. In contrast to dielectrics, where the charges are fixed in the molecules, the charged species in conductors can translate over distances large as compared with the intermolecular separations. In metals and semiconductors, the continuous current is carried by the electrons, whereas in electrolyte solutions, and in certain solids termed superionic conductors, the current is carried by ions. An example of the latter is silver iodide. The dimensions of the AgI unit cell are determined entirely by interionic contacts between neighboring I^- ions. Within this structure, the smaller Ag^+ ions are free to move through the interstitial spaces between the I^- ions. The Ag^+ ions carry the current. In the case of conducting materials, we identify the charge carriers, be they electrons or ions, as *free* charges.

Each molecule in a dielectric material occupies a space that is impenetrable to any other molecule. From the point of view of an observer stationed on a given molecule, the ups and downs of the local charge density due to the molecular packing become smeared out with increasing distance, so that the distant local net charge density appears to be zero. At the atomic dimension, atoms and molecules are mostly empty space, so it is a fair approximation to regard a molecule as a cavity in an otherwise continuous dielectric. At the center of the cavity, a point dipole is placed so as to represent the molecular dipole induced by the external field. If the dielectric is polar, a permanent electric dipole is also present. This molecular model consisting of a dipoles in an empty cavity immersed in an otherwise continuous dielectric is known as the *semi-continuum model.*

Below, we shall begin by introducing the macroscopic theory of polarization. We shall then use the theory to derive Born's equation (Max Born, German–English, 1882–1970) for the heat of solution of an ion. The next task will be to examine the theoretical basis for the various electromagnetic equations of state that have been advanced to represent the density dependence of the electric permittivity. This will entail calculation of the local electric field. The local electric field concept will then lead us to a theory for the effect of the solvent on the absorption spectra of solutes (solvatochromism), as well as a theory for the heat of solution of polar molecules.

3.2 The Polarization and the Displacement

According to Eq. (1.11.1) of Chap. 1, the electrostatic potential at a point b in a vacuum due to a volume distribution, $\rho(\mathbf{r}_a)$, of free charge centered on point a is

$$\Phi(\mathbf{r}_b) = \frac{1}{4\pi\varepsilon_o}\int\frac{\rho(\mathbf{r}_a)d^3r_a}{|\mathbf{r}_b - \mathbf{r}_a|} = \frac{1}{4\pi\varepsilon_o}\int\frac{\rho(\mathbf{r}_a)}{r_{ab}}d^3r_a \tag{3.2.1}$$

In Eq. (3.2.1), we have used the definitions, $\mathbf{r}_{ab} = \mathbf{r}_b - \mathbf{r}_a$, and $r_{ab} = |\mathbf{r}_b - \mathbf{r}_a|$. On the basis of this statement of the electrostatic potential due to a *free* charge density, we will proceed below to derive an equation for the electrostatic potential which includes both free charges and also bound charges in the form of electric dipoles. We will then introduce a new vector, called the *displacement*, which when combined with the Gauss' divergence theorem will permit us to arrive at an equation for the electrostatic potential that applies in the case of a continuous dielectric material containing both free charge and bound charge.

We start with the result derived in Chap. 1 for the electrostatic potential of an isolated *point* electric dipole. According to Eq. (1.9.2), the potential at b due to a point electric dipole located at a is given by

$$\Phi(\mathbf{r}_b) = \frac{\mathbf{p}_a\cdot\mathbf{r}_{ab}}{4\pi\varepsilon_o r_{ab}^3} = \frac{\mathbf{p}_a\cdot(\mathbf{r}_b - \mathbf{r}_a)}{4\pi\varepsilon_o|\mathbf{r}_b - \mathbf{r}_a|^3} \tag{3.2.2}$$

Equation (3.2.2) can be generalized so that it represents the electrostatic potential at b due to an electric dipole *distribution* which is centered on point a. The following mathematics will prove helpful:

$$\begin{aligned}\nabla_b \frac{1}{r_{ab}} &= \left(\hat{\mathbf{e}}_x \frac{\partial}{\partial x_b} + \hat{\mathbf{e}}_y \frac{\partial}{\partial y_b} + \hat{\mathbf{e}}_z \frac{\partial}{\partial z_b}\right)\left(\frac{1}{r_{ab}}\right) \\ &= \hat{\mathbf{e}}_x \frac{\partial}{\partial x_b}\left(\frac{1}{r_{ab}}\right) + \hat{\mathbf{e}}_y \frac{\partial}{\partial y_b}\left(\frac{1}{r_{ab}}\right) + \hat{\mathbf{e}}_z \frac{\partial}{\partial z_b}\left(\frac{1}{r_{ab}}\right) \\ &= -\left(\frac{1}{r_{ab}^2}\right)\left(\hat{\mathbf{e}}_x \frac{\partial r_{ab}}{\partial x_b} + \hat{\mathbf{e}}_y \frac{\partial r_{ab}}{\partial y_b} + \hat{\mathbf{e}}_z \frac{\partial r_{ab}}{\partial z_b}\right) \\ &= -\frac{\nabla_b r_{ab}}{r_{ab}^2} \end{aligned} \tag{3.2.3}$$

By referring to Eq. (1.3.5) of Chap. 1, we can evaluate $\nabla_b r_{ab}$ as follows:

$$\begin{aligned}\nabla_b r_{ab} &= \left[\hat{\mathbf{e}}_x \frac{\partial}{\partial x_b} + \hat{\mathbf{e}}_y \frac{\partial}{\partial y_b} + \hat{\mathbf{e}}_z \frac{\partial}{\partial z_b}\right]\left[(x_b - x_a)^2 + (y_b - y_a)^2 + (z_b - z_a)^2\right]^{1/2} \\ &= \frac{1}{2}\left[\frac{2(x_b - x_a)\hat{\mathbf{e}}_x + 2(y_b - y_a)\hat{\mathbf{e}}_y + 2(z_b - z_a)\hat{\mathbf{e}}_z}{\left((x_b - x_a)^2 + (y_b - y_a)^2 + (z_b - z_a)^2\right)^{1/2}}\right] \\ &= \frac{\mathbf{r}_{ab}}{r_{ab}} \end{aligned} \tag{3.2.4}$$

The combination of Eqs. (3.2.3) and (3.2.4) leads to the simple result

$$\nabla_b \frac{1}{r_{ab}} = -\frac{\mathbf{r}_{ab}}{r_{ab}^3} \tag{3.2.5}$$

If Eq. (3.2.5) is substituted into Eq. (3.2.2), the electrostatic potential associated with a point dipole can be written as

$$\Phi(\mathbf{r}_b) = -\frac{1}{4\pi\varepsilon_o}\mathbf{p}_a \cdot \nabla_b\left(\frac{1}{r_{ab}}\right) \tag{3.2.6}$$

Because $\mathbf{r}_a$ and $\mathbf{r}_b$ appear with opposite signs in the definition of r_{ab}, the two possible gradients of $(1/r_{ab})$ are related by

$$\nabla_b\left(\frac{1}{r_{ab}}\right) = -\nabla_a\left(\frac{1}{r_{ab}}\right) \tag{3.2.7}$$

Upon substitution of Eq. (3.2.7) into Eq. (3.2.6), one obtains for a point dipole,

$$\Phi(\mathbf{r}_b) = \frac{1}{4\pi\varepsilon_o}\mathbf{p}_a \cdot \nabla_a\left(\frac{1}{r_{ab}}\right) \tag{3.2.8}$$

The volume density of dipole moment at the point a located at $\mathbf{r}_a$ is termed the *polarization* and is given the symbol $\mathbf{P}(\mathbf{r}_a)$. If the volume element associated with point a is d^3r_a, then the dipole moment contained within this volume is $\mathbf{P}(\mathbf{r}_a)d^3r_a$. The electrostatic potential at point b due to the total dipole moment within the volume surrounding point a is then given by the integral

$$\Phi(\mathbf{r}_b) = \frac{1}{4\pi\varepsilon_o}\int d^3r_a\mathbf{P}(\mathbf{r}_a)\cdot\nabla_a\left(\frac{1}{r_{ab}}\right) \tag{3.2.9}$$

Next refer to the vector identity

$$\nabla\cdot(\Phi\mathbf{A}) = \mathbf{A}\cdot\nabla\Phi + \Phi\nabla\cdot\mathbf{A} \tag{3.2.10a}$$

which can be found in the Mathematical Methods Appendix. With $\mathbf{A} = \mathbf{P}(\mathbf{r}_a)$ and $\Phi = 1/r_{ab}$, the identity predicts,

$$\mathbf{P}(\mathbf{r}_a)\cdot\nabla_a\left(\frac{1}{r_{ab}}\right) = \nabla_a\cdot\left(\frac{\mathbf{P}(\mathbf{r}_a)}{r_{ab}}\right) - \left(\frac{1}{r_{ab}}\right)\nabla_a\cdot\mathbf{P}(\mathbf{r}_a) \tag{3.2.10b}$$

Substitution of Eq. (3.2.10b) into Eq. (3.2.9) results in the sum of integrals,

$$\Phi(\mathbf{r}_b) = \frac{1}{4\pi\varepsilon_o}\int d^3r_a\nabla_a\cdot\left(\frac{\mathbf{P}(\mathbf{r}_a)}{r_{ab}}\right) - \frac{1}{4\pi\varepsilon_o}\int d^3r_a\left(\frac{1}{r_{ab}}\right)\nabla_a\cdot\mathbf{P}(\mathbf{r}_a) \tag{3.2.11}$$

A differential element of the surface area surrounding the volume element, d^3r_a, can be written in the form

$$d\mathbf{S}_a = dS_a\hat{\mathbf{n}}_a \tag{3.2.12}$$

where $\hat{\mathbf{n}}_a$ is the unit normal vector drawn outward from the surface. If we use the Gauss–Green divergence theorem, which can be found in the Mathematical Methods Appendix, the first volume integral on the right hand side of Eq. (3.2.11) can be replaced by a surface integral so that the equation reads,

$$\Phi(\mathbf{r}_b) = \frac{1}{4\pi\varepsilon_o}\int dS_a\hat{\mathbf{n}}_a\cdot\left(\frac{\mathbf{P}(\mathbf{r}_a)}{r_{ab}}\right) - \frac{1}{4\pi\varepsilon_o}\int d^3r_a\left(\frac{1}{r_{ab}}\right)\nabla_a\cdot\mathbf{P}(\mathbf{r}_a) \tag{3.2.13}$$

The ends of some of the dipoles will rest on the element of surface area, dS_a. These charged ends constitute a surface charge density, $\sigma_P(\mathbf{r}_a)$, given by

$$\sigma_P(\mathbf{r}_a) = \mathbf{P}(\mathbf{r}_a)\cdot\hat{\mathbf{n}}_a \tag{3.2.14}$$

Although the material in the volume element, dr_a^3, is electrically neutral overall, the dipole ends have opposite charge, and any spatial non-uniformity in their distribution will be sensed as an apparent net charge density by an observer passing from one point to the next. According to the second integral in Eq. (3.2.13), this apparent volume charge density, which we denote by $\rho_P(\mathbf{r}_a)$, is given by

$$\rho_P(\mathbf{r}_a) = -\nabla_a \cdot \mathbf{P}(\mathbf{r}_a) \tag{3.2.15}$$

By employing Eqs. (3.2.14) and (3.2.15), Eq. (3.2.13) can be rewritten as

$$\Phi(\mathbf{r}_b) = \frac{1}{4\pi\varepsilon_o}\int dS_a \cdot \left(\frac{\sigma_P(\mathbf{r}_a)}{r_{ab}}\right) + \frac{1}{4\pi\varepsilon_o}\int d^3r_a\left(\frac{\rho_P(\mathbf{r}_a)}{r_{ab}}\right) \tag{3.2.16}$$

As mentioned, the contributions made by the polarization to the surface charge density, $\sigma_P(\mathbf{r}_a) = \mathbf{P}(\mathbf{r}_a)\cdot\hat{\mathbf{n}}_a$, are limited to those dipoles positioned in the volume so that one end of the dipole lies on the surface bounding d^3r_a. Because the surface-to-volume ratio decreases with increasing radius, r_{ab}, the contribution to the potential due to the volume charge density will ultimately dominate the contribution due to the surface charge density. Hence, in Eq. (3.2.16) we can ignore $\sigma_P(\mathbf{r}_a)$ in comparison to $\rho_P(\mathbf{r}_a)$ and write simply,

$$\Phi(\mathbf{r}_b) = \frac{1}{4\pi\varepsilon_o}\int d^3r_a\left(\frac{\rho_P(\mathbf{r}_a)}{r_{ab}}\right) \tag{3.2.17}$$

The role played by $\rho_P(\mathbf{r}_a)$ in Eq. (3.2.17) in determining the electric field can be understood by comparing it with the effect of free charge existing alone. According to Eq. (1.12.1) of Chap. 1, the integral of the electric field over a surface in the case of free charge in vacuum is given by

$$\int \mathbf{E}(\mathbf{r}_a)\cdot d\mathbf{S}_a = (1/\varepsilon_o)\int \rho(\mathbf{r}_a)d^3r_a \tag{3.2.18}$$

where $\rho(\mathbf{r}_a)$ is the density of the free charge. To extend this equation to apply in the case of both free and bound charge, we need to take into account the contribution to the electric field made by the polarization density. This is done by adding $\rho_P(\mathbf{r}_a) = -\nabla_a \cdot \mathbf{P}(\mathbf{r}_a)$ from Eq. (3.2.15) to the integrand on right hand side of Eq. (3.2.18). The result is

$$\begin{aligned}\int \mathbf{E}(\mathbf{r}_a)\cdot d\mathbf{S}_a &= (1/\varepsilon_o)\int [\rho(\mathbf{r}_a) + \rho_P(\mathbf{r}_a)]\, d^3r_a \\ &= (1/\varepsilon_o)\int [\rho(\mathbf{r}_a) - \nabla_a \cdot \mathbf{P}(\mathbf{r}_a)]\, d^3r_a\end{aligned} \tag{3.2.19}$$

Upon application of the Gauss–Green divergence theorem, which can be found in the Mathematical Methods Appendix, the surface integral on the left can be rewritten as a volume integral. Equation (3.2.19) then reads,

$$\int [\nabla_a \cdot \mathbf{E}(\mathbf{r}_a)] d^3 r_a = (1/\varepsilon_o) \int [\rho(\mathbf{r}_a) - \nabla_a \cdot \mathbf{P}(\mathbf{r}_a)]\, d^3 r_a \tag{3.2.20}$$

The terms in Eq. (3.2.20) can be regrouped so that

$$\int \{\nabla_a \cdot [\varepsilon_o \mathbf{E}(\mathbf{r}_a) + \mathbf{P}(\mathbf{r}_a)]\}\, d^3 r_a = \int \rho(\mathbf{r}_a) d^3 r_a \tag{3.2.21}$$

If Eq. (3.2.21) is to be true regardless of the choice of the differential volume element, $d^3 r_a$, then the integrands on either side must be equal. This condition is satisfied if

$$\nabla_a \cdot [\varepsilon_o \mathbf{E}(\mathbf{r}_a) + \mathbf{P}(\mathbf{r}_a)] = \rho(\mathbf{r}_a) \tag{3.2.22}$$

We next define the *displacement* vector, **D**, by

$$\mathbf{D} = \varepsilon_o \mathbf{E} + \mathbf{P} \tag{3.2.23}$$

If reference to the specific location, $\mathbf{r}_a$, is suppressed, then Eqs. (3.2.22) and (3.2.23) can be combined to read,

$$\nabla \cdot \mathbf{D} = \rho \tag{3.2.24}$$

According to Eq. (3.2.24), the source of **D** is to be found exclusively in the free charge density, ρ. Equation (3.2.24) constitutes one of the fundamental laws of electromagnetism and is known as Gauss' Law.

In the case of a continuous, isotropic dielectric where the polarization depends linearly upon the electric field, the two vectors are connected by the phenomenological relation,

$$\mathbf{P} = \varepsilon_o \chi_e \mathbf{E} \tag{3.2.25}$$

where the coefficient, χ_e, known as the electric susceptibility, is material specific. If we substitute Eq. (3.2.25) into Eq. (3.2.23), the displacement takes the form,

$$\mathbf{D} = \varepsilon_o \mathbf{E} + \mathbf{P} = \varepsilon_o \mathbf{E} + \varepsilon_o \chi_e \mathbf{E} = \varepsilon_o (1 + \chi_e) \mathbf{E} \tag{3.2.26}$$

Now define the electric permittivity, ε, of the dielectric by writing Eq. (3.2.26) as

$$\mathbf{D} = \varepsilon \mathbf{E} \tag{3.2.27}$$

By virtue of Eq. (3.2.26), ε has the definition,

$$\varepsilon = \varepsilon_r \varepsilon_o \tag{3.2.28}$$

where

$$\varepsilon_r = 1 + \chi_e \tag{3.2.29}$$

The quantity, ε_r, is known as the relative permittivity, and is unity in vacuum, where $\chi_e = 0$. On the basis of the definition of ε in Eq. (3.2.28), Eq. (3.2.27) becomes

$$\mathbf{D} = \varepsilon_r \varepsilon_o \mathbf{E} \tag{3.2.30}$$

We recall from the theory of electrostatics in Chap. 1 that

$$\mathbf{E} = -\nabla\Phi \tag{3.2.31}$$

is the equation for the electric field in terms of the potential. By using Eqs. (3.2.30) and (3.2.31), Eq. (3.2.24) can then be rewritten in the form

$$\nabla^2\Phi = -\rho/\varepsilon_r\varepsilon_o \tag{3.2.32}$$

We remarked in Chap. 1 that Eq. (3.2.32) is known as Poisson's equation.

3.3 Coulomb's Law for Free Charges in a Continuous Dielectric

In analogy with the Coulomb force acting between two charges in vacuum, the corresponding force, $\mathbf{F}_{ab}$, acting between two charges, q_a and q_b, immersed in a dielectric is

$$\mathbf{F}_{ab} = \frac{q_a q_b}{4\pi\varepsilon r_{ab}^2}\hat{\mathbf{r}}_{ab} \tag{3.3.1}$$

Example Born Charging Formula.

We consider an ion in solution to be a sphere of radius R immersed in a continuous dielectric. The solvation free energy is reckoned as the difference between the energy of the charged sphere in the dielectric and a charged sphere in vacuum. The difference can be evaluated by first considering the energy required to bring an uncharged sphere to its full charge. This is calculated as the integral of the differential of the work, dW_ε, done when an additional charge in the amount, dq', is moved from infinity to the surface of the sphere with charge q'. According to Eq. (3.3.1), the differential of the magnitude of the force in the case where $r_{ab} = r$, $q_a = q'$, and $q_b = dq'$ is

$$dF = \frac{q'dq'}{4\pi\varepsilon r^2} \tag{3.3.2}$$

The differential of the charging energy is

$$d^2W_\varepsilon = -dFdr = -\frac{q'dq'}{4\pi\varepsilon r^2}dr \tag{3.3.3}$$

Integration from $r = \infty$ to $r = R$ gives

$$dW_\varepsilon = -\int_\infty^R \frac{q'dq'}{4\pi\varepsilon r^2}dr = \frac{q'dq'}{4\pi\varepsilon R} \tag{3.3.4}$$

A second integration over dq' from zero to $q' = q$ gives,

$$W_\varepsilon = \frac{1}{4\pi\varepsilon R}\int_0^q q'dq' = \frac{q^2}{8\pi\varepsilon R} \tag{3.3.5}$$

Let the sphere represent an ion with electrovalence, Z_i. The charge on the ion is then $q = Z_i e$, where e is the magnitude of the charge of the electron. Applied to the ion, the work calculated in Eq. (3.3.5) is

$$W_\varepsilon = \frac{Z_i^2 e^2}{8\pi\varepsilon R_i} \tag{3.3.6}$$

If the ion is charged in vacuum, ε in Eq. (3.3.6) must be replaced with ε_o. The corresponding work of charging is

$$W_{\varepsilon_o} = \frac{Z_i^2 e^2}{8\pi\varepsilon_o R_i} \tag{3.3.7}$$

The energy required to transfer the sphere from vacuum to the dielectric is the difference between Eqs. (3.3.6) and (3.3.7), namely,

$$W_\varepsilon - W_{\varepsilon_o} = \frac{Z_i^2 e^2}{8\pi\varepsilon R_i} - \frac{Z_i^2 e^2}{8\pi\varepsilon_o R_i} \tag{3.3.8}$$

The electric work measured at constant temperature and pressure is equal to the Gibbs free energy. If ε is determined at constant temperature and pressure, the work in Eq. (3.3.8) can be identified with the Gibbs free energy of solvation, $\Delta_{solv}G^o(i)$, namely,

$$\Delta_{solv}G^o(i) = \frac{Z_i^2 e^2}{8\pi\varepsilon R_i} - \frac{Z_i^2 e^2}{8\pi\varepsilon_o R_i} \tag{3.3.9}$$

After factoring Eq. (3.3.9) and introducing Eq. (3.2.28), we have

$$\Delta_{solv}G^{o}(i) = -\frac{Z_i^2 e^2}{8\pi\,\varepsilon_o R_i}\left[1 - \frac{1}{\varepsilon_r}\right] \tag{3.3.10}$$

Equation (3.3.10) is due to Max Born (German, English 1882–1970). The value of $\Delta_{solv}G^{o}(i)$ can be rendered in molar units by multiplying Eq. (3.3.10) by the Avogadro number. The relative permittivity of water at 298.15 K is $\varepsilon_r = 78.54$, so in the case of an aqueous solution,

$$\Delta_{solv}G^{o}(i) = -\frac{Z_i^2}{R_i} \times (6.86 \times 10^4\,\text{kJ mol}^{-1}\,\text{pm}) \tag{3.3.11}$$

Notice that $\Delta_{solv}G^{o}(i) < 0$, which indicates that the ion in solution is energetically favored as compared to the ion in vacuum. That said, Eq. (3.3.11) is an idealization, which ignores chemical bonding effects. Its accuracy can be improved, however, if its application is restricted to the comparison of ions having the same electrovalence. For example, in the case Na^+ and K^+, we can write

$$\Delta_{solv}G^{o}(Na^{+}) - \Delta_{solv}G^{o}(K^{+}) = -\left(\frac{1}{R_{Na^{+}}} - \frac{1}{R_{K^{+}}}\right) \times (6.86 \times 10^4\,\text{kJ mol}^{-1}\,\text{pm}) \tag{3.3.12}$$

After the introduction of $R_{Na^{+}} = 185.8$ pm and $R_{K^{+}} = 227.2$ pm, the numerical evaluation of Eq. (3.3.12) gives

$$\begin{aligned}\Delta_{solv}G^{o}(Na^{+}) - \Delta_{solv}G^{o}(K^{+}) &= -\left(\frac{1}{185.8} - \frac{1}{227.2}\right)\\ &\quad \times (6.86 \times 10^4\,\text{kJ mol}^{-1}\,\text{pm})\\ &= -67.3\,\text{kJ/mol}\end{aligned} \tag{3.3.13}$$

which serves as a reasonable estimate.

By virtue of the appearance of ε_r in Eq. (3.3.10), $\Delta_{solv}G^{o}(i)$ is temperature dependent. This temperature dependence permits the heat of solution, $\Delta_{solv}H^{o}(i)$, to be calculated by substituting $\Delta_{solv}G^{o}(i)$ into the Gibbs–Helmholtz equation, which is given by Eq. (2.5.20). Used to calculate $\Delta_{solv}H^{o}(i)$, Eq. (2.5.20) reads,

$$\Delta_{solv}H^{o}(i) = -T^2\left(\frac{\partial(\Delta_{solv}G^{o}(i)/T)}{\partial T}\right) \tag{3.3.14}$$

Upon substitution of Eq. (3.3.10) into Eq. (3.3.14) followed by computation of the indicated derivative, the heat of solution is

$$\Delta_{solv}H^{o}(i) = \frac{Z_i^2 e^2 T^2}{8\pi\,\varepsilon_o R_i \varepsilon_r^2}\left(\frac{\partial \varepsilon_r}{\partial T}\right) \tag{3.3.15}$$

In the case of most solvents capable of dissolving ions, $(\partial \varepsilon_r/\partial T)$ is negative. Thus, according to the Born model $\Delta_{solv}H^o(i)$ is exothermic.

3.4 Electric Field and Potential Due to a Distribution of Free Charges in a Dielectric

According to Eq. (3.2.1), the electrostatic potential at an arbitrary point $\mathbf{r}$ in a dielectric of permittivity, ε, due to a distribution, $\rho(\mathbf{r}_a)$, of free charges located at a point a can be calculated in the form,

$$\Phi(\mathbf{r}) = \frac{1}{4\pi\varepsilon}\int \frac{\rho(\mathbf{r}_a)d^3r_a}{|\mathbf{r}-\mathbf{r}_a|} \tag{3.4.1}$$

The corresponding electric field is

$$\mathbf{E}(\mathbf{r}) = -\nabla\Phi(\mathbf{r}) \tag{3.4.2}$$

Upon substitution of Eq. (3.4.1) into Eq. (3.4.2) and evaluation of the gradient using the analogous derivation for a point charge in Sect. 1.6 of Chap. 1, the corresponding electric field is

$$\mathbf{E}(\mathbf{r}) = \frac{1}{4\pi\varepsilon}\int \rho(\mathbf{r}_a)\frac{(\mathbf{r}-\mathbf{r}_a)}{|\mathbf{r}-\mathbf{r}_a|^3}d^3r_a \tag{3.4.3}$$

3.5 Lorentz Local Electric Field and the Clausius–Mossotti Dielectric Equation of State

A mechanical equation of state for a material under isothermal conditions links the density of the material to the ambient pressure. By analogy, a so-called electromagnetic equation of state links the isothermal electric permittivity of a substance to its density. In a remarkable coincidence of eponymous association, the Lorenz–Lorentz equation for the isothermal electric permittivity was proposed by the Danish physicist, L. V. Lorenz (1829–1899) in 1869 and again a year later by the Dutch physicist H. A. Lorentz (1853–1928).

Because two molecules cannot occupy the same space at the same time, Lorenz and Lorentz assumed that a molecule in the dielectric carves out for itself a sphere of occupation. See Fig. 3.1.

In the Lorenz–Lorentz model, the sphere of occupation of a molecule is assumed to be so small that it does not distort the strength or direction of an external electric field that permeates the surrounding material. Since atoms and molecules are mostly

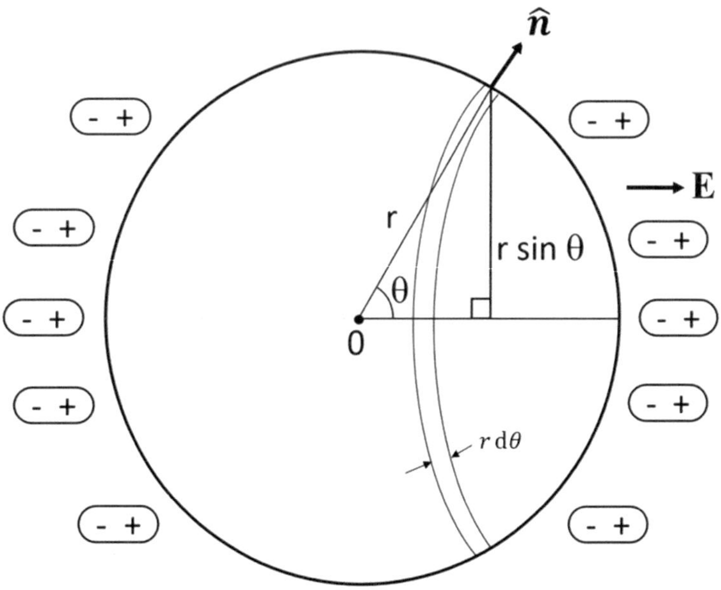

Fig. 3.1 Empty sphere of radius, r, in a continuous dielectric. The presence of the sphere is assumed to have no effect on the direction of the electric field lines associated with the applied electric field, **E**, which are everywhere parallel and taken to be in the horizontal direction. The orientations of the polarized molecular dipoles with ends touching the surface of the sphere are parallel to **E**. The radius, r, makes an angle θ with the horizontal. The outward unit normal vector to a spherical band of width $rd\theta$ on the surface of the sphere is $\hat{\mathbf{n}}$

empty space, the relative permittivity inside the sphere can be taken as unity. Although the molecules, which are contiguous to the surface of the sphere cannot enter it, their dipoles will, nonetheless, orient themselves in response to the applied field so that poles having all the same sign come to rest on the surface of one hemisphere of the cavity, while poles of the opposite sign come to rest on the surface of the opposite hemisphere. If the bulk polarization is **P**, then according to Eq. (3.2.14), the dipolar orientation creates on the sphere an effective surface charge density, σ_P, given by

$$\sigma_P = \mathbf{P} \cdot \hat{\mathbf{n}} \tag{3.5.1}$$

For ease in calculation, the origin of coordinates is located at the center of the sphere. As shown in Fig. 3.1, the unit normal vector, $\hat{\mathbf{n}}$, points in the positive radial direction. As the external polarization **P** is assumed to be parallel to **E**, the angle between $\hat{\mathbf{n}}$ and **P** is thus θ. Hence, $\mathbf{P} \cdot \hat{\mathbf{n}} = P\cos\theta$. On the basis of Eq. (3.5.1), the angular dependence of the charge density at the surface of the sphere is

$$\sigma_P = P\cos\theta \tag{3.5.2}$$

The radius of the sphere being, r, a circular band of polarization charge density lying on the surface of the sphere is located at a height $r\sin\theta$ above the horizontal axis. This height serves as the radius of the band. The circumference of the band is thus $2\pi r\sin\theta$. Lying as it is on the surface of the sphere, the band has an infinitesimal width equal to $rd\theta$. Thus the area of the band is the product of its circumference times its width or $2\pi r\sin\theta rd\theta$. The charge on the band equals the area times the charge density, which is $2\pi r^2\sigma_P\sin\theta d\theta = 2\pi r^2 P\cos\theta\sin\theta d\theta$. According to Coulomb's law, this charge produces at the center of the empty sphere a radially directed electric field of magnitude,

$$\frac{2\pi r^2 P\cos\theta\sin\theta d\theta}{4\pi\varepsilon_o r^2}$$

The component of this electric field which is parallel to the horizontal direction is

$$\frac{2\pi r^2 P\cos\theta\sin\theta d\theta}{4\pi\varepsilon_o r^2}\cdot\cos\theta$$

The total horizontal electric field produced at the origin of the sphere due to the surface charge density can be calculated by integration over the range of the polar angle, θ. The result is

$$\frac{2\pi r^2 P}{4\pi\varepsilon_o r^2}\int_o^{\pi}\cos^2\theta\sin\theta d\theta = \frac{P}{2\varepsilon_o}\left[-\frac{\cos^3\theta}{3}\right]_0^{\pi} = \frac{P}{2\varepsilon_o}\left(\frac{1}{3}+\frac{1}{3}\right) = \frac{P}{3\varepsilon_o} \quad (3.5.3)$$

In the Lorenz–Lorentz model, the local electric field, F, which is sensed at the center of the sphere is the sum of the applied field, E, and the field given by Eq. (3.5.3). Thus

$$F = E + \frac{P}{3\varepsilon_o} \quad (3.5.4)$$

We can solve this equation for

$$P = 3\varepsilon_o(F - E) \quad (3.5.5a)$$

With reference to Eqs. (3.2.26) and (3.2.30), the displacement can be written in scalar form as

$$D = \varepsilon_r\varepsilon_o E = \varepsilon_o E + P \quad (3.5.5b)$$

Upon substitution of Eq. (3.5.5a) into the far right hand side of Eq. (3.5.5b), the result is

$$\varepsilon_r \varepsilon_o E = \varepsilon_o E + 3\varepsilon_o (F - E) \tag{3.5.5c}$$

which can then be solved for F, where

$$F = \left(\frac{\varepsilon_r + 2}{3} \right) E \tag{3.5.6}$$

This representation of F is known as the Lorentz local field and is sometimes denoted by F_L. Because $\varepsilon_r > 1$, the value of F, as given by Eq. (3.5.6), is greater than the value of the applied electric field, E. The local electric field induces a dipole moment in any molecule on which it acts. In the case of a non-polar dielectric, the induced dipoles constitute the principal response. If the polarizability of a molecule is α, then the dipole moment, p, induced in the molecule is given by

$$p = \alpha F \tag{3.5.7}$$

If there are N such molecules per unit volume, then the macroscopic polarization is

$$P = N\alpha F \tag{3.5.8}$$

Substitute Eq. (3.5.6) into Eq. (3.5.8) to obtain,

$$P = N\alpha \left(\frac{\varepsilon_r + 2}{3} \right) E \tag{3.5.9}$$

Equation (3.5.9) can be substituted into the right hand side Eq. (3.5.5b) to obtain the result,

$$\varepsilon_r \varepsilon_o E = \varepsilon_o E + N\alpha \left(\frac{\varepsilon_r + 2}{3} \right) E \tag{3.5.10}$$

which upon cancellation of E and algebraic rearrangement, takes the form,

$$\frac{\varepsilon_r - 1}{\varepsilon_r + 2} = \frac{N\alpha}{3\varepsilon_o} \tag{3.5.11}$$

Equation (3.5.11) gives the relative permittivity, ε_r, of a non-polar dielectric in terms of its molecular number density, N, and the polarizability α, of the molecules, which it contains. As such, Eq. (3.5.11) qualifies as an electromagnetic equation of state. Due to the fact that the sole response of the material to the applied electric field is induced polarization, it is applicable only to non-polar dielectrics. It was introduced independently by Rudolf Clausius (German, 1822–1888) and O. F. Mossotti (Italian, 1791–1863) and is known universally as the Clausius–Mossotti formula. Equation (3.5.11) can be used to represent the optical properties of a material by replacing ε_r by the square of the refractive index, n^2, and noting that α becomes a function of the wavelength of the light. With these replacements, Eq. (3.5.11) is known as the Lorenz–Lorentz equation.

3.6 Boundary Conditions in a Dielectric

In the case of a heterogeneous chemical mixture, there are interfaces that separate the coexisting phases. Each coexisting phase has a value of the relative permittivity, ε_r, unique to that phase. To calculate the electrical properties in such cases, one must know how the displacement changes when crossing the interface between adjacent phases. See Fig. 3.2.

To derive the boundary condition satisfied by the displacement, we start by computing the volume integral of Eq. (3.2.24), which is

$$\int d^3r\nabla \cdot \mathbf{D} = \int d^3r\rho \tag{3.6.1a}$$

The Gauss-Green divergence theorem from the Mathematical Methods Appendix can be used to evaluate left hand side of Eq. (3.6.1a). The result is

$$\int d^3r\nabla \cdot \mathbf{D} = \int d\mathbf{S} \cdot \mathbf{D} \tag{3.6.1b}$$

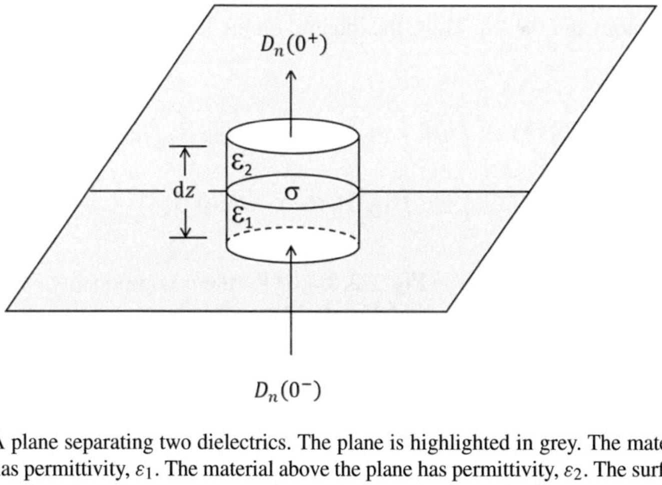

Fig. 3.2 A plane separating two dielectrics. The plane is highlighted in grey. The material below the plane has permittivity, ε_1. The material above the plane has permittivity, ε_2. The surface charge density at the interface between the two planes is σ. The normal component of the displacement vector at the surface just below the plane is $D_n(0^-)$. The normal component of the displacement vector at the surface just above the plane is $D_n(0^+)$. The white ovals below and above the plane represent the bottom and top, respectively, of a pillbox. The pillbox contains an element of the plane having charge density, σ. The height of the pillbox is dz

where $d\mathbf{S}$ is an element of the area of the planar surface in Fig. 3.2. Combine Eqs.(3.6.1a) and (3.6.1b) to obtain

$$\int d\mathbf{S}\cdot\mathbf{D} = \int d^3r\,\rho \tag{3.6.2}$$

The volume of integration for the integral on the right-hand side of Eq. (3.6.2) is the interior of the pillbox as shown in Fig. 3.2. Written in terms of a free charge volume density, ρ, the charge in the pillbox is ρd^3r, where $d^3r = dxdydz$. The volume charge density has its origin, of course, in the surface charge density, σ, at the junction between the two dielectrics. The relation between the two charge densities is $\sigma = \int \rho dz$. As a consequence, the integral on the right hand side of Eq. (3.6.1a) can be simplified to read,

$$\int d^3r\rho = \int dx \int dy \int dz\,\rho = \int dx \int dy\,\sigma \tag{3.6.3}$$

We now turn to the integral on the left-hand side of Eq. (3.6.2). The z-direction is perpendicular to the surface of the plane bisecting the pillbox. The height of the pillbox is dz while an differential element of the area of its base is $dS = dxdy$. The outward normal (not shown) associated with the base of the pillbox is pointed in the $-z$ direction, while the outward normal (not shown) to the lid of the pillbox is pointed in the $+z$ direction. We assume that the direction of $\mathbf{D}$ within the pillbox is everywhere parallel to the z-axis. There is thus no flux of $\mathbf{D}$ through the sides of the pillbox. Hence, as indicated on the left hand side of Eq. (3.6.2), if we integrate over the entire surface of the pillbox where $\mathbf{D}$ is non-zero, the sole contributions come from the bottom and the lid. Thus, the integral on the left hand side of Eq. (3.6.2) is just

$$\begin{aligned}\int d\mathbf{S}\cdot\mathbf{D} &= \int dx \int dy\,D_n(0^+) - \int dx \int dy\,D_n(0^-)\\ &= \int dx \int dy\big(D_n(0^+) - D_n(0^-)\big)\end{aligned} \tag{3.6.4}$$

In Eq. (3.6.4) and as shown in Fig. 3.2, $D_n(0^+)$ is the z- component of $\mathbf{D}$ evaluated on the surface element, dS, at a height infinitesimally above $z = 0$, while $D_n(0^-)$ represents the z- component of $\mathbf{D}$ evaluated on dS at a height infinitesimally below $z = 0$.

Upon combining Eqs. (3.6.3) and (3.6.4), Eq. (3.6.2) reads,

$$\int dx \int dy\big(D_n(0^+) - D_n(0^-)\big) = \int dx \int dy\,\sigma \tag{3.6.5}$$

from which it is possible to conclude that

$$D_n(0^+) - D_n(0^-) = \sigma \tag{3.6.6}$$

According to Eq. (3.6.6), if the surface of contact between two materials contains a non-zero surface charge density, σ, as shown Fig. 3.2, then the value of D_n changes discontinuously upon crossing the surface; otherwise, D_n is continuous.

It is possible to convert Eq. (3.6.6) into a condition satisfied by the potential Φ. By combining Eqs. (3.2.27) and (3.2.31), we have

$$\mathbf{D} = -\varepsilon \nabla \Phi \tag{3.6.7}$$

Referring again to Fig. 3.2, if we designate $\hat{\mathbf{n}}$ as the unit vector perpendicular to the plane, then according to Eq. (3.6.7), the normal component of the displacement vector, D_n, is given by

$$D_n = -\varepsilon \nabla \Phi \cdot \hat{\mathbf{n}} = -\varepsilon \frac{\partial \Phi}{\partial n} \tag{3.6.8}$$

where n, not be confused with the refractive index, designates the coordinate parallel to $\hat{\mathbf{n}}$. In rectangular coordinates, n, is ordinarily one of (x, y, z). If the permittivity of the material in the half-space defined by $n > 0$ is ε_2, while the permittivity of the material in the half-space, $n < 0$, is ε_1, then by virtue of Eq. (3.6.8) we can write Eq. (3.6.6) in the form,

$$\varepsilon_2 \left(\frac{\partial \Phi}{\partial n} \right)_{0^+} - \varepsilon_1 \left(\frac{\partial \Phi}{\partial n} \right)_{0^-} = -\sigma \tag{3.6.9}$$

It is possible to derive a second boundary condition satisfied by Φ. To obtain this equation, we begin by forming the curl of both sides of Eq. (3.2.31). Using a vector identity from the Mathematical Methods Appendix, we find that

$$\nabla \times \mathbf{E} = -\nabla \times \nabla \Phi = 0 \tag{3.6.10}$$

At the junction between the two dielectrics, we choose a path of integration consisting of a narrow rectangle. See Fig. 3.3.

One long side of the rectangle lies in the material with permittivity, ε_1, while the other long side lies in the material with permittivity ε_2. The two short sides of the rectangle are both perpendicular to the surface that contacts the two dielectrics. By virtue of Stokes theorem (Sir George G. Stokes, British, (1819–1903), which can be found in the Mathematical Methods Appendix, the integral over the surface $d\mathbf{S}$ of the rectangle is related to the integral around its perimeter by

$$\int d\mathbf{S} \cdot (\nabla \times \mathbf{E}) = \oint d\mathbf{l} \cdot \mathbf{E} \tag{3.6.11}$$

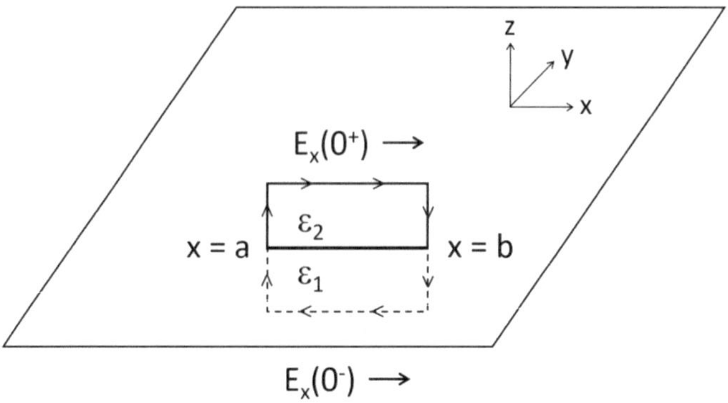

Fig. 3.3 Rectangle with two long sides extending from $x = a$ to $x = b$ parallel to the plane with two short sides perpendicular to the plane. The material below the plane has permittivity, ε_1, while the material above the plane has permittivity, ε_2. The x- component of the electric field in the space above the plane is $E_x(0^+)$ while that below the plane is $E_x(0^-)$. The direction of the path of integration around the rectangle is indicated by the arrows

where $d\mathbf{l}$ is an element of length of the perimeter. According to Eq. (3.6.10), the integrand on the left hand side of Eq. (3.6.11) is zero. Moreover, if the short sides are of infinitesimal length, only the long sides of the rectangle contribute to the line integral on the right hand side of Eq. (3.6.11). The long sides of the rectangle are assumed to be parallel to the x-axis and to extend from $x = a$ to $x = b$. The line integral is then

$$\oint d\mathbf{l} \cdot \mathbf{E} = \int_a^b dx E_x(0^+) + \int_b^a dx E_x(0^-) = 0 \tag{3.6.12}$$

where 0^+ and 0^- have the same meanings as in Eq. (3.6.4). Equation (3.6.12) can be rewritten as

$$\int_a^b dx\big(E_x(0^+) - E_x(0^-)\big) = 0 \tag{3.6.13}$$

Since a and b are arbitrary, we can conclude

$$E_x(0^+) = E_x(0^-) \tag{3.6.14}$$

The path of integration in Eq. (3.6.12) need not be parallel to the x-axis. It can be in any direction lying in the x, y-plane. Let the coordinate parallel to an arbitrary tangent to the surface be τ. We can then rewrite Eq. (3.6.14) as

$$E_\tau(0^+) = E_\tau(0^-) \tag{3.6.15}$$

In terms of τ, the tangential electric field and the electrostatic potential are related by

$$\Phi = -\int E_\tau(0^+)d\tau = -\int E_\tau(0^-)d\tau \tag{3.6.16}$$

By combining Eqs. (3.6.15) and (3.6.16), we obtain

$$\Phi(0^+) = \Phi(0^-) \tag{3.6.17}$$

Equations (3.6.9) and (3.6.17) are the boundary conditions that must be satisfied by the solutions to Laplace's equation, $\nabla^2\Phi = 0$, in cases where materials with different values of the electric permittivity join.

3.7 Solution to Laplace's Equation in Cartesian Coordinates

Consider a dielectric with permittivity, ε, sandwiched between two planes formed by parallel metal plates separated by a distance, d. This arrangement of conductors and dielectric forms a parallel plate capacitor. See Fig. 3.4.

Let the normal to the plates be the x-axis. The surface charge density is $+\sigma$ on the plate at $x = 0^+$ and $-\sigma$ on the plate at $x = d^-$. The density of free charges within the dielectric is zero; hence, Eq. (3.2.32) for the electrostatic potential, Φ, between the plates assumes the form of Laplace's equation,

$$\nabla^2\Phi = 0 \tag{3.7.1}$$

Equation (3.7.1) can be written in Cartesian coordinates as

$$\frac{\partial^2\Phi}{\partial x^2} + \frac{\partial^2\Phi}{\partial y^2} + \frac{\partial^2\Phi}{\partial z^2} = 0 \tag{3.7.2}$$

(See the Mathematical Methods Appendix). If the plates extend to infinity in the y and z-directions, then the lines of electric force are everywhere parallel to the x-direction, and Eq. (3.7.2) simplifies to

$$\frac{\partial^2\Phi}{\partial x^2} = 0 \tag{3.7.3}$$

The first integral of Eq. (3.7.3) is

$$\frac{\partial\Phi}{\partial x} = C_1 \tag{3.7.4a}$$

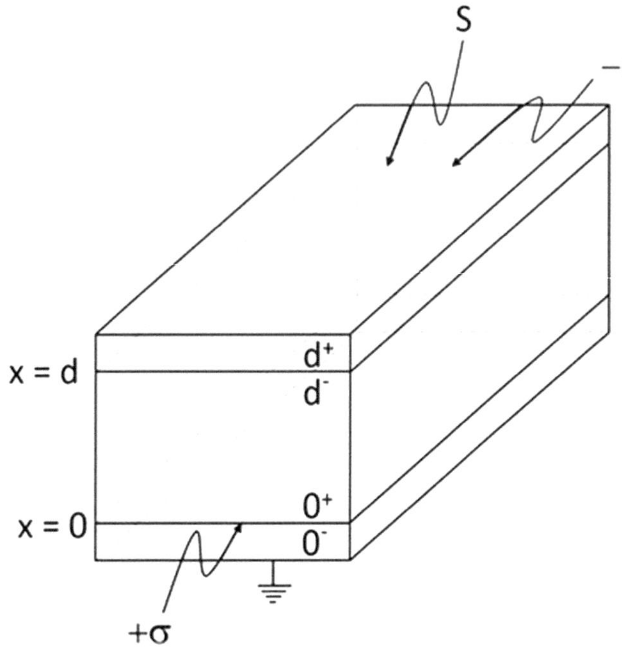

Fig. 3.4 Parallel plate capacitor. Two identical metallic plates have area S and are separated in the x-direction by the distance, d. The dielectric between the plates occupies the space lying between $x = 0^+$ to $x = d^-$. The surfaces labeled $x = 0^-$ and $x = d^+$ are located entirely within the material of the plates. The surface charge density on the top plate is $-\sigma$, whereas on the bottom plate it is $+\sigma$

where C_1 is a constant of integration. The integral of Eq. (3.7.4a) is

$$\Phi(x) = C_1 x + C_2 \tag{3.7.4b}$$

where C_2 is a second constant of integration. Since there is no natural zero for the electrostatic potential, we can assume that the plate at $x = 0$ is grounded. This latter statement means that

$$\Phi(0) = 0 \tag{3.7.5a}$$

The plate at $x = d$ is assumed to carry a voltage, $-V$, which implies that

$$\Phi(d) = -V \tag{3.7.5b}$$

Upon substitution of Eq. (3.7.4b) into Eq. (3.7.5a), we find

$$C_2 = 0 \tag{3.7.6a}$$

Upon substitution of Eq. (3.7.6a) into Eq. (3.7.4b) and the result into Eq. (3.7.5b), we find

$$C_1 = -\frac{V}{d} \tag{3.7.6b}$$

When both of Eqs. (3.7.6a) and (3.76b) are substituted into Eq. (3.7.4b), the electrostatic potential is found to be

$$\Phi(x) = -\frac{V}{d}x \tag{3.7.7}$$

Equation (3.7.7) is the solution to Laplace's equation subject to the boundary conditions in Eqs. (3.7.5a) and (3.7.5b).

Equation (3.7.7) can be used to calculate the surface charge density on each of the plates in Fig. 3.4. The electric permittivity of the dielectric material between the plates is ε. The point, $x = 0^+$, coincides with the top surface of the bottom plate, while the point at $x = 0^-$ lies in the bottom plate just below its boundary with the dielectric. Since the x-direction is normal to the plates, Eq. (3.6.9) becomes

$$\varepsilon_2\left(\frac{\partial\Phi}{\partial x}\right)_{0^+} - \varepsilon_1\left(\frac{\partial\Phi}{\partial x}\right)_{0^-} = -\sigma \tag{3.7.8}$$

Since the potential is constant and zero everywhere within the bottom plate, we have $(\partial\Phi/\partial x)_{0^-} = 0$; hence, with $\varepsilon_2 = \varepsilon$, Eq. (3.7.8) simplifies to

$$\varepsilon\left(\frac{\partial\Phi}{\partial x}\right)_{0^+} = -\sigma \tag{3.7.9}$$

Upon substitution of Eq. (3.7.7) into Eq. (3.7.9), one obtains,

$$\sigma = \frac{\varepsilon V}{d} \tag{3.7.10}$$

which is the charge density on the bottom plate. To calculate the charge density on the top plate, we evaluate Eq. (3.6.9) at the top plate. The result is

$$\varepsilon_2\left(\frac{\partial\Phi}{\partial x}\right)_{d^+} - \varepsilon_1\left(\frac{\partial\Phi}{\partial x}\right)_{d^-} = -\sigma \tag{3.7.11}$$

The point, $x = d^+$, lies wholly within the top plate where the potential is uniform; hence, $(\partial\Phi/\partial x)_{d^+} = 0$. The point at $x = d^-$ coincides with the inside surface of the top plate. With $\varepsilon_1 = \varepsilon$, Eq. (3.7.11) simplifies to

$$\varepsilon\left(\frac{\partial\Phi}{\partial x}\right)_{d^-} = \sigma \tag{3.7.12}$$

Upon substitution of Eq. (3.7.7) into Eq. (3.7.12), we find

$$\sigma = -\frac{\varepsilon V}{d} \tag{3.7.13}$$

for the charge density on the top plate.

If S is the surface area of the plates, then the magnitude of the total charge, $q = \sigma S$, on the bottom plate is

$$q = \frac{\varepsilon S}{d} V \tag{3.7.14}$$

The capacitance, C, of the capacitor formed by the plates and the dielectric is defined by

$$q = CV \tag{3.7.15}$$

Hence, by comparison of Eq. (3.7.14) with Eq. (3.7.15), we obtain for the capacitance,

$$C = \frac{\varepsilon S}{d} \tag{3.7.16}$$

Note that C depends upon the geometry of the capacitor in the form of the ratio of the surface area, S, of the plates and the distance, d, between them. Given a set of fixed dimensions, the capacitance increases linearly with ε.

The magnitude of the electric field within the dielectric is

$$E = -\frac{\partial \Phi}{\partial x} \tag{3.7.17}$$

Substitute Eq. (3.7.7) into Eq. (3.7.17) to obtain

$$E = \frac{V}{d} \tag{3.7.18}$$

According to Eq. (3.2.27), the magnitude of the electric displacement is

$$D = \frac{\varepsilon V}{d} \tag{3.7.19}$$

Both E and D are uniform and parallel to the positive x-direction. Both are positive, because the negatively charged plate of the capacitor is on the top.

3.8 Solution to Laplace's Equation for a Spherical Cavity in a Continuous Dielectric Subject to a Uniform External Electric Field

In this section, we improve on the Lorenz–Lorentz model of a dielectric by dropping the assumption that the sphere of occupation of a molecule does not distort the applied electric field.

As mentioned, an isolated molecule that finds itself between the plates of a parallel plate capacitor will suffer a distortion of its internal charge distribution due to the action of the electric field. The atomic nuclei will be displaced toward the negative plate while the atomic and molecular electrons will be displaced toward the positive plate. In the case of the rare gases, and symmetrical molecules, such as CCl_4, SF_6, and benzene, charge displacement is the principal response of the molecule to the applied electric field. In these materials, the induced electric dipole moment vector, **p**, is proportional to the applied electric field vector, **E**, namely,

$$\mathbf{p} = \alpha\mathbf{E} \tag{3.8.1}$$

where α is the molecular polarizability. By contrast, in the case of polar molecules, such as H_2O, HCl, and bromobenzene, there is in addition to the polarizability also a permanent molecular electric dipole moment that exists in the absence of an applied field. As mentioned, this permanent dipole will rotate so as to be as parallel as possible to the direction of the applied field. The total dipole moment in the case of a polar molecule is then the vector sum of its permanent dipole and its induced dipole.

Because no two molecules can occupy the same space, the space of a given molecule cannot be invaded by any of its neighbors. Nonetheless, a given molecule is still subject to the electric fields produced by the dipole moments on its neighbors. As we have seen in analyzing the Lorenz–Lorentz model, the combination of the applied field, **E**, and the fields due to neighboring molecules produces a local electric field, **F**, which is sensed by the molecule in question. Sufficiently far away from the given molecule, the electric dipole distributions on the neighboring molecules appear to smooth out forming a uniform continuum. Under these circumstances, it is not unreasonable to assume that a given molecule can be regarded as forging for itself a cavity in an otherwise continuous dielectric. The coexistence of a molecular cavity with an otherwise continuous dielectric is known as the "semi-continuum" model.

In contrast to Lorentz, Lars Onsager (Norwegian, American, 1903–1976) stressed the idea that the presence of a molecular cavity should distort the electric field lines in its vicinity thus altering the value of the local electric field [1]. We now pursue this idea. In reprising Onsager's theory, we will assume the cavity to be a sphere of radius, a. The relative permittivity of the space inside the cavity will be denoted by $\varepsilon^{(i)}$. The permittivity of the material outside the cavity will be denoted by $\varepsilon^{(o)}$. Later in the calculation, we will take $\varepsilon^{(i)} = \varepsilon_o$, which is appropriate to a vacuum, and

$\varepsilon^{(o)} = \varepsilon_r \varepsilon_o$, which is appropriate to a continuous dielectric. Assuming the central molecule to be polar, we will endow the cavity with both induced and permanent electric dipoles.

The empty cavity is assumed to lie between the plates of a parallel plate capacitor. See Fig. 3.5.

A normal to the positively charged plate defines the Cartesian $+z$ direction. When a voltage is applied to the capacitor, the electric field far from the cavity is parallel to this direction and is given by

$$\mathbf{E} = E_o \hat{\mathbf{e}}_z \tag{3.8.2}$$

where E_o is the magnitude of the field. As shown in Fig. 3.6, because $\varepsilon^{(i)} < \varepsilon^{(o)}$, the cavity distorts the uniformity of the field lines in its immediate presence.

In the absence of free charge in the dielectric, the electrostatic potential, Φ, raised by the external electric field is given by Laplace's equation. In spherical polar coordinates, Eq. (3.7.1) takes the form,

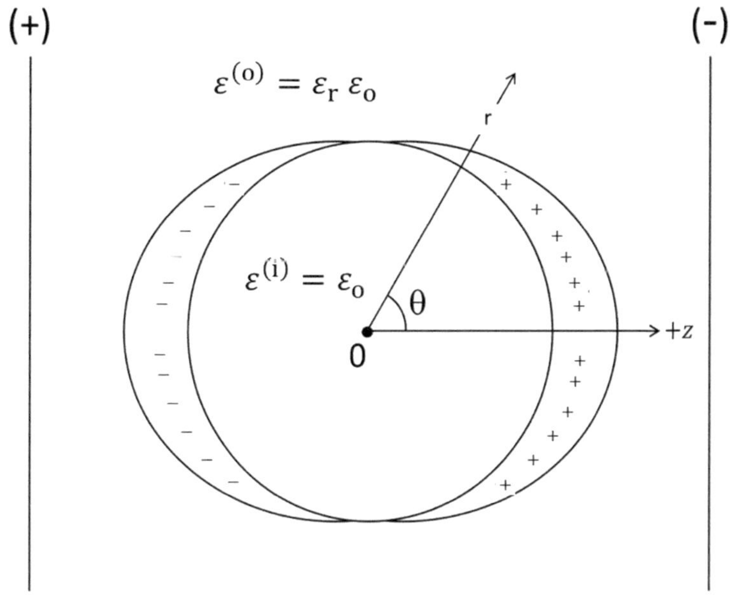

Fig. 3.5 Empty cavity in an otherwise continuous dielectric located between the parallel plates of a capacitor. The center of the cavity is at 0. The $+z$-axis points in the horizontal direction, from left to right. The applied electric field is parallel to this axis. The radius, r, locating an arbitrary point in the dielectric makes an angle, θ, with the $+z$-axis. Due to the polarization raised by the action of the applied field on the dielectric material contiguous with the surface of the cavity, there is a net positive surface charge density on the right hemisphere and a net negative charge density of equal magnitude on the left hemisphere

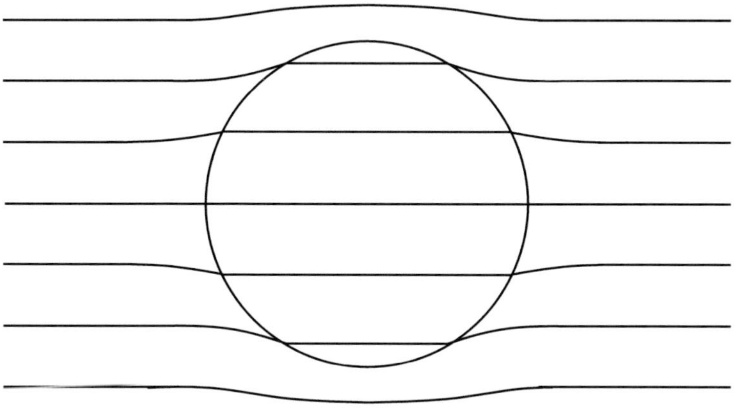

Fig. 3.6 Field lines in the vicinity of an empty cavity in an otherwise uniform dielectric

$$\frac{1}{r^2}\frac{\partial}{\partial r}\left(r^2\frac{\partial\Phi}{\partial r}\right)+\frac{1}{r^2\sin\theta}\frac{\partial}{\partial\theta}\left(\sin\theta\frac{\partial\Phi}{\partial\theta}\right)+\frac{1}{r^2\sin^2\theta}\left(\frac{\partial^2\Phi}{\partial\phi^2}\right)=0 \tag{3.8.3}$$

Equation (3.8.3) can be found derived in the Mathematical Methods Appendix. By taking advantage of the symmetry about the z-axis, it is possible to simplify Eq. (3.8.3). Let the angle of rotation about the z-axis be ϕ. By virtue of Fig. 3.5, it is clear that neither the parallel plate capacitor nor the spherical cavity is changed by such a rotation. As a consequence, the electrostatic potential can be represented as $\Phi = \Phi(r, \theta)$, which is independent of ϕ. With this simplification, Eq. (3.8.3) becomes

$$\frac{1}{r^2}\frac{\partial}{\partial r}\left(r^2\frac{\partial\Phi}{\partial r}\right)+\frac{1}{r^2\sin\theta}\frac{\partial}{\partial\theta}\left(\sin\theta\frac{\partial\Phi}{\partial\theta}\right)=0 \tag{3.8.4}$$

Although Eq. (3.8.4) is still a partial differential equation, it can be separated into two ordinary differential equations by assuming that Φ can be factored in the form

$$\Phi(r,\theta)=R(r)\Theta(\theta) \tag{3.8.5}$$

where $R(r)$ is a function of r alone and $\Theta(\theta)$ is a function of θ alone. To check that this factorization is feasible, we substitute Eq. (3.8.5) into Eq. (3.8.4) to obtain

$$\frac{\Theta(\theta)}{r^2}\frac{\partial}{\partial r}\left(r^2\frac{\partial R(r)}{\partial r}\right)+\frac{R(r)}{r^2\sin\theta}\frac{\partial}{\partial\theta}\left(\sin\theta\frac{\partial\Theta(\theta)}{\partial\theta}\right)=0 \tag{3.8.6}$$

Next, divide Eq. (3.8.6) through by $R(r)\Theta(\theta)$. The result is

$$\frac{1}{r^2R(r)}\frac{\partial}{\partial r}\left(r^2\frac{\partial R(r)}{\partial r}\right)+\frac{1}{r^2\Theta(\theta)\sin\theta}\frac{\partial}{\partial\theta}\left(\sin\theta\frac{\partial\Theta(\theta)}{\partial\theta}\right)=0 \tag{3.8.7}$$

Upon multiplication of Eq. (3.8.7) through by r^2, one obtains,

$$\frac{1}{R(r)}\frac{\partial}{\partial r}\left(r^2\frac{\partial R(r)}{\partial r}\right)+\frac{1}{\Theta(\theta)\sin\theta}\frac{\partial}{\partial\theta}\left(\sin\theta\frac{\partial\Theta(\theta)}{\partial\theta}\right)=0 \tag{3.8.8}$$

Equation (3.8.8) is now in a form termed, "variables separated". That is to say, the first term on the left is a function of r alone, while the second term is a function of θ alone. Consider a path in (r,θ) space with r varying while θ is fixed. Along this path, the second term on the left in Eq. (3.8.8) is necessarily a constant. Likewise, if we consider a path in (r,θ) space where r is fixed and θ is varying, the first term on the left is fixed. If Eq. (3.8.8) is to be satisfied for values r and θ varying simultaneously, the terms on opposite sides of the plus sign must have the same constant magnitude with one being positive while the other is negative. It is convenient to let that constant be $\ell(\ell+1)$, where for reasons which will become clear, ℓ will need to be an integer. The first term on the left hand side of Eq. (3.8.8) can then be written as

$$\frac{1}{R(r)}\frac{\partial}{\partial r}\left(r^2\frac{\partial R(r)}{\partial r}\right)=\ell(\ell+1) \tag{3.8.9}$$

Equation (3.8.9) is termed the radial equation. Given Eq. (3.8.9), the second term on the left in Eq. (3.8.8) is necessarily

$$\frac{1}{\Theta(\theta)\sin\theta}\frac{\partial}{\partial\theta}\left(\sin\theta\frac{\partial\Theta(\theta)}{\partial\theta}\right)=-\,\ell(\ell+1) \tag{3.8.10}$$

Equation (3.8.10) is termed the angular equation.

Upon multiplication of both sides of Eq. (3.8.9) by $R(r)$, the radial equation assumes the form,

$$\frac{\partial}{\partial r}\left(r^2\frac{\partial R(r)}{\partial r}\right)-\ell(\ell+1)R(r)=0 \tag{3.8.11}$$

By trial and error, one finds that Eq. (3.8.11) has two solutions. These are

$$R(r)=r^{\ell} \tag{3.8.12}$$

and

$$R(r)=r^{-(\ell+1)} \tag{3.8.13}$$

Since Eq. (3.8.11) is linear in $R(r)$, its general solution is formed by a linear combination of the two solutions, r^{ℓ} and $r^{-(\ell+1)}$. That is,

$$R_{\ell}(r)=A_{\ell}r^{\ell}+B_{\ell}r^{-(\ell+1)} \tag{3.8.14}$$

where A_ℓ and B_ℓ are constants of integration.

By multiplication of both sides of Eq. (3.8.10) by $\Theta(\theta) sin\theta$, the angular equation can be recast in the form,

$$\frac{\partial}{\partial\theta}\left(\sin\theta\frac{\partial\Theta(\theta)}{\partial\theta}\right)+\ell(\ell+1)\Theta(\theta)\sin\theta=0 \tag{3.8.15}$$

There are two types of solutions to Eq. (3.8.15). One is of no physical significance [2]. The other is

$$\Theta_\ell(\theta)=P_\ell(\cos\theta) \tag{3.8.16}$$

where $P_\ell(\cos\theta)$ is the same Legendre polynomial, which we encountered in Sect. 1.14 of Chap. 1 while calculating the electrostatic potentials associated with the linear electric dipole and the linear electric quadrupole. As we saw there, the first three Legendre polynomials, $\{P_\ell(\cos\theta)\}$, are

$$P_o(\cos\theta)=1 \tag{3.8.17a}$$

$$P_1(\cos\theta)=\cos\theta \tag{3.8.17b}$$

$$P_2(\cos\theta)=\frac{1}{2}\left(3\cos^2\theta-1\right) \tag{3.8.17c}$$

Since $\theta = 0$ and $\theta = 2\pi$ locate the same point on the sphere, the function, $\Theta_\ell(\theta)$, must be periodic with period 2π. This periodicity is guaranteed by the fact that each of Eqs. (3.8.17) is either a constant or involves a power of $\cos\theta$. According to Eqs. (3.8.5), (3.8.14) and (3.8.16), the special solution to Eq. (3.8.4) for a fixed value of ℓ is

$$\Phi_\ell(r,\theta)=R_\ell(r)\Theta_\ell(\theta)=\left(A_\ell r^\ell+B_\ell r^{-(\ell+1)}\right)P_\ell(\cos\theta) \tag{3.8.18}$$

Since for each integral value of ℓ, Eq. (3.8.18) serves as a special solution of Eq. (3.8.4), the general solution can be constructed as a sum of these special solutions. Hence, the general solution for the potential, $\Phi^{(i)}(r,\theta)$, inside the cavity reads,

$$\Phi^{(i)}(r,\theta)=\sum_{\ell=0}^{\infty}\left(A_\ell^{(i)}r^\ell+B_\ell^{(i)}r^{-(\ell+1)}\right)P_\ell(\cos\theta)\quad\text{for}\quad 0\le r\le a \tag{3.8.19}$$

The general solution for the potential, $\Phi^{(o)}(r,\theta)$, outside the cavity is

$$\Phi^{(o)}(r,\theta)=\sum_{\ell=0}^{\infty}\left(A_\ell^{(o)}r^\ell+B_\ell^{(o)}r^{-(\ell+1)}\right)P_\ell(\cos\theta)\quad\text{for}\quad a<r<\infty \tag{3.8.20}$$

Compared to the size of the cavity, which is on the molecular scale, the separation between the plates of the capacitor is on the macroscopic scale. This allows us to extend the range of r over which the exterior solution applies so as to include $r = \infty$. We now specify the boundary conditions that will permit us to determine the allowed values of ℓ and the corresponding values of the interior constants of integration, $A_\ell^{(i)}$ and $B_\ell^{(i)}$, and the exterior constants of integration $A_\ell^{(o)}$ and $B_\ell^{(o)}$.

The boundary conditions take the form:

$$\Phi^{(i)}(0, \theta) = 0 \tag{3.8.21a}$$

$$\Phi^{(i)}(a, \theta) = \Phi^{(o)}(a, \theta) \tag{3.8.21b}$$

$$\varepsilon^{(i)}\left(\frac{\partial \Phi^{(i)}}{\partial r}\right)_{r=a} = \varepsilon^{(o)}\left(\frac{\partial \Phi^{(o)}}{\partial r}\right)_{r=a} \tag{3.8.21c}$$

$$\lim_{r\to\infty} \Phi^{(o)}(r, \theta) = -E_o z = -E_o r \cos\theta \tag{3.8.21d}$$

Since there is no absolute zero for the electrostatic potential, we let $\Phi^{(i)}$ be zero at the origin, $r = 0$. This accounts for Eq. (3.8.21a). Equations (3.8.21b) and (3.8.21c) have their origin in Sect. 3.6. In analogy with Eq. (3.6.17), we require the electrostatic potential to be continuous across the boundary between the cavity and the surrounding dielectric. This accounts for Eq. (3.8.21b). Since there is no free surface charge density at the boundary between the cavity and the dielectric, Eq. (3.6.9) applies with zero on its right hand side. This accounts for Eq. (3.8.21c). Finally, the electric field at great distances from the cavity is not influenced by its presence and is uniform. As shown in Sect. 3.7, the electric field satisfies $E_o = -(\partial \Phi^{(o)}/\partial z)$, in this case, where E_o is a constant. This accounts for Eq. (3.8.21d).

We now turn to the task of using Eqs. (3.8.21a)–(3.8.21d) to determine the constants of integration. In Eq. (3.8.19), $r^{-(\ell+1)}$ becomes infinite when $r = 0$. To avoid this singularity, we set

$$B_\ell^{(i)} = 0 \quad \text{for all values of } \ell \tag{3.8.22a}$$

With Eq. (3.8.22a) established, Eq. (3.8.19) evaluated at $r = 0$ reads $\Phi^{(i)}(0, \theta) = A_0^{(i)}$. To satisfy Eq. (3.8.21a), we set

$$A_0^{(i)} = 0 \tag{3.8.22b}$$

When Eqs. (3.8.22a) and (3.8.22b) are substituted into Eq. (3.8.19), the sum simplifies to

$$\Phi^{(i)}(r, \theta) = \sum_{\ell=1}^{\infty} A_\ell^{(i)} r^\ell P_\ell(\cos\theta) \tag{3.8.23}$$

Next, we turn our attention to the exterior solution. Since $P_1(\cos\theta) = \cos\theta$, Eq. (3.8.20) satisfies Eq. (3.13.21d) if

$$A_1^{(o)} = -E_o \tag{3.8.24a}$$

and

$$A_\ell^{(o)} = 0, \quad \text{when } \ell = 0 \quad \text{and} \quad \ell \geq 2 \tag{3.8.24b}$$

After incorporating Eqs. (3.8.24a) and (3.8.24b), Eq. (3.8.20) can be rewritten in the form

$$\Phi^{(o)}(r,\theta) = -E_o r\cos\theta + \sum_{\ell=0}^{\infty} B_\ell^{(o)} r^{-(\ell+1)} P_\ell(\cos\theta) \tag{3.8.25}$$

Now consider the boundary conditions that must be satisfied by Eqs. (3.8.23) and (3.8.25) at $r = a$. Substitute Eq. (3.8.23) into the left had side of Eq. (3.8.21b) and (3.8.25) into the right hand side. The result is

$$\sum_{\ell=1}^{\infty} A_\ell^{(i)} a^\ell P_\ell(\cos\theta) = -E_o a\cos\theta + \sum_{\ell=0}^{\infty} B_\ell^{(o)} a^{-(\ell+1)} P_\ell(\cos\theta) \tag{3.8.26}$$

The sums on either side of Eq. (3.8.26) must be equal term by term. Because there is no term involving $\ell = 0$ on the left hand side of Eq. (3.8.26), we must have

$$B_0^{(o)} = 0 \quad \text{when } \ell = 0 \tag{3.8.27a}$$

Next, recalling that $P_1(\cos\theta) = \cos\theta$, we equate coefficients of $P_\ell(\cos\theta)$ with $\ell = 1$ on either side of Eq. (3.8.26) to obtain

$$A_1^{(i)} a = -E_o a + B_1^{(0)} a^{-2} \quad \text{when } \ell = 1 \tag{3.8.27b}$$

Upon equating the coefficients of $P_\ell(\cos\theta)$ with $\ell \geq 2$, on either side of Eq. (3.8.26), the result is

$$A_\ell^{(i)} a^\ell = B_\ell^{(o)} a^{-(\ell+1)} \quad \text{when } \ell \geq 2 \tag{3.8.27c}$$

Next, substitute Eq. (3.8.23) into the left hand side of Eqs. (3.8.21c) and (3.8.25) into the right hand side of Eq. (3.8.21c). The result is

$$\varepsilon^{(i)} \sum_{\ell=1}^{\infty} A_\ell^{(i)} \ell a^{\ell-1} P_\ell(\cos\theta) = -\varepsilon^{(o)} E_o \cos\theta - \varepsilon^{(o)} \sum_{\ell=1}^{\infty} (\ell+1) B_\ell^{(o)} a^{-(\ell+2)} P_\ell(\cos\theta) \tag{3.8.28}$$

Because $B_0^{(o)} = 0$, we have not included a term in $\ell = 0$ in the sum on the right hand side of Eq. (3.8.28). After equating the coefficients of $P_\ell(\cos\theta)$ with $\ell = 1$ on either side of Eq. (3.8.28), we find

$$\varepsilon^{(i)} A_1^{(i)} = -\,\varepsilon^{(o)} E_o - 2\varepsilon^{(o)} B_1^{(o)} a^{-3}, \quad \ell = 1 \tag{3.8.29a}$$

After equating coefficients of $P_\ell(\cos\theta)$ with $\ell \geq 2$, we have

$$\varepsilon^{(i)} A_\ell^{(i)} \ell a^{\ell-1} = -\,\varepsilon^{(o)} (\ell + 1) B_\ell^{(o)} a^{-(\ell+2)}, \quad \ell \geq 2 \tag{3.8.29b}$$

Equations (3.8.27b) and (3.8.29a) can be solved algebraically for $A_1^{(i)}$ and $B_1^{(o)}$. The results are

$$A_1^{(i)} = -\left(\frac{3\varepsilon^{(o)}}{\varepsilon^{(i)} + 2\varepsilon^{(o)}}\right) E_o \tag{3.8.30a}$$

and

$$B_1^{(o)} = \left(\frac{\varepsilon^{(i)} - \varepsilon^{(o)}}{\varepsilon^{(i)} + 2\varepsilon^{(o)}}\right) E_o a^3 \tag{3.8.30b}$$

Next, rearrange Eq. (3.8.27c) for $B_\ell^{(o)}$ so that it reads,

$$B_\ell^{(o)} = a^{(2\ell+1)} A_\ell^{(i)} \tag{3.8.31a}$$

We can solve Eq. (3.8.29b) for $B_\ell^{(o)}$ by writing,

$$B_\ell^{(o)} = -\frac{\varepsilon^{(i)}}{\varepsilon^{(o)}}\left(\frac{\ell}{\ell + 1}\right) a^{(2\ell+1)} A_\ell^{(i)} \tag{3.8.31b}$$

When Eqs. (3.8.31a) and (3.8.31b) are equated, the result is

$$(\ell + 1)\varepsilon^{(o)} + \ell\varepsilon^{(i)} = 0 \tag{3.8.31c}$$

There is no solution to Eq. (3.8.31c) for all $\ell \geq 2$ and fixed non-zero values of $\varepsilon^{(o)}$ and $\varepsilon^{(i)}$, so we must conclude that

$$A_\ell^{(i)} = B_\ell^{(o)} = 0, \quad \ell \geq 2 \tag{3.8.32}$$

When Eqs. (3.8.27a), (3.8.30a), (3.8.30b), and (3.8.32) are substituted into Eqs. (3.8.23) and (3.8.25), the results are

$$\Phi^{(i)}(r, \theta) = -\left(\frac{3\varepsilon^{(o)}}{2\varepsilon^{(o)} + \varepsilon^{(i)}}\right) E_o r \cos\theta \tag{3.8.33}$$

and

$$\Phi^{(o)}(r,\theta) = -E_o a\left(\frac{\varepsilon^{(o)} - \varepsilon^{(i)}}{2\varepsilon^{(o)} + \varepsilon^{(i)}}\right)\left(\frac{a}{r}\right)^2 \cos\theta - E_o r\cos\theta \tag{3.8.34}$$

respectively. Equation (3.8.33) gives the electrostatic potential sensed in the empty cavity as a result of the direct action of the applied electric field and the polarization which it raises in the material surrounding the cavity. Equation (3.8.33) will prove useful in deriving an electromagnetic equation of state from the semi-continuum model. Equation (3.8.34) gives the electrostatic potential exterior to the cavity.

3.9 Solution to Laplace's Equation for a Point Dipole at the Center of a Spherical Cavity in a Continuous Dielectric

In this model, a polarized molecule is represented as a point dipole at the center of an empty cavity in an otherwise continuous dielectric. The dipole acts on the surrounding dielectric in the absence of an external field, E_o. The point dipole induces a polarization in the molecules outside the cavity. This exterior polarization in turn makes a contribution to the electrostatic potential at the position of the cavity dipole. See Fig. 3.7.

In the figure, the point dipole lies along the z-axis. As before, we seek the solution to Eq. (3.8.4) for Φ. The search begins with the general solutions given by Eqs. (3.8.19) and (3.8.20) with Eqs. (3.8.21b) and (3.8.21c) serving to specify the potential at the boundary between the cavity and the surrounding dielectric. Boundary conditions, Eqs. (3.8.21a) and (3.8.21d), however, will need to be replaced in order to take into account the presence of the point dipole.

From the aspect of an observer interior to the cavity, the internal potential has the same form as that of a point dipole embedded in a material having permittivity, $\varepsilon^{(i)}$. That is to say, the behavior of $\Phi^{(i)}(r,\theta)$ as $r \to 0$ can be expressed as the limit,

$$\lim_{r\to 0}\left[\Phi^{(i)}(r,\theta) - \frac{p\cos\theta}{4\pi\varepsilon^{(i)}r^2}\right] = 0 \tag{3.9.1}$$

where p is the magnitude of the electric dipole moment. One notices that with $\varepsilon^{(i)} = \varepsilon_o$, the second term in Eq. (3.9.1) becomes identical with the right hand side of Eq. (1.7.18) in Chap. 1. Equation (3.9.1) replaces Eq. (3.8.21a) as the interior boundary condition.

Far from the cavity, the boundary between the cavity and the surrounding dielectric has little effect on the potential. As $r \to \infty$, the outside potential needs to reflect only the presence of a dipole at the origin. To meet this requirement, the outside potential assumes the asymptotic form,

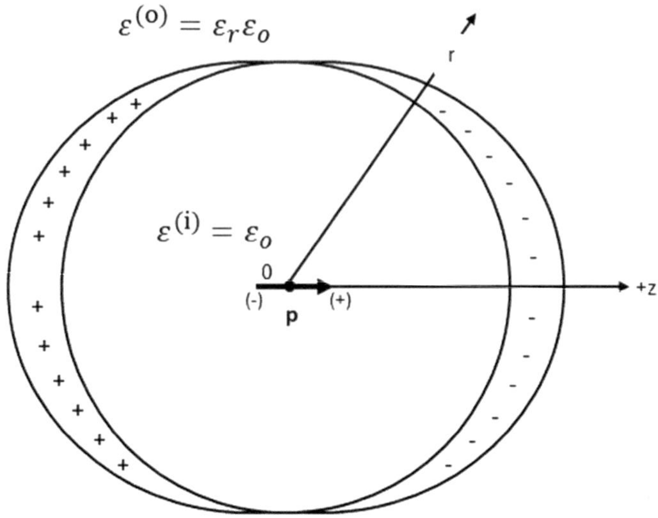

Fig. 3.7 Spherical cavity containing a central point electric dipole, **p**, in an otherwise continuous dielectric. The electric permittivity interior to the cavity is $\varepsilon^{(i)} = \varepsilon_o$ and exterior to the cavity it is $\varepsilon^{(o)} = \varepsilon_r\varepsilon_o$. The positive z-direction is collinear with the dipole. As in Fig. 3.5, the radius, r, makes an angle of θ with respect to the z-direction

$$\lim_{r\to\infty} \Phi^{(o)}(r,\theta) \to \frac{\cos\theta}{r^2} \tag{3.9.2}$$

Equation (3.9.2) replaces Eq. (3.8.21d) as the exterior boundary condition.

Near $r = 0$, it is useful to write the general solution to the interior problem, Eq. (3.8.19), in the expanded form,

$$\Phi^{(i)}(r,\theta) = \sum_{\ell=0}^{\infty} A_\ell^{(i)} r^\ell P_\ell(\cos\theta) + \frac{B_0^{(i)}}{r} + \frac{B_1^{(i)}\cos\theta}{r^2} + \sum_{\ell=2}^{\infty} \frac{B_\ell^{(i)}}{r^{\ell+1}} P_\ell(\cos\theta) \tag{3.9.3}$$

Upon substitution of Eq. (3.9.3) into Eq. (3.9.1), the result is

$$\lim_{r\to 0}\left[\sum_{\ell=0}^{\infty} A_\ell^{(i)} r^\ell P_\ell(\cos\theta) + \frac{B_0^{(i)}}{r} + \frac{B_1^{(i)}\cos\theta}{r^2} + \sum_{\ell=2}^{\infty} \frac{B_\ell^{(i)}}{r^{\ell+1}} P_\ell(\cos\theta) - \frac{p\cos\theta}{4\pi\varepsilon^{(i)} r^2}\right] = 0 \tag{3.9.4}$$

The last term on the left hand side of Eq. (3.9.4) is a function of r^{-2}. To achieve the zero value of the left hand side of Eq. (3.9.4) as $r \to 0$, the last term must be balanced by one or more of terms involving $1/r^2$, which precede it. In the case of terms involving r^ℓ, $\ell \geq 0$, all go to zero as $r \to 0$. Hence, except for the necessity,

$A_0^{(i)} = 0$, our proposed balancing condition places no restriction on any of the $A_\ell^{(i)}$. By contrast, the out of balance term with coefficient, $B_0^{(i)}$, as well as the out of balance terms involving $B_\ell^{(i)}$ for $\ell \geq 2$, must all be zero. The specific $1/r^2$ balance is achieved by setting $B_1^{(i)} = p/4\pi\varepsilon^{(i)}$. Hence, we conclude that balance is achieved if,

$$A_0^{(i)} = 0 \tag{3.9.5a}$$

$$B_0^{(i)} = 0 \tag{3.9.5b}$$

$$B_1^{(i)} = \frac{p}{4\pi\varepsilon^{(i)}} \tag{3.9.5c}$$

and

$$B_\ell^{(i)} = 0 \quad \text{for all } \ell \geq 2 \tag{3.9.5d}$$

When Eqs. (3.9.5) are substituted into Eq. (3.9.3), the result is

$$\Phi^{(i)}(r,\theta) = \sum_{\ell=1}^{\infty} A_\ell^{(i)} r^\ell P_\ell(\cos\theta) + \frac{p\cos\theta}{4\pi\varepsilon^{(i)} r^2} \tag{3.9.6}$$

In order to convert the general solution to the exterior problem, Eq. (3.8.20), to a form consistent with the boundary condition in Eq. (3.9.2), it is useful to write the former in the expanded form:

$$\Phi^{(o)}(r,\theta) = \sum_{\ell=0}^{\infty} A_\ell^{(o)} r^\ell P_\ell(\cos\theta) + \frac{B_0^{(o)}}{r} + \sum_{\ell=1}^{\infty} B_\ell^{(o)} r^{-(\ell+1)} P_\ell(\cos\theta) \tag{3.9.7}$$

If $\Phi^{(o)}(r,\theta)$ is to fall off as fast as r^{-2} or faster as $r \to \infty$, Eq. (3.9.7) can contain no constant term; neither any terms proportional to r^ℓ, $\ell \geq 1$; nor any term in r^{-1}. These requirements are satisfied, if

$$A_\ell^{(o)} = 0 \quad \text{for all values of } \ell \tag{3.9.8a}$$

and

$$B_0^{(o)} = 0 \tag{3.9.8b}$$

When these limitations are incorporated, Eq. (3.9.7) reads

$$\Phi^{(o)}(r,\theta) = \sum_{\ell=1}^{\infty} \frac{B_\ell^{(o)}}{r^{(\ell+1)}} P_\ell(\cos\theta) \tag{3.9.9}$$

To obtain equations involving the constants of integration in Eqs. (3.9.6) and (3.9.9), we substitute these two equations into Eq. (3.8.21b). The result is

$$\sum_{\ell=1}^{\infty} A_{\ell}^{(i)} a^{\ell} P_{\ell}(\cos\theta) + \frac{p\cos\theta}{4\pi\varepsilon^{(i)}a^2} = \sum_{\ell=1}^{\infty} \frac{B_{\ell}^{(o)}}{a^{(\ell+1)}} P_{\ell}(\cos\theta) \tag{3.9.10}$$

Equation (3.9.10) can then be expanded to read,

$$\begin{aligned} & A_1^{(i)} a P_1(\cos\theta) + \sum_{\ell=2}^{\infty} A_{\ell}^{(i)} a^{\ell} P_{\ell}(\cos\theta) + \frac{p\cos\theta}{4\pi\varepsilon^{(i)}a^2} \\ & = \frac{B_1^{(o)}}{a^2} P_1(\cos\theta) + \sum_{\ell=2}^{\infty} \frac{B_{\ell}^{(o)}}{a^{(\ell+1)}} P_{\ell}(\cos\theta) \end{aligned} \tag{3.9.11}$$

Matching terms in $P_{\ell}(\cos\theta)$ on either side if Eq. (3.9.11), and noting that $P_1(\cos\theta) = \cos\theta$, we conclude

$$A_1^{(i)} a + \frac{p}{4\pi\varepsilon^{(i)}a^2} = \frac{B_1^{(o)}}{a^2} \tag{3.9.12a}$$

and

$$A_{\ell}^{(i)} a^{\ell} = B_{\ell}^{(o)} a^{-(\ell+1)} \quad \text{for all } \ell \geq 2 \tag{3.9.12b}$$

To obtain equations involving the remaining constants of integration in Eqs. (3.9.6) and (3.9.9) substitute Eqs. (3.9.6) and (3.9.9) into Eq. (3.8.21c). The result is

$$\begin{aligned} & \varepsilon^{(i)} \sum_{\ell=1}^{\infty} \ell a^{(\ell-1)} A_{\ell}^{(i)} P_{\ell}(\cos\theta) - \varepsilon^{(i)} \left(\frac{2p\cos\theta}{4\pi\varepsilon^{(i)}a^3} \right) \\ & = -\varepsilon^{(o)} \sum_{\ell=1}^{\infty} (\ell+1) B_{\ell}^{(o)} a^{-(\ell+2)} P_{\ell}(\cos\theta) \end{aligned} \tag{3.9.13}$$

Terms involving $P_{\ell}(\cos\theta)$ with like values of ℓ can be equated. The results are

$$\varepsilon^{(i)} A_1^{(i)} - \frac{2p}{4\pi a^3} = -2\varepsilon^{(o)} B_1^{(o)} a^{-3} \tag{3.9.14a}$$

and

$$\varepsilon^{(i)} \ell a^{(\ell-1)} A_{\ell}^{(i)} = -\varepsilon^{(o)} (\ell+1) B_{\ell}^{(o)} a^{-(\ell+2)} \quad \text{for } \ell \geq 2 \tag{3.9.14b}$$

We can solve Eqs. (3.9.12a) and (3.9.14a) algebraically to obtain

$$A_1^{(i)} = \frac{2p}{4\pi a^3}\left[\frac{\varepsilon^{(i)} - \varepsilon^{(o)}}{\varepsilon^{(i)}(2\varepsilon^{(o)} + \varepsilon^{(i)})}\right] \tag{3.9.15a}$$

and

$$B_1^{(o)} = \frac{3p}{4\pi}\frac{1}{(2\varepsilon^{(o)} + \varepsilon^{(i)})} \tag{3.9.15b}$$

Rewrite Eq. (3.9.12b) as

$$B_\ell^{(o)} = A_\ell^{(i)} a^{(2\ell+1)} \tag{3.9.16a}$$

and Eq. (3.9.14b) as

$$B_\ell^{(o)} = -\frac{\varepsilon^{(i)}}{\varepsilon^{(o)}}\left(\frac{\ell}{\ell+1}\right)a^{(2\ell+1)}A_\ell^{(i)} \tag{3.9.16b}$$

If we eliminate $B_\ell^{(o)}$ between Eqs. (3.9.16a) and (3.9.16b), the result is

$$(\ell+1)\varepsilon^{(o)} + \ell\varepsilon^{(i)} = 0 \tag{3.9.16c}$$

There is no solution to Eq. (3.9.16c) for all $\ell \geq 2$ and fixed non-zero values of $\varepsilon^{(o)}$ and $\varepsilon^{(i)}$, so we must conclude that

$$A_\ell^{(i)} = 0 \quad \text{for } \ell \geq 2 \tag{3.9.17a}$$

and

$$B_\ell^{(o)} = 0 \quad \text{for } \ell \geq 2 \tag{3.9.17b}$$

Upon substitution of Eq. (3.9.17a) into Eq. (3.9.6), one obtains,

$$\Phi^{(i)}(r,\theta) = A_1^{(i)} r\cos\theta + \frac{p}{4\pi\varepsilon^{(i)}}\frac{\cos\theta}{r^2} \tag{3.9.18}$$

Upon substitution of Eqs. (3.9.17b) into Eq. (3.9.9), the result is

$$\Phi^{(o)}(r,\theta) = \frac{B_1^{(o)}\cos\theta}{r^2} \tag{3.9.19}$$

Substitute Eq. (3.9.15a) into Eq. (3.9.18) with the result,

$$\Phi^{(i)}(r,\theta) = \frac{2p}{4\pi\varepsilon^{(i)}a^2}\left(\frac{\varepsilon^{(i)} - \varepsilon^{(o)}}{2\varepsilon^{(o)} + \varepsilon^{(i)}}\right)\left(\frac{r}{a}\right)\cos\theta + \frac{p}{4\pi\varepsilon^{(i)}}\frac{\cos\theta}{r^2} \tag{3.9.20}$$

Finally, substitute Eq. (3.9.15b) into Eq. (3.9.19) with the result,

$$\Phi^{(o)}(r,\theta) = \frac{3p}{4\pi\varepsilon^{(i)}}\left(\frac{1}{2\varepsilon^{(o)}+\varepsilon^{(i)}}\right)\frac{\cos\theta}{r^2} \tag{3.9.21}$$

Equation (3.9.20) represents the electrostatic potential inside the cavity due to the central electric dipole plus the external polarization which it induces. Equation (3.9.21) represents the electrostatic potential outside the cavity.

3.10 The Onsager Local Electric Field and the Bottcher–Onsager Equation of State for a Non-polar Dielectric

Onsager used the semi-continuum model to derive an electromagnetic equation of state for materials containing molecules manifesting both permanent and induced dipoles [1]. Such a material qualifies as a polar dielectric. Subsequently, Bottcher (C. J. F. Bottcher, Dutch, 1915–2008) showed that by restricting Onsager's model to induced polarization only, a simple, but effective, equation of state for non-polar dielectrics resulted [3]. The Onsager model is based upon the superposition of the solutions to Laplace's equation [2] for the two cases treated above, namely: (1) an empty cavity in a dielectric subjected to uniform external field and (2) a point dipole at the center of a cavity in a dielectric in the absence of an external field. Below, we use this model to derive Bottcher's simple formula for a non-polar dielectric.

The total potential, $\Phi_{tot}^{(i)}(r,\theta)$, inside the cavity in Bottcher's case is obtained by summing Eqs. (3.8.33) and (3.9.20) with the result,

$$\Phi_{tot}^{(i)}(r,\theta) = -E_o\left(\frac{3\varepsilon^{(o)}}{2\varepsilon^{(o)}+\varepsilon^{(i)}}\right)r\cos\theta + \frac{2p}{4\pi\varepsilon^{(i)}a^2}\left(\frac{\varepsilon^{(i)}-\varepsilon^{(o)}}{2\varepsilon^{(o)}+\varepsilon^{(i)}}\right)\left(\frac{r}{a}\right)\cos\theta + \frac{p}{4\pi\varepsilon^{(i)}}\frac{\cos\theta}{r^2} \tag{3.10.1}$$

If the cavity interior is a vacuum, then $\varepsilon^{(i)} = \varepsilon_o$, where ε_o is the permittivity of free space. For the surrounding dielectric continuum, the permittivity is $\varepsilon^{(o)} = \varepsilon_r\varepsilon_o$, where ε_r is the relative permittivity of the dielectric. With these substitutions, Eq. (3.10.1) becomes

$$\Phi_{tot}^{(i)}(r,\theta) = -E_o\left(\frac{3\varepsilon_r}{2\varepsilon_r+1}\right)r\cos\theta - \frac{2p}{4\pi\varepsilon_o a^2}\left(\frac{\varepsilon_r-1}{2\varepsilon_r+1}\right)\left(\frac{r}{a}\right)\cos\theta + \frac{p}{4\pi\varepsilon_o}\frac{\cos\theta}{r^2} \tag{3.10.2}$$

The local electric field, **F**, is the negative gradient of $\Phi_{tot}^{(i)}(r,\theta)$,

$$\mathbf{F} = -\nabla \Phi_{tot}^{(i)} = -\hat{\mathbf{e}}_r \frac{\partial \Phi_{tot}^{(i)}}{\partial r} - \hat{\mathbf{e}}_\theta \frac{\partial \Phi_{tot}^{(i)}}{r \partial \theta} - \hat{\mathbf{e}}_\phi \frac{\partial \Phi_{tot}^{(i)}}{r \sin \theta \partial \phi} \tag{3.10.3}$$

The gradient operator in Eq. (3.10.3) is derived in the Mathematical Methods Appendix. The last derivative on the right hand side of Eq. (3.10.3) is zero because $\Phi_{tot}^{(i)} = \Phi_{tot}^{(i)}(r, \theta)$ is independent of the angle, ϕ. Upon substitution of Eq. (3.10.2) into Eq. (3.10.3), the result is

$$\begin{aligned} \mathbf{F} = & E_o \left(\frac{3\varepsilon_r}{2\varepsilon_r + 1} \right) \cos \theta \hat{\mathbf{e}}_r + \frac{2p}{4\pi \varepsilon_o a^2} \left(\frac{\varepsilon_r - 1}{2\varepsilon_r + 1} \right) \left(\frac{1}{a} \right) \cos \theta \hat{\mathbf{e}}_r \\ & + \frac{2p}{4\pi \varepsilon_o} \frac{\cos \theta}{r^3} \hat{\mathbf{e}}_r - E_o \left(\frac{3\varepsilon_r}{2\varepsilon_r + 1} \right) \sin \theta \hat{\mathbf{e}}_\theta \\ & - \frac{2p}{4\pi \varepsilon_o a^2} \left(\frac{\varepsilon_r - 1}{2\varepsilon_r + 1} \right) \left(\frac{1}{a} \right) \sin \theta \hat{\mathbf{e}}_\theta + \frac{p}{4\pi \varepsilon_o} \frac{\sin \theta}{r^3} \hat{\mathbf{e}}_\theta \end{aligned} \tag{3.10.4}$$

The induced molecular electric dipole moment will be in the same direction as the applied electric field, which according to Eq. (3.8.2), is parallel to the z-axis. In spherical polar coordinates, this direction corresponds to $\theta = 0$. Along the $\theta = 0$ direction, $\hat{\mathbf{e}}_r$ is parallel to $\hat{\mathbf{e}}_z$. Upon setting $\theta = 0$ and replacing $\hat{\mathbf{e}}_r$ by $\hat{\mathbf{e}}_z$, Eq. (3.10.4) reads,

$$\mathbf{F} = E_o \left(\frac{3\varepsilon_r}{2\varepsilon_r + 1} \right) \hat{\mathbf{e}}_z + \frac{2p}{4\pi \varepsilon_o a^2} \left(\frac{\varepsilon_r - 1}{2\varepsilon_r + 1} \right) \left(\frac{1}{a} \right) \hat{\mathbf{e}}_z + \frac{2p}{4\pi \varepsilon_o r^3} \hat{\mathbf{e}}_z \tag{3.10.5}$$

The third term on the right hand side of Eq. (3.10.5) is proportional to $1/r^3$. This term, which diverges when $r = 0$, represents the action of the point dipole on itself. As in the evaluation of the classical radius of the electron in Chap. 1, which led to Eq. (1.2.8), this self-action has a questionable physical meaning, so we drop the last term in Eq. (3.10.5). Accordingly, the magnitude, F, of the local electric field is

$$F = E_o \left(\frac{3\varepsilon_r}{2\varepsilon_r + 1} \right) + \frac{p}{4\pi \varepsilon_o a^3} \left(\frac{2\varepsilon_r - 2}{2\varepsilon_r + 1} \right) \tag{3.10.6}$$

The first term on the right of Eq. (3.10.6) represents the direct effect of the applied electric field. The second term, which is called the "reaction" field, represents the interaction of the surrounding dielectric with the induced dipole at the center of the cavity. Together, the two terms constitute what is known as the Onsager local electric field.

If we make the assumption that the molecular cavities fill all space, we can use Eq. (3.10.6) to recover the value of F derived by Lorentz. The calculation proceeds as follows: Let N be the number of molecules per unit volume, then if the molecular cavities fill all space, the volume per cavity is, $1/N$, which can be related to the cavity radius, a, by virtue of

$$\frac{1}{N} = \frac{4\pi}{3}a^3 \tag{3.10.7}$$

Since E_o is the magnitude of the applied electric field, the displacement, D, and the polarization, P, satisfy

$$D = \varepsilon_r\varepsilon_o E_o = \varepsilon_o E_o + P \tag{3.10.8}$$

We can solve Eq. (3.10.8) for

$$P = (\varepsilon_r - 1)\varepsilon_o E_o \tag{3.10.9}$$

If p is the magnitude of the electric dipole moment on a single molecule, then the polarization is

$$P = Np \tag{3.10.10}$$

When Eqs. (3.10.9) and (3.10.10) are combined, the result is

$$p = \frac{(\varepsilon_r - 1)\varepsilon_o}{N}E_o \tag{3.10.11}$$

Upon substitution of Eq. (3.10.11) into Eq. (3.10.6), we find

$$F = E_o\left(\frac{3\varepsilon_r}{2\varepsilon_r + 1}\right) + \frac{2(\varepsilon_r - 1)^2 E_o}{(2\varepsilon_r + 1)}\frac{1}{4\pi a^3 N} \tag{3.10.12}$$

Equation (3.10.7) can be used to simplify the second term on the right hand side of Eq. (3.10.12). The result is

$$F = \frac{3\varepsilon_r E_o}{(2\varepsilon_r + 1)} + \frac{2(\varepsilon_r - 1)^2 E_o}{3(2\varepsilon_r + 1)} \tag{3.10.13}$$

If the two terms on the right of Eq. (3.10.13) are written with a common denominator, one obtains

$$F_L = \left(\frac{\varepsilon_r + 2}{3}\right)E_o \tag{3.10.14}$$

We recognize, F_L, as the Lorentz local electric field as it appears Eq. (3.5.6), where it was derived on the basis of the Lorenz–Lorentz model.

Equation (3.10.6) expresses the local field, F, as the sum of a term involving the external electric field, E_o, and term involving the molecular dipole moment, p, at the center of the cavity. In the case of a non-polar material, the magnitude of the dipole depends solely on the polarizability. Being proportional to the local electric field, we can write it in the form,

$$p = \alpha F \tag{3.10.15}$$

According to Eq. (3.10.6), F depends upon p. To make Eq. (3.10.15) consistent with Eq. (3.10.6), we must have

$$p = \alpha F = \alpha E_o\left(\frac{3\varepsilon_r}{2\varepsilon_r + 1}\right) + \frac{\alpha p}{4\pi\varepsilon_o a^3}\left(\frac{2\varepsilon_r - 2}{2\varepsilon_r + 1}\right) \tag{3.10.16}$$

which can be solved for p to obtain

$$p = \frac{\alpha E_o\left(\frac{3\varepsilon_r}{2\varepsilon_r+1}\right)}{1 - \frac{\alpha}{2\pi\varepsilon_o a^3}\left(\frac{\varepsilon_r-1}{2\varepsilon_r+1}\right)} \tag{3.10.17}$$

Equation (3.10.17) is the starting point for our derivation of the Bottcher - Onsager equation of state for a non-polar dielectric.

If Eq. (3.10.17) is set equal to (3.10.11) and the common factor of E_o cancelled, the result is

$$\frac{(\varepsilon_r - 1)\varepsilon_o}{N} = \frac{\alpha\left(\frac{3\varepsilon_r}{2\varepsilon_r+1}\right)}{1 - \frac{\alpha}{2\pi\varepsilon_o a^3}\left(\frac{\varepsilon_r-1}{2\varepsilon_r+1}\right)} \tag{3.10.18}$$

After cross multiplication, we can rewrite Eq. (3.10.18) as

$$\frac{(2\varepsilon_r + 1)(\varepsilon_r - 1)}{3N\varepsilon_r} = \frac{(\alpha/\varepsilon_o)}{1 - \frac{(\alpha/\varepsilon_o)}{2\pi a^3}\left(\frac{\varepsilon_r-1}{2\varepsilon_r+1}\right)} \tag{3.10.19}$$

The reciprocal of Eq. (3.10.19) is

$$\frac{3N\varepsilon_r}{(2\varepsilon_r + 1)(\varepsilon_r - 1)} = \frac{1}{(\alpha/\varepsilon_o)} - \left(\frac{1}{2\pi a^3}\right)\left(\frac{\varepsilon_r - 1}{2\varepsilon_r + 1}\right) \tag{3.10.20}$$

Equation (3.10.20) is often referred to as the Bottcher–Onsager formula in honor of Lars Onsager, who created the theory, and C. J. F. Bottcher, who first used it to determine the values of α and a for molecules of various gases and liquids by fitting Eq. (3.10.20) to experimental data in the form of ε_r as a function of N [3]. An example is to be seen in Fig. 3.8, where measurements of ε_r vs density made by Danforth [4] on liquid carbon disulfide have been plotted.

In the case of the data in Fig. 3.8, ε_r was determined as a function of the density by varying the pressure [4].

At optical frequencies,

$$n^2 = \varepsilon_r \tag{3.10.21}$$

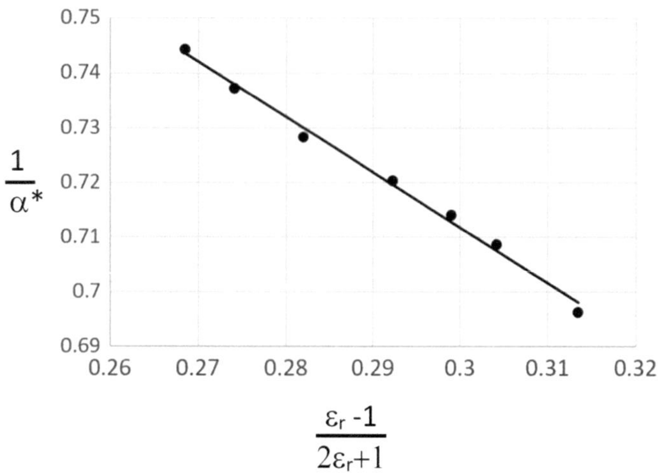

Fig. 3.8 Relative permittivity data for carbon disulfide plotted in Bottcher form with $\frac{1}{\alpha^*}$ versus $\left(\frac{\varepsilon_r-1}{2\varepsilon_r+1}\right)$ where by comparison with Eq. (3.10.20), $\frac{1}{\alpha^*} = \frac{3N\varepsilon_r}{(2\varepsilon_r+1)(\varepsilon_r-1)}$. The units of the $1/\alpha^*$ entries on the ordinate are 10^{28} m^{-3}. The entries on the abscissa are dimensionless. The quantities, (α/ε_o) and $\left(\frac{1}{2\pi a^3}\right)$, in Eq. (3.10.20) are treated as adjustable parameters. The slope and intercept of the line are consistent with $a = 0.246$ nm and $\alpha = 8.72 \times 10^{-40}$ C^2 m/N, respectively

where n is the refractive index. Upon substitution of Eq. (3.10.21) into Eq. (3.10.20), the result is

$$\frac{3Nn^2}{(2n^2+1)(n^2-1)} = \frac{1}{(\alpha/\varepsilon_o)} - \left(\frac{1}{2\pi a^3}\right)\left(\frac{n^2-1}{2n^2+1}\right) \tag{3.10.22}$$

Equation (3.10.22) applies when the wavelength of the light is larger than the cavity radius. For light in the visible region, this is ordinarily the case. When liquids are involved, the values of α and a can be evaluated by fitting Eq. (3.10.22) to measurements of n vs N, both determined as functions of the temperature [5, 6].

It is possible to recover the Clausius–Mossotti equation from the Bottcher–Onsager formula by using Eq. (3.10.7) to replace a on the right hand side of Eq. (3.10.20). The immediate result is

$$\frac{3N\varepsilon_r}{(2\varepsilon_r+1)(\varepsilon_r-1)} = \frac{1}{(\alpha/\varepsilon_o)} - \frac{2}{3}\left(\frac{\varepsilon_r-1}{2\varepsilon_r+1}\right)N \tag{3.10.23a}$$

If the second term on the right hand side of Eq. (3.10.23a) is transferred to the left hand side, one obtains after finding a common denominator,

$$\frac{N}{(2\varepsilon_r+1)}\left[\frac{9\varepsilon_r+2(\varepsilon_r-1)^2}{3(\varepsilon_r-1)}\right] = \frac{1}{(\alpha/\varepsilon_o)} \tag{3.10.23b}$$

After expanding the square of the binomial and factoring the results, Eq. (3.10.23b) reads

$$\frac{N}{(2\varepsilon_r + 1)}\left[\frac{2\varepsilon_r^2 + 5\varepsilon_r + 2}{3(\varepsilon_r - 1)}\right] = N\left[\frac{(\varepsilon_r + 2)}{3(\varepsilon_r - 1)}\right] = \frac{1}{(\alpha/\varepsilon_o)} \tag{3.10.23c}$$

Upon rearrangement, the final result is

$$\frac{\varepsilon_r - 1}{\varepsilon_r + 2} = \frac{N\alpha}{3\varepsilon_o} \tag{3.10.24}$$

which is the same as Eq. (3.5.11).

If d is the mass density, and M is the molar mass, then the number density is

$$N = \frac{N_A d}{M} \tag{3.10.25}$$

where N_A is the Avogadro number. Upon substitution of Eq. (3.10.25) into Eq. (3.10.24), the result is

$$\frac{M}{d}\left(\frac{\varepsilon_r - 1}{\varepsilon_r + 1}\right) = \frac{N_A\alpha}{3\varepsilon_o} \tag{3.10.26}$$

where the dimensionless product, $N_A\alpha$, is known as the molar polarizability. Equation (3.10.26) is the Clausius–Mossotti equation written in a form which is useful for comparison with experiment.

3.11 Solvatochromism

A molecule in a low pressure gas absorbs light at frequencies that are characteristic of the free molecule. If the molecule is in solution, however, its absorption frequencies are shifted generally to lower values as compared to the same transitions in the gas phase. The magnitude of the shift towards the red depends upon the polarity of the solvent. The role of the solvent in determining the red shift is termed *solvatochromism*.

A light wave consists of an electric field vector and a magnetic field vector at right angles to one another, the two oscillating in both time and space. The direction of propagation of the wave is normal to the plane of the field vectors. Upon interaction with a molecule, the electric vector of the light wave induces an oscillating electric dipole moment in the molecule. If the frequency of the light is resonant with one of the molecular absorption frequencies, an electronic transition occurs and the molecule absorbs energy.

Although quantum theory is required to describe the absorption process in detail, the general mathematical structure of the quantum mechanical result can be duplicated in classical mechanics by assuming that the electron is harmonically bound to the nucleus. In this classical model, an electron exposed to an electromagnetic wave responds as if it were a damped harmonic oscillator. The damping represents the energy loss of the electron due to various radiative and non-radiative processes. This idea is due to Lorentz and was subsequently used by Bayless to describe solvatochromism [7]. Bayless adopted the Bottcher–Onsager model to represent the local electric field associated with the light wave interacting with a molecule in solution.

If the mass of the electron is m, and the charge of the electron is e, the classical equation of motion for a damped harmonic oscillator is

$$m\frac{d^2x}{dt^2} + m\gamma\frac{dx}{dt} + m\omega_o^2 x = eF \tag{3.11.1}$$

where t is the time, x is the displacement of the electron away from equilibrium, ω_o is the resonance frequency of the oscillation, γ is the damping coefficient, $m\omega_o^2$ is the force constant governing the harmonic restoring force, and F is the local value of the electric field sensed by the molecule.

The light wave interacts with the molecule by inducing an instantaneous electric dipole moment of magnitude, $p = f_o ex$, where f_o is termed the "oscillator strength". The oscillator strength can be assumed to be proportional to the number of electrons in the molecule that can absorb light with frequency ω_o, but in point of fact, its most important role is to account for the physical limitations inherent in using Eq. (3.11.1) to describe what is basically a quantum phenomenon. If Eq. (3.10.6) is chosen to represent the local electric field acting on the electron, then one writes,

$$F = E_o\left(\frac{3\varepsilon_r}{2\varepsilon_r + 1}\right) + \frac{f_o ex}{4\pi\varepsilon_o a^3}\left(\frac{2\varepsilon_r - 2}{2\varepsilon_r + 1}\right) \tag{3.11.2}$$

where E_o is the electric field of the light wave. Upon combining Eqs. (3.11.1) and (3.11.2) and grouping together all of the terms proportional to x, the result is

$$m\frac{d^2x}{dt^2} + m\gamma\frac{dx}{dt} + m\left[\omega_o^2 - \frac{f_o e^2}{4\pi\varepsilon_o m a^3}\left(\frac{2\varepsilon_r - 2}{2\varepsilon_r + 1}\right)\right]x = eE_o\left(\frac{3\varepsilon_r}{2\varepsilon_r + 1}\right) \tag{3.11.3}$$

At optical frequencies, $\varepsilon_r = n^2$, where n is the refractive index. With this replacement, Eq. (3.11.3) becomes

$$m\frac{d^2x}{dt^2} + m\gamma\frac{dx}{dt} + m\left[\omega_o^2 - \frac{f_o e^2}{4\pi\varepsilon_o m a^3}\left(\frac{2n^2 - 2}{2n^2 + 1}\right)\right]x = eE_o\left(\frac{3n^2}{2n^2 + 1}\right) \tag{3.11.4}$$

Upon comparison of Eqs. (3.11.1) and (3.11.4), we can identify the quantity within the brackets in Eq. (3.11.4) as the square of an effective oscillator frequency, ω, where

$$\omega^2 = \omega_o^2 - \frac{f_o e^2}{4\pi \varepsilon_o m a^3}\left(\frac{2n^2 - 2}{2n^2 + 1}\right) \tag{3.11.5a}$$

Equation (3.11.5a) can be written in the factored form

$$(\omega_o + \omega)(\omega_o - \omega) = \frac{f_o e^2}{4\pi \varepsilon_o m a^3}\left(\frac{2n^2 - 2}{2n^2 + 1}\right) \tag{3.11.5b}$$

The magnitude of the solvatochromic shift is defined by

$$\Delta\omega = \omega_o - \omega \tag{3.11.5c}$$

Because $\Delta\omega$ is ordinarily much smaller than ω_o, we can replace $\omega_o + \omega$ in the first parentheses on the left in Eq. (3.11.5b) by $2\omega_o$. After making this change and substituting Eq. (3.11.5c), Eq. (3.11.5b) reads,

$$\Delta\omega = \frac{f_o e^2}{4\pi \varepsilon_o \omega_o m a^3}\left(\frac{n^2 - 1}{2n^2 + 1}\right) \tag{3.11.6}$$

Both $\Delta\omega$ and ω_o are angular frequencies. We can rewrite Eq. (3.11.6) in terms of circular frequencies $\Delta\nu$ and ν_o by making the identifications

$$\Delta\omega = 2\pi\,\Delta\nu \tag{3.11.7a}$$

and

$$\omega_o = 2\pi\,\nu_o \tag{3.11.7b}$$

With these substitutions, Eq. (3.11.6) becomes

$$\Delta\nu = \frac{f_o e^2}{16\pi^3 \varepsilon_o \nu_o m a^3}\left(\frac{n^2 - 1}{2n^2 + 1}\right) \tag{3.11.8}$$

According to Eq. (3.11.8), the magnitude of the red shift depends upon the properties of the solvent through the factor involving the refractive index, n.

Of course, the local electric field expressed by Eq. (3.11.2) is not the only interaction of the optically active electron with its surroundings. Other phenomena, including dispersion forces, hydrogen bonding, and diffusion can play a role. These interactions are sometimes sufficient to change the red shift predicted by the cavity model into a blue shift. Such effects have been extensively investigated [8–10]. Nonetheless, Eq. (3.11.8) describes the basic phenomenon.

3.12 Lorenz–Lorentz Equation and Landolt's Rule

To a fair approximation, each atom and/or localized chemical bond in a molecule makes a separate and independent contribution to the total polarizability of the molecule. These distinctive contributions are called atomic polarizabilities. In the case of pi-bonded systems, where the π-electrons are delocalized, the π-system must be assigned its own polarizability. The values assigned to the atom/bond polarizabilities prove to be transferable from one molecular species to another. This allows the polarizability of a given molecule to be calculated on the basis of its molecular formula as a sum of atomic and bond polarizabilities. The principle of transferability and additivity is sometimes termed Landolt's rule (Hans Heinrich Landolt, Swiss, 1831–1910). At wavelengths in the visible or ultraviolet where the molecule absorbs light, the molecular polarizability becomes a function of the wavelength and the concepts of transferability and additivity break down. Hence, to apply Landolt's rule with accuracy, the molecular polarizabilities are measured at the 589.3 nm wavelength of the Na D-line where most covalently bonded molecules do not absorb.

The density of a pure substance and the polarizability of its molecules serve to determine its refractive index. Substitute Eq. (3.10.21) into Eq. (3.10.26) to obtain the result,

$$\frac{M}{d}\left(\frac{n^2-1}{n^2+1}\right)=\frac{N_A\alpha}{3\varepsilon_o} \tag{3.12.1a}$$

which is known as the Lorenz–Lorentz equation. Landolt's rule polarizabilities are almost exclusively rendered in cgs units which means that Eq. (3.12.1a) requires modification. If M is the molar mass, d is the density in g/cm^3, and n is the refractive index, then in cgs units, Eq. (3.12.1a) reads

$$\frac{M}{d}\left(\frac{n^2-1}{n^2+1}\right)=\frac{4\pi}{3}N_A\alpha_{cgs} \tag{3.12.1b}$$

where N_A is the Avogadro number, α_{cgs}, is the cgs molecular polarizability in the units of cm^3.

Landolt's rule asserts that in the case of a molecule made up of atoms, A and B, it is possible to define atom or bond "molar refractions" expressed in the form,

$$R(A)=\frac{4\pi}{3}N_A\alpha(A) \tag{3.12.2a}$$

and

$$R(B)=\frac{4\pi}{3}N_A\alpha(B) \tag{3.12.2b}$$

respectively. If the molecule has formula, A_xB_y, its molar refraction, $R(A_xB_y)$, is calculated as the stoichiometrically weighted sum,

$$R(A_xB_y) = xR(A) + yR(B) \tag{3.12.3}$$

where according to Eq. (3.12.1b)

$$R(A_xB_y) = \frac{4\pi}{3}N_A\alpha(A_xB_y) = \frac{M}{d}\left(\frac{n^2-1}{n^2+1}\right) \tag{3.12.4}$$

In Eq. (3.12.4), M is the molar mass of A_xB_y, n is the refractive index of the material, and d is its mass density. Table 3.1 lists some bond and atom molar refractions [11].

Example According to Table 3.1, $R(\text{C}) = 2.591$, $R(\text{H}) = 1.028$, $R(=\text{O}) = 2.122$, and $R(\text{OH}) = 2.553$, respectively, each rendered in the units of cm^3/mol. Based upon these atom/bond refractions, the molar refraction of propionic acid, C_2H_5COOH, is

Table 3.1 Landolt's rule atom and bond molar refractions

Bond/atom	R (cm^3/mol)
C	2.591
H	1.028
=O	2.122
–O–	1.643
OH	2.553
F	0.81
Cl	5.844
Br	8.741
I	13.954
C_6H_5	25.463
$C_{10}H_7$	43.00
–S	7.729
=S	7.921
CN	5.459
N (primary aliphatic)	2.376
N (secondary aliphatic)	2.582
Ethylenic bond	1.575
Acetylenic bond	1.977
N (aromatic)	3.550
Four membered ring	0.317
Three membered ring	0.614

See Ref. [11] at the end of this chapter

rendered as.

$$R(C_3H_5O(OH)) = 3R(C) + 5R(H) + R(=O) + R(OH)$$
$$= 3(2.591) + 5(1.028) + 2.122 + 2.553 = 17.588\,\text{cm}^3/\text{mol} \tag{3.12.5}$$

If the calculated value of $R(C_3H_5COOH)$ is combined with the molecular weight, $M = 74.08$ g/mol, and mass density, $d = 0.9930\,\text{g/cm}^3$, one can solve Eq. (3.12.4) for the refractive index. The result differs from the experimental value by about 8%.

The value of $R(C_3H_5COOH)$ can also be substituted into the left hand side of Eq. (3.12.4), and the result used to evaluate the molecular polarizability:

$$\alpha_{C_2H_5COOH} = \frac{3R}{4\pi N_A} = \frac{3(17.588\,\text{cm}^3/\text{mol})}{4\pi\left(6.022\times 10^{23}/\text{mol}\right)} = 6.972\times 10^{-24}\,\text{cm}^3 \tag{3.12.6}$$

Although cgs units are ordinarily used when applying Landolt's rule, it is possible to convert the molar refraction to MKSQ units. In cgs units, we write,

$$R_{cgs}(\text{cm}^3/\text{mol}) = \frac{4\pi}{3}N_A\alpha_{cgs} \tag{3.12.7a}$$

According to Eq. (3.12.1a), the MKSQ equivalent of Eq. (3.12.7a) is

$$R_{MKSQ}(\text{m}^3/\text{mol}) = \frac{N_A\alpha_{MKSQ}}{3\varepsilon_o} \tag{3.12.7b}$$

The relation between R_{cgs} and R_{MKSQ} is

$$R_{MKSQ}(\text{m}^3/\text{mol}) = R_{cgs}(\text{cm}^3/\text{mol}) \times \frac{\text{m}^3}{10^6\text{cm}^3} \tag{3.12.7c}$$

When Eqs. (3.12.7a) through (3.12.7c) are combined, one obtains

$$\alpha_{MKSQ} = (10^{-6}\text{m}^3/\text{cm}^3) \times 4\pi\,\varepsilon_o\alpha_{cgs} \tag{3.12.8}$$

where the units of α_{MKSQ} are C^2m^2/J. Equation (3.12.8) can be used to convert molecular polarizabilities from cgs to MKSQ units.

3.13 Onsager Theory of Polar Dielectrics

3.13.1 *Induced Dipole* ($\mathbf{p}_\alpha$)*, Reaction Field* ($\mathbf{R}_\alpha$)*, and Directing Field* ($\mathbf{F}_\alpha$)

The electric field expressed by Eq. (3.10.6) can be divided into two parts. Rendered in vector notation, the first term on the right is the electric field, $\mathbf{F}_c$, produced at the center of the cavity due to the external electric field. Known as the cavity field,

$$\mathbf{F}_c = \mathbf{E}_o\left(\frac{3\varepsilon_r}{2\varepsilon_r + 1}\right) \tag{3.13.1}$$

The second term can be written as

$$\mathbf{R}_\alpha = \frac{\mathbf{p}_\alpha}{4\pi\varepsilon_o a^3}\left(\frac{2\varepsilon_r - 2}{2\varepsilon_r + 1}\right) \tag{3.13.2}$$

and is termed the reaction field. The reaction field arises in the response of the dielectric outside the cavity to the electric dipole, $\mathbf{p}_\alpha$, induced at the center of the cavity by the external field. We use the subscript, α, to refer to the molecular polarizability and to stress that $\mathbf{p}_\alpha$ is an *induced* dipole. Neither it nor the reaction field, $\mathbf{R}_\alpha$, would exist in the absence of the external electric field.

We can rewrite Eq. (3.13.2) in the form

$$\mathbf{R}_\alpha = f\mathbf{p}_\alpha \tag{3.13.3a}$$

where

$$f = \frac{1}{4\pi\varepsilon_o a^3}\left(\frac{2\varepsilon_r - 2}{2\varepsilon_r + 1}\right) \tag{3.13.3b}$$

The sum of the cavity field, $\mathbf{F}_c$, and the reaction field, $\mathbf{R}_\alpha$, is

$$\mathbf{F}_\alpha = \mathbf{F}_c + \mathbf{R}_\alpha \tag{3.13.4}$$

where $\mathbf{F}_\alpha$ is termed the "directing" field. Because $\mathbf{p}_\alpha$ forms in response to $\mathbf{F}_\alpha$,

$$\mathbf{p}_\alpha = \alpha\mathbf{F}_\alpha \tag{3.13.5}$$

where the multiplicative coefficient, α, is of course, the molecular polarizability. The combination of Eqs. (3.13.3a) and (3.13.5) has the result

$$\mathbf{R}_\alpha = f\alpha\mathbf{F}_\alpha \tag{3.13.6}$$

Substitution of Eq. (3.13.6) into Eq. (3.13.4) leads to

$$\mathbf{F}_{\alpha} = \mathbf{F}_{c} + f\alpha\mathbf{F}_{\alpha} \tag{3.13.7}$$

which can be solved for $\mathbf{F}_{\alpha}$ in terms of $\mathbf{F}_{c}$.

$$\mathbf{F}_{\alpha} = \frac{\mathbf{F}_{c}}{1 - f\alpha} \tag{3.13.8a}$$

Upon insertion of Eq. (3.13.1), Eq. (3.13.8a) reads,

$$\mathbf{F}_{\alpha} = \frac{1}{1 - f\alpha}\left(\frac{3\varepsilon_r}{2\varepsilon_r + 1}\right)\mathbf{E}_{o} \tag{3.13.8b}$$

3.13.2 Permanent Electric Dipole ($\boldsymbol{\mu}$), Reaction Field ($\mathbf{R}_{\mu}$), and Local Field ($\mathbf{F}_{\mu}$)

Next consider the case where the external field is zero, but there is a *permanent* electric dipole, $\boldsymbol{\mu}$, at the center of the cavity. In the calculation which follows, this permanent dipole, $\boldsymbol{\mu}$, replaces $\mathbf{E}_{o}$ as a source of the electric field that induces polarization in the dielectric. Closely associated with the permanent electric dipole is the reaction field, $\mathbf{R}_{\mu}$.

Since all molecules are polarizable to a greater or lesser extent, a molecule with a permanent electric dipole moment will also have a polarizability α. The reaction field, $\mathbf{R}_{\mu}$, associated with the permanent electric dipole induces a dipole moment, $\alpha\mathbf{R}_{\mu}$, in the molecule in the cavity. The total electric dipole moment, $\mathbf{p}_{\mu}$, at the center of the cavity thus has a permanent part, $\boldsymbol{\mu}$, and an induced part $\alpha\mathbf{R}_{\mu}$. The sum is

$$\mathbf{p}_{\mu} = \boldsymbol{\mu} + \alpha\mathbf{R}_{\mu} \tag{3.13.9}$$

In analogy with Eq. (3.13.3a), the reaction field associated with $\mathbf{p}_{\mu}$ is

$$\mathbf{R}_{\mu} = f\mathbf{p}_{\mu} = f\boldsymbol{\mu} + f\alpha\mathbf{R}_{\mu} \tag{3.13.10}$$

where we have used Eq. (3.13.9). Solving Eq. (3.13.10) for $\mathbf{R}_{\mu}$, we find

$$\mathbf{R}_{\mu} = \frac{f\boldsymbol{\mu}}{1 - f\alpha} \tag{3.13.11}$$

Since there is no external electric field, the local electric field, $\mathbf{F}_\mu$, is identical with the reaction field, $\mathbf{R}_\mu$; hence, according to Eq. (3.13.11),

$$\mathbf{F}_\mu = \mathbf{R}_\mu = \frac{f\boldsymbol{\mu}}{1 - f\alpha} \tag{3.13.12}$$

In the case where there is an external field and also a permanent dipole $\boldsymbol{\mu}$ at the center of the cavity, the local field, $\mathbf{F}_\alpha$, given by Eq. (3.13.8b) serves as a "directing field" acting on the permanent dipole $\boldsymbol{\mu}$ so as to orient it in the direction parallel to the external field. Opposing the effect of $\mathbf{F}_\alpha$ are the frequent inter-molecular collisions which tend to randomize the orientation of the dipole. Because of these frequent collisions, only the average orientation of the dipole lies along the direction of $\mathbf{F}_\alpha$. This average, which is temperature dependent, can be calculated using statistical mechanics.

According to Eq. (1.8.9) of Chap. 1, the energy $u(\theta)$ of interaction of $\mathbf{F}_\alpha$ with $\boldsymbol{\mu}$ is given by

$$u(\theta) = -\boldsymbol{\mu} \cdot \mathbf{F}_\alpha = -\mu F_\alpha \cos\theta \tag{3.13.13}$$

where θ is the angle between $\boldsymbol{\mu}$ and $\mathbf{F}_\alpha$. By analogy with Eq. (2.9.22) of Chap. 2, the probability, $P(\theta)d\theta d\phi$, that we can find $\boldsymbol{\mu}$ within the differential solid angle, $d\Omega = \sin\theta d\theta d\phi$, defined by the polar angle, θ, and the azimuthal angle, ϕ, is

$$P(\theta)d\theta d\phi = \frac{\exp(-u(\theta)/k_B T)\sin\theta d\theta d\phi}{Z} \tag{3.13.14}$$

where by comparison with Eq. (2.9.20), the canonical partition function is

$$Z = \int_0^{2\pi} d\phi \int_0^{\pi} d\theta \sin\theta \exp(-u(\theta)/k_B T) \tag{3.13.15}$$

Equations (3.13.14) and (3.13.15) can be combined to calculate the average orientation of the dipole. Let the direction of the applied electric field, $\mathbf{E}_o$, define the z-axis. The x and y components of $\boldsymbol{\mu}$ are then given by $\mu_x = \mu\sin\theta\cos\phi$ and $\mu_y = \mu\sin\theta\sin\phi$, respectively. It can be shown that both of these components average to zero. As noted above, the differential solid angle in polar coordinates is $d\Omega = \sin\theta d\theta d\phi$ (see the Mathematical Methods Appendix). With $\beta = 1/k_B T$, the angle average of μ_x is

$$\begin{aligned}\langle \mu_x \rangle &= \int d\Omega P(\theta) \mu \sin\theta \cos\phi \\ &= \frac{\mu}{Z} \int_0^{\pi} d\theta \sin^2\theta e^{-\beta u(\theta)} \int_0^{2\pi} d\phi \cos\phi \\ &= \frac{\mu}{Z} \int_0^{\pi} d\theta \sin^2\theta e^{-\beta u(\theta)} [\sin\phi]_0^{2\pi} \\ &= \frac{\mu}{Z} \int_0^{\pi} d\theta \sin^2\theta e^{-\beta u(\theta)} [0 - 0] = 0 \end{aligned} \tag{3.13.16a}$$

Next, the angle average of μ_y is

$$\begin{aligned}\langle \mu_y \rangle &= \frac{\mu}{Z} \int_0^{\pi} d\theta \sin^2\theta e^{-\beta u(\theta)} \int_0^{2\pi} d\phi \sin\phi \\ &= \frac{\mu}{Z} \int_0^{\pi} d\theta \sin^2\theta e^{-\beta u(\theta)} [-\cos\phi]_0^{2\pi} \\ &= \frac{\mu}{Z} \int_0^{\pi} d\theta \sin^2\theta e^{-\beta u(\theta)} [-1 + 1] = 0 \end{aligned} \tag{3.13.16b}$$

By contrast $\langle \mu_z \rangle$ is non-zero. The z-component of $\boldsymbol{\mu}$ is $\mu_z = \mu \cos\theta$. Its average is given by

$$\langle \mu_z \rangle = \mu \langle \cos\theta \rangle = \frac{\mu}{Z} \int_0^{\pi} d\phi \int_0^{2\pi} d\theta e^{-\beta u(\theta)} \sin\theta \cos\theta \tag{3.13.17}$$

To evaluate $\langle \mu_z \rangle$, we shall need to calculate the canonical partition function, Z. Upon substitution of Eq. (3.13.13) into Eq. (3.13.15), we obtain

$$Z = \int_0^{2\pi} d\phi \int_0^{\pi} d\theta \sin\theta \exp(\mu F_\alpha \cos\theta / k_B T) \tag{3.13.18a}$$

Evaluation of the integral over ϕ gives

$$Z = 2\pi \int_0^{\pi} d\theta \sin\theta \exp(\mu F_\alpha \cos\theta / k_B T) \tag{3.13.18b}$$

The remaining integral can be evaluated by recognizing that $\sin\theta d\theta = -d(\cos\theta)$. Equation (3.13.18b) then becomes

$$Z = -2\pi \int_0^{\pi} d(\cos\theta) \exp(\mu F_\alpha \cos\theta / k_B T) \tag{3.13.18c}$$

Let $a = \mu F_\alpha / k_B T$ and $x = \cos\theta$ then Eq. (3.13.18c) reads

$$Z = -2\pi \int_{+1}^{-1} dx \exp(ax) \tag{3.13.18d}$$

where we have noted that when $\theta = 0$, $x = +1$ and when $\theta = \pi$, $x = -1$. Computation of the integral gives

$$Z = -2\pi \left[\frac{\exp(-a) - \exp(a)}{a} \right] \tag{3.13.18e}$$

Equation (3.13.18e) can be rewritten as

$$Z = 4\pi \frac{\sinh(a)}{a} \tag{3.13.19}$$

According to Eq. (3.13.13), $e^{-\beta u(\theta)} = e^{\mu F_\alpha \cos\theta / k_B T}$. This substitution permits Eq. (3.13.17) to be written as

$$\langle \mu_z \rangle = \mu \langle \cos\theta \rangle = \int_0^{\pi} d\phi \int_0^{2\pi} d\theta \sin\theta \frac{\exp(\mu F_\alpha \cos\theta / k_B T)}{Z} \mu \cos\theta \tag{3.13.20}$$

After evaluation of the integral over ϕ and substitution of $\sin\theta d\theta = -d(\cos\theta)$, Eq. (3.13.20) becomes

$$\langle \mu_z \rangle = -\frac{2\pi\mu}{Z} \int_0^{\pi} d(\cos\theta) \left[\cos\theta \exp(\mu F_\alpha \cos\theta / k_B T) \right] \tag{3.13.21a}$$

Again let $a = \mu F_\alpha / k_B T$, $x = \cos\theta$, and $dx = d(\cos\theta)$. Equation (3.13.21a) can then be rewritten as

$$\langle \mu_z \rangle = \frac{2\pi \mu}{Z} \int_{-1}^{+1} dx\left[x \exp(ax)\right] \tag{3.13.21b}$$

The integral in Eq. (3.13.21b) can be done by parts. The first integration yields

$$\int_{-1}^{+1} dx\left[x \exp(ax)\right] = \left[\frac{xe^{ax}}{a}\right]_{-1}^{+1} - \frac{1}{a} \int_{-1}^{+1} dx \exp(ax) \tag{3.13.21c}$$

After the second integration, the result is

$$\begin{aligned} \int_{-1}^{+1} dx\left[x \exp(ax)\right] &= \left[\frac{xe^{ax}}{a}\right]_{-1}^{+1} - \frac{1}{a}\left[\frac{e^{ax}}{a}\right]_{-1}^{+1} = \frac{e^{a} + e^{-a}}{a} - \frac{e^{a} - e^{-a}}{a^2} \\ &= \frac{2\cosh(a)}{a} - \frac{2\sinh(a)}{a^2} \end{aligned} \tag{3.13.21d}$$

Upon substitution of Eq. (3.13.21d) into Eq. (3.13.21b), one obtains,

$$\langle \mu_z \rangle = \frac{4\pi \mu}{Z}\left[\frac{\cosh(a)}{a} - \frac{\sinh(a)}{a^2}\right] \tag{3.13.22}$$

Substitution of Eq. (3.13.19) into Eq. (3.13.22) gives

$$\langle \mu_z \rangle = \frac{4\pi \mu}{4\pi \frac{\sinh(a)}{a}}\left[\frac{\cosh(a)}{a} - \frac{\sinh(a)}{a^2}\right] \tag{3.13.23}$$

After algebraic simplification, Eq. (3.13.23) can be written as

$$\langle \mu_z \rangle = \mu L(a) \tag{3.13.24}$$

where

$$L(a) = \left[\coth(a) - \frac{1}{a}\right] \tag{3.13.25}$$

is termed the Langevin function (Paul Langevin, French, 1872–1946).

In most experimental situations, $a \ll 1$. To simply the Langevin function to apply to this case, we construct the common denominator,

$$L(a) = \frac{\cosh(a)}{\sinh(a)} - \frac{1}{a} = \frac{a\cosh(a) - \sinh(a)}{a\sinh(a)} \tag{3.13.26}$$

When $a \ll 1$, approximate $\sinh(a)$ by

$$\sinh(a) = a + \frac{a^3}{3!} + \cdots \tag{3.13.27a}$$

and $\cosh(a)$ by

$$\cosh(a) = 1 + \frac{a^2}{2!} + \cdots \tag{3.13.27b}$$

With these series approximations, the Langevin function becomes

$$L(a) \simeq \frac{a\left(1 + \frac{a^2}{2!}\right) - \left(a + \frac{a^3}{3!}\right)}{a\left(a + \frac{a^3}{3!}\right)} \simeq \frac{\left(\frac{a^3}{2} - \frac{a^3}{6}\right)}{a^2} \tag{3.13.28a}$$

In the fraction that constitutes the right hand side of Eq. (3.13.28a), the numerator contains factors of order a^3, while in the denominator, the leading factor is of order a^2. Cancellation of common factors between the numerator and the denominator indicates that to lowest order, $L(a)$ is proportional to a. The final result reads,

$$L(a) = \frac{a}{3} \tag{3.13.28b}$$

Upon substitution of $a = \mu F_\alpha / k_B T$, Eq. (3.13.28b) becomes

$$L(a) = \frac{\mu F_\alpha}{3 k_B T} \tag{3.13.29}$$

Substitution of Eq. (3.13.29) into Eq. (3.13.24) gives

$$\langle \mu_z \rangle = \frac{\mu^2 F_\alpha}{3 k_B T} \tag{3.13.30}$$

Hence, by virtue of Eqs. (3.13.16a), (3.13.16b), and (3.13.30), $\boldsymbol{\mu}$ has on average only a z-component; hence, the average value of the vector $\mathbf{F}_\mu$ in Eq. (3.13.12) is

$$F_\mu = R_\mu = \frac{f \langle \mu_z \rangle}{1 - f\alpha} \tag{3.13.31}$$

After substitution of Eq. (3.13.30), Eq. (3.13.31) reads

$$F_\mu = \frac{f}{(1 - f\alpha)} \cdot \frac{\mu^2 F_\alpha}{3 k_B T} \tag{3.13.32}$$

Recognizing that $\mathbf{E}_o$ has only a z- component, we can replace F_α in Eq. (3.13.32) by the scalar form of Eq. (3.13.8b). The result is

$$F_\mu = \frac{f}{(1-f\alpha)^2} \cdot \frac{\mu^2}{3k_B T} \cdot \frac{3\varepsilon_r}{2\varepsilon_r + 1} E_o \tag{3.13.33}$$

At this point, the Onsager theory of polar dielectrics can be summarized as follows: The applied electric field, E_o, is distorted in the vicinity of the cavity, so that the sensible field at the center of the cavity is $F_c = (3\varepsilon_r/(2\varepsilon_r + 1))E_o$. In Sect. 3.13.1, we found that F_c induces an electric dipole moment in the polarizable molecule at the center of the cavity. The field of this induced dipole polarizes the dielectric exterior to the cavity. This induced polarization is the source of a reaction field, R_α, that is sensed by the dipole at the center of the cavity. The sum, $F_\alpha = F_c + R_\alpha$, constitutes a contribution to the local field, which has its origin in the induced polarization. In Sect. 3.13.2, we considered the effect of a permanent electric dipole located at the center of the cavity. Although restricted in its orientation by thermal fluctuations, this dipole responds by becoming partially aligned with F_α. This partial orientation is the source of an electric field which polarizes the surrounding dielectric. The polarization in the surrounding dielectric is the source of a field, R_μ, which is sensed at the center of the cavity. The field $R_\mu = F_\mu$ constitutes the contribution of the permanent dipole to the field sensed at the center of the cavity. Below, we will note that the total local electric field at the center of the cavity is the sum, $F = F_\alpha + F_\mu$.

3.13.3 Superposition of an Induced Dipole and a Permanent Dipole

The total local electric field, $\mathbf{F}$, in a polar material is the sum of $\mathbf{F}_\alpha$, whose origin is to be found in the induced dipoles and $\mathbf{F}_\mu$, whose origin is to be found in the permanent dipoles. Because the proximate cause of both $\mathbf{F}_\alpha$ and $\mathbf{F}_\mu$ is the external field, which has only a z-component, we focus on the component of $\mathbf{F}$ in the z-direction; namely,

$$F = F_\alpha + F_\mu \tag{3.13.34}$$

Upon substitution of Eq. (3.13.33) and the scalar form of Eq. (3.13.8b) into. Equation (3.13.34), the result is

$$F = \frac{1}{(1-f\alpha)}\left(\frac{3\varepsilon_r}{2\varepsilon_r + 1}\right)E_o + \frac{f}{(1-f\alpha)^2} \cdot \frac{\mu^2}{3k_B T} \cdot \left(\frac{3\varepsilon_r}{2\varepsilon_r + 1}\right)E_o \tag{3.13.35}$$

Factor Eq. (3.13.35) to obtain,

$$F = \left[1 + \frac{f}{(1-f\alpha)} \cdot \frac{\mu^2}{3k_B T} \cdot\right] \frac{1}{(1-f\alpha)}\left(\frac{3\varepsilon_r}{2\varepsilon_r + 1}\right)E_o \tag{3.13.36}$$

The total macroscopic polarization, P, has a component, P_α, associated with the induced molecular electric dipoles and a component, P_μ, associated with the permanent molecular dipoles. The total macroscopic polarization is thus

$$P = P_\alpha + P_\mu \tag{3.13.37}$$

The source of P_α is the total local electric field, F. The magnitude of the induced electric dipole moment per molecule is αF. If N is the number of molecules per unit volume, then

$$P_\alpha = N\alpha F \tag{3.13.38}$$

With the use of Eq. (3.13.36), Eq. (3.13.38) takes the form,

$$P_\alpha = N\alpha\left[1 + \frac{f}{(1-f\alpha)}\cdot\frac{\mu^2}{3k_BT}\right]\frac{1}{(1-f\alpha)}\left(\frac{3\varepsilon_r}{2\varepsilon_r+1}\right)E_o \tag{3.13.39}$$

The source of P_μ is the directing field, F_α, the field that serves to orient the permanent dipole. Given that there are N molecules per unit volume and the average projection of a permanent molecular electric dipole onto z-direction is $\langle\mu_z\rangle$, we have

$$P_\mu = N\langle\mu_z\rangle \tag{3.13.40}$$

The scalar form of Eq. (3.13.8b) gives F_α. Substitution of F_α into Eq. (3.13.30) leads to the result,

$$\langle\mu_z\rangle = \left(\frac{\mu^2}{3k_BT}\right)\frac{1}{(1-f\alpha)}\left(\frac{3\varepsilon_r}{2\varepsilon_r+1}\right)E_o \tag{3.13.41}$$

Upon substitution of Eq. (3.13.41), Eq. (3.13.40) becomes

$$P_\mu = N\left[\left(\frac{\mu^2}{3k_BT}\right)\frac{1}{(1-f\alpha)}\left(\frac{3\varepsilon_r}{2\varepsilon_r+1}\right)\right]E_o \tag{3.13.42}$$

Now substitute Eqs. (3.13.39) and (3.13.42) into Eq. (3.13.37) to obtain for the total polarization,

$$\begin{aligned} P = {} & N\left[1 + \frac{f}{(1-f\alpha)}\cdot\frac{\mu^2}{3k_BT}\right]\frac{\alpha}{(1-f\alpha)}\left(\frac{3\varepsilon_r}{2\varepsilon_r+1}\right)E_o \\ & + N\left[\left(\frac{\mu^2}{3k_BT}\right)\frac{1}{(1-f\alpha)}\left(\frac{3\varepsilon_r}{2\varepsilon_r+1}\right)\right]E_o \end{aligned} \tag{3.13.43a}$$

Next factor Eq. (3.13.43a) to arrive at

$$P = \left(\frac{N}{(1-f\alpha)}\right)\left[\alpha + \frac{f\alpha}{(1-f\alpha)}\cdot\frac{\mu^2}{3k_BT} + \frac{\mu^2}{3k_BT}\right]\left(\frac{3\varepsilon_r}{2\varepsilon_r+1}\right)E_o \quad (3.13.43b)$$

Now combine terms within the bracket that involve $\mu^2/3k_BT$. The immediate result is

$$P = \left(\frac{N}{(1-f\alpha)}\right)\left[\alpha + \frac{\mu^2}{3k_BT}\left(\frac{f\alpha+1-f\alpha}{1-f\alpha}\right)\right]\left(\frac{3\varepsilon_r}{2\varepsilon_r+1}\right)E_o \quad (3.13.43c)$$

Finally, perform the indicated subtraction within the bracket to obtain,

$$P = \left(\frac{N}{(1-f\alpha)}\right)\left[\alpha + \frac{1}{3k_BT}\left(\frac{\mu^2}{1-f\alpha}\right)\right]\left(\frac{3\varepsilon_r}{2\varepsilon_r+1}\right)E_o \quad (3.13.44)$$

Equations (3.2.26) and (3.2.30) can be combined to express **P** in the form,

$$\mathbf{P} = \mathbf{D} - \varepsilon_o\mathbf{E} = \varepsilon_o(\varepsilon_r - 1)\mathbf{E} \quad (3.13.45a)$$

By noting that the external electric field, $\mathbf{E}_o$, has the same meaning as, **E** we can use Eq. (3.13.45a) in scalar form to replace P on the left hand side of Eq. (3.13.44). The result is

$$\varepsilon_o(\varepsilon_r - 1)E_o = \left(\frac{N}{1-f\alpha}\right)\left[\alpha + \frac{1}{3k_BT}\left(\frac{\mu^2}{1-f\alpha}\right)\right]\left(\frac{3\varepsilon_r}{2\varepsilon_r+1}\right)E_o \quad (3.13.45b)$$

After cancellation of E_o from either side of Eq. (3.13.45b), and multiplication by $(2\varepsilon_r + 1)/3\varepsilon_r\varepsilon_o$, we obtain

$$\frac{(\varepsilon_r - 1)(2\varepsilon_r + 1)}{3\varepsilon_r} = \left(\frac{N}{\varepsilon_o(1-f\alpha)}\right)\left[\alpha + \frac{1}{3k_BT}\left(\frac{\mu^2}{1-f\alpha}\right)\right] \quad (3.13.46)$$

which is Onsager's electromagnetic equation of state for a polar dielectric consisting a pure chemical substance.

In the case of a multicomponent solution containing "c" components, the right hand side can be augmented by writing it as a sum of the contributions from each component. Equation (3.13.46) then becomes an electromagnetic equation of state for the solution as a whole, which reads,

$$\frac{(\varepsilon_r - 1)(2\varepsilon_r + 1)}{3\varepsilon_r} = \sum_{k=1}^{c}\left(\frac{N_k}{\varepsilon_o(1-f_k\alpha_k)}\right)\left[\alpha_k + \frac{1}{3k_BT}\left(\frac{\mu_k^2}{1-f_k\alpha_k}\right)\right] \quad (3.13.47a)$$

where

$$f_k = \frac{1}{4\pi\varepsilon_o a_k^3}\left(\frac{2\varepsilon_r - 2}{2\varepsilon_r + 1}\right) \quad (3.13.47b)$$

The special case of a non-polar dielectric consisting of a single component is obtained by setting $\mu = 0$ in Eq. (3.13.46). The result is

$$\frac{(\varepsilon_r - 1)(2\varepsilon_r + 1)}{3\varepsilon_r} = \left(\frac{N\alpha/\varepsilon_o}{(1 - f\alpha)}\right) \tag{3.13.48}$$

Upon suppressing the subscript, k, in Eq. (3.13.47b) and substituting the result into Eq. (3.13.48) and rearranging, the Bottcher - Onsager formula

$$\frac{3N\varepsilon_r}{(2\varepsilon_r + 1)(\varepsilon_r - 1)} = \frac{1}{(\alpha/\varepsilon_o)} - \left(\frac{1}{2\pi a^3}\right)\left(\frac{\varepsilon_r - 1}{2\varepsilon_r + 1}\right) \tag{3.13.49}$$

is the result. Equation (3.13.49) is seen to be the same as Eq. (3.10.20).

The model, upon which Eqs. (3.13.47) and (3.13.49) are based, can be summarized as follows: Each molecule in a fluid occupies a space that cannot be occupied by another molecule. In essence, each molecule carves out a cavity for itself. From the point of view of the molecule in its cavity, the intermolecular packing of the surrounding molecules produces ups and downs in the relative permittivity. At great distances from the cavity, these oscillations become smeared out, so that the surrounding material appears to be a continuous dielectric. When a uniform electric field is applied to the material, the presence of the cavity has the effect of distorting the field in its proximity. The distortion of this direct field by the cavity causes the electrostatic potential at the center of the cavity to be different from that in the surrounding dielectric. This distortion in the potential is the source of a local electric field sensed at the center of the cavity. The response of the electronic structure of the molecule to the local field can be modeled by placing at the center of the cavity an induced dipole in the case of a non-polar dielectric and also a permanent dipole in the case of a polar dielectric. Both are sources of a reaction field that contributes to the polarization of the surrounding dielectric. The value of the local field is adjusted to be self-consistent with the sum of values of the direct field and the reaction fields. Calculation of the Onsager dielectric equation of state for a polar material then proceeds by making the local field consistent with the constitutive relations governing a macroscopic dielectric. Equations (3.13.47a) and (3.13.47b) have been used extensively to determine the polarizabilities and cavity radii that apply in the case of both inorganic [6] and organic molecules [12].

3.14 Application of the Onsager Local Electric Field Concept to Solution Thermodynamics

The infinite dilution enthalpy of a solute dissolving in a liquid solvent can be written in phenomenological terms as

$$\Delta H_{sol'n} = \Delta H_{sep} + \Delta H_{cav} + \Delta H_{int} \tag{3.14.1}$$

In Eq. (3.14.1), $\Delta H_{sol'n}$ is the enthalpy of solution, ΔH_{sep} is the enthalpy required to separate the condensed solute phase into individual molecules, ΔH_{cav} is the enthalpy needed to form a cavity in the solvent to accommodate a solute molecule, and ΔH_{int} is the enthalpy of interaction of the solute molecule with the surrounding solvent molecules.

As a first approximation, ΔH_{sep} can be set equal to the heat of vaporization in the case of a liquid solute and the heat of sublimation in the case of a solid solute. The enthalpy of formation of the cavity, ΔH_{cav}, can be evaluated by using the concept of surface tension to calculate the energy required to create a free spherical surface in the solvent. The last contribution, ΔH_{int}, is based upon a calculation of the electrostatic work, W, associated with the equilibrium between a molecular dipole and its reaction field as treated in the semi-continuum model of dielectrics [13]. We shall see that the work, W, depends upon the solvent static relative permittivity, ε_r, and the solvent refractive index, n. Since both of these material parameters are ordinarily measured at constant temperature and pressure, W, can be identified with the Gibbs free energy. The associated enthalpy, ΔH_{int}, can then be calculated by substituting W into the Gibbs–Helmholtz equation, Eq. (2.5.20), in Chap. 2. In terms of W, the appropriate formula is

$$\Delta H_{int} = -T^2\left(\frac{\partial(W/T)}{\partial T}\right) \tag{3.14.2}$$

We shall see that the electrostatic energy, W, in Eq. (3.14.2) is the dipolar analog of the Born charging energy used to calculate the enthalpy of solvation of a free charge as it was derived in Sect. 3.3.

In the case of a dipole moment of size, $d\mu$, interacting with its reaction field, R_μ, the differential of the work, dW, can be calculated by writing Eq. (1.8.9) in Chap. 1 in differential form as

$$dW = -R_\mu d\mu \tag{3.14.3}$$

Since the final value of the dipole moment is μ, we can represent the differential of μ as $d\mu = \mu d\lambda$, where λ is a dimensionless variable that runs from zero to one. By virtue of Eq. (3.13.11), the final value of the reaction field, R_μ, is proportional to μ; hence, its instantaneous value is $R_\mu\lambda$. In terms of λ, Eq. (3.14.3) can be written as

$$dW = -\lambda R_\mu \mu d\lambda \tag{3.14.4a}$$

The integral of Eq. (3.14.4a) is

$$W = -\mu R_\mu \int_0^1 \lambda d\lambda = -\mu R_\mu\left[\frac{\lambda^2}{2}\right]_0^1 \tag{3.14.4b}$$

which upon evaluation at the limits of integration takes the form,

$$W = -\frac{1}{2}R_{\mu}\mu \tag{3.14.5}$$

Next, we invoke the scalar form of Eq. (3.13.11),

$$R_{\mu} = \frac{f\mu}{1 - f\alpha} \tag{3.14.6a}$$

where we recall that f is defined by Eq. (3.13.3b) as

$$f = \frac{1}{4\pi\varepsilon_o a^3}\left(\frac{2\varepsilon_r - 2}{2\varepsilon_r + 1}\right) \tag{3.14.6b}$$

Rewrite Eq. (3.14.6a) as

$$R_{\mu} = \frac{\mu}{\frac{1}{f} - \alpha} \tag{3.14.7a}$$

Upon substitution of Eq. (3.14.6b), Eq. (3.14.7a) becomes

$$R_{\mu} = \frac{\mu}{\frac{4\pi\varepsilon_o a^3(2\varepsilon_r+1)}{(2\varepsilon_r-2)} - \alpha} \tag{3.14.7b}$$

Assuming that the molecular cavities fill all space, we can rearrange Eq. (3.10.7) to read

$$4\pi a^3 = \frac{3}{N} \tag{3.14.7c}$$

We assume that at wavelengths sufficiently far away from where the solute absorbs, the polarizability is the same as the static polarizability. On the basis of this assumption, we can set $\varepsilon_r = n^2$ in Eq. (3.10.24), and solve the result for α:

$$\alpha = \frac{3\varepsilon_o}{N}\left(\frac{n^2 - 1}{n^2 + 2}\right) \tag{3.14.7d}$$

Substitute Eq. (3.14.7c) and (3.14.7d) into Eq. (3.14.7b). The result is

$$R_{\mu} = \frac{\mu}{\frac{3\varepsilon_o}{N}\frac{(2\varepsilon_r+1)}{(2\varepsilon_r-2)} - \frac{3\varepsilon_o}{N}\left(\frac{n^2-1}{n^2+2}\right)} \tag{3.14.7e}$$

After construction of a common denominator, Eq. (3.14.7e) reads

$$R_{\mu} = \frac{(2\varepsilon_r - 2)(n^2 + 2)}{(2\varepsilon_r + 1)(n^2 + 2) - (2\varepsilon_r - 2)(n^2 - 1)}\left(\frac{N\mu}{3\varepsilon_o}\right) \tag{3.14.7g}$$

Multiply out the denominator and cancel like terms having opposite signs. The result is

$$R_\mu = \frac{2(\varepsilon_r - 1)(n^2 + 2)}{3(2\varepsilon_r + n^2)} \left(\frac{N\mu}{3\varepsilon_o} \right) \tag{3.14.8}$$

Upon substitution of Eq. (3.14.8) into Eq. (3.14.5), we obtain the final result [13],

$$W = -\frac{(\varepsilon_r - 1)(n^2 + 2)}{(2\varepsilon_r + n^2)} \left(\frac{N\mu^2}{9\varepsilon_o} \right) \tag{3.14.9}$$

Because $\varepsilon_r = 1$ describes free space, the energy required to solvate the dipole in vacuum is zero. Hence, the vacuum formation energy does not have to be subtracted from Eq. (3.14.9), and the equation by itself accounts for the energy released upon transfer of the dipole from vacuum to an empty cavity surrounded by a continuous dielectric. This equation and its variations have found wide application, including the calculation of the cohesive energy of liquids and solids [13] and the calculation of the effects of solvation on the position of chemical equilibrium [14].

The Onsager reaction field concept has proved most useful. Among its various applications has been the calculation of the solvation energy of the free electrons produced by the exposure of dielectric liquids to ionizing radiation [15–17] and its use as the foundation for theories of the intermolecular forces [18–20].

3.15 Debye Electromagnetic Equation of State for Polar Dielectrics

The Debye theory (P. J. W. Debye, Dutch, American, 1884–1966) of polar dielectrics, which preceded Onsager's theory, relies upon the Lorentz local electric field, F_L, as expressed by Eq. (3.10.14) to calculate the macroscopic polarization. If in Eq. (3.13.30) we replace F_α by F_L, then

$$\langle \mu_z \rangle = \frac{\mu^2 F_L}{3k_B T} \tag{3.15.1}$$

The contribution to the macroscopic polarization due to permanent dipoles is as before,

$$P_\mu = N\langle \mu_z \rangle \tag{3.15.2}$$

Substitute Eq. (3.15.1) into Eq. (3.15.2) to obtain

$$P_\mu = \frac{N\mu^2 F_L}{3k_B T} \tag{3.15.3}$$

The contribution to the macroscopic polarization due to induced dipoles is

$$P_\alpha = N\alpha F_L \tag{3.15.4}$$

The total macroscopic polarization, P, is the sum

$$P = P_\alpha + P_\mu \tag{3.15.5}$$

Upon substitution of Eqs. (3.15.3) and (3.15.4) into Eq. (3.15.5), the result is

$$P = N\left(\alpha + \frac{\mu^2}{3k_B T}\right)F_L \tag{3.15.6}$$

According to Eq. (3.10.14),

$$F_L = \left(\frac{\varepsilon_r + 2}{3}\right)E_o \tag{3.15.7}$$

Upon substitution of Eq. (3.15.7) into Eq. (3.15.6), we obtain

$$P = N\left(\alpha + \frac{\mu^2}{3k_B T}\right)\left(\frac{\varepsilon_r + 2}{3}\right)E_o \tag{3.15.8a}$$

We can combine Eq. (3.15.8a) with Eq. (3.13.45a) with **P** replaced by P and **E** replaced by E_o. The result is

$$\varepsilon_o(\varepsilon_r - 1)E_o = N\left(\alpha + \frac{\mu^2}{3k_B T}\right)\left(\frac{\varepsilon_r + 2}{3}\right)E_o \tag{3.15.8b}$$

Finally, upon cancellation of E_o from both sides of Eq. (3.15.8b), and rearranging terms, we obtain

$$\frac{\varepsilon_r - 1}{\varepsilon_r + 2} = \frac{N}{3\varepsilon_o}\left(\alpha + \frac{\mu^2}{3k_B T}\right) \tag{3.15.9}$$

which is known as the Debye electromagnetic equation of state for polar dielectrics.

If we set $\mu = 0$ in Eq. (3.15.9), we recover the Clausius-Mosotti equation,

$$\frac{\varepsilon_r - 1}{\varepsilon_r + 2} = \frac{N\alpha}{3\varepsilon_o} \tag{3.15.10}$$

which is an electromagnetic equation of state for non-polar dielectrics.

3.16 Discussion

With the definition of the electrostatic potential, Gauss' Law, Eq. (3.2.24) takes the form of Poisson's equation, Eq. (3.2.32). If the density of free charges is zero, Eq. (3.2.32) becomes Laplace's equation, Eq. (3.7.1). Laplace's equation is a partial differential equation, which is said to be "homogeneous" because the right hand side is zero. Both Laplace's equation and Poisson's equation require the statement of boundary conditions for their complete solution. A differential equation with associated boundary conditions is said to constitute a "boundary value" problem. If the boundaries have sufficient symmetry, the partial differential equation can be reduced to one or more ordinary differential equations. Equation (3.7.3) is the result of a one-dimensional reduction of Laplace's equation. The associated boundary conditions are stated by Eqs. (3.7.5a) and (3.7.5b).

Another example of a boundary problem is Eq. (3.8.3), which is Laplace's equation in spherical polar coordinates. It becomes independent of the angle, ϕ, in the case considered where the boundaries have rotational symmetry about the z-axis. This simplification has been applied in Eq. (3.8.4). The latter can be solved by the general method known as "separation of variables". In the case of Eq. (3.8.4), the separation technique involves writing the electrostatic potential as a product of a function of r alone and a function of θ alone. This strategy permits Eq. (3.8.4) to be replaced by the two ordinary differential equations, which are Eqs. (3.8.11) and (3.8.15). The associated boundary conditions are stated in Eqs. (3.8.21a)–(3.8.21d).

Whereas the molecular polarizability and permanent electric dipole moment can be calculated from first principles by quantum mechanics [19, 20], there is no equivalent quantum calculation for the cavity radius. The best that can be done is to estimate the radius on the basis of the atomic pair correlation function of statistical mechanics [18], or by substituting the quantum calculated values of the polarizability and permanent dipole moment into Eq. (3.13.46) and fitting the result to data in the form of ε_r versus N in order to determine f. The value of the radius can then be determined from the value of f by solving Eq. (3.13.3b) for a [21].

Some limitations of the Onsager model include the neglect of hydrogen bonding in the case of solvents containing hydrogen, oxygen, nitrogen, and fluorine, and the neglect of the dispersion force which accounts for the attractive part of the cohesive energies in rare gas liquids and molecular solids [22]. Nonetheless, in cases where fully molecular theories are out of reach, reaction field models can be used, as in Sect. 3.14, to relate molecular properties to macroscopic observables.

References

1. L. Onsager, Electric moments of molecules in liquids. J. Am. Chem. Soc. **58**, 1486–1493 (1936)
2. G. Arfken, *Mathematical Methods for Physicists*, 3rd edn (Academic Press, New York, 1985), pp.701–706
3. C.J.F. Bottcher, *Theory of Electric Polarization* (Elsevier Publishing, Co., Amsterdam, 1952), pp.210–212
4. W.E. Danforth, The dielectric constant of liquids under high pressure. Phys. Rev. **38**, 1224–1235 (1931)
5. W.H. Orttung, Polarizability and apparent radius of water from refractive index data. J. Phys. Chem. **67**, 503–504 (1963)
6. H.R. Petty, J.A. Crumb, V.E. Anderson, E.T. Arakawa, J.K. Baird, New differential Bottcher Onsager method used to determine polarizability and apparent radius of the tungstosilicate anion. J. Phys. Chem. **81**, 696–703 (1977)
7. N.S. Bayless, The effect of the electrostatic polarization of the solvent on electronic absorption spectra in solution. J. Chem. Phys. **18**, 292–296 (1950)
8. J. Ray, N. Panja, P.K. Nandi, J.J. Martin, W.E. Jones Jr., Spectroscopic and ab initio study of an intramolecular charge transfer (ICT) rhodanine derivative. J. Mol. Struct. **874**, 121–127 (2008)
9. A. Kawski, P. Bojarski, B. Kuklinski, Estimation of ground—and excited—state dipole moments of Nile Red dye from solvatochromic effect on absorption and fluorescence spectra. Chem. Phys. Lett. **463**, 410–412 (2008)
10. A. Marini, A. Munoz-Losa, A. Biancardi, B. Mennucci, What is solvatochromism? J. Phys. Chem. B **114**, 17128–17135 (2010)
11. R.C. Weast (ed.), *CRC Handbook of Chemistry and Physics*, 64th edn (CRC Press, Boca Raton, FL 1983). p. E-363
12. W.H. Orttung, Polarizability and apparent radius of glycine from refractive index data. J. Phys. Chem. **67**, 1911–1915 (1963)
13. H.P. Diogo, M.E. Minas da Piedade, J.J. Moura Ramos, J.A. Simoni, J.A. Martinho Simoes, Intermolecular forces in solution and lattice energies in ionic crystals. J. Chem. Ed. **70**, A227–A233 (1993)
14. C.H. Cortney, V.V. Krishnan, Keto-enol tautomerization of acetylacetone in mixed solvents by NMR spectroscopy. A physical chemistry experiment of the application of the Onsager–Kirkwood model for solvation thermodynamics. J. Chem. Ed. **97**, 825–830 (2020)
15. J.K. Baird, Continuum dielectric model for an electron in a nonpolar fluid. J. Phys. Chem. **79**, 2862–2866 (1975)
16. J.K. Baird, Dielectric screening and the electron scattering length in liquid argon. Phys. Rev. A **32**, 1235–1236 (1985)
17. J.K. Baird, D.P. Russell, Physical interpretation of the disappearance of dielectric screening in the case of electrons in nonpolar fluids. Phys. Rev. A **39**, 4295–4297 (1989)
18. B. Linder, Reaction field techniques and their applications to intermolecular forces. Adv. Chem. Phys. **12**, 225–283 (1967)
19. S.W. Harrison, H.-J. Nolte, D.L. Beveridge, Free energy of charge distributions in a spheroidal cavity in a polarizable dielectric continuum. J. Phys. Chem. **80**, 2580–2585 (1976)
20. J. Tomasi, B. Mennucci, R. Cammi, Quantum mechanical continuum solvation models. Chem. Rev. **105**, 2999–3093 (2005)
21. Y. Luo, H. Agren, K.V. Mikkelsen, Unique determination of the cavity radius in Onsager reaction field theory. Chem. Phys. Lett. **52**, 145–150 (1997)
22. N.W. Ashcroft, N.D. Mermin, *Solid State Physics* (Saunders College, Philadelphia, PA, 1976), pp. 404–408

Recommended Additional Reading

1. C.J.F. Bottcher, *Theory of Polarization*, vol. 1, 2nd edn. Dielectrics and Static Fields (Elsevier Publishing Co., New York, 1973)
2. A. Hinchliffe, R.W. Munn, *Molecular Electromagnetism* (Wiley, Chichester, West Sussex, UK, 1985)
3. D.J. Griffiths, *Introduction to Electrodynamics*, 4th edn (Pearson Education Ltd., India, 2013)

Chapter 4
Molecular Theory of Non-polar Dielectrics

In this chapter, we reprise Kirkwood's model for a non-polar dielectric material. In the Kirkwood model, a molecule in the dielectric is no longer represented by a cavity and its neighbors are no longer represented by a surrounding continuous dielectric. Rather, the individual molecules are treated all on the same basis, being represented as point polarizable electric dipoles in a vacuum. The point electric dipoles interact pairwise by way of the electric dipole–dipole energy formula derived in Chap. 1. With the separation between two interacting dipoles fixed, the effect of the remaining dipoles is replaced by a Boltzmann weighted average over their various possible positions and orientations. This thermal average cannot be calculated exactly but is represented by an approximation that is best described mathematically as replacing the average of a product of two quantities by the product of the two averages. This approximation is nonetheless effective, and leads directly to the Clausius–Mossotti equation.

4.1 The Dipolar Electric Field, the Transfer Matrix, and the Displacement

In the derivations of the Clausius–Mossotti and Bottcher–Onsager electromagnetic equations of state in the previous chapter, a molecule was assumed to behave as if it occupied a cavity in an otherwise continuous dielectric. In the same year that Onsager published his semi-continuum theory, J. G. Kirkwood (American, 1901–1959) introduced a discrete molecular model for a non-polar dielectric, wherein both the central molecule and the surrounding molecules were treated on equal footing as point polarizable electric dipoles.

In Kirkwood's model [1], the point polarizable molecular electric dipoles are positioned in a vacuum between the charged plates of a parallel plate capacitor. Except for the central molecule, which is assumed to occupy a small volume of

J. K. Baird, *Electromagnetism for Chemistry*,
https://doi.org/10.1007/978-3-031-88508-2_4

atomic dimensions, all the other dipoles can occupy positions anywhere between the plates. According to Gauss' Law, Eq. (3.2.24) of Chap. 3, the free charges on the capacitor plates produce a displacement, $\mathbf{D}$, in the vacuum between them. The electric field, $\mathbf{E}_o$, in the vacuum is related to the displacement according to $\mathbf{D} = \varepsilon_o \mathbf{E}_o$. The local field acting on a dipole is the sum of $\mathbf{D}/\varepsilon_o$ and the ensemble average of the electric fields produced by the various spatial configurations of the other electric dipoles. By combining this value of the local field with the constitutive relations governing dielectrics, Kirkwood was able to derive the Clausius–Mossotti equation. Kirkwood's calculation is a good example of how statistical mechanics can be combined with electrostatics to forge a link between molecular properties and the macroscopic properties of a substance. Making these connections is certainly one of the goals of chemistry.

The theory [1] starts with the formula for the electric field evaluated at position b due to a point dipole located at position a. That field is

$$\mathbf{E}(\mathbf{r}_b) = -\frac{1}{4\pi\varepsilon_o}\left[\frac{\mathbf{p}_a}{r_{ab}^3} - \frac{3(\mathbf{p}_a \cdot \mathbf{r}_{ab})}{r_{ab}^5}\mathbf{r}_{ab}\right] \tag{4.1.1}$$

which we recognize as Eq. (1.9.11) of Chap. 1. To simplify the notation, we suppress explicit reference to a and b, and rewrite Eq. (4.1.1) as

$$\mathbf{E}(\mathbf{r}) = -\frac{1}{4\pi\varepsilon_o}\left[\frac{\mathbf{p}}{r^3} - \frac{3(\mathbf{p} \cdot \mathbf{r})}{r^5}\mathbf{r}\right] = -\frac{1}{4\pi\varepsilon_o}\left[\frac{\mathbf{p}}{r^3} - \frac{3\mathbf{p} \cdot \mathbf{rr}}{r^5}\right] \tag{4.1.2}$$

where $\mathbf{r}$ is the vector distance between the reference dipole and the point in space where the electric field is being sensed. We will have more to say about the dyadic symbol, $\mathbf{rr}$, below.

In Cartesian coordinates, the dipole vector, $\mathbf{p}$, has the representation,

$$\mathbf{p} = p_x\hat{\mathbf{e}}_x + p_y\hat{\mathbf{e}}_y + p_z\hat{\mathbf{e}}_z \tag{4.1.3a}$$

In Cartesian coordinates, the unit vectors satisfy

$$(\hat{\mathbf{e}}_x \cdot \hat{\mathbf{e}}_x) = (\hat{\mathbf{e}}_y \cdot \hat{\mathbf{e}}_y) = (\hat{\mathbf{e}}_z \cdot \hat{\mathbf{e}}_z) = 1 \tag{4.1.3b}$$

$$(\hat{\mathbf{e}}_x \cdot \hat{\mathbf{e}}_y) = (\hat{\mathbf{e}}_x \cdot \hat{\mathbf{e}}_z) = (\hat{\mathbf{e}}_y \cdot \hat{\mathbf{e}}_z) = 0 \tag{4.1.3c}$$

It is now worthwhile to introduce a new mathematical object termed the tensor. The product of a tensor times a vector is another vector. The basis for tensor multiplication is the *unit tensor*, $\overleftrightarrow{\mathbf{I}}$. The product of $\overleftrightarrow{\mathbf{I}}$ and a vector equals the original vector, and the product of a vector with $\overleftrightarrow{\mathbf{I}}$ is also the original vector. The multiplication is commutative. In terms of the Cartesian unit vectors, $\overleftrightarrow{\mathbf{I}}$ can be written as

$$\overleftrightarrow{\mathbf{I}} = \hat{\mathbf{e}}_x\hat{\mathbf{e}}_x + \hat{\mathbf{e}}_y\hat{\mathbf{e}}_y + \hat{\mathbf{e}}_z\hat{\mathbf{e}}_z \tag{4.1.4}$$

where the notation, $\hat{\mathbf{e}}_i\hat{\mathbf{e}}_i$, $(i = x, y, z;\ j = x, y, z)$ is meant to indicate multiplication in the same sense as scalar numbers are multiplied together. Equation (4.1.4) is said to be the *dyadic* representation of $\overleftrightarrow{\mathbf{I}}$. The dot product rules in Eqs. (4.1.3b) and (4.1.3c) form the basis for calculating the product of a vector, such as $\mathbf{p}$, with the tensor $\overleftrightarrow{\mathbf{I}}$. Using Eqs. (4.1.3) and (4.1.4), we can compute

$$\begin{aligned}
\mathbf{p}\cdot\overleftrightarrow{\mathbf{I}} &= (p_x\hat{\mathbf{e}}_x + p_y\hat{\mathbf{e}}_y + p_z\hat{\mathbf{e}}_z)\cdot(\hat{\mathbf{e}}_x\hat{\mathbf{e}}_x + \hat{\mathbf{e}}_y\hat{\mathbf{e}}_y + \hat{\mathbf{e}}_z\hat{\mathbf{e}}_z)\\
&= p_x(\hat{\mathbf{e}}_x\cdot\hat{\mathbf{e}}_x)\hat{\mathbf{e}}_x + p_x(\hat{\mathbf{e}}_x\cdot\hat{\mathbf{e}}_y)\hat{\mathbf{e}}_y + p_x(\hat{\mathbf{e}}_x\cdot\hat{\mathbf{e}}_z)\hat{\mathbf{e}}_z\\
&\quad + p_y(\hat{\mathbf{e}}_y\cdot\hat{\mathbf{e}}_x)\hat{\mathbf{e}}_x + p_y(\hat{\mathbf{e}}_y\cdot\hat{\mathbf{e}}_y)\hat{\mathbf{e}}_y + p_y(\hat{\mathbf{e}}_y\cdot\hat{\mathbf{e}}_z)\hat{\mathbf{e}}_z\\
&\quad + p_z(\hat{\mathbf{e}}_z\cdot\hat{\mathbf{e}}_x)\hat{\mathbf{e}}_x + p_z(\hat{\mathbf{e}}_z\cdot\hat{\mathbf{e}}_y)\hat{\mathbf{e}}_y + p_z(\hat{\mathbf{e}}_z\cdot\hat{\mathbf{e}}_z)\hat{\mathbf{e}}_z\\
&= p_x\hat{\mathbf{e}}_x + p_y\hat{\mathbf{e}}_y + p_z\hat{\mathbf{e}}_z\\
&= \mathbf{p}
\end{aligned} \tag{4.1.5}$$

The symbol $\mathbf{rr}$, in Eq. (4.1.2) also constitutes a tensor. Its dyadic form is

$$\begin{aligned}
\mathbf{rr} &= xx\hat{\mathbf{e}}_x\hat{\mathbf{e}}_x + xy\hat{\mathbf{e}}_x\hat{\mathbf{e}}_y + xz\hat{\mathbf{e}}_x\hat{\mathbf{e}}_z + yx\hat{\mathbf{e}}_y\hat{\mathbf{e}}_x + yy\hat{\mathbf{e}}_y\hat{\mathbf{e}}_y + yz\hat{\mathbf{e}}_y\hat{\mathbf{e}}_z\\
&\quad + zx\hat{\mathbf{e}}_z\hat{\mathbf{e}}_x + zy\hat{\mathbf{e}}_z\hat{\mathbf{e}}_y + zz\hat{\mathbf{e}}_z\hat{\mathbf{e}}_z
\end{aligned} \tag{4.1.6}$$

Equations (4.1.5) and (4.1.6) permit Eq. (4.1.2) to be factored and written in dyadic form as

$$\mathbf{E}(\mathbf{r}) = -\frac{\mathbf{p}}{4\pi\varepsilon_o r^3}\cdot\left[\overleftrightarrow{\mathbf{I}} - \frac{3\mathbf{rr}}{r^2}\right] \tag{4.1.7}$$

The unit tensor, $\overleftrightarrow{\mathbf{I}}$, and the tensor, $\mathbf{rr}$, both have matrix representations. Consider first the unit tensor, which has the matrix representation,

$$\overleftrightarrow{\mathbf{I}} = \begin{bmatrix} 1\,0\,0\\ 0\,1\,0\\ 0\,0\,1 \end{bmatrix} \tag{4.1.8}$$

Consider the dot product, $\mathbf{p}\cdot\overleftrightarrow{\mathbf{I}}$, with $\mathbf{p}$ on the left of $\overleftrightarrow{\mathbf{I}}$. To compute this dot product by matrix multiplication, we represent $\mathbf{p}$ as a row vector. In matrix notation, the multiplication then reads,

$$\mathbf{p}\cdot\overleftrightarrow{\mathbf{I}} = \begin{bmatrix} p_x\ p_y\ p_z \end{bmatrix}\begin{bmatrix} 1\,0\,0\\ 0\,1\,0\\ 0\,0\,1 \end{bmatrix} = \begin{bmatrix} p_x\ p_y\ p_z \end{bmatrix} = \mathbf{p} \tag{4.1.9}$$

If $\mathbf{p}$ appears on the right, the rules of matrix multiplication require $\mathbf{p}$ to be represented by a column vector, and the multiplication reads,

$$\overleftrightarrow{\mathbf{I}} \cdot \mathbf{p} = \begin{bmatrix} 1 & 0 & 0 \\ 0 & 1 & 0 \\ 0 & 0 & 1 \end{bmatrix} \begin{bmatrix} p_x \\ p_y \\ p_z \end{bmatrix} = \begin{bmatrix} p_x \\ p_y \\ p_z \end{bmatrix} = \mathbf{p} \tag{4.1.10}$$

$\overleftrightarrow{\mathbf{I}}$ is said to be a symmetric tensor, because its elements satisfy $I_{ij} = I_{ji}$. Dot products on the left and right of a symmetric tensor give the same result. This is confirmed by our calculations in Eqs. (4.1.9) and (4.1.10), where

$$\mathbf{p} \cdot \overleftrightarrow{\mathbf{I}} = \overleftrightarrow{\mathbf{I}} \cdot \mathbf{p} \tag{4.1.11}$$

Next consider the matrix representation of $\mathbf{rr}$, which is

$$\mathbf{rr} = \begin{bmatrix} x^2 & xy & xz \\ yx & y^2 & yz \\ zx & zy & z^2 \end{bmatrix} \tag{4.1.12}$$

Like $\overleftrightarrow{\mathbf{I}}$, the tensor $\mathbf{rr}$ is symmetric; hence,

$$\mathbf{p} \cdot \mathbf{rr} = \mathbf{rr} \cdot \mathbf{p} \tag{4.1.13}$$

Because both $\overleftrightarrow{\mathbf{I}}$ and $\mathbf{rr}$ are symmetric tensors, Eq. (4.1.7) can be rewritten as

$$\mathbf{E}(\mathbf{r}) = -\frac{1}{4\pi\varepsilon_o r^3}\left[\overleftrightarrow{\mathbf{I}} - \frac{3\mathbf{rr}}{r^2}\right] \cdot \mathbf{p} \tag{4.1.14}$$

Equation (4.1.14) can be used to rewrite Eq. (4.1.1) in tensor form as

$$\mathbf{E}(\mathbf{r}_b) = -\frac{1}{4\pi\varepsilon_o r_{ab}^3}\left[\overleftrightarrow{\mathbf{I}} - \frac{3\mathbf{r}_{ab}\mathbf{r}_{ab}}{r_{ab}^2}\right] \cdot \mathbf{p}_a \tag{4.1.15}$$

where $\mathbf{r}_{ab} = \mathbf{r}_a - \mathbf{r}_b$ and $r_{ab} = |\mathbf{r}_a - \mathbf{r}_b|$.

Define the tensor,

$$\overleftrightarrow{\mathbf{T}}_{ab} = \frac{1}{4\pi\varepsilon_o r_{ab}^3}\left[\overleftrightarrow{\mathbf{I}} - \frac{3\mathbf{r}_{ab}\mathbf{r}_{ab}}{r_{ab}^2}\right] \tag{4.1.16}$$

Using Eqs. (4.1.8) and (4.1.12), we can express Eq. (4.1.16) in matrix form as

$$\overleftrightarrow{\mathbf{T}}_{ab} = \frac{1}{4\pi\varepsilon_o r_{ab}^3}\begin{bmatrix} 1-\frac{3x_{ab}^2}{r_{ab}^2} & -\frac{3x_{ab}y_{ab}}{r_{ab}^2} & -\frac{3x_{ab}z_{ab}}{r_{ab}^2} \\ -\frac{3y_{ab}x_{ab}}{r_{ab}^2} & 1-\frac{3y_{ab}^2}{r_{ab}^2} & -\frac{3y_{ab}z_{ab}}{r_{ab}^2} \\ -\frac{3z_{ab}x_{ab}}{r_{ab}^2} & -\frac{3z_{ab}y_{ab}}{r_{ab}^2} & 1-\frac{3z_{ab}^2}{r_{ab}^2} \end{bmatrix} \tag{4.1.17}$$

By virtue of $x_{ab} = -x_{ba}$, $y_{ab} = -y_{ba}$, $z_{ab} = -z_{ba}$, we have $r_{ab} = r_{ba}$; hence, $\overleftrightarrow{\mathbf{T}}_{ab} = \overleftrightarrow{\mathbf{T}}_{ba}$ This permits Eq. (4.1.15) to be rewritten as

$$\mathbf{E}(\mathbf{r}_a) = -\overleftrightarrow{\mathbf{T}}_{ab}\cdot\mathbf{p}_b \tag{4.1.18}$$

According to Eq. (4.1.18), the operation with the tensor $\overleftrightarrow{\mathbf{T}}_{ab}$ on a dipole vector $\mathbf{p}_b$ produces another vector, the electric field, $\mathbf{E}(\mathbf{r}_a)$. Because $\overleftrightarrow{\mathbf{T}}_{ab}$ has a matrix representation and functions to determine the electric field at point a due to a dipole at point b, it is said to be the *transfer matrix*. Because $\overleftrightarrow{\mathbf{T}}_{ab}$ has two subscripts, it constitutes a "second rank" tensor.

We next show that the dyadic representation of $\nabla\nabla\left(\frac{1}{r}\right)$ leads to an additional simplification of Eq. (4.1.16). It should be stressed that in writing $\nabla\nabla\left(\frac{1}{r}\right)$, there is no dot missing between the two ∇ operators. The calculation of this dyad begins with Eq. (1.6.12) of Chap. 1, which reads,

$$\nabla\left(\frac{1}{r}\right) = -\frac{\mathbf{r}}{r^3} \tag{4.1.19a}$$

Equation (4.1.19a) shows that the gradient of a scalar is a vector. A second application of ∇ is obtained by exploiting Eq. (4.1.19a). The result is

$$\nabla\nabla\left(\frac{1}{r}\right) = \nabla\left(\frac{-\mathbf{r}}{r^3}\right) = -\left(\frac{1}{r^3}\right)\nabla\mathbf{r} - \mathbf{r}\nabla\left(\frac{1}{r^3}\right) \tag{4.1.19b}$$

where the calculus rule for differentiation of a product has been applied. Calculation of the derivative, $\nabla\mathbf{r}$, in Eq. (4.1.19b) proceeds in dyadic form as follows:

$$\begin{aligned}\nabla\mathbf{r} &= \left(\hat{\mathbf{e}}_x\frac{\partial}{\partial x} + \hat{\mathbf{e}}_y\frac{\partial}{\partial y} + \hat{\mathbf{e}}_z\frac{\partial}{\partial z}\right)(x\hat{\mathbf{e}}_x + y\hat{\mathbf{e}}_y + z\hat{\mathbf{e}}_z)\\ &= \frac{\partial x}{\partial x}\hat{\mathbf{e}}_x\hat{\mathbf{e}}_x + \frac{\partial y}{\partial x}\hat{\mathbf{e}}_x\hat{\mathbf{e}}_y + \frac{\partial z}{\partial x}\hat{\mathbf{e}}_x\hat{\mathbf{e}}_z + \frac{\partial x}{\partial y}\hat{\mathbf{e}}_y\hat{\mathbf{e}}_x\\ &+ \frac{\partial y}{\partial y}\hat{\mathbf{e}}_y\hat{\mathbf{e}}_y + \frac{\partial z}{\partial y}\hat{\mathbf{e}}_y\hat{\mathbf{e}}_z + \frac{\partial x}{\partial z}\hat{\mathbf{e}}_z\hat{\mathbf{e}}_x + \frac{\partial y}{\partial z}\hat{\mathbf{e}}_z\hat{\mathbf{e}}_y + \frac{\partial z}{\partial z}\hat{\mathbf{e}}_z\hat{\mathbf{e}}_z\\ &= \hat{\mathbf{e}}_x\hat{\mathbf{e}}_x + \hat{\mathbf{e}}_y\hat{\mathbf{e}}_y + \hat{\mathbf{e}}_z\hat{\mathbf{e}}_z = \overleftrightarrow{\mathbf{I}}\end{aligned} \tag{4.1.19c}$$

The final result in Eq. (4.1.19c) follows because x, y, and z are independent variables. If we recall Eq. (1.6.10) in Chap. 1, which reads,

$$\nabla r = \frac{\mathbf{r}}{r} \tag{4.1.19d}$$

then the remaining derivative in Eq. (4.1.19b) is

$$\nabla\left(\frac{1}{r^3}\right) = -3\frac{\nabla r}{r^4} = -3\frac{\mathbf{r}}{r^5} \tag{4.1.19e}$$

Substitution of Eqs. (4.1.19c) and (4.1.19e) into Eq. (4.1.19b) gives

$$\nabla\nabla\left(\frac{1}{r}\right) = -\left(\frac{1}{r^3}\right)\nabla\mathbf{r} - \mathbf{r}\nabla\left(\frac{1}{r^3}\right) = -\left(\frac{\overleftrightarrow{\mathbf{I}}}{r^3}\right) + 3\frac{\mathbf{rr}}{r^5} \tag{4.1.20}$$

Equation (4.1.20) shows that the gradient of a vector is a tensor. If we replace r by $r_{ab} = |\mathbf{r}_a - \mathbf{r}_b|$, then upon comparing, Eq. (4.1.20) with (4.1.16), we have

$$\frac{1}{4\pi\varepsilon_o}\nabla_a\nabla_a\left(\frac{1}{r_{ab}}\right) = -\left(\frac{1}{4\pi\varepsilon_o r_{ab}^3}\right)\left[\overleftrightarrow{\mathbf{I}} - 3\frac{\mathbf{r}_{ab}\mathbf{r}_{ab}}{r_{ab}^2}\right] = -\overleftrightarrow{\mathbf{T}}_{ab} \tag{4.1.21}$$

The symmetric nature of Eq. (4.1.21) can now be made complete by noting that because of the relation, $r_{ab} = |\mathbf{r}_a - \mathbf{r}_b|$, we can write

$$\nabla_a\left(\frac{1}{r_{ab}}\right) = -\nabla_b\left(\frac{1}{r_{ab}}\right) \tag{4.1.22}$$

Equation (4.1.22) permits Eq. (4.1.21) to be restated in the form,

$$\frac{1}{4\pi\varepsilon_o}\nabla_a\nabla_a\left(\frac{1}{r_{ab}}\right) = -\frac{1}{4\pi\varepsilon_o}\nabla_a\nabla_b\left(\frac{1}{r_{ab}}\right) = -\overleftrightarrow{\mathbf{T}}_{ab} \tag{4.1.23}$$

According to Eq. (4.1.23), $\overleftrightarrow{\mathbf{T}}_{ab}$ can rewritten as a symmetric tensor in the dyadic form,

$$\overleftrightarrow{\mathbf{T}}_{ab} = \frac{1}{4\pi\varepsilon_o}\nabla_a\nabla_b\left(\frac{1}{r_{ab}}\right) \tag{4.1.24}$$

Using Eq. (4.1.24), the electric field in Eq. (4.1.18) assumes the form,

$$\mathbf{E}(\mathbf{r}_a) = -\frac{1}{4\pi\varepsilon_o}\nabla_a\nabla_b\left(\frac{1}{r_{ab}}\right)\cdot\mathbf{p}_b \tag{4.1.25}$$

If we consider a system of just two dipoles, then in the Kirkwood model, the local electric field, $\mathbf{F}_a$, at point a is the sum of the applied electric field, $\mathbf{E}_o = \mathbf{D}/\varepsilon_o$, and the electric field $\mathbf{E}_a = \mathbf{E}(\mathbf{r}_a)$, due to the dipole at b:

$$\mathbf{F}_a = \frac{\mathbf{D}}{\varepsilon_o} + \mathbf{E}_a \tag{4.1.26}$$

Now assume that there are N' identical dipoles making up the dielectric. According to Eq. (4.1.18), each dipole makes a contribution $\mathbf{E}_a = -\overset{\leftrightarrow}{\mathbf{T}}_{ab} \cdot \mathbf{p}_b$ to the local electric field. Taking into account all N' dipoles by summing over b, Eqs. (4.1.18) and (4.1.26) can be combined to read,

$$\mathbf{F}_a = \frac{\mathbf{D}}{\varepsilon_o} - \sum_{b=1}^{N'} \overset{\leftrightarrow}{\mathbf{T}}_{ab} \cdot \mathbf{p}_b \tag{4.1.27}$$

If the dipole at a is polarizable, then

$$\mathbf{p}_a = \alpha \mathbf{F}_a \tag{4.1.28}$$

where α is the molecular polarizability. When Eq. (4.1.27) is substituted into Eq. (4.1.28), the result is

$$\mathbf{p}_a = \alpha \mathbf{F}_a = \alpha \left(\frac{\mathbf{D}}{\varepsilon_o} \right) - \alpha \sum_{b=1}^{N'} \overset{\leftrightarrow}{\mathbf{T}}_{ab} \cdot \mathbf{p}_b \tag{4.1.29}$$

which can be rearranged to read

$$\mathbf{p}_a + \alpha \sum_{b=1}^{N'} \overset{\leftrightarrow}{\mathbf{T}}_{ab} \cdot \mathbf{p}_b = \alpha (\mathbf{D}/\varepsilon_o) \tag{4.1.30}$$

Equation (4.1.30) serves to express the dipole moment,$\mathbf{p}_a$, in terms of the transfer matrix, $\overset{\leftrightarrow}{\mathbf{T}}_{ab}$, and the displacement, $\mathbf{D}$. Since the dipoles are exerting forces on one another, their positions are not entirely random. The next task is to exploit the pairwise interaction of the dipoles in order to compute the ensemble average of their various spatial configurations.

4.2 The Potential of Mean Force

According to Eq. (1.9.12) of Chap. 1, the potential energy of interaction of two point dipoles, $\mathbf{p}_i$ and $\mathbf{p}_j$ separated by a distance, r_{ij}, is

$$U_e(\mathbf{r}_{ij}) = \frac{1}{4\pi \varepsilon_o} \left[\frac{\mathbf{p}_i \cdot \mathbf{p}_j}{r_{ij}^3} - 3 \frac{(\mathbf{p}_i \cdot \mathbf{r}_{ij})(\mathbf{p}_j \cdot \mathbf{r}_{ij})}{r_{ij}^5} \right] \tag{4.2.1}$$

Because $\mathbf{r}_{ij} = -\mathbf{r}_{ji}$ and $r_{ij} = |\mathbf{r}_i - \mathbf{r}_j| = r_{ji}$, Eq. (4.2.1) indicates that $U_e(\mathbf{r}_{ij}) = U_e(\mathbf{r}_{ji})$. For the sake of clarity, when introducing the potential of mean force, we will confine our attention to the energy of interaction of four point dipoles. The four dipoles $\mathbf{p}_1$, $\mathbf{p}_2$, $\mathbf{p}_3$, and $\mathbf{p}_4$, are located in space by the vector coordinates $\mathbf{r}_1$, $\mathbf{r}_2$, $\mathbf{r}_3$, and $\mathbf{r}_4$, respectively. The values of the four sets of spatial coordinates are said to describe a "configuration". According to Eq. (4.2.1), the energy of interaction of any two dipoles, for example $\mathbf{p}_1$ and $\mathbf{p}_2$, is completely independent of the locations of dipoles, $\mathbf{p}_3$, and $\mathbf{p}_4$. As such, Eq. (4.2.1) is said to describe a "pair potential". Having assumed pairwise interactions, the total potential energy, $U_4(\mathbf{r}_1, \mathbf{r}_2, \mathbf{r}_3, \mathbf{r}_4)$, of the four dipoles can be represented as the sum of the pair potentials, $U_e(\mathbf{r}_{ij})$. In the case of the four dipoles, this sum is

$$U_4(\mathbf{r}_1, \mathbf{r}_2, \mathbf{r}_3, \mathbf{r}_4) = \frac{1}{2} \sum_{i=1,\neq j}^{4} \sum_{j=1,\neq i}^{4} U_e(\mathbf{r}_{ij}) \tag{4.2.2}$$

Equation (4.2.2) contains a factor of 1/2 to avoid counting $U_e(\mathbf{r}_{ij})$ and $U_e(\mathbf{r}_{ji})$ as separate interactions.

The restriction, $i \neq j$, in Eq. (4.2.2) removes terms which involve a dipole interacting with itself; that is to say, it removes the self-energy from consideration. On the basis of Eq. (4.2.2), $U_4(\mathbf{r}_1, \mathbf{r}_2, \mathbf{r}_3, \mathbf{r}_4)$ is said to be a *pairwise additive* potential energy.

If Eq. (2.9.33a) of Chap. 2 is specialized to four interacting dipoles, the configuration integral, Z_4, is

$$Z_4 = \int d^3r_1 \int d^3r_2 \int d^3r_3 \int d^3r_4 \exp(\ \beta U_4(\mathbf{r}_1, \mathbf{r}_2, \mathbf{r}_3, \mathbf{r}_4)) \tag{4.2.3}$$

where $\beta = 1/k_B T$, and the range of integration is assumed to include all possible configurations in the ensemble. According to Eq. (2.9.22) of Chap. 2, the probability of finding the four dipoles at positions, $\mathbf{r}_1$, $\mathbf{r}_2$, $\mathbf{r}_3$, and $\mathbf{r}_4$ is

$$P(\mathbf{r}_1, \mathbf{r}_2, \mathbf{r}_3, \mathbf{r}_4) = \frac{\exp(-\beta U_4(\mathbf{r}_1, \mathbf{r}_2, \mathbf{r}_3, \mathbf{r}_4))}{Z_4} \tag{4.2.4}$$

The probability, $P(\mathbf{r}_1, \mathbf{r}_2)$, of finding the dipoles $\mathbf{p}_1$ and $\mathbf{p}_2$, at $\mathbf{r}_1$ and $\mathbf{r}_2$, respectively, without reference to the positions of $\mathbf{p}_3$, and $\mathbf{p}_4$ is defined as

$$P(\mathbf{r}_1, \mathbf{r}_2) = \frac{\int d^3r_3 \int d^3r_4 \exp(-\beta U_4(\mathbf{r}_1, \mathbf{r}_2, \mathbf{r}_3, \mathbf{r}_4))}{Z_4} \tag{4.2.5}$$

The ensemble average value $\langle f(\mathbf{r}_1, \mathbf{r}_2)\rangle$ of an arbitrary function, $f(\mathbf{r}_1, \mathbf{r}_2)$, of $\mathbf{r}_1$ and $\mathbf{r}_2$ is given by

$$\langle f(\mathbf{r}_1, \mathbf{r}_2)\rangle = \int d^3r_1 \int d^3r_2 P(\mathbf{r}_1, \mathbf{r}_2) f(\mathbf{r}_1, \mathbf{r}_2) \tag{4.2.6}$$

At this point, it will prove worthwhile to define a potential of mean force, $W_2(\mathbf{r}_1, \mathbf{r}_2)$, by the exponential relation,

$$\exp[-\beta W_2(\mathbf{r}_1, \mathbf{r}_2)] = \frac{V^2 \int d^3r_3 \int d^3r_4 \exp(-\beta U_4(\mathbf{r}_1, \mathbf{r}_2, \mathbf{r}_3, \mathbf{r}_4))}{\int d^3r_1 \int d^3r_2 \int d^3r_3 \int d^3r_4 \exp(-\beta U_4(\mathbf{r}_1, \mathbf{r}_2, \mathbf{r}_3, \mathbf{r}_4))} \tag{4.2.7}$$

where V is the volume of the dielectric [2]. By use of Eq. (4.2.3), Eq. (4.2.7) can be rewritten in terms of Z_4 as

$$\exp[-\beta W_2(\mathbf{r}_1, \mathbf{r}_2)] = \frac{V^2 \int d^3r_3 \int d^3r_4 \exp(-\beta U_4(\mathbf{r}_1, \mathbf{r}_2, \mathbf{r}_3, \mathbf{r}_4))}{Z_4} \tag{4.2.8}$$

To understand the physical meaning of $W_2(\mathbf{r}_1, \mathbf{r}_2)$, apply the operator ∇_1 to both sides of Eq. (4.2.8), where ∇_1 operates exclusively on $\mathbf{r}_1$. Since the denominator on the right hand side of Eq. (4.2.8) is just a constant, application of ∇_1 to the right hand side of Eq. (4.2.8) involves differentiation of the numerator only. We write

$$\nabla_1 \exp[-\beta W_2(\mathbf{r}_1, \mathbf{r}_2)] = \frac{V^2 \int d^3r_3 \int d^3r_4 \nabla_1 \exp(-\beta U_4(\mathbf{r}_1, \mathbf{r}_2, \mathbf{r}_3, \mathbf{r}_4))}{Z_4} \tag{4.2.9a}$$

Upon performing the indicated differentiations, Eq. (4.2.9a) can be written as

$$\begin{aligned} &\exp[-\beta W_2(\mathbf{r}_1, \mathbf{r}_2)](-\beta)\nabla_1 W_2(\mathbf{r}_1, \mathbf{r}_2) \\ &= \frac{V^2}{Z_4} \int d^3r_3 \int d^3r_4 \exp(-\beta U_4(\mathbf{r}_1, \mathbf{r}_2, \mathbf{r}_3, \mathbf{r}_4))(-\beta)\nabla_1 U_4(\mathbf{r}_1, \mathbf{r}_2, \mathbf{r}_3, \mathbf{r}_4) \end{aligned} \tag{4.2.9b}$$

On the left hand side of Eq. (4.2.9b), replace $\exp[-\beta W_2(\mathbf{r}_1, \mathbf{r}_2)]$ by its definition in Eq. (4.2.8). The result is

$$\begin{aligned} &(-\beta)\nabla_1 W_2(\mathbf{r}_1, \mathbf{r}_2)\left[\frac{V^2}{Z_4} \int d^3r_3 \int d^3r_4 \exp(-\beta U_4(\mathbf{r}_1, \mathbf{r}_2, \mathbf{r}_3, \mathbf{r}_4))\right] \\ &= \frac{V^2}{Z_4} \int d^3r_3 \int d^3r_4 \exp(-\beta U_4(\mathbf{r}_1, \mathbf{r}_2, \mathbf{r}_3, \mathbf{r}_4))(-\beta)\nabla_1 U_4(\mathbf{r}_1, \mathbf{r}_2, \mathbf{r}_3, \mathbf{r}_4) \end{aligned} \tag{4.2.9c}$$

Solve Eq. (4.2.9c), for $-\nabla_1 W_2(\mathbf{r}_1, \mathbf{r}_2)$, to obtain,

$$-\nabla_1 W_2(\mathbf{r}_1, \mathbf{r}_2) = \frac{\int d^3r_3 \int d^3r_4 \exp(-\beta U_4(\mathbf{r}_1, \mathbf{r}_2, \mathbf{r}_3, \mathbf{r}_4))(-1)\nabla_1 U_4(\mathbf{r}_1, \mathbf{r}_2, \mathbf{r}_3, \mathbf{r}_4)}{\int d^3r_3 \int d^3r_4 \exp(-\beta U_4(\mathbf{r}_1, \mathbf{r}_2, \mathbf{r}_3, \mathbf{r}_4))} \tag{4.2.10}$$

The name, "potential of mean force", is based upon Eq. (4.2.10). The force $\mathbf{F}_1(\mathbf{r}_1|\mathbf{r}_2, \mathbf{r}_3, \mathbf{r}_4)$ on dipole $\mathbf{p}_1$ at $\mathbf{r}_1$ due to dipoles $\mathbf{p}_2$, $\mathbf{p}_3$, and $\mathbf{p}_4$ fixed at locations, $\mathbf{r}_2$, $\mathbf{r}_3$, and $\mathbf{r}_4$, respectively, is $\mathbf{F}_1(\mathbf{r}_1|\mathbf{r}_2, \mathbf{r}_3, \mathbf{r}_4) = -\nabla_1 U_4(\mathbf{r}_1, \mathbf{r}_2, \mathbf{r}_3, \mathbf{r}_4)$. This describes the force associated with a particular configuration. The right hand side of Eq. (4.2.10) is a mean force, $\langle \mathbf{F}_1(\mathbf{r}_1|\mathbf{r}_2)\rangle$. The mean force is $\mathbf{F}_1(\mathbf{r}_1|\mathbf{r}_2, \mathbf{r}_3, \mathbf{r}_4)$ evaluated at $\mathbf{r}_1$ with $\mathbf{r}_2$ fixed, and $\mathbf{r}_3$ and $\mathbf{r}_4$ thermally averaged over all positions consistent with a fixed value for $\mathbf{r}_2$. Stated succinctly, $\langle \mathbf{F}_1(\mathbf{r}_1|\mathbf{r}_2)\rangle = -\nabla_1 W(\mathbf{r}_1, \mathbf{r}_2)$ is the mean force on $\mathbf{p}_1$ due to $\mathbf{p}_2$ when its position is fixed and $\mathbf{p}_3$ and $\mathbf{p}_4$ are located at their average positions.

Two previous equations can now be rewritten in terms of the potential of mean force. By virtue of Eq. (4.2.5) and (4.2.8), one obtains,

$$\exp[-\beta W_2(\mathbf{r}_1, \mathbf{r}_2)] = V^2 P(\mathbf{r}_1, \mathbf{r}_2) \tag{4.2.11}$$

On the basis of Eq. (4.2.6) and (4.2.11), the ensemble average of $f(\mathbf{r}_1, \mathbf{r}_2)$ can be rewritten as

$$\langle f(\mathbf{r}_1, \mathbf{r}_2)\rangle = \frac{1}{V^2}\int d^3 r_1 \int d^3 r_2 f(\mathbf{r}_1, \mathbf{r}_2)\exp[-\beta W_2(\mathbf{r}_1, \mathbf{r}_2)] \tag{4.2.12}$$

4.3 Statistical Averages in the Absence of Correlation

The various possible configurations of the four dipoles occur with probabilities given by Eq. (4.2.4). Because experiments on macroscopic samples of matter determine average properties, no single configuration is accessible to experiment. Only the thermal average polarization can be measured. If $\mathbf{p}_i$ is the instantaneous value of the dipole moment on the i-th molecule, then by using Eq. (4.1.30), its ensemble average value $\langle \mathbf{p}_i\rangle$ satisfies,

$$\langle \mathbf{p}_i\rangle + \alpha \sum_{\substack{j=1 \\ j\neq i}}^{N'} \left\langle \overleftrightarrow{\mathbf{T}}_{ij}\cdot \mathbf{p}_j\right\rangle = \alpha(\mathbf{D}/\varepsilon_o) \tag{4.3.1}$$

The tensor product, $\overleftrightarrow{\mathbf{T}}_{ij}\cdot \mathbf{p}_j$, is a vector. Its average value is $\left\langle \overleftrightarrow{\mathbf{T}}_{ij}\cdot \mathbf{p}_j\right\rangle$. Based upon this average, we can define the quantity,

$$\langle \mathbf{n}_i\rangle = \sum_{\substack{j=1 \\ j\neq i}}^{N'} \left[\left\langle \overleftrightarrow{\mathbf{T}}_{ij}\cdot \mathbf{p}_j\right\rangle - \left\langle \overleftrightarrow{\mathbf{T}}_{ij}\right\rangle\cdot\langle \mathbf{p}_i\rangle\right] \tag{4.3.2}$$

Equation (4.3.2) serves as a measure of the correlation between the tensor, $\overleftrightarrow{\mathbf{T}}_{ij}$, and the vector, $\mathbf{p}_j$, the correlation being known as the covariance. Equation (4.3.2) is constructed as follows: The average value of $\overleftrightarrow{\mathbf{T}}_{ij}$ is $\left\langle \overleftrightarrow{\mathbf{T}}_{ij} \right\rangle$, and the average value of $\mathbf{p}_j$ is $\langle \mathbf{p}_j \rangle$. The projection of $\langle \mathbf{T}_{ij} \rangle$ onto $\langle \mathbf{p}_j \rangle$ is $\langle \mathbf{T}_{ij} \rangle \cdot \langle \mathbf{p}_i \rangle$, where it need not be the same as the average value of their projection, which is $\langle \mathbf{T}_{ij} \cdot \mathbf{p}_j \rangle$. That is to say, the product of the averages need not be the same as the average of the product. If such is the case, then the correlation, $\langle \mathbf{n}_i \rangle \neq 0$ [1]. In terms of $\langle \mathbf{n}_i \rangle$, Eq. (4.3.1) can be rewritten in the form,

$$\langle \mathbf{p}_i \rangle + \alpha \sum_{\substack{j=1 \\ j \neq i}}^{N'} \left\langle \overleftrightarrow{\mathbf{T}}_{ij} \right\rangle \cdot \langle \mathbf{p}_j \rangle + \alpha \langle \mathbf{n}_i \rangle = \alpha (\mathbf{D}/\varepsilon_o) \tag{4.3.3}$$

Equation (4.3.3) is not easily solved unless $\langle \mathbf{n}_i \rangle = 0$. Although a tempting assumption, there is no a priori evidence that $\langle \mathbf{n}_i \rangle = 0$. Nonetheless, for the moment we will proceed with $\langle \mathbf{n}_i \rangle = 0$, so that Eq. (4.3.3) simplifies to read,

$$\langle \mathbf{p}_i \rangle + \alpha \sum_{\substack{j=1 \\ j \neq i}}^{N'} \left\langle \overleftrightarrow{\mathbf{T}}_{ij} \right\rangle \cdot \langle \mathbf{p}_j \rangle = \alpha (\mathbf{D}/\varepsilon_o) \tag{4.3.4a}$$

In a uniform dielectric, the mean value of $\mathbf{p}_i$ is a constant independent of subscript i. This allows us to suppress the subscript i on $\langle \mathbf{p}_i \rangle$ and replace it simply by $\langle \mathbf{p} \rangle$. Nevertheless, the average value, $\left\langle \overleftrightarrow{\mathbf{T}}_{ij} \right\rangle$, of the transfer matrix, $\overleftrightarrow{\mathbf{T}}_{ij}$, is still a function of the distance between the dipoles $\mathbf{p}_i$ and $\mathbf{p}_j$, so we will still need to maintain the ij subscript notation. With these ideas taken into account, Eq. (4.3.4a) now reads,

$$\langle \mathbf{p} \rangle + \alpha \sum_{\substack{j=1 \\ j \neq i}}^{N'} \left\langle \overleftrightarrow{\mathbf{T}}_{ij} \right\rangle \cdot \langle \mathbf{p} \rangle = \alpha (\mathbf{D}/\varepsilon_o) \tag{4.3.4b}$$

When $i = j$ is omitted in the sum in Eq. (4.3.4b), there remain $N' - 1$ identical pairings of i and j. Consequently, Eq. (4.3.4b) assumes the simpler form,

$$\langle \mathbf{p} \rangle + \alpha (N' - 1) \left\langle \overleftrightarrow{\mathbf{T}}_{ij} \right\rangle \cdot \langle \mathbf{p} \rangle = \alpha (\mathbf{D}/\varepsilon_o) \tag{4.3.5}$$

The next task is to evaluate $\langle \mathbf{T}_{ij} \rangle$.

4.4 Evaluation of the Mean Value of the Transfer Matrix

4.4.1 Introduction of the Potential of Mean Force

If we replace a and b with i and j, respectively, the average value of Eq. (4.1.24) can be written as

$$\left\langle \overleftrightarrow{\mathbf{T}}_{ij} \right\rangle = \frac{1}{4\pi\varepsilon_o}\left\langle \nabla_i \nabla_j \left(\frac{1}{r_{ij}}\right)\right\rangle \tag{4.4.1}$$

Now with

$$f(\mathbf{r}_i, \mathbf{r}_j) = \frac{1}{4\pi\varepsilon_o}\nabla_i \nabla_j \left(\frac{1}{r_{ij}}\right) \tag{4.4.2}$$

Equation (4.2.12) can be used to write Eq. (4.4.1) in the form,

$$\left\langle \mathbf{T}_{ij} \right\rangle = \frac{1}{4\pi\varepsilon_o V^2}\int d^3 r_i \int d^3 r_j \nabla_i \nabla_j \left(\frac{1}{r_{ij}}\right)\exp\left[-\beta W_2(\mathbf{r}_i, \mathbf{r}_j)\right] \tag{4.4.3}$$

In order to rule out the interaction of the central dipole with itself, that is $r_{ij} = 0$, Kirkwood [1] excluded from the volume of integration an atomic sized sphere of volume, v, occupied by the central dipole. See Fig. 4.1.

With the volume of exclusion, v, taken into account, Eq. (4.4.3) can be rewritten as

$$\left\langle \mathbf{T}_{ij} \right\rangle = \frac{1}{4\pi\varepsilon_o V^2}\int\limits_{V-v} d^3 r_i \int\limits_{V-v} d^3 r_j \nabla_i \nabla_j \left(\frac{1}{r_{ij}}\right)\exp\left[-\beta W_2(\mathbf{r}_i, \mathbf{r}_j)\right] \tag{4.4.4}$$

where V is the volume of the entire sample. As an aid to the evaluation of the integrals in Eq. (4.4.4), note that by virtue of Eqs. (4.2.1) and (4.2.7), the potential of mean force does not depend upon $\mathbf{r}_i$ and $\mathbf{r}_j$ separately, but rather just on the magnitude of their difference, r_{ij}. This restriction can be taken into account by writing,

$$W_2(\mathbf{r}_i, \mathbf{r}_j) = W_2\left(\left|\mathbf{r}_i - \mathbf{r}_j\right|\right) \tag{4.4.5}$$

Let,

$$\mathbf{r} = \mathbf{r}_{ij} = \mathbf{r}_i - \mathbf{r}_j \tag{4.4.6a}$$

This variable substitution permits us to switch from the independent variables of integration,$\mathbf{r}_i$ and $\mathbf{r}_j$, to the independent variables $\mathbf{r}_i$ and $\mathbf{r}$. This change leads to the following: If $\mathbf{r} = (x, y, z)$, $\mathbf{r}_i = (x_i, y_i, z_i)$, and $\mathbf{r}_j = (x_j, y_j, z_j)$, then $x = x_i - x_j$, etc. According to Eq. (4.4.6a), $\nabla_i = \nabla$. If x and x_i are considered to be independent,

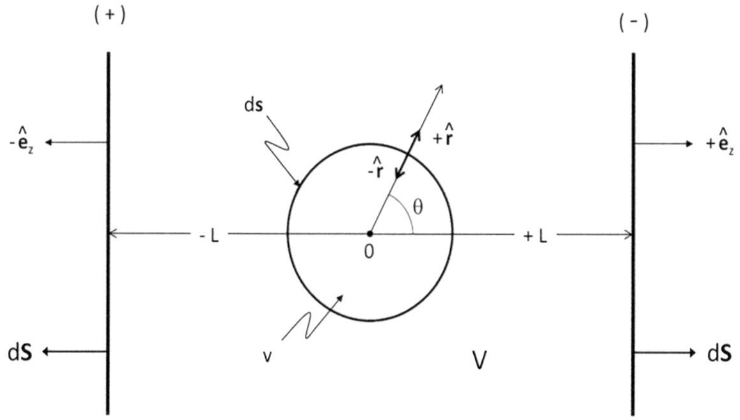

Fig. 4.1 Spherical volume of exclusion about the central dipole. The volume of the entire sample is V. The volume occupied by the central dipole is v. The dielectric is contained within a parallel plate capacitor. The outward drawn unit normal to the positive plate is $-\hat{\mathbf{e}}_z$. The outward drawn unit normal to the negative plate is $+\hat{\mathbf{e}}_z$. The sphere of exclusion is centered on the point positioned at the origin, 0, which is located a distance, L, from each plate. The outward drawn unit normal to the surface of the sphere is $+\hat{\mathbf{r}}$. The inward drawn unit normal to this surface is $-\hat{\mathbf{r}}$. The radius, r, makes an angle, θ, with respect to the horizontal. The vector element of the surface area of the sphere, denoted by $d\mathbf{s}$, points in the $+\hat{\mathbf{r}}$ direction. A vector element of the surface area of a plate is $d\mathbf{S}$

then x varies only with x_j so $dx = -dx_j$, which accounts for $d^3r_j = -d^3r$, while $\partial/\partial x_j = -\partial/\partial x$, etc. which accounts for $\nabla_j = -\nabla$. After making these changes in variables, Eq. (4.4.4) becomes,

$$\langle \mathbf{T}_{ij} \rangle = \frac{1}{4\pi\varepsilon_o V^2} \int\limits_{V-v} d^3r_i \int\limits_{V-v} d^3r \left[-\nabla\nabla\left(\frac{1}{r}\right) \right] \exp[-\beta W_2(r)] \qquad (4.4.6b)$$

Since the integrand is independent of $\mathbf{r}_i$, the integral with respect to d^3r_i over the domain, $V - v$, is simply,

$$\int\limits_{V-v} d^3r_i = V - v \qquad (4.4.6c)$$

Substitution of Eq. (4.4.6c) into Eq. (4.4.6b) leads to the result,

$$\langle \mathbf{T}_{ij} \rangle = \frac{(V-v)}{4\pi\varepsilon_o V^2} \int\limits_{V-v} d^3r \left[-\nabla\nabla\left(\frac{1}{r}\right) \right] \exp[-\beta W_2(r)] \qquad (4.4.6d)$$

Because $v \ll V$, the difference $V - v \approx V$. This allows Eq. (4.4.6d) to be rewritten as,

$$\langle \mathbf{T}_{ij} \rangle = \frac{1}{4\pi\varepsilon_o V} \int_{V-v} d^3r \left[-\nabla\nabla\left(\frac{1}{r}\right) \right] \exp[-\beta W_2(r)] \qquad (4.4.6e)$$

By addition and subtraction of $\frac{1}{4\pi\varepsilon_o V}\int_{V-v} d^3r\left[\nabla\nabla\left(\frac{1}{r}\right)\right]$ on the right hand side of Eq. (4.4.6e), one obtains the sum of integrals,

$$-\langle \mathbf{T}_{ij} \rangle = \frac{1}{4\pi\varepsilon_o V} \int_{V-v} d^3r \left[\nabla\nabla\left(\frac{1}{r}\right) \right] \left(e^{-\beta W_2(r)} - 1\right) + \frac{1}{4\pi\varepsilon_o V} \int_{V-v} d^3r \left[\nabla\nabla\left(\frac{1}{r}\right) \right] \qquad (4.4.7)$$

Upon definition of the tensor,

$$\overset{\leftrightarrow}{\mathbf{S}}_1 = \frac{1}{4\pi\varepsilon_o V} \int_{V-v} d^3r \left[\nabla\nabla\left(\frac{1}{r}\right) \right] \left(e^{-\beta W_2(r)} - 1\right) \qquad (4.4.8)$$

and the tensor,

$$\overset{\leftrightarrow}{\mathbf{S}}_2 = \frac{1}{4\pi\varepsilon_o V} \int_{V-v} d^3r \left[\nabla\nabla\left(\frac{1}{r}\right) \right] \qquad (4.4.9)$$

Equation (4.4.7) can be rewritten as

$$-\left\langle \overset{\leftrightarrow}{\mathbf{T}}_{ij} \right\rangle = \overset{\leftrightarrow}{\mathbf{S}}_1 + \overset{\leftrightarrow}{\mathbf{S}}_2 \qquad (4.4.10)$$

Equation (4.4.10) expresses the ensemble average of the transfer matrix in terms of the integrals, $\overset{\leftrightarrow}{\mathbf{S}}_1$ and $\overset{\leftrightarrow}{\mathbf{S}}_2$.

4.4.2 Evaluation of $\overset{\leftrightarrow}{\mathbf{S}}_1$ as Defined by Eq. (4.4.8)

Equation (4.1.20) can be rewritten as

$$\nabla\nabla\left(\frac{1}{r}\right) = -\left(\frac{1}{r^3}\right)\left[\overset{\leftrightarrow}{\mathbf{I}} - 3\frac{\mathbf{rr}}{r^2}\right] \qquad (4.4.11a)$$

which when substituted into Eq. (4.4.8) gives

$$\overset{\leftrightarrow}{\mathbf{S}}_1 = \frac{1}{4\pi\varepsilon_o V} \int_{V-v} d^3r \left(\frac{-1}{r^3}\right)\left[\overset{\leftrightarrow}{\mathbf{I}} - 3\frac{\mathbf{rr}}{r^2}\right]\left(e^{-\beta W_2(r)} - 1\right) \qquad (4.4.11b)$$

The geometry in Fig. 4.1 suggests spherical polar coordinates. In spherical polar coordinates, $d^3r = r^2 dr \sin\theta d\theta d\phi$. With this substitution, Eq. (4.4.11b) becomes

$$\overleftrightarrow{\mathbf{S}}_1 = \frac{1}{4\pi\varepsilon_o V}\int\limits_{V-v} drr^2 \int\limits_0^{2\pi} d\phi \int\limits_0^{\pi} d\theta \sin\theta\left(\frac{-1}{r^3}\right)\left[\overleftrightarrow{\mathbf{I}} - 3\frac{\mathbf{rr}}{r^2}\right]\left(e^{-\beta W_2(r)} - 1\right) \tag{4.4.12}$$

Define

$$\overleftrightarrow{\mathbf{S}}_{1A} = \frac{1}{4\pi\varepsilon_o V}\int\limits_{V-v} drr^2\left(\frac{-1}{r^3}\right)\left(e^{-\beta W_2(r)} - 1\right)\int\limits_0^{2\pi} d\phi \int\limits_0^{\pi} d\theta \sin\theta\, \overleftrightarrow{\mathbf{I}} \tag{4.4.13}$$

and

$$\overleftrightarrow{\mathbf{S}}_{1B} = \frac{1}{4\pi\varepsilon_o V}\int\limits_{V-v} drr^2\left(\frac{-1}{r^3}\right)\left(e^{-\beta W_2(r)} - 1\right)\int\limits_0^{2\pi} d\phi \int\limits_0^{\pi} d\theta \sin\theta \frac{\mathbf{rr}}{r^2} \tag{4.4.14}$$

Equation (4.4.12) can then be written as

$$\overleftrightarrow{\mathbf{S}}_1 = \overleftrightarrow{\mathbf{S}}_{1A} - 3\overleftrightarrow{\mathbf{S}}_{1B} \tag{4.4.15}$$

We recall that the elements of $\overleftrightarrow{\mathbf{I}}$ are either zero or unity. Hence, the integrand in Eq. (4.4.13) is independent of the angles θ and ϕ. The angular part of the integral is then simply,

$$\begin{aligned}\int\limits_0^{2\pi} d\phi \int\limits_0^{\pi} d\theta \sin\theta &= \int\limits_0^{2\pi} d\phi[-\cos\theta]_0^{\pi} = 2\pi[-\cos\theta]_0^{\pi} = 2\pi[-(-1) - (-1)(1)] \\ &= 4\pi \end{aligned} \tag{4.4.16}$$

Into Eq. (4.4.13), substitute Eq. (4.4.16). The result is

$$\overleftrightarrow{\mathbf{S}}_{1A} = \frac{\overleftrightarrow{\mathbf{I}}}{\varepsilon_o V}\int\limits_{V-v} drr^2\left(\frac{-1}{r^3}\right)\left(e^{-\beta W_2(r)} - 1\right) \tag{4.4.17}$$

Next turn to the evaluation of $\overleftrightarrow{\mathbf{S}}_{1B}$ in Eq. (4.4.14). The components of the vector, $\mathbf{r}$, in spherical polar coordinates are

$$x = r\sin\theta\cos\phi \tag{4.4.18a}$$

$$y = r \sin\theta \sin\phi \tag{4.4.18b}$$

$$z = r \cos\theta \tag{4.4.18c}$$

In terms of the Cartesian unit vectors, $\hat{\mathbf{e}}_x$, $\hat{\mathbf{e}}_y$, $\hat{\mathbf{e}}_z$, the vector $\mathbf{r}$ reads,

$$\mathbf{r} = x\hat{\mathbf{e}}_x + y\hat{\mathbf{e}}_y + z\hat{\mathbf{e}}_z \tag{4.4.19a}$$

Upon using Eqs. (4.4.18), Eq. (4.4.19a) assumes the form,

$$\mathbf{r} = r \sin\theta \cos\phi \hat{\mathbf{e}}_x + r \sin\theta \sin\phi \hat{\mathbf{e}}_y + r \cos\theta \hat{\mathbf{e}}_z \tag{4.4.19b}$$

Using Eq. (4.4.19b), the dyad, $\mathbf{rr}$, in Eq. (4.1.6) is then

$$\begin{aligned}\mathbf{rr} = r^2[&\sin^2\theta\cos^2\phi(\hat{\mathbf{e}}_x\hat{\mathbf{e}}_x) + \sin^2\theta\cos\phi\sin\phi(\hat{\mathbf{e}}_x\hat{\mathbf{e}}_y) + \sin\theta\cos\theta\cos\phi(\hat{\mathbf{e}}_x\hat{\mathbf{e}}_z)\\ &+ \sin^2\theta\cos\phi\sin\phi(\hat{\mathbf{e}}_y\hat{\mathbf{e}}_x) + \sin^2\theta\sin^2\phi(\hat{\mathbf{e}}_y\hat{\mathbf{e}}_y) + \sin\theta\cos\theta\sin\phi(\hat{\mathbf{e}}_y\hat{\mathbf{e}}_z)\\ &+ \sin\theta\cos\theta\cos\phi(\hat{\mathbf{e}}_z\hat{\mathbf{e}}_x) + \sin\theta\cos\theta\sin\phi(\hat{\mathbf{e}}_z\hat{\mathbf{e}}_y) + \cos^2\theta(\hat{\mathbf{e}}_z\hat{\mathbf{e}}_z)]\end{aligned} \tag{4.4.20}$$

On the basis of Eq. (4.4.20), the dyad, $\mathbf{rr}$, can be written in matrix form as

$$\mathbf{rr} = r^2\begin{bmatrix} \sin^2\theta\cos^2\phi & \sin^2\theta\sin\phi\cos\phi & \sin\theta\cos\theta\cos\phi \\ \sin^2\theta\sin\phi\cos\phi & \sin^2\theta\sin^2\phi & \sin\theta\cos\theta\sin\phi \\ \sin\theta\cos\theta\cos\phi & \sin\theta\cos\theta\sin\phi & \cos^2\theta \end{bmatrix} \tag{4.4.21}$$

Upon use of Eq. (4.4.21), the integral over angles on the right hand side of Eq. (4.4.14) can be written as

$$\begin{aligned}\int_0^{2\pi} d\phi \int_0^{\pi} d\theta \sin\theta \mathbf{rr} = r^2 &\int_0^{2\pi} d\phi \int_0^{\pi} d\theta \sin\theta \\ &\times \begin{bmatrix} \sin^2\theta\cos^2\phi & \sin^2\theta\sin\phi\cos\phi & \sin\theta\cos\theta\cos\phi \\ \sin^2\theta\sin\phi\cos\phi & \sin^2\theta\sin^2\phi & \sin\theta\cos\theta\sin\phi \\ \sin\theta\cos\theta\cos\phi & \sin\theta\cos\theta\sin\phi & \cos^2\theta \end{bmatrix}\end{aligned} \tag{4.4.22}$$

Because $\mathbf{rr}$ is a symmetric tensor, we need consider only the three integrals involving the diagonal elements and three of the integrals on one side of the diagonal. Recalling that in spherical polar coordinates, the differential of the solid angle is $d\phi \sin\theta d\theta$, the integrals over the diagonal element, D_{xx}, is

$$D_{xx} = \int_0^{2\pi} d\phi \int_0^{\pi} d\theta \sin\theta \cdot \sin^2\theta \cos^2\phi = \int_0^{2\pi} d\phi \cos^2\phi \int_0^{\pi} d\theta \sin\theta \sin^2\theta \tag{4.4.23a}$$

The integral over ϕ is evaluated as follows:

$$\int_0^{2\pi} d\phi \cos^2\phi = \int_0^{2\pi} d\phi \frac{1}{2}[1+\cos(2\phi)] = \left[\frac{\phi}{2}\right]_0^{2\pi} + \frac{1}{2}\left[\frac{\sin(2\phi)}{2}\right]_0^{2\pi} = \pi \tag{4.4.23b}$$

The integral over θ is computed as,

$$\begin{aligned} \int_0^{\pi} d\theta \sin\theta \sin^2\theta &= \int_0^{\pi} -d(\cos\theta)\left[1-\cos^2\theta\right] = -[\cos\theta]_0^{\pi} + \left[\frac{\cos^3\theta}{3}\right]_0^{\pi} \\ &= -[-1-1] + \frac{1}{3}\left[(-1)^3 - 1\right] = \frac{6}{3} - \frac{2}{3} = \frac{4}{3} \end{aligned} \tag{4.4.23c}$$

Substitute Eqs. (4.4.23b) and (4.4.23c) into (4.4.23a) to obtain

$$D_{xx} = \int_0^{2\pi} d\phi \int_0^{\pi} d\theta \sin\theta \cdot \sin^2\theta \cos^2\phi = \frac{4\pi}{3} \tag{4.4.23d}$$

The next diagonal element is

$$D_{yy} = \int_0^{2\pi} d\phi \int_0^{\pi} d\theta \sin\theta \cdot \sin^2\theta \sin^2\phi = \int_0^{2\pi} d\phi \sin^2\phi \int_0^{\pi} d\theta \sin\theta \sin^2\theta \tag{4.4.23e}$$

The integral over ϕ is evaluated as follows:

$$\int_0^{2\pi} d\phi \sin^2\phi = \int_0^{2\pi} d\phi \frac{1}{2}[1-\cos(2\phi)] = \left[\frac{\phi}{2}\right]_0^{2\pi} - \frac{1}{2}\left[\frac{\sin(2\phi)}{2}\right]_0^{2\pi} = \pi \tag{4.4.23f}$$

The integral over θ in Eq. (4.4.23e) is given by Eq. (4.4.23c). Hence, by combining Eqs. (4.4.23c) and (4.4.23f), Eq. (4.4.23e) reads,

$$D_{yy} = \int_0^{2\pi} d\phi \int_0^{\pi} d\theta \sin\theta \cdot \sin^2\theta \sin^2\phi = \frac{4\pi}{3} \tag{4.4.23g}$$

The remaining diagonal element is

$$D_{zz} = \int_0^{2\pi} d\phi \int_0^{\pi} d\theta \sin\theta \cdot \cos^2\theta = 2\pi \int_0^{\pi} -d(\cos\theta)\cos^2\theta = -2\pi\left[\frac{\cos^3\theta}{3}\right]_0^{\pi}$$
$$= -\frac{2\pi}{3}\left[(-1)^3 - 1\right] = \frac{4\pi}{3} \tag{4.4.23h}$$

We now turn to the off-diagonal elements. The first is

$$D_{xy} = \int_0^{2\pi} d\phi \int_0^{\pi} d\theta \sin\theta \cdot \sin^2\theta \sin\phi\cos\phi = \int_0^{2\pi} d\phi \sin\phi\cos\phi \int_0^{\pi} d\theta \sin\theta \cdot \sin^2\theta \tag{4.4.23i}$$

The integral over ϕ proceeds as follows:

$$\int_0^{2\pi} d\phi \sin\phi\cos\phi = \int_0^{2\pi} d(\sin\phi)\sin\phi = \left[\frac{\sin^2\phi}{2}\right]_0^{2\pi} = \frac{\sin^2(2\pi)}{2} - \frac{\sin^2(0)}{2} = 0 \tag{4.4.23j}$$

Upon substituting Eq. (4.4.23j), Eq. (4.4.23i) reads,

$$D_{xy} = \int_0^{2\pi} d\phi \int_0^{\pi} d\theta \sin\theta \cdot \sin^2\theta \sin\phi\cos\phi = 0 \tag{4.4.23k}$$

We now turn to the off-diagonal element,

$$D_{xz} = \int_0^{2\pi} d\phi \int_0^{\pi} d\theta \sin\theta \cdot \sin\theta\cos\theta\cos\phi = \int_0^{2\pi} d\phi \cos\phi \int_0^{\pi} d\theta \cos\theta \cdot \sin^2\theta \tag{4.4.23l}$$

The integral over ϕ is

$$\int_0^{2\pi} d\phi \cos\phi = [\sin\phi]_0^{2\pi} = 0 \tag{4.4.23m}$$

Upon substitution of Eq. (4.4.23 m), Eq. (4.4.23l) reads,

$$D_{xz} = \int_0^{2\pi} d\phi \int_0^{\pi} d\theta \sin\theta \cdot \sin\theta \cos\theta \cos\phi = 0 \tag{4.4.23n}$$

Finally, we consider

$$D_{yz} = \int_0^{2\pi} d\phi \int_0^{\pi} d\theta \sin\theta \cdot \sin\theta \cos\theta \sin\phi = \int_0^{2\pi} d\phi \sin\phi \int_0^{\pi} d\theta \sin\theta \cdot \sin\theta \cos\theta \tag{4.4.23o}$$

The integral over ϕ is

$$\int_0^{2\pi} d\phi \sin\phi = [-\cos\phi]_0^{2\pi} = 0 \tag{4.4.23p}$$

Upon substitution of Eq. (4.4.23p), Eq. (4.423o) reads,

$$D_{yz} = \int_0^{2\pi} d\phi \int_0^{\pi} d\theta \sin\theta \cdot \sin\theta \cos\theta \sin\phi = 0 \tag{4.4.23q}$$

By virtue of Eqs. (4.4.23a) through (4.4.23q), Eq. (4.4.22) can be written as,

$$\int_0^{2\pi} d\phi \int_0^{\pi} d\theta \sin\theta \mathbf{rr} = r^2 \begin{bmatrix} \left(\frac{4\pi}{3}\right) & 0 & 0 \\ 0 & \left(\frac{4\pi}{3}\right) & 0 \\ 0 & 0 & \left(\frac{4\pi}{3}\right) \end{bmatrix} \tag{4.4.24a}$$

We note that Eq. (4.4.24a) can be simplified to read,

$$\int_0^{2\pi} d\phi \int_0^{\pi} d\theta \sin\theta \mathbf{rr} = \frac{4\pi}{3} r^2 \begin{bmatrix} 1 & 0 & 0 \\ 0 & 1 & 0 \\ 0 & 0 & 1 \end{bmatrix} = \frac{4\pi}{3} r^2 \overleftrightarrow{\mathbf{I}} \tag{4.4.24b}$$

where we used Eq. (4.1.8). Upon substitution of Eq. (4.4.24b) into Eq. (4.4.14), the result is

$$\overleftrightarrow{\mathbf{S}}_{1B} = \frac{\overleftrightarrow{\mathbf{I}}}{3\varepsilon_o V} \int_{V-v} dr r^2 \left(\frac{-1}{r^3}\right) \left(e^{-\beta W_2(r)} - 1\right) \tag{4.4.25}$$

When Eqs. (4.4.17) and (4.4.25) are substituted into Eq. (4.4.15), we find that

$$\overleftrightarrow{\mathbf{S}}_1 = \overleftrightarrow{\mathbf{S}}_{1A} - 3\overleftrightarrow{\mathbf{S}}_{1B} = 0 \tag{4.4.26}$$

4.4.3 Evaluation of $\overleftrightarrow{\mathbf{S}}_2$ as Defined by Eq. (4.4.9)

The sphere of exclusion associated with the central dipole has the microscopic volume, v.

The surface area of this sphere is $s = 4\pi r^2$. Outside the sphere of exclusion, the dielectric extends to macroscopic dimensions. As illustrated in Fig. 4.1, the dielectric ends only when it comes into contact with one of the identical parallel metal plates of the capacitor bounding it. The surface area of each plate is S. The plates are regarded to be infinite in extent, so that edge effects are not a consideration.

To evaluate $\overleftrightarrow{\mathbf{S}}_2$ in Eq. (4.4.9), we will need the tensor version of the Gauss divergence theorem. If the tensor $\nabla\nabla f(r)$ is defined throughout the volume $V - v$, which is itself bounded by the two surfaces of area, s and S, the theorem states that

$$\int\limits_{V-v} d^3r \nabla\nabla f = \int\limits_{s} d\mathbf{s}\nabla f + \int\limits_{S} d\mathbf{S}\nabla f \tag{4.4.27}$$

where $d\mathbf{s}$ is the outward drawn normal to the surface of the sphere, whereas $d\mathbf{S}$ is outward drawn normal to the surface of a plate, and it is assumed that both plates are taken into account in the integration. If we apply this tensor divergence theorem to the integrand in Eq. (4.4.9), the result is

$$\overleftrightarrow{\mathbf{S}}_2 = \frac{1}{4\pi\varepsilon_o V}\int\limits_{V-v} d^3r\left[\nabla\nabla\left(\frac{1}{r}\right)\right] = \frac{1}{4\pi\varepsilon_o V}\int\limits_{s} \nabla\left(\frac{1}{r}\right)d\mathbf{s} + \frac{1}{4\pi\varepsilon_o V}\int\limits_{S} \nabla\left(\frac{1}{r}\right)d\mathbf{S} \tag{4.4.28}$$

In reading Eq. (4.4.28), one may imagine both $\nabla\left(\frac{1}{r}\right)d\mathbf{s}$ and $\nabla\left(\frac{1}{r}\right)d\mathbf{S}$ written as dyads. Define

$$\overleftrightarrow{\mathbf{S}}_{2A} = \frac{1}{4\pi\varepsilon_o V}\int\limits_{s} \nabla\left(\frac{1}{r}\right)d\mathbf{s} \tag{4.4.29}$$

and

$$\overleftrightarrow{\mathbf{S}}_{2B} = \frac{1}{4\pi\varepsilon_o V}\int\limits_{S} \nabla\left(\frac{1}{r}\right)d\mathbf{S} \tag{4.4.30}$$

so that

$$\overleftrightarrow{\mathbf{S}}_2 = \overleftrightarrow{\mathbf{S}}_{2A} + \overleftrightarrow{\mathbf{S}}_{2B} \tag{4.4.31}$$

We turn first to the evaluation of the surface integral appearing in the formula for $\overleftrightarrow{\mathbf{S}}_{2A}$ in Eq. (4.4.29). The computation $\nabla(1/r)$ in Cartesian coordinates proceeds as usual:

$$\begin{aligned}\nabla\left(\frac{1}{r}\right) &= \nabla\left(\frac{1}{\sqrt{x^2+y^2+z^2}}\right) = \left(\hat{\mathbf{e}}_x\frac{\partial}{\partial x} + \hat{\mathbf{e}}_y\frac{\partial}{\partial y} + \hat{\mathbf{e}}_z\frac{\partial}{\partial z}\right)\left(x^2+y^2+z^2\right)^{-1/2} \\ &= \left(-\frac{1}{2}\right)\frac{2x\hat{\mathbf{e}}_x + 2y\hat{\mathbf{e}}_y + 2z\hat{\mathbf{e}}_z}{\left(x^2+y^2+z^2\right)^{3/2}} = -\frac{x\hat{\mathbf{e}}_x + y\hat{\mathbf{e}}_y + z\hat{\mathbf{e}}_z}{\left(x^2+y^2+z^2\right)^{3/2}}\end{aligned} \tag{4.4.32a}$$

Equation (4.4.32a) can also be written as

$$\nabla\left(\frac{1}{r}\right) = -\frac{\mathbf{r}}{r^3} \tag{4.4.32b}$$

Upon substitution of Eq. (4.4.32b) into Eq. (4.4.29), the result is

$$\overleftrightarrow{\mathbf{S}}_{2A} = \frac{1}{4\pi\varepsilon_o V}\int_s \nabla\left(\frac{1}{r}\right)d\mathbf{s} = -\frac{1}{4\pi\varepsilon_o V}\int_s \frac{\mathbf{r}}{r^3}d\mathbf{s} \tag{4.4.32c}$$

From the point of view of the interior of the spherical surface bounding the volume, v, the outward unit normal is $\hat{\mathbf{r}}$. But in Eq. (4.4.28), the volume of integration is the volume, $V - v$, which is external to the sphere. With reference to the interior of $V - v$, the outward normal on the spherical surface bounding v is $-\hat{\mathbf{r}}$ where $-\hat{\mathbf{r}} = -\mathbf{r}/r$. The corresponding element of surface area is $d\mathbf{s} = d\Omega r^2(-\hat{\mathbf{r}})$ where in spherical polar coordinates, $d\Omega = \sin\theta d\theta d\phi$. With these substitutions, Eq. (4.4.32c) reads

$$\begin{aligned}\overleftrightarrow{\mathbf{S}}_{2A} &= \frac{1}{4\pi\varepsilon_o V}\int_s \nabla\left(\frac{1}{r}\right)d\mathbf{s} = -\frac{1}{4\pi\varepsilon_o V}\int_s \frac{\mathbf{r}}{r^3}d\mathbf{s} \\ &= \frac{-1}{4\pi\varepsilon_o V}\int_s d\Omega r^2(-\hat{\mathbf{r}})\left(\frac{\mathbf{r}}{r^3}\right) = \frac{r^2}{4\pi\varepsilon_o V r^4}\int_0^{2\pi} d\phi\int_0^{\pi} d\theta\,\sin\theta\,\mathbf{rr} \\ &= \frac{1}{4\pi\varepsilon_o V r^2}\int_0^{2\pi} d\phi\int_0^{\pi} d\theta\,\sin\theta\,\mathbf{rr}\end{aligned} \tag{4.4.32d}$$

where we have used, $\hat{\mathbf{r}} = \mathbf{r}/r$. Equation (4.4.32d) can be evaluated by substituting Eq. (4.4.24b) for the double integral of the dyad, $\mathbf{rr}$. The result is

$$\overleftrightarrow{\mathbf{S}}_{2A} = \left(\frac{1}{3V\varepsilon_o}\right)\overleftrightarrow{\mathbf{I}} \tag{4.4.33}$$

The integral, $\overleftrightarrow{\mathbf{S}}_{2B}$, in Eq. (4.4.30) involves the outer boundary of the dielectric which consists of the plates of the capacitor. Let the origin of coordinates for this integral be the center of the dielectric. If the separation of the two plates is $2L$, a point on the surface of one plate is located by the coordinates $(x, y, +L)$, while a point on the opposite plate is located by the coordinates $(x, y, -L)$. See Fig. 4.1. The outward drawn normal to the plate at $z = +L$ is $\hat{\mathbf{e}}_z$. A differential element of surface area on this plate is $d\mathbf{S} = \hat{\mathbf{e}}_z dxdy$. The outward drawn normal to the plate at $z = -L$ is $-\hat{\mathbf{e}}_z$. On this plate, a differential element of surface area is $d\mathbf{S} = -\hat{\mathbf{e}}_z dxdy$. Hence, Eq. (4.4.30) can be broken up into two integrals,

$$\overleftrightarrow{\mathbf{S}}_{2B} = \frac{1}{4\pi\varepsilon_o V}\int_S \nabla\left(\frac{1}{r}\right)d\mathbf{S} = \frac{1}{4\pi\varepsilon_o V}\int_S dxdy(\hat{\mathbf{e}}_z)\nabla\left(\frac{1}{r}\right)_{z=+L} + \frac{1}{4\pi\varepsilon_o V}\int_S dxdy(-\hat{\mathbf{e}}_z)\nabla\left(\frac{1}{r}\right)_{z=-L} \tag{4.4.34a}$$

Evaluation of Eq. (4.4.32a) at coordinates $(x, y, +L)$ leads to

$$\nabla\left[\left(\frac{1}{r}\right)\right]_{z=L} = \nabla\left(\frac{1}{\sqrt{x^2+y^2+z^2}}\right)_{z=L} = -\frac{x\hat{\mathbf{e}}_x + y\hat{\mathbf{e}}_y + L\hat{\mathbf{e}}_z}{\left(x^2+y^2+L^2\right)^{3/2}} \tag{4.4.34b}$$

Evaluation of Eq. (4.4.32a) at coordinates $(x, y, -L)$ leads to

$$\nabla\left[\left(\frac{1}{r}\right)\right]_{z=-L} = -\frac{x\hat{\mathbf{e}}_x + y\hat{\mathbf{e}}_y - L\hat{\mathbf{e}}_z}{\left(x^2+y^2+L^2\right)^{3/2}} \tag{4.4.34c}$$

Substitute Eq. (4.4.34b) into the first term in Eq. (4.4.34a) and (4.4.34c) into the second term. With $d\mathbf{S} = \hat{\mathbf{e}}_z dxdy$ representing the area in the first term and $d\mathbf{S} = -\hat{\mathbf{e}}_z dxdy$ representing the area in the second term, the result is

$$\begin{aligned}\overleftrightarrow{\mathbf{S}}_{2B} &= \frac{1}{4\pi\varepsilon_o V}\int_S \nabla\left(\frac{1}{r}\right)d\mathbf{S}\\ &= \frac{-1}{4\pi\varepsilon_o V}\int dx\int dy\frac{x(+\hat{\mathbf{e}}_z)\hat{\mathbf{e}}_x + y(+\hat{\mathbf{e}}_z)\hat{\mathbf{e}}_y + L(+\hat{\mathbf{e}}_z)\hat{\mathbf{e}}_z}{\left(x^2+y^2+L^2\right)^{3/2}}\\ &\quad + \frac{-1}{4\pi\varepsilon_o V}\int dx\int dy\frac{x(-\hat{\mathbf{e}}_z)\hat{\mathbf{e}}_x + y(-\hat{\mathbf{e}}_z)\hat{\mathbf{e}}_y + (-L)(-\hat{\mathbf{e}}_z)\hat{\mathbf{e}}_z}{\left(x^2+y^2+L^2\right)^{3/2}}\end{aligned} \tag{4.4.34d}$$

When summed, the two integrals on the right hand side of Eq. (4.4.34d) yield the net result,

$$\overleftrightarrow{\mathbf{S}}_{2B} = -\frac{1}{4\pi\varepsilon_o V}\int dx\int dy\frac{2L\hat{\mathbf{e}}_z\hat{\mathbf{e}}_z}{\left(x^2+y^2+L^2\right)^{3/2}} \tag{4.4.34e}$$

To evaluate the integral in Eq. (4.4.34e), we introduce plane polar coordinates, (ρ, ϕ), which are defined in terms of the Cartesian coordinates by

$$x = \rho \cos \phi \tag{4.4.35a}$$

$$y = \rho \sin \phi \tag{4.4.35b}$$

where the radius, ρ, satisfies

$$\rho^2 = x^2 + y^2 \tag{4.4.35c}$$

To avoid the consideration of edge effects, we have assumed the area of each plate to be infinite. This allows the extent of the radius to be $0 \leq \rho \leq \infty$. The domain of the angle, ϕ, is $0 \leq \phi \leq 2\pi$. The differential area in plane polar coordinates is $\rho d\phi d\rho$.

After substitution of Eqs. (4.4.35c) into Eq. (4.4.34e) and noting the limits of the ranges of the plane polar coordinates, we have

$$\overleftrightarrow{\mathbf{S}}_{2B} = -\frac{1}{4\pi\varepsilon_o V}\int_0^{\infty} d\rho\rho \int_0^{2\pi} d\phi \frac{2L\hat{\mathbf{e}}_z\hat{\mathbf{e}}_z}{\left(\rho^2 + L^2\right)^{3/2}} \tag{4.4.36a}$$

Integration over ϕ gives

$$\overleftrightarrow{\mathbf{S}}_{2B} = -\frac{2\pi L\hat{\mathbf{e}}_z\hat{\mathbf{e}}_z}{4\pi\varepsilon_o V}\int_0^{\infty} d\rho \frac{2\rho}{\left(\rho^2 + L^2\right)^{3/2}} \tag{4.4.36b}$$

Evaluation of the remaining integral over ρ yields

$$\overleftrightarrow{\mathbf{S}}_{2B} = -\frac{2\pi L\hat{\mathbf{e}}_z\hat{\mathbf{e}}_z}{4\pi\varepsilon_o V}\left[\frac{\left(\rho^2 + L^2\right)^{-1/2}}{(-1/2)}\right]_0^{\infty} = -\frac{2\pi L\hat{\mathbf{e}}_z\hat{\mathbf{e}}_z}{4\pi\varepsilon_o V}\left(\frac{2}{L}\right) \tag{4.4.36c}$$

After performing the indicated cancellations, one obtains,

$$\overleftrightarrow{\mathbf{S}}_{2B} = -\frac{\hat{\mathbf{e}}_z\hat{\mathbf{e}}_z}{\varepsilon_o V} \tag{4.4.37}$$

When Eq. (4.4.33) and (4.4.37) are substituted into Eq. (4.4.31), the result is

$$\overleftrightarrow{\mathbf{S}}_2 = \frac{\overleftrightarrow{\mathbf{I}}}{3\varepsilon_o V} - \frac{\hat{\mathbf{e}}_z\hat{\mathbf{e}}_z}{\varepsilon_o V} \tag{4.4.38}$$

4.4.4 Formula for the Mean Value of the Transfer Matrix

Having evaluated $\overleftrightarrow{\mathbf{S}}_1$ and $\overleftrightarrow{\mathbf{S}}_2$, we can substitute Eqs. (4.4.26) and (4.4.38) into Eq. (4.4.10) to obtain,

$$-\left\langle \overleftrightarrow{\mathbf{T}}_{ij} \right\rangle = \overleftrightarrow{\mathbf{S}}_1 + \overleftrightarrow{\mathbf{S}}_2 = 0 + \left(\frac{1}{3\varepsilon_o V} \right) \left(\overleftrightarrow{\mathbf{I}} - 3\hat{\mathbf{e}}_z \hat{\mathbf{e}}_z \right) \tag{4.4.39}$$

or equivalently,

$$\left\langle \overleftrightarrow{\mathbf{T}}_{ij} \right\rangle = -\left(\frac{1}{3\varepsilon_o V} \right) \left(\overleftrightarrow{\mathbf{I}} - 3\hat{\mathbf{e}}_z \hat{\mathbf{e}}_z \right) \tag{4.4.40}$$

Equation (4.4.40) confirms our earlier assumption (Eq. (4.3.3) through Eq. (4.3.5)) that $\left\langle \overleftrightarrow{\mathbf{T}}_{ij} \right\rangle$ is independent of the indices, i and j.

4.5 The Clausius–Mossotti Equation

For a macroscopic sample of matter, the number of molecules, N', in the dielectric greatly exceeds unity. This permits Eq. (4.3.5) to be approximated as

$$\langle \mathbf{p} \rangle + \alpha N' \left\langle \overleftrightarrow{\mathbf{T}}_{ij} \right\rangle \cdot \langle \mathbf{p} \rangle = \alpha (\mathbf{D}/\varepsilon_o) \tag{4.5.1}$$

Now replace $\left\langle \overleftrightarrow{\mathbf{T}}_{ij} \right\rangle$ in Eq. (4.5.1) by Eq. (4.4.40). The result is

$$\langle \mathbf{p} \rangle - \alpha \left(\frac{N'}{3\varepsilon_o V} \right) \left(\overleftrightarrow{\mathbf{I}} - 3\hat{\mathbf{e}}_z \hat{\mathbf{e}}_z \right) \cdot \langle \mathbf{p} \rangle = \alpha (\mathbf{D}/\varepsilon_o) \tag{4.5.2a}$$

Upon substituting Eq. (4.1.4) for $\overleftrightarrow{\mathbf{I}}$, Eq. (4.5.2a) reads

$$\langle \mathbf{p} \rangle - \alpha \left(\frac{N'}{3\varepsilon_o V} \right) (\hat{\mathbf{e}}_x \hat{\mathbf{e}}_x + \hat{\mathbf{e}}_y \hat{\mathbf{e}}_y - 2\hat{\mathbf{e}}_z \hat{\mathbf{e}}_z) \cdot \langle \mathbf{p} \rangle = \alpha (\mathbf{D}/\varepsilon_o) \tag{4.5.2b}$$

In a parallel plate capacitor, the displacement is everywhere in the z-direction, so

$$\mathbf{D} = D\hat{\mathbf{e}}_z \tag{4.5.2c}$$

As $\mathbf{D}$ includes the effects of polarization in the dielectric, the ensemble average dipole moment is necessarily parallel to $\hat{\mathbf{e}}_z$; hence,

$$\langle \mathbf{p} \rangle = p\hat{\mathbf{e}}_z \tag{4.5.2d}$$

Substitute Eqs. (4.5.2c) and (4.5.2d) into Eq. (4.5.2b). The result is

$$\hat{\mathbf{e}}_z p - \alpha\left(\frac{N'}{3\varepsilon_o V}\right)\left[\hat{\mathbf{e}}_x\hat{\mathbf{e}}_x + \hat{\mathbf{e}}_y\hat{\mathbf{e}}_y - 2\hat{\mathbf{e}}_z\hat{\mathbf{e}}_z\right] \cdot \hat{\mathbf{e}}_z p = \alpha\hat{\mathbf{e}}_z(D/\varepsilon_o) \tag{4.5.2e}$$

In Eq. (4.5.2e), the dot product of $\hat{\mathbf{e}}_z p$ with the dyad in the square bracket can be expanded to read,

$$\hat{\mathbf{e}}_z p - \alpha\left(\frac{N'}{3\varepsilon_o V}\right)\left[\hat{\mathbf{e}}_x(\hat{\mathbf{e}}_x \cdot \hat{\mathbf{e}}_z) + \hat{\mathbf{e}}_y(\hat{\mathbf{e}}_y \cdot \hat{\mathbf{e}}_z) - 2\hat{\mathbf{e}}_z(\hat{\mathbf{e}}_z \cdot \hat{\mathbf{e}}_z)\right]p = \alpha\hat{\mathbf{e}}_z(D/\varepsilon_o) \tag{4.5.2f}$$

Upon substitution of the unit vector dot product identities,

$$(\hat{\mathbf{e}}_x \cdot \hat{\mathbf{e}}_z) = 0 \tag{4.5.3a}$$

$$(\hat{\mathbf{e}}_y \cdot \hat{\mathbf{e}}_z) = 0 \tag{4.5.3b}$$

$$(\hat{\mathbf{e}}_z \cdot \hat{\mathbf{e}}_z) = 1 \tag{4.5.3c}$$

Equation (4.5.2f) simplifies to read,

$$p\hat{\mathbf{e}}_z + \alpha\left(\frac{2N'}{3\varepsilon_o V}\right)p\hat{\mathbf{e}}_z = \alpha\hat{\mathbf{e}}_z(D/\varepsilon_o) \tag{4.5.4a}$$

Next, cancel $\hat{\mathbf{e}}_z$ from either side of Eq. (4.5.4a) and solve the result for p:

$$p = \frac{\alpha(D/\varepsilon_o)}{\left(1 + \frac{2N'\alpha}{3\varepsilon_o V}\right)} \tag{4.5.4b}$$

If the N' molecules in the dielectric are confined to a volume, V, the macroscopic polarization is $P = N'p/V$. When combined with Eq. (4.5.4b), the polarization satisfies the equation,

$$P = \frac{N'p}{V} = \frac{D}{\left(1 + \frac{2N'\alpha}{3\varepsilon_o V}\right)}\left(\frac{N'\alpha}{\varepsilon_o V}\right) \tag{4.5.4c}$$

According to Sect. 3.2 of Chap. 3, the displacement is given by

$$D = \varepsilon_o E + P \tag{4.5.5a}$$

and also by

$$D = \varepsilon_r \varepsilon_o E \tag{4.5.5b}$$

Equation (4.5.5b) can be rearranged to read

$$E = \frac{D}{\varepsilon_r \varepsilon_o} \tag{4.5.5c}$$

which when substituted into Eq. (4.5.5a) gives

$$D = \frac{D}{\varepsilon_r} + P \tag{4.5.5d}$$

When solved for P, Eq. (4.5.5d) reads

$$P = \left(1 - \frac{1}{\varepsilon_r}\right) D \tag{4.5.5e}$$

Eliminate P between Eq. (4.5.4c) and (4.5.5e). The result is

$$\frac{D}{\left(1 + \frac{2N'\alpha}{3\varepsilon_o V}\right)} \left(\frac{N'\alpha}{\varepsilon_o V}\right) = \left(1 - \frac{1}{\varepsilon_r}\right) D \tag{4.5.6a}$$

Upon cancellation of D from either side of Eq. (4.5.6a), the resulting equation can be rearranged to read,

$$\frac{(N'\alpha/\varepsilon_o V)}{\left[1 + \frac{2}{3}\left(\frac{N'\alpha}{\varepsilon_o V}\right)\right]} = \frac{(\varepsilon_r - 1)}{\varepsilon_r} \tag{4.5.6b}$$

Finally, solve Eq. (4.5.6b) for $(N'\alpha/\varepsilon_o V)$. The result is

$$\frac{\varepsilon_r - 1}{\varepsilon_r + 2} = \frac{1}{3}\left(\frac{N'\alpha}{V\varepsilon_o}\right) \tag{4.5.6c}$$

The number density, N, is defined by

$$N = \frac{N'}{V} \tag{4.5.6d}$$

Upon substitution of Eq. (4.5.6d) into Eq. (4.5.6c), the result is

$$\frac{\varepsilon_r - 1}{\varepsilon_r + 2} = \frac{N\alpha}{3\varepsilon_o} \tag{4.5.7}$$

Equation (4.5.7) can be recognized as the Clausius–Mossotti equation. The latter was previously derived in Chap. 3 on the basis of both the Lorenz–Lorentz continuous dielectric model [see Eq. (3.5.11)] and the Onsager semi-continuum model [(Eq. 3.10.24)].

4.6 Discussion

The free charges on the metal capacitor plates are the source of the electric field that polarizes the dielectric. These charges determine the displacement, **D**, which by virtue of Gauss' Law has the same value in the absence of the dielectric as it does in the presence of the dielectric. It is the electric field, **E**, that changes when the dielectric is inserted. In the absence of the dielectric, the value of the macroscopic electric field is $\mathbf{E_0} = \mathbf{D}/\varepsilon_o$, and the value of the local electric field, $\mathbf{F}_a$, is the same. In the presence of the dielectric, $\mathbf{F}_a = (\mathbf{D}/\varepsilon_o) + \mathbf{E}_a$, where $\mathbf{E}_a$ is the contribution due to the induced dipoles.

Unlike the radial electric field acting between point charges, the electric field at point b due to a dipole at point a, as given by Eq. (4.1.15), has components which are not parallel to the vector, $\mathbf{r}_{ab}$, connecting the two points. As a consequence, the dipole is said to be the source of a non-central electric field. This non-central characteristic has its origin in the tensor, $\overset{\leftrightarrow}{\mathbf{T}}_{ab}$. The latter is defined by Eq. (4.1.16).

According to Eq. (4.3.1), the contribution to the average polarization at i due to a dipole at j is determined by the ensemble average of the product of the matrix of the transfer tensor, $\overset{\leftrightarrow}{\mathbf{T}}_{ab}$, and the dipole moment at j. By setting $\langle \mathbf{n}_i \rangle = 0$ as mentioned in the text below Eq. (4.3.3), we are ignoring the effect of molecular clusters, which involve dipole–dipole correlations. This assumption is not unique to dielectric theory but appears elsewhere when statistical mechanics is applied to systems of interacting molecules. Clusters should be rare when the molecular number density is low. This presumably accounts for the good agreement between experiment and the Clausius–Mossotti equation in the case of non-polar gases at low pressure.

The Kirkwood assumption $\langle \mathbf{n}_i \rangle = 0$ is a mathematical approximation. By contrast, in the Onsager model, the assumption of a semi-continuum is a physical approximation. The purpose of both approximations is to cope with the interaction of the exterior molecules with the central molecule. Although the two may not otherwise be connected, it is amusing to note that if one lets the cavity radius a in the semi-continuum model satisfy the condition $(4\pi/3)a^3 = 1/N$, the Onsager theory and the Kirkwood theory both lead to the Clausius–Mossotti formula.

The transfer matrix method forms the basis for still more advanced theories of the permittivity of non-polar fluids. Particularly readable versions can be found in Refs. [3, 4]. The transfer matrix method has also been exploited in order to represent the intermolecular forces in condensed matter [5, 6].

References

1. J.G. Kirkwood, On the theory of dielectric polarization. J. Chem. Phys. **4**, 592–601 (1936)
2. D.A. McQuarrie, *Statistical Mechanics* (Harper and Rowe, New York, 1976)
3. A. Ishihara, Theory of the dielectric constant of a nonpolar fluid. J. Chem. Phys. **38**, 2437–2444 (1963)
4. S. Kielich, General molecular theory and electric field effects in isotropic dielectrics, in *Specialist Periodical Reports, Dielectrics and Related Molecular Processes*, ed. By M. Davies, vol. 1 (The Chemical Society, London, UK, 1972)
5. B. Linder, Reaction field techniques and their applications to intermolecular forces. Adv. Chem. Phys. **12**, 225–283 (1967)
6. T. Keyes, B.M. Ladanyi, The internal field problem in depolarized light scattering. Adv. Chem. Phys. **56**, 411–465 (1984)

Recommended Additional Reading

1. C.J.F. Bottcher, *Theory of Polarization*, vol. 1. Dielectrics in Static Fields (Elsevier Publishing Co., New York, 1973)

Chapter 5
Plasmas and the Poisson–Boltzmann Equation

A plasma is a spatial distribution of charges which are in thermal equilibrium or steady state with one another. Examples include the ions in an electrolyte solution, and the ions and electrons in a steady state electric discharge. In a plasma, charges of opposite sign tend to drift toward one another under the action of their mutual electric fields. If this were the sole force acting, the oppositely charged ions would recombine and the spatial distribution of charges would collapse. As realized by Debye, however, the drift of charges of opposite signs towards one another increases the local charge density. As the second law of thermodynamics favors a uniform distribution of charge, the local increase in charge density adjusts by partially diffusing back into the surrounding material. The result is an equilibrium charge distribution in which each charge is surrounded by a radially decreasing atmosphere of opposite charge. Consistent with the charge neutrality condition, the integral of the charge distribution over the entire sample is zero. The determination of the electrostatic potential requires a method of solution of the Poisson equation which maintains self-consistency. This is achieved by setting the potential of mean force acting on a given ion equal to the charge on the ion times the electrostatic potential evaluated at the position of the ion. The result is the so-called Poisson–Boltzmann equation. In this chapter, we use the Poisson-Boltzmann equation to solve for the electrostatic potential in the case of a charged flat plate dipped in an electrolyte solution. The result is subsequently used to calculate the electrostatic force of repulsion between two like charged parallel plates. This case is relevant to the problem of colloidal charged particle coagulation. In another example, we solve the Poisson–Boltzmann equation in cylindrical polar coordinates for the case of a charged rod in an electrolyte solution. This result improves upon the single rod model for a charged DNA strand considered in Chap. 1.

J. K. Baird, *Electromagnetism for Chemistry*,
https://doi.org/10.1007/978-3-031-88508-2_5

5.1 Introduction

At elevated temperatures which occur, for example, in electric discharges or in stars, the atoms and molecules are dissociated into ions and electrons. At room temperature, cases of ionization include aqueous solutions of strong electrolytes and also metals, where the charges are the ion cores and the conduction electrons. At either extreme of temperature, however, the ionized state is said to constitute a *plasma.*

As in the case of a chemical reaction, the equilibrium state of the ions in an electrolyte solution involves a balance between a forward rate and a reverse rate. As every cation attracts anions and every anion attracts cations, the electrostatic forces are always conspiring to achieve ion recombination. If this were the sole force acting, then opposite charged ions would recombine, and the electrolyte plasma would collapse. In a plasma, the ions are free to move by diffusion, however. Diffusion acts in a sense opposite to the electrostatic attractive force driving recombination. If, for example, the concentration of anions around a given cation begins to build up, the anion concentration will depart from the uniform distribution that maximizes the entropy. To adjust for this imbalance, the local excess of anions will begin to diffuse back into the bulk of the solution. Equilibrium occurs when the rate of drift of the anions toward the cation is balanced by the rate diffusion of anions away from the cation. In the resulting equilibrium state, every positive ion is surrounded by a stationary atmosphere of negative ions, while every negative ion is surrounded by a stationary atmosphere of positive ions. The theory, which describes this state of equilibrium, exploits electrostatics to account for the charge density and statistical mechanics to account for the equilibrium between drift and diffusion.

5.2 Energy Stored in an Electrostatic Charge Distribution

Consider an ionic system consisting of three free charges, q_1, q_2, and q_3, embedded in a dielectric liquid having relative permittivity, ε_r. An example might be water containing a single molecular unit of the strong electrolyte, $CaCl_2$. The ions are located by the vectors, $\mathbf{r}_1$, $\mathbf{r}_2$, and $\mathbf{r}_3$, respectively. The electrostatic potential energy of the system of ions in this configuration is

$$U_3(\mathbf{r}_1, \mathbf{r}_2, \mathbf{r}_3) = \frac{q_1 q_2}{4\pi\varepsilon_r\varepsilon_o|\mathbf{r}_1 - \mathbf{r}_2|} + \frac{q_1 q_3}{4\pi\varepsilon_r\varepsilon_o|\mathbf{r}_1 - \mathbf{r}_3|} + \frac{q_2 q_3}{4\pi\varepsilon_r\varepsilon_o|\mathbf{r}_2 - \mathbf{r}_3|} \tag{5.2.1}$$

which when written as a sum is

$$U_3 = \frac{1}{2}\sum_{i=1}^{3}\sum_{j\neq i=1}^{3} \frac{q_i q_j}{4\pi\varepsilon_r\varepsilon_o\left|\mathbf{r}_i - \mathbf{r}_j\right|} \tag{5.2.2}$$

Equation (5.2.2) is a double sum over both i and j. As mentioned in Sect. 1.5 of Chap. 1, the factor of $(1/2)$ corrects the double sum, which counts q_i interacting with q_j and q_j interacting with q_i as separate interactions. The restriction, $i \neq j$, removes from consideration infinite self-energy terms. The electrostatic potential at an arbitrary point located by the vector $\mathbf{r}$ due to the three ions is

$$\Phi(\mathbf{r}) = \sum_{j=1}^{3} \frac{q_j}{4\pi \varepsilon_r \varepsilon_o |\mathbf{r} - \mathbf{r}_j|} \tag{5.2.3}$$

We can combine Eqs. (5.2.2) and (5.2.3) to obtain for the electrostatic energy,

$$U_3 = \frac{1}{2} \sum_{i=1}^{3} q_i \Phi(\mathbf{r}_i) \tag{5.2.4}$$

If instead of consisting of point charges, the charge is distributed continuously according to the charge density, $\rho(\mathbf{r})$, then Eq. (5.2.4) assumes the form,

$$U_e = \frac{1}{2} \int d^3 r \rho(\mathbf{r}) \Phi(\mathbf{r}) \tag{5.2.5}$$

where the subscript, e, on U identifies it as the electrostatic energy. As shown in Sect. 3.2 of Chap. 3, Gauss' Law for the displacement is

$$\nabla \cdot \mathbf{D} = \rho. \tag{5.2.6}$$

where ρ is the density of free charge. Since the displacement and the electric field are related by

$$\mathbf{D} = \varepsilon_r \varepsilon_o \mathbf{E}. \tag{5.2.7}$$

Equation (5.2.6) can be rewritten as

$$\nabla \cdot \mathbf{E} = \rho / \varepsilon_r \varepsilon_o. \tag{5.2.8}$$

Replace ρ in Eq. (5.2.5) using Eq. (5.2.8). The result is

$$U_e = \frac{\varepsilon_r \varepsilon_o}{2} \int d^3 r \Phi(\mathbf{r}) (\nabla \cdot \mathbf{E}(\mathbf{r})) \tag{5.2.9a}$$

One may find the following identity in the Mathematical Methods Appendix,

$$\nabla \cdot (\mathbf{E}\Phi) = \mathbf{E} \cdot \nabla \Phi + \Phi \nabla \cdot \mathbf{E}. \tag{5.2.9b}$$

Upon use of Eq. (5.2.9b), Eq. (5.2.9a) can be transformed to read,

$$U_e = \frac{\varepsilon_r \varepsilon_o}{2}\left[-\int d^3r \mathbf{E}(\mathbf{r}) \cdot \nabla \Phi(\mathbf{r}) + \int d^3r \nabla \cdot (\mathbf{E}(\mathbf{r})\Phi(\mathbf{r})) \right] \tag{5.2.9c}$$

The second integral on the right hand side of Eq. (5.2.9c) can be evaluated using the Gauss-Green divergence theorem (see the Mathematical Methods Appendix). The result is

$$U_e = \frac{\varepsilon_r \varepsilon_o}{2}\left[-\int d^3r \mathbf{E}(\mathbf{r}) \cdot \nabla \Phi(\mathbf{r}) + \int d\mathbf{S} \cdot (\mathbf{E}(\mathbf{r})\Phi(\mathbf{r})) \right] \tag{5.2.9d}$$

If the boundaries are sufficiently distant such that the ratio of the surface area to the volume goes to zero, the electric field can be considered to vanish on the boundary with the result,

$$\int d\mathbf{S} \cdot (\mathbf{E}(\mathbf{r})\Phi(\mathbf{r})) = 0 \tag{5.2.9e}$$

The electric field and the electrostatic potential are related by

$$\mathbf{E} = -\nabla \Phi. \tag{5.2.9f}$$

Upon substitution of Eqs. (5.2.9e) and (5.2.9f) into Eq. (5.2.9d), the result is

$$U_e = \frac{\varepsilon_r \varepsilon_o}{2} \int d^3r |\nabla \Phi(\mathbf{r})|^2 \tag{5.2.10}$$

Equation (5.2.10) provides a general formula for calculating the energy stored in an electric field where the boundaries make no contribution.

5.3 Derivation of the Poisson–Boltzmann Equation from Statistical Mechanics

Equation (5.2.1) gives the electrostatic energy of interaction of three ions imbedded in a dielectric. The ions are assumed to be free charges, which are capable of moving to positions of mutual thermal equilibrium. According to Sect. 4.2 of Chap. 4, the *configuration integral* in the canonical ensemble for this case is

$$Z_3 = \int d^3r_1 \int d^3r_2 \int d^3r_3 \exp(-\beta U_3(\mathbf{r}_1, \mathbf{r}_2, \mathbf{r}_3)) \tag{5.3.1a}$$

where $\beta = 1/k_B T$. The probability of finding the three charges in any given configuration is

$$P(\mathbf{r}_1, \mathbf{r}_2, \mathbf{r}_2) = \frac{\exp(-\beta U_3(\mathbf{r}_1, \mathbf{r}_2, \mathbf{r}_2))}{Z_3} \tag{5.3.1b}$$

Now assume that q_1 is fixed at position, $\mathbf{r}_1$, while q_2 and q_3 are allowed to adopt configurations consistent with thermal equilibrium. According to Sect. 4.2 of Chap. 4, the potential of mean force, $W_1(\mathbf{r}_1)$, at $\mathbf{r}_1$ is defined by

$$\exp(-\beta W_1(\mathbf{r}_1)) = \frac{V \int d^3r_2 \int d^3r_3 \exp(-\beta U_3(\mathbf{r}_1, \mathbf{r}_2, \mathbf{r}_3))}{Z_3}. \tag{5.3.2}$$

For the sake of generality, replace the coordinate, $\mathbf{r}_1$ in Eq. (5.3.2) by the generic coordinate, $\mathbf{r}$, so that Eq. (5.3.2) reads,

$$\exp(-\beta W_1(\mathbf{r})) = \frac{V \int d^3r_2 \int d^3r_3 \exp(-\beta U_3(\mathbf{r}, \mathbf{r}_2, \mathbf{r}_3))}{Z_3} \tag{5.3.3}$$

where $U_3(\mathbf{r}, \mathbf{r}_2, \mathbf{r}_3)$ is given by Eq. (5.2.1) with $\mathbf{r}_1 = \mathbf{r}$. Equation (5.3.3) will find use below.

Consider a system of three fixed point charges located at coordinates, $\mathbf{r}_1$, $\mathbf{r}_2$, and $\mathbf{r}_3$. Let $\Phi(\mathbf{r}|\mathbf{r}_1, \mathbf{r}_2, \mathbf{r}_3)$ be the electrostatic potential evaluated at $\mathbf{r}$ and $\rho(\mathbf{r}|\mathbf{r}_1, \mathbf{r}_2, \mathbf{r}_3)$ be the corresponding charge density. The Poisson's equation then reads,

$$\nabla^2\Phi(\mathbf{r}|\mathbf{r}_1, \mathbf{r}_2, \mathbf{r}_3) = -\frac{\rho(\mathbf{r}|\mathbf{r}_1, \mathbf{r}_2, \mathbf{r}_3)}{\varepsilon_r\varepsilon_o} \tag{5.3.4}$$

Because charges are fixed at $\mathbf{r}_1$, $\mathbf{r}_2$, and $\mathbf{r}_3$, it is understood that $\rho(\mathbf{r}|\mathbf{r}_1, \mathbf{r}_2, \mathbf{r}_3)$ is zero except when $\mathbf{r}$ is equal to $\mathbf{r}_1$, $\mathbf{r}_2$, or $\mathbf{r}_3$. The thermal average of Eq. (5.3.4) is denoted as

$$\nabla^2\langle\Phi(\mathbf{r}|\mathbf{r}_1, \mathbf{r}_2, \mathbf{r}_3)\rangle = -\frac{\langle\rho(\mathbf{r}|\mathbf{r}_1, \mathbf{r}_2, \mathbf{r}_3)\rangle}{\varepsilon_r\varepsilon_o} \tag{5.3.5}$$

According to results of Chap. 4, Sect. 4.2 and Eq. (5.3.1b), $\nabla^2\langle\Phi(\mathbf{r}|\mathbf{r}_1, \mathbf{r}_2, \mathbf{r}_3)\rangle$, is given by the configuration integral,

$$\begin{aligned}\nabla^2\langle\Phi(\mathbf{r}|\mathbf{r}_1, \mathbf{r}_2, \mathbf{r}_3)\rangle &= \int d^3r_1 \int d^3r_2 \int d^3r_3 \nabla^2\Phi(\mathbf{r}|\mathbf{r}_1, \mathbf{r}_2, \mathbf{r}_3) P(\mathbf{r}_1, \mathbf{r}_2, \mathbf{r}_2)\\ &= \frac{1}{Z_3}\int d^3r_1 \int d^3r_2 \int d^3r_3 \nabla^2\Phi(\mathbf{r}|\mathbf{r}_1, \mathbf{r}_2, \mathbf{r}_3)\\ &\quad \times \exp(-\beta U_3(\mathbf{r}_1, \mathbf{r}_2, \mathbf{r}_3))\end{aligned} \tag{5.3.6}$$

where the integration is over $\mathbf{r}_1$, $\mathbf{r}_2$, and $\mathbf{r}_3$ with $\mathbf{r}$ serving as a parameter.

Since we are treating q_1 as a point charge, its contribution to the macroscopic charge density can be represented as the product, $q_1\delta(\mathbf{r} - \mathbf{r}_1)$, where $\delta(\mathbf{r} - \mathbf{r}_1)$ is the Dirac delta function. See the Mathematical Methods Appendix for the properties of the delta function. If each of the three charges are represented in this way, the volume density of charge at **r** is given by

$$\rho(\mathbf{r}|\mathbf{r}_1, \mathbf{r}_2, \mathbf{r}_3) = q_1\delta(\mathbf{r}-\mathbf{r}_1) + q_2\delta(\mathbf{r}-\mathbf{r}_2) + q_3\delta(\mathbf{r}-\mathbf{r}_3) \tag{5.3.7}$$

Recognizing that the thermal average, $\langle\rho(\mathbf{r}|\mathbf{r}_1, \mathbf{r}_2, \mathbf{r}_3)\rangle$, involves integration over the three variables $\mathbf{r}_1$, $\mathbf{r}_2$, and $\mathbf{r}_3$, which will disappear from the final result, we write $\rho(\mathbf{r}) \equiv \langle\rho(\mathbf{r}|\mathbf{r}_1, \mathbf{r}_2, \mathbf{r}_3)\rangle$. In analogy with Eq. (5.3.6), $\langle\rho(\mathbf{r}|\mathbf{r}_1, \mathbf{r}_2, \mathbf{r}_3)\rangle$ on the right hand side of Eq. (5.3.5) reads,

$$\rho(\mathbf{r}) = \frac{\int d^3r_1 \int d^3r_2 \int d^3r_3\, \rho(\mathbf{r}|\mathbf{r}_1, \mathbf{r}_2, \mathbf{r}_3)\exp(-\beta U_3(\mathbf{r}_1, \mathbf{r}_2, \mathbf{r}_3))}{Z_3} \tag{5.3.8}$$

Upon substitution of Eq. (5.3.7) into Eq. (5.3.8), the result is

$$\rho(\mathbf{r}) = \frac{1}{Z_3}\left\{\int d^3r_1 \int d^3r_2 \int d^3r_3\big[q_1\delta(\mathbf{r}-\mathbf{r}_1) + q_2\delta(\mathbf{r}-\mathbf{r}_2) + q_3\delta(\mathbf{r}-\mathbf{r}_3)\big]\right.$$
$$\left.\times\exp[-\beta U_3(\mathbf{r}_1, \mathbf{r}_2, \mathbf{r}_3)]\right\} \tag{5.3.9}$$

As an example of the contribution of a delta function to $\rho(\mathbf{r})$ in Eq. (5.3.9), we evaluate explicitly the integral involving $q_1\delta(\mathbf{r}-\mathbf{r}_1)$. In computing the indicated integral, we will exploit the following properties of the delta function,

$$\delta(\mathbf{r}-\mathbf{r}_1) = \delta(\mathbf{r}_1-\mathbf{r}) \tag{5.3.10a}$$

and

$$\int d^3r_1\, \delta(\mathbf{r}_1-\mathbf{r}) f(\mathbf{r}_1) = f(\mathbf{r}) \tag{5.3.10b}$$

With the aid of Eqs. (5.3.10a) and (5.3.10b), we'll define the first integral of Eq. (5.3.9) as I_1, which can be evaluated as follows:

$$\begin{aligned} I_1 &= \frac{1}{Z_3}\int d^3r_1 \int d^3r_2 \int d^3r_3\, q_1\delta(\mathbf{r}-\mathbf{r}_1)\exp(-\beta U_3(\mathbf{r}_1, \mathbf{r}_2, \mathbf{r}_3)) \\ &= \frac{1}{Z_3}\int d^3r_2 \int d^3r_3 \int d^3r_1\, q_1\delta(\mathbf{r}_1-\mathbf{r})\exp(-\beta U_3(\mathbf{r}_1, \mathbf{r}_2, \mathbf{r}_3)) \\ &= \frac{q_1}{Z_3}\int d^3r_2 \int d^3r_3 \exp(-\beta U_3(\mathbf{r}, \mathbf{r}_2, \mathbf{r}_3)) \end{aligned} \tag{5.3.11a}$$

If we multiply and divide Eq. (5.3.11a) by the volume, V, and refer to Eq. (5.3.3), we have

$$I_1 = \left(\frac{q_1}{V}\right)\frac{V}{Z_3}\int d^3r_2 \int d^3r_3 \exp(-\beta U_3(\mathbf{r}, \mathbf{r}_2, \mathbf{r}_3)) = \left(\frac{q_1}{V}\right)\exp(-\beta W_1(\mathbf{r})) \tag{5.3.11b}$$

Using Eq. (5.3.11b) as a template, the sum of the three integrals in Eq. (5.3.9) can be written as follows:

$$\begin{aligned}\rho(\mathbf{r}) = &\left(\frac{q_1}{V}\right)\frac{V}{Z_3}\int d^3r_2 \int d^3r_3 \exp(-\beta U_3(\mathbf{r}, \mathbf{r}_2, \mathbf{r}_3)) \\ &+ \left(\frac{q_2}{V}\right)\frac{V}{Z_3}\int d^3r_1 \int d^3r_3 \exp(-\beta U_3(\mathbf{r}_1, \mathbf{r}, \mathbf{r}_3)) \\ &+ \left(\frac{q_3}{V}\right)\frac{V}{Z_3}\int d^3r_1 \int d^3r_2 \exp(-\beta U_3(\mathbf{r}_1, \mathbf{r}_2, \mathbf{r}))\end{aligned} \tag{5.3.12}$$

Again upon reference to Eq. (5.3.3), Eq. (5.3.12) becomes

$$\rho(\mathbf{r}) = \left(\frac{q_1}{V}\right)\exp(-\beta W_1(\mathbf{r})) + \left(\frac{q_2}{V}\right)\exp(-\beta W_2(\mathbf{r})) + \left(\frac{q_3}{V}\right)\exp(-\beta W_3(\mathbf{r})) \tag{5.3.13}$$

As the variables $\mathbf{r}_1$, $\mathbf{r}_2$, and $\mathbf{r}_3$ have disappeared upon integration, we write $\Phi(\mathbf{r}) \equiv \langle\Phi(\mathbf{r}|\mathbf{r}_1, \mathbf{r}_2, \mathbf{r}_3)\rangle$. After introducing this notation in Eq. (5.3.5), and Eq. (5.3.13) for the charge density, we obtain

$$\begin{aligned}\nabla^2\Phi(\mathbf{r}) = -\frac{1}{\varepsilon_r\varepsilon_o}\Big[&\left(\frac{q_1}{V}\right)\exp(-\beta W_1(\mathbf{r})) + \left(\frac{q_2}{V}\right)\exp(-\beta W_2(\mathbf{r})) \\ &+ \left(\frac{q_3}{V}\right)\exp(-\beta W_3(\mathbf{r}))\Big]\end{aligned} \tag{5.3.14}$$

Equation (5.3.14) is known as the Poisson–Boltzmann equation.

5.4 Poisson–Boltzmann Equation and the Debye–Hückel Approximation

In the case of an electrolyte solution containing N' ions, the charge density in Eq. (5.3.13) can be generalized to read,

$$\rho(\mathbf{r}) = \frac{1}{V}\sum_{j=1}^{N'} q_j \exp\big(-\beta W_j(\mathbf{r})\big) \quad j = 1, 2, 3, \ldots, N' \tag{5.4.1}$$

On the basis of Eq. (5.4.1), Eq. (5.3.14) then becomes,

$$\nabla^2\Phi(\mathbf{r}) = -\frac{1}{\varepsilon_r\varepsilon_o V}\sum_{j=1}^{N'} q_j \exp\big(-\beta W_j(\mathbf{r})\big) \tag{5.4.2}$$

The general solution to Eq. (5.4.2) is not known. To evaluate the sum and achieve closure, P. J. W. Debye and Erich Hückel (German, 1896–1980) suggested that the potential of mean force, $W_j(\mathbf{r})$, could be represented in terms of the local electrostatic potential, $\Phi(\mathbf{r})$, by the equations [1, 2],

$$W_j(\mathbf{r}) = q_j\Phi(\mathbf{r}) \quad j = 1, 2, \ldots, N' \tag{5.4.3}$$

Equation (5.4.3) is said to be a *closure* approximation. Upon substitution of Eq. (5.4.3) into Eq. (5.4.2), the result is the self-consistent Poisson–Boltzmann equation,

$$\nabla^2\Phi(\mathbf{r}) = -\frac{1}{\varepsilon_r\varepsilon_o V}\sum_{j=1}^{N'} q_j \exp\big(-\beta q_j(\mathbf{r})\Phi(\mathbf{r})\big) \tag{5.4.4}$$

5.5 Solution of the Poisson–Boltzmann Equation in Rectangular Coordinates

5.5.1 Statement of the Boundary Value Problem

Consider a charged flat metal plate of infinite extent suspended in an electrolyte solution [3]. The ions in the solution that have a charge opposite in sign to that of the plate will be attracted to the plate, while the ions having a charge with the same sign as the plate will be repelled. Hence, in the immediate vicinity of the plate, there will be a layer of solution rich in counter ions. The charge distribution will resume uniformity only at distances far enough from the plate that the influence of the plate on the electrolyte disappears. The spatial dependence of the charge distribution can be obtained by rendering Eq. (5.4.4) in rectangular coordinates.

Begin with the Poisson equation written in the form,

$$\nabla^2\Phi(\mathbf{r}) = -\frac{\rho(\mathbf{r})}{\varepsilon_r\varepsilon_o} \tag{5.5.1}$$

In rectangular coordinates, the left hand side of Eq. (5.5.1) is

$$\nabla^2\Phi(x,y,z) = \frac{\partial^2\Phi(x,y,z)}{\partial x^2} + \frac{\partial^2\Phi(x,y,z)}{\partial y^2} + \frac{\partial^2\Phi(x,y,z)}{\partial z^2} \tag{5.5.2}$$

(see the Mathematical Methods Appendix). If the x-axis is perpendicular to the plate and the plate is assumed to be infinite in extent in the other two directions, Eq. (5.5.2) simplifies to

$$\nabla^2\Phi(x, y, z) = \frac{\partial^2\Phi(x, y, z)}{\partial x^2} \tag{5.5.3}$$

In this case, Eq. (5.5.1) becomes the ordinary differential equation,

$$\frac{d^2\Phi(x)}{dx^2} = -\frac{\rho(x)}{\varepsilon_r\varepsilon_o} \tag{5.5.4}$$

For an electrolyte made up of two different ions having charges, $q_1 = Z_1 e$ and $q_2 = Z_2 e$, respectively, the thermally averaged charge density in Eq. (5.3.13), $\rho(x)$, assumes the form,

$$\rho(x) = \frac{1}{V}\left[N_1 Z_1 e \exp(-\beta Z_1 e\Phi(x)) + N_2 Z_2 e \exp(-\beta Z_2 e\Phi(x))\right] \tag{5.5.5}$$

where N_1 is the number of ions with electrovalence, Z_1, and N_2 is the number of ions with electrovalence, Z_2. When Eq. (5.5.5) is substituted into the right hand side of Eq. (5.5.4), the result is the one-dimensional Poisson-Boltzmann equation,

$$\frac{d^2\Phi(x)}{dx^2} = -\frac{1}{\varepsilon_r\varepsilon_o V}\left[N_1 Z_1 e \exp(-\beta Z_1 e\Phi(x)) + N_2 Z_2 e \exp(-\beta Z_2 e\Phi(x))\right] \tag{5.5.6}$$

Equation (5.5.6) can be solved exactly, if we assume that

$$Z_1 = Z > 0 \tag{5.5.7a}$$

$$Z_2 = -Z \tag{5.5.7b}$$

Some chemical terminology can be adopted at this point. An alkali metal chloride electrolyte such as NaCl, where $Z_1 = +1$ and $Z_2 = -1$, is termed a 1:1 electrolyte. An alkaline earth chloride, such as $CaCl_2$, where $Z_1 = +2$ and $Z_2 = -1$, is termed a 2:1 electrolyte. An alkali sulfate electrolyte, such as K_2SO_4, is termed a 1:2 electrolyte, etc. Equations (5.5.7) describe a $Z : Z$ electrolyte. In the case of a $Z : Z$ electrolyte, electroneutrality requires,

$$N_1 = N_2 \equiv N \tag{5.5.8}$$

Upon substitution of Eqs. (5.5.7a), (5.5.7b), and (5.5.8) into Eq. (5.5.6), the result is

$$\frac{d^2\Phi(x)}{dx^2} = -\frac{NZe}{\varepsilon_r\varepsilon_o V}\left[\exp(-\beta Ze\Phi(x)) - \exp(+\beta Ze\Phi(x))\right] \tag{5.5.9a}$$

With the identification,

$$\sinh(\beta Ze\Phi(x)) = \frac{\exp(\beta Ze\Phi(x)) - \exp(-\beta Ze\Phi(x))}{2} \tag{5.5.9b}$$

Equation (5.5.9a) can be rewritten as

$$\frac{d^2\Phi(x)}{dx^2} = \frac{2NZe}{\varepsilon_r\varepsilon_o V}\sinh(\beta Ze\Phi(x)) \tag{5.5.10}$$

The appropriate boundary conditions can be stated as follows: At the surface of the plate, where $x = 0$, we assign a constant value to the electrostatic potential; hence,

$$\Phi(0) = \Phi_o \tag{5.5.11a}$$

where for convenience, $\Phi_o > 0$. Far from the plate at $x = \infty$, the electrolyte is undisturbed by the presence of the plate. At infinity, the electrostatic potential is zero; hence,

$$\lim_{x\to\infty}\Phi(x) = 0 \tag{5.5.11b}$$

Also, at $x = \infty$, both types of ions are uniformly distributed, which makes the local value of the electric field zero, so

$$\lim_{x\to\infty}E(x) = \lim_{x\to\infty}\left(-\frac{d\Phi(x)}{dx}\right) = 0 \tag{5.5.11c}$$

Equations (5.5.10) and (5.5.11a) through (5.5.11c) complete the statement of a boundary value problem.

5.5.2 Solution to the Boundary Value Problem

The first step in the integration of Eq. (5.5.10) is to multiply both sides by $2(d\Phi(x)/dx)$. The result is

$$2\left(\frac{d\Phi(x)}{dx}\right)\left(\frac{d^2\Phi(x)}{dx^2}\right) = \frac{4NZe}{\varepsilon_r\varepsilon_o V}\sinh(\beta Ze\Phi(x))\left(\frac{d\Phi(x)}{dx}\right) \tag{5.5.12a}$$

Equation (5.5.12a) can then be rewritten as

$$\frac{d}{dx}\left(\frac{d\Phi(x)}{dx}\right)^2 = \frac{4NZe}{\varepsilon_r\varepsilon_o V}\sinh(\beta Ze\Phi(x))\left(\frac{d\Phi(x)}{dx}\right) \tag{5.5.12b}$$

Multiply Eq. (5.5.12b) through by dx. This permits Eq. (5.5.12b) to be written in integral form as

$$\begin{aligned}\int\left[\frac{d}{dx}\left(\frac{d\Phi(x)}{dx}\right)^2\right]dx &= \frac{4NZe}{\varepsilon_r\varepsilon_o V}\int\left[\sinh(\beta Ze\Phi(x))\left(\frac{d\Phi(x)}{dx}\right)\right]dx\\ &= \frac{4NZe}{\varepsilon_r\varepsilon_o V}\int\sinh(\beta Ze\Phi(x))d\Phi(x)\end{aligned} \tag{5.5.13a}$$

Evaluation of the integrals on each side of Eq. (5.5.13a) leads to

$$\left(\frac{d\Phi(x)}{dx}\right)^2 = \left(\frac{4NZe}{\varepsilon_r\varepsilon_o V}\right)\frac{\cosh(\beta Ze\Phi(x))}{\beta Ze} + A \tag{5.5.13b}$$

where A is a constant of integration. After cancellation of common factors,

Equation (5.5.13b) reads

$$\left(\frac{d\Phi(x)}{dx}\right)^2 = \frac{4N}{\varepsilon_r\varepsilon_o V\beta}\cosh(\beta Ze\Phi(x)) + A \tag{5.5.14}$$

Equation (5.5.14) is said to be the first integral of Eq. (5.5.10).

To proceed further, it is worthwhile to introduce the dimensionless potential, y, defined by

$$y = \beta Ze\Phi \tag{5.5.15a}$$

and the dimensionless distance, ξ, defined by

$$\xi = \alpha x \tag{5.5.15b}$$

where the parameter, α, is given by

$$\alpha^2 = \frac{2NZ^2e^2\beta}{\varepsilon_r\varepsilon_o V} \tag{5.5.15c}$$

In these new variables, $d\Phi$ becomes,

$$d\Phi = dy/\beta Ze \tag{5.5.16a}$$

while dx is

$$dx = d\xi/\alpha \tag{5.5.16b}$$

On the basis of Eqs. (5.5.16a) and (5.5.16b),

$$\left(\frac{d\Phi(x)}{dx}\right)^2 = \frac{\alpha^2}{\beta^2 Z^2 e^2}\left(\frac{dy}{d\xi}\right)^2 \tag{5.5.17a}$$

Using Eqs. (5.5.15a) and (5.5.17a), we can rewrite Eq. (5.5.14) in the form

$$\frac{\alpha^2}{\beta^2 Z^2 e^2}\left(\frac{dy}{d\xi}\right)^2 = \frac{4N}{\varepsilon_r \varepsilon_o V \beta}\cosh(y) + A \tag{5.5.17b}$$

The derivative square, $(dy/d\xi)^2$, can be isolated on the left by writing. Equation (5.5.17b) in the form,

$$\left(\frac{dy}{d\xi}\right)^2 = \frac{2}{\alpha^2}\left(\frac{2NZ^2 e^2 \beta}{\varepsilon_r \varepsilon_o V}\right)\cosh(y) + \frac{\beta^2 Z^2 e^2 A}{\alpha^2} \tag{5.5.17c}$$

Define a new constant of integration, A', as

$$A' = \frac{\beta^2 Z^2 e^2 A}{\alpha^2} \tag{5.5.17d}$$

Use Eq. (5.5.15c) and Eq. (5.5.17d) to rewrite Eq. (5.5.17c) as

$$\left(\frac{dy}{d\xi}\right)^2 = 2\cosh(y) + A' \tag{5.5.18}$$

In terms of the variables defined in Eqs. (5.5.15a) through (5.5.15c), the boundary conditions in Eqs. (5.5.11a) through (5.5.11c) can be rewritten as

$$y(0) = \beta Z e \Phi_o \tag{5.5.19a}$$

$$\lim_{\xi\to\infty} y(\xi) = 0 \tag{5.5.19b}$$

$$\lim_{\xi\to\infty}\left(\frac{dy(\xi)}{d\xi}\right) = 0 \tag{5.5.19c}$$

respectively. Equations (5.5.19a) through (5.5.19c) express the boundary conditions in dimensionless variables.

When applied to Eq. (5.5.18), Eqs. (5.5.19b) and (5.519c) combine to produce the result,

$$\lim_{\xi\to\infty}\left(\frac{dy(\xi)}{d\xi}\right)^2 = \lim_{\xi\to\infty}\left[2\cosh(y)+A'\right] = 2\cosh(0)+A' = 2+A' = 0 \tag{5.5.20a}$$

which can be solved for A' to obtain

$$A' = -2 \tag{5.5.20b}$$

When Eq. (5.5.20b) is substituted into Eq. (5.5.18), the result is

$$\left(\frac{dy}{d\xi}\right)^2 = 2\cosh(y) - 2 \tag{5.5.21a}$$

Because the electrostatic potential, $\Phi(x)$, decreases from $\Phi_o > 0$ at $x = 0$ to zero at $x = \infty$, both $(d\Phi(x)/dx)$ and $(dy(\xi)/d\xi)$ are negative; hence, the appropriate root of Eq. (5.5.21a) is

$$\left(\frac{dy}{d\xi}\right) = -\left[2\cosh(y) - 2\right]^{1/2} \tag{5.5.21b}$$

Next use

$$\cosh(y) = \left(\frac{e^y + e^{-y}}{2}\right) \tag{5.5.21c}$$

to simplify Eq. (5.5.21b) as follows:

$$\begin{aligned}\left(\frac{dy}{d\xi}\right) &= -\left[2\left(\frac{e^y+e^{-y}}{2}\right)-2\right]^{1/2} = -\left[e^y - 2 + e^{-y}\right]^{1/2}\\ &= -2\left[\frac{(e^{y/2})^2 - 2 + (e^{-y/2})^2}{4}\right]^{1/2}\\ &= -2\left[\frac{(e^{y/2})^2 - 2(e^{y/2})(e^{-y/2}) + (e^{-y/2})^2}{4}\right]^{1/2}\\ &= -2\left[\left(\frac{e^{y/2}-e^{-y/2}}{2}\right)^2\right]^{1/2}\\ &= -2\frac{\left(e^{y/2}-e^{-y/2}\right)}{2}\end{aligned} \tag{5.5.21d}$$

Now use

$$\sinh(y/2) = \left(\frac{e^{y/2} - e^{-y/2}}{2}\right) \tag{5.5.21e}$$

to write Eq. (5.5.21d) in the form,

$$\left(\frac{dy}{d\xi}\right) = -2\sinh(y/2) \tag{5.5.22a}$$

In Eq. (5.5.22a), the variables are separable. This permits the equation to be written as

$$\frac{d(y/2)}{\sinh(y/2)} = -d\xi \tag{5.5.22b}$$

If we use

$$\operatorname{csch}(y/2) = \frac{1}{\sinh(y/2)} \tag{5.5.22c}$$

then Eq. (5.5.22b) reads

$$\operatorname{csch}(y/2)d(y/2) = -d\xi \tag{5.5.22d}$$

In integral form, Eq. (5.5.22d) reads,

$$-\int d\xi + \ln B' = \int \operatorname{csch}(y/2)(dy/2) \tag{5.5.22d}$$

where $\ln B'$ is a constant of integration. After using the tabulated integral in the Mathematical Appendix,

$$\int \operatorname{csch}(x)dx = \ln[\tanh(x/2)] \tag{5.5.22e}$$

Equation (5.5.22d) reads,

$$-\xi + \ln B' = \ln\left[\tanh(y/4)\right] \tag{5.5.22f}$$

Exponentiating both sides of Eq. (5.5.22f) leads to

$$B'e^{-\xi} = \left[\tanh(y/4)\right] \tag{5.5.23a}$$

To simplify notation, we introduce a new variable, $z = y(0)$. Equation (5.5.19a) can then be written in the form,

$$z = y(0) = \beta Ze\Phi_o \tag{5.5.23b}$$

If we evaluate the left of Eq. (5.5.23a) at $\xi = 0$ and use Eq. (5.5.23b) to evaluate the right in terms of z, the result is

$$B' = [\tanh(z/4)] \tag{5.5.23c}$$

Substitute Eq. (5.5.23c) into Eq. (5.5.23a), to obtain

$$e^{-\xi} = \frac{\tanh(y/4)}{\tanh(z/4)} \tag{5.5.24}$$

Equation (5.5.24) can be transformed to a more useful form by using the definition

$$\tanh(x/4) = \frac{e^{x/4} - e^{-x/4}}{e^{x/4} + e^{-x/4}} = \frac{e^{-x/4}}{e^{-x/4}}\left(\frac{e^{x/2} - 1}{e^{x/2} + 1}\right) = \left(\frac{e^{x/2} - 1}{e^{x/2} + 1}\right) \tag{5.5.25a}$$

Application of Eq. (5.5.25a) to both the numerator and denominator of Eq. (5.5.24) permits the latter equation to be written in the form

$$e^{-\xi} = \left(\frac{e^{y/2} - 1}{e^{y/2} + 1}\right)\left(\frac{e^{z/2} + 1}{e^{z/2} - 1}\right) \tag{5.5.25b}$$

Now define the constant

$$\frac{1}{\gamma} = \left(\frac{e^{z/2} + 1}{e^{z/2} - 1}\right) \tag{5.5.25c}$$

so that Eq. (5.5.25b) reads,

$$e^{-\xi} = \frac{1}{\gamma}\left(\frac{e^{y/2} - 1}{e^{y/2} + 1}\right) \tag{5.5.25d}$$

Equation (5.5.25d) can now be solved for $e^{y/2}$ by cross multiplying and rearranging as

$$\gamma e^{-\xi} e^{y/2} + \gamma e^{-\xi} = e^{y/2} - 1 \tag{5.5.25e}$$

and upon isolating $e^{y/2}$

$$\gamma e^{-\xi} e^{y/2} - e^{y/2} = -1 - \gamma e^{-\xi} \tag{5.5.25f}$$

Equation (5.5.25f), can be solved for $e^{y/2}$ as follows,

$$e^{y/2} = \frac{1 + \gamma e^{-\xi}}{1 - \gamma e^{-\xi}} \tag{5.5.25g}$$

Taking the logarithm of both sides of Eq. (5.5.25 g) yields

$$y = 2\left[\frac{1 + \gamma e^{-\xi}}{1 - \gamma e^{-\xi}}\right] \tag{5.5.26}$$

Equation (5.5.26) is the general solution to the Poisson–Boltzmann equation for a charged flat plate immersed in an electrolyte solution.

With the use of Eqs. (5.5.15a) and (5.5.15b) with $\beta = 1/k_B T$, we can express Eq. (5.5.26) in dimensioned variables. The result is

$$\Phi(x) = 2\left(\frac{k_B T}{Ze}\right)\ln\left(\frac{1 + \gamma e^{-\alpha x}}{1 - \gamma e^{-\alpha x}}\right) \tag{5.5.27}$$

Upon combining Eqs. (5.5.23b) and (5.5.25c),

$$\gamma = \left(\frac{\exp(Ze\Phi_o/2k_B T) - 1}{\exp(Ze\Phi_o/2k_B T) + 1}\right) \tag{5.5.28}$$

Equations (5.5.27) and (5.5.28) constitute the solution to Eq. (5.5.10) written in terms of dimensioned variables.

5.5.3 A Useful Approximate Solution

Sufficiently far away from the plate, Eq. (5.5.27) takes on a particularly simple form [3]. To derive this form, we make use of the series approximation,

$$\ln\left(\frac{1 + \lambda}{1 - \lambda}\right) \simeq 2\lambda + \cdots \tag{5.5.29a}$$

which applies when $\lambda \ll 1$. In the limit that x is large and $\gamma e^{-\alpha x} \ll 1$, we can let $\lambda = \gamma e^{-\alpha x}$ in Eq. (5.5.29a) and use the result to replace the logarithm on the right hand side of Eq. (5.5.27) by

$$\ln\left(\frac{1 + \gamma e^{-\alpha x}}{1 - \gamma e^{-\alpha x}}\right) \simeq 2\gamma e^{-\alpha x} \tag{5.5.29b}$$

Upon substitution of Eq. (5.5.29b) into Eq. (5.5.27), the result is

$$\Phi(x) = 4\gamma\left(\frac{k_B T}{Ze}\right)e^{-\alpha x} \tag{5.5.30}$$

For the purpose of continued discussion, we consider the reciprocal of Eq. (5.5.25c),

$$\gamma = \left(\frac{e^{z/2} - 1}{e^{z/2} + 1}\right) \tag{5.5.31}$$

There are two interesting limiting cases of Eq. (5.5.30), which depend upon the value of z [3].

Case I: We assume $\Phi_o \ll k_B T/Ze$, so that $z \ll 1$. Using $k_B = 1.38 \times 10^{-23}$J/K, we find in the case $Z = 1$ and $T = 300$ K, that these conditions place an upper limit on Φ_o which is

$$\Phi_o \ll \frac{(1.38 \times 10^{-23}\text{J/K})(300\text{ K})(\text{C} \cdot \text{V/J})}{1.6 \times 10^{-19}\text{C}} = 26 \times 10^{-3}\text{ V} = 26\,\text{mV} \quad (5.5.32a)$$

By employing the expansion,

$$e^{\varsigma} = 1 + \varsigma + (\varsigma^2/2!) + \cdots \quad (5.5.32b)$$

we can simplify Eq. (5.5.31) for the case of small z. If we let $\varsigma = z/2$ and retain only the first two terms in Eq. (5.5.32b), then Eq. (5.5.31) simplifies to read,

$$\gamma = \left(\frac{e^{z/2} - 1}{e^{z/2} + 1}\right) \simeq \frac{1 + (z/2) - 1}{1 + (z/2) + 1} \simeq \frac{(z/2)}{2} = \frac{z}{4} = \frac{Ze\Phi_o}{4k_B T} \quad (5.5.32c)$$

Upon substitution of Eq. (5.5.32c), Eq. (5.5.30) becomes

$$\Phi(x) = \Phi_o e^{-\alpha x} \quad (5.5.32d)$$

Equation (5.5.32d) applies under conditions where x is large, and $\Phi_o \ll k_B T/Ze$. In this case, if one observes the plate from the vantage point of large x, the potential on the surface of the plate at $x = 0$ appears to be Φ_o [3].

Case II: We assume $\Phi_o \gg k_B T/Ze$, which corresponds to $z \gg 1$. When $Z = 1$, and $T = 300$ K, K, this requirement is satisfied by $\Phi_o \gg 26$ mV. According to Eq. (5.5.31), in the limit $z \gg 1$,

$$\gamma = \left(\frac{e^{z/2} - 1}{e^{z/2} + 1}\right) \simeq \frac{e^{z/2}}{e^{z/2}} = 1 \quad (5.5.33a)$$

With the use of Eq. (5.5.33a), Eq. (5.5.30) assumes the form

$$\Phi(x) = \left(\frac{4k_B T}{Ze}\right) e^{-\alpha x} \quad (5.5.33b)$$

Equation (5.5.33b) applies in the limit of large x and $\Phi_o \gg k_B T/Ze$. In this case, if one observes the plate from the vantage point of large x, the potential on the surface of the plate at $x = 0$ appears to be $(4k_B T/Ze)$, which is about 100 mV when $Z = 1$ and $T = 300$ K[4].

5.5.4 Charge Density on the Surface of the Plate

As shown in Eq. (3.6.6) in Chap. 3, the surface charge density, σ, at the boundary between two materials is determined by the difference in the normal components of the displacements in the two materials, specifically,

$$D_n(0^+) - D_n(0^-) = \sigma \tag{5.5.34a}$$

If we let $x = 0^-$ lie in the metal plate, then we have $D_n(0^-) = 0$, because the electric field in a perfect conductor is zero. The point $x = 0^+$ then lies in the electrolyte where the electric field is non-zero. Hence, Eq. (5.5.34a) simplifies to read

$$D_n(0^+) = \sigma \tag{5.5.34b}$$

According to Eqs. (5.2.7) and (5.2.9f), the normal component of the displacement is related to the gradient of the potential by

$$D_n(0^+) = \varepsilon_r\varepsilon_o E_n(0^+) = -\,\varepsilon_r\varepsilon_o\left[\frac{\partial\Phi(x)}{\partial x}\right]_{x=0^+} \tag{5.5.34c}$$

Upon substitution of Eq. (5.5.34c) into Eq. (5.5.34b), the result is

$$\sigma = -\,\varepsilon_r\varepsilon_o\left[\frac{\partial\Phi(x)}{\partial x}\right]_{x=0^+} \tag{5.5.34d}$$

In the case of a flat plate in an electrolyte, the derivative on the right hand side of Eq. (5.5.34d) can be evaluated using Eq. (5.5.27):

$$\begin{aligned}\frac{\partial\Phi(x)}{\partial x} &= 2\left(\frac{k_BT}{Ze}\right)\frac{\partial}{\partial x}\left[\ln\left(1+\gamma e^{-\alpha x}\right) - \ln\left(1-\gamma e^{-\alpha x}\right)\right]\\ &= 2\left(\frac{k_BT}{Ze}\right)\left[\frac{-\alpha\gamma e^{-\alpha x}}{(1+\gamma e^{-\alpha x})} - \frac{(-\alpha)(-\gamma)e^{-\alpha x}}{(1-\gamma e^{-\alpha x})}\right]\end{aligned} \tag{5.5.34e}$$

When evaluated at $x = 0$, which corresponds to the surface of the plate, Eq. (5.5.34e) becomes

$$\begin{aligned}\left[\frac{\partial\Phi(x)}{\partial x}\right]_{x=0} &= -\,2\alpha\gamma\left(\frac{k_BT}{Ze}\right)\left[\frac{1}{(1+\gamma)} + \frac{1}{(1-\gamma)}\right]\\ &= -\,4\alpha\left(\frac{k_BT}{Ze}\right)\left[\frac{\gamma}{\left(1-\gamma^2\right)}\right]\end{aligned} \tag{5.5.34f}$$

Upon substitution of Eq. (5.5.34f) into Eq. (5.5.34d), the surface charge density is

$$\sigma = 4\alpha\varepsilon_r\varepsilon_o\left(\frac{k_BT}{Ze}\right)\left[\frac{\gamma}{1-\gamma^2}\right] \tag{5.5.35}$$

In Eq. (5.5.7a), Z was chosen to be positive. Hence, using the definition, $z = \beta Ze\Phi_o$, and referencing Eq. (5.5.31), we can conclude that if $\Phi_o < 0$, then $-1 < \gamma < 0$. Equation (5.5.35) predicts for this case that $\sigma < 0$. On the other hand, when $\Phi_o > 0$, then $0 < \gamma < 1$, and Eq. (5.5.35) predicts that $\sigma > 0$.

5.6 Mutual Repulsion of Like Charged Flat Plates Separated by an Electrolyte

We can apply the results of Sect. 5.5 to the problem posed by the electrostatic energy associated with the interaction of two flat plates. The geometry is shown in Fig. 5.1.

The electrostatic potential raised by two parallel, like charged, flat plates dipped in an electrolyte is denoted by $\Phi^{(2)}(x)$, where superscript (2) specializes the meaning of Φ to a the problem of two plates. The exact solution to the Poisson-Boltzmann equation for $\Phi^{(2)}(x)$ in the case of the arrangement in Fig. 5.1 is not known. Nonetheless,

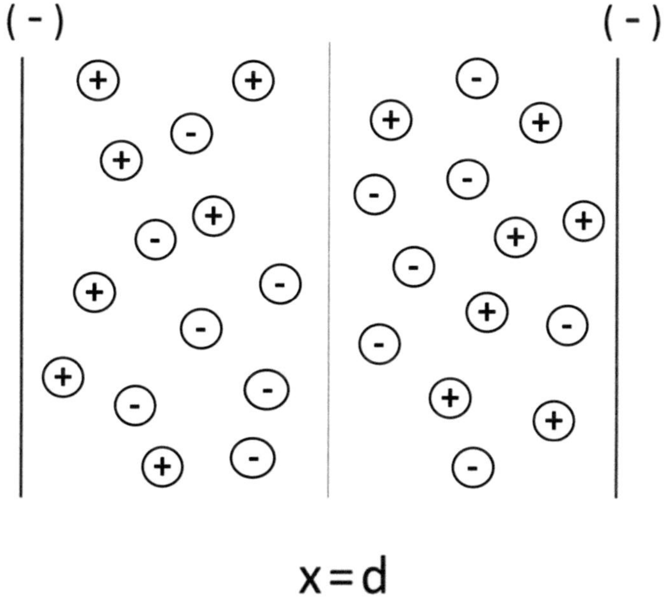

Fig. 5.1 A 1:1 strong electrolyte confined between two negatively charged flat plates. The plates are separated by the distance, $2d$. The center line is located at $x = d$

an approximate solution is feasible. As shown in the figure, the plates are separated by the distance, $2d$. If d is sufficiently large, then at the position, $x = d$, half-way between them, it is possible to approximate $\Phi^{(2)}(x)$ evaluated at $x = d$ by $2\Phi(d)$, where $\Phi(x)$ is the potential produced by a single isolated plate. Letting Eq. (5.5.30) represent $\Phi(x)$, we have for $\Phi^{(2)}(d)$ [3]:

$$\Phi^{(2)}(d) = 8\gamma\left(\frac{k_B T}{Ze}\right)e^{-\alpha d} \tag{5.6.1}$$

The electric field at a point, x, between the plates is given by

$$E(x) = -\frac{\partial \Phi^{(2)}(x)}{\partial x} \tag{5.6.2}$$

If the space between the plates has cross section, A, then the charge in a planar layer of electrolyte having thickness, dx, is given by $\rho(x)Adx$, where $\rho(x)$ is the charge density. The electric field, $E(x)$, produced by the plates exerts an electrostatic force, $\rho(x)Adx \cdot E(x)$, on this charge, which is given by

$$\rho(x)Adx \cdot E(x) = -\rho(x)Adx\left(\frac{\partial \Phi^{(2)}(x)}{\partial x}\right) \tag{5.6.3a}$$

The electrostatic force on the charge in the volume Adx creates a local pressure difference, $\Delta P(x) = P(x + dx) - P(x)$ acting perpendicular to the cross section, A. Known as the plasma pressure, it resists the volume decrease associated with the narrowing of the separation between the plates. The resulting force balance is

$$\Delta P(x)A = [P(x + dx) - P(x)]A = -\rho(x)Adx\left(\frac{\partial \Phi^{(2)}(x)}{\partial x}\right) \tag{5.6.3b}$$

Expand $P(x + dx)$ in a Taylor series. The result is

$$P(x + dx) = P(x) + \left(\frac{\partial P(x)}{\partial x}\right)dx \tag{5.6.3c}$$

Substitute Eq. (5.6.3c) into Eq. (5.6.3b) to obtain

$$\left[P(x) + \left(\frac{\partial P(x)}{\partial x}\right)dx - P(x)\right]A = -\rho(x)Adx\left(\frac{\partial \Phi^{(2)}(x)}{\partial x}\right) \tag{5.6.3d}$$

with the result,

$$\left(\frac{\partial P(x)}{\partial x}\right) = -\rho(x)\left(\frac{\partial \Phi^{(2)}(x)}{\partial x}\right) \tag{5.6.4}$$

At the separation, $x = d$, where the electrostatic potential is $\Phi^{(2)}(d)$, the ambient pressure is $P(d)$. By contrast, when the plates are infinitely far apart, which corresponds to $x = \infty$, the potential is $\Phi^{(2)}(\infty) = 0$, and the pressure is $P(\infty) = 0$. Taking these limits into account, we can multiply both sides of Eq. (5.6.4) by dx and integrate

$$P(d) = \int_{P(\infty)}^{P(d)} dP = \int_{0}^{\Phi^{(2)}(d)} (-\rho(\Phi))d\Phi \tag{5.6.5a}$$

According to Eq. (5.5.5), $\rho(x)$ is a function of $\Phi(x)$ which allows us to write $(-\rho(\Phi))$ as the integrand in Eq. (5.6.5a) with Φ regarded to be the independent variable. Equation (5.5.5) can be rearranged to read,

$$\rho(x) = \frac{1}{V}\left[N_2Z_2e\exp(-\beta Z_2e\Phi(x)) + N_1Z_1e\exp(-\beta Z_1e\Phi(x))\right] \tag{5.6.5b}$$

In the case that $Z_1 = Z$, $Z_2 = -Z$, and $N_1 = N_2 \equiv N$, the charge density in Eq. (5.6.5b) becomes

$$\begin{aligned}\rho(x) &= -\frac{NZe}{V}\left[\exp(\beta Ze\Phi(x)) - \exp(-\beta Ze\Phi(x))\right] \\ &= -\frac{2NZe}{V}\sinh(\beta Ze\Phi(x))\end{aligned} \tag{5.6.5c}$$

Upon substitution of Eq. (5.6.5c), the integral on the far right of Eq. (5.6.5a), becomes

$$\begin{aligned}\int_{0}^{\Phi^{(2)}(d)} (-\rho(\Phi))d\Phi &= 2\left(\frac{NZe}{V}\right)\left[\int_{0}^{\Phi^{(2)}(d)} \sinh(\beta Ze\Phi)d\Phi\right] \\ &= 2\left(\frac{N}{\beta V}\right)\left[\int_{0}^{\Phi^{(2)}(d)} \sinh(\beta Ze\Phi)d(\beta Ze\Phi)\right] \\ &= 2\left(\frac{N}{\beta V}\right)[\cosh(\beta Ze\Phi)]_0^{\Phi^{(2)}(d)} \\ &= 2\left(\frac{N}{\beta V}\right)\left[\cosh(\beta Ze\Phi^{(2)}(d)) - 1\right]\end{aligned} \tag{5.6.5d}$$

Upon substitution of Eq. (5.6.5d) into Eq. (5.6.5a), we obtain for the pressure at $x = d$,

$$P(d) = 2\left(\frac{N}{\beta V}\right)\left[\cosh\left(\beta Ze\Phi^{(2)}(d)\right) - 1\right] \tag{5.6.6}$$

A simple form of Eq. (5.6.6) can be derived in the limit $\beta Ze\Phi^{(2)}(d) \ll 1$. This limit is encountered if the plates are sufficiently far apart. Start by replacing the hyperbolic cosine by its definition. Equation (5.6.6) then reads,

$$\begin{aligned} P(d) &= 2\left(\frac{N}{\beta V}\right)\left[\frac{\exp\left(\beta Ze\Phi^{(2)}(d)\right) + \exp\left(-\beta Ze\Phi^{(2)}(d)\right)}{2} - 1\right] \\ &= 2\left(\frac{N}{\beta V}\right)\left[\frac{\exp\left(\beta Ze\Phi^{(2)}(d)\right) + \exp\left(-\beta Ze\Phi^{(2)}(d)\right) - 2}{2}\right] \end{aligned} \tag{5.6.7a}$$

Let $\varsigma = \beta Ze\Phi^{(2)}(d)$. With this identification, Eq. (5.6.7a) reads

$$P(d) = 2\left(\frac{N}{\beta V}\right)\left[\frac{\exp(\varsigma) + \exp(-\varsigma) - 2}{2}\right] \tag{5.6.7b}$$

Insert the $e^{\varsigma} \approx 1 + \varsigma + \varsigma^2/2$ into Eq. (5.6.7b), so that it reads

$$P(d) \simeq 2\left(\frac{N}{\beta V}\right)\left[\frac{1 + \varsigma + (1/2)\varsigma^2 + 1 + (-\varsigma) + (1/2)(-\varsigma)^2 - 2}{2}\right] = \left(\frac{N}{\beta V}\right)\varsigma^2 \tag{5.6.7c}$$

Upon substituting $\varsigma = \beta Ze\Phi^{(2)}$, Eq. (5.6.7c) reads

$$P(d) \simeq \left(\frac{N}{\beta V}\right)\left(\beta Ze\Phi^{(2)}(d)\right)^2 \tag{5.6.7d}$$

Use Eq. (5.6.1) to replace $\Phi^{(2)}(d)$ in Eq. (5.6.7d). With the identification, $\beta = 1/k_B T$, Eq. (5.6.7d) reads

$$P(d) \simeq \left(\frac{Nk_B T}{V}\right)\left[\frac{Ze(8\gamma)(k_B T/Ze)e^{-\alpha d}}{k_B T}\right]^2 \tag{5.6.7e}$$

Upon simplification of Eq. (5.6.7e), the final result for the pressure at the mid-point between the two plates is

$$P(d) = 64\left(\frac{Nk_B T}{V}\right)\gamma^2 e^{-2\alpha d} \tag{5.6.8}$$

The pressure, $P(x)$, being the force per unit area, the work required to bring one plate from $x = \infty$ to a separation $x = d$ against the opposing plasma pressure is given by W, where

$$W = -\int_{\infty}^{d} P(x)A dx \tag{5.6.9a}$$

and A is the cross-sectional area. In order to evaluate the pressure at position, x, we replace d by x in Eq. (5.6.8). The integral in Eq. (5.6.9a) then takes the form,

$$W = -64A\left(\frac{Nk_BT}{V}\right)\gamma^2\int_{\infty}^{d} e^{-2\alpha x}dx = -64A\left(\frac{Nk_BT}{V}\right)\gamma^2\left[\frac{e^{-2\alpha x}}{-2\alpha}\right]_{\infty}^{d} \tag{5.6.9b}$$

After evaluation of the integrated term in Eq. (5.6.9b) at its limits, the final formula for the work exerted on a plate is

$$W = 32A\left(\frac{Nk_BT}{V\alpha}\right)\gamma^2 e^{-2\alpha d} \tag{5.6.10}$$

The work, W, required to bring two plates from $x = \infty$ to a separation $x = d$ is twice that calculated in Eq. (5.6.10); hence, for a pair of plates,

$$W_{pair} = 64A\left(\frac{Nk_BT}{V\alpha}\right)\gamma^2 e^{-2\alpha d} \tag{5.6.11}$$

Equation (5.6.11) is a measure of the electrical work stored in a pair of flat plates. Assuming platelet shaped colloidal particles, Eq. (5.6.11) can serve as a measure of the electrostatic energy stored in a colloidal suspension. As can be seen in Eq. (5.5.15c), $\alpha^2 = 2(N/V)(Z^2e^2/\varepsilon_r\varepsilon_o k_BT)$. Thus, Eq. (5.6.11) shows that the work depends upon the ion concentration, N/V; the temperature, T; the ion electrovalence, Z; the relative permittivity of the solvent, ε_r; plus the plate cross sectional area, A, and the relative separation, d, respectively. By virtue of Eqs. (5.5.23b) and (5.5.25c), it also depends upon the electrostatic potential, Φ_o, on the surface of the plates. In colloid science, Eq. (5.6.11) serves as an estimate of the energy barrier opposing the agglomeration of charged colloidal particles.

5.7 Solution to the Poisson–Boltzmann Equation in Cylindrical Polar Coordinates

5.7.1 Statement of the Boundary Value Problem

In Chap. 1, we calculated the electric field in the vicinity of an infinitely long linear charge distribution immersed in a vacuum. This model was intended to represent a hypothetical isolated DNA strand. We mentioned that the model was an inadequate representation of a DNA strand in solution. There was no mention of the solvent, nor the counter ions that are necessary to satisfy the charge neutrality principle required of condensed matter. In the model described below, we correct these omissions. For the sake of generality, we will consider the interaction of two parallel charged cylinders immersed in an electrolyte solution. See Fig. 5.2. This is the parallel cylinder analog of the parallel plate model analyzed in Sect. 5.6.

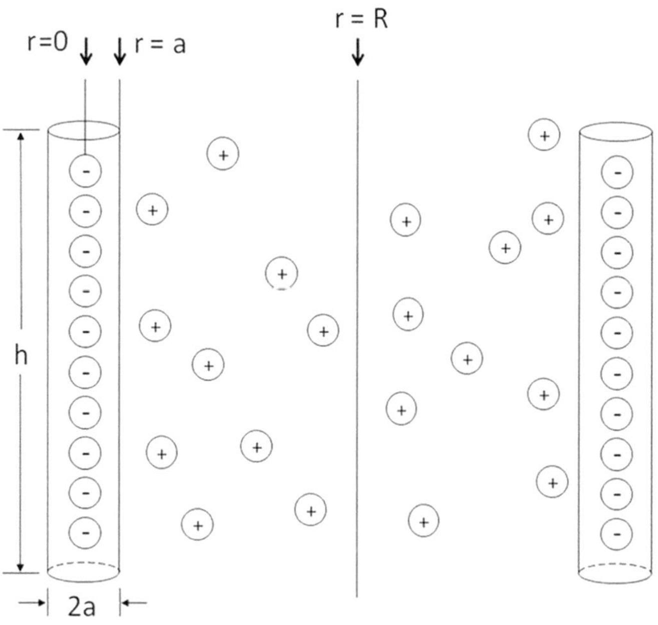

Fig. 5.2 Identical parallel charged rods of length, h and diameter, $2a$, separated by the distance, $2R$. The radial distance, r, is measured in the direction perpendicular to the long axis of the rod on the left. The midpoint between the two rods corresponds to $r = R$. As the rods are supposed to represent linear chain molecules, h is much larger than a. The rods bear fixed negative charges, while the surrounding solution contains positive ions. Given the assumed wide separation between the rods, they neither repel nor attract one another

Our DNA strand will be represented as a long, uniformly charged, right circular cylinder immersed in a solvent, which has relative permittivity ε_r and contains counter ions. Let the total charge on the strand be q_1. Each counter ion carries the charge, q_2. Electrical neutrality is maintained if $q_1 + N_2 q_2 = 0$, where N_2 is the number of counter ions surrounding the strand. The bulk number density of counter ions is $n_2 = N_2/V$, where the volume of the surrounding solvent is V.

To take into account the cylindrical geometry, we will need to convert Eq. (5.4.4) to cylindrical polar coordinates. As a consequence of the infinitely long, cylindrical symmetry, the electrostatic potential in the surrounding medium will be independent of the axial and angular coordinates and will depend solely on the radial distance, r. Thus, when writing out the Laplacian of the electrostatic potential in cylindrical coordinates, only the derivatives with respect to r in $\nabla^2\Phi(r)$ will survive (See the Mathematical Methods Appendix). Under these conditions, the Poisson–Boltzmann equation, Eq. (5.4.4), is [5]:

$$\frac{1}{r}\frac{\partial}{\partial r}\left(r\frac{\partial\Phi(r)}{\partial r}\right) = -\frac{n_2 q_2}{\varepsilon_r\varepsilon_o}\exp(-\beta q_2\Phi(r)) \tag{5.7.1}$$

where we recall that $\beta = 1/k_B T$. Consider for the moment a cylinder of length, h, and radius a. Excluding the circular cross section ends, the cylinder has area is $A = 2\pi ah$. The hypothesis of a uniform charge can be met if the surface charge density on the cylinder is set equal to $q_1/2\pi ah$. In this cylinder model, the distance between the DNA strand and any neighboring DNA strand dissolved in the solution will is set equal to $2R$. See Fig. 5.2. The axes of the two DNA strands are assumed to be parallel. The strand on the left dominates the volume of the cylindrical shell having radius $a \leq r \leq R$. Since we have taken into account charge balance, the net charge within the larger cylindrical volume, $0 \leq r \leq R$, must be zero.

The electrostatic potential, $\Phi(r)$, associated with a cylinder satisfies boundary conditions at $r = a$ and $r = R$. To establish these boundary conditions, we turn to Gauss' law,

$$\nabla\cdot\mathbf{D} = \rho \tag{5.7.2a}$$

where ρ is the density of counter ions in the solution. The counter ions occupy the space $a \leq r \leq R$. The total charge in this space is $N_2 q_2$. As pointed out, charge balance requires, $N_2 q_2 = -q_1$. The volume integral of ρ over the space $a \leq r \leq R$ necessarily equals $-q_1$. According to Eq. (5.7.2a), $-q_1$ satisfies,

$$\int d^3r\,\nabla\cdot\mathbf{D} = \int d^3r\,\rho = -q_1 \tag{5.7.2b}$$

where the volume of integration extends over $a \leq r \leq R$. By use of the Gauss-Green divergence theorem, the volume integral on the left in Eq. (5.7.2b) can be replaced by a surface integral (see Mathematical Methods Appendix), so that

$$\int \mathbf{D} \cdot d\mathbf{A} = -q_1 \tag{5.7.2c}$$

where $d\mathbf{A}$ is a differential of the area on the surface of the inner cylinder of radius, a. According to Eqs. (5.2.7) and (5.2.9f), the electric field and the displacement are related by

$$\mathbf{D} = \varepsilon_r \varepsilon_o \mathbf{E} = -\varepsilon_r \varepsilon_o \nabla \Phi. \tag{5.7.2d}$$

Because the charged cylinders on the left and right are identical, the electric field **E**, and hence the displacement **D**, are both zero at a point midway between the two cylinders. As such, there is no contribution to the surface integral in Eq. (5.7.2c) from the cylindrical surface at $r = R$. Upon substitution of Eq. (5.7.2d) into Eq. (5.7.2c), integration over the cylindrical surface area, $2\pi ah$, proceeds as follows:

$$\begin{aligned} -q_1 &= \int \mathbf{D} \cdot d\mathbf{A} = \varepsilon_r \varepsilon_o \int \mathbf{E} \cdot d\mathbf{A} \cdot = -\varepsilon_r \varepsilon_o \int \nabla \Phi \cdot d\mathbf{A} \\ &= -\varepsilon_r \varepsilon_o 2\pi ah(-\hat{\mathbf{r}} \cdot \nabla \Phi) = 2\pi ah \varepsilon_r \varepsilon_o \left[\frac{\partial \Phi}{\partial r} \right]_{r=a} \end{aligned} \tag{5.7.2e}$$

where $-\hat{\mathbf{r}}$ is the unit vector drawn outward from the solution, and $\partial\Phi/\partial r$ is the radial component of the gradient. We can rewrite Eq. (5.7.2e) in the form,

$$a \left[\frac{\partial \Phi}{\partial r} \right]_{r=a} + \frac{q_1}{2\pi h \varepsilon_r \varepsilon_o} = 0 \tag{5.7.3a}$$

Let r increase in the positive sense, that is into the solution away from our reference cylinder on the left in Fig. 5.2. Along this path, the potential in the solution is decreasing. At $r = R$ it reaches a minimum and then starts to increase again due to the charge on the cylinder at $r = 2R$. By symmetry, at the midpoint, $r = R$, the gradient of the potential is necessarily zero; hence, at $r = R$, there is a second boundary condition given by

$$\left[\frac{\partial \Phi}{\partial r} \right]_{r=R} = 0 \tag{5.7.3b}$$

This is equivalent to noting, as above, that the electric field at $r = R$ is zero. Equations (5.7.1) and (5.7.3) define a boundary value problem.

5.7.2 Solution to the Boundary Value Problem

To integrate Eq. (5.7.1), we begin with the variable substitution,

$$r = e^u \tag{5.7.4a}$$

The differential of Eq. (5.7.4a) is

$$dr = e^u du \tag{5.7.4b}$$

Substitute Eqs. (5.7.4a) and (5.7.4b) into Eq. (5.7.1). The result is

$$\frac{1}{e^u}\frac{\partial}{e^u \partial u}\left[e^u\left(\frac{1}{e^u}\right)\frac{\partial \Phi(u)}{\partial u}\right] = -\frac{n_2 q_2}{\varepsilon_r \varepsilon_o}\exp(-\beta q_2 \Phi(u)) \tag{5.7.4c}$$

Upon cancellation of the factors of e^u within the square bracket, Eq. (5.7.4c) can be rewritten as

$$e^{-2u}\left(\frac{d^2\Phi(u)}{du^2}\right) = -\frac{n_2 q_2}{\varepsilon_r \varepsilon_o}\exp(-\beta q_2 \Phi(u)) \tag{5.7.4d}$$

Equation (5.7.4d) is a non-linear, second order ordinary differential equation. Many such equations are not integrable analytically. Remarkably, Eq. (5.7.4d) is not one of them! Introduce the dependent variable, $v = v(u)$, which is defined by [5]

$$-\beta q_2 \Phi = v - 2u \tag{5.7.5a}$$

The derivative of Eq. (5.7.5a) with respect to u is

$$-\beta q_2 \frac{d\Phi}{du} = \frac{dv}{du} - 2 \tag{5.7.5b}$$

The derivative of Eq. (5.7.5b) with respect to u is then,

$$\frac{d^2\Phi}{du^2} = -\frac{1}{\beta q_2}\frac{d^2 v}{du^2} \tag{5.7.5c}$$

Now exponentiate Eq. (5.7.5a). The result is

$$\exp(-\beta q_2 \Phi) = \exp(v)\cdot\exp(-2u) \tag{5.7.5d}$$

Upon substitution of Eqs. (5.7.5c) and (5.7.5d) into Eq. (5.7.4d), one obtains

$$e^{-2u}\left(\frac{-1}{\beta q_2}\frac{d^2 v}{du^2}\right) = -\frac{n_2 q_2}{\varepsilon_r \varepsilon_o}e^v \cdot e^{-2u} \tag{5.7.6a}$$

which can be rearranged to read

$$\frac{d^2v}{du^2} = \frac{\beta n_2 q_2^2}{\varepsilon_r \varepsilon_o} e^v \tag{5.7.6b}$$

Define

$$\kappa^2 = \frac{\beta n_2 q_2^2}{\varepsilon_r \varepsilon_o} \tag{5.7.6c}$$

and substitute Eq. (5.7.6c) into Eq. (5.7.6b). The result is

$$\frac{d^2v}{du^2} = \kappa^2 e^v \tag{5.7.8}$$

Equation (5.7.8) completes the variable transformation of Eq. (5.7.1) to a form that can be integrated by elementary methods.

Start by multiplying both sides of Eq. (5.7.8) by $2(dv/du)$ to obtain,

$$2\frac{dv}{du}\frac{d^2v}{du^2} = 2\kappa^2 e^v \frac{dv}{du} \tag{5.7.9a}$$

Equation (5.7.9a) can be rewritten in the form,

$$\frac{d}{du}\left(\frac{dv}{du}\right)^2 = 2\kappa^2 \frac{d}{du} e^v \tag{5.7.9b}$$

The first integral of Eq. (5.7.9b) is

$$\left(\frac{dv}{du}\right)^2 = 2\kappa^2 e^v + 4C \tag{5.7.9c}$$

where $4C$ is a constant of integration. Upon factoring out $4C$ from the right of Eq. (5.7.9c), the result is

$$\left(\frac{dv}{du}\right)^2 = 4C\left[\frac{\kappa^2 e^v}{2C} + 1\right] \tag{5.7.9d}$$

Equation (5.7.9d) has two roots, which are

$$\left(\frac{dv}{du}\right) = \pm 2\sqrt{C}\left[\frac{\kappa^2 e^v}{2C} + 1\right]^{1/2} \tag{5.7.9e}$$

respectively. The variables in Eq. (5.7.9e) can be separated with the result,

$$\frac{dv}{\left[\frac{\kappa^2 e^v}{2C} + 1\right]^{1/2}} = \pm 2\sqrt{C}du \qquad (5.7.10)$$

Further progress can be made by defining a new variable, x, where

$$x = \frac{\kappa e^{v/2}}{\sqrt{2C}} \qquad (5.7.11a)$$

According to Eq. (5.7.11a), the differential of this new variable is

$$dx = \frac{\kappa e^{v/2}}{\sqrt{2C}} \cdot \frac{dv}{2} = x\frac{dv}{2} \qquad (5.7.11b)$$

Equation (5.7.11b) can then be rearranged to read,

$$dv = \frac{2}{x}dx \qquad (5.7.11c)$$

Substitution of Eqs. (5.7.11a) and (5.7.11c) into the left hand side of Eq. (5.7.10) transforms it to read,

$$\frac{dx}{x\sqrt{x^2+1}} = \pm\sqrt{C}du \qquad (5.7.11d)$$

The integral of the right hand side of Eq. (5.7.11d) is $\pm\sqrt{C}u$. The integral of the left hand side is [See the Mathematical Appendix],

$$\int \frac{dx}{x\sqrt{x^2+1}} = -\operatorname{csch}^{-1}(x) \qquad (5.7.11e)$$

The integral of Eq. (5.7.11d) can thus be written in the form,

$$-\operatorname{csch}^{-1}(x) = \pm\sqrt{C}u \pm \sqrt{C}\ln A. \qquad (5.7.11f)$$

where $\pm\sqrt{C}\ln A$ is a constant of integration. Upon multiplication of both sides of Eq. (5.7.11f) by (-1) and use of Eq. (5.7.4a) to substitute $\ln r$ for u, Eq. (5.7.11f) reads

$$\operatorname{csch}^{-1}(x) = \mp\sqrt{C}\ln r \mp \sqrt{C}\ln A = \mp\sqrt{C}\ln(Ar) \qquad (5.7.11g)$$

According to Eq. (5.7.11 g)

$$x = \operatorname{csch}\left(\mp\sqrt{C}\ln(Ar)\right) \qquad (5.7.11h)$$

The reciprocal of Eq. (5.7.11 h) is

$$\frac{1}{x} = \sinh\left(\mp\sqrt{C}\ln(Ar)\right) \tag{5.7.11i}$$

Upon substitution of Eq. (5.7.11a) for x, the final result is

$$\frac{\sqrt{2C}}{\kappa}e^{-v/2} = \sinh\left(\mp\sqrt{C}\ln(Ar)\right) \tag{5.7.11j}$$

The argument of the hyperbolic sine on the right hand side of Eq. (5.7.11j) must be consistent with the positive definite nature of $\left(\sqrt{2C}/\kappa\right)e^{-v/2}$ on the left hand side. Because the hyperbolic sine is an odd function of its argument, that is $\sinh(-\omega) = -\sinh(\omega)$, we are obliged to choose $+\sqrt{C}\ln(Ar)$, with the proviso that the argument, Ar, of the logarithm is greater than unity (i.e., $Ar > 1$). With this choice, the natural logarithm of Eq. (5.7.11j) reads,

$$-\frac{1}{2}v + \ln\left(\frac{\sqrt{2C}}{\kappa}\right) = \ln\left[\sinh\left(\sqrt{C}\ln(Ar)\right)\right] \tag{5.7.12a}$$

Equation (5.7.12a) can be rearranged to read,

$$v = -2\ln\left[\left(\frac{\kappa}{\sqrt{2C}}\right)\sinh\left(\sqrt{C}\ln(Ar)\right)\right] \tag{5.7.12b}$$

Substitute Eq. (5.7.4a) into Eq. (5.7.5a) to obtain

$$\Phi = -\frac{1}{\beta q_2}(v - 2u) = -\frac{1}{\beta q_2}(v - 2\ln r) \tag{5.7.12c}$$

The electrostatic potential can be obtained as a function of r by substituting v as given by Eq. (5.7.12b) into the far right of Eq. (5.7.12c). The result is

$$\Phi(r) = -\frac{1}{\beta q_2}\left\{-2\ln\left[\frac{\kappa}{\sqrt{2C}}\sinh\left(\sqrt{C}\ln(Ar)\right)\right] - 2\ln r\right\} \tag{5.7.12c}$$

Equation (5.7.12c) can in turn be rewritten as the difference of two logarithms involving squares, where

$$\Phi(r) = -\frac{1}{\beta q_2}\left\{-\ln\left[\frac{\kappa}{\sqrt{2C}}\sinh\left(\sqrt{C}\ln(Ar)\right)\right]^2 - \ln(r^2)\right\} \tag{5.7.12d}$$

Upon combination of the logarithms in Eq. (5.7.12d), the result is

$$\Phi(r) = \frac{1}{\beta q_2} \ln\left[\frac{\kappa^2 r^2}{2C} \sinh^2\left(\sqrt{C}\ln(Ar)\right)\right] \tag{5.7.13}$$

Equation (5.7.13) is the general solution to the original Poisson–Boltzmann equation defined by Eq. (5.7.1).

Equations satisfied by the constants of integration, C and A in Eq. (5.7.13) can be found by substituting Eq. (5.7.13) into the boundary conditions expressed by Eqs. (5.7.3a) and (5.7.3b). The result is a pair of transcendental equations that must be solved numerically [5]. We shall assume that the solutions to these transcendental equations have been found, and proceed to use Eq. (5.7.13) to calculate the electrostatic contribution to the internal energy of a charged rod and the surrounding electrolyte solution.

5.8 Electrostatic Contribution to the Free Energy

The energy stored in the electrostatic field [5] can be calculated by evaluating Eq. (5.2.10), which we copy below:

$$U_e = \frac{\varepsilon_r \varepsilon_o}{2} \int d^3 r |\nabla\Phi|^2 \tag{5.8.1}$$

In cylindrical polar coordinates, the volume of the integration in Eq. (5.8.1) is defined by the limits,

$$\begin{aligned} 0 &\le z \le h \\ a &\le r \le R \\ 0 &\le \phi \le 2\pi \end{aligned} \tag{5.8.2a}$$

which describe the space occupied by the electrolyte, which surrounds the rod. In cylindrical polar coordinates, the differential volume element is

$$d^3 r = r dr d\phi dz \tag{5.8.2b}$$

(see the Mathematical Methods Appendix). Recalling that the cylindrical symmetry restricts $\Phi(r)$ to being a function of r alone, we note that $\nabla\Phi(r) = \hat{\mathbf{r}}(\partial\Phi(r)/\partial r)$, so that $|\nabla\Phi|^2 = (\hat{\mathbf{r}}\cdot\hat{\mathbf{r}})(\partial\Phi(r)/\partial r)^2 = (\partial\Phi(r)/\partial r)^2$. Upon incorporating this result plus Eqs. (5.8.2a) and (5.8.2b), Eq. (5.8.1) reads,

$$U_e = \frac{\varepsilon_r \varepsilon_o}{2} \int_0^h dz \int_0^{2\pi} d\phi \int_a^R dr\, r \left(\frac{\partial\Phi(r)}{\partial r}\right)^2 \tag{5.8.3}$$

The derivative that appears in Eq. (5.8.3) is evaluated by differentiating Eq. (5.7.13). The first step is to write,

$$\frac{\partial\Phi(r)}{\partial r} = \frac{1}{\beta q_2}\frac{\partial}{\partial r}\ln\left[\frac{\kappa^2 r^2}{2C}\sinh^2\left(\sqrt{C}\ln(Ar)\right)\right]$$
$$= \frac{1}{\beta q_2}\left\{\frac{\frac{2\kappa^2 r}{2C}\sinh^2\left(\sqrt{C}\ln(Ar)\right) + \frac{\kappa^2 r^2}{2C}\cdot 2\sinh\left(\sqrt{C}\ln(Ar)\right)\cdot\cosh\left(\sqrt{C}\ln(Ar)\right)\cdot\frac{A\sqrt{C}}{Ar}}{\frac{\kappa^2 r^2}{2C}\sinh^2\left(\sqrt{C}\ln(Ar)\right)}\right\} \tag{5.8.4a}$$

After cancellation of common factors, Eq. (5.8.4a) reads

$$\frac{\partial\Phi(r)}{\partial r} = \frac{2}{\beta q_2}\left\{\frac{\sinh^2\left(\sqrt{C}\ln(Ar)\right) + \sqrt{C}\sinh\left(\sqrt{C}\ln(Ar)\right)\cdot\cosh\left(\sqrt{C}\ln(Ar)\right)}{r\sinh^2\left(\sqrt{C}\ln(Ar)\right)}\right\} \tag{5.8.4b}$$

In Eq. (5.8.4b), divide the numerator by the denominator to obtain,

$$\frac{\partial\Phi(r)}{\partial r} = \frac{2}{\beta q_2 r}\left\{1 + \sqrt{C}\coth\left(\sqrt{C}\ln(Ar)\right)\right\} \tag{5.8.5}$$

Now substitute Eq. (5.8.5). into Eq. (5.8.3) to obtain

$$U_e = \frac{\varepsilon_r\varepsilon_o}{2}\left(\frac{2}{\beta q_2}\right)^2\int_0^h dz\int_0^{2\pi} d\phi\int_a^R dr\, r\left(\frac{1}{r^2}\right)\left\{1 + \sqrt{C}\coth\left(\sqrt{C}\ln(Ar)\right)\right\}^2 \tag{5.8.6a}$$

The integrand in Eq. (5.8.6a) is independent of ϕ and z, so the double integral, $\int_0^h dz\int_0^{2\pi} d\phi = 2\pi h$. Equation (5.8.6a) can then be rewritten as

$$U_e = \frac{\varepsilon_r\varepsilon_o}{2}\left(\frac{2}{\beta q_2}\right)^2 2\pi h\int_a^R dr\left(\frac{1}{r}\right)\left\{1 + \sqrt{C}\coth\left(\sqrt{C}\ln(Ar)\right)\right\}^2 \tag{5.8.6b}$$

To perform the remaining integral over the domain of r, let

$$x = \ln(Ar) \tag{5.8.7a}$$

The differential of Eq. (5.8.7a) is

$$dx = d\ln(Ar) = \frac{Adr}{Ar} = \frac{dr}{r} \tag{5.8.7b}$$

With this change in variables, the lower limit on the integral in Eq. (5.8.6b) is replaced by $\ln(Aa)$ while the upper limit is replaced by $\ln(AR)$. Define

$$\theta = \sqrt{C} \tag{5.8.7c}$$

so that the hyperbolic cotangent in the integrand of Eq. (5.8.6b) can be written as

$$\coth\left(\sqrt{C}\ln(Ar)\right) = \coth(\theta x) \tag{5.8.7d}$$

When Eqs. (5.8.7a) through (5.8.7d) are employed, the integral in Eq. (5.8.6b) reads

$$\begin{aligned} U_e &= \frac{\varepsilon_r \varepsilon_o}{2}\left(\frac{2}{\beta q_2}\right)^2 2\pi h \int\limits_{\ln(Aa)}^{\ln(AR)} dx[1 + \theta\coth(\theta x)]^2 \\ &= \frac{\varepsilon_r \varepsilon_o}{2}\left(\frac{2}{\beta q_2}\right)^2 2\pi h \int\limits_{\ln(Aa)}^{\ln(AR)} dx\left[1 + 2\theta\coth(\theta x) + \theta^2\coth^2(\theta x)\right] \end{aligned} \tag{5.8.8}$$

We can rewrite Eq. (5.8.8) in terms of a sum of integrals as

$$U_e = \frac{\varepsilon_r \varepsilon_o}{2}\left(\frac{2}{\beta q_2}\right)^2 2\pi h(I_1 + I_2 + I_3) \tag{5.8.9}$$

I_1, I_2 and I_3, are defined and evaluated as follows:

$$I_1 = \int\limits_{\ln(Aa)}^{\ln(AR)} dx = \ln(AR) - \ln(Aa) = \ln\left(\frac{R}{a}\right) \tag{5.8.10a}$$

$$\begin{aligned} I_2 &= 2\theta \int\limits_{\ln(Aa)}^{\ln(AR)} \coth(\theta x)dx = 2\int\limits_{\ln(Aa)}^{\ln(AR)} \frac{\cosh(\theta x)}{\sinh(\theta x)}d(\theta x) = 2\ln[\sinh(\theta x)]_{\ln(Aa)}^{\ln(AR)} \\ &= 2\ln[\sinh(\theta AR)] - 2\ln[\sinh(\theta Aa)] = 2\ln\left(\frac{\sinh(\theta AR)}{\sinh(\theta Aa)}\right) \end{aligned} \tag{5.8.10b}$$

and finally

$$I_3 = \theta^2 \int\limits_{\ln(Aa)}^{\ln(AR)} \coth^2(\theta x)dx \tag{5.8.10c}$$

To evaluate the integral in Eq. (5.8.10c), compute first

$$\frac{d}{dy}\coth(y) = \frac{d}{dy}\left(\frac{\cosh(y)}{\sinh(y)}\right) = \frac{\sinh^2(y) - \cosh^2(y)}{\sinh^2(y)} = 1 - \coth^2(y) \quad (5.8.11a)$$

Rearrange Eq. (5.8.11a) to read,

$$\coth^2(y) = 1 - \frac{d}{dy}\coth(y) \quad (5.8.11b)$$

Let $y = \theta x$, and substitute Eq. (5.8.11b) into Eq. (5.8.10c) to obtain,

$$\begin{aligned} I_3 &= \theta \int_{\ln(Aa)}^{\ln(AR)} \coth^2(\theta x) d(\theta x) = \theta \int_{\ln(Aa)}^{\ln(AR)} d(\theta x)\left[1 - \frac{d\coth(\theta x)}{d(\theta x)}\right] \\ &= \theta[\theta x]_{\ln(Aa)}^{\ln(AR)} - \theta[\coth(\theta x)]_{\ln(Aa)}^{\ln(AR)} \end{aligned} \quad (5.8.12a)$$

The final result for I_3 is

$$I_3 = \theta^2 \ln\left(\frac{R}{a}\right) - \theta\coth(\theta\ln(AR)) + \theta\coth(\theta\ln(Aa)) \quad (5.8.12b)$$

Upon substitution of Eqs. (5.8.10a), (5.8.10b), and (5.8.12b) into Eq. (5.8.9), we obtain

$$\begin{aligned} U_e &= \pi h\varepsilon_r\varepsilon_o\left(\frac{2k_BT}{q_2}\right)^2 \\ &\cdot\left[\ln\left(\frac{R}{a}\right) + 2\ln\left(\frac{\sinh(\theta AR)}{\sinh(\theta Aa)}\right)\right. \\ &\left. + \theta^2\ln\left(\frac{R}{a}\right) - \theta\coth(\theta\ln(AR)) + \theta\coth(\theta\ln(Aa))\right] \end{aligned} \quad (5.8.13)$$

where we used $\beta = 1/k_BT$. Equation (5.8.7c) can be used to re-express Eq. (5.8.13) in terms of the original constant of integration, C. The result is

$$\begin{aligned} U_e &= \pi h\varepsilon_r\varepsilon_o\left(\frac{2k_BT}{q_2}\right)^2 \\ &\cdot\left[(1 + C)\ln\left(\frac{R}{a}\right) + \ln\left[\left(\frac{\sinh(\sqrt{C}AR)}{\sinh(\sqrt{C}Aa)}\right)^2\right]\right. \\ &\left. - \sqrt{C}\coth\left(\sqrt{C}\ln(AR)\right) + \sqrt{C}\coth\left(\sqrt{C}\ln(Aa)\right)\right] \end{aligned} \quad (5.8.14)$$

Equation (5.8.14) represents the energy change in passing from the unionized state where the macromolecule is neutral to the ionized state where the macro-ion has charge, q_1, and each counter ion has charge, q_2. If the process of charging is carried out reversibly at constant pressure and temperature, the electrostatic contribution to the internal energy, U_e, can be identified as the Gibbs energy, ΔG_{ion}, of charging. The enthalpy of charging ΔH_{ion} is then given by the Gibbs–Helmholtz equation,

$$\Delta H_{ion} = -T^2\left[\frac{\partial(\Delta G_{ion}/T)}{\partial T}\right]_P = -T^2\left[\frac{\partial(U_e/T)}{\partial T}\right]_P \tag{5.8.15}$$

Evaluation of the temperature derivative in Eq. (5.8.15) can be expected to be elaborate, because ε_r is a function of temperature, and both C and A are functions of ε_r.

References

1. S.A. Rice, M. Nagasawa, *Polyelectrolyte Solutions* (Academic Press, New York, 1961), pp. 99–119
2. D.A. McQuarrie, *Statistical Mechanics* (Harper and Row, New York, 1973), pp. 326–339
3. D.J. Shaw, *Introduction to Colloid and Surface Chemistry*, 2nd edn. (Butterworths, London, 1970). (Chapter 7)
4. D.H. Everett, *Basic Principles of Colloid Science* (Royal Society of Chemistry, London, UK, 1988), p. 131
5. R.M. Fuoss, A. Katchalsky, S. Lifson, The potential of an infinite rod-like molecule and the distribution of counter ions. Proc. Nat. Acad. Sci. **37**, 579 (1951)

Recommended Additional Reading

1. Y. Levin, Electrostatic correlations. Rep. Prog. Phys. **65**, 1577–1632 (2002)

Chapter 6
Debye-Hückel Theory of Strong Electrolytes

In this chapter, the Poisson–Boltzmann equation is solved in spherical polar coordinates using what is known as the Debye–Hückel approximation. The goal is to calculate the ion activity coefficients, whose concentration dependence lies behind the non-ideal behavior of solutions of electrolytes. Formulae for the ion activity coefficients are given in the both the Debye–Hückel limiting form, which applies rigorously in the limit of infinitely dilute solution, and also in a practical form which applies in the case of moderately concentrated solutions. The Debye–Hückel theory of activity coefficients is applicable to both equilibrium phenomena and dynamic phenomena. Examples of equilibrium phenomena include the solubility of an ionic solid in water and the analysis of the effect of a strong electrolyte on the ionization of a weak acid. In the case of dynamic phenomena, the theory can be used to calculate: (1) the effect of the ion concentration on the rate coefficient for an ionic reaction, and (2) the effect of the ionization of a weak electrolyte on its conductivity.

6.1 Introduction

Sodium chloride is completely ionized in aqueous solution. As such, it is said to be a strong electrolyte. By contrast, acetic acid is only partially ionized and is said to be a weak electrolyte. Weak and strong electrolytes respond differently upon dilution. In the case of strong electrolytes, dilution does not increase the number of charge carriers, whereas in the case of weak electrolytes, the degree of dissociation into ions increases strongly upon dilution. The increase in ionization of weak electrolytes upon dilution is known as Ostwald's dilution law (F. W. Ostwald, German, 1853–1932). In the case of both weak and strong electrolytes, the interionic interactions between the ions can be taken into account by using the Debye–Hückel theory.

J. K. Baird, *Electromagnetism for Chemistry*,
https://doi.org/10.1007/978-3-031-88508-2_6

Prior to the Debye–Hückel theory, there were many attempts to understand the properties of electrolytes [1, 2]. As proposed by Svante Arrhenius (Swedish 1859–1927), the interaction between ions in an aqueous solution was electrostatic. This was emphasized by the other founders of physical chemistry, namely J. H. van't Hoff (Dutch 1852–1911), W. Nernst (German 1864–1941), and W. Ostwald. Although it was generally agreed that the distance between the ions played a role, the structure of an electrolyte solution was thought to be that of an expanded crystal with solvent serving as a dielectric filling the space between the ions. In this expanded crystal model, if there are C ions distributed uniformly per unit volume in the solution, then each ion can be regarded as commanding a sphere of influence of radius, R, where $(4\pi R^3/3)C = 1$. According to Coulomb's law, the inter-ionic electrostatic energy acting between the ions should vary with distance in proportion to R^{-1}. By virtue of the sphere of influence theory, R^{-1} should then be proportional to $C^{1/3}$. Thus, any average property of an electrolyte solution could be expected to be a function of $C^{1/3}$. Experiment, on the other hand, showed that the equilibrium properties of strong electrolytes exhibited a $C^{1/2}$ dependence [2].

According to Debye, the Coulomb field of an ion draws in toward itself a current of ions of the opposite sign, called "counter ions". The inward flowing current causes the local concentration of counter ions in the vicinity of the central ion to exceed the counter ion concentration in the bulk of the solution. The result is a local gradient in the chemical potential. In order to return the chemical potential to the uniform value characteristic of the bulk, the counter ions diffuse away from the central ion. At steady state, the inward and outward ion fluxes balance. This serves to establish an equilibrium ion distribution as specified by a Boltzmann factor. This equilibrium distribution is known as the Debye interionic atmosphere. We have already encountered this concept in Chap. 5 in the case of charged flat plates and cylindrical macro-ions in contact with electrolytes.

6.2 Thermal Average Electrostatic Potential of a Charge of Finite Size

Consider three ions represented by the point charges, q_1, q_2, and q_3 located at positions in the solution specified by the vectors, $\mathbf{r}_1$, $\mathbf{r}_2$, and $\mathbf{r}_3$, respectively. The electrostatic potential energy of this system of ions is

$$U_3(\mathbf{r}_1, \mathbf{r}_2, \mathbf{r}_3) = \frac{q_1 q_2}{4\pi\varepsilon_r\varepsilon_o|\mathbf{r}_1 - \mathbf{r}_2|} + \frac{q_1 q_3}{4\pi\varepsilon_r\varepsilon_o|\mathbf{r}_1 - \mathbf{r}_3|} + \frac{q_2 q_3}{4\pi\varepsilon_r\varepsilon_o|\mathbf{r}_2 - \mathbf{r}_3|} \tag{6.2.1}$$

By virtue of Eq. (2.9.33a) of Chap. 2, the configuration integral is

$$Z_3 = \int d^3r_1 \int d^3r_2 \int d^3r_3 \exp(-\beta U_3(\mathbf{r}_1, \mathbf{r}_2, \mathbf{r}_3)) \tag{6.2.2}$$

where, as before, $\beta = 1/k_B T$, k_B is Boltzmann's constant, and T is the Kelvin temperature. If the ions are dissolved in a volume, V, then according to Eq. (2.9.35) of Chap. 2, the Helmholtz free energy, A, evaluated for the three charges is

$$A = A^{id} - k_B T \ln\left(\frac{Z_3}{V^3}\right) \tag{6.2.3}$$

where A^{id} is the Helmholtz energy in the case of an ideal solution which is assumed to be formed by the ions in the absence of their charges. With reference to Eqs. (6.2.1) and (6.2.2), the ideal state is described by $U_3(\mathbf{r}_1, \mathbf{r}_2, \mathbf{r}_3) \equiv 0$, for which $Z_3 = V^3$.

On the basis of Eq. (6.2.3), the free energy per unit charge on ion "1" is

$$\psi_1 = \frac{\partial A}{\partial q_1} \tag{6.2.4a}$$

Substitute Eq. (6.2.3) into Eq. (6.2.4a). Because A^{id} and V are independent of charge, the result is

$$\psi_1 = -k_B T \frac{\partial \ln Z_3}{\partial q_1} = -\frac{1}{\beta}\left(\frac{1}{Z_3}\right)\frac{\partial Z_3}{\partial q_1} \tag{6.2.4b}$$

Explicit differentiation of Eq. (6.2.2) has the result,

$$\frac{\partial Z_3}{\partial q_1} = \int d^3 r_1 \int d^3 r_2 \int d^3 r_3 \left(-\beta \frac{\partial U_3}{\partial q_1}\right) \exp(-\beta U_3(\mathbf{r}_1, \mathbf{r}_2, \mathbf{r}_3)) \tag{6.2.5a}$$

Substitute Eq. (6.2.5a) into the far right hand side of (6.2.4b). The result is

$$\psi_1 = \frac{\partial A}{\partial q_1} = -\frac{1}{\beta}\left(\frac{1}{Z_3}\right) \int d^3 r_1 \int d^3 r_2 \int d^3 r_3 \left(-\beta \frac{\partial U_3}{\partial q_1}\right) \exp(-\beta U_3(\mathbf{r}_1, \mathbf{r}_2, \mathbf{r}_3)) \tag{6.2.5b}$$

On the basis of Eq. (6.2.1), the charge derivative of the potential energy is

$$\frac{\partial U_3}{\partial q_1} = \frac{q_2}{4\pi\varepsilon_r\varepsilon_o|\mathbf{r}_1 - \mathbf{r}_2|} + \frac{q_3}{4\pi\varepsilon_r\varepsilon_o|\mathbf{r}_1 - \mathbf{r}_3|} \tag{6.2.5c}$$

Upon substitution of Eq. (6.2.5c) into Eq. (6.2.5b), the result is

$$\psi_1 = \left(\frac{1}{Z_3}\right) \int d^3 r_1 \int d^3 r_2 \int d^3 r_3 \left[\frac{q_2}{4\pi\varepsilon_r\varepsilon_o|\mathbf{r}_1 - \mathbf{r}_2|} + \frac{q_3}{4\pi\varepsilon_r\varepsilon_o|\mathbf{r}_1 - \mathbf{r}_3|}\right] \times \exp(-\beta U_3(\mathbf{r}_1, \mathbf{r}_2, \mathbf{r}_3)) \tag{6.2.5d}$$

The local electrostatic potential, $\Phi_{loc}(\mathbf{r}|\mathbf{r}_1, \mathbf{r}_2, \mathbf{r}_3)$, at an arbitrary point in the solution is

$$\Phi_{loc}(\mathbf{r}|\mathbf{r}_1, \mathbf{r}_2, \mathbf{r}_3) = \frac{q_1}{4\pi\varepsilon_r\varepsilon_o|\mathbf{r}-\mathbf{r}_1|} + \frac{q_2}{4\pi\varepsilon_r\varepsilon_o|\mathbf{r}-\mathbf{r}_2|} + \frac{q_3}{4\pi\varepsilon_r\varepsilon_o|\mathbf{r}-\mathbf{r}_3|} \tag{6.2.6a}$$

Following the logical pattern pursued in Sect. 5.3 of Chap. 5, the ensemble average electrostatic potential has the integral definition,

$$\Phi(\mathbf{r}) = \frac{1}{Z_3}\int d^3r_1 \int d^3r_2 \int d^3r_3 \Phi_{loc}(\mathbf{r}|\mathbf{r}_1, \mathbf{r}_2, \mathbf{r}_3)\exp(-\beta U_3(\mathbf{r}_1, \mathbf{r}_2, \mathbf{r}_3)) \tag{6.2.6b}$$

Substitute Eq. (6.2.6a) into Eq. (6.2.6b) to obtain

$$\begin{aligned}\Phi(\mathbf{r}) =& \frac{1}{Z_3}\int d^3r_1 \int d^3r_2 \int d^3r_3 \\ &\times \left[\left(\frac{q_1}{4\pi\varepsilon_r\varepsilon_o|\mathbf{r}-\mathbf{r}_1|} + \frac{q_2}{4\pi\varepsilon_r\varepsilon_o|\mathbf{r}-\mathbf{r}_2|} + \frac{q_3}{4\pi\varepsilon_r\varepsilon_o|\mathbf{r}-\mathbf{r}_3|}\right)\right. \\ &\left.\exp(-\beta U_3(\mathbf{r}_1, \mathbf{r}_2, \mathbf{r}_3))\right]\end{aligned} \tag{6.2.6c}$$

Next, define

$$\begin{aligned}\psi_1(\mathbf{r}|\mathbf{r}_2, \mathbf{r}_3) &= \Phi_{loc}(\mathbf{r}|\mathbf{r}_1, \mathbf{r}_2, \mathbf{r}_3) - \frac{q_1}{4\pi\varepsilon_r\varepsilon_o|\mathbf{r}-\mathbf{r}_1|} \\ &= \frac{q_2}{4\pi\varepsilon_r\varepsilon_o|\mathbf{r}-\mathbf{r}_2|} + \frac{q_3}{4\pi\varepsilon_r\varepsilon_o|\mathbf{r}-\mathbf{r}_3|}\end{aligned} \tag{6.2.7a}$$

which is the electrostatic potential at $\mathbf{r}$ in the absence of q_1. The ensemble average of $\psi_1(\mathbf{r}|\mathbf{r}_2, \mathbf{r}_3)$ is

$$\psi_1(\mathbf{r}) = \frac{1}{Z_3}\int d^3r_1 \int d^3r_2 \int d^3r_3 \psi_1(\mathbf{r}|\mathbf{r}_2, \mathbf{r}_3)\exp(-\beta U_3(\mathbf{r}_1, \mathbf{r}_2, \mathbf{r}_3)) \tag{6.2.7b}$$

Substitute the far right hand side of Eq. (6.2.7a) into Eq. (6.2.7b). The result is

$$\begin{aligned}\psi_1(\mathbf{r}) =& \frac{1}{Z_3}\int d^3r_1 \int d^3r_2 \int d^3r_3 \left[\frac{q_2}{4\pi\varepsilon_r\varepsilon_o|\mathbf{r}-\mathbf{r}_2|} + \frac{q_3}{4\pi\varepsilon_r\varepsilon_o|\mathbf{r}-\mathbf{r}_3|}\right] \\ &\times \exp(-\beta U_3(\mathbf{r}_1, \mathbf{r}_2, \mathbf{r}_3))\end{aligned} \tag{6.2.7c}$$

By comparison of Eq. (6.2.6c) with Eq. (6.2.7c), we can conclude that

$$\begin{aligned}\psi_1(\mathbf{r}) = \Phi(\mathbf{r}) &- \frac{1}{Z_3}\int d^3r_1 \int d^3r_2 \int d^3r_3 \left[\frac{q_1}{4\pi\varepsilon_r\varepsilon_o|\mathbf{r}-\mathbf{r}_1|}\right] \\ &\times \exp(-\beta U_3(\mathbf{r}_1, \mathbf{r}_2, \mathbf{r}_3))\end{aligned} \tag{6.2.8a}$$

Let the ion with charge, q_1, be a sphere of radius, a, with center located at $\mathbf{r}_1$. We propose to evaluate Eq. (6.2.8a) at some point located at $\mathbf{r} = \mathbf{r}_1 + \mathbf{a}$ on the surface of the sphere where the radius vector is $\mathbf{a}$. The required integral is

$$\begin{aligned}\psi_1(\mathbf{r}_1 + \mathbf{a}) &= \Phi(\mathbf{r}_1 + \mathbf{a}) \\ &\quad - \frac{1}{Z_3}\int d^3r_1 \int d^3r_2 \int d^3r_3 \left[\frac{q_1}{4\pi\varepsilon_r\varepsilon_o|(\mathbf{r}_1 + \mathbf{a}) - \mathbf{r}_1|}\right] \\ &\quad \times \exp(-\beta U_3(\mathbf{r}_1, \mathbf{r}_2, \mathbf{r}_3)) \end{aligned} \tag{6.2.8b}$$

which can be rewritten as

$$\begin{aligned}\psi_1(\mathbf{r}_1 + \mathbf{a}) = \Phi(\mathbf{r}_1 + \mathbf{a}) - \frac{1}{Z_3}\int d^3r_1 \int d^3r_2 \int d^3r_3 \left[\frac{q_1}{4\pi\varepsilon_r\varepsilon_o|\mathbf{a}|}\right] \\ \times \exp(-\beta U_3(\mathbf{r}_1, \mathbf{r}_2, \mathbf{r}_3)) \end{aligned} \tag{6.2.8c}$$

Recognizing that $a = |\mathbf{a}|$, Eq. (6.2.8c) can be simplified to read,

$$\psi_1(\mathbf{r}_1 + \mathbf{a}) = \Phi(\mathbf{r}_1 + \mathbf{a}) - \left[\frac{q_1}{4\pi\varepsilon_r\varepsilon_o a}\right]\frac{\int d^3r_1 \int d^3r_2 \int d^3r_3 \exp(-\beta U_3(\mathbf{r}_1, \mathbf{r}_2, \mathbf{r}_3))}{Z_3} \tag{6.2.8d}$$

Use of Eq. (6.2.2) allows Eq. (6.2.8d) to be written in the simple form,

$$\psi_1(\mathbf{r}_1 + \mathbf{a}) = \Phi(\mathbf{r}_1 + \mathbf{a}) - \frac{q_1}{4\pi\varepsilon_r\varepsilon_o a} \tag{6.2.8e}$$

Finally, we assume that q_1 is located at the origin of coordinates, which implies that $\mathbf{r}_1 = 0$. This permits Eq. (6.2.8e) to be written as

$$\psi_1(\mathbf{a}) = \Phi(\mathbf{a}) - \frac{q_1}{4\pi\varepsilon_r\varepsilon_o a} \tag{6.2.9}$$

The results of this section can be summarized as follows: According to Eq. (6.2.6c), $\Phi(\mathbf{r})$ is the thermal average electrostatic potential evaluated at the point, $\mathbf{r}$. By comparison, $\psi_1(\mathbf{r})$, as defined by Eq. (6.2.8a), is the thermal average potential with q_1 removed. Finally $\psi_1(\mathbf{a})$, as in Eq. (6.2.9), is the potential sensed by q_1 due to all of the other ions.

In the following Section, the Poisson–Boltzmann equation will be expressed in terms of spherical polar coordinates. In Sect. 6.4, a boundary value problem appropriate to spherical ions will be introduced, and the Poisson–Boltzmann equation will be solved approximately for $\Phi(\mathbf{r})$. With an analytic form for $\Phi(\mathbf{r})$ then available, it will be shown in Sect. 6.5 how the action of q_1 on itself can be removed from $\Phi(\mathbf{r})$. It will then be possible to give $\psi_1(\mathbf{a})$ a thermodynamic interpretation in terms of the Helmholtz free energy [3, 4].

6.3 Poisson–Boltzmann Equation in Spherical Polar Coordinates

The Poisson–Boltzmann equation in the Debye–Hückel approximation, Eq. (5.4.4), is

$$\nabla^2\Phi(\mathbf{r}) = -\frac{1}{\varepsilon_r\varepsilon_o V}\sum_{j=1}^{N'} q_j \exp\bigl(-\beta q_j\Phi(\mathbf{r})\bigr) \tag{6.3.1}$$

where N' is the total number of ions of all kinds in the electrolyte solution. In the case of an ion atmosphere which is spherically symmetric, $\Phi(\mathbf{r})$ is independent of the polar and azimuthal angles, θ, and ϕ, respectively. Hence, only the radial part of ∇^2 survives. This permits Eq. (6.3.1) to be replaced by the ordinary differential equation (see the Mathematical Methods Appendix),

$$\frac{1}{r^2}\frac{\partial}{\partial r}\left(r^2\frac{\partial\Phi(r)}{\partial r}\right) = -\frac{1}{\varepsilon_r\varepsilon_o V}\sum_{j=1}^{N'} q_j \exp\bigl(-\beta q_j\Phi(r)\bigr) \tag{6.3.2}$$

Unfortunately, Eq. (6.3.2) has no known exact analytical solution in closed form. If it is assumed that $\beta q_j\Phi(r) \ll 1$, the series expansion of the exponential in Eq. (6.3.2) can be exploited. Upon retaining terms up to first order in $\Phi(r)$, the result is

$$\exp\bigl(-\beta q_j\Phi(r)\bigr) \approx 1 - \beta q_j\Phi(r) \tag{6.3.3a}$$

After Eq. (6.3.3a) has been substituted, Eq. (6.3.2) reads,

$$\frac{1}{r^2}\frac{\partial}{\partial r}\left(r^2\frac{\partial\Phi(r)}{\partial r}\right) = -\frac{1}{\varepsilon_r\varepsilon_o V}\sum_{j=1}^{N'} q_j + \frac{1}{\varepsilon_r\varepsilon_o V}\sum_{j=1}^{N'} q_j^2\beta\,\Phi(r) \tag{6.3.3b}$$

Because the solution is overall electrically neutral,

$$\sum_{j=1}^{N'} q_j = 0 \tag{6.3.3c}$$

After Eq. (6.3.3c) is substituted into Eq. (6.3.3b), we have

$$\frac{1}{r^2}\frac{\partial}{\partial r}\left(r^2\frac{\partial\Phi(r)}{\partial r}\right) = \frac{\beta}{\varepsilon_r\varepsilon_o V}\left[\sum_{j=1}^{N'} q_j^2\right]\Phi(r) \tag{6.3.3d}$$

Equation (6.3.3d) can be summarized by writing it in the form,

$$\frac{1}{r^2}\frac{\partial}{\partial r}\left(r^2\frac{\partial \Phi(r)}{\partial r}\right) - \kappa^2\Phi(r) = 0 \tag{6.3.4a}$$

where

$$\kappa^2 = \frac{\beta}{\varepsilon_r\varepsilon_o V}\left[\sum_{i=1}^{N'} q_i^2\right] = \frac{1}{\varepsilon_r\varepsilon_o V k_B T}\left[\sum_{i=1}^{N'} q_i^2\right] \tag{6.3.4b}$$

The units of κ^2 in Eq. (6.3.4b) are m^{-2}. The reciprocal of $1/\kappa$ has the units of meters, and is known as the Debye length. Equation (6.3.4a) is known as the Debye–Hückel equation.

6.4 Solution to the Debye–Hückel Equation

To aid us in the solution to Eq. (6.3.4a), each ion will be assumed to be surrounded by a spherical volume of radius, a, where a is the ionic radius. The spherical model serves to divide the domain, $0 < r \le \infty$, into two regions. The value of κ as defined by Eq. (6.3.4b) determines the free charge density in the exterior region, which corresponds to $a < r \le \infty$. In this region, Eq. (6.3.4a) assumes the form,

$$\frac{1}{r^2}\frac{d}{dr}\left(r^2\frac{d\Phi_{\text{ext}}(r)}{dr}\right) - \kappa^2\Phi_{\text{ext}}(r) = 0 \tag{6.4.1a}$$

where $\Phi(r) = \Phi_{\text{ext}}(r)$ is the exterior potential. The interior region corresponds to $0 < r \le a$. In this region, $\Phi(r) = \Phi_{\text{int}}(r)$, where $\Phi_{\text{int}}(r)$ is the potential interior to the sphere. In the interior, all free charge density other than that of the central ion is assumed to be absent; hence, $\kappa = 0$. In the space, $0 < r \le a$, Eq. (6.3.4a) assumes the form,

$$\frac{1}{r^2}\frac{d}{dr}\left(r^2\frac{d\Phi_{\text{int}}(r)}{dr}\right) = 0 \tag{6.4.1b}$$

The zero of the potential is arbitrary. A convenient zero is expressed by the limit,

$$\lim_{r\to\infty} \Phi_{\text{ext}}(r) = 0 \tag{6.4.2a}$$

The electric field at the surface of the sphere containing charge, q_1, is $E_1(a) = -(d\Phi_{\text{int}}/dr)_{r=a} = q_1/(4\pi\varepsilon_r\varepsilon_o a^2)$; hence, at this location, Eqs. (3.2.27) and (3.2.28) of Chap. 3 give for the displacement,

$$D(a) = \varepsilon_r\varepsilon_o E_1(a) = -\varepsilon_r\varepsilon_o\left(\frac{d\Phi_{\text{int}}}{dr}\right)_{r=a} = \frac{q_1}{4\pi a^2} \tag{6.4.2b}$$

Whereas κ is non-zero in Eq. (6.4.1a) ($r > a$), it is zero in Eq. (6.4.1b) ($r < a$). If ρ is the local volume charge density on the boundary of the sphere, then on the basis of dimensional analysis, the surface charge density on the boundary is $\sigma = \rho dr$ When the thickness of the boundary layer vanishes, $dr \rightarrow 0$, it makes $\sigma = 0$. According to Sect. 3.6 of Chap. 3, $\sigma = 0$ implies that the displacement is continuous across the boundary. Taking $\sigma = 0$ and assuming that on the molecular scale the appropriate value of ε_r is that of vacuum, i.e., $\varepsilon_r = 1$, then

$$\left(\frac{d\Phi_{\text{int}}(r)}{dr}\right)_{r=a} = \left(\frac{d\Phi_{\text{ext}}(r)}{dr}\right)_{r=a} \tag{6.4.2c}$$

At the surface of the sphere, the tangential component of the electric field must be continuous. This implies that

$$\Phi_{\text{int}}(a) = \Phi_{\text{ext}}(a) \tag{6.4.2d}$$

Equations (6.4.1a) and (6.4.1b) with boundary conditions given by Eqs. (6.4.2c) through (6.4.2d) serve to define a boundary value problem.

To find the general solution to Eq. (6.4.1a), we let

$$\Phi_{\text{ext}}(r) = \frac{u(r)}{r} \tag{6.4.3a}$$

Using Eq. (6.4.3a), the derivative of $\Phi_{ext}(r)$ is

$$\frac{d\Phi_{\text{ext}}(r)}{dr} = -\frac{u(r)}{r^2} + \frac{1}{r}\frac{du(r)}{dr} \tag{6.4.3b}$$

Next, multiply both sides of Eq. (6.4.3b) by r^2 to get,

$$r^2\frac{d\Phi_{\text{ext}}(r)}{dr} = -u(r) + r\frac{du(r)}{dr} \tag{6.4.3c}$$

Upon substitution of Eqs. (6.4.3a) and (6.4.3c) into Eq. (6.4.1a), the result is

$$\frac{1}{r^2}\frac{d}{dr}\left(-u(r) + r\frac{du(r)}{dr}\right) - \kappa^2\frac{u(r)}{r} = 0 \tag{6.4.3d}$$

Calculation using the derivative operator outside the parentheses in Eq. (6.4.3d) produces the result,

$$\frac{1}{r^2}\left(-\frac{du(r)}{dr} + \frac{du(r)}{dr} + r\frac{d^2u(r)}{dr^2}\right) - \kappa^2\frac{u(r)}{r} = 0 \tag{6.4.3e}$$

Cancellation of terms with opposite signs in Eq. (6.4.3e) yields,

$$\frac{1}{r}\frac{d^2u(r)}{dr^2} - \kappa^2\frac{u(r)}{r} = 0 \tag{6.4.3f}$$

Finally, upon multiplication of both sides by r, we have

$$\frac{d^2u(r)}{dr^2} - \kappa^2 u(r) = 0 \tag{6.4.4}$$

which is an ordinary differential equation with constant coefficients.

The general solution to Eq. (6.4.4) is

$$u(r) = A_{\text{ext}}e^{-\kappa r} + B_{\text{ext}}e^{\kappa r} \tag{6.4.5}$$

where A_{ext} and B_{ext} are constants of integration. Substitute Eq. (6.4.5) into Eq. (6.4.3a). The result is

$$\Phi_{\text{ext}}(r) = A_{\text{ext}}\frac{e^{-\kappa r}}{r} + B_{\text{ext}}\frac{e^{\kappa r}}{r} \tag{6.4.6}$$

Equation (6.4.6) is the general solution to Eq. (6.4.1a).

We now turn to the solution of the differential equation in Eq. (6.4.1b). Upon multiplication of both sides of this equation by r^2, the result is the ordinary differential equation,

$$\frac{d}{dr}\left(r^2\frac{d\Phi_{\text{int}}(r)}{dr}\right) = 0 \tag{6.4.7a}$$

The first integral of Eq. (6.4.7a) is

$$r^2\frac{\partial\Phi_{\text{int}}}{\partial r} = -A_{\text{int}} \tag{6.4.7b}$$

where $-A_{\text{int}}$ is a constant of integration. Divide both sides of Eq. (6.4.7b) by r^2 to obtain

$$\frac{d\Phi_{\text{int}}}{dr} = -\frac{A_{\text{int}}}{r^2} \tag{6.4.7c}$$

The integral of Eq. (6.4.7c) is

$$\Phi_{\text{int}}(r) = \frac{A_{\text{int}}}{r} + B_{\text{int}} \tag{6.4.8}$$

where B_{int} is a second constant of integration. Equation (6.4.8) is the general solution to Eq. (6.4.1b).

The right hand side of Eq. (6.4.2b) states that $(d\Phi_{\text{int}}(r)/dr)_{r=a} = -q_1/4\pi\varepsilon_r\varepsilon_o a^2$. According to Eq. (6.4.7c), this requirement implies that

$$A_{\text{int}} = \frac{q_1}{4\pi\varepsilon_r\varepsilon_o} \tag{6.4.9}$$

Upon substitution of Eq. (6.4.9), Eq. (6.4.8) reads

$$\Phi_{\text{int}}(r) = \frac{q_1}{4\pi\varepsilon_r\varepsilon_o r} + B_{\text{int}} \tag{6.4.10}$$

To determine the remaining constants of integration, we refer to Eqs. (6.4.2a), (6.4.2c), and (6.4.2d). According to Eq. (6.4.2a), the external potential is zero at $r = \infty$. This requirement is met if in Eq. (6.4.6), $B_{\text{ext}} = 0$. We thus have

$$\Phi_{\text{ext}}(r) = A_{\text{ext}}\frac{e^{-\kappa r}}{r} \tag{6.4.11}$$

Equations (6.4.10) and (6.4.11) must satisfy (6.4.2d), which implies

$$\frac{q_1}{4\pi\varepsilon_r\varepsilon_o a} + B_{\text{int}} = A_{\text{ext}}\frac{e^{-\kappa a}}{a} \tag{6.4.12}$$

Next, Eqs. (6.4.10) and (6.4.11) must satisfy Eq. (6.4.2c). To evaluate Eq. (6.4.2c), we will require the derivative of Eq. (6.4.10), which is

$$\frac{d\Phi_{\text{int}}(r)}{dr} = -\frac{q_1}{4\pi\varepsilon_r\varepsilon_o r^2} \tag{6.4.13a}$$

and the derivative of Eq. (6.4.11) which is

$$\frac{d\Phi_{\text{ext}}(r)}{dr} = A_{\text{ext}}\left[-\kappa\frac{e^{-\kappa r}}{r} - \frac{e^{-\kappa r}}{r^2}\right] \tag{6.4.13b}$$

When Eqs. (6.4.13a) and (6.4.13b) are substituted into Eq. (6.4.2c), the result is

$$\frac{-q_1}{4\pi\varepsilon_r\varepsilon_o a^2} = \left[-\frac{\kappa e^{-\kappa a}}{a} - \frac{e^{-\kappa a}}{a^2}\right]A_{\text{ext}} = -\left[\frac{\kappa a + 1}{a^2}\right]e^{-\kappa a}A_{\text{ext}} \tag{6.4.14a}$$

Equation (6.4.14a) can be solved for A_{ext} with the result,

$$A_{\text{ext}} = \frac{q_1/4\pi\varepsilon_r\varepsilon_o}{(1+\kappa a)}e^{\kappa a} \tag{6.4.14b}$$

Substitute Eq. (6.4.14b) into Eq. (6.4.11) to obtain,

$$\Phi_{\text{ext}}(r) = \frac{q_1}{4\pi \varepsilon_r \varepsilon_o r} \frac{e^{-\kappa(r-a)}}{(1+\kappa a)} \tag{6.4.15}$$

To evaluate the remaining constant associated with the interior solution, substitute Eq. (6.4.14b) into Eq. (6.4.12) to obtain

$$\frac{q_1}{4\pi \varepsilon_r \varepsilon_o a} + B_{\text{int}} = \frac{(q_1/4\pi \varepsilon_r \varepsilon_o)}{(1+\kappa a)} e^{\kappa a} \left(\frac{e^{-\kappa a}}{a} \right) \tag{6.4.16a}$$

Equation (6.4.16a) simplifies to read

$$\frac{q_1}{4\pi \varepsilon_r \varepsilon_o a} + B_{\text{int}} = \frac{(q_1/4\pi a \varepsilon_r \varepsilon_o)}{(1+\kappa a)} \tag{6.4.16b}$$

which when solved for B_{int} gives

$$B_{\text{int}} = \frac{q_1}{4\pi a \varepsilon_r \varepsilon_o} \left[\frac{1}{(1+\kappa a)} - 1 \right] = \frac{-q_1}{4\pi a \varepsilon_r \varepsilon_o} \frac{\kappa a}{(1+\kappa a)} \tag{6.4.16c}$$

Upon substitution of Eq. (6.4.16c) into Eq. (6.4.10), the result is

$$\Phi_{\text{int}}(r) = \frac{q_1}{4\pi \varepsilon_r \varepsilon_o r} - \frac{q_1}{4\pi \varepsilon_r \varepsilon_o a} \frac{\kappa a}{(1+\kappa a)} \tag{6.4.17}$$

Equations (6.4.15) and (6.4.17) serve to solve the boundary value problem.

In Sect. 6.5, below, we will show that Eq. (6.4.17) can be combined with Eq. (6.2.9) for $\psi_1(\mathbf{a})$ so as to evaluate the Helmholtz free energy [3, 4].

6.5 Evaluation of the Helmholtz Free Energy

For the three ions considered above, the total differential of the Helmholtz free energy with the charges treated as independent variables can be written as

$$dA = \left(\frac{\partial A}{\partial q_1} \right) dq_1 + \left(\frac{\partial A}{\partial q_2} \right) dq_2 + \left(\frac{\partial A}{\partial q_3} \right) dq_3 \tag{6.5.1}$$

As a generalization of Eq. (6.2.4a), we can define

$$\psi_j = \left(\frac{\partial A}{\partial q_j} \right), \quad j = 1, 2, 3 \tag{6.5.2a}$$

Equation (6.5.1) then reads

$$dA = \psi_1 dq_1 + \psi_2 dq_2 + \psi_3 dq_3 \tag{6.5.2b}$$

According to Eqs. (6.5.1), (6.5.2a), and (6.5.2b), $\psi_1 dq_1$ is the change in the Helmholtz free energy associated with an infinitesimal change, dq_1, in the charge, q_1, on the first ion with all the charges on the other ions fixed. Given that $\Phi(\mathbf{r})$ takes into account the equilibrium effects of all of the charges, we can interpret the potential $\psi_1(\mathbf{a})$ in Eq. (6.2.9) as the potential sensed by q_1 when it is fixed and all the other charges are at their equilibrium positions. The electrostatic energy associated with changing the charge on the first ion by from q_1 to $q_1 + dq_1$ is $\psi_1(\mathbf{a})dq_1$. Because the problem is spherically symmetric, the surface potential, $\Phi(\mathbf{a})$, can be equated to $\Phi_{\text{int}}(a)$. Upon combining all of these observations, we can rewrite Eq. (6.2.9) in the form,

$$\psi_1 = \Phi_{\text{int}}(a) - \frac{q_1}{4\pi \varepsilon_r \varepsilon_o a} \tag{6.5.3}$$

If we evaluate Eq. (6.4.17) at $r = a$ and substitute the result into Eq. (6.5.3), we find

$$\psi_1 = \frac{q_1}{4\pi \varepsilon_r \varepsilon_o a} - \frac{q_1}{4\pi \varepsilon_r \varepsilon_o a} \frac{\kappa a}{(1 + \kappa a)} - \frac{q_1}{4\pi \varepsilon_r \varepsilon_o a} \tag{6.5.4a}$$

or simply

$$\psi_1 = -\frac{q_1}{4\pi \varepsilon_r \varepsilon_o a} \frac{\kappa a}{(1 + \kappa a)} \tag{6.5.4b}$$

Generalizing Eq. (6.5.4b) to the j-th ion, we write it as

$$\psi_j = -\frac{q_j}{4\pi \varepsilon_r \varepsilon_o a} \frac{\kappa a}{(1 + \kappa a)} \tag{6.5.5}$$

According to Eq. (6.5.5), ψ_j depends upon κ. Equation (6.3.4b) in turn indicates that κ depends upon the charges of all of the ions. We stress this link by writing Eq. (6.5.2b) in the general form,

$$dA = \sum_{j=1}^{N'} \psi_j(q_1', q_2', \ldots, q_N')\, dq_j' \tag{6.5.6}$$

To integrate Eq. (6.5.6) and obtain the Helmholtz energy, we will start by considering an electrolyte solution with each ion replaced by an uncharged sphere. If all of the non-ideality of an electrolyte solution can be assigned to the electrostatic interactions between the ion charges, then a liquid containing uncharged spheres represents a thermodynamically ideal solution. The effect of the charges in establishing the free energy of the non-ideal solution can then be calculated by increasing each charge in Eq. (6.5.6) uniformly from $q_j' = 0$ to a final value of $q_j' = q_j$. This charging process is the analog of the Born charging process for a single ion in a dielectric as was encountered in Sect. 3.3 of Chap. 3. Generalizing the Born process, we require the instantaneous magnitudes of the various charges to be proportional to their final values. This uniform proportion requirement is achieved by letting $q_j' = \lambda q_j$, and $dq_j' = q_j d\lambda$, where λ is the coefficient of proportionality. The starting condition corresponds to $(\lambda = 0, q_j' = 0)$, whereas the final condition corresponds to $(\lambda = 1, q_j' = q_j)$. In the context of ionic solutions, this is known as the Güntelberg–Müller charging procedure. Equation (6.5.6) can rewritten in the form

$$dA = \sum_{j=1}^{N'} \psi_j(\lambda q_1, \lambda q_2, ..., \lambda q_N)\, q_j d\lambda \tag{6.5.7a}$$

With the use of Eq. (6.5.5), Eq. (6.5.7a) reads

$$dA = \sum_{j=1}^{N'} \frac{-\lambda q_j}{4\pi\varepsilon_r\varepsilon_o a} \cdot \frac{\kappa(\lambda q_1, \lambda q_2, ..., \lambda q_N)a}{1 + \kappa(\lambda q_1, \lambda q_2, ..., \lambda q_N)a}(q_j d\lambda) \tag{6.5.7b}$$

where upon referring to Eq. (6.3.4b), we have noted that $\kappa = \kappa(\lambda q_1, \lambda q_2, ..., \lambda q_N)$ is a function of the instantaneous values all of the charges. According to Eq. (6.3.4b), $\kappa = \kappa(\lambda q_1, \lambda q_2, ..., \lambda q_N)$ is given explicitly by

$$\begin{aligned}\kappa^2(\lambda q_1, \lambda q_2, ..., \lambda q_N) &= \frac{1}{\varepsilon_r\varepsilon_o V k_B T}\left[\sum_{i=1}^{N'} (\lambda q_i)^2\right] \\ &= \lambda^2 \cdot \frac{1}{\varepsilon_r\varepsilon_o V k_B T}\left[\sum_{i=1}^{N'} q_i^2\right] = \lambda^2\kappa^2(q_1, q_2, ..., q_n) \equiv \lambda^2\kappa^2\end{aligned} \tag{6.5.8a}$$

The square root of Eq. (6.5.8a) is

$$\kappa(\lambda q_1, \lambda q_2, ..., \lambda q_N) = \lambda\kappa(q_1, q_2, ..., q_n) \equiv \lambda\kappa \tag{6.5.8b}$$

Upon substitution of Eq. (6.5.8b) into Eq. (6.5.7b), we obtain a differential equation for A as a function of λ, which reads

$$dA = \sum_{j=1}^{N'} \frac{-q_j^2}{4\pi\varepsilon_r\varepsilon_o a} \frac{\lambda^2\kappa a}{(1 + \lambda\kappa a)} d\lambda \tag{6.5.9}$$

In the limit $\lambda = 0$, the ions are uncharged, and $A = A^{id}$. The integral of Eq. (6.5.9) can then be written with A^{id} serving as the constant of integration. The result is

$$A = A^{id} - \sum_{j=1}^{N'} \frac{q_j^2 \kappa}{4\pi \varepsilon_r \varepsilon_o} \int_0^1 \frac{\lambda^2}{(1 + \lambda \kappa a)} d\lambda \tag{6.5.10a}$$

The remaining integral on the right hand side of Eq. (6.5.10a) is known and is given by:

$$\int \frac{x^2 dx}{A + Bx} = \frac{1}{B^3} \left[\frac{1}{2}(A + Bx)^2 - 2A(A + Bx) + A^2 \ln(A + Bx) \right] \tag{6.5.10b}$$

To make use of Eq. (6.5.10b), we let $x = \lambda$, $A = 1$ and $B = \kappa a$. When these substitutions are made in Eq. (6.5.10b), the integral in Eq. (6.5.10a) can be evaluated with the result,

$$A = A^{id} - \sum_{j=1}^{N} \frac{q_j^2 \kappa}{4\pi \varepsilon_r \varepsilon_o} \frac{1}{(\kappa a)^3} \left[\frac{1}{2}(1 + \kappa a \lambda)^2 - 2(1 + ka\lambda) + \ln(1 + \kappa a \lambda) \right]_0^1 \tag{6.5.10c}$$

After rearranging the terms in the bracket and taking into account the limits, Eq. (6.5.10c) reads,

$$A = A^{id} - \sum_{j=1}^{N'} \frac{q_j^2 \kappa}{4\pi \varepsilon_r \varepsilon_o} \frac{1}{(\kappa a)^3} \left[\left(\ln(1 + \kappa a) - 2(1 + \kappa a) + \frac{1}{2}(1 + \kappa a)^2 \right) - \left(-2 + \frac{1}{2} \right) \right] \tag{6.5.11a}$$

Expansion of the squared binomial within the bracket in Eq. (6.5.11a) yields,

$$A = A^{id} - \sum_{j=1}^{N'} \frac{q_j^2 \kappa}{4\pi \varepsilon_r \varepsilon_o} \frac{1}{(\kappa a)^3} \left[\ln(1 + \kappa a) - 2 - 2\kappa a + \frac{1}{2} + \kappa a + \frac{1}{2}(\kappa a)^2 + \frac{3}{2} \right] \tag{6.5.11b}$$

which leads finally to

$$A = A^{id} - \frac{1}{4\pi \varepsilon_r \varepsilon_o} \sum_{j=1}^{N'} q_j^2 \kappa \left(\frac{1}{(\kappa a)^3} \right) \left[\ln(1 + \kappa a) - \kappa a + \frac{1}{2}(\kappa a)^2 \right] \tag{6.5.11c}$$

According to Eq. (6.3.4b),

$$\sum_{i=1}^{N'} q_i^2 = \varepsilon_r \varepsilon_o V k_B T \kappa^2 \tag{6.5.12a}$$

Upon substitution of Eq. (6.5.12a) into Eq. (6.5.11c), one obtains

$$A = A^{id} - \frac{1}{4\pi\varepsilon_r\varepsilon_o}\left(\varepsilon_r\varepsilon_o V k_B T \kappa^2\right)\left(\frac{\kappa}{(\kappa a)^3}\right)\left[\ln(1+\kappa a) - \kappa a + \frac{\kappa^2 a^2}{2}\right] \quad (6.5.12b)$$

After cancellation of common factors in Eq. (6.5.12b), the solution to Eq. (6.5.9) assumes the form,

$$A = A^{id} - \frac{1}{4\pi a^3}(V k_B T)\left[\ln(1+\kappa a) - \kappa a + \frac{\kappa^2 a^2}{2}\right] \quad (6.5.13)$$

The second term in Eq. (6.5.13) corrects the ideal solution model for the electrostatic energy stored in the interactions among the charges.

6.6 Ion Chemical Potentials

If we make the variable substitution,

$$x = \kappa a \quad (6.6.1a)$$

then Eq. (6.5.13) can be rewritten in the form

$$A = A^{id} - \frac{1}{4\pi a^3}(V k_B T)\left[\ln(1+x) - x + \frac{x^2}{2}\right] \quad (6.6.1b)$$

Consider the electrolyte $CaCl_2$. The electrovalence of Ca^{2+} is $Z = +2$, and the electrovalence of Cl^- is $Z = -1$. In general we shall represent the chemical formula for such an electrolyte by writing $\left(M^{Z_M}\right)_{\nu_M}\left(X^{Z_X}\right)_{\nu_X}$, where ν_M and ν_X are the respective stoichiometric coefficients, and Z_M and Z_X are the respective electrovalences. If there are N of these molecular units dissolved in the solution, then the number of cations is $N_M = \nu_M N$, while the number of anions is $N_X = \nu_X N$. In the uncharged state, the chemical potential of the cation is

$$\mu_M^{id} = \frac{\partial A^{id}}{\partial N_M} \quad (6.6.2a)$$

In the charged state, it is

$$\mu_M = \frac{\partial A}{\partial N_M} \quad (6.6.2b)$$

Using the chain rule, differentiation of Eq. (6.6.1b) with respect to N_M has the result,

$$\frac{\partial A}{\partial N_M} = \frac{\partial A^{id}}{\partial N_M} - \frac{1}{4\pi a^3}(Vk_BT)\frac{\partial}{\partial x}\left[\ln(1+x) - x + \frac{x^2}{2}\right]\left(\frac{\partial x}{\partial \kappa}\right)\left(\frac{\partial \kappa}{\partial N_M}\right) \quad (6.6.3a)$$

where

$$\frac{\partial}{\partial x}\left[\ln(1+x) - x + \frac{x^2}{2}\right] = \frac{1}{1+x} - 1 + x = \frac{x^2}{1+x} \quad (6.6.3b)$$

By virtuc of Eq. (6.6.1a)

$$\frac{\partial x}{\partial \kappa} = a \quad (6.6.3c)$$

Now, substitute Eqs. (6.6.3b) and (6.6.3c) into Eq. (6.6.3a). The result is

$$\frac{\partial A}{\partial N_M} = \frac{\partial A^{id}}{\partial N_M} - \frac{1}{4\pi a^3}(Vk_BT)\left(\frac{ax^2}{1+x}\right)\left(\frac{\partial \kappa}{\partial N_M}\right) \quad (6.6.4a)$$

Because there are only two species of ions, the sum of the squares of the charges is

$$\sum_{i=1}^{N'} q_i^2 = N_M q_M^2 + N_X q_X^2 \quad (6.6.4b)$$

where $N' = N_M + N_X$. Use of Eq. (6.6.4b) permits Eq. (6.3.4b) to be written as

$$\kappa^2 = \frac{1}{(\varepsilon_r \varepsilon_o V k_B T)}\left[N_M q_M^2 + N_X q_X^2\right] \quad (6.6.5a)$$

The derivative of κ^2 with respect to N_M is then

$$\frac{\partial \kappa^2}{\partial N_M} = 2\kappa \frac{\partial \kappa}{\partial N_M} = \frac{q_M^2}{\varepsilon_r \varepsilon_o V k_B T} \quad (6.6.5b)$$

Equation (6.6.5b) can be solved for $\partial\kappa/\partial N_M$ with the result,

$$\frac{\partial \kappa}{\partial N_M} = \frac{q_M^2}{2\kappa(\varepsilon_r \varepsilon_o V k_B T)} \quad (6.6.5c)$$

Now substitute Eq. (6.6.5c) into Eq. (6.6.4a) with x replaced by κa. The result is

$$\frac{\partial A}{\partial N_M} = \frac{\partial A^{id}}{\partial N_M} - \frac{1}{4\pi a^3}(Vk_BT)\left(\frac{\kappa^2 a^2}{1+\kappa a}\right)\left(\frac{a q_M^2}{2\kappa \varepsilon_r \varepsilon_o V k_B T}\right) \quad (6.6.6)$$

After algebraic simplification and use of the definitions in Eqs. (6.6.2a) and (6.6.2b), Eq. (6.6.6) reads,

$$\mu_M = \mu_M^{id} - \frac{q_M^2}{8\pi\varepsilon_r\varepsilon_o a}\left(\frac{\kappa a}{1+\kappa a}\right) \tag{6.6.7}$$

The chemical potential for the anion, X^{Z_X}, is obtained from Eq. (6.6.7) simply by replacing M everywhere by X. The anion chemical potential is thus,

$$\mu_X = \mu_X^{id} - \frac{q_X^2}{8\pi\varepsilon_r\varepsilon_o a}\left(\frac{\kappa a}{1+\kappa a}\right) \tag{6.6.8}$$

6.7 Ideal Solution Limit and the Geometric Mean Activity Coefficient

The chemical potential of the cation in the binary electrolyte with stoichiometry, $M_{\nu_M}^{Z_M}X_{\nu_X}^{Z_X}$, can be represented by Eq. (2.7.2) as derived in Chap. 2, namely,

$$\mu_M = \mu_M^o + k_B T \ln a_M \tag{6.7.1a}$$

where μ_M^o is the chemical potential in the standard state, which we shall take to be the uncharged ion at a concentration of $m^o = 1$ molal. The activity, a_M, is related to the molal concentration, m_M of the cation by Eq. (2.7.3a) of Chap. 2, which we shall write in the form,

$$a_M = \gamma_M (m_M/m^o) \tag{6.7.1b}$$

where γ_M is the cation activity coefficient. When Eq. (6.7.1b) is substituted into Eq. (6.7.1a), the result is

$$\mu_M = \mu_M^o + k_B T \ln\left[\gamma_M (m_M/m^o)\right] \tag{6.7.1c}$$

We can expand Eq. (6.7.1c) to read,

$$\mu_M = \mu_M^o + k_B T \ln \gamma_M + k_B T \ln(m_M/m^o) \tag{6.7.1d}$$

The cation is said to behave ideally in the limit $\gamma_M \to 1$. In this limit, Eq. (6.7.1d) becomes,

$$\mu_M^{id} = \mu_M^o + k_B T \ln(m_M/m^o) \tag{6.7.2a}$$

Note that $\mu_M^{id} = \mu_M^o$ in the standard state where $m_M = m^o$. Equation (6.7.2a) permits Eq. (6.7.1d) to be rewritten as

$$\mu_M = \mu_M^{id} + k_B T \ln \gamma_M \tag{6.7.2b}$$

Likewise, the chemical potential of the anion species in $M_{\nu_M}^{Z_M} X_{\nu_X}^{Z_X}$ is given by

$$\mu_X = \mu_X^o + k_B T \ln a_X \tag{6.7.3a}$$

The activity, a_X, of the anion is related to its molal concentration, m_X, by

$$a_X = \gamma_X (m_X / m^o) \tag{6.7.3b}$$

where γ_X is the anion activity coefficient. Substitute Eq. (6.7.3b) into (6.7.3a) to obtain

$$\mu_X = \mu_X^o + k_B T \ln\left[\gamma_X (m_X / m^o)\right] \tag{6.7.3c}$$

which can be rewritten as

$$\mu_X = \mu_X^o + k_B T \ln \gamma_X + k_B T \ln\left(m_X / m^o\right) \tag{6.7.3d}$$

In an ideal solution, $\gamma_X \to 1$, and the analog of Eq. (6.7.2a) is

$$\mu_X^{id} = \mu_X^o + k_B T \ln\left(m_X / m^o\right) \tag{6.7.4a}$$

Equation (6.7.4a) permits Eq. (6.7.3d) to be written as

$$\mu_X = \mu_X^{id} + k_B T \ln \gamma_X \tag{6.7.4b}$$

If we compare Eq. (6.7.2b) with Eq. (6.6.7), we can conclude that

$$k_B T \ln \gamma_M = -\frac{q_M^2}{8\pi \varepsilon_r \varepsilon_o a}\left[\frac{\kappa a}{(1 + \kappa a)}\right] \tag{6.7.5a}$$

If we compare Eq. (6.7.4b) with Eq. (6.6.8), we obtain

$$k_B T \ln \gamma_X = -\frac{q_X^2}{8\pi \varepsilon_r \varepsilon_o a}\left[\frac{\kappa a}{(1 + \kappa a)}\right] \tag{6.7.5b}$$

Although Eqs. (6.7.5a) and (6.7.5b) provide theoretical expressions for the cation and anion activity coefficients individually, it is impossible to measure either experimentally, because there is no property of the solution where one acts in isolation from the other. Hence, for the purposes of comparing the theory with experiment, we shall need to construct an equation for the chemical potential of the electrolyte as a whole. That chemical potential takes the form,

$$\mu = \nu_M \mu_M + \nu_X \mu_X \tag{6.7.6a}$$

Substitute Eqs. (6.7.1c) and (6.7.3c), into Eq. (6.7.6a). The result is

$$\mu = \left(\nu_M \mu_M^o + \nu_X \mu_X^o\right) + \nu_M k_B T \ln\left[\gamma_M (m_M/m^o)\right] + \nu_X k_B T \ln\left[\gamma_X (m_X/m^o)\right] \tag{6.7.6b}$$

The chemical potential of the electrolyte in the standard state is

$$\mu^o = \nu_M \mu_M^o + \nu_X \mu_X^o \tag{6.7.6c}$$

Substitute Eq. (6.7.6c) into Eq. (6.7.6b) to obtain

$$\mu = \mu^o + \nu_M k_B T \ln\left[\gamma_M (m_M/m^o)\right] + \nu_X k_B T \ln\left[\gamma_X (m_X/m^o)\right] \tag{6.7.7a}$$

which can be rewritten as

$$\mu = \mu^o + k_B T \ln\left[(\gamma_M^{\nu_M} \gamma_X^{\nu_X})(m_M^{\nu_M} m_X^{\nu_X})/(m^o)^{\nu_M+\nu_X}\right] \tag{6.7.7b}$$

If the molality of the electrolyte is m, then by virtue of the stoichiometry, the molalities of the individual ions are

$$m_M = \nu_M m \tag{6.7.8a}$$

and

$$m_X = \nu_X m \tag{6.7.8b}$$

Equations (6.7.8a) and (6.7.8b) permit Eq. (6.7.7b) to be rewritten as

$$\mu = \mu^o + k_B T \ln\left[(\gamma_M^{\nu_M} \gamma_X^{\nu_X})\right] + k_B T \ln\left[(\nu_M^{\nu_M} \nu_X^{\nu_X})(m/m^o)^{\nu_M+\nu_X}\right] \tag{6.7.9}$$

Define the stoichiometric coefficient relations,

$$\nu = \nu_M + \nu_X \tag{6.7.10a}$$

and

$$\nu_\pm^\nu = (\nu_M)^{\nu_M} (\nu_X)^{\nu_X} \tag{6.7.10b}$$

Define the geometric mean activity coefficient, $\gamma_\pm$, by

$$\gamma_{\pm}^{\nu} = (\gamma_M)^{\nu_M} (\gamma_X)^{\nu_X} \tag{6.7.11}$$

Substitute Eqs. (6.7.10a), (6.7.10b) and (6.7.11) into Eq. (6.7.9). The result is

$$\mu = \mu^o + k_B T \ln(\gamma_{\pm})^{\nu} + k_B T\left[(\nu_{\pm})^{\nu} (m/m^o)^{\nu}\right] \tag{6.7.12}$$

Equation (6.7.12) is a result that can be compared with experiment [5].

6.8 Debye–Hückel Theory of the Geometric Mean Activity Coefficient

Equation (6.7.11) can be written in logarithmic form as

$$\nu k_B T \ln \gamma_{\pm} = \nu_M k_B T \ln \gamma_M + \nu_X k_B T \ln \gamma_X \tag{6.8.1}$$

Upon substitution of Eqs. (6.7.5a) and (6.7.5b) into Eq. (6.8.1), the result is

$$\nu k_B T \ln \gamma_{\pm} = -\frac{\nu_M q_M^2}{8\pi \varepsilon_r \varepsilon_o}\left(\frac{\kappa}{1+\kappa a}\right) - \frac{\nu_X q_X^2}{8\pi \varepsilon_r \varepsilon_o}\left(\frac{\kappa}{1+\kappa a}\right) \tag{6.8.2a}$$

which simplifies to

$$\nu k_B T \ln \gamma_{\pm} = -\frac{\left(\nu_M q_M^2 + \nu_X q_X^2\right)}{8\pi \varepsilon_r \varepsilon_o}\left(\frac{\kappa}{1+\kappa a}\right) \tag{6.8.2b}$$

By virtue of charge neutrality,

$$\nu_M q_M + \nu_X q_X = 0 \tag{6.8.3a}$$

which can be rearranged to read

$$\nu_M q_M = -\nu_X q_X \tag{6.8.3b}$$

Multiply both sides of Eq. (6.8.3b) by q_M. The result is

$$\nu_M q_M^2 = -\nu_X q_M q_X \tag{6.8.3c}$$

Multiply both sides of Eq. (6.8.3b) by q_X. The result is

$$\nu_X q_X^2 = -\nu_M q_M q_X \tag{6.8.3d}$$

Equations (6.8.3c) and (6.8.3d) permit us to simplify the sum in parentheses in the numerator of Eq. (6.8.2b) to read

$$\nu_M q_M^2 + \nu_X q_X^2 = -\nu_M q_M q_X - \nu_X q_M q_X = -(\nu_M + \nu_X) q_M q_X = -\nu q_M q_X \tag{6.8.4a}$$

where we have made use of Eq. (6.7.10a). Because $q_M > 0$ and $q_X < 0$, we have $-q_M q_X = |q_M q_X|$. This permits Eq. (6.8.4a) to be rewritten as

$$\nu_M q_M^2 + \nu_X q_X^2 = \nu |q_M q_X| \tag{6.8.4b}$$

Substitute Eq. (6.8.4b) into Eq. (6.8.2b). After cancellation of a factor of ν from either side, the result is

$$k_B T \ln \gamma_\pm = -\frac{|q_M q_X|}{8\pi \varepsilon_r \varepsilon_o}\left(\frac{\kappa}{1+\kappa a}\right) \tag{6.8.5}$$

Now, $q_M = Z_M e$ and $q_X = Z_X e$, where Z_M and Z_X are the respective charge numbers, and e is the magnitude of the charge of the electron. Upon using these identities to replace the corresponding charges in Eq. (6.8.5), our final result is

$$k_B T \ln(\gamma_\pm) = -\frac{|Z_M Z_X| e^2}{8\pi \varepsilon_r \varepsilon_o}\left(\frac{\kappa}{1+\kappa a}\right) \tag{6.8.6}$$

6.9 Geometric Mean Activity Coefficient with Concentrations in Molar Units

If Eq. (6.8.6) is to be compared with experimental data, it should be converted from molecular units to molar units. Tradition dictates that we first convert from natural logarithms to decadic logarithms. Using the equivalence, $\ln \gamma_\pm = 2.303 \log \gamma_\pm$, Eq. (6.8.6) reads,

$$\log(\gamma_\pm) = -\frac{e^2}{(2.303)(8\pi)\varepsilon_r \varepsilon_o k_B T}\left(\frac{\kappa}{1+\kappa a}\right)|Z_M Z_X| \tag{6.9.1}$$

Using the relations between the electrovalences and the charges, $q_M = Z_M e$ and $q_X = Z_X e$, we can rewrite Eq. (6.6.5a) in the form,

$$\kappa^2 = \frac{1}{\varepsilon_r \varepsilon_o k_B T}\left[\frac{N_M}{V}(Z_M e)^2 + \frac{N_X}{V}(Z_X e)^2\right] \tag{6.9.2a}$$

The number densities can be converted to molar units by use of

$$\frac{N_M}{V(\mathrm{m}^3)} = N_A c_M \ (\mathrm{mol/L}) \times 10^3 \ \mathrm{L/m}^3 \tag{6.9.2b}$$

and

$$\frac{N_X}{V\ (\mathrm{m}^3)} = N_A c_X \ (\mathrm{mol/L}) \times 10^3 \ \mathrm{L/m}^3 \tag{6.9.2c}$$

where c_M and c_X are the molar concentrations of the cation and the anion, respectively, and N_A is the Avogadro number. Upon substitution of Eqs. (6.9.2b) and (6.9.2c), Eq. (6.9.2a) becomes

$$\kappa^2 = \frac{1000 N_A e^2}{\varepsilon_r \varepsilon_o k_B T}\left(c_M Z_M^2 + c_X Z_X^2\right) \tag{6.9.2d}$$

which by tradition is always rewritten as

$$\kappa^2 = \frac{2000 N_A e^2}{\varepsilon_r \varepsilon_o k_B T}\left(\frac{1}{2}\right)\left(c_M Z_M^2 + c_X Z_X^2\right) \tag{6.9.2e}$$

For a binary electrolyte, the ionic strength, I, is defined by

$$I = \frac{1}{2}\sum_{s=1}^{2} Z_s^2 c_s \tag{6.9.2f}$$

Now substitute Eq. (6.9.2f) into Eq. (6.9.2e) and take the square root. The result is

$$\kappa = \left(\frac{2000 N_A e^2}{\varepsilon_r \varepsilon_o k_B T}\right)^{1/2} \sqrt{I} \tag{6.9.3a}$$

Use Eq. (6.9.3a) to write the quantity in parenthesis in Eq. (6.9.1) as

$$\left(\frac{\kappa}{1+\kappa a}\right) = \frac{\left(\frac{2000 N_A e^2}{\varepsilon_r \varepsilon_o k_B T}\right)^{1/2}\sqrt{I}}{1 + a\left(\frac{2000 N_A e^2}{\varepsilon_r \varepsilon_o k_B T}\right)^{1/2}\sqrt{I}} \tag{6.9.3b}$$

Define the coefficients,

$$A = \frac{1}{2.303}\left(\frac{e^2}{8\pi\varepsilon_r \varepsilon_o k_B T}\right)\left(\frac{2000 N_A e^2}{\varepsilon_r \varepsilon_o k_B T}\right)^{1/2} = \frac{1}{2.303}\left(\frac{e^2}{8\pi\varepsilon_r \varepsilon_o k_B T}\right)B \tag{6.9.4a}$$

where

$$B = \left(\frac{2000 N_A e^2}{\varepsilon_r \varepsilon_o k_B T} \right)^{1/2} \tag{6.9.4b}$$

Substitute Eq. (6.9.3b) into Eq. (6.9.1) and use Eqs. (6.9.4a) and (6.9.4b) to simplify. One finds

$$\log \gamma_{\pm} = -\frac{A|Z_M Z_X|\sqrt{I}}{1 + aB\sqrt{I}} \tag{6.9.5}$$

which is our final result.

Equation (6.9.5) is found to agree best with experiment in the limit of dilute solutions where $aB\sqrt{I} \ll 1$. In this limit, Eq. (6.9.5) takes the simple form,

$$\log \gamma_{\pm} = -A|Z_M||Z_X|\sqrt{I} \tag{6.9.6}$$

When the solvent is water at 25 °C, $A = 0.5092\ \mathrm{L}^{1/2}\ \mathrm{mol}^{-1/2}$. Equation (6.9.6) is known as the Debye–Hückel limiting law. See Fig. 6.1.

Note that Eq. (6.9.6) predicts that the activity coefficient data for 1:1 electrolytes and 2:2 electrolytes are distinct, falling on lines differing in slope by a factor of four. By contrast the activity coefficient data for 1:2 electrolytes and 2:1 electrolytes are predicted to fall on the same line. It is one of the successes of the theory that these trends are in complete agreement with experiments involving dilute solutions.

The activity coefficient of a single ion can be calculated by referring to the appropriate choice between Eqs. (6.7.5a) and (6.7.5b). For example, for the cation, $q_M = Z_M e$, and Eq. (6.7.5a) reads,

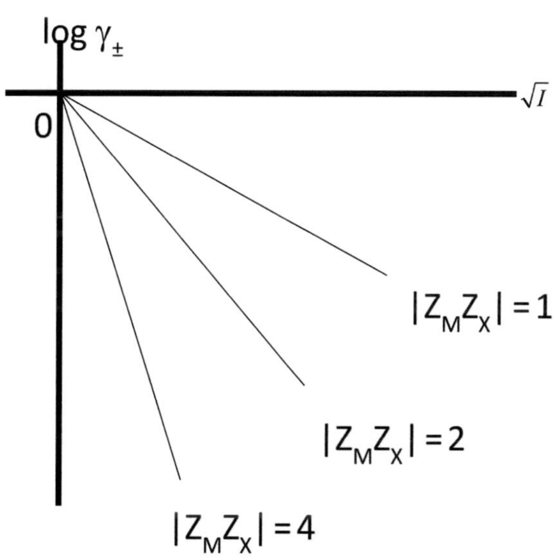

Fig. 6.1 Plot of Eq. (6.9.6) in the form, $\log \gamma_{\pm}$ versus $\sqrt{I}$ for different values of the electrovalence product, $|Z_M Z_X|$

$$k_B T \ln \gamma_M = -\frac{Z_M^2 e^2}{8\pi \varepsilon_r \varepsilon_o}\left(\frac{\kappa}{1+\kappa a}\right) \tag{6.9.7}$$

When Eq. (6.9.3b) is substituted into Eq. (6.9.7), the coefficients defined by Eqs. (6.9.4a) and (6.9.4b) can be made to reappear. These identifications permit Eq. (6.9.7) to be rewritten as

$$\log \gamma_M = -\frac{Z_M^2 A\sqrt{I}}{1+aB\sqrt{I}} \tag{6.9.8a}$$

The same arguments apply in the case of the anion, where the activity coefficient is given by

$$\log \gamma_X = -\frac{Z_X^2 A\sqrt{I}}{1+aB\sqrt{I}} \tag{6.9.8b}$$

In the limit, $aB\sqrt{I} \ll I$, Eqs. (6.9.8a) and (6.9.8b) read

$$\log \gamma_M = -Z_M^2 A\sqrt{I} \tag{6.9.9a}$$

and

$$\log \gamma_X = -Z_X^2 A\sqrt{I} \tag{6.9.9b}$$

respectively.

In deriving Eq. (6.9.5), the cation and anion were tacitly assumed to have the same value for the radius, a. As this is seldom the case, a compromise is needed when applying Eq. (6.9.5) to fit experimental data. As the contact distance between two ions in a crystal can be determined by X-ray diffraction, the contact distance is often substituted in the place of a. For some electrolytes, aB is of the order of unity in the units of $(\text{L/mol})^{1/2}$, so that in dilute solutions, $aB\sqrt{I}$ can be replaced simply by $\sqrt{I}$ so long as I is expressed in mol/L. This allows us to ignore unit cancellation, so that Eq. (6.9.5) reads simply,

$$\log \gamma_\pm = -\frac{A|Z_M Z_X|\sqrt{I}}{1+\sqrt{I}} \tag{6.9.10}$$

The standard state, $m = m^o = 1$ molal, is said to be "hypothetical," in the sense that most actual solutions at a concentration of $m = 1$ molal fail to behave ideally. Nonetheless, the chemical potential, μ, of the solute can be determined as a function of concentration, m, by any one of a number of methods, which include vapor pressure, boiling point elevation, freezing point depression, and electrochemical cell potential. As we saw in Chap. 2, if data in the form of μ versus $\log(m/m^o)$ are collected at sufficiently dilute concentrations, $m \ll m^o$, where $\gamma_\pm$ is essentially unity, then the solution behaves ideally, and the data will fall on a straight line. When the line is extrapolated to a concentration of $m = m^o = 1$ molal, the value of the intercept on the chemical potential axis corresponds to μ^o.

It should be noted that the charge density, ρ, which appears in Gauss' Law, $\nabla \cdot \mathbf{D} = \rho$, is the charge per unit volume. This is the origin of the definition of I in terms of volumetric concentrations in Eq. (6.9.2f). The goal of thermodynamics is to describe equilibrium in terms of independent variables, such as temperature, pressure, and some measure of concentration, such as the molality, which is temperature and pressure independent. As often occurs, molality gets substituted for molarity in Eq. (6.9.2f), which means that the temperature dependence of I gets lost. In practice, this technicality is of little consequence in the case of aqueous solutions, so long as thc data refer to a temperature where the density of water is close to 1 g/cm^3.

In the case of a 1:1 electrolyte, the factor of 1/2 in Eq. (6.9.2f) makes the ionic strength equal to the molality, that is $I = (1/2)[(+1)^2 m + (-1)^2 m] = m$. In cases of solutions of several electrolytes, the sum in Eq. (6.9.2f) runs over all ions present in the solution. Thus, to apply Eq. (6.9.5) to an electrolyte $M_{\nu_M}^{Z_M} X_{\nu_X}^{Z_X}$, it should be remembered that the ionic strength, I, includes contributions from all of the ions in the solution.

6.10 Applications

6.10.1 *Solubility of CaF$_2$*

Calcium fluoride is a weakly soluble ionic compound that dissolves in water to form a strong electrolyte. The dissolution reaction is

$$CaF_2(s) \rightarrow Ca^{2+}(aq) + 2F^-(aq) \tag{6.10.1}$$

The solubility product, $K_{sp} = 1.6 \times 10^{-10}$ is defined as the ratio

$$K_{sp} = \frac{(a_{Ca^{2+}})(a_{F^-})^2}{a_{CaCl_2}} \tag{6.10.2}$$

As shown in Chap. 2, Sect. 2.6, the activity of any solid is unity at $P_0 = 1$ bar pressure; hence, Eq. (6.10.2) simplifies to

$$K_{sp} = (a_{Ca^{2+}})(a_{F^-})^2 \tag{6.10.3}$$

If m is the molality of the electrolyte, then the Ca^{2+} concentration is $m_{Ca^{2+}} = m$, while the F^- concentration is $m_{F^-} = 2m$. According to Eq. (6.7.1b),

$$a_{Ca^{2+}} = \gamma_{Ca^{2+}}(m_{Ca^{2+}}/m^o) = \gamma_{Ca^{2+}}(m/m^o) \tag{6.10.4a}$$

and according to Eq. (6.7.3b),

$$a_{F^-} = \gamma_{F^-}(m_{F^-}/m^o) = \gamma_{F^-}(2m/m^o) \tag{6.10.4b}$$

Substitute Eqs. (6.10.4a) and (6.10.4b) into Eq. (6.10.3) to obtain the solubility product in the form,

$$\begin{aligned} K_{sp} &= \gamma_{Ca^{2+}}(m/m^o)\left[\gamma_{F^-}(2m/m^o)\right]^2 \\ &= 4\gamma_{Ca^{2+}}\gamma_{F^-}^2(m/m^o)^3 \end{aligned} \tag{6.10.5}$$

Given the stoichiometric coefficients, the geometric mean activity coefficient, as defined in Sect. 6.7 of this chapter, assumes the form,

$$\gamma_\pm^3 = \gamma_{Ca^{2+}}\gamma_{F^-}^2 \tag{6.10.6}$$

Substitute Eq. (6.10.6) into Eq. (6.10.5) to obtain

$$K_{sp} = 4\gamma_\pm^3(m/m^o)^3 \tag{6.10.7}$$

Solved for m/m^o, Eq. (6.10.7) reads,

$$\frac{m}{m^o} = \frac{1}{\gamma_\pm}\left(\frac{K_{sp}}{4}\right)^{1/3} \tag{6.10.8}$$

To calculate m using Eq. (6.10.8), we need first to evaluate γ_+. In principle, Eq. (6.9.10) could be substituted for $\gamma_\pm$, but since I depends upon concentration, this leads to an equation that must be solved numerically. Short of a computer-based calculation, we can, nonetheless, obtain a good estimate of the value of m by exploiting the method of successive approximations.

As a zeroth order approximation, we set the ionic strength to zero. According to Eq. (6.9.10), under this condition $\gamma_\pm = 1$. With $\gamma_\pm = 1$, Eq. (6.10.8) gives for the solubility in the zeroth approximation,

$$\frac{m}{m^o} = \left(\frac{K_{sp}}{4}\right)^{1/3} = \left(\frac{1.6 \times 10^{-10}}{4}\right)^{1/3} = 3.42 \times 10^{-4} \tag{6.10.9}$$

To launch the next approximation, estimate the value of I using the result, $m = 3.42 \times 10^{-4}$ molal obtained in Eq. (6.10.9). At such a low concentration, there is little distinction between molarity, c_s, and molality, m_s, of ion species, s, so that Eq. (6.9.2f) can be replaced by

$$I = \frac{1}{2}\sum_{s=1}^{2} Z_s^2 m_s \tag{6.10.10a}$$

Equation (6.10.10a) reads

$$I = \frac{1}{2}\sum_{s=1}^{2} Z_s^2 m_s = \frac{1}{2}\left(Z_M^2 \nu_M m + Z_X^2 \nu_X m\right) \tag{6.10.10b}$$

The right hand has been specialized to account for a binary electrolyte where, $n = 2$, and Eqs. (6.7.8a) and (6.7.8b) apply. For calcium fluoride, $Z_M = +2, Z_X = -1$ and $\nu_M = 1$, $\nu_X = 2$, and Eq. (6.10.10b) becomes,

$$I = \frac{1}{2}\left[(+2)^2(1)m + (-1)^2(2)m\right] = 3m \tag{6.10.10c}$$

Inserting the zeroth approximation $m = 3.42 \times 10^{-4}$ molal into Eq. (6.10.10c), we obtain the first order correction to the ionic strength, which is

$$I = 3m = 3\left(3.42 \times 10^{-4}\right) = 1.03 \times 10^{-3}\text{molal} \tag{6.10.10d}$$

Substitution of Eq. (6.10.10d) into Eq. (6.9.10) with $A = 0.5092\,\text{L}^{1/2}\,\text{mol}^{-1/2}$ gives for the log of the geometric mean activity coefficient,

$$\log\gamma_\pm = -\frac{0.5092|(+2)(-1)|\sqrt{1.03 \times 10^{-3}}}{1 + \sqrt{1.03 \times 10^{-3}}} = -3.16 \times 10^{-2} \tag{6.10.11a}$$

The antilog of Eq. (6.10.11a) gives for the geometric mean activity coefficient,

$$\gamma_\pm = 0.930 \tag{6.10.11b}$$

Substitution of Eq. (6.10.11b) into Eq. (6.10.8) leads to the first order estimate of the solubility,

$$\frac{m}{m^o} = \left(\frac{1}{0.930}\right)\left(\frac{1.6 \times 10^{-10}}{4}\right)^{1/3} = 3.68 \times 10^{-4} \tag{6.10.12}$$

The improvement in the solubility estimate in going from the zeroth order to the first order is thus

$$\frac{(3.68 \times 10^{-4}) - (3.42 \times 10^{-4})}{(3.42 \times 10^{-4})} \times 100\% = 8\% \tag{6.10.13}$$

If one proceeds to second order, a slightly better estimate of this percentage change can be achieved.

6.10.2 The "Salting in" Effect

The percent ionization of a weak acid or a weak base can be increased by the addition of a strong electrolyte. This is known as "salting in."

Consider the ionization of 0.1 molal acetic acid. The ionization equilibrium is

$$HA + H_2O \rightarrow H_3O^+ + A^- \tag{6.10.14}$$

where HA is acetic acid and A^- is the acetate ion. The acid ionization constant is

$$K_a = \frac{a_{H_3O^+}a_{A^-}}{a_{HA}a_{H_2O}} \simeq \frac{a_{H_3O^+}a_{A^-}}{a_{HA}} \tag{6.10.15}$$

where $K_a = 1.754 \times 10^{-5}$. In dilute solution, $a_{H_2O} \simeq 1$, which accounts for its disappearance from the far right hand side of Eq. (6.10.15). If an exact value is required for a_{H_2O}, one can be determined by measuring the water vapor pressure over the solution.

The concentration of hydronium ion and acetate ion are equal by virtue of the principle of electrical neutrality. If m is the molality of HA as prepared, and y the change in molality due to ionization, then the concentrations of the two ion species are both equal to y, and their corresponding activities are

$$a_{H_3O^+} = \gamma_{H_3O^+}(y/m^o) \tag{6.10.16a}$$

and

$$a_{A^-} = \gamma_{A^-}(y/m^o) \tag{6.10.16b}$$

The activity of the undissociated acid, HA, is unknown, but given that it is a neutral species, its interactions with the surrounding molecules and ions will likely be much less than the long range Coulomb interactions between the ions, H_3O^+ and A^-. On the basis of this argument, the activity coefficient of the dilute undissociated acid can be taken to be $\gamma_{HA} = 1$. Since the molality of the undissociated acid is $(m - y)$, its activity is thus approximately,

$$a_{HA} = (m - y)/m^o \tag{6.10.17a}$$

The product of the ion activity coefficients serves to define the geometric mean activity coefficient as

$$\gamma_\pm^2 = \gamma_{H_3O^+}\gamma_{A^-} \tag{6.10.17b}$$

Into Eq. (6.10.15), substitute Eqs. (6.10.16a) and (6.10.16b) followed by Eqs. (6.10.17a) and Eq. (6.10.17b). The result is

$$K_a = \frac{\gamma_{H_3O^+}\gamma_{A^-}y^2}{m^o(m-y)} = \frac{\gamma_\pm^2 y^2}{m^o(m-y)} \tag{6.10.18}$$

The value of $K_a = 1.754 \times 10^{-5}$ is small enough that the ionized fraction of the acid is slight; hence, we can introduce the zeroth order approximation, $(m-y) \approx m$. At this same level of approximation, we can also set $\gamma_\pm = 1$. Equation (6.10.18) then becomes

$$K_a = \frac{y^2}{m^o m} \tag{6.10.19a}$$

Equation (6.10.19a) is readily solved for y. Recalling that $m = 0.1$ molal, we find that

$$y = \sqrt{m^o m K_a} = \sqrt{(0.1)(1.754 \times 10^{-5})} = 1.324 \times 10^{-3} \tag{6.10.19b}$$

Now suppose that sufficient sodium chloride is added to make the solution 0.01 molal in the strong electrolyte. All of the ions in the solution must be taken into account in calculating the ionic strength, so we write

$$I = \frac{1}{2}\sum_{s=1}^{4} Z_s^2 m_s \tag{6.10.20a}$$

We set $m_{Na^+} = 0.01 = m_{Cl^-} = 0.01$ and use the estimate in Eq. (6.10.19b) to evaluate $m_{H_3O^+} = m_{A^-} = 1.324 \times 10^{-3}$. Substitution into Eq. (6.10.20a) gives the result,

$$\begin{aligned} I &= \frac{1}{2}\big[(+1)^2(0.01) + (-1)^2(0.01) + (+1)^2(1.324 \times 10^{-3}) \\ &\quad +(-1)^2(1.324 \times 10^{-3})\big] \\ &= 1.1324 \times 10^{-2} \end{aligned} \tag{6.10.20b}$$

According to Eq. (6.9.10), the value of the log of the geometric mean activity coefficient is

$$\log \gamma_\pm = -\frac{0.5092|(+1)(-1)|\sqrt{1.1324 \times 10^{-2}}}{1 + \sqrt{1.1324 \times 10^{-2}}} = -4.897 \times 10^{-2} \tag{6.10.20c}$$

which when solved for the activity coefficient gives

$$\gamma_\pm = 0.8934 \tag{6.10.21}$$

To begin the first order level of approximation, we set $(m-y) = 0.1 - 0.00132 = 0.09868$. When $(m-y) = 0.09868$, $m^o = 1$, and $\gamma_\pm = 0.8934$ are substituted into

Eq. (6.10.18), we have as the first order approximation,

$$K_a = \frac{\gamma_{\pm}^2 y^2}{m^o(m - y)} = \frac{(0.8934)^2 y^2}{0.09868} \tag{6.10.22}$$

When solved for y, the result is

$$y = \frac{\sqrt{0.09868 \times 1.754 \times 10^{-5}}}{0.8934} = 1.473 \times 10^{-3} \tag{6.10.23}$$

The addition of 0.01 molal NaCl thus enhances the ionization by

$$\frac{(1.473 \times 10^{-3}) - (1.324 \times 10^{-3})}{(1.324 \times 10^{-3})} \times 100\% = 11\% \tag{6.10.24}$$

In this case, the extent of "salting-in" is 11%.

6.10.3 Primary Kinetic Salt Effect

When two ions undergo an electron transfer reaction, the ions pass through a charged transition state before forming the products. Since charge must be conserved, the charge of the transition state is necessarily the sum of the charges on the two reactants. As every degree of freedom possessed by the transition state is assumed to be at equilibrium except for that corresponding to the reaction coordinate, the transition state can be expected to be surrounded by an equilibrium ion atmosphere consistent with its total charge.

Denote the reacting ions by $(1)^{Z_1}$ and $(2)^{Z_2}$, respectively, where Z_1 and Z_2 are the corresponding electrovalences. The charged transition state is denoted by $\otimes^{(Z_1+Z_2)}$, where by conservation of charge, its electrovalence is $(Z_1 + Z_2)$. The stoichiometry of the reaction, including the assumed equilibrium between the reactants and the transition state is

$$(1)^{Z_1} + (2)^{Z_2} \rightleftarrows \otimes^{(Z_1+Z_2)} \rightarrow \text{products} \tag{6.10.25}$$

The activities, a_1, a_2 and $a_\otimes$ of the corresponding charged species are related to the equilibrium constant, $K_\otimes$, for formation of the transition state by

$$K_\otimes = \frac{a_\otimes}{a_1 a_2} \tag{6.10.26}$$

If the solution is sufficiently dilute, we need not make any distinction between molality and molarity. In such case, the activities can be written in terms of the corresponding molar concentrations, c_1, c_2 and $c_\otimes$, as

$$a_1 = \gamma_1(c_1/c^o) \tag{6.10.27a}$$

$$a_2 = \gamma_2(c_2/c^o) \tag{6.10.27b}$$

and

$$a_\otimes = \gamma_\otimes(c_\otimes/c^o) \tag{6.10.27c}$$

where $c^o = 1.0$ M is the concentration in the standard state, and γ_1, γ_2 and $\gamma_\otimes$ are the respective activity coefficients. Upon substitution of Eqs. (6.10.27a), (6.10.27b), and (6.10.27c) into Eq. (6.10.26), one obtains

$$K_\otimes = \frac{\gamma_\otimes}{\gamma_1\gamma_2}\frac{(c_\otimes/c^o)}{(c_1/c^o)(c_2/c^o)} \tag{6.10.28a}$$

Equation (6.10.28a) can be solved for the transition state concentration,

$$c_\otimes = \left(\frac{K_\otimes}{c^o}\right)\left(\frac{\gamma_1\gamma_2}{\gamma_\otimes}\right)c_1c_2 \tag{6.10.28b}$$

According to transition state theory, the instantaneous value, r, of the rate of reaction is given by

$$r = \nu c_\otimes \tag{6.10.28c}$$

where ν is the frequency which determines the rate of decay of the transition state into the products. We can substitute Eq. (6.10.28b) into Eq. (6.10.28c) to obtain for the rate

$$r = \nu\left(\frac{K_\otimes}{c^o}\right)\left(\frac{\gamma_1\gamma_2}{\gamma_\otimes}\right)c_1c_2 \tag{6.10.29}$$

In the laboratory, the instantaneous reaction rate is given by

$$r = k_{obs}c_1c_2 \tag{6.10.30}$$

where k_{obs} is the observed value of the rate constant. Set Eq. (6.10.29) equal to (6.10.30) and cancel the product, c_1c_2, from either side. The result is

$$k_{obs} = \nu\left(\frac{K_\otimes}{c^o}\right)\left(\frac{\gamma_1\gamma_2}{\gamma_\otimes}\right) \tag{6.10.31}$$

Equation (6.10.31) expresses the observed rate constant in terms of the relevant properties of the reacting ions and the transition state.

In the hypothetical state of infinite dilution, the ionic strength will be zero, and the activity coefficients will satisfy, $\gamma_1 = \gamma_2 = \gamma_\otimes = 1$. On the basis of Eq. (6.10.31), we can define a zero ionic strength rate coefficient,k_{obs}^0, by

$$k_{obs}^0 = \nu\left(\frac{K_\otimes}{c^o}\right) \tag{6.10.32}$$

Using Eq. (6.10.32), Eq. (6.10.31) can then be rewritten in the form,

$$k_{obs} = \left(\frac{\gamma_1\gamma_2}{\gamma_\otimes}\right)k_{obs}^0 \tag{6.10.33}$$

Equation (6.10.33) can be represented in logarithmic form as

$$\log k_{obs} = \log k_{obs}^0 + \log\gamma_1 + \log\gamma_2 - \log\gamma_\otimes \tag{6.10.34}$$

If the solution is sufficiently dilute so that the Debye–Hückel limiting law applies, then equations of the form of Eqs. (6.9.9a) and (6.9.9b) give for the individual ion activity coefficients,

$$\log\gamma_1 = -AZ_1^2\sqrt{I} \tag{6.10.35a}$$

$$\log\gamma_2 = -AZ_2^2\sqrt{I} \tag{6.10.35b}$$

$$\log\gamma_\otimes = -A(Z_1 + Z_2)^2\sqrt{I} \tag{6.10.35c}$$

where by conservation of charge, the electrovalence of the transition state is $Z_1 + Z_2$. Substitute Eqs. (6.10.35a) through (6.10.35c) into Eq. (6.10.34) to obtain,

$$log(k_{obs}) = log(k_{obs}^0) - AZ_1^2\sqrt{I} - AZ_1^2\sqrt{I} + A(Z_1 + Z_2)^2\sqrt{I} \tag{6.10.36a}$$

If the product, $A\sqrt{I}$, is factored from the right hand side of Eq. (6.10.36a), the result is

$$log(k_{obs}) = log(k_{obs}^0) + \left[-Z_1^2 - Z_1^2 + (Z_1 + Z_2)^2\right]A\sqrt{I} \tag{6.10.36b}$$

Finally, upon squaring the binomial and canceling terms of opposite sign, we arrive at

$$log(k_{obs}) = log(k_{obs}^0) + 2AZ_1Z_2\sqrt{I} \tag{6.10.37}$$

Equation (6.10.37) indicates that when $log(k_{obs})$ is plotted versus $\sqrt{I}$, a straight line should result. The slope of the line is preportional to Z_1Z_2 while the intercept is

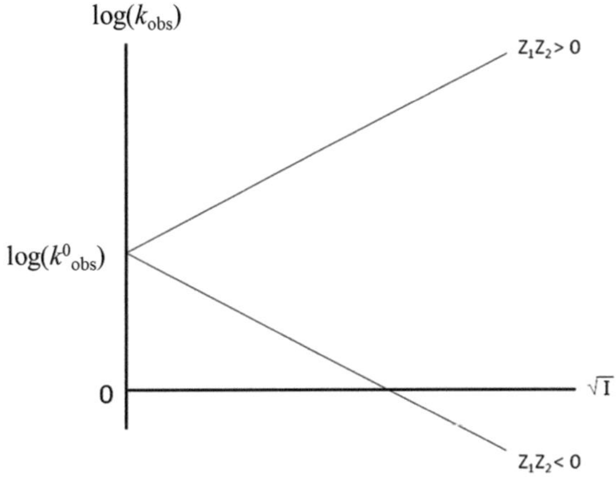

Fig. 6.2 Plot of $\log(k_{obs})$ versus $\sqrt{I}$ according to Eq. (6.10.37). The intercept labeled, $\log(k_{obs}^0)$, on the $\log(k_{obs})$ axis is the rate constant in the limit of zero ionic strength

equal to $log(k_{obs}^0)$. The ion atmosphere serves to weaken the Coulomb force acting between the reactants. According to Eq. (6.10.37), this weakening has the effect of accelerating the rate of reaction when Z_1 and Z_2 have the same sign and decelerating the rate of reaction when Z_1 and Z_2 have opposite signs. See Fig. 6.2.

If the solution is a mixture of ions of differing electrovalences, the sign of the product, Z_1Z_2, can serve as a clue to the identity of the two ions which are responsible for the reaction. The value of k_{obs}^0 provides information on the rate of reaction in the absence of Debye – Hückel screening. Equation (6.10.37) is said to describe the "primary kinetic salt effect".

6.10.4 Ostwald's Dilution Law

Wilhelm Ostwald (German, 1853–1932) noticed that the conductivity of a solution of a weak electrolyte increased markedly upon dilution. S. Arrhenius (Swedish, 1859–1927) had previously shown that the molar conductance, Λ_V, of a weak acid in a volume, V, of solution was proportional to the degree of dissociation, α, of the acid, where $0 < \alpha < 1$. The coefficient of proportionality between Λ_V and α was, Λ_∞, the molar conductance extrapolated to infinite dilution. As the conductance strongly increased as the acid concentration decreased, it was difficult to determine the ionization constant of the weak electrolyte.

Consider the ionization of a weak electrolyte, AB,

$$AB \rightarrow A^+ + B^- \tag{6.10.38}$$

If K is the equilibrium constant for the ionization reaction, then the law of mass action states,

$$K = \frac{a_{A^+}a_{B^-}}{a_{AB}} \tag{6.10.39}$$

In terms of the corresponding molar concentrations m_{A^+}, m_{B^-}, and m_{AB}, and activity coefficients, γ_{A^+}, γ_{B^+}, γ_{AB}, the activities, a_{A^+}, a_{B^-}, a_{AB} are given by

$$a_{A^+} = \gamma_{A^+}(m_{A^+}/m^o) \tag{6.10.40a}$$

$$a_{B^-} = \gamma_{B^-}(m_{B^-}/m^o) \tag{6.10.40b}$$

$$a_{AB} = \gamma_{AB}(m_{AB}/m^o) \tag{6.10.40c}$$

where $m^o = 1$ molal, and m_{A^+}, m_{B^-}, m_{AB} are molalities. Upon substitution of Eqs. (6.10.40a), (6.10.40b), and (6.10.40c) into Eq. (6.10.39), the result is

$$K = \frac{\gamma_{A^+}(m_{A^+}/m^o)\gamma_{B^-}(m_{B^-}/m^o)}{\gamma_{AB}(m_{AB}/m^o)} \tag{6.10.41}$$

The geometric mean activity coefficient is

$$\gamma_\pm^2 = \gamma_{A^+}\gamma_{B^-} \tag{6.10.42}$$

Upon substitution of Eq. (6.10.42) into Eq. (6.10.41), the result is

$$K = \frac{\gamma_\pm^2}{\gamma_{AB}} \frac{m_{A^+}m_{B^-}}{m_{AB}m^o} \tag{6.10.43}$$

In terms of the fraction, α, of the weak electrolyte concentration, m_{AB}, the unionized concentration is $(1-\alpha)m_{AB}$, and the individual ion concentrations are

$$m_{A^+} = \alpha m_{AB} \tag{6.10.44a}$$

$$m_{B^-} = \alpha m_{AB} \tag{6.10.44b}$$

Substitute Eqs. (6.10.44a) and (6.10.44b) into Eq. (6.10.43) to obtain

$$K = \frac{\gamma_\pm^2}{\gamma_{AB}} \cdot \frac{\alpha^2(m_{AB})^2}{(1-\alpha)m_{AB}m_o} \simeq \gamma_\pm^2 \frac{\alpha^2 m_{AB}}{(1-\alpha)m^o} \tag{6.10.45a}$$

where in the case of the neutral species, *AB*, we have again represented its activity coefficient by unity. The logarithm of Eq. (6.10.45a) reads,

$$\log K = 2\log\gamma_\pm + \log\frac{\alpha^2}{(1-\alpha)} + \log\left(m_{AB}/m^o\right) \tag{6.10.45b}$$

Because we are considering a solution under conditions of increasing dilution, where $I \ll 1$, we can approximate Eq. (6.9.5) by writing it as

$$\log \gamma_{\pm} = -A|Z_{A^+}||Z_{B^-}|\sqrt{I} = -A\sqrt{I} \tag{6.10.46a}$$

where $Z_{A^+} = 1$ and $Z_{B^-} = -1$. The ionic strength calculated on the basis of the ion concentrations in Eqs. (6.10.44a) and (6.10.44b) is

$$I = \frac{1}{2}\left[Z_{A^+}^2 m_{A^+} + Z_{B^-}^2 m_{B^-}\right] = \frac{1}{2}\left[(+1)^2 \alpha m_{AB} + (-1)^2 \alpha m_{AB}\right] = \alpha m_{AB} \tag{6.10.46b}$$

Upon substitution of Eq. (6.10.46b), Eq. (6.10.46a) reads,

$$\log \gamma_{\pm} = -A\sqrt{\alpha m_{AB}} \tag{6.10.46c}$$

Now substitute Eq. (6.10.46c) into Eq. (6.10.45b) with the result,

$$\log K = -2A\sqrt{\alpha m_{AB}} + \log \frac{\alpha^2}{(1-\alpha)} + \log\left(m_{AB}/m^o\right) \tag{6.10.46d}$$

Rearranged, Eq. (6.10.46d) reads

$$\log \frac{\alpha^2}{(1-\alpha)} - 2A\sqrt{\alpha m_{AB}} = \log K - \log\left(m_{Ab}/m^o\right) \tag{6.10.47a}$$

As a measure to the fraction ionized, we use the Arrhenius law,

$$\alpha = \frac{\Lambda}{\Lambda_\infty} \tag{6.10.47b}$$

where Λ is the molar conductance at the finite concentration, m_{AB}, while Λ_∞ is the molar conductance at infinite dilution. Upon substitution of Eq. (6.10.47b), Eq. (6.10.47a) reads,

$$\log\left[\frac{(\Lambda/\Lambda_\infty)^2}{1-(\Lambda/\Lambda_\infty)}\right] - 2A\sqrt{(\Lambda/\Lambda_\infty)m_{AB}} = \log K - \log\left(m_{AB}/m^o\right) \tag{6.10.48}$$

The left hand side of Eq. (6.10.48) can be plotted as a function of $\log(m_{AB}/m^o)$. The result is a straight line with intercept given by $\log K$. The conductivity method thus suffices to determine the ionization constant of a weak electrolyte, which by other methods is often difficult to determine.

6.11 Conclusions

In this chapter we have used spherical polar coordinates to solve the Debye–Hückel equation for the case of dilute electrolytes. Assuming that the ions of the electrolyte form an ideal solution when uncharged, we have used the Güntelberg–Müller charging process to calculate the Helmholtz free energy of the solution when the ions are fully charged. From the Helmholtz energy, we have computed the chemical potential. By comparing this representation of the chemical potential with that defined in thermodynamics, we were able to obtain a formula for the geometric mean activity coefficient of the electrolyte. The formula for the geometric mean activity coefficient was used to demonstrate the salting in effect of an inert electrolyte on a weakly soluble ionic solid, the effect on an inert electrolyte on the rate of a reaction involving ions, and the effect of an inert electrolyte on the ionization of a weak acid.

References

1. G.N. Lewis, M. Randall as revised by K.S. Pitzer, L. Brewer, *Thermodynamics* (McGraw-Hill, New York, 1961), chap. 23
2. L. Hoddeson, E. Braun, J. Teichmann, S. Weart, *Out of the Crystal Maze* (Oxford UP, Oxford, UK, 1992), p.490
3. S.A. Rice, M. Nagasawa, *Polyelectrolyte Solutions* (Academic Press, New York, 1961), pp.99–119
4. D.A. McQuarrie, *Statistical Mechanics* (Harper and Row, New York, 1973), pp.326–339
5. M.C. Gupta, *Statistical Thermodynamics* (Wiley, New York, 1990), chap. 19

Recommended Additional Reading

1. S.E. Wood, R. Battino, *Themodynamics of Chemical Systems* (Cambridge University Press, Cambridge, UK, 1990)
2. D. Kondepudi, I. Prigogine, *Modern Thermodynamics*, 2nd edn. (Wiley, Chichester, W. Sussex, UK, 2015)
3. J.W. Moore, R.G. Pearson, *Kinetics and Mechanism* (Wiley, New York, 1981)

Chapter 7
Electromagnetic Waves

In this chapter, we introduce Ampere's law of magnetism and Faraday's law of electromagnetic induction. When Ampere's law and Faraday's law are combined with Gauss' law, $\nabla \cdot \mathbf{D} = \rho$, and the equation $\nabla \cdot \mathbf{B} = 0$, expressing the continuity of the magnetic induction, the resulting set is known as Maxwell's equations. This powerful combination of equations can be used to determine the space and time dependence of the electromagnetic field under any set of circumstances. To apply Maxwell's equations to matter requires that they be supplemented by experimentally determined constitutive relations, one linking the electric displacement to the electric field and the other linking the magnetic induction to the magnetic field. With these relations specified, we solve the four Maxwell equations for the case of a plane electromagnetic wave propagating through a dielectric medium. To include the frequency dependence of the electric permittivity of the material through which the wave is propagating in our calculation, we exploit the method of Fourier transforms. The Fourier transform pair of equations permits us to derive the *convolution theorem*, and give mathematical expression to the principle of causality, which asserts that an effect cannot precede its cause. The law of conservation of electromagnetic energy is derived. This law leads to the identification of the *Poynting vector* that governs the propagation of electromagnetic energy through a dispersive medium. On the basis of a model for the frequency response of a medium, we use the *Poynting vector* concept to construct a purely electromagnetic proof of the Beer–Lambert Law of spectrophotometry.

7.1 Introduction

So far, we have encountered phenomena which are static or are sufficiently time independent that the theory of electrostatics suffices for their description. Our objective in this chapter is to take into account true time dependence. In addition to electric charge distributions, we will encounter electric current distributions. The latter

J. K. Baird, *Electromagnetism for Chemistry*,
https://doi.org/10.1007/978-3-031-88508-2_7

are the sources of magnetic fields. The complete description of the charge distributions, the current distributions, and the simultaneous electric and magnetic fields, which they generate, will require us to introduce in addition to Gauss' Law, three new vector differential equations. One of these is a continuity equation, which is similar to Gauss' law in the sense that it governs the continuity of the magnetic field lines. The remaining two equations, named for Ampere (Andre Marie Ampere, French, 1775–1836) and Faraday (Michael Faraday, English, 1791–1867), respectively, specify the coupling between the electric field and the magnetic field. When the electric and magnetic fields are time varying, electromagnetic waves are possible. Following Maxwell (James Clerk Maxwell, Scottish, 1831–1879), we will show that these waves travel through empty space at the speed of light.

The main goal of this chapter is to derive the equations that govern the propagation of electromagnetic waves through matter. One of our results will be the Beer–Lambert law, which forms the basis for applications of spectrophotometry in chemistry [1]. The wave equations will also prove useful in the next chapter where they serve as the basis for the theory of light scattering.

In the case of the static phenomena, which we have considered so far, the displacement, **D**, has its source in the density of free charge, ρ, as expressed by Gauss' law,

$$\nabla \cdot \mathbf{D} = \rho \tag{7.1.1}$$

The corresponding electric field satisfies Eq. (7.1.1) in the form,

$$\nabla \cdot \mathbf{E} = \rho / \varepsilon \tag{7.1.2}$$

where ε is the electric permittivity, which is defined by

$$\varepsilon = \varepsilon_r \varepsilon_o \tag{7.1.3}$$

where ε_o is the permittivity of free space, and ε_r is the relative permittivity of the material.

It is possible to render Eq. (7.1.2) in an integral form. If both sides of Eq. (7.1.2) are integrated over the volume containing the charge density, the result is

$$\int d^3 r \nabla \cdot \mathbf{E} = \frac{1}{\varepsilon_r \varepsilon_o} \int d^3 r \rho = \frac{q}{\varepsilon_r \varepsilon_o} \tag{7.1.4a}$$

where Eq. (7.1.3) has been used, and where $q = \int d^3 r \rho$ is the total charge within the volume. By virtue of the Gauss-Green divergence theorem [See the Mathematical Methods Appendix],

$$\int d\mathbf{S} \cdot \mathbf{E} = \int d^3 r \nabla \cdot \mathbf{E} \tag{7.1.4b}$$

where $d\mathbf{S}$ is an element of surface area surrounding the volume. Upon combining Eqs. (7.1.4a) and (7.1.4b), we arrive at

$$\int d\mathbf{S} \cdot \mathbf{E} = \frac{q}{\varepsilon_r \varepsilon_o} \tag{7.1.5}$$

Equation (7.1.5) is the integral form of Gauss' law. It serves as the basis for the idea of conservation of electric field lines. It states that $q/\varepsilon_r\varepsilon_o$ is the total number of electric field lines emanating from the volume containing the total charge, q.

It is now possible to compare electrostatic and magnetostatic phenomena with respect to continuity of their respective field lines. We use the symbol $\mathbf{B}$ to stand for the magnetic induction (sometimes referred to as the magnetic flux density). The magnetic induction is related to the magnetic field, $\mathbf{H}$, by

$$\mathbf{B} = \mu \mathbf{H} \tag{7.1.6}$$

where

$$\mu = \mu_r \mu_o \tag{7.1.7}$$

In Eq. (7.1.6), μ is the magnetic permeability of the material, μ_r is the relative magnetic permeability of the material, and $\mu_o = 4\pi \times 10^{-7}$ N/A^2 is the permeability of free space.

By contrast to lines of electric displacement that start on positive charges and end on negative charges, lines of magnetic induction pass continuously through a material without interruption. This is a consequence of the fact that an isolated magnetic pole, which is the analog of the isolated charge, q, in Eq. (7.1.5), does not exist. There are, of course north poles, and south poles, but unlike free charges, a north pole never appears in the absence of a south pole and vice versa. Due to the absence of free magnetic poles, the magnetic analog of Eq. (7.1.5) is

$$\int \mathbf{B} \cdot d\mathbf{S} = 0 \tag{7.1.8}$$

By virtue of the Gauss-Green divergence theorem,

$$\int \mathbf{B} \cdot d\mathbf{S} = \int d^3 r \nabla \cdot \mathbf{B} \tag{7.1.9}$$

Since in Eq. (7.1.9), the volume element in the integral and the surface that surrounds it are arbitrary, we can conclude that

$$\nabla \cdot \mathbf{B} = 0 \tag{7.1.10}$$

Stated alone, Eqs. (7.1.2) and (7.1.10) would imply that **E** and **B** are independent. That is true if both **E** and **B** do not depend upon the time. In such case, the electrostatics and magnetostatics are independent.

7.2 Time Dependent Phenomena

7.2.1 *Faraday's Law*

The flux Φ_m of magnetic field lines crossing a surface is defined as

$$\Phi_m = \int \mathbf{B} \cdot d\mathbf{S} \tag{7.2.1}$$

Faraday passed a loop of wire through the field lines of a magnet. He found that as long as the loop was moving, a voltage difference appeared between the ends of the loop. He realized that in the moving frame of the loop, the magnetic field lines appeared to be time dependent. See Fig. 7.1.

When Φ_m was increasing with time, the observed voltage difference was positive, whereas when it was decreasing, the observed voltage difference was negative. If ΔV is the voltage difference, then Faraday's observations can be summarized by the equation,

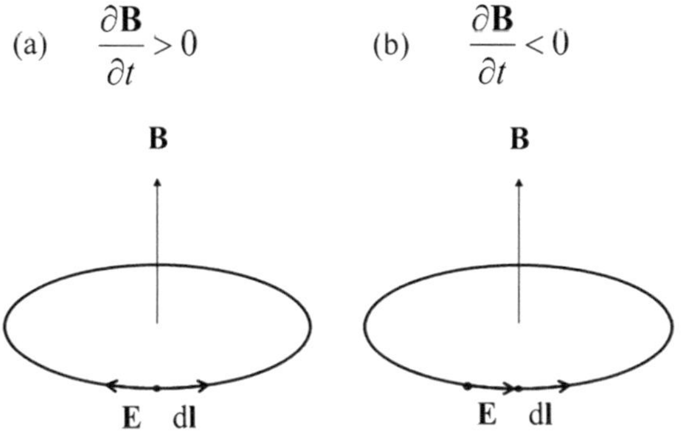

Fig. 7.1 **a** A magnetic induction increasing with time, $(\partial \mathbf{B}/\partial t) > 0$, acting perpendicular to a current loop produces an electric field, **E**, in the clockwise direction. The clockwise direction is opposite to the differential element of circumference, $d\mathbf{l}$, that defines the positive direction of circulation around the loop. **b** A magnetic induction decreasing with time, $(\partial \mathbf{B}/\partial t) < 0$, acting perpendicular to a current loop produces an electric field, **E**, in the counter clockwise direction. The direction of the electric field is parallel to the direction of the differential element of circumference, $d\mathbf{l}$, that defines the positive direction around the loop

$$\frac{\partial \Phi_m}{\partial t} = \Delta V \tag{7.2.2}$$

If $\mathbf{E}$ is the electric field, then $\int d\mathbf{l} \cdot \mathbf{E}$ is tangent to the wire. The loop integral traversing the entire circumference of the wire is

$$\Delta V = -\oint d\mathbf{l} \cdot \mathbf{E} \tag{7.2.3}$$

In Eq. (7.2.3), as in Fig. 7.1, $d\mathbf{l}$ is an differential element of the length of the wire, and the positive direction of integration is assumed to be counter clockwise. When Eqs. (7.2.1) and (7.2.3) are substituted into Eq. (7.2.2), the result is

$$-\oint d\mathbf{l} \cdot \mathbf{E} = \int \frac{\partial \mathbf{B}}{\partial t} \cdot d\mathbf{S} \tag{7.2.4a}$$

where $d\mathbf{S}$ is an element of the surface area bounded by the contour of the wire. By use of Stokes' theorem [See the Mathematical Methods Appendix], the integral on the left of Eq. (7.2.4a) can be written as a surface integral,

$$\oint d\mathbf{l} \cdot \mathbf{E} = \int (\nabla \times \mathbf{E}) \cdot d\mathbf{S} \tag{7.2.4b}$$

Upon substitution of Eq. (7.2.4b) into the left hand side of Eq. (7.2.4a), the result is

$$-\int (\nabla \times \mathbf{E}) \cdot d\mathbf{S} = \int \frac{\partial \mathbf{B}}{\partial t} \cdot d\mathbf{S} \tag{7.2.4c}$$

Upon equating the integrands on either side Eq. (7.2.4c), we obtain

$$\nabla \times \mathbf{E} = -\frac{\partial \mathbf{B}}{\partial t} \tag{7.2.5}$$

Equation (7.2.5) is known as Faraday's law of induction.

7.2.2 Ampere's Law

The experiments of Ampere and Oersted (Hans Christian Oersted, Danish, 1777–1851) indicated that when a wire carrying a current was brought near a compass needle, the needle was deflected. The direction of the deflection depended upon the direction of the current.

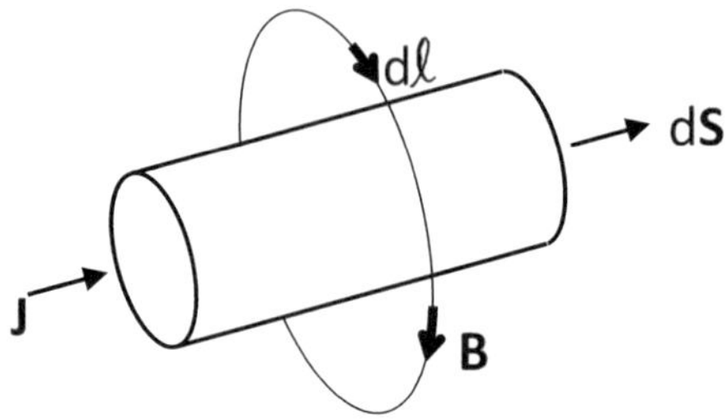

Fig. 7.2 A line of magnetic induction encircling a section of a wire of differential cross-sectional area, $d\mathbf{S}$, through which a steady current density, $\mathbf{J}$, is passing. A differential element of the length of the closed path followed by the induction, $\mathbf{B}$, is denoted by $d\ell$

The current carried by the wire produces lines of magnetic induction encircling it. See Fig. 7.2.

According to Ampere, the loop integral of **B** around any path completely enclosing the wire is related to the current, I, in the wire by

$$\oint (\mathbf{B}/\mu) \cdot d\mathbf{l} = I \tag{7.2.6a}$$

where $d\mathbf{l}$ is the vector differential element of length $d\ell$ of the encircling path, and μ is the magnetic permeability of the material surrounding the wire. The current density, $\mathbf{J}$ (A/m^2), in the wire, and the current, I, are related by the surface integral,

$$I = \int d\mathbf{S} \cdot \mathbf{J} \tag{7.2.6b}$$

where $d\mathbf{S}$ is an element of the cross-sectional area of the wire. Upon equating Eqs. (7.2.6a) and (7.2.6b), the result is

$$\oint (\mathbf{B}/\mu) \cdot d\mathbf{l} = \int d\mathbf{S} \cdot \mathbf{J} \tag{7.2.6c}$$

By virtue of Stokes' theorem [See the Mathematical Methods Appendix], the left hand side of Eq. (7.2.6c) can be written as

$$\oint (\mathbf{B}/\mu) \cdot d\mathbf{l} = \int d\mathbf{S} \cdot (\nabla \times \mathbf{B}/\mu) \tag{7.2.6d}$$

When combined, Eqs. (7.2.6c) and (7.2.6d) imply that

$$\int d\mathbf{S} \cdot (\nabla \times \mathbf{B}/\mu) = \int d\mathbf{S} \cdot \mathbf{J} \tag{7.2.6e}$$

Upon equating the integrands in Eq. (7.2.6e) and use of Eq. (7.1.6), the result is

$$\nabla \times \mathbf{H} = \mathbf{J} \tag{7.2.7}$$

Equation (7.2.7) is known as Ampere's law.

Maxwell suggested that Eq. (7.2.7) was incomplete when time dependent phenomena were under consideration. To the right hand side of Eq. (7.2.7), Maxwell added the time derivative, $\partial \mathbf{D}/\partial t$, to obtain,

$$\nabla \times \mathbf{H} = \mathbf{J} + \frac{\partial \mathbf{D}}{\partial t} \tag{7.2.8}$$

He then showed that Eqs. (7.1.1), (7.1.10), (7.2.5), and (7.2.8) were the necessary and sufficient conditions for the self-consistent determination of the electric and magnetic fields. From these equations, he was able to predict the existence of electromagnetic waves and calculate the speed of light in vacuum.

7.2.3 *Electrostatics and Magnetostatics*

The situation where $\partial \mathbf{B}/\partial t = 0$ and ρ is time independent constitutes a special case. In this static case, the electric field is determined by

$$\nabla \cdot \mathbf{E} = \rho/\varepsilon \tag{7.2.9a}$$

and according to Eq. (7.2.5)

$$\nabla \times \mathbf{E} = 0 \tag{7.2.9b}$$

This pair of equations governs the subject of electrostatics.

When $\partial \mathbf{D}/\partial t = 0$ and $\mathbf{J}$ is steady, then according to Eqs. (7.1.6), (7.1.10) and (7.2.8), the magnetic field is determined by

$$\nabla \times \mathbf{H} = \mathbf{J} \tag{7.2.10a}$$

with

$$\nabla \cdot \mathbf{H} = 0 \tag{7.2.10b}$$

This second pair of equations is said to define the subject of magnetostatics. By virtue of the pair of Eqs. (7.2.9) and the pair of Eqs. (7.2.10), the electric and magnetic fields in the static case are specified independently of one another.

7.2.4 Maxwell's Equations

With the inclusion of $\partial \mathbf{D}/\partial t$ in Eq. (7.2.8), the mathematical basis for understanding the electromagnetic phenomena that occur in material media of importance to chemistry is now complete. Considered as a group, the following are known as Maxwell's equations:

$$\nabla \cdot \mathbf{D} = \rho \tag{7.2.11a}$$

$$\nabla \times \mathbf{E} = -\frac{\partial \mathbf{B}}{\partial t} \tag{7.2.11b}$$

$$\nabla \times \mathbf{H} = \mathbf{J} + \frac{\partial \mathbf{D}}{\partial t} \tag{7.2.11c}$$

$$\nabla \cdot \mathbf{B} = 0 \tag{7.2.11d}$$

As mentioned above, Eq. (7.2.11a) is known as Gauss' Law, Eq. (7.2.11b) is known as Faraday's Law, and Eq. (7.2.11c) is known as Ampere's Law. The outlier is Eq. (7.1.11d) which has no official name. The reader can assign their own name to this equation!

We now turn to the constitutive relations. These equations summarize the response of matter to the fields. Of the four vector fields, **D**, **E**, **B**, and **H**, mentioned in Eqs. (7.2.11), it is **E** and **B** that take precedence, because they determine the electric and magnetic forces, respectively, which act on a charged particle. In particular, consider a particle with charge, q, proceeding through space with velocity, **v**. In the presence of an electric field, **E**, the electric force on the particle is $q\mathbf{E}$, while in the presence of a magnetic induction, **B**, the magnetic force is $q(\mathbf{v} \times \mathbf{B})$. The former is known as the Coulomb force. The latter is known as the Lorentz force.

As we have seen, an applied electric field polarizes a dielectric material. In materials that respond linearly, the polarization, **P**, is proportional to the applied electric field, **E**, according to

$$\mathbf{P} = \varepsilon_o \chi_e \mathbf{E} \tag{7.2.12}$$

where χ_e is termed the electric susceptibility. The displacement **D** is related to **P** and **E** by

$$\mathbf{D} = \varepsilon_o \mathbf{E} + \mathbf{P} \tag{7.2.13}$$

If we substitute Eq. (7.2.12) into Eq. (7.2.13), the result is

$$\mathbf{D} = \varepsilon_o (1 + \chi_e) \mathbf{E} \tag{7.2.14}$$

If we write

$$\mathbf{D} = \varepsilon \mathbf{E} \tag{7.2.15}$$

where ε is the electric permittivity of the material, Eq. (7.2.15) is the constitutive relation for the electric field. Upon comparing Eqs. (7.2.14) and (7.2.15), ε depends upon the electric susceptibility, χ_e, according to

$$\varepsilon = \varepsilon_o(1 + \chi_e) \tag{7.2.16}$$

By comparison of Eqs. (7.1.3) and (7.2.16), the relative permittivity satisfies,

$$\varepsilon_r = 1 + \chi_e \tag{7.2.17}$$

The values of the relative permittivity, ε_r, of most chemical substances lie in the range of 1–40. Exceptions are the highly polar, hydrogen bonded liquids, such as hydrogen peroxide and water. For example, near room temperature, water exhibits a relative permittivity of $\varepsilon_r = 80$.

The magnetic equivalent of Eqs. (7.2.12) through (7.2.17) can be constructed by recognizing, that the fundamental field is **B** whereas **H** is a derived field. By convention, the magnetization, **M**, i.e. the magnetic dipole moment per unit volume, is related to the magnetic field, **H**, by

$$\mathbf{M} = \chi_m \mathbf{H} \tag{7.2.18}$$

where χ_m is the magnetic susceptibility. The magnetic field, **H**, in turn is related to the induction, **B**, and the magnetization, **M**, by

$$\mathbf{H} = \frac{1}{\mu_o}\mathbf{B} - \mathbf{M} \tag{7.2.19}$$

The appearance of **B** and $1/\mu_o$ on the right hand side of Eq. (7.2.19), like **E** and ε_o on the right hand side of Eq. (7.2.13), reminds us that **B** is fundamental in the sense that it determines the force acting on moving charges. Upon solving for **B**, Eq. (7.2.19) reads

$$\mathbf{B} = \mu_o \mathbf{H} + \mu_o \mathbf{M} \tag{7.2.20}$$

If we substitute Eq. (7.2.18) into Eq. (7.2.20). The result is

$$\mathbf{B} = \mu_o(1 + \chi_m)\mathbf{H} \tag{7.2.21}$$

According to Eq. (7.1.6),

$$\mathbf{B} = \mu \mathbf{H} \tag{7.2.22}$$

Equation (7.2.22) is the constitutive relation for the magnetic field. Equations (7.2.21) and (7.2.22) are made compatible if

$$\mu = \mu_o(1 + \chi_m) \tag{7.2.23}$$

By comparing Eq. (7.1.7) with Eq. (7.2.23), the relative magnetic permeability is given by

$$\mu_r = 1 + \chi_m \tag{7.2.24}$$

Generally speaking, there are three classes of magnetic materials. In the first class are the *diamagnetic* materials. All of the electrons in the atoms and or molecules in diamagnetic materials have their spins paired. For every spin up electron there is a spin down electron. The weak magnetism that they exhibit is due to electron orbital motion, which opposes the action of any applied magnetic field. The corresponding magnetic susceptibilities are thus negative with values in the range $\chi_m \sim -10^{-9}$ to $-\ 10^{-5}$. The rare gases are an example. Because the susceptibility is negative, a diamagnetic material is deflected to weakest region of an applied magnetic field. The effect is slight, however, because χ_m is small. The second class of magnetic materials are said to be *paramagnetic*. Paramagnetic materials consist of atoms or molecules with one or more of their electron spins unpaired. The magnetic susceptibilities of paramagnetic materials are of the order of $\chi_m \sim 10^{-5}$ to 10^{-3}. Gadolinium sulfate is an example. In $(Gd)_2(SO_4)_3$, Gd^{2+} has seven unpaired electrons. The electron orbits in these materials still exhibit diamagnetism, but because the diamagnetism is orders of magnitude smaller than the paramagnetism, the effect of the diamagnetism is seldom noticeable. Because the net susceptibility is positive, paramagnetic materials are attracted to the strongest region of an applied magnetic field. In the third class are the *ferromagnetic* materials. Examples include elemental iron, cobalt, and nickel and some of their oxides and alloys. Like the paramagnetic materials, ferromagnetic materials consist of atoms with unpaired electron spins. In contrast to the paramagnetic materials where the source of the magnetism are the atoms responding individually to an applied magnetic field, in ferromagnetism, groups of atoms with spins pointed in the same direction respond collectively. The groups of polarized atoms are said to form magnetic domains. If a strong enough applied magnetic field is applied, the domains can grow to macroscopic size. As a consequence, ferromagnetic materials can exhibit susceptibilities, χ_m, of the order of 1000 or more. Ferromagnetic materials are attracted to the strongest region of an applied magnetic field.

Helmholtz (Hermann Ludwig von Helmholtz, German, 1821–1894) showed that a vector field is completely determined if both its divergence and its curl are specified. Given the constitutive relation, Eq. (7.2.15), linking **D** to **E**, Eqs. (7.2.11a) and (7.2.11b) fulfill the conditions of Helmholtz's theorem. Given the constitutive relation, Eq. (7.2.22), linking **B** and **H**, Eqs. (7.2.11c) and (7.2.11d) satisfy the conditions of Helmholtz's theorem. Hence, these four vector equations serve to specify both **E** and **B**.

7.3 The Continuity Equations

7.3.1 General Conservation Law in Differential Form

Consider parallel planes having cross sectional area, A, with normal in the $+x$-direction. The planes, which are a distance dx apart, enclose a volume equal to Adx. See Fig. 7.3.

The cross-sectional area of each plane is A. We assume that charge flows into the volume at x and departs at $x + dx$. At time, t, the current entering the plane at x is $J(x, t)A$, and that departing the plate at $x + dx$ is $J(x + dx, t)A$. In the space between the two planes, some of the charge disappears by charge neutralization at the rate, $(\partial Q(x, t)/\partial t)Adx < 0$. At time, t, let $\rho(x, t)$ be the volume charge density at a point, x. The net rate of change of charge within the volume is $(\partial\rho/\partial t)Adx$ is equal to the rate of loss, $(\partial Q/\partial t)Adx$, plus the net rate of flow, $J(x, t)A - J(x + dx, t)A$. The required balance equation is

$$\frac{\partial\rho(x,t)}{\partial t}Adx = \frac{\partial Q(x,t)}{\partial t}Adx + J(x,t)A - J(x+dx,t)A \tag{7.3.1a}$$

Expand $J(x + dx, t)$ in a two-term Taylor series about x. The result is

$$J(x+dx,t) = J(x,t) + \frac{\partial J(x,t)}{\partial x}(x + dx - x) \tag{7.3.1b}$$

Substitute Eq. (7.3.1b) into Eq. (7.3.1a) to obtain,

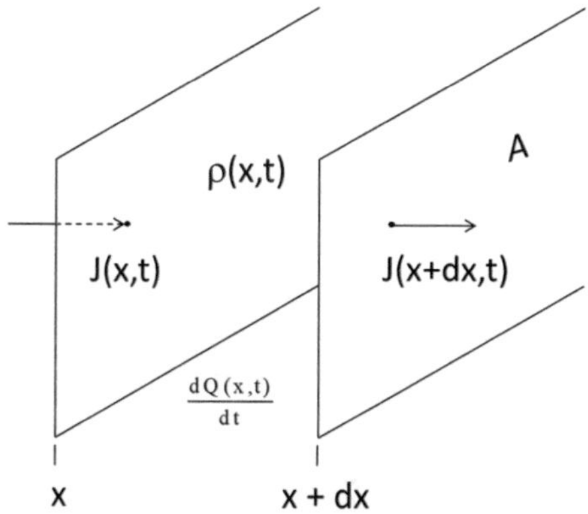

Fig. 7.3 Parallel planes spaced by a distance, dx, sandwiching a volume where the charge density is, $\rho(x, t)$, at position, x, and time, t. A current density, $J(x, t)$, enters the volume through the plane at coordinate, x, while a current density, $J(x+dx, t)$, exits the volume through the plane at coordinate, $x + dx$. Between the planes, there is a sink for charge density $Q(x, t)$, which changes at the rate, $\partial Q(x, t)/\partial t < 0$

$$\frac{\partial \rho(x,t)}{\partial t} A dx = \frac{\partial Q(x,t)}{\partial t} A dx - \frac{\partial J(x,t)}{\partial x} A dx \tag{7.3.1c}$$

Upon cancellation of Adx from either side and rearrangement, Eq. (7.3.1c) reads

$$\frac{\partial \rho(x,t)}{\partial t} + \frac{\partial J(x,t)}{\partial x} = \frac{\partial Q(x,t)}{\partial t} \tag{7.3.2}$$

Equation (7.3.2) is said to be a continuity equation.

7.3.2 Conservation of Charge

The Maxwell equations can be shown to conserve charge as illustrated in Eq. (7.3.2). To prove this assertion, we compute the divergence of both sides of Eq. (7.2.11c). The result is

$$\nabla \cdot (\nabla \times \mathbf{H}) = \nabla \cdot \mathbf{J} + \frac{\partial (\nabla \cdot \mathbf{D})}{\partial t} \tag{7.3.3}$$

By virtue of the vector identity $\nabla \cdot (\nabla \times \mathbf{H}) = 0$ [See the Mathematical Methods Appendix], the left hand side of Eq. (7.3.3) is zero. We can simplify the right hand side of Eq. (7.3.3) by substituting Eq. (7.2.11a). Equation (7.3.3) then reads

$$\nabla \cdot \mathbf{J} + \frac{\partial \rho}{\partial t} = 0 \tag{7.3.4}$$

Equation (7.3.4) is the vector analog of the scalar continuity relation in Eq. (7.3.2). The term, $\nabla \cdot \mathbf{J}$, accounts for the net rate of charge accumulation due to the spatial dependence of the current density, $\mathbf{J}$, while $\frac{\partial \rho}{\partial t}$ is the time rate of change of the buildup of charge density within the volume. Since charge neutralization is not under consideration in Maxwell's equations, there is no equivalent in Eq. (7.3.4) of the sink term, $\frac{\partial Q(x,t)}{\partial t}$, in Eq. (7.3.2).

7.3.3 Conservation of Electromagnetic Energy

Energy is stored in the electromagnetic field both in the presence of matter and in the absence of matter. The flux of electromagnetic energy has a magnitude and direction, which is known as the *Poynting vector* (John Henry Poynting, English, 1852–1914). The *Poynting vector* can be constructed by multiplying Eq. (7.2.11b) through by $\mathbf{H}$ and Eq. (7.2.11c) through by $\mathbf{E}$ and then subtracting the latter from the former. The result is

$$\mathbf{H}\cdot(\nabla\times\mathbf{E})-\mathbf{E}\cdot(\nabla\times\mathbf{H})=-\mathbf{H}\cdot\frac{\partial\mathbf{B}}{\partial t}-\mathbf{E}\cdot\frac{\partial\mathbf{D}}{\partial t}-\mathbf{J}\cdot\mathbf{E}\tag{7.3.5a}$$

The left hand side Eq. (7.3.5a) can be simplified by using the vector calculus identity,

$$\nabla\cdot(\mathbf{E}\times\mathbf{H})=\mathbf{H}\cdot(\nabla\times\mathbf{E})-\mathbf{E}\cdot(\nabla\times\mathbf{H})\tag{7.3.5b}$$

[See the Mathematical Methods Appendix]. Upon substituting Eq. (7.3.5b) into Eq. (7.3.5a) and transferring terms from right to left, the result is

$$\nabla\cdot(\mathbf{E}\times\mathbf{H})+\mathbf{H}\cdot\frac{\partial\mathbf{B}}{\partial t}+\mathbf{E}\cdot\frac{\partial\mathbf{D}}{\partial t}=-\mathbf{J}\cdot\mathbf{E}\tag{7.3.5c}$$

The sum of the second and third terms on the left hand side of Eq. (7.3.5c) can be rewritten as

$$\begin{aligned}\mathbf{H}\cdot\frac{\partial\mathbf{B}}{\partial t}+\mathbf{E}\cdot\frac{\partial\mathbf{D}}{\partial t}&=\mu\mathbf{H}\cdot\frac{\partial\mathbf{H}}{\partial t}+\varepsilon\mathbf{E}\cdot\frac{\partial\mathbf{E}}{\partial t}=\frac{\partial}{\partial t}\left[\frac{1}{2}\mu(\mathbf{H}\cdot\mathbf{H})+\frac{1}{2}\varepsilon(\mathbf{E}\cdot\mathbf{E})\right]\\&=\frac{\partial}{\partial t}\left[\frac{1}{2}(\mathbf{H}\cdot\mathbf{B}+\mathbf{E}\cdot\mathbf{D})\right]\end{aligned}\tag{7.3.5d}$$

where we have used the constitutive relations Eqs. (7.2.15) and (7.2.22). Upon substitution of Eq. (7.3.5d) into the left hand side of Eq. (7.3.5c), the result is

$$\nabla\cdot(\mathbf{E}\times\mathbf{H})+\frac{\partial}{\partial t}\left[\frac{1}{2}(\mathbf{H}\cdot\mathbf{B}+\mathbf{E}\cdot\mathbf{D})\right]=-\mathbf{J}\cdot\mathbf{E}\tag{7.3.6}$$

By analogy with $\rho(x, t)$ in Eq. (7.3.2), we can identify the electromagnetic energy density, ρ_E, as

$$\rho_E=\frac{1}{2}(\mathbf{H}\cdot\mathbf{B}+\mathbf{E}\cdot\mathbf{D})\tag{7.3.7}$$

By analogy with $J(x, t)$ in Eq. (7.3.2), we can define the energy flux, **N**, associated with the electromagnetic field as

$$\mathbf{N}=\mathbf{E}\times\mathbf{H}\tag{7.3.8}$$

With these identifications, Eq. (7.3.6) reads

$$\frac{\partial\rho_E}{\partial t}+\nabla\cdot\mathbf{N}=-\mathbf{J}\cdot\mathbf{E}\tag{7.3.9}$$

The term $-\mathbf{J} \cdot \mathbf{E}$ on the right hand side of Eq. (7.3.9) is the net loss of energy per unit volume per unit time and plays the role of $\partial Q/\partial t$ in Eq. (7.3.2). The identification of $-\mathbf{J} \cdot \mathbf{E}$ with an energy loss rate can be understood on the basis of dimensional argument. Let the voltage applied to a material be V, and the current in the material be I. If the material has thickness, L, and cross section, A, then the electric field is $E = V/L$, and the current density is $J = I/A$. The energy lost per unit volume per second is thus $JE = IV/AL$, where IV is the power, and AL is the volume.

Equation (7.3.9) is the continuity equation for the electromagnetic energy density. It accounts for the requirement that energy should be conserved both locally and instantaneously

7.4 Maxwell's Equations in Empty Space

In the absence of free charges, currents, and in the presence of vacuum or optically linear materials, the Maxwell equations, Eqs. (7.2.11a)–(7.2.11d) simplify considerably. Under these restricted conditions, they become

$$\nabla \cdot \mathbf{D} = 0 \tag{7.4.1a}$$

$$\nabla \times \mathbf{E} = -\frac{\partial \mathbf{B}}{\partial t} \tag{7.4.1b}$$

$$\nabla \times \mathbf{H} = \frac{\partial \mathbf{D}}{\partial t} \tag{7.4.1c}$$

and

$$\nabla \cdot \mathbf{B} = 0 \tag{7.4.1d}$$

In the absence of material media, the constitutive relations in Eqs. (7.2.15) and (7.2.22) simplify to

$$\mathbf{D} = \varepsilon_o \mathbf{E} \tag{7.4.2}$$

and

$$\mathbf{B} = \mu_o \mathbf{H} \tag{7.4.3}$$

respectively. We will now show that Eqs. (7.4.1a) through (7.4.1d) have plane wave solutions.

7.5 Plane Waves

We propose to show that **E** and **H** form a plane wave propagating in the direction of the wave vector, **k**, with angular frequency, ω, and speed, c, where c is the vacuum speed of light. Consider an electric field, $\mathbf{E}(\mathbf{r}, t)$, evaluated at position, **r**, and time, t, with components in the three Cartesian directions,

$$\mathbf{E}(\mathbf{r}, t) = E_1(\mathbf{r}, t)\hat{\mathbf{e}}_1 + E_2(\mathbf{r}, t)\hat{\mathbf{e}}_2 + E_3(\mathbf{r}, t)\hat{\mathbf{e}}_3 \tag{7.5.1a}$$

where $\hat{\mathbf{e}}_1$, $\hat{\mathbf{e}}_2$, $\hat{\mathbf{e}}_3$ are the three mutually perpendicular Cartesian unit vectors. To represent a wave, each component of $\mathbf{E}(\mathbf{r}, t)$ must vary sinusoidally in both space and time. This condition can be met by writing the components in the form of complex valued exponentials. For example, the first component of $\mathbf{E}(\mathbf{r}, t)$ is

$$E_1(\mathbf{r}, t) = E_1(x_1, x_2, x_3, t) = E_1^o \exp[i(k_1x_1 + k_2x_2 + k_3x_3) - i\omega t] \tag{7.5.1b}$$

where x_1, x_2, and x_3 are the three components of **r**, $i = \sqrt{-1}$, E_1^o is a constant, and $E_1(\mathbf{r}, t)$ is pointed in the x_1-direction. For the sake of generality, we allow the electric field to have components in all three perpendicular directions, so that the constant vector amplitude, $\mathbf{E}^o$ takes the form,

$$\mathbf{E}^o = E_1^o\hat{\mathbf{e}}_1 + E_2^o\hat{\mathbf{e}}_2 + E_3^o\hat{\mathbf{e}}_3 \tag{7.5.1c}$$

In terms of Eq. (7.5.1c), $\mathbf{E}(\mathbf{r}, t)$, can be written concisely as

$$\mathbf{E}(\mathbf{r}, t) = \mathbf{E}(x_1, x_2, x_3, t) = \mathbf{E}^o \exp[i(k_1x_1 + k_2x_2 + k_3x_3) - i\omega t] \tag{7.5.2}$$

To show that Eq. (7.5.2) satisfies the Maxwell equation represented by Eq. (7.4.1b), we will need to compute $\nabla \times \mathbf{E}$. To evaluate this vector quantity, certain space derivatives will be involved. For example, the derivative of $E_2(x_1, x_2, x_3, t)$ with respect to the coordinate, x_1, is

$$\begin{aligned}\frac{\partial}{\partial x_1}E_2(x_1, x_2, x_3, t) &= E_2^o\frac{\partial}{\partial x_1}\exp[i(k_1x_1 + k_2x_2 + k_3x_3) - i\omega t]\\ &= (ik_1)E_2^o \exp[i(k_1x_1 + k_2x_2 + k_3x_3) - i\omega t]\\ &= ik_1E_2(x_1, x_2, x_3, t)\end{aligned} \tag{7.5.3a}$$

Written as a matrix with the proviso that it be expanded in minors of the first row, the curl of **E** is

$$\nabla \times \mathbf{E} = \begin{vmatrix} \hat{\mathbf{e}}_1 & \hat{\mathbf{e}}_2 & \hat{\mathbf{e}}_3 \\ \frac{\partial}{\partial x_1} & \frac{\partial}{\partial x_2} & \frac{\partial}{\partial x_3} \\ E_1 & E_2 & E_3 \end{vmatrix}$$

$$
\begin{aligned}
&= \hat{\mathbf{e}}_1\left(\frac{\partial E_3}{\partial x_2} - \frac{\partial E_2}{\partial x_3}\right) + \hat{\mathbf{e}}_2\left(\frac{\partial E_1}{\partial x_3} - \frac{\partial E_3}{\partial x_1}\right) + \hat{\mathbf{e}}_3\left(\frac{\partial E_2}{\partial x_1} - \frac{\partial E_1}{\partial x_2}\right)\\
&= \hat{\mathbf{e}}_1(ik_2E_3 - ik_3E_2) + \hat{\mathbf{e}}_2(ik_3E_1 - ik_1E_3) + \hat{\mathbf{e}}_3(ik_1E_2 - ik_2E_1)\\
&= i\begin{vmatrix} \hat{\mathbf{e}}_1 & \hat{\mathbf{e}}_2 & \hat{\mathbf{e}}_3 \\ k_1 & k_2 & k_3 \\ E_1 & E_2 & E_3 \end{vmatrix} = i(\mathbf{k}\times\mathbf{E})
\end{aligned} \tag{7.5.3b}
$$

where, as we shall see, the vector $\mathbf{k} = k_1\hat{\mathbf{e}} + k_2\hat{\mathbf{e}}_2 + k_3\hat{\mathbf{e}}_3$ determines the direction of propagation of the wave. With intermediate steps removed, Eq. (7.5.3b) reads,

$$\nabla \times \mathbf{E} = i\mathbf{k}\times\mathbf{E} \tag{7.5.4}$$

The corresponding time dependent magnetic field vector is

$$\mathbf{H}(\mathbf{r},t) = H_1(\mathbf{r},t)\hat{\mathbf{e}}_1 + H_2(\mathbf{r},t)\hat{\mathbf{e}}_2 + H_3(\mathbf{r},t)\hat{\mathbf{e}}_3 \tag{7.5.5a}$$

where in analogy with Eq. (7.5.1b), the first component varies with space and time as

$$
\begin{aligned}
H_1(\mathbf{r},t) &= H_1(x_1,x_2,x_3,t)\\
&= H_1^o \exp[i(k_1x_1 + k_2x_2 + k_3x_3) - i\omega t]
\end{aligned} \tag{7.5.5b}
$$

In Eq. (7.5.5b), H_1^o, is a constant. The constant vector amplitude of the magnetic field is

$$\mathbf{H}^o = H_1^o\hat{\mathbf{e}}_1 + H_2^o\hat{\mathbf{e}}_2 + H_3^o\hat{\mathbf{e}}_3 \tag{7.5.5c}$$

With these identifications, Eq. (7.5.5a) can be written concisely as

$$\mathbf{H}(\mathbf{r},t) = \mathbf{H}(x_1,x_2,x_3,t) = \mathbf{H}^o \exp[i(k_1x_1 + k_2x_2 + k_3x_3) - i\omega t] \tag{7.5.6}$$

By virtue of Eq. (7.4.3), the magnetic induction can be written as

$$\mathbf{B}(x_1,x_2,x_3,t) = \mu_o\mathbf{H}(x_1,x_2,x_3) = \mu_o\mathbf{H}^o \exp[i(k_1x_1 + k_2x_2 + k_3x_3) - i\omega t] \tag{7.5.7}$$

To show that Eq. (7.5.7) satisfies Eq. (7.4.1b), we shall need the time derivative,

$$
\begin{aligned}
\frac{\partial \mathbf{B}}{\partial t} &= \mu_o\frac{\partial \mathbf{H}}{\partial t}\\
&= \mu_o\mathbf{H}^o \exp[i(k_1x_1 + k_2x_2 + k_3x_3)]\frac{\partial(e^{-i\omega t})}{\partial t}\\
&= -i\omega\mu_o\mathbf{H}^o \exp[i(k_1x_1 + k_2x_2 + k_3x_3) - \omega t]
\end{aligned}
$$

$$= -i\omega\mu_o\mathbf{H} \tag{7.5.8}$$

Upon substitution of Eqs. (7.5.4) and (7.5.8), Eq. (7.4.1b) becomes

$$\mathbf{k} \times \mathbf{E} = \omega\mu_o\mathbf{H} \tag{7.5.9}$$

We now turn to the solution of Eq. (7.4.1c). Because Eqs. (7.5.2) and (7.5.6) have the same functional form, we can replace **E** by **H** in Eq. (7.5.4) and write the result as

$$\nabla \times \mathbf{H} = i\mathbf{k} \times \mathbf{H} \tag{7.5.10}$$

By reference to Eq. (7.5.8), the analogous time derivative of the electric field is

$$\frac{\partial \mathbf{D}}{\partial t} = \varepsilon_o \frac{\partial \mathbf{E}}{\partial t} = -i\omega\varepsilon_o\mathbf{E} \tag{7.5.11}$$

Upon substitution of Eqs. (7.5.10) and (7.5.11) into Eq. (7.4.1c), the result is

$$\mathbf{k} \times \mathbf{H} = -\omega\varepsilon_o\mathbf{E} \tag{7.5.12}$$

To show that Eq. (7.5.2) is consistent with Eq. (7.4.1a), we will need the divergence of the electric field. This will require the calculation of

$$\begin{aligned}
\nabla \cdot \mathbf{E}(\mathbf{r}, t) &= \frac{\partial}{\partial x_1} E_1(x_1, x_2, x_3, t) + \frac{\partial}{\partial x_2} E_2(x_1, x_2, x_3, t) + \frac{\partial}{\partial x_3} E_3(x_1, x_2, x_3, t) \\
&= ik_1E_1 + ik_2E_2 + ik_3E_3 \\
&= i\mathbf{k} \cdot \mathbf{E}(\mathbf{r}, t)
\end{aligned} \tag{7.5.13}$$

where we have applied Eq. (7.5.3a) for the case in which the coordinates have the same index. When we combine Eq. (7.4.1a) with Eq. (7.4.2) and Eq. (7.5.13), the result is

$$\nabla \cdot \mathbf{D} = \varepsilon_o \nabla \cdot \mathbf{E} = \varepsilon_o i\mathbf{k} \cdot \mathbf{E} = 0 \tag{7.5.14}$$

According to Eq. (7.5.14), the **k**-vector is perpendicular to the electric field vector, **E**.

To show that Eq. (7.5.6) is consistent with Eq. (7.4.1d), we shall need the divergence of the magnetic field, which is

$$\begin{aligned}
\nabla \cdot \mathbf{H}(\mathbf{r}, t) &= \frac{\partial}{\partial x_1} H_1(x_1, x_2, x_3, t) + \frac{\partial}{\partial x_2} H_2(x_1, x_2, x_3, t) + \frac{\partial}{\partial x_3} H_3(x_1, x_2, x_3, t) \\
&= ik_1H_1 + ik_2H_2 + ik_3H_3 \\
&= i\mathbf{k} \cdot \mathbf{H}
\end{aligned} \tag{7.5.15}$$

If we combine Eqs. (7.4.1d), (7.4.3), and (7.5.15), the result is

$$\nabla \cdot \mathbf{B} = \mu_o \nabla \cdot \mathbf{H} = \mu_o i\mathbf{k} \cdot \mathbf{H} = 0 \tag{7.5.16}$$

Equation (7.5.16) proves that the **k**-vector is perpendicular to the magnetic field vector, **H**. Hence, the vectors, **E** and **H**, are both normal to **k**, the vector that determines the direction of propagation.

To find a formula for the speed of advance of this disturbance, we will need the vector cross product of **k** with both sides of Eq. (7.5.9). The result is

$$\mathbf{k} \times (\mathbf{k} \times \mathbf{E}) = \omega\mu_o(\mathbf{k} \times \mathbf{H}) \tag{7.5.17a}$$

The left hand side of Eq. (7.5.17a) can be expanded using the vector identity,

$$\mathbf{A} \times (\mathbf{B} \times \mathbf{C}) = (\mathbf{A} \cdot \mathbf{C})\mathbf{B} - (\mathbf{A} \cdot \mathbf{B})\mathbf{C} \tag{7.5.17b}$$

[See the Mathematical Methods Appendix]. According to Eq. (7.5.17b), the expansion of the left hand side Eq. (7.5.17a) yields

$$\mathbf{k} \times (\mathbf{k} \times \mathbf{E}) = (\mathbf{k} \cdot \mathbf{E})\mathbf{k} - (\mathbf{k} \cdot \mathbf{k})\mathbf{E} \tag{7.5.17c}$$

According to Eq. (7.5.14), $(\mathbf{k} \cdot \mathbf{E}) = 0$. Because $\mathbf{k} \cdot \mathbf{k} = k^2$, Eq. (7.5.17c) can be rewritten as

$$\mathbf{k} \times (\mathbf{k} \times \mathbf{E}) = -k^2\mathbf{E} \tag{7.5.17d}$$

The combination of Eqs. (7.5.17a) and (7.5.17d) leads to

$$-k^2\mathbf{E} = \omega\mu_o(\mathbf{k} \times \mathbf{H}) \tag{7.5.17e}$$

Upon substitution of Eq. (7.5.12) for $\mathbf{k} \times \mathbf{H}$, Eq. (7.5.17e) reads

$$-k^2\mathbf{E} = -\omega^2\mu_o\varepsilon_o\mathbf{E} \tag{7.5.17f}$$

Cancel **E** from both sides of Eq. (7.4.17f). The result is

$$k^2 = \omega^2\varepsilon_o\mu_o \tag{7.5.17g}$$

The square root of Eq. (7.5.17g) is

$$k = \omega\sqrt{\varepsilon_o\mu_o} \tag{7.5.18}$$

The magnitude of the wave vector is related to the wavelength, λ, by

$$k = \frac{2\pi}{\lambda} \tag{7.5.19a}$$

while the angular frequency, ω, is related to the circular frequency, ν, by

$$\omega = 2\pi\nu \tag{7.5.19b}$$

The vacuum speed of light, c, is given by

$$c = \frac{\omega}{k} = \lambda\nu \tag{7.5.20}$$

On the basis of Eqs. (7.5.18) and (7.5.20), we can conclude that

$$c = \frac{1}{\sqrt{\varepsilon_o \mu_o}} \tag{7.5.21}$$

which is the speed of advance of an electromagnetic wave in vacuum.

To show that **E** and **H** are perpendicular to each other, we form the dot product of **E** with both sides of Eq. (7.5.9). The result is

$$\omega\mu_o(\mathbf{E}\cdot\mathbf{H}) = \mathbf{E}\cdot(\mathbf{k}\times\mathbf{E}) = \mathbf{k}\cdot(\mathbf{E}\times\mathbf{E}) = 0 \tag{7.5.22}$$

To get the result in Eq. (7.5.22), we have rearranged the middle term using the vector identity, $\mathbf{A}\cdot(\mathbf{B}\times\mathbf{C}) = \mathbf{B}\cdot(\mathbf{C}\times\mathbf{A})$. [See the Mathematical Methods Appendix]. In summary, Eq. (7.5.22) proves that $\mathbf{E}\cdot\mathbf{H} = 0$. According to Eq. (7.5.14), $\mathbf{k}\cdot\mathbf{E} = 0$. According to Eq. (7.5.16), $\mathbf{k}\cdot\mathbf{H} = 0$. Hence, **E**, and **H** define a plane whose normal is **k**, the vector which defines the direction of propagation.

We have now shown that Eqs. (7.5.2) and (7.5.6) solve Eqs. (7.4.1) through (7.4.3) and describe a plane wave propagating in the **k**-direction with the speed, $c = 1/\sqrt{\varepsilon_o\mu_o}$. This plane wave solution would not exist if Maxwell had not added the term, $\partial\mathbf{D}/\partial t$, to the right hand side of Eq. (7.2.8).

7.6 Fourier Transforms and the Dielectric Response Function

In Sects. 7.4 and 7.5 above, we have described an electromagnetic wave propagating in free space. The relative electric permittivity was $\varepsilon_r = 1$, and the relative magnetic permeability was $\mu_r = 1$. The energy in the wave traveled without gain or loss. By contrast, in chemical spectroscopy, the medium of propagation is capable of absorbing energy from the electromagnetic wave. In the visible and ultraviolet regions of the spectrum, ε_r is a function of the frequency while μ_r maintains a constant value close to unity. In the analysis in this chapter, we shall assume the frequencies to be in the visible and ultraviolet bands with $\mu_r = 1$.

In the case of a material absorbing in the ultraviolet and/or the visible, the permittivity is a complex valued function of the frequency. We represent the complex value of the permittivity by $\tilde{\varepsilon}(\omega) = \varepsilon_o \tilde{\varepsilon}_r(\omega)$, where the frequency is ω. Starting with this introduction, the "tilde" notation will be used to represent a complex valued function. The instantaneous values of the fields, **D** and **E**, are separated by a time delay so that they no longer track each other as indicated by Eq. (7.2.15). To cope with the time delays involved, we shall switch the emphasis from time to frequency. This will require the transformation of functions from the time domain to the frequency domain and back, a task which is accomplished by the Fourier transform theorem (Jean-Baptiste Joseph Fourier, French, 1768–1830). The theorem states that if $\mathbf{E}(t)$ is the instantaneous value of the electric field, then the corresponding electric field in the frequency domain, $\tilde{\mathbf{E}}(\omega)$, is a complex valued function given by the integral,

$$\tilde{\mathbf{E}}(\omega) = \int_{-\infty}^{+\infty} dt\, \mathbf{E}(t) e^{i\omega t} \tag{7.6.1a}$$

Conversely, $\mathbf{E}(t)$, can be retrieved as a function of the time by computing the integral

$$\mathbf{E}(t) = \frac{1}{2\pi} \int_{-\infty}^{+\infty} d\omega\, \tilde{\mathbf{E}}(\omega) e^{-i\omega t} \tag{7.6.1b}$$

Equation (7.6.1b) is said to be the "Fourier integral representation" of $\mathbf{E}(t)$, while $\tilde{\mathbf{E}}(\omega)$ in Eq. (7.6.1a) is said to be the "Fourier transform" of $\mathbf{E}(t)$. In what follows, we will need other Fourier transform pairs. In the case of the displacement, the transform pair is

$$\tilde{\mathbf{D}}(\omega) = \int_{-\infty}^{+\infty} dt\, \mathbf{D}(t) e^{i\omega t} \tag{7.6.2a}$$

and

$$\mathbf{D}(t) = \frac{1}{2\pi} \int_{-\infty}^{+\infty} d\omega\, \tilde{\mathbf{D}}(\omega) e^{-i\omega t} \tag{7.6.2b}$$

In the case of the magnetic field, we shall need to introduce

$$\tilde{\mathbf{H}}(\omega) = \int_{-\infty}^{+\infty} dt \mathbf{H}(t) e^{i\omega t} \tag{7.6.3a}$$

and

$$\mathbf{H}(t) = \frac{1}{2\pi} \int_{-\infty}^{+\infty} d\omega \tilde{\mathbf{H}}(\omega) e^{-i\omega t} \tag{7.6.3b}$$

In the case of the permittivity, we shall make use of

$$\tilde{\varepsilon}(\omega) = \int_{-\infty}^{+\infty} dt \varepsilon(t) e^{i\omega t} \tag{7.6.4a}$$

and

$$\varepsilon(t) = \frac{1}{2\pi} \int_{-\infty}^{+\infty} d\omega \tilde{\varepsilon}(\omega) e^{-i\omega t} \tag{7.6.4b}$$

As defined by Eq. (7.6.4b), $\varepsilon(t)$ is a time dependent quantity referred to as the *permittivity response function*. In general, $\tilde{\varepsilon}(\omega)$, $\tilde{\mathbf{E}}(\omega)$, $\tilde{\mathbf{D}}(\omega)$, and $\tilde{\mathbf{H}}(\omega)$ are complex valued functions of ω. Experiments that determine the time response of the polarization are sensitive to $\varepsilon(t) = \varepsilon_o \varepsilon_r(t)$. By contrast, experiments involving spectroscopy are sensitive to its transform, $\tilde{\varepsilon}(\omega) = \varepsilon_o \tilde{\varepsilon}_r(\omega)$.

In the time domain, the displacement, $\mathbf{D}(t)$, and the electric field, $\mathbf{E}(t)$, are related to the response function, $\varepsilon(t)$, by the integral expression,

$$\mathbf{D}(t) = \int_{-\infty}^{+\infty} d\tau \varepsilon(t-\tau) \mathbf{E}(\tau) \tag{7.6.5}$$

Equation (7.6.5) replaces Eq. (7.2.15). In mathematical parlance, the integral in Eq. (7.6.5) is said to be the *convolution* of $\varepsilon(t)$ and $\mathbf{E}(t)$. [See the Mathematical Methods Appendix].

Equation (7.6.5) replaces $\mathbf{D} = \varepsilon \mathbf{E}$, a relation which implies that if ε is a constant, then $\mathbf{D}$ and $\mathbf{E}$ share the same time dependence. By contrast, Eq. (7.6.5) assumes that at time τ the material is acted upon by an electric field, $\mathbf{E}(\tau)$, which creates some electric polarization. As represented in Eq. (7.6.5), the displacement, $\mathbf{D}(t)$, measured at the time, $t > \tau$, is the sum (that is the integral) of the displacements that have developed over the elapsed times, $(t-\tau)$, due to all preceding electrical field stimuli stretching back to $\tau = -\infty$. The particular dynamics of the development of the polarization, which are specific to the material, are included in the time dependence of the response function, $\varepsilon(t-\tau)$. These effects, which have their origin in the intermolecular interactions, impede the development of the polarization and prevent the material from responding instantaneously to $\mathbf{E}(\tau)$. Although the domain of the integral extends all the way from $\tau=-\infty$ to $\tau=+\infty$, it is recognized that $\varepsilon(t-\tau)$ is necessarily zero when $\tau>t$. This restriction avoids any contribution to the value of $\mathbf{D}(t)$ from values of $\mathbf{E}(\tau)$ that occur at future at times, $\tau>t$. If it were not for this restriction, part of the contribution to $\mathbf{D}(t)$, would precede its cause, $\mathbf{E}(\tau)$, a situation which is said to violate the *principle of causality*.

Equation (7.6.5) is an integral relation connecting the time domain functions $\mathbf{D}(t)$, $\mathbf{E}(t)$, and $\varepsilon(t)$. A relation among the corresponding transforms, $\tilde{\mathbf{D}}(\omega)$, $\tilde{\mathbf{E}}(\omega)$, and $\tilde{\varepsilon}(\omega)$ in the frequency domain can be derived by replacing $\varepsilon(t-\tau)$ in Eq. (7.6.5) by its Fourier integral representation,

$$\varepsilon(t-\tau)=\frac{1}{2\pi}\int_{-\infty}^{+\infty} d\omega\,\varepsilon(\omega)e^{-i\omega(t-\tau)} \tag{7.6.6a}$$

Equation (7.6.6a) is the same as Eq. (7.6.4b) but with $t-\tau$ substituted for t. Upon substitution of Eq. (7.6.6a) into the right hand side of Eq. (7.6.5), the result is

$$\mathbf{D}(t)=\int_{-\infty}^{+\infty} d\tau\varepsilon(t-\tau)\mathbf{E}(\tau)=\int_{-\infty}^{+\infty} d\tau\left[\frac{1}{2\pi}\int_{-\infty}^{+\infty} d\omega\tilde{\varepsilon}(\omega)e^{-i\omega(t-\tau)}\right]\mathbf{E}(\tau) \tag{7.6.6b}$$

Upon reordering the sequence of integrations, Eq. (7.6.6b) reads,

$$\mathbf{D}(t)=\frac{1}{2\pi}\int_{-\infty}^{+\infty} d\omega\tilde{\varepsilon}(\omega)e^{-i\omega t}\left[\int_{-\infty}^{+\infty} d\tau e^{i\omega\tau}\mathbf{E}(\tau)\right] \tag{7.6.6c}$$

Next replace the integral within the brackets in Eq. (7.6.6c) using Eq. (7.6.1a). The final result is

$$\mathbf{D}(t)=\frac{1}{2\pi}\int_{-\infty}^{+\infty} d\omega\tilde{\varepsilon}(\omega)\tilde{\mathbf{E}}(\omega)e^{-i\omega t} \tag{7.6.6d}$$

If we equate the Fourier integral representation of $\mathbf{D}(t)$ in Eq. (7.6.2b) to its representation in Eq. (7.6.6d), we have

$$\mathbf{D}(t)=\frac{1}{2\pi}\int_{-\infty}^{+\infty} d\omega\tilde{\mathbf{D}}(\omega)e^{-i\omega t}=\frac{1}{2\pi}\int_{-\infty}^{+\infty} d\omega\tilde{\varepsilon}(\omega)\tilde{\mathbf{E}}(\omega)e^{-i\omega t} \tag{7.6.6e}$$

Equating integrands of the integrals in Eq. (7.6.6e), we conclude that

$$\tilde{\mathbf{D}}(\omega) = \tilde{\varepsilon}(\omega)\tilde{\mathbf{E}}(\omega) \tag{7.6.7}$$

A medium, for which Eq. (7.6.7) applies, is said to be *dispersive*. That means that waves with different frequencies proceed with different speeds. Equation (7.6.7) also allows for a medium where a part or all of the electromagnetic energy passing through the medium is absorbed. Such a medium is said to be *dissipative*. In the case of such a medium, Eq. (7.6.7) replaces Eq. (7.2.15).

7.7 Plane Wave Solutions to Maxwell's Equations in a Dispersive Medium

To solve Maxwell's equations in a dispersive medium, we start with the plane wave function,

$$e^{i(\mathbf{k}\cdot\mathbf{r}-\omega t)} = \exp[i(k_1x_1 + k_2x_2 + k_3x_3) - i\omega t] \tag{7.7.1}$$

to represent the space and time dependence of the fields. According to the Fourier transform theorem [See the Mathematical Methods Appendix], any arbitrary time varying electric field, $\mathbf{E}(\mathbf{r}, t)$, propagating as part of a plane wave can be represented as a sum of frequency components, $\tilde{\mathbf{E}}(\omega)e^{i(\mathbf{k}\cdot\mathbf{r}-\omega t)}$, where $\tilde{\mathbf{E}}(\omega)$ is perpendicular to $\mathbf{k}$. Since the frequency is a continuous variable, the sum of components can be written as the integral,

$$\mathbf{E}(\mathbf{r}, t) = \frac{1}{2\pi}\int_{-\infty}^{+\infty} d\omega\,\tilde{\mathbf{E}}(\omega)e^{i(\mathbf{k}\cdot\mathbf{r}-\omega t)} \tag{7.7.2}$$

To satisfy Faraday's law as given by Eq. (7.4.1b), we shall need to compute the curl of Eq. (7.7.2). The result is

$$\nabla\times\mathbf{E}(\mathbf{r}, t) = \frac{1}{2\pi}\int_{-\infty}^{+\infty} d\omega\,\nabla\times\left[\tilde{\mathbf{E}}(\omega)e^{i(\mathbf{k}\cdot\mathbf{r}-\omega t)}\right] = \frac{1}{2\pi}\int_{-\infty}^{+\infty} d\omega(i\mathbf{k}\times\tilde{\mathbf{E}}(\omega))e^{i(\mathbf{k}\cdot\mathbf{r}-\omega t)} \tag{7.7.3}$$

Equation (7.7.3) is the analog of Eq. (7.5.4). According to the Fourier theorem, the magnetic field in a dispersive medium is

$$\mathbf{H}(\mathbf{r}, t) = \frac{1}{2\pi}\int_{-\infty}^{+\infty} d\omega\,\tilde{\mathbf{H}}(\omega)e^{i(\mathbf{k}\cdot\mathbf{r}-\omega t)} \tag{7.7.4}$$

To satisfy Faraday's law, we shall need the time derivative,

$$\frac{\partial}{\partial t}\mathbf{H}(\mathbf{r},t) = \frac{1}{2\pi}\int_{-\infty}^{+\infty} d\omega\,\tilde{\mathbf{H}}(\omega)\frac{\partial}{\partial t}e^{i(\mathbf{k}\cdot\mathbf{r}-\omega t)} = \frac{1}{2\pi}\int_{-\infty}^{+\infty} d\omega\,\tilde{\mathbf{H}}(\omega)(-i\omega)e^{i(\mathbf{k}\cdot\mathbf{r}-\omega t)} \tag{7.7.5}$$

Multiply Eq. (7.7.5) by μ_o and add the result to Eq. (7.7.3). This combination of equations reads,

$$\nabla\times\mathbf{E}(\mathbf{r},t) + \mu_o\frac{\partial}{\partial t}\mathbf{H}(\mathbf{r},t) = \frac{1}{2\pi}\int_{-\infty}^{+\infty} d\omega\left[(i\mathbf{k}\times\tilde{\mathbf{E}}(\omega)) + \mu_o\tilde{\mathbf{H}}(\omega)(-i\omega)\right]e^{i(\mathbf{k}\cdot\mathbf{r}-\omega t)} \tag{7.7.6}$$

According to Eqs. (7.4.1b) and (7.4.3), the left hand side of Eq. (7.7.6) is zero. This condition implies that the integrand in Eq. (7.7.6) should satisfy,

$$\mathbf{k}\times\tilde{\mathbf{E}}(\omega) = \omega\mu_o\tilde{\mathbf{H}}(\omega) \tag{7.7.7}$$

Equation (7.7.7) is the analog of Eq. (7.5.9). We thus conclude that Faraday's Law is satisfied by Eq. (7.7.7).

To satisfy Ampere's Law, Eq. (7.4.1c), we shall need to compute,

$$\nabla\times\mathbf{H}(\mathbf{r},t) = \frac{1}{2\pi}\int_{-\infty}^{+\infty} d\omega\nabla\times\left(\tilde{\mathbf{H}}(\omega)e^{i(\mathbf{k}\cdot\mathbf{r}-\omega t)}\right) = \frac{1}{2\pi}\int_{-\infty}^{+\infty} d\omega(i\mathbf{k}\times\tilde{\mathbf{H}}(\omega))e^{i(\mathbf{k}\cdot\mathbf{r}-\omega t)} \tag{7.7.8}$$

Equation (7.7.8) is the analog of Eq. (7.5.10). The displacement in a dispersive medium is

$$\mathbf{D}(\mathbf{r},t) = \frac{1}{2\pi}\int_{-\infty}^{+\infty} d\omega\tilde{\mathbf{D}}(\omega)e^{i(\mathbf{k}\cdot\mathbf{r}-\omega t)} \tag{7.7.9}$$

To satisfy Ampere's law, we shall also need the time derivative,

$$\begin{aligned}\frac{\partial}{\partial t}\mathbf{D}(\mathbf{r},t) &= \frac{1}{2\pi}\int_{-\infty}^{+\infty} d\omega\tilde{\mathbf{D}}(\omega)\frac{\partial}{\partial t}e^{i(\mathbf{k}\cdot\mathbf{r}-\omega t)} = \frac{1}{2\pi}\int_{-\infty}^{+\infty} d\omega\tilde{\mathbf{D}}(\omega)(-i\omega)e^{i(\mathbf{k}\cdot\mathbf{r}-\omega t)} \\ &= \frac{1}{2\pi}\int_{-\infty}^{+\infty} d\omega\tilde{\varepsilon}(\omega)\tilde{\mathbf{E}}(\omega)(-i\omega)e^{i(\mathbf{k}\cdot\mathbf{r}-\omega t)}\end{aligned} \tag{7.7.10}$$

where in the last step we used Eq. (7.6.7). If we subtract Eq. (7.7.10) from Eq. (7.7.8), the result is

$$\begin{aligned}\nabla \times \mathbf{H}(\mathbf{r},t) - \frac{\partial}{\partial t}\mathbf{D}(\mathbf{r},t) &= \frac{1}{2\pi}\int_{-\infty}^{+\infty} d\omega\big[(i\mathbf{k}\times\tilde{\mathbf{H}}(\omega)) - \tilde{\mathbf{D}}(\omega)(-i\omega)\big]e^{i(\mathbf{k}\cdot\mathbf{r}-\omega t)} \\ &= \frac{1}{2\pi}\int_{-\infty}^{+\infty} d\omega\big[(i\mathbf{k}\times\tilde{\mathbf{H}}(\omega)) - \tilde{\varepsilon}(\omega)\tilde{\mathbf{E}}(\omega)(-i\omega)\big]e^{i(\mathbf{k}\cdot\mathbf{r}-\omega t)} \\ &= 0 \end{aligned} \tag{7.7.11}$$

where by virtue of Eq. (7.4.1c), the left hand side of Eq. (7.7.11) is zero. Hence, the right-hand side is equivalent to

$$\mathbf{k}\times\tilde{\mathbf{H}}(\omega) = -\omega\tilde{\varepsilon}(\omega)\tilde{\mathbf{E}}(\omega) \tag{7.7.12}$$

Equation (7.7.12) is the analog of Eq. (7.5.12). We thus conclude that Ampere's Law is satisfied by Eq. (7.7.12).

To get the relation between k and ω for a dispersive medium, we shall need the cross product, $\mathbf{k}\times(\mathbf{k}\times\tilde{\mathbf{E}}(\omega))$. Using Eqs. (7.5.17c) and (7.5.17d), the result is

$$\mathbf{k}\times(\mathbf{k}\times\tilde{\mathbf{E}}(\omega)) = (\mathbf{k}\cdot\tilde{\mathbf{E}}(\omega))\mathbf{k} - (\mathbf{k}\cdot\mathbf{k})\tilde{\mathbf{E}}(\omega) = -k^2\tilde{\mathbf{E}}(\omega) \tag{7.7.13}$$

We can use Eq. (7.7.7) to express the cross product of $\mathbf{k}$ with $\mathbf{k}\times\tilde{\mathbf{E}}(\omega)$. The result is

$$\mathbf{k}\times(\mathbf{k}\times\tilde{\mathbf{E}}(\omega)) = \omega\mu_o(\mathbf{k}\times\tilde{\mathbf{H}}(\omega)) \tag{7.7.14a}$$

Equation (7.7.12) reads, $\mathbf{k}\times\tilde{\mathbf{H}}(\omega) = -\omega\tilde{\varepsilon}(\omega)\tilde{\mathbf{E}}(\omega)$. With the use of this result, Eq. (7.7.14a) reads

$$\mathbf{k}\times(\mathbf{k}\times\tilde{\mathbf{E}}(\omega)) = -\omega^2\mu_o\tilde{\varepsilon}(\omega)\tilde{\mathbf{E}}(\omega) \tag{7.7.14b}$$

Upon comparison of Eqs. (7.7.13) and (7.7.14b), we conclude that

$$k^2\tilde{\mathbf{E}}(\omega) = \omega^2\mu_o\tilde{\varepsilon}(\omega)\tilde{\mathbf{E}}(\omega) \tag{7.7.15}$$

Cancellation of $\tilde{\mathbf{E}}(\omega)$ from either side of Eq. (7.7.15) leads to the result,

$$k^2 = \omega^2\mu_o\tilde{\varepsilon}(\omega) \tag{7.7.16}$$

We can define a complex valued relative permittivity, $\tilde{\varepsilon}_r(\omega)$, according to

$$\tilde{\varepsilon}(\omega) = \varepsilon_o\tilde{\varepsilon}_r(\omega) \tag{7.7.17}$$

Since ε_o is a real valued constant, $\tilde{\varepsilon}_r(\omega)$ has the same frequency dependence as $\tilde{\varepsilon}(\omega)$.

By virtue of its dependence on $\tilde{\varepsilon}(\omega)$, k in Eq. (7.7.16) is in general a complex valued function of the frequency. We denote this behavior by writing it as $\tilde{k}(\omega)$, so that Eq. (7.7.16) now reads,

$$\tilde{k}^2(\omega) = \omega^2 \mu_o \varepsilon_o \tilde{\varepsilon}_r(\omega) \tag{7.7.18}$$

Recalling that the vacuum speed of light is given by Eq. (7.5.21), we can rewrite Eq. (7.7.18) as

$$\tilde{k}(\omega) = \frac{\omega}{c}\sqrt{\tilde{\varepsilon}_r(\omega)} \tag{7.7.19}$$

One can consider the case where $\tilde{\varepsilon}_r(\omega)$ is real and independent of frequency. In this circumstance, Eq. (7.7.19) reads $k = (\omega/c)\sqrt{\varepsilon_r}$. The corresponding wave speed is $v = (\omega/k) = c/\sqrt{\varepsilon_r}$. The refractive index, n, is defined by $n = c/v$; hence, $n = \sqrt{\varepsilon_r}$. In materials, $\varepsilon_r > 1$ which explains why electromagnetic waves in a dielectric travel slower than in vacuum.

7.8 Poynting Vector in a Dispersive Medium

7.8.1 *Time Average of the Poynting Vector*

For a plane wave traveling in the x-direction, the Poynting vector [2], $\mathbf{N}(x, t)$, is,

$$\mathbf{N}(x, t) = \mathbf{E}(x, t) \times \mathbf{H}(x, t) \tag{7.8.1}$$

If the signal propagating in the x-direction is made up of a single wave of fixed frequency, i.e. it is monochromatic, then the electric field associated with the wave is given by

$$\begin{aligned} \mathbf{E}(x, t) &= \mathbf{E}_o \exp\left[i(\tilde{k}(\omega)x - \omega t\right] \\ &= \overline{\mathbf{E}}(x)e^{-i\omega t} \end{aligned} \tag{7.8.2a}$$

where

$$\overline{\mathbf{E}}(x) = \mathbf{E}_o e^{i\tilde{k}(\omega)x} \tag{7.8.2b}$$

As defined by Eqs. (7.8.2a) and (7.8.2b), the electric field is a complex valued function, which can be written as the sum of its real and imaginary parts [2] in the form,

$$\mathbf{E}(x,t) = \mathrm{Re}(\mathbf{E}(x,t)) + i\ \mathrm{Im}(\mathbf{E}(x,t)) \tag{7.8.3a}$$

The complex valued amplitude function, $\overline{\mathbf{E}}(x)$, in Eq. (7.8.2b) can in turn be expressed as the sum of its real and imaginary parts as

$$\overline{\mathbf{E}}(x) = \mathrm{Re}\,(\overline{\mathbf{E}}(x)) + i\ \mathrm{Im}(\overline{\mathbf{E}}(x)) \tag{7.8.3b}$$

The complex valued time function in Eq. (7.8.2a) can be expanded in terms of its real and imaginary parts as,

$$e^{-i\omega t} = \cos(\omega t) - i\sin(\omega t) \tag{7.8.3c}$$

The system of Eqs. (7.8.2) and (7.8.3) can be reassembled to express $\mathbf{E}(x,t)$ in the form,

$$\begin{aligned}\mathbf{E}(x,t) &= \overline{\mathbf{E}}(x)e^{-i\omega t}\\ &= \left[\mathrm{Re}(\overline{\mathbf{E}}(x)) + i\,\mathrm{Im}(\overline{\mathbf{E}}(x))\right][\cos(\omega t) - i\sin(\omega t)]\\ &= \left[\mathrm{Re}(\overline{\mathbf{E}}(x))\cos(\omega t) + \mathrm{Im}(\overline{\mathbf{E}}(x))\sin(\omega t)\right]\\ &\quad + i\left[\mathrm{Im}(\overline{\mathbf{E}}(x))\cos(\omega t) - \mathrm{Re}(\overline{\mathbf{E}}(x))\sin(\omega t)\right]\end{aligned} \tag{7.8.3d}$$

Upon comparison of Eqs. (7.8.3a) and (7.8.3d), we conclude the

$$\mathrm{Re}(\mathbf{E}(x,t)) = \mathrm{Re}(\overline{\mathbf{E}}(x))\cos(\omega t) + \mathrm{Im}(\overline{\mathbf{E}}(x))\sin(\omega t) \tag{7.8.4}$$

The magnetic field associated with the wave is

$$\mathbf{H}(x,t) = \mathbf{H}_o \exp\left[i(\tilde{k}(\omega)x - \omega t)\right] = \overline{\mathbf{H}}(x)e^{-i\omega t} \tag{7.8.5a}$$

where

$$\overline{\mathbf{H}}(x) = \mathbf{H}_o e^{i\tilde{k}(\omega)x} \tag{7.8.5b}$$

If in Eqs. (7.8.3), we replace $\mathbf{E}$ everywhere by $\mathbf{H}$, and follow the same steps as those leading to Eq. (7.8.4), we find for the real part of the magnetic field,

$$\mathrm{Re}(\mathbf{H}(x,t)) = \mathrm{Re}(\overline{\mathbf{H}}(x))\cos(\omega t) + \mathrm{Im}(\overline{\mathbf{H}}(x))\sin(\omega t) \tag{7.8.6}$$

Since $\mathbf{N}(x,t)$ is measurable, it has to be real, so we write it in terms of the real parts of $\mathbf{E}(x,t)$ and $\mathbf{H}(x,t)$ as [2]

$$\mathbf{N}(x,t) = \mathrm{Re}(\mathbf{E}(x,t)) \times \mathrm{Re}(\mathbf{H}(x,t)) \tag{7.8.7}$$

Upon substitution of Eqs. (7.8.4) and (7.8.6) into Eq. (7.8.7), $\mathbf{N}(x, t)$ can be represented as the product

$$\begin{aligned}\mathbf{N}(x, t) = & [\mathrm{Re}(\overline{\mathbf{E}}(x))\cos(\omega t) + \mathrm{Im}(\overline{\mathbf{E}}(x))\sin(\omega t)] \\ & \times [\mathrm{Re}(\overline{\mathbf{H}}(x))\cos(\omega t) + \mathrm{Im}(\overline{\mathbf{H}}(x))\sin(\omega t)]\end{aligned} \tag{7.8.8a}$$

Perform the indicated multiplications so that Eq. (7.8.8a) reads

$$\begin{aligned}\mathbf{N}(x, t) = & \left[\mathrm{Re}(\overline{\mathbf{E}}(x)) \times \mathrm{Re}(\overline{\mathbf{H}}(x))\right]\cos^2(\omega t) \\ & + \left[\mathrm{Im}(\overline{\mathbf{E}}(x)) \times \mathrm{Im}(\overline{\mathbf{H}}(x))\right]\sin^2(\omega t) \\ & + \left[\mathrm{Im}(\overline{\mathbf{E}}(x)) \times \mathrm{Re}(\overline{\mathbf{H}}(x))\right]\sin(\omega t)\cos(\omega t) \\ & + \left[\mathrm{Re}(\overline{\mathbf{E}}(x)) \times \mathrm{Im}(\overline{\mathbf{H}}(x))\right]\cos(\omega t)\sin(\omega t)\end{aligned} \tag{7.8.8b}$$

Of physical interest is the time average of $\mathbf{N}(x, t)$, which we designate by the notation,

$$\begin{aligned}\langle\mathbf{N}(x)\rangle = & \left[\mathrm{Re}(\overline{\mathbf{E}}(x)) \times \mathrm{Re}(\overline{\mathbf{H}}(x))\right]\left\langle\cos^2(\omega t)\right\rangle \\ & + \left[\mathrm{Im}(\overline{\mathbf{E}}(x)) \times \mathrm{Im}(\overline{\mathbf{H}}(x))\right]\left\langle\sin^2(\omega t)\right\rangle \\ & + \left[\mathrm{Im}(\overline{\mathbf{E}}(x)) \times \mathrm{Re}(\overline{\mathbf{H}}(x))\right]\langle\sin(\omega t)\cos(\omega t)\rangle \\ & + \left[\mathrm{Re}(\overline{\mathbf{E}}(x)) \times \mathrm{Im}(\overline{\mathbf{H}}(x))\right]\langle\cos(\omega t)\sin(\omega t)\rangle\end{aligned} \tag{7.8.8c}$$

Given that the period of oscillation is the inverse of the circular frequency, $1/\nu$, and $\omega = 2\pi\nu$, the period of oscillation can be represented by $2\pi/\omega$. The time average of $\cos^2(\omega t)$ over one period is then

$$\begin{aligned}\left\langle\cos^2(\omega t)\right\rangle &= \left(\frac{1}{2\pi/\omega}\right)\int_0^{2\pi/\omega} dt\,\cos^2(\omega t) = \frac{\omega}{2\pi}\int_0^{2\pi/\omega} dt\left[\frac{1+\cos(2\omega t)}{2}\right] \\ &= \frac{\omega}{2\pi}\left\{\left[\frac{t}{2}\right]_0^{2\pi/\omega} + \frac{1}{2}\left[\frac{\sin(2\omega t)}{2\omega}\right]_0^{2\pi/\omega}\right\} = \left(\frac{\omega}{2\pi}\right)\frac{2\pi/\omega}{2} = \frac{1}{2}\end{aligned} \tag{7.8.8d}$$

Similarly,

$$\begin{aligned}\left\langle\sin^2(\omega t)\right\rangle &= \left(\frac{1}{2\pi/\omega}\right)\int_0^{2\pi/\omega} dt\,\sin^2(\omega t) = \frac{\omega}{2\pi}\int_0^{2\pi/\omega} dt\left[\frac{1-\cos(2\omega t)}{2}\right] \\ &= \frac{\omega}{2\pi}\left\{\left[\frac{t}{2}\right]_0^{2\pi/\omega} - \frac{1}{2}\left[\frac{\sin(2\omega t)}{2\omega}\right]_0^{2\pi/\omega}\right\} = \left(\frac{\omega}{2\pi}\right)\frac{2\pi/\omega}{2} = \frac{1}{2}\end{aligned} \tag{7.8.8e}$$

By contrast $\langle\sin(\omega t)\cos(\omega t)\rangle$ is

$$\langle \sin(\omega t)\cos(\omega t)\rangle = \frac{1}{2\pi/\omega}\int_0^{2\pi/\omega} dt \sin(\omega t)\cos(\omega t)$$
$$= \frac{\omega}{2\pi}\left[\frac{\sin^2(\omega t)}{2\omega}\right]_0^{2\pi/\omega} = 0 \qquad (7.8.8f)$$

Upon substitution of Eqs. (7.8.8d), (7.8.8e), and (7.8.8f) into Eq. (7.8.8c), the result is

$$\langle \mathbf{N}(x)\rangle = \left(\frac{1}{2}\right)\left[\mathrm{Re}(\overline{\mathbf{E}}(x)) \times \mathrm{Re}(\overline{\mathbf{H}}(x))\right] + \left(\frac{1}{2}\right)\left[\mathrm{Im}(\overline{\mathbf{E}}(x)) \times \mathrm{Im}(\overline{\mathbf{H}}(x))\right] \quad (7.8.9)$$

There is a simpler way of writing Eq. (7.8.9) [2]. Consider the complex conjugate, $\mathbf{H}^*(x, t)$, of the magnetic field, $\mathbf{H}(x, t)$. According to Eq. (7.8.5a),

$$\mathbf{H}^*(x, t) = \overline{\mathbf{H}}^*(x)e^{i\omega t} = \left[\mathrm{Re}(\overline{\mathbf{H}}(x)) - i\ \mathrm{Im}(\overline{\mathbf{H}}(x))\right]e^{i\omega t} \qquad (7.8.10a)$$

According to the second line of Eq. (7.8.2a),

$$\mathbf{E}(x, t) = \overline{\mathbf{E}}(x)e^{-i\omega t} = \left[\mathrm{Re}(\overline{\mathbf{E}}(x)) + i\mathrm{Im}(\overline{\mathbf{E}}(x))\right]e^{-i\omega t} \qquad (7.8.10b)$$

Use Eqs. (7.8.10a) and (7.8.10b) to compute the vector cross product,

$$\begin{aligned}\mathbf{E}(x, t) \times \mathbf{H}^*(x, t) &= \left[\mathrm{Re}(\overline{\mathbf{E}}(x)) + i\mathrm{Im}(\overline{\mathbf{E}}(x))\right]e^{-i\omega t} \times \left[\mathrm{Re}(\overline{\mathbf{H}}(x)) - i\mathrm{Im}(\overline{\mathbf{H}}(x))\right]e^{i\omega t} \\ &= \left[\mathrm{Re}(\overline{\mathbf{E}}(x)) \times \mathrm{Re}(\overline{\mathbf{H}}(x)) + \mathrm{Im}(\overline{\mathbf{E}}(x)) \times \mathrm{Im}(\overline{\mathbf{H}}(x))\right] \\ &\quad + i\left[\mathrm{Im}(\overline{\mathbf{E}}(x)) \times \mathrm{Re}(\overline{\mathbf{H}}(x)) - \mathrm{Re}(\overline{\mathbf{E}}(x)) \times \mathrm{Im}(\overline{\mathbf{H}}(x))\right]\end{aligned} \qquad (7.8.10c)$$

The real part of Eq. (7.8.10c) is

$$\mathrm{Re}\big(\mathbf{E}(x, t) \times \mathbf{H}^*(x, t)\big) = \left[\mathrm{Re}(\overline{\mathbf{E}}(x)) \times \mathrm{Re}(\overline{\mathbf{H}}(x)) + \mathrm{Im}(\overline{\mathbf{E}}(x)) \times \mathrm{Im}(\overline{\mathbf{H}}(x))\right] \qquad (7.8.10d)$$

By comparing Eq. (7.8.10d) with Eq. (7.8.9), the simple expression

$$\langle \mathbf{N}(x)\rangle = \frac{1}{2}\mathrm{Re}\big(\mathbf{E}(x, t) \times \mathbf{H}^*(x, t)\big) \qquad (7.8.11)$$

follows [2].

7.8.2 Effect of Dispersion

To take into account the dispersive/dissipative effect of the medium on wave propagation, we shall need plane wave Fourier integral representations for both $\mathbf{E}(x, t)$ and $\mathbf{H}(x, t)$. The appropriate formulae are

$$\mathbf{E}(x,t) = \frac{1}{2\pi}\int_{-\infty}^{+\infty} d\omega\, \tilde{\mathbf{E}}(\omega) e^{i(\tilde{k}x-\omega t)} \tag{7.8.12a}$$

and

$$\mathbf{H}(x,t) = \frac{1}{2\pi}\int_{-\infty}^{+\infty} d\omega\, \tilde{\mathbf{H}}(\omega) e^{i(\tilde{k}x-\omega t)} \tag{7.8.12b}$$

where the complex valued wave vector, $\tilde{k} = \tilde{k}(\omega)$, in Eq. (7.7.19) takes into account the dispersion/dissipation. To simplify our analysis, we shall consider a single frequency component selected from the integrals in Eqs. (7.8.12). This selection procedure is the analog of choosing a single term from a sum. For the case of a wave propagating in the x-direction according to Eqs. (7.8.12a) and (7.8.12b), a single component with frequency, ω, takes the form of Eqs. (7.8.13a) and (7.8.13b):

$$\mathbf{E}(x,t) = \tilde{\mathbf{E}}(\omega) e^{i\tilde{k}x} e^{-i\omega t} \tag{7.8.13a}$$

and

$$\mathbf{H}(x,t) = \tilde{\mathbf{H}}(\omega) e^{i\tilde{k}x} e^{-i\omega t} \tag{7.8.13b}$$

Because propagation is confined to the x-direction, the $\mathbf{k}$, that appears in Eqs. (7.7.2) and (7.7.4), has only an x-component, which we represent by $\tilde{k} = \tilde{k}(\omega)$. Recognizing that $\tilde{k}$ can be complex valued, the complex conjugate of Eq. (7.8.13b) is

$$\mathbf{H}^*(x,t) = \tilde{\mathbf{H}}^*(\omega) e^{-i\tilde{k}^* x} e^{i\omega t} \tag{7.8.13c}$$

Using Eqs. (7.8.13a) and (7.8.13c), the vector cross product needed to evaluate Eq. (7.8.11) is

$$\mathbf{E}(x,t) \times \mathbf{H}^*(x,t) = \tilde{\mathbf{E}}(\omega) e^{i\tilde{k}x} e^{-i\omega t} \times \tilde{\mathbf{H}}^*(\omega) e^{-i\tilde{k}^* x} e^{i\omega t} \tag{7.8.14a}$$

Because $\tilde{k}$ can be complex valued, the same is true of $\mathbf{k}$, so the complex conjugate of Eq. (7.7.7) takes the form,

$$\tilde{\mathbf{H}}^*(\omega) = \frac{1}{\omega\mu_o}\left(\tilde{\mathbf{k}}^* \times \tilde{\mathbf{E}}^*(\omega)\right) \tag{7.8.14b}$$

Upon insertion of Eq. (7.8.14b) into Eq. (7.8.14a), the resulting equation is

$$\begin{aligned}\mathbf{E}(x,t) \times \mathbf{H}^*(x,t) &= \tilde{\mathbf{E}}(\omega)e^{i\tilde{k}x}e^{-i\omega t} \times \frac{1}{\mu_o\omega}\left(\tilde{\mathbf{k}}^* \times \tilde{\mathbf{E}}^*(\omega)\right)e^{-i\tilde{k}^*x}e^{i\omega t} \\ &= \frac{1}{\mu_o\omega}\left[\tilde{\mathbf{E}}(\omega) \times \left(\tilde{\mathbf{k}}^* \times \tilde{\mathbf{E}}^*(\omega)\right)\right]e^{i(\tilde{k}-\tilde{k}^*)x} \end{aligned} \tag{7.8.14c}$$

Upon use of Eq. (7.5.17b) with $\mathbf{A} = \tilde{\mathbf{E}}(\omega)$, $\mathbf{B} = \mathbf{k}^*$, and $\mathbf{C} = \tilde{\mathbf{E}}^*(\omega)$, Eq. (7.8.14c) becomes,

$$\mathbf{E}(x,t) \times \mathbf{H}^*(x,t) = \frac{1}{\mu_o\omega}\left(\left(\tilde{\mathbf{E}}(\omega) \cdot \tilde{\mathbf{E}}^*(\omega)\right)\tilde{\mathbf{k}}^* - \left(\tilde{\mathbf{E}}(\omega) \cdot \tilde{\mathbf{k}}^*\right)\tilde{\mathbf{E}}^*(\omega)\right)e^{i(\tilde{k}-\tilde{k}^*)x} \tag{7.8.14d}$$

Because a plane wave is being considered,

$$\left(\tilde{\mathbf{E}}(\omega) \cdot \tilde{\mathbf{k}}^*\right) = 0 \tag{7.8.14e}$$

Substitute Eq. (7.8.14e) into Eq. (7.8.14d). The result is

$$\begin{aligned}\mathbf{E}(x,t) \times \mathbf{H}^*(x,t) &= \frac{1}{\mu_o\omega}\left[\left(\tilde{\mathbf{E}}(\omega) \cdot \tilde{\mathbf{E}}^*(\omega)\right)\tilde{\mathbf{k}}^*\right]e^{i(\tilde{k}-\tilde{k}^*)x} \\ &= \frac{1}{\mu_o\omega}\left(\tilde{\mathbf{k}}^*\left|\tilde{\mathbf{E}}(\omega)\right|^2\right)e^{-2\mathrm{Im}(\tilde{k}x)} \end{aligned} \tag{7.8.14f}$$

where

$$\tilde{\mathbf{E}}(\omega) \cdot \tilde{\mathbf{E}}^*(\omega) = \left|\tilde{\mathbf{E}}(\omega)\right|^2 \tag{7.8.14g}$$

and

$$\tilde{k} - \tilde{k}^* = [\mathrm{Re}(\tilde{k}) + i\mathrm{Im}(\tilde{k})] - [\mathrm{Re}(\tilde{k}) - i\mathrm{Im}(\tilde{k})] = 2i\mathrm{Im}(\tilde{k}) \tag{7.8.14h}$$

Substitute Eq. (7.8.14f) into Eq. (7.8.11). The result is

$$\langle\mathbf{N}(x)\rangle = \frac{1}{2}\left(\frac{\mathrm{Re}(\tilde{\mathbf{k}}^*)}{\mu_o\omega}\right)\left|\tilde{\mathbf{E}}(\omega)\right|^2 e^{-2\mathrm{Im}(\tilde{k})x} \tag{7.8.15}$$

On the basis of Eq. (7.8.15), it is clear that the magnitude of $\langle\mathbf{N}(x)\rangle$ decreases with increasing thickness, x, of the material. We will show that Eq. (7.8.15) can be used to derive the Beer–Lambert law and evaluate the optical density in spectrophotometry.

7.8.3 Propagation in Vacuum

In vacuum, $\text{Re}(\tilde{k}) = \text{Re}(\tilde{k}^*)$ and $\text{Im}(\tilde{k}) = 0$; hence, $\tilde{\mathbf{k}} = \tilde{\mathbf{k}}^* = \text{Re}(\tilde{k})\hat{\mathbf{k}}$, where $\hat{\mathbf{k}}$ is a unit vector in the direction of $\tilde{\mathbf{k}}$. We note further that $\text{Re}(\tilde{k}) = k = \omega/c$, so we have finally, $\text{Re}(\tilde{\mathbf{k}}^*) = (\omega/c)\hat{\mathbf{k}}$. With these simplifications, Eq. (7.8.15) can be applied to empty space. The result is

$$\langle \mathbf{N} \rangle = \frac{1}{2}\left(\frac{\omega/c}{\mu_o \omega}\right)\left|\tilde{\mathbf{E}}(\omega)\right|^2 \hat{\mathbf{k}} \tag{7.8.16a}$$

Cancellation of ω between the numerator and the denominator and use of Eq. (7.5.21) leads to the result,

$$\langle \mathbf{N} \rangle = \frac{1}{2}\left(\frac{\sqrt{\varepsilon_o \mu_o}}{\mu_o}\right)\left|\tilde{\mathbf{E}}(\omega)\right|^2 \hat{\mathbf{k}} = \frac{1}{2}\left(\sqrt{\frac{\varepsilon_o}{\mu_o}}\right)\left|\tilde{\mathbf{E}}(\omega)\right|^2 \hat{\mathbf{k}} \tag{7.8.16b}$$

The factor $\sqrt{\mu_o/\varepsilon_o}$ is known as the "impedance of free space", Z_o:

$$Z_o = \sqrt{\frac{\mu_o}{\varepsilon_o}} \tag{7.8.17}$$

Upon substitution of Eq. (7.8.17) into Eq. (7.8.16b), the magnitude, $\langle N \rangle$, of the time average of the Poynting vector is found to be

$$\langle N \rangle = \frac{1}{2Z_o}\left|\tilde{\mathbf{E}}(\omega)\right|^2 \tag{7.8.18}$$

Equation (7.8.18) applies in the case of the propagation of an electromagnetic energy through free space.

7.9 Material Response to an Electromagnetic Wave

We now turn our attention to the forces in matter which resist the formation of electric polarization. In the visible or ultraviolet range of frequencies, the electric field of an electromagnetic wave couples to the atomic and molecular electrons. In condensed matter, the electron responds to the local electric field value, $F(t)$. The response of the electron is governed by quantum mechanics. Prior to the development of the quantum theory, however, it was assumed by Lorentz and others that an electron in an atom was harmonically bound. As discussed in Chap. 3, Sect. 3.11, the electron in this model was represented as having mass, m, charge, $-e$, and natural oscillation frequency, ω_o. On the basis of this model, the displacement, $x(t)$, of the electron from its equilibrium position was governed by the damped harmonic oscillator equation, which can be written in slightly altered form as

$$m\frac{d^2x}{dt^2} + 2m\gamma\frac{dx}{dt} + m\omega_o^2 x = -eF(t) \tag{7.9.1}$$

where t is the time, and e is the magnitude of the charge of the electron. The rate of energy loss due to dissipative forces in the medium is taken into account by inclusion of the semi-empirical damping term, $2m\gamma(dx/dt)$, where 2γ serves as the coefficient of friction. The restoring force associated with the atomic binding of the electron is represented as $m\omega_o^2 x$.

After the success of the Bohr theory of the hydrogen atom in 1913 and the development of the wave mechanics which followed, it became abundantly clear that electrons were not bound harmonically. By a twist of fate, however, the solutions to the appropriate quantum theory equations for the time dependence of the electron motion happened to have a mathematical form which was similar to the solutions of Eq. (7.9.1). So, for the moment, we will go with history and adopt Eq. (7.9.1). The equation recommends itself, because solving it requires less effort than solving the corresponding quantum theory equation. In a subsequent chapter, we will have no choice but to use the quantum theory, and the true story will then emerge.

If the local electric field is taken to have the form,

$$F(t) = \tilde{F}(\omega)e^{-i\omega t} \tag{7.9.2}$$

where the complex valued amplitude, $\tilde{F}(\omega)$, depends upon the light frequency, the corresponding solution of Eq. (7.9.1) has the form,

$$x(t) = \tilde{x}(\omega)e^{-i\omega t} \tag{7.9.3}$$

where the amplitude $\tilde{x}(\omega)$ is a complex valued function of ω. When Eqs. (7.9.2) and (7.9.3) are substituted into Eq. (7.9.1), the result is

$$-m\omega^2\tilde{x}(\omega)e^{-i\omega t} - 2i\gamma m\omega\tilde{x}(\omega)e^{-i\omega t} + m\omega_o^2\tilde{x}(\omega)e^{-i\omega t} = -e\tilde{F}(\omega)e^{-i\omega t} \tag{7.9.4a}$$

Upon cancellation of $e^{-i\omega t}$ from both sides, the left hand side of Eq. (7.9.4a) can be factored to read,

$$m\tilde{x}(\omega)\left[-\omega^2 - 2i\gamma\omega + \omega_o^2\right] = -e\tilde{F}(\omega) \tag{7.9.4b}$$

When Eq. (7.9.4b) is solved algebraically for the amplitude,$\tilde{x}(\omega)$, the result is

$$\tilde{x}(\omega) = \frac{-(e/m)}{\left(\omega_o^2 - \omega^2\right) - 2i\gamma\omega} \cdot \tilde{F}(\omega) \tag{7.9.4c}$$

The next step is to rationalize Eq. (7.9.4c) by multiplying the numerator and denominator of the fraction by the complex conjugate of the denominator. This leads to

$$\tilde{x}(\omega) = \frac{-(e/m)\tilde{F}(\omega)}{\left[\left(\omega_o^2 - \omega^2\right) - 2i\gamma\omega\right]} \cdot \frac{\left[\left(\omega_o^2 - \omega^2\right) + 2i\gamma\omega\right]}{\left[\left(\omega_o^2 - \omega^2\right) + 2i\gamma\omega\right]} \tag{7.9.4d}$$

Upon multiplication of the factors in the denominator, Eq. (7.9.4d) can be written as

$$\tilde{x}(\omega) = \frac{-(e/m)\tilde{F}(\omega)\left[\left(\omega_o^2 - \omega^2\right) + 2i\gamma\omega\right]}{\left[\left(\omega_o^2 - \omega^2\right)^2 + 4\gamma^2\omega^2\right]} \tag{7.9.5}$$

The Fourier frequency component, $\tilde{p}(\omega)$, of the induced electric dipole moment is

$$\tilde{p}(\omega) = -e\tilde{x}(\omega) = \tilde{\alpha}(\omega)\tilde{F}(\omega) \tag{7.9.6a}$$

where $\tilde{\alpha}(\omega)$ is the frequency dependent polarizability, and e is the magnitude of the electron charge. Solution of Eq. (7.9.6a) for $\tilde{\alpha}(\omega)$ gives

$$\tilde{\alpha}(\omega) = \frac{-e\tilde{x}(\omega)}{\tilde{F}(\omega)} \tag{7.9.6b}$$

Upon substitution of Eq. (7.9.5) into Eq. (7.9.6b), we obtain for the polarizability,

$$\tilde{\alpha}(\omega) = \frac{\left(\omega_o^2 - \omega^2\right)(e^2/m)}{\left[\left(\omega_o^2 - \omega^2\right)^2 + 4\gamma^2\omega^2\right]} + i\frac{2\gamma\omega(e^2/m)}{\left[\left(\omega_o^2 - \omega^2\right)^2 + 4\gamma^2\omega^2\right]} \tag{7.9.7}$$

If we represent $\tilde{\alpha}(\omega)$ in complex form as

$$\tilde{\alpha}(\omega) = \mathrm{Re}[\tilde{\alpha}(\omega)] + i\mathrm{Im}[\tilde{\alpha}(\omega)] = \alpha'(\omega) + i\alpha''(\omega) \tag{7.9.8}$$

then upon comparison with Eq. (7.9.7), we have

$$\mathrm{Re}[\tilde{\alpha}(\omega)] = \alpha'(\omega) = \frac{\left(\omega_o^2 - \omega^2\right)(e^2/m)}{\left[\left(\omega_o^2 - \omega^2\right)^2 + 4\gamma^2\omega^2\right]} \tag{7.9.9a}$$

and

$$\mathrm{Im}[\tilde{\alpha}(\omega)] = \alpha''(\omega) = \frac{2\gamma\omega(e^2/m)}{\left[\left(\omega_o^2 - \omega^2\right)^2 + 4\gamma^2\omega^2\right]} \tag{7.9.9b}$$

Equations (7.9.9a) and (7.9.9b) form the basis for the Lorentz representation of the frequency response of a bound electron in a molecule subject to the oscillating electric field of a light wave. We are now in position to use the theory of electromagnetism to derive one of the most important relations in spectrochemical analysis, the Beer–Lambert law.

7.10 Beer–Lambert Law

Named for August Beer (German, 1825–1863) and Johann Heinrich Lambert (Swiss, 1728–1777), the Beer–Lambert law relates the concentration of a solute to the optical density of the solution that contains it. The optical density is defined as $\log(I(0)/I(b))$ where $I(0)$ is the intensity of the light incident on the sample, and $I(b)$ is the intensity of the light that has passed through a thickness b of the sample.

Using traditional notation, we let $\bar{c}$ be the concentration of the solute and define a as an empirical parameter accounting for the absorption of the light by the solution. The Beer–Lambert Law states that $I(0)$ and $I(b)$ are related by the equation,

$$I(b) = I(0)e^{-ab\bar{c}} \tag{7.10.1}$$

where a is ordinarily a function of the wavelength of the light.

We will show that the validity of Eq. (7.10.1) can be established on the basis of electromagnetic theory. Let $\langle N(x)\rangle$ be the magnitude of the vector, $\langle \mathbf{N}(x)\rangle$. In Eq. (7.8.15) set $x = b$ and identify $I(b)$ with $\langle N(b)\rangle$, so that

$$I(b) = \langle N(b)\rangle = \frac{1}{2}\left(\frac{\mathrm{Re}\left(\tilde{k}^*\right)}{\mu_o\omega}\right)\left|\tilde{\mathbf{E}}(\omega)\right|^2 e^{-2\mathrm{Im}(\tilde{k})b} \tag{7.10.2a}$$

where $\tilde{k}^*$ is the magnitude of the vector, $\tilde{\mathbf{k}}^*$. Next, make the identification,

$$I(0) = \frac{1}{2}\left(\frac{\mathrm{Re}\left(\tilde{k}^*\right)}{\mu_o\omega}\right)\left|\tilde{\mathbf{E}}(\omega)\right|^2 \tag{7.10.2b}$$

Let

$$2\mathrm{Im}(\tilde{k}) = a\bar{c}, \tag{7.10.2c}$$

Upon combining Eqs. (7.10.2a), (7.10.2b), and (7.10.2c), we conclude that

$$I(b) = I(0)\exp\left(-2\mathrm{Im}(\tilde{k})b\right) \tag{7.10.3}$$

Equation (7.10.3) satisfactorily accounts for the exponential form of the Beer–Lambert law as expressed by Eq. (7.10.1).

To complete the proof of the identity of Eqs. (7.10.1) and (7.10.3), the theory of dielectrics can be used to demonstrate the dependence of $\mathrm{Im}(\tilde{k})$ on $\bar{c}$ and also to determine the the dependence of a on the wavelength. Both results follow on the basis of the Clausius–Mossotti equation, which we write in frequency dependent form as

$$\frac{\tilde{\varepsilon}_r(\omega)-1}{\tilde{\varepsilon}_r(\omega)+2}=\frac{N\tilde{\alpha}(\omega)}{3\varepsilon_o} \tag{7.10.4a}$$

The number of molecules per unit volume is N. According to Eq. (7.10.4a), the frequency dependence of $\tilde{\varepsilon}_r(\omega)$ has its origin in the complex valued molecular polarizability, $\tilde{\alpha}(\omega)$.

To apply the Clausius–Mossotti formula to a solution consisting of a mixture of solvent, "A", and a solute, "B", Eq. (7.10.4a) needs to be extended to read,

$$\frac{\tilde{\varepsilon}_r(\omega)-1}{\tilde{\varepsilon}_r(\omega)+2}=\frac{1}{3\varepsilon_o}(c_A\tilde{\alpha}_A(\omega)+c_B\tilde{\alpha}_B(\omega)) \tag{7.10.4b}$$

In Eq. (7.10.4b), $c_A=N_A/V$ and $c_B=N_B/V$ where N_A and N_B are the numbers of molecules of "A" and "B", respectively, and V is the volume. With A as the solvent and B as the solute, the complex valued solvent and solute molecular polarizabilities are $\tilde{\alpha}_A(\omega)$ and $\tilde{\alpha}_B(\omega)$, respectively.

To achieve the most accurate determination of the concentration of a solute by optical density measurement, the solvent is ordinarily chosen to be transparent in the frequency range where the solute absorbs. The polarizability of a solvent molecule under such conditions is purely real. In this case, we can set $\mathrm{Im}[\tilde{\alpha}_A(\omega)]=0$; hence,

$$\tilde{\alpha}_A(\omega)=\mathrm{Re}[\tilde{\alpha}_A(\omega)] \tag{7.10.5}$$

If B is to absorb radiation, the polarizability, $\tilde{\alpha}_B(\omega)$, must have a non-zero imaginary part, so we write

$$\tilde{\alpha}_B(\omega)=\mathrm{Re}[\tilde{\alpha}_B(\omega)]+i\mathrm{Im}[\tilde{\alpha}_B(\omega)] \tag{7.10.6}$$

According to Eq. (7.9.9a),

$$\mathrm{Re}[\tilde{\alpha}_B(\omega)]=\frac{\left(\omega_o^2-\omega^2\right)(e^2/m)}{\left[\left(\omega_o^2-\omega^2\right)^2+4\gamma^2\omega^2\right]} \tag{7.10.7a}$$

and Eq. (7.9.9b),

$$\mathrm{Im}[\tilde{\alpha}_B(\omega)]=\frac{2\gamma\omega(e^2/m)}{\left[\left(\omega_o^2-\omega^2\right)^2+4\gamma^2\omega^2\right]} \tag{7.10.7b}$$

Near resonance where $\omega = \omega_o$, a comparison of Eqs. (7.10.7a) and (7.10.7b) indicates that $\text{Im}[\tilde{\alpha}_B(\omega)] \gg \text{Re}[\tilde{\alpha}_B(\omega)]$. Equation (7.10.6) can then be simplified to read

$$\tilde{\alpha}_B(\omega) \approx i\text{Im}[\tilde{\alpha}_B(\omega)] \tag{7.10.8}$$

On the basis of Eqs. (7.10.5) and (7.10.8), the sum of the terms enclosed by the parentheses on the right hand side of Eq. (7.10.4b) can be replaced by

$$c_A\tilde{\alpha}_A(\omega) + c_B\tilde{\alpha}_B(\omega) = c_A\text{Re}[\tilde{\alpha}_A(\omega)] + ic_B\text{Im}[\tilde{\alpha}_B(\omega)] \tag{7.10.9}$$

To simplify the subsequent algebra, let

$$x = \left(\frac{c_A}{3\varepsilon_o}\right)\text{Re}[\tilde{\alpha}_A(\omega)] \tag{7.10.10a}$$

and

$$y = \left(\frac{c_B}{3\varepsilon_o}\right)\text{Im}[\tilde{\alpha}_B(\omega)] \tag{7.10.10b}$$

then Eq. (7.10.4b) can be transformed to read

$$\frac{\tilde{\varepsilon}_r(\omega) - 1}{\tilde{\varepsilon}_r(\omega) + 2} = x + iy \tag{7.10.10c}$$

Define z as

$$z = x + iy \tag{7.10.11}$$

then Eq. (7.10.10c) assumes the simplified form,

$$\frac{\tilde{\varepsilon}_r(\omega) - 1}{\tilde{\varepsilon}_r(\omega) + 2} = z \tag{7.10.12}$$

Upon simplifying the notation so that $\tilde{\varepsilon}_r = \tilde{\varepsilon}_r(\omega)$, Eq. (7.10.12) can rearranged to read,

$$\tilde{\varepsilon}_r - 1 = z(\tilde{\varepsilon}_r + 2) \tag{7.10.13a}$$

or

$$\tilde{\varepsilon}_r(1 - z) = (1 + 2z) \tag{7.10.13b}$$

We solve Eq. (7.10.13b) for $\tilde{\varepsilon}_r$ and for future reference, we take the square root. The result is

$$\sqrt{\tilde{\varepsilon}_r} = \frac{(1+2z)^{1/2}}{(1-z)^{1/2}} \tag{7.10.14}$$

Substitution of Eq. (7.10.14) into Eq. (7.7.19), then gives for $\tilde{k}(\omega)$,

$$\tilde{k}(\omega) = \left(\frac{\omega}{c}\right) f(z) \tag{7.10.15}$$

where

$$f(z) = \sqrt{\tilde{\varepsilon}_r} = \frac{(1+2z)^{1/2}}{(1-z)^{1/2}} \tag{7.10.16}$$

In Eq. (7.10.16), $f(z)$ is a function of the complex variable, z. If the mixture of the solute B in solvent A is dilute, then $c_B \ll c_A$. According to Eqs. (7.10.10a) and (7.10.10b), this implies that $y \ll x$. This condition permits $f(z)$ to be expanded in a Taylor series about a point, $z_o = x$, on the real axis of the complex plane where $y = 0$. Retaining only the first two terms, the Taylor series representation of $f(z)$ is

$$f(z) = f(z_o) + f'(z_o)(z - z_o) + \cdots \tag{7.10.17}$$

See Fig. 7.4 for an example of a locus of $f(z)$ as defined in Eq. (7.10.17).

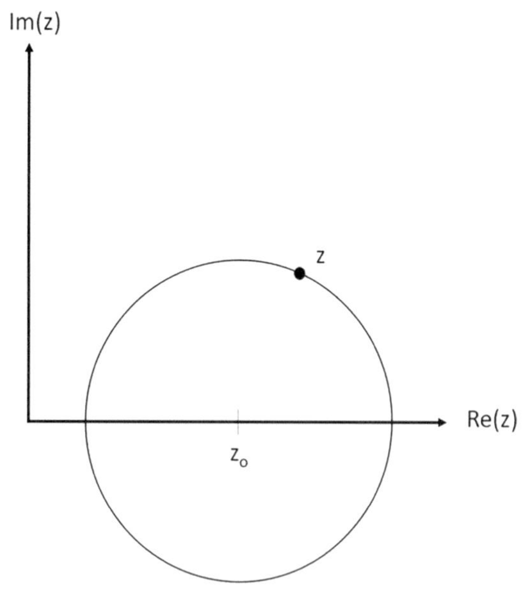

Fig. 7.4 Locus of $f(z)$ in the complex plane for a fixed value of $|z - z_o|$, where z_o is purely real

Upon introducing $z_o = x$ and $z = x + iy$, Eq. (7.10.17) reads,

$$f(z) = f(x) + if'(x)y + \cdots \tag{7.10.18}$$

Substitute Eq. (7.10.18) into Eq. (7.10.15). The result is

$$\tilde{k}(\omega) = \left(\frac{\omega}{c}\right)\left[f(x) + if'(x)y\right] \tag{7.10.19}$$

Evaluated at $z = x$, Eq. (7.10.16) reads,

$$f(x) = \frac{(1+2x)^{1/2}}{(1-x)^{1/2}} \tag{7.10.20}$$

Use Eq. (7.10.20) to compute $f'(x)$. The calculation proceeds as follows:

$$f'(x) = \frac{(1/2)(1+2x)^{-1/2}(2)(1-x)^{1/2} - (1+2x)^{1/2}(1/2)(1-x)^{-1/2}(-1)}{(1-x)} \tag{7.10.21a}$$

Upon cancelation of numerical and algebraic factors and transfer of binomials with negative exponents to the denominator, Eq. (7.10.21a) becomes

$$f'(x) = \frac{1}{(1+2x)^{1/2}(1-x)^{1/2}} + \frac{(1+2x)^{1/2}}{2(1-x)^{3/2}} \tag{7.10.21b}$$

Equation (7.10.21b) can be written over a common denominator to read,

$$f'(x) = \frac{2(1-x) + (1+2x)}{2(1+2x)^{1/2}(1-x)^{3/2}} \tag{7.10.21c}$$

from which the final result follows:

$$f'(x) = \frac{3}{2(1+2x)^{1/2}(1-x)^{3/2}} \tag{7.10.22}$$

According to Eq. (7.10.19),

$$\mathrm{Im}\left[\tilde{k}(\omega)\right] = \left(\frac{\omega}{c}\right)f'(x)y \tag{7.10.23a}$$

Substitute Eq. (7.10.22) into Eq. (7.10.23a). The result is

$$\mathrm{Im}\left[\tilde{k}(\omega)\right] = \left(\frac{\omega}{2c}\right)\frac{3y}{(1+2x)^{1/2}(1-x)^{3/2}} \tag{7.10.23b}$$

The real part of $\tilde{\alpha}_A(\omega)$ determines the solvent refractive index. The solvent refractive index, n_A, satisfies the Lorenz–Lorentz equation. Equation (7.10.10a) permits the Lorenz–Lorentz equation for the solvent to be written in the form,

$$\frac{n_A^2 - 1}{n_A^2 + 2} = \left(\frac{c_A}{3\varepsilon_o}\right)\mathrm{Re}[\tilde{\alpha}_A(\omega)] = x \tag{7.10.24}$$

Equation (7.10.24) can be used to express the denominator in the fraction on the right hand side of Eq. (7.10.23b) in terms of the measurable quantity, n_A. For this purpose, we evaluate first

$$1 + 2x = 1 + \frac{2n_A^2 - 2}{n_A^2 + 2} = \frac{3n_A^2}{n_A^2 + 2} \tag{7.10.25a}$$

and then

$$1 - x = 1 - \frac{n_A^2 - 1}{n_A^2 + 2} = \frac{3}{n_A^2 + 2} \tag{7.10.25b}$$

Upon use of Eqs. (7.10.25a) and (7.10.25b), the denominator in Eq. (7.10.23b) becomes simply,

$$(1 + 2x)^{1/2}(1 - x)^{3/2} = \left[\frac{3n_A^2}{n_A^2 + 2}\right]^{1/2}\left[\frac{3}{n_A^2 + 2}\right]^{3/2} = \frac{9n_A}{\left(n_A^2 + 2\right)^2} \tag{7.10.25c}$$

Substitute Eqs. (7.10.10b) and (7.10.25c) into Eq. (7.10.23b). The result is

$$\mathrm{Im}\left[\tilde{k}(\omega)\right] = \left(\frac{\omega}{2c}\right)(3)\left(\frac{\mathrm{Im}[\tilde{\alpha}_B(\omega)]}{3\varepsilon_o}\right)c_B\frac{\left(n_A^2 + 2\right)^2}{9n_A} \tag{7.10.26a}$$

In Eq. (7.10.26a), the pure solvent speed of light, $c_m = c/n_A$, where c is the speed of light in vacuum. With this substitution, Eq. (7.10.26a) reads

$$\mathrm{Im}\left[\tilde{k}(\omega)\right] = \left(\frac{\omega}{2n_A c_m}\right)(3)\left(\frac{\mathrm{Im}[\tilde{\alpha}_B(\omega)]}{3\varepsilon_o}\right)c_B\frac{\left(n_A^2 + 2\right)^2}{9n_A} \tag{7.10.26b}$$

Upon regrouping and cancellation of common factors, Eq. (7.10.26b) reads,

$$\mathrm{Im}\left[\tilde{k}(\omega)\right] = \left(\frac{\omega}{2c_m}\right)\left(\frac{n_A^2 + 2}{3n_A}\right)^2\left(\frac{\mathrm{Im}[\tilde{\alpha}_B(\omega)]}{\varepsilon_o}\right)c_B \tag{7.10.26c}$$

In the context where A is the solvent and B is the solute, we can replace $\overline{c}$ in Eq. (7.10.1) by c_B. Equation (7.10.26c) then provides a theoretical interpretation for the empirical quantity $(1/2)a\overline{c}$, as defined in Eq. (7.10.2c).

The Beer–Lambert law in its empirical decadic form reads,

$$I(b) = I(0)10^{-\{\varepsilon\}[B]\ell} \tag{7.10.27}$$

where $[B]$ is the molar solute concentration,$\{\varepsilon\}$ is the molar absorptivity, and ℓ is the thickness of the sample in centimeters. On the basis of Eq. (7.10.27), the optical density is defined as

$$OD = \log\left(\frac{I(0)}{I(b)}\right) = \{\varepsilon\}[B]\ell \tag{7.10.28}$$

Equation (7.10.3) can be used to evaluate the optical density if the form,

$$OD = \log\left(\frac{I(0)}{I(b)}\right) = 2\mathrm{Im}(\tilde{k})b \cdot \log e = 2(0.4343)\mathrm{Im}(\tilde{k})b \tag{7.10.29}$$

When the optical density is eliminated between Eqs. (7.10.28) and (7.10.29), the result is

$$\{\varepsilon\}[B]\ell = 2(0.4343)\mathrm{Im}(\tilde{k})b \tag{7.10.30}$$

Upon substitution of Eq. (7.10.26c) into the right hand side of Eq. (7.10.30), one obtains

$$\{\varepsilon\}[B]\ell = (0.4343)\left(\frac{\omega}{c_m}\right)\left(\frac{n_A^2+2}{3n_A}\right)^2\left(\frac{\mathrm{Im}[\tilde{\alpha}_B(\omega)]}{\varepsilon_o}\right) \cdot c_B b \tag{7.10.31}$$

Equation (7.10.31) includes quantities having a variety of units.

On the right side of Eq. (7.10.4b), the units of ε_o, are C^2/J m, the units of the number density, c_B, are m^{-3}, while the units of $\mathrm{Im}[\tilde{\alpha}_B(\omega)]$ are $C^2\ m^2/J$. The combination, $\mathrm{Im}[\tilde{\alpha}_B(\omega)]c_B/\varepsilon_o$, is dimensionless. The remaining quantities on the right hand side of Eq. (7.10.31) are ω with units of s^{-1}, c_m with units of m/s, b with units of m, and n_A which is dimensionless. The combination $\omega b/c_m$ is thus dimensionless. Hence, the right hand side of Eq. (7.10.31) is dimensionless. We now turn to the left hand side of Eq. (7.10.31). On the left hand side, $[B]$ is in the units of mol/L, ℓ is ordinarily in the units of cm, and the molar absorptivity, $\{\varepsilon\}$, has the units, L/mol cm. The product, $\{\varepsilon\}[B]\ell$, is dimensionless. Hence, both sides of Eq. (7.10.31) are dimensionless as required by the nature of this equation.

In order to solve Eq. (7.10.31) for $\{\varepsilon\}$, we shall need to cancel the product, $[B]\ell$, on the left hand side against $c_B b$ on the right. Since the units of these two products are different, we make use of the following equivalencies:

$$[B] = \frac{c_B}{N_A} \cdot \frac{\mathrm{m}^3}{10^3\,\mathrm{L}} \tag{7.10.32a}$$

where N_A is the Avogadro number. The two measures of the cell thickness are related by

$$\ell = b \cdot \frac{10^2\,\mathrm{cm}}{\mathrm{m}} \tag{7.10.32b}$$

Upon substitution of Eqs. (7.10.32a) and (7.10.32b) into Eq. (7.10.31), the immediate result is

$$\{\varepsilon\}\left(\frac{c_B}{N_A} \cdot \frac{\mathrm{m}^3}{10^3\mathrm{L}}\right)\left(b \cdot \frac{10^2\,\mathrm{cm}}{\mathrm{m}}\right) = (0.4343)\left(\frac{\omega}{c}\right)\left(\frac{n_A^2+2}{3n_A}\right)^2 \cdot \frac{\mathrm{Im}[\tilde{\alpha}_B(\omega)]}{\varepsilon_o} \cdot c_B b \tag{7.10.33}$$

The product, $c_B b$, which appears on both sides of Eq. (7.10.34), can be cancelled from either side with the final result,

$$\{\varepsilon\} = (0.4343)N_A\left(\frac{\omega}{c_m}\right)\left(\frac{\mathrm{m}}{10^2\,\mathrm{cm}}\right)\left(\frac{n_A^2+2}{3n_A}\right)^2\left(\frac{\mathrm{Im}[\tilde{\alpha}_B(\omega)]}{\varepsilon_o}\right)\left(\frac{10^3\,\mathrm{L}}{\mathrm{m}^3}\right) \tag{7.10.34}$$

In writing Eq. (7.10.34), the conversion factor $(10^3\,\mathrm{L}/\mathrm{m}^3)$ has been grouped with $\mathrm{Im}[\tilde{\alpha}_B(\omega)]/\varepsilon_o$, which has the units of m^3, while the conversion factor, $(\mathrm{m}/10^2\,\mathrm{cm})$ has been grouped with ω/c_m, which has the units of m^{-1}.

In optical density measurements, the spectrophotometer is tuned to one of the absorption frequencies, $\omega = \omega_o$, of the solute, B. The chosen absorption frequency corresponds to one of the spectral lines which maximizes $\mathrm{Im}[\tilde{\alpha}_B(\omega)]$. In the Lorentz model, $\mathrm{Im}[\tilde{\alpha}_B(\omega)]$ is given by Eq. (7.10.7b). Thus, with the use of Eq. (7.10.7b), Eq. (7.10.34) assumes the final form,

$$\{\varepsilon\} = (0.4343)N_A\left(\frac{\omega}{c_m}\right)\left(\frac{\mathrm{m}}{10^2\,\mathrm{cm}}\right)\left(\frac{n_A^2+2}{3n_A}\right)^2\left(\frac{1}{\varepsilon_o}\right)$$
$$\left[\frac{2\gamma\omega(e^2/m)}{\left(\omega_o^2-\omega^2\right)^2+4\gamma^2\omega^2}\right]\left(\frac{10^3\,\mathrm{L}}{\mathrm{m}^3}\right) \tag{7.10.35}$$

Intuitively, one would expect the molar absorptivity, $\{\varepsilon\}$, in Eq. (7.10.35) to depend upon the optical properties of both the solvent and the solute. The solvent dependence enters through the factor $\left((n_A^2+2)/3n_A\right)^2$, which is a function of the solvent refractive index, n_A. Starting from a minimum, when $n_A = 1$, the solvent factor, $\left((n_A^2+2)/3n_A\right)^2$ increases with increasing values of n_A. This implies that in cases where one has a choice of solvents, the solvent with the larger refractive index can produce the larger absorptivity. The solute dependence appears within the square bracket in

Eq. (7.10.35). The solute properties are not totally immune to the choice of the solvent, however. As pointed out in Sect. 3.11 of Chap. 3, the dielectric property of the solvent can shift the absorption frequency of a solute. It also can have an effect on the damping constant, γ. The factor in brackets in Eq. (7.10.35) has a maximum at the absorption frequency ω_o. Plotted as a function of ω, this factor assumes a shape which is termed a "Lorentzian".

7.11 Discussion

Equation (7.10.1) can also be derived by assigning to the solute an effective interaction cross section that governs its absorption of the incident radiation [1, 3]. This approach is attractive because the derivation of Eq. (7.10.1) requires only a few lines. Although laborious, the longer derivation above, which is based upon Maxwell's equations and the Lorentz model of the electron, has the advantage that it predicts a frequency dependence for $\{\varepsilon\}$ and also clearly distinguishes between the separate effects of the solvent and solute.

References

1. D.A. Skoog, D.M. West, F.J. Holler, S.R. Crouch, *Fundamentals of Analytical Chemistry* (Thomson, Brooks and Cole, Belmont, CA, 2004)
2. J.C. Slater, N.H. Frank, *Electromagnetism* (McGraw-Hill, New York, 1947), pp. 102–103
3. P. Norman, K. Ruud, T. Saue, *Principles and Practices of Molecular Properties (Theory and Simulations)* (Wiley, Hoboken, NJ, 2018)

Recommended Additional Reading

1. A. Hinchliffe, R. W. Munn, *Molecular Electromagnetism* (Wiley, Chichester, W. Sussex, UK, 1985)
2. D.J. Griffiths, *Introduction to Electrodynamics*, 4th edn. (Pearson Education, India, 2013)

Chapter 8
Light Scattering

This chapter begins with a recap of the free space Maxwell's equations, the density of the electromagnetic energy, and the Poynting vector for the energy flux. In order to separate the Maxwell equations, a vector potential is introduced in addition to the usual scalar potential. Fluctuations in the electric permittivity are identified as the cause of the scattering of the incident light. It is then shown that both the vector potential and the scalar potential can be combined in terms of a Hertz potential, which can be used to calculate the electric and magnetic fields that determine the intensity of the scattered light. The Hertz potential is shown to satisfy an inhomogeneous wave equation, which is solved by the Green's function method. The Rayleigh ratio is introduced in order to quantify the wavelength and angular dependence of the scattering. The turbidity follows as the angle average of the Rayleigh ratio. The theory of fluctuations, that was introduced in Chap. 2, is then invoked in order to identify the specific material properties associated with the scattering. Specific systems analyzed for their light scattering powers include the ideal gas, a real fluid with a liquid–vapor critical point, a binary liquid mixture with a critical point of solution, and finally a solution where the solute is a macromolecule.

8.1 Introduction

Atoms and molecules interact with electromagnetic radiation over a wide range of frequencies extending from radio waves to X-rays. Depending on the wavelength, the radiation can be refracted, reflected, scattered, or absorbed. Absorption was considered in Chap. 7. Historically, the scattering of visible light from aerosol particles was first demonstrated experimentally in 1869 by John Tyndall (Irish, 1820–1893). The phenomenon was subsequently analyzed theoretically by Rayleigh in 1871 using

J. K. Baird, *Electromagnetism for Chemistry*,
https://doi.org/10.1007/978-3-031-88508-2_8

a mechanical model of light, which was current at the time, and then analyzed again by him in 1881 and 1889 using Maxwell's theory of electromagnetic radiation. In this chapter, we shall concentrate on the elastic scattering of visible light. This phenomenon is known as Rayleigh scattering [1, 2].

Rayleigh showed that the intensity of scattered light is inversely proportional to the fourth power of the wavelength. From this result, he was able to explain why the sky is blue and the sunset is orange. He began by observing that light coming from the Sun contains a distribution of wavelengths. Sunlight entering the atmosphere is scattered by density fluctuations and suspended dust particles. Because the scattering power of a medium depends upon the wavelength, λ, as $1/\lambda^4$, the blue wavelengths are preferentially scattered. An observer looking at the sky in a direction away from the Sun, sees only the scattered light. By contrast, an observer who is looking directly at the Sun during sunset sees predominantly transmitted light, which is orange, the blue component having been scattered out of his field of vision. Given the many previous generations of humans who have marveled at these colors without understanding their origin, Rayleigh's theory is all the more remarkable.

When visible light interacts with matter, the electric field of the light induces oscillating electric dipoles in the molecules. These oscillating dipoles reradiate the light into all directions. The degree to which visible light interacts with matter is determined by the electric permittivity. Scattering is due to spatial variations in the permittivity. Marian Smoluchowski (Polish, 1872–1917) in 1908 and Albert Einstein (German, Swiss, American, 1879–1955) in 1910 realized that the fluctuations in the density in the case of a fluid and fluctuations in concentration in the case of a liquid solution are responsible for the variations in the permittivity. They showed that the magnitude of these fluctuations could be calculated using a combination of classical thermodynamics and statistical mechanics.

Our main interest in this chapter will be the solution to Maxwell's equations of electromagnetism in order to obtain the intensity and angular distribution of the scattered light. To achieve this goal, we will proceed as follows: (1) review the results of Chap. 7 for the balance and flow of energy associated with an electromagnetic field (Poynting's vector), (2) introduce the magnetic vector potential, and derive a wave equation satisfied by the electric and magnetic fields of the scattered light, and (3) solve for the scattered electric and magnetic fields using the Green's function technique. At the end of the chapter, we will incorporate electric permittivity fluctuations into the theory. This may seem like a big program, but it will be broken down into parts. The mathematical and physical reasoning involved in each stage the calculation will be revealed and explained.

8.2 Electric and Magnetic Constitutive Relations and Maxwell's Equations

In Chap. 3, we introduced the constitutive relations satisfied by the electric field, **E**, polarization density, **P**, and the displacement, **D**. We summarize these equations below:

$$\mathbf{P} = \varepsilon_o \chi_e \mathbf{E} \tag{8.2.1a}$$

$$\mathbf{D} = \varepsilon_o \mathbf{E} + \mathbf{P} \tag{8.2.1b}$$

$$\mathbf{D} = \varepsilon \mathbf{E} \tag{8.2.1c}$$

$$\varepsilon = \varepsilon_o (1 + \chi_e) = \varepsilon_r \varepsilon_o \tag{8.2.1d}$$

The corresponding constitutive relations satisfied by the magnetic field, **H**, the magnetization, **M**, and the magnetic induction, **B**, were introduced in Chap. 7. They are:

$$\mathbf{M} = \chi_m \mathbf{H} \tag{8.2.2a}$$

$$\mathbf{H} = \frac{1}{\mu_o}\mathbf{B} - \mathbf{M} \tag{8.2.2b}$$

$$\mathbf{B} = \mu \mathbf{H} \tag{8.2.2c}$$

$$\mu = \mu_o (1 + \chi_m) = \mu_r \mu_o \tag{8.2.2d}$$

In Chap. 7 we encountered the Maxwell equations, which are summarized below:

$$\nabla \cdot \mathbf{D} = \rho \qquad \text{(Gauss)} \tag{8.2.3}$$

$$\nabla \times \mathbf{H} = \mathbf{J} + \frac{\partial \mathbf{D}}{\partial t} \qquad \text{(Ampere)} \tag{8.2.4}$$

$$\nabla \times \mathbf{E} = -\frac{\partial \mathbf{B}}{\partial t} \qquad \text{(Faraday)} \tag{8.2.5}$$

$$\nabla \cdot \mathbf{B} = 0 \tag{8.2.6}$$

The mobile charges in any system consist of free electrons and/or free ions. As a reminder, ρ in Eq. (8.2.3) is the volume density of the free charges, while **J** in

Eq. (8.2.4) is the current density associated with these free charges when they are moving.

8.3 Energy Density and the Poynting Vector

In this section, we review the results of Chap. 7 concerning energy balance in the electromagnetic field. The energy density, ρ_E, is

$$\rho_E = \frac{1}{2}(\mathbf{H} \cdot \mathbf{B} + \mathbf{E} \cdot \mathbf{D}) \tag{8.3.1}$$

The Poynting vector, $\mathbf{N}$, for the energy flux is defined by

$$\mathbf{N} = \mathbf{E} \times \mathbf{H} \tag{8.3.2}$$

With these identifications, the continuity equation for energy conservation takes the form,

$$\frac{\partial \rho_E}{\partial t} + \nabla \cdot \mathbf{N} = -\mathbf{J} \cdot \mathbf{E} \tag{8.3.3}$$

where $\mathbf{J}$ is the current density in the material through which the radiation is traveling, and $\mathbf{E}$ is the electric field.

8.4 Vector and Scalar Potentials and the Separation of Maxwell's Equations

8.4.1 Vector and Scalar Potentials

Equations (8.2.3) through (8.2.6) are four coupled first order partial differential equations. In this section, we condense these four equations for $\mathbf{D}$, $\mathbf{E}$, $\mathbf{B}$, and $\mathbf{H}$ into two uncoupled second order partial differential equations for the scalar potential Φ and vector potential $\mathbf{A}$.

In electrostatics, there is only one vector field to consider, which is the electric field $\mathbf{E}$. The electric field can be calculated as the gradient of a scalar potential. When time dependence is under consideration, we will denote this scalar potential temporarily by Φ'. To take into account magnetic effects, a second potential is required. This additional potential is a vector, which we shall temporarily call, $\mathbf{A}'$. In both the static and time dependent cases, the magnetic induction is calculated as the curl,

$$\mathbf{B} = \nabla \times \mathbf{A}' \tag{8.4.1}$$

In the time varying case, $\mathbf{E}$ depends upon both Φ' and $\mathbf{A}'$. This assertion can be demonstrated. If Eq. (8.4.1) is substituted into Eq. (8.2.5), the result is

$$\nabla \times \mathbf{E} = -\frac{\partial(\nabla \times \mathbf{A}')}{\partial t} \tag{8.4.2a}$$

which can be rewritten as

$$\nabla \times \left(\mathbf{E} + \frac{\partial \mathbf{A}'}{\partial t}\right) = 0 \tag{8.4.2b}$$

Since Φ' is a scalar valued function, we can invoke the vector operator identity [See the Mathematical Methods Appendix],

$$\nabla \times \nabla \Phi' = 0 \tag{8.4.2c}$$

This identity permits us to define Φ' so that Eq. (8.4.2b) is equivalent to

$$-\nabla \Phi' = \mathbf{E} + \frac{\partial \mathbf{A}'}{\partial t} \tag{8.4.2d}$$

This proves our assertion that $\mathbf{E}$ can be expressed in terms of Φ' and $\mathbf{A}'$.

8.4.2 The Concept of Gauge Invariance

Equation (8.4.2d) can be rearranged to read

$$\mathbf{E} = -\nabla \Phi' - \frac{\partial \mathbf{A}'}{\partial t} \tag{8.4.3}$$

Next, we use Eqs. (8.2.1c) and (8.2.2c) to write Eq. (8.2.4) in the form,

$$\nabla \times \left(\frac{\mathbf{B}}{\mu}\right) = \mathbf{J} + \varepsilon \frac{\partial \mathbf{E}}{\partial t} \tag{8.4.4a}$$

If we substitute Eq. (8.4.1) into the left hand side of Eqs. (8.4.4a) and (8.4.3) into the right hand side, the result is

$$\nabla \times \left(\frac{\nabla \times \mathbf{A}'}{\mu}\right) = \mathbf{J} - \varepsilon \nabla\left(\frac{\partial \Phi'}{\partial t}\right) - \varepsilon\left(\frac{\partial^2 \mathbf{A}'}{\partial t^2}\right) \tag{8.4.4b}$$

Use the vector operator identity,

$$\nabla \times (\nabla \times \mathbf{A}') = \nabla(\nabla \cdot \mathbf{A}') - \nabla^2 \mathbf{A}' \tag{8.4.4c}$$

[See the Mathematical Methods Appendix] to expand the left hand side of Eq. (8.4.4b), so that the whole equation reads

$$\frac{1}{\mu}\nabla(\nabla\cdot\mathbf{A}') - \frac{1}{\mu}\nabla^2\mathbf{A}' = \mathbf{J} - \varepsilon\nabla\left(\frac{\partial\Phi'}{\partial t}\right) - \varepsilon\left(\frac{\partial^2\mathbf{A}'}{\partial t^2}\right) \tag{8.4.4d}$$

If Eq. (8.4.4d) is multiplied through by $-\mu$, it can be rearranged to read,

$$\nabla^2\mathbf{A}' - \varepsilon\mu\frac{\partial^2\mathbf{A}'}{\partial t^2} = -\mu\mathbf{J} + \mu\nabla\left(\varepsilon\frac{\partial\Phi'}{\partial t} + \frac{1}{\mu}\nabla\cdot\mathbf{A}'\right) \tag{8.4.4e}$$

If it were not for the term in parentheses on the right, Eq. (8.4.4e) would be a recognizable second order partial differential equation for the vector field, $\mathbf{A}'$. Fortunately, it is possible to transform $\mathbf{A}'$ in a way which does make the quantity in parentheses in Eq. (8.4.4e) zero and at the same time preserves Eq. (8.4.1). To do this, we introduce the scalar valued function, Λ, called the "gauge" function. Like all scalar valued functions, it satisfies, the vector operator identity

$$\nabla\times\nabla\Lambda = 0 \tag{8.4.5a}$$

Introduce a new vector potential, $\mathbf{A}$, which satisfies

$$\mathbf{A}' = \mathbf{A} - \nabla\Lambda \tag{8.4.5b}$$

By virtue of Eq. (8.4.5b), the magnetic flux density in Eq. (8.4.1) is

$$\mathbf{B} = \nabla\times\mathbf{A}' = \nabla\times(\mathbf{A} - \nabla\Lambda) = (\nabla\times\mathbf{A}) - (\nabla\times\nabla\Lambda) = \nabla\times\mathbf{A} \tag{8.4.5c}$$

where $\nabla\times(\nabla\Lambda) = 0$ by virtue of it being a vector identity. Because $\mathbf{B}$ is measurable, its value should be independent of the way in which it is calculated. We see that the value of $\mathbf{B}$ is the same when calculated as $\nabla\times\mathbf{A}$ or when calculated as $\nabla\times\mathbf{A}'$. This is due to the fact that $\mathbf{A}$ and $\mathbf{A}'$ differ by $\nabla\Lambda$, where Λ is an arbitrary scalar function, which because the curl of $\nabla\Lambda$ is zero, makes no contribution to the value of $\mathbf{B}$. Being insensitive to the to the choice of Λ, $\mathbf{B}$ is said to be "gauge invariant", where Λ is the gauge function. We remove the dependence upon Φ' in Eq. (8.4.4e), by supposing that

$$\frac{1}{\mu}\nabla\cdot\mathbf{A}' + \varepsilon\frac{\partial\Phi'}{\partial t} = 0 \tag{8.4.6a}$$

Below, we show that it is possible to derive a wave equation, which in principle can always be solved so as to determine Λ. Upon multiplication through by μ and introduction of the speed of light, c_m, in the medium,

$$c_m = \frac{1}{\sqrt{\varepsilon\mu}} \tag{8.4.6b}$$

Equation (8.4.6a) reads,

$$\nabla \cdot \mathbf{A}' + \frac{1}{c_m^2}\frac{\partial \Phi'}{\partial t} = 0 \tag{8.4.6c}$$

Equation (8.4.6c) is satisfied if we introduce a new scalar function Φ that is related to Φ' according to

$$\Phi' = \Phi + \frac{\partial \Lambda}{\partial t} \tag{8.4.6d}$$

Upon substitution of Eqs. (8.4.5b) and (8.4.6d) into Eq. (8.4.6c), we obtain

$$\nabla \cdot \mathbf{A} - \nabla^2 \Lambda + \frac{1}{c_m^2}\frac{\partial \Phi}{\partial t} + \frac{1}{c_m^2}\frac{\partial^2 \Lambda}{\partial t^2} = 0 \tag{8.4.6e}$$

Upon rearrangement, Eq. (8.4.6e) becomes

$$\nabla \cdot \mathbf{A} + \frac{1}{c_m^2}\frac{\partial \Phi}{\partial t} = \nabla^2 \Lambda - \frac{1}{c_m^2}\frac{\partial^2 \Lambda}{\partial t^2} \tag{8.4.6f}$$

Since Λ is an arbitrary scalar function, we can impose on it the condition,

$$\nabla^2 \Lambda - \frac{1}{c_m^2}\frac{\partial^2 \Lambda}{\partial t^2} = 0 \tag{8.4.6g}$$

Equation (8.4.6g) is a wave equation, which is said to be *homogeneous*, because the right hand side is zero. As such, its solutions can always be found. Consequently, it is safe to conclude that

$$\nabla \cdot \mathbf{A} + \frac{1}{c_m^2}\frac{\partial \Phi}{\partial t} = 0 \tag{8.4.6h}$$

Eq. (8.4.6h) is said to define the Lorenz gauge.

8.4.3 Wave Equations that Replace Maxwell's Equations

With $\mathbf{A}'$ and Φ' chosen so that Eq. (8.4.6a) is satisfied, Eq. (8.4.4e) simplifies to

$$\nabla^2 \mathbf{A}' - \frac{1}{c_m^2}\frac{\partial^2 \mathbf{A}'}{\partial t^2} = -\mu \mathbf{J} \tag{8.4.7}$$

Upon substitution of Eq. (8.4.5b), Eq. (8.4.7) reads,

$$\nabla^2(\mathbf{A} - \nabla\Lambda) - \frac{1}{c_m^2}\frac{\partial^2(\mathrm{A} - \nabla\Lambda)}{\partial t^2} = -\mu\mathbf{J} \tag{8.4.8a}$$

which can be rewritten as

$$\nabla^2\mathbf{A} - \frac{1}{c_m^2}\frac{\partial^2\mathbf{A}}{\partial t^2} - \nabla\left(\nabla^2\Lambda - \frac{1}{c_m^2}\frac{\partial^2\Lambda}{\partial t^2}\right) = -\mu\mathbf{J} \tag{8.4.8b}$$

By virtue of Eq. (8.4.6g), Eq. (8.4.8b) becomes

$$\nabla^2\mathbf{A} - \frac{1}{c_m^2}\frac{\partial^2\mathbf{A}}{\partial t^2} = -\mu\mathbf{J} \tag{8.4.9}$$

The wave equations, Eqs. (8.4.7) and (8.4.9), are mathematically exactly the same. They are thus independent of the choice of Λ so long as Λ satisfies Eq. (8.4.6g). Equation (8.4.5c) shows that the magnetic induction is gauge invariant. When combined, Eqs. (8.4.7) and (8.4.9) show that the wave equation that determines the vector potential is also gauge invariant. Since the right hand side of Eq. (8.4.9) is non-zero, it is said to be an *inhomogeneous* wave equation. The product, $-\mu\mathbf{J}$, is called the source term.

We now derive a wave equation satisfied by Φ, which is consistent with the Lorenz gauge relation, Eq. (8.4.6h). Use Eq. (8.2.1c) to rewrite Eq. (8.2.3) in the form

$$\nabla\cdot\mathbf{E} = \rho/\varepsilon \tag{8.4.10}$$

When Eq. (8.4.3) is substituted into Eq. (8.4.10), the result is

$$-\nabla^2\Phi' - \frac{\partial\left(\nabla\cdot\mathbf{A}'\right)}{\partial t} = \frac{\rho}{\varepsilon} \tag{8.4.11a}$$

Substitute Eqs. (8.4.5b) and (8.4.6d) into Eq. (8.4.11a), so that it reads

$$-\nabla^2\Phi - \nabla^2\left(\frac{\partial\Lambda}{\partial t}\right) - \frac{\partial(\nabla\cdot\mathbf{A})}{\partial t} + \frac{\partial\left(\nabla^2\Lambda\right)}{\partial t} = \frac{\rho}{\varepsilon} \tag{8.4.11b}$$

In the second term of Eq. (8.4.11b), commute ∇^2 and $\partial/\partial t$ so that the result is

$$-\nabla^2\Phi - \frac{\partial\left(\nabla^2\Lambda\right)}{\partial t} - \frac{\partial(\nabla\cdot\mathbf{A})}{\partial t} + \frac{\partial\left(\nabla^2\Lambda\right)}{\partial t} = \frac{\rho}{\varepsilon} \tag{8.4.11c}$$

Upon cancellation of terms of opposite sign in Eq. (8.4.11c), the result is

$$-\nabla^2\Phi - \frac{\partial(\nabla\cdot\mathbf{A})}{\partial t} = \frac{\rho}{\varepsilon} \tag{8.4.11d}$$

If we substitute the Lorenz gauge relation, Eq. (8.4.6h), into Eq. (8.4.11d), we obtain

$$\nabla^2\Phi - \frac{1}{c_m^2}\frac{\partial^2\Phi}{\partial t^2} = -\frac{\rho}{\varepsilon} \tag{8.4.12}$$

Equation (8.4.12) is an inhomogeneous wave equation satisfied by the scalar potential, Φ.

Equation (8.4.9) involves $\mathbf{A}$ alone, while Eq. (8.4.12) involves Φ alone. As such, they are said to be uncoupled. The two, second order differential equations, Eqs. (8.4.9) and (8.4.12) replace the four, first order Maxwell differential equations, which are Eqs. (8.2.3) through (8.2.6). The solutions to Eqs. (8.4.9) and (8.4.12) can be used to determine the fields, $\mathbf{E}$ and $\mathbf{B}$.

8.5 Solution to the Wave Equation in Terms of the Green's Function

Equation (8.4.9) is solved by the vector valued function, $\mathbf{A}(\mathbf{R}, t)$ which has the three components, $(A_1(\mathbf{R}, t), A_2(\mathbf{R}, t), A_3(\mathbf{R}, t))$. Equation (8.4.9) can thus be decomposed into three independent scalar wave equations, one for each vector component. These equations are

$$\nabla^2 A_i - \frac{1}{c_m^2}\frac{\partial^2 A_i}{\partial t^2} = -\mu J_i \quad (i = x, y, z) \tag{8.5.1}$$

Equations (8.5.1) are examples of the general scalar wave equation,

$$\nabla^2\psi(\mathbf{R}, t) - \frac{1}{c_m^2}\frac{\partial^2\psi(\mathbf{R}, t)}{\partial t^2} = -q(\mathbf{R}, t) \tag{8.5.2}$$

where $\psi(\mathbf{R}, t)$ stands for one of the A_i $(i = 1, 2, 3)$, while $-q(\mathbf{R}, t)$ represents any of the corresponding components, $-\mu J_i$ $(i = 1, 2, 3)$. In our case, the goal is to consider a plane wave incident on a target and then solve Eq. (8.5.2) for the outgoing scattered wave. We take the origin of coordinates to be located at a point somewhere within the target. The scattered wave amplitude, $\psi(\mathbf{R}, t)$ is observed at the point, $\mathbf{R}$, and the time, t. If $\mathbf{R}$ describes a point outside the target, then $-q(\mathbf{R}, t) = 0$; otherwise, $-q(\mathbf{R}, t) \neq 0$.

We will use a method of solution first introduced by George Green (British, 1793–1841). The goal will be to determine $\psi(\mathbf{R}, t)$ at the arbitrary point, $\mathbf{R}$, and time, t, by exploiting the solution to a simpler companion wave problem. In the case of this companion problem, the wave emanates from an instantaneous point source, as opposed to the enduring, distributed source, $-q(\mathbf{R}, t)$, in Eq. (8.5.2). Specifically, the wave equation for the companion problem is

$$\nabla_o^2 G(\mathbf{R}, t; \mathbf{R}_o, t_o) - \frac{1}{c_m^2}\frac{\partial^2 G(\mathbf{R}, t; \mathbf{R}_o, t_o)}{\partial t_o^2} = -\delta(\mathbf{R}_o - \mathbf{R})\delta(t_o - t) \tag{8.5.3}$$

The delta function, $\delta(\mathbf{R}_o - \mathbf{R})$, on the right hand side of Eq. (8.5.3) represents a point source. The second delta function, $\delta(t_o - t)$, indicates that the point source acts instantaneously at $t = t_o$. [For a fuller description of the delta function, see the Mathematical Methods Appendix]. The solution, $G(\mathbf{R}, t; \mathbf{R}_o, t_o)$, to Eq. (8.5.3) represents a signal generated at the point $\mathbf{R}$ at time t. The signal is subsequently observed at time, t_o, at the position, $\mathbf{R}_o$. The solution, $G(\mathbf{R}, t; \mathbf{R}_o, t_o)$, is known as Green's function. [See the Mathematical Methods Appendix for a simple example of a Green's function]. Green's theorem asserts that the solution, $\psi(\mathbf{R}, t)$, to Eq. (8.5.2) can be written as an integral involving $G(\mathbf{R}, t; \mathbf{R}_o, t_o)$. The procedure for solving Eq. (8.5.3) for $G(\mathbf{R}, t; \mathbf{R}_o, t_o)$ is known as Green's problem.

The proof begins by replacing $\mathbf{R}$ as the independent variable in Eq. (8.5.2) by $\mathbf{R}_o$ and the time, t, by t_o. This is motivated by the fact that $\delta(\mathbf{R}_o - \mathbf{R})$ and $\delta(t_o - t)$, which constitute the ultimate source of the disturbance, are both even functions of their arguments. [See the Mathematical Methods Appendix]. That is to say, $\delta(\mathbf{R}_o-\mathbf{R})$ does not change sign when $\mathbf{R}$ and $\mathbf{R}_o$ are interchanged, nor does $\delta(t_o - t)$ change sign when t and t_o are interchanged. After the proposed substitution of variables, Eq. (8.5.2) reads,

$$\nabla_o^2 \psi(\mathbf{R}_o, t_o) - \frac{1}{c_m^2}\frac{\partial^2 \psi(\mathbf{R}_o, t_o)}{\partial t_o^2} = -q(\mathbf{R}_o, t_o) \tag{8.5.4a}$$

We next multiply both sides of Eq. (8.5.4a) by $G = G(\mathbf{R}, t; \mathbf{R}_o, t_o)$. The result is

$$G\nabla_o^2 \psi - \frac{1}{c_m^2} G\frac{\partial^2 \psi}{\partial t_o^2} = -Gq(\mathbf{R}_o, t_o) \tag{8.5.4b}$$

Next, multiply Eq. (8.5.3) through by $\psi = \psi(\mathbf{R}_o, t_o)$. That result is

$$\psi\nabla_o^2 G - \frac{1}{c_m^2}\psi\frac{\partial^2 G}{\partial t_o^2} = -\psi\delta(\mathbf{R} - \mathbf{R}_o)\delta(t - t_o) \tag{8.5.4c}$$

Subtract Eq. (8.5.4c) from Eq. (8.5.4b) and integrate both sides of the resulting equation over the volume differential element, d^3R_o, and over the time differential element, dt_o. The integral of the difference between the two equations is

$$\int_0^t dt_o \int d^3R_o \delta(\mathbf{R} - \mathbf{R}_o)\delta(t - t_o)\psi(\mathbf{R}_o, t_o)$$

$$
\begin{aligned}
&-\int_0^t dt_o \int d^3R_o G(\mathbf{R}, t; \mathbf{R}_o, t_o) q(\mathbf{R}_o, t_o) \\
&= \int_0^t dt_o \int d^3R_o \left[G\nabla_o^2\psi - \psi\nabla_o^2 G\right] \\
&-\frac{1}{c_m^2}\int_0^t dt_o \int d^3R_o \left[G\frac{\partial^2\psi}{\partial t_o^2} - \psi\frac{\partial^2 G}{\partial t_o^2}\right]
\end{aligned} \tag{8.5.4d}
$$

The even nature of the delta functions permits Eq. (8.5.4d) to be rewritten as

$$
\begin{aligned}
&\int_0^t dt_o \int d^3R_o \delta(\mathbf{R}_o - \mathbf{R})\delta(t_o - t)\psi(\mathbf{R}_o, t_o) \\
&-\int_0^t dt_o \int d^3R_o G(\mathbf{R}, t; \mathbf{R}_o, t_o) q(\mathbf{R}_o, t_o) \\
&= \int_0^t dt_o \int d^3R_o \left[G\nabla_o^2\psi - \psi\nabla_o^2 G\right] \\
&-\frac{1}{c_m^2}\int_0^t dt_o \int d^3R_o \left[G\frac{\partial^2\psi}{\partial t_o^2} - \psi\frac{\partial^2 G}{\partial t_o^2}\right]
\end{aligned} \tag{8.5.4e}
$$

The range of the spatial integration is assumed to extend over the entire volume of the source, while the time covers the range, $0 \le t_o \le t$. Because of the delta functions, the first integral on the left hand side of Eq. (8.5.4e) is just $\psi(\mathbf{R}, t)$; hence, Eq. (8.5.4d) can be rewritten as

$$
\begin{aligned}
&\psi(\mathbf{R}, t) - \int_0^t dt_o \int d^3R_o G(\mathbf{R}, t; \mathbf{R}_o, t_o) q(\mathbf{R}_o, t_o) \\
&= \int dt_o \int d^3R_o \left[G\nabla_o^2\psi - \psi\nabla_o^2 G\right] - \frac{1}{c_m^2}\int_0^t dt_o \int d^3R_o \left[G\frac{\partial^2\psi}{\partial t_o^2} - \psi\frac{\partial^2 G}{\partial t_o^2}\right]
\end{aligned} \tag{8.5.5}
$$

The first integral on the right hand side of Eq. (8.5.5) can be evaluated by exploiting the following vector identity [See the Mathematical Methods Appendix],

$$
\nabla \cdot (\phi\mathbf{A}) = \phi\nabla \cdot \mathbf{A} + \mathbf{A} \cdot \nabla\phi \tag{8.5.6a}
$$

where **A** is an arbitrary vector, and ϕ is an arbitrary scalar. Equation (8.5.6a) can then be rearranged to read,

$$\phi\nabla\cdot\mathbf{A}=\nabla\cdot(\phi\mathbf{A})-\mathbf{A}\cdot\nabla\phi \tag{8.5.6b}$$

In Eq. (8.5.6b), replace ϕ by G and **A** by $\nabla_o\psi$, and write $\nabla_o^2=\nabla_o\cdot\nabla_o$, so that expand $G\nabla_o^2\psi$ is expanded in the form,

$$G\nabla_o^2\psi=G\nabla_o\cdot(\nabla_o\psi)=\nabla_o\cdot(G\nabla_o\psi)-\nabla_o\psi\cdot\nabla_oG \tag{8.5.7a}$$

Likewise, in Eq. (8.5.6b) replace ϕ by ψ and **A** by ∇_oG and write for $\psi\nabla_o^2G$ in the form,

$$\psi\nabla_o^2G=\psi\nabla_o\cdot(\nabla_oG)=\nabla_o\cdot(\psi\nabla_oG)-\nabla_oG\cdot\nabla_o\psi \tag{8.5.7b}$$

The difference between Eqs. (8.5.7a) and (8.5.7b) is

$$\begin{aligned}G\nabla_o^2\psi-\psi\nabla_o^2G&=\nabla_o\cdot(G\nabla_o\psi)-\nabla_o\cdot(\psi\nabla_oG)\\&=\nabla_o\cdot(G\nabla_o\psi-\psi\nabla_oG)\end{aligned} \tag{8.5.8}$$

We now turn to the second integral on the right hand side of Eq. (8.5.5). The integrand can be written as

$$G\frac{\partial^2\psi}{\partial t_o^2}-\psi\frac{\partial^2G}{\partial t_o^2}=\frac{\partial}{\partial t_o}\left[G\frac{\partial\psi}{\partial t_o}-\psi\frac{\partial G}{\partial t_o}\right] \tag{8.5.9}$$

Evaluate the right hand side of Eq. (8.5.5) using Eqs. (8.5.8) and (8.5.9). The result is

$$\begin{aligned}&\psi(\mathbf{R},t)-\int_0^t dt_o\int d^3R_oG(\mathbf{R},t;\mathbf{R}_o,t_o)q(\mathbf{R}_o,t_o)\\&=\int_0^t dt_o\int d^3R_o\nabla_o\cdot[G\nabla_o\psi-\psi\nabla_oG]\\&\quad-\frac{1}{c^2}\int d^3R_o\int_0^t dt_o\frac{\partial}{\partial t_o}\left[G\frac{\partial\psi}{\partial t_o}-\psi\frac{\partial G}{\partial t_o}\right]\end{aligned} \tag{8.5.10}$$

The space part of the first double integral on the right hand side of Eq. (8.5.10) can be converted to a surface integral using the Gauss–Green divergence theorem [See the Mathematical Methods Appendix]. The time part of the second double integral on the right hand side of Eq. (8.5.10) can be computed directly. Equation (8.5.10) can be rewritten as

$$\psi(\mathbf{R}, t) - \int_0^t dt_o \int d^3R_o G(\mathbf{R}, t; \mathbf{R}_o, t_o) q(\mathbf{R}_o, t_o)$$
$$= \int_0^t dt_o \int d\mathbf{S}_o \cdot [G\nabla_o\psi - \psi\nabla_o G]$$
$$- \frac{1}{c^2}\int d^3R_o \int_0^t dt_o \frac{\partial}{\partial t_o}\left[G\frac{\partial\psi}{\partial t_o} - \psi\frac{\partial G}{\partial t_o}\right] \quad (8.5.11)$$

The integral over $d\mathbf{S}_o$ on the right hand side of Eq. (8.5.11) extends over the entire surface surrounding the volume of interest. The boundary conditions associated with the companion Eq. (8.5.3) are still at our disposal. The surface integral on the right hand side of Eq. (8.5.11) can be made to disappear if we choose for Green's problem the boundary conditions on $\mathbf{S}_o$ to be,

$$G = 0 \quad (8.5.12a)$$

and

$$\nabla_o G = 0 \quad (8.5.12b)$$

After computation of the time integral, Eq. (8.5.11) then reads,

$$\psi(\mathbf{R}, t) - \int_0^t dt_o \int d^3R_o G(\mathbf{R}, t; \mathbf{R}_o, t_o) q(\mathbf{R}_o, t_o)$$
$$= -\frac{1}{c^2}\int d^3R_o \left[G\frac{\partial\psi}{\partial t_o} - \psi\frac{\partial G}{\partial t_o}\right]_{t_o=0}^{t_o=t} \quad (8.5.12c)$$

We can assume for the companion problem, that the initial time values of Green's function are

$$G(\mathbf{R}_o, 0) = 0 \quad (8.5.13a)$$

and

$$\left[\frac{\partial G(\mathbf{R}_o, t_o)}{\partial t_o}\right]_{t_o=0} = 0 \quad (8.5.13b)$$

We also require,

$$\psi(\mathbf{R}_o, t_o \geq t) = 0 \quad (8.5.14a)$$

and

$$\left[\frac{\partial \psi(\mathbf{R}_o, t_o)}{\partial t_o}\right]_{t_o \geq t} = 0 \tag{8.5.14b}$$

Equations (8.5.14a) and (8.5.14b) state that values of the signal, $\psi(\mathbf{R}_o, t)$ and its time derivative, cannot depend upon conditions in the source which pertain at times, $t_o \geq t$. The fact that the time, t_o, cannot be later than t is a statement of the Principle of Causality.

When Eqs. (8.5.13a) through (8.5.14b) are substituted into Eq. (8.5.12c), the result is

$$\psi(\mathbf{R}, t) = \int_0^t dt_o \int d^3R_o G(\mathbf{R}, t; \mathbf{R}_o, t_o) q(\mathbf{R}_o, t_o) \tag{8.5.15}$$

We have now shown that Eq. (8.5.15) is the solution to Eq. (8.5.2), subject to Eqs. (8.5.14a) and (8.5.14b).

8.6 The Green's Function Solution for the Scattered Radiation

To evaluate Eq. (8.5.15), we need to solve the companion problem, Eq. (8.5.3), subject to the boundary conditions expressed by Eqs. (8.5.12a) and (8.5.12b) and the initial conditions expressed by Eqs. (8.5.13a) and (8.5.13b). As mentioned, the task of solving Eq. (8.5.3) is sometimes known as Green's problem. As we have noted, the delta functions in a source term, $\delta(\mathbf{R} - \mathbf{R}_o)\delta(t - t_o)$, are even functions of their arguments. This permits us to write, $G(\mathbf{R}, t; \mathbf{R}_o, t_o) = G(\mathbf{R}_o, t_o; \mathbf{R}, t)$. This interchangeability implies that the derivatives of $G(\mathbf{R}, t; \mathbf{R}_o, t_o)$ with respect to $\mathbf{R}$ are the same as the derivatives with respect to $\mathbf{R}_o$, while the derivatives with respect to t are the same as derivatives with respect t_o. Hence, Eq. (8.5.3) can be rewritten as,

$$\nabla^2 G(\mathbf{R}, t; \mathbf{R}_o, t_o) - \frac{1}{c_m^2}\frac{\partial^2 G(\mathbf{R}, t; \mathbf{R}_o, t_o)}{\partial t^2} = -\delta(\mathbf{R} - \mathbf{R}_o)\delta(t - t_o) \tag{8.6.1}$$

The signal travels at the speed, c_m, from the source point, $\mathbf{R}_o$, to the point of observation at $\mathbf{R}$. Hence, the signal cannot depend upon $\mathbf{R}$ and $\mathbf{R}_o$ independently but only upon the difference. $\mathbf{r} = \mathbf{R} - \mathbf{R}_o$. This permits Green's function to be represented as $G(\mathbf{r}, t)$. The latter serves as the solution to Eq. (8.6.1) written as

$$\nabla^2 G(\mathbf{r}, t) - \frac{1}{c_m^2}\frac{\partial^2 G(\mathbf{r}, t)}{\partial t^2} = -\delta(\mathbf{r})\delta(t - t_o) \tag{8.6.2}$$

We can remove the explicit dependence upon time in Eq. (8.6.2) by introducing the time Fourier transform, $\tilde{G}(\mathbf{r}, \omega)$, defined by

$$\tilde{G}(\mathbf{r}, \omega) = \int_{-\infty}^{+\infty} dt \exp(i\omega t) G(\mathbf{r}, t) \tag{8.6.3}$$

The inverse of Eq. (8.6.3) is

$$G(\mathbf{r}, t) = \frac{1}{2\pi} \int_{-\infty}^{+\infty} d\omega \, \exp(-i\omega t) \tilde{G}(\mathbf{r}, \omega) \tag{8.6.4}$$

Equation (8.6.4) is said to be the Fourier integral representation of $G(\mathbf{r}, t)$. We shall also need the Fourier transform of the time delta function, which is

$$\exp(i\omega\, t_o) = \int_{-\infty}^{+\infty} dt \exp(i\omega\, t)\, \delta(t - t_o) \tag{8.6.5}$$

The inverse of Eq. (8.6.5) is

$$\delta(t - t_o) = \frac{1}{2\pi} \int_{-\infty}^{+\infty} d\omega \, \exp(-i\omega\, (t - t_o)) \tag{8.6.6}$$

which is known as the Fourier integral representation of the delta function. [See the Mathematical Methods Appendix]. The Fourier integral representation in Eq. (8.6.4) can be used to suppress the explicit time dependence in Eq. (8.6.2) by replacing $G(\mathbf{r}, t)$ by $\tilde{G}(\mathbf{r}, \omega)$. The second time derivative of Eq. (8.6.4) is

$$\begin{aligned} \frac{\partial^2 G(\mathbf{r}, t)}{\partial t^2} &= \frac{1}{2\pi} \int_{-\infty}^{+\infty} d\omega \left[\frac{\partial^2}{\partial t^2} \exp(-i\omega t) \right] \tilde{G}(\mathbf{r}, \omega) \\ &= \frac{1}{2\pi} \int_{-\infty}^{+\infty} d\omega \left[(-\omega^2) \exp(-i\omega t) \right] \tilde{G}(\mathbf{r}, \omega) \end{aligned} \tag{8.6.7a}$$

Equations (8.6.4), (8.6.6) and (8.6.7a) can be substituted into Eq. (8.6.2) with the result,

$$\frac{1}{2\pi} \int_{-\infty}^{+\infty} d\omega \exp(-i\omega\, t) \left[\nabla^2 \tilde{G}(\mathbf{r}, \omega) + \frac{\omega^2}{c_m^2} \tilde{G}(\mathbf{r}, \omega) \right]$$

$$= -\delta(\mathbf{r})\frac{1}{2\pi}\int_{-\infty}^{+\infty} d\omega \exp(-i\omega(t-t_o)) \tag{8.6.7b}$$

Because the delta function, $\delta(\mathbf{r})$, on the right hand side of Eq. (8.6.7b) is frequency independent, it can be moved inside the integral. This permits the integral on the right hand side of, Eq. (8.6.7b) to be written as

$$\frac{1}{2\pi}\int_{-\infty}^{+\infty} d\omega \exp(-i\omega\, t)\left[\nabla^2\tilde{G}(\mathbf{r},\omega) + \frac{\omega^2}{c^2}\tilde{G}(\mathbf{r},\omega)\right]$$

$$= -\frac{1}{2\pi}\int_{-\infty}^{+\infty} d\omega \exp(-i\omega\, t)\left[\exp(i\omega\, t_o)\,\delta(\mathbf{r})\right] \tag{8.6.7c}$$

If the left and right hand sides of Eq. (8.6.7c) are to be equal, the terms in brackets on opposite sides must be equal. Hence, upon introducing

$$k^2 = \frac{\omega^2}{c_m^2} \tag{8.6.7d}$$

on the left hand side of Eq. (8.6.7c), we have

$$\nabla^2\tilde{G}(\mathbf{r},\omega) + k^2\tilde{G}(\mathbf{r},\omega) = -\delta(\mathbf{r})\exp(i\omega\, t_o) \tag{8.6.7e}$$

Equation (8.6.7e) is a partial differential equation to be solved for $\tilde{G}(\mathbf{r},\omega)$.

The Fourier transform, $\widehat{G}(\mathbf{r},\omega)$, of the Green's function represents the amplitude of a wave scattered by a delta function point source located a distance, $\mathbf{r}$, away from the point of observation. As the scattering from a point source is isotropic, the scattered signal will depend solely upon the scalar distance $r = |\mathbf{R} - \mathbf{R_o}|$. Being independent of the spherical polar angles, the Laplacian takes on the simple form, $\nabla^2 G = (1/r)\left[d^2(rG)/dr^2\right]$, which involves r alone. [See the Mathematical Methods Appendix]. With this substitution, Eq. (8.6.7e) becomes

$$\frac{1}{r}\frac{d^2\left(r\tilde{G}(r,\omega)\right)}{dr^2} + k^2\tilde{G}(r,\omega) = -\delta(\mathbf{r})\exp(i\omega t_o) \tag{8.6.8}$$

At the observation point where, $\mathbf{r} = (\mathbf{R} - \mathbf{R_o}) > 0$, the delta function on the right hand side of Eq. (8.6.8) is zero. In this case, Eq. (8.6.8) takes on the simplified homogeneous form,

$$\frac{d^2\left(r\tilde{G}\right)}{dr^2} + k^2\left(r\tilde{G}\right) = 0 \tag{8.6.9a}$$

Equation (8.6.9a) is a constant coefficient ordinary differential equation, which determines the function, $r\tilde{G}$. The general solution to Eq. (8.6.9a) is

$$r\tilde{G} = A\exp(ikr) + B\exp(-ikr) \tag{8.6.9b}$$

where A and B are constants of integration. The wave cannot be traveling in the $+k$ and the $-k$ directions at the same time. As the $+k$ direction refers to an outgoing wave, we are obliged to set $B = 0$. With this simplification, Eq. (8.6.9b) can then be rewritten as

$$\tilde{G} = A\frac{\exp(ikr)}{r} \tag{8.6.9c}$$

To take into account waves in the source region where $\mathbf{r} = 0$, we will need to retain the delta function in Eq. (8.6.7e). The effect of the delta function is taken into account by integrating both sides of Eq. (8.6.7e) over an infinitesimal volume, δv, centered on the origin. We write

$$\int_{\delta v} d^3r\nabla^2\tilde{G}(\mathbf{r},\omega) + \lim_{\delta v\to 0} k^2\int_{\delta v} d^3r\tilde{G}(\mathbf{r},\omega) = -\exp(i\omega t_o)\int_{\delta v} d^3r\delta(\mathbf{r}) \tag{8.6.10a}$$

After performance of the indicated integrals, Eq. (8.6.10a) will serve as a second boundary conditions to be satisfied by Eq. (8.6.9b).

The first integral in Eq. (8.6.10a) can be evaluated using the Gauss–Green divergence theorem. [See the Mathematical Methods Appendix]. We let $d\mathbf{S} = dS\hat{\mathbf{r}}$ be an element of the surface area of the infinitesimal sphere centered on the origin where $\hat{\mathbf{r}}$ is a unit vector in the radial direction. The Gauss–Green integration proceeds as follows:

$$\begin{aligned}\int d^3r\nabla^2\tilde{G}(r,\omega) &= \int d^3r\nabla\cdot\left(\nabla\tilde{G}(r,\omega)\right) = \int d\mathbf{S}\cdot\nabla\tilde{G}(r,\omega)\\ &= \int dS\left[\hat{r}\cdot\nabla\tilde{G}(r,\omega)\right] = \int dS\frac{d\tilde{G}(r,\omega)}{dr}\\ &= 4\pi r^2\frac{d\tilde{G}(r,\omega)}{dr}\end{aligned} \tag{8.6.10b}$$

Because $\tilde{G}(r,\omega)$ is independent of the polar angles, the surface integral in the middle line of Eq. (8.6.10b) has been replaced in the last line by the surface area, $4\pi r^2$, times the radial component, $\frac{d\tilde{G}(r,\omega)}{dr}$, of the gradient, $\nabla\tilde{G}(\mathbf{r},\omega)$. The second integral in Eq. (8.6.10a) is calculated by noting that according to Eq. (8.6.9c), $\tilde{G}$ is proportional to $1/r$, whereas $d^3r = r^2\sin\theta d\theta d\phi$. Since their product is proportional to r, which goes to zero as $\delta v \to 0$, we can ignore the second integral in Eq. (8.6.10a). The third integral in Eq. (8.6.10a) takes the form,

$$-\exp(i\omega\, t_o)\int d^3r\delta(\mathbf{r}) = -\exp(i\omega t_o) \tag{8.6.10c}$$

where we have noted that the volume integral, $\int d^3r\delta(\mathbf{r})$, of the delta function is unity. [See the Mathematical Methods Appendix]. Upon combining the results of the three integrations, we can write Eq. (8.6.10a) as

$$4\pi r^2 \frac{d\tilde{G}(r,\omega)}{dr} = -\exp(i\omega t_o) \tag{8.6.10d}$$

Equation (8.6.10d) serves as a boundary condition that must be satisfied by $\tilde{G}(r,\omega)$ at the origin. Since the wavelength, λ, of the scattered radiation is large compared with the radius of the infinitesimal sphere surrounding the origin, we are safe in letting $k = 2\pi/\lambda = 0$ in this region. The appropriate value of the constant of integration, A, is then found by setting $k = 0$ in Eq. (8.6.9c) and substituting the result into Eq. (8.6.10d). The calculation proceeds as follows:

$$4\pi r^2 \frac{d\tilde{G}(r,\omega)}{dr} = 4\pi r^2 \frac{d}{dr}\left(\frac{A}{r}\right) = -4\pi r^2 \frac{A}{r^2} = -4\pi A \tag{8.6.10e}$$

We thus find that

$$A = \frac{1}{4\pi}\exp(i\omega t_o) \tag{8.6.10f}$$

To obtain the final form of $\tilde{G}(r,\omega)$, substitute Eq. (8.6.10f) into Eq. (8.6.9c). The result is

$$\tilde{G}(r,\omega) = \frac{\exp(ikr)\exp(i\omega t_o)}{4\pi r} \tag{8.6.11a}$$

To construct $G(r,t)$ from its transform, $\tilde{G}(r,\omega)$, substitute Eq. (8.6.11a) into Eq. (8.6.4). That result is

$$G(r,t) = \left(\frac{1}{4\pi r}\right)\left(\frac{1}{2\pi}\right)\int_{-\infty}^{+\infty} d\omega \exp(-i\omega t)\exp[i(kr + \omega t_o)] \tag{8.6.11b}$$

Next group together the two exponentials within the integrand to obtain

$$G(r,t) = \left(\frac{1}{4\pi r}\right)\left(\frac{1}{2\pi}\right)\int_{-\infty}^{+\infty} d\omega \exp[-i\omega(t - (t_o + kr/\omega))] \tag{8.6.11c}$$

Substitute c_m for ω/k in the exponential, where c_m is the speed of light in the material. The result is

$$G(r,t) = \left(\frac{1}{4\pi r}\right)\left\{\frac{1}{2\pi}\int_{-\infty}^{+\infty} d\omega \exp[-i\omega(t-(t_o+r/c_m))]\right\} \tag{8.6.11d}$$

By comparison with Eq. (8.6.6), the factor in braces on the right hand side of Eq. (8.6.11d) is recognizable as the Fourier integral representation of a delta function with argument, $t-(t_o+(r/c_m))$. Hence, Eq. (8.6.11d) can be rewritten as

$$G(r,t) = \frac{\delta[t-(t_o+(r/c_m))]}{4\pi r} \tag{8.6.11e}$$

Upon substitution of $r = |\mathbf{R}-\mathbf{R_o}|$, Eq. (8.6.11e) becomes

$$G(\mathbf{R},t;\mathbf{R}_o,t_o) = \frac{\delta\left[t-\left(t_o+\left(|\mathbf{R}-\mathbf{R_o}|/c_m\right)\right)\right]}{4\pi|\mathbf{R}-\mathbf{R_o}|} \tag{8.6.11f}$$

Because the delta function is an even function of its argument [See the Mathematical Methods Appendix], Eq. (8.6.11f) can be rewritten as

$$G(\mathbf{R},t;\mathbf{R}_o,t_o) = \frac{\delta\left[\left(t_o+\left(|\mathbf{R}-\mathbf{R_o}|/c_m\right)\right)-t\right]}{4\pi|\mathbf{R}-\mathbf{R_o}|} \tag{8.6.11g}$$

Upon regrouping of terms, Eq. (8.6.11g) reads,

$$G(\mathbf{R},t;\mathbf{R}_o,t_o) = \frac{\delta\left[t_o-\left(t-\left(|\mathbf{R}-\mathbf{R}_o|/c_m\right)\right)\right]}{4\pi|\mathbf{R}-\mathbf{R}_o|} \tag{8.6.12}$$

Equation (8.6.12) is the solution to Green's problem for the scattered wave. Due to the presence of the delta function, the Green's function in Eq. (8.6.12) is zero except when its argument is zero where it is infinite. Such a function can have no experimental significance. Its usefulness is restricted to those cases where it appears in the integrand of an integral. This application occurs when we substitute Eq. (8.6.12) into Eq. (8.5.15) to obtain,

$$\psi(\mathbf{R},t) = \int d^3R_o \int_0^t dt_o \frac{\delta\left[t_o-\left(t-\left(|\mathbf{R}-\mathbf{R}_o|/c_m\right)\right)\right]}{4\pi|\mathbf{R}-\mathbf{R}_o|} q(\mathbf{R}_o,t_o) \tag{8.6.13}$$

Equation (8.6.13) gives the amplitude, $\psi(\mathbf{R},t)$, of the wave that results from the isotropic scattering from a source having strength, $q(\mathbf{R}_o,t_o)$, at location, $\mathbf{R}_o$, and time, t_o.

8.7 Electric and Magnetic Fields of the Scattered Light

In this short section, we will determine the form assumed by the Maxwell equations when they are used to describe the electric and magnetic fields associated with the scattered light. Our derivation follows those provided by Berne and Pecora [1] and also Chu [2]. It is to be noted that Chu based his treatment on the 1967 MIT thesis of J. B. Lastovka [3]. In referring to the fields, the subscript, I, stands for "incident", whereas the subscript, S, stands for "scattered". The total electric field, $\mathbf{E}$, is thus the sum of the electric field, $\mathbf{E}_I$, of the incident light and the electric field, $\mathbf{E}_s$, of the scattered light:

$$\mathbf{E} = \mathbf{E}_I + \mathbf{E}_S \tag{8.7.1a}$$

Likewise, for the displacement

$$\mathbf{D} = \mathbf{D}_I + \mathbf{D}_S \tag{8.7.1b}$$

As pointed out in the introduction to this chapter, light scattering is due to fluctuations in the electric permittivity, ε. If the spatial average of the permittivity is $\langle\varepsilon\rangle$, and $\Delta\varepsilon$ is the fluctuation, then

$$\varepsilon = \langle\varepsilon\rangle + \Delta\varepsilon \tag{8.7.1c}$$

By extension of Eq. (8.2.1c), the total displacement in the medium is related to the total electric field by

$$\mathbf{D} = \mathbf{D}_I + \mathbf{D}_S = \varepsilon\mathbf{E} = (\langle\varepsilon\rangle + \Delta\varepsilon)(\mathbf{E}_I + \mathbf{E}_S) \tag{8.7.1d}$$

The fluctuation, $\Delta\varepsilon$, is assumed to be small as compared with $\langle\varepsilon\rangle$, namely $\Delta\varepsilon \ll \varepsilon$. The scattered electric field and the scattered displacement are both small as compared with their incident counterparts; that is, $\mathbf{E}_S \ll \mathbf{E}_I$ and $\mathbf{D}_S \ll \mathbf{D}_I$. We are then justified in multiplying out the binomial on the right hand side of Eq. (8.7.1d), retaining only those terms which are zeroth order (the product of two large terms) and first order (the product of a large term and a small term). The result is

$$\mathbf{D}_I + \mathbf{D}_S \simeq \langle\varepsilon\rangle\mathbf{E}_I + \Delta\varepsilon\mathbf{E}_I + \langle\varepsilon\rangle\mathbf{E}_S \tag{8.7.1e}$$

By pairing quantities of like order of magnitude, we can equate quantities on the left hand side of Eq. (8.7.1e) to quantities on the right to obtain

$$\mathbf{D}_I = \langle\varepsilon\rangle\mathbf{E}_I \tag{8.7.1f}$$

and

$$\mathbf{D}_S = \Delta\varepsilon\mathbf{E}_I + \langle\varepsilon\rangle\mathbf{E}_S \tag{8.7.1g}$$

If the scattered wave is observed in a region of space where there are no free charges, then in Eq. (8.2.3), $\rho = 0$, and in Eq. (8.2.4), $\mathbf{J} = 0$. In this case, the scattered fields satisfy the simplified Maxwell equations,

$$\nabla \cdot \mathbf{D}_S = 0 \tag{8.7.2}$$

$$\nabla \cdot \mathbf{B}_S = 0 \tag{8.7.3}$$

$$\nabla \times \mathbf{E}_S = -\frac{\partial \mathbf{B}_S}{\partial t} \tag{8.7.4}$$

$$\nabla \times \mathbf{H}_S = \frac{\partial \mathbf{D}_S}{\partial t} \tag{8.7.5}$$

8.8 Wave Equation for the Scalar Potential Describing the Scattered Fields

Assuming for the moment that we are observing the scattered wave while it is still traversing the material, we can use Eq. (8.2.2c) to link $\mathbf{B}_S$ to $\mathbf{H}_S$. The result,$\mathbf{B}_S = \mu\mathbf{H}_S$, can be substituted into Eq. (8.7.5) to obtain

$$\nabla \times \left(\frac{\mathbf{B}_S}{\mu}\right) = \frac{\partial \mathbf{D}_S}{\partial t} \tag{8.8.1a}$$

Substitute Eq. (8.7.1g) into the right hand side of Eq. (8.8.1a). The result is

$$\nabla \times \left(\frac{\mathbf{B}_S}{\mu}\right) = \langle\varepsilon\rangle\frac{\partial \mathbf{E}_S}{\partial t} + \Delta\varepsilon\frac{\partial \mathbf{E}_I}{\partial t} \tag{8.8.1b}$$

Multiplication of both sides of Eq. (8.8.1b) by μ yields

$$\nabla \times \mathbf{B}_S = \mu\langle\varepsilon\rangle\frac{\partial \mathbf{E}_S}{\partial t} + \mu\Delta\varepsilon\frac{\partial \mathbf{E}_I}{\partial t} \tag{8.8.1c}$$

By virtue of Eq. (8.4.5c), we can represent the scattered magnetic flux density in terms of a scattered field vector potential, $\mathbf{A}_S$, as

$$\mathbf{B}_S = \nabla \times \mathbf{A}_S \tag{8.8.2}$$

According to Eqs. (8.7.4) and (8.8.2), the scattered electric field satisfies,

$$\nabla \times \mathbf{E}_S = -\frac{\partial(\nabla \times \mathbf{A}_S)}{\partial t} \tag{8.8.3a}$$

or simply,

$$\nabla \times \left[\mathbf{E}_S + \frac{\partial \mathbf{A}_S}{\partial t}\right] = 0 \tag{8.8.3b}$$

We now introduce a scalar potential, Φ_S. By analogy with Eq. (8.4.3), Φ_S is related to $\mathbf{E}_S$ and $\mathbf{A}_S$ according to

$$\mathbf{E}_S = -\nabla\Phi_S - \frac{\partial \mathbf{A}_S}{\partial t} \tag{8.8.4}$$

Using Eq. (8.7.1g), the divergence of $\mathbf{D}_S$ is

$$\nabla \cdot \mathbf{D}_S = \langle\varepsilon\rangle\nabla \cdot \mathbf{E}_S + \Delta\varepsilon(\nabla \cdot \mathbf{E}_I) \tag{8.8.5a}$$

On the right hand side of Eq. (8.8.5a), substitute Eq. (8.8.4) for the scattered electric field. Taking note of Eq. (8.7.2), the result is

$$\nabla \cdot \mathbf{D}_S = -\langle\varepsilon\rangle\nabla \cdot \left[\nabla\Phi_S + \frac{\partial \mathbf{A}_S}{\partial t}\right] + \Delta\varepsilon(\nabla \cdot \mathbf{E}_I) = 0 \tag{8.8.5b}$$

The bracketed quantity in Eq. (8.8.5b) can be written out, so that Eq. (8.8.5b) reads

$$-\langle\varepsilon\rangle\nabla^2\Phi_S - \langle\varepsilon\rangle\frac{\partial(\nabla \cdot \mathbf{A}_S)}{\partial t} + \Delta\varepsilon(\nabla \cdot \mathbf{E}_I) = 0 \tag{8.8.5c}$$

In terms of the potentials representing the scattered fields, the Lorenz gauge Eq. (8.4.6h) assumes the form,

$$\nabla \cdot \mathbf{A}_S + \frac{1}{c_m^2}\frac{\partial \Phi_S}{\partial t} = 0 \tag{8.8.6}$$

Use Eq. (8.8.6) to replace $\nabla \cdot \mathbf{A}_S$ in Eq. (8.8.5c). The result is

$$-\langle\varepsilon\rangle\nabla^2\Phi_S + \frac{\langle\varepsilon\rangle}{c_m^2}\frac{\partial^2\Phi_S}{\partial t^2} + \Delta\varepsilon(\nabla \cdot \mathbf{E}_I) = 0 \tag{8.8.7a}$$

After dividing Eq. (8.8.7a) through by $-\langle\varepsilon\rangle$, one obtains,

$$\nabla^2\Phi_S - \frac{1}{c_m^2}\frac{\partial^2\Phi_S}{\partial t^2} = \frac{\Delta\varepsilon}{\langle\varepsilon\rangle}(\nabla \cdot \mathbf{E}_I) \tag{8.8.7b}$$

On the basis of Eq. (8.2.1d), we can assume that the electric susceptibility, χ_e, is related to the average electric permittivity by the equation,

$$\langle \varepsilon \rangle = \varepsilon_o (1 + \langle \chi_e \rangle) \tag{8.8.8}$$

where $\langle \chi_e \rangle$ is the mean susceptibility, and ε_o is the permittivity of free space. By virtue of Eq. (8.2.1d), the spatial fluctuation in the susceptibility, $\Delta \chi_e$, is related to the spatial fluctuation in the permittivity, $\Delta \varepsilon$, by

$$\Delta \varepsilon = \varepsilon_o \Delta \chi_e \tag{8.8.9}$$

Next, substitute Eqs. (8.8.8) and (8.8.9) into the right hand side of Eq. (8.8.7b) to obtain

$$\nabla^2 \Phi_S - \frac{1}{c_m^2} \frac{\partial^2 \Phi_S}{\partial t^2} = \left(\frac{1}{\varepsilon_o} \right) \frac{\varepsilon_o \Delta \chi_e}{(1 + \langle \chi_e \rangle)} (\nabla \cdot \mathbf{E}_I) \tag{8.8.10}$$

To simplify the right hand side of Eq. (8.8.10), define the effective field,

$$\mathbf{F} = \frac{\varepsilon_o \Delta \chi_e}{(1 + \langle \chi_e \rangle)} \mathbf{E}_I \tag{8.8.11}$$

Upon substitution of Eq. (8.8.11) into Eq. (8.8.10), the result is

$$\nabla^2 \Phi_S - \frac{1}{c_m^2} \frac{\partial^2 \Phi_S}{\partial t^2} = \frac{1}{\varepsilon_o} \nabla \cdot \mathbf{F} \tag{8.8.12}$$

Equation (8.8.12) is the wave equation satisfied by the scattered wave scalar potential [1]. It replaces Eq. (8.4.12) in cases where the potential is due to a source term that involves changes in the electric polarization in a dielectric medium.

8.9 Wave Equation for the Hertz Potential for Scattered Light

A vector such as $\mathbf{F}$, is uniquely determined if its divergence $\nabla \cdot \mathbf{F}$ and its curl $\nabla \times \mathbf{F}$ are specified. This statement is known as Helmholtz's theorem. The source term on the right hand side of the wave equation in Eq. (8.8.12) is the divergence, $\nabla \cdot \mathbf{F}$. Missing from this equation is any mention of $\nabla \times \mathbf{F}$. Both $\nabla \cdot \mathbf{F}$ and $\nabla \times \mathbf{F}$, must be known in order to calculate the scattered fields, E_S and $\mathbf{B}_S$. In order to recover $\mathbf{F}$ in its entirety, we replace the scalar potential, Φ_S, with a vector potential, $\mathbf{\Pi}_S$, whose source term is $\mathbf{F}$ itself. This new vector is known as the Hertz potential (Heinrich Rudolph Hertz, German, 1857–1894). To make the scalar, Φ_S, consistent with the vector, $\mathbf{\Pi}_S$, we write

$$\Phi_S = -\nabla \cdot \mathbf{\Pi}_S \tag{8.9.1}$$

Equation (8.8.6) is the Lorenz gauge condition satisfied by the scattered potentials, Φ_S and $\mathbf{A}_S$, which was used to derive the wave equation, Eq. (8.8.12), satisfied by Φ_S. Equation (8.8.6) will also serve as the basis for deriving the wave equation satisfied by $\mathbf{\Pi}_S$. Substitute Eq. (8.9.1) into Eq. (8.8.6). The result is

$$\nabla \cdot \mathbf{A}_S + \frac{1}{c_m^2}\frac{\partial(-\nabla \cdot \mathbf{\Pi}_S)}{\partial t} = 0 \tag{8.9.2a}$$

Factor out the $\nabla\cdot$ operator so that Eq. (8.9.2a) reads

$$\nabla \cdot \left[\mathbf{A}_S - \frac{1}{c_m^2}\frac{\partial \mathbf{\Pi}_S}{\partial t}\right] = 0 \tag{8.9.2b}$$

Equation (8.9.2b) is satisfied if

$$\mathbf{A}_S = \frac{1}{c_m^2}\frac{\partial \mathbf{\Pi}_S}{\partial t} \tag{8.9.3}$$

According to Eqs. (8.8.2) and (8.9.3), the magnetic induction is

$$\mathbf{B}_S = \nabla \times \mathbf{A}_S = \nabla \times \left[\frac{1}{c_m^2}\frac{\partial \mathbf{\Pi}_S}{\partial t}\right] \tag{8.9.4}$$

or

$$\mathbf{B}_S = \frac{1}{c_m^2}\frac{\partial(\nabla \times \mathbf{\Pi}_S)}{\partial t} \tag{8.9.5}$$

Now form the curl of Eq. (8.9.5). The result is

$$\nabla \times \mathbf{B}_S = \frac{1}{c_m^2}\frac{\partial(\nabla \times \nabla \times \mathbf{\Pi}_S)}{\partial t} \tag{8.9.6a}$$

Apply Eq. (8.2.1c) to represent the scattered field, $\mathbf{D}_S$, and substitute the result into Eq. (8.8.1a) to obtain,

$$\nabla \times \left(\frac{\mathbf{B}_S}{\mu}\right) = \frac{\partial(\varepsilon \mathbf{E}_S)}{\partial t} \tag{8.9.6b}$$

Multiply Eq. (8.9.6b) through by μ and rearrange the result to read,

$$\nabla \times \mathbf{B}_S = \mu\varepsilon\frac{\partial \mathbf{E}_S}{\partial t} = \frac{1}{c_m^2}\frac{\partial \mathbf{E}_S}{\partial t} \tag{8.9.6c}$$

where we have used $c_m^2 = 1/\mu\varepsilon$ to identify the speed of light in the medium. The combination of Eqs. (8.9.6a) and (8.9.6c) yields the result,

$$\frac{1}{c_m^2}\frac{\partial \mathbf{E}_S}{\partial t} = \frac{1}{c_m^2}\frac{\partial(\nabla \times \nabla \times \mathbf{\Pi}_S)}{\partial t} \tag{8.9.6d}$$

According to Eq. (8.9.6d)

$$\mathbf{E}_S = \nabla \times \nabla \times \mathbf{\Pi}_S \tag{8.9.7}$$

Equations (8.9.5) and (8.9.7) express the scattered fields, $\mathbf{B}_S$ and $\mathbf{E}_S$, in terms of the Hertz potential. Substitute Eq. (8.9.1) into Eq. (8.8.12) to obtain

$$\nabla^2(-\nabla \cdot \mathbf{\Pi}_S) - \frac{1}{c^2}\frac{\partial^2(-\nabla \cdot \mathbf{\Pi}_S)}{\partial t^2} = \frac{1}{\varepsilon_o}\nabla \cdot \mathbf{F} \tag{8.9.8a}$$

If we multiply both sides of Eq. (8.9.8a) by (− 1) and factor out the operator, $\nabla\cdot$, the result is

$$\nabla \cdot \left[\nabla^2 \mathbf{\Pi}_S - \frac{1}{c_m^2}\frac{\partial^2 \mathbf{\Pi}_S}{\partial t^2}\right] = \nabla \cdot \left[-\frac{\mathbf{F}}{\varepsilon_o}\right] \tag{8.9.8b}$$

Equation (8.9.8b) is satisfied if

$$\nabla^2 \mathbf{\Pi}_S - \frac{1}{c_m^2}\frac{\partial^2 \mathbf{\Pi}_S}{\partial t^2} = -\frac{\mathbf{F}}{\varepsilon_o} \tag{8.9.9}$$

In Eq. (8.8.12), the source term that determines Φ_S involves only $\nabla\cdot\mathbf{F}$. By contrast, in Eq. (8.9.9), the source term that determines $\mathbf{\Pi}_S$ involves $\mathbf{F}$, itself. The solution to Eq. (8.9.9) provides sufficient information to determine $\nabla \times \mathbf{\Pi}_S$. The curl of $\mathbf{\Pi}_S$ will form the basis for calculating the strengths of the scattered electric and magnetic fields.

8.10 Green's Function Solution to the Wave Equation for the Hertz Potential

Equations (8.4.9) and (8.9.9) are both vector wave equations. Like the vector potential, $\mathbf{A}_S$, the Hertz vector, $\mathbf{\Pi}_S$, has three components. Exploiting the analogy between the vector potential $\mathbf{A}_S$ and the Hertz potential $\mathbf{\Pi}_S$, Eqs. (8.5.1) imply that each component of the Hertz vector satisfies a scalar wave equation. If $\psi(\mathbf{R}, t)$ in Eq. (8.5.2) is set equal to one of the components, $\Pi_{s,i}(\mathbf{R}, t)$ $(i = 1, 2, 3)$ of $\mathbf{\Pi}_S$, Eq. (8.6.13) provides a template for the solution to the wave equation for that component. This assertion is illustrated by writing,

$$\Pi_{s,i}(\mathbf{R},t)=\frac{1}{\varepsilon_o}\int d^3R_o\int_0^t dt_o\frac{\delta[t_o-(t-(|\mathbf{R}-\mathbf{R}_o|/c_m))]}{4\pi|\mathbf{R}-\mathbf{R}_o|}F_i(\mathbf{R}_o,t_o)\quad(i=1,2,3)\tag{8.10.1}$$

where $F_i(\mathbf{R}_o,t_o)(i=1,2,3)$ is a component of $\mathbf{F}(\mathbf{R}_o,t_o)$. The three equations summarized by Eq. (8.10.1) can be written as one vector equation:

$$\mathbf{\Pi}_S(\mathbf{r},t)=\frac{1}{4\pi\varepsilon_o}\int d^3R_o\int_0^t dt_o\frac{\delta\left[t_o-\left(t-\left(|\mathbf{R}-\mathbf{R}_o|/c_m\right)\right)\right]}{|\mathbf{R}-\mathbf{R}_o|}\mathbf{F}(\mathbf{R}_o,t_o)\tag{8.10.2}$$

Equation (8.10.2) is the Green's function solution to the wave equation, Eq. (8.9.9), for the $\mathbf{\Pi}_S$. In the following sections, we will evaluate the integrals on the right hand side of Eq. (8.10.2) and use the result to calculate the electric and magnetic fields associated with the scattered radiation.

8.11 Evaluation of the Hertz Potential

In order to make use of Eq. (8.10.2), we need to understand how the direction of the scattered electric field vector, $\mathbf{E}_S$, depends upon the orientation of the Hertz vector potential $\mathbf{\Pi}_S(\mathbf{r},t)$. Figure 8.1 is a guide to the geometry of the scattering event.

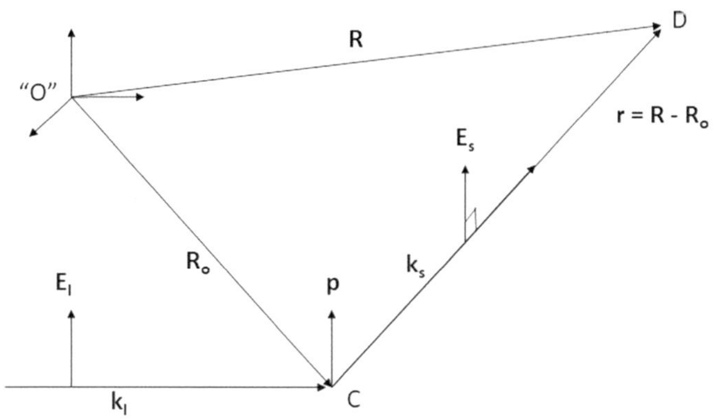

Fig. 8.1 The geometry associated with the scattering of a plane wave. The origin of coordinates is at "O". The vector distance from "O" to the scattering center at C is $\mathbf{R}_o$, while the vector distance from "O" to the point at **D** where the scattered wave is detected is **R**. The difference, $\mathbf{r}=\mathbf{R}-\mathbf{R}_o$, is the vector distance from C to D. The electric vector of the incident light is $\mathbf{E}_I$, and its direction of propagation toward the scattering center at C is given by the waver vector, $\mathbf{k}_I$. The electric vector of the scattered light is $\mathbf{E}_S$. Its wave vector $\mathbf{k}_S$ is the direction of propagation along the line CD. The electric dipole induced by $\mathbf{E}_I$ in the scattering center material is **p**

Substitution Eq. (8.10.2) into the right hand side of Eq. (8.9.7) yields,

$$\begin{aligned}\mathbf{E}_S &= \nabla \times \nabla \times \mathbf{\Pi}_S \\ &= \frac{1}{4\pi\varepsilon_o}\nabla_\mathbf{R} \times \nabla_\mathbf{R} \\ &\times \int d^3R_o \int_0^t dt_o \frac{\delta\left[t_o - \left(t - |\mathbf{R}-\mathbf{R}_o|/c_m\right)\right]}{|\mathbf{R}-\mathbf{R}_o|}\mathbf{F}(\mathbf{R}_o, t_o)\end{aligned} \tag{8.11.1}$$

Here "$\nabla_\mathbf{R}\times$" indicates that the curl operator differentiates the coordinates of the vector, $\mathbf{R}$. The quantity, $|\mathbf{R}-\mathbf{R}_o|$, can be written in the form of the square root of a polynomial:

$$|\mathbf{R}-\mathbf{R}_o| = [(\mathbf{R}-\mathbf{R}_o)\cdot(\mathbf{R}-\mathbf{R}_o)]^{1/2} = \left[R^2 - 2\mathbf{R}\cdot\mathbf{R}_o + R_o^2\right]^{1/2} \tag{8.11.2a}$$

The vector $\mathbf{R}$ locates a point in the laboratory where the scattered radiation is being observed. By contrast, $\mathbf{R}_o$ locates a point confined to the scattering region. Ordinarily, the distance, R is much greater than the distance, R_o. Hence, because $R_o^2 \ll 2(\mathbf{R}\cdot\mathbf{R}_o)$, we can drop R_o^2 in Eq. (8.11.2a) in comparison with $2(\mathbf{R}\cdot\mathbf{R}_o)$ and write the result as

$$|\mathbf{R}-\mathbf{R}_o| \simeq \left[R^2 - 2\mathbf{R}\cdot\mathbf{R}_o\right]^{1/2} = R\left[1 - 2(\mathbf{R}\cdot\mathbf{R}_o)/R^2\right]^{1/2} \tag{8.11.2b}$$

Set $x = -2(\mathbf{R}\cdot\mathbf{R}_o)/R^2$ and use the binomial theorem in the form, $(1+x)^{1/2} = 1 + (1/2)x + \cdots$ to approximate the right hand side of Eq. (8.11.2b) as

$$|\mathbf{R}-\mathbf{R}_o| \simeq R\left[1 + (1/2)(-2(\mathbf{R}\cdot\mathbf{R}_o))/R^2\right] = R\left[1 - (\mathbf{R}\cdot\mathbf{R}_o)/R^2\right] \tag{8.11.2c}$$

Bring the factor of R back inside the bracket in Eq. (8.11.2c) and introduce the unit vector, $\hat{\mathbf{R}}$, where

$$\hat{\mathbf{R}} = \frac{\mathbf{R}}{R} \tag{8.11.2d}$$

then Eq. (8.11.2c) reads,

$$|\mathbf{R}-\mathbf{R}_o| = R - \frac{\mathbf{R}_o\cdot\mathbf{R}}{R} = R - \mathbf{R}_o\cdot\hat{\mathbf{R}} \tag{8.11.3}$$

We can now substitute Eq. (8.11.3) into the argument of the delta function in Eq. (8.11.1). In the denominator of the integrand in Eq. (8.11.1), we approximate $|\mathbf{R}-\mathbf{R}_o|$ simply by R. With these two simplifications, Eq. (8.11.1) becomes

$$\mathbf{E}_S(\mathbf{R}, t) = \nabla_{\mathbf{R}} \times \nabla_{\mathbf{R}} \times \int d^3R_o \int_0^t dt_o \frac{\delta\left[t_o - \left(t - \left(R - \mathbf{R}_o \cdot \hat{\mathbf{R}}\right)/c_m\right)\right]}{4\pi\varepsilon_o R}\mathbf{F}(\mathbf{R}_o, t_o) \tag{8.11.4}$$

Equation (8.8.11) evaluated at time, t_o, and position, $\mathbf{R}_o$, reads,

$$\mathbf{F}(\mathbf{R}_o, t_o) = \frac{\varepsilon_o \Delta\chi_e(\mathbf{R}_o)}{(1 + \langle\chi_e\rangle)}\mathbf{E}_I(\mathbf{R}_o, t_o) \tag{8.11.5}$$

By writing, $\Delta\chi_e = \Delta\chi_e(\mathbf{R}_o)$, we are assuming that the electric susceptibility fluctuates only in space but not in time. The average value $\langle\chi_e\rangle$ is considered to be independent of space and time. When Eq. (8.11.5) is substituted into Eq. (8.11.4), the result is

$$\mathbf{E}_S(\mathbf{R}, t) = \nabla_{\mathbf{R}} \times \nabla_{\mathbf{R}} \times \int d^3R_o^2 \int_0^t dt_o \frac{\delta\left[t_o - \left(t - \left(R - \mathbf{R}_o \cdot \hat{\mathbf{R}}\right)/c_m\right)\right]}{4\pi\varepsilon_o R} \frac{\varepsilon_o \Delta\chi_e(\mathbf{R}_o)}{(1 + \langle\chi_e\rangle)}\mathbf{E}_I(\mathbf{R}_o, t_o) \tag{8.11.6}$$

Next, redefine ε in Eq. (8.2.1d) as $\varepsilon = \varepsilon_o(1 + \langle\chi_e\rangle)$. Use this relation to rewrite Eq. (8.11.6) in the form,

$$\mathbf{E}_S(\mathbf{R}, t) = \nabla_{\mathbf{R}} \times \nabla_{\mathbf{R}} \times \int d^3R_o \int_0^t dt_o \frac{\delta\left[t_o - \left(t - (R - \mathbf{R}_o \cdot \hat{\mathbf{R}})/c_m\right)\right]}{4\pi R}\left(\frac{\varepsilon_o}{\varepsilon}\right)\Delta\chi_e(\mathbf{R}_o)\mathbf{E}_I(\mathbf{R}_o, t_o) \tag{8.11.7}$$

The electric field of the incident light wave at $\mathbf{R}_o$ and t_o is

$$\mathbf{E}_I(\mathbf{R}_o, t_o) = \mathbf{E}_I^o \exp[i(\mathbf{k}_I \cdot \mathbf{R}_o - \omega_I t_o)] \tag{8.11.8}$$

where ω_I is the angular frequency of the incident light, and $\mathbf{E}_I^o$ points in the positive z-direction. When Eq. (8.11.8) is substituted into Eq. (8.11.7), the result is

$$\mathbf{E}_S(\mathbf{R}, t) = \nabla_{\mathbf{R}} \times \nabla_{\mathbf{R}} \times \int d^3R_o \int_0^t dt_o \frac{\delta\left[t_o - \left(t - (R - \mathbf{R}_o \cdot \hat{\mathbf{R}})/c_m\right)\right]}{4\pi R}\left(\frac{\varepsilon_o}{\varepsilon}\right)\Delta\chi_e(\mathbf{R}_o)\mathbf{E}_I^o \times \exp[i(\mathbf{k}_I \cdot \mathbf{R}_o - \omega_I t_o)] \tag{8.11.9a}$$

Integration over the delta function with respect to dt_o has the effect in Eq. (8.11.9a) of using $\left(t - (R - \mathbf{R}_o \cdot \hat{\mathbf{R}})/c_m\right)$ to replace t_o in $\exp[i(\mathbf{k}_I \cdot \mathbf{R}_o - \omega_I t_o)]$. With this done, Eq. (8.11.9a) reads,

$$\begin{aligned}\mathbf{E}_S(\mathbf{R}, t) &= \nabla_{\mathbf{R}} \times \nabla_{\mathbf{R}} \\ &\times \int d^3R_o \left(\frac{\varepsilon_o}{4\pi\varepsilon R}\right) \Delta\chi_e(\mathbf{R}_o) \mathbf{E}_I^o \\ &\exp\left[i\left(\mathbf{k}_I \cdot \mathbf{R}_o - \omega_I\left(t - \frac{R}{c_m} + \frac{\hat{\mathbf{R}} \cdot \mathbf{R}_o}{c_m}\right)\right)\right]\end{aligned} \tag{8.11.9b}$$

In Eq. (8.11.9b), we can factor the exponential so as to group together terms involving a dot product with $\mathbf{R}_o$. The result is

$$\begin{aligned}\mathbf{E}_S(\mathbf{R}, t) &= \nabla_{\mathbf{R}} \times \nabla_{\mathbf{R}} \\ &\times \int d^3R_o \left(\frac{\varepsilon_o}{4\pi\varepsilon R}\right) \Delta\chi_e(\mathbf{R}_o) \mathbf{E}_I^o \\ &\exp\left[i\left(\mathbf{k}_I - \frac{\omega_I}{c_m}\hat{\mathbf{R}}\right) \cdot \mathbf{R}_o\right] \exp\left[i\left(\frac{\omega_I R}{c_m} - \omega_I t\right)\right]\end{aligned} \tag{8.11.9c}$$

In Eq. (8.11.9c), light is scattered by the fluctuation, $\Delta\chi_e(\mathbf{R}_o)$, centered on $\mathbf{R}_o$. The light, which is scattered from the position, $\mathbf{R}_o$, travels in the direction of the wave, vector, $\mathbf{k}_S$, arriving at the point, $\mathbf{R}$, in the laboratory where it is observed. The scattered wave vector, $\mathbf{k}_S$, is parallel to the $\mathbf{R} - \mathbf{R}_o$ direction. Because $\mathbf{R}$ is a macroscopic distance, whereas $\mathbf{R}_o$ is of the order of the wave length of the light, the inequality, $R \gg R_o$, holds, which makes $\mathbf{k}_S$ essentially parallel to $\mathbf{R}$. If the scattering is elastic, no energy is transferred from the light to the material or vice versa. As a consequence, the frequency of the scattered light is the same as the frequency of the incident light, and the magnitude of the scattered wave vector, $\mathbf{k}_S$, is the same as the magnitude of the incident wave vector, $\mathbf{k}_I$, so $k_I = k_S$. The angular frequency of the incident light satisfies, $\omega_I = c_m k_I$; here, $k_I = 2\pi/\lambda$ where λ is the wavelength of the light. The above observations are summarized by the equation,

$$\mathbf{k}_S = k_S\hat{\mathbf{R}} = k_I\hat{\mathbf{R}} = \left(\frac{\omega_I}{c_m}\right)\hat{\mathbf{R}} \tag{8.11.10a}$$

where $\hat{\mathbf{R}}$ is the unit vector in the direction of $\mathbf{R}$. Hence, in Eq. (8.11.9c), we can replace $(\omega_I/c_m)\hat{\mathbf{R}}$ in the first exponential by $k_S\hat{\mathbf{R}}$, and set $k_S = \omega_I/c_m$ into the second. The result is

$$\begin{aligned}\mathbf{E}_S(\mathbf{R}, t) &= \nabla_{\mathbf{R}} \times \nabla_{\mathbf{R}} \\ &\times \mathbf{E}_I^o \left(\frac{\varepsilon_o}{4\pi\varepsilon R}\right) \int d^3R_o \Delta\chi_e(\mathbf{R}_o) \\ &\exp[i(\mathbf{k}_I - \mathbf{k}_S) \cdot \mathbf{R}_o] \exp[i(k_S R - \omega_I t)]\end{aligned} \tag{8.11.10b}$$

In Eq. (8.11.10b), the quantity, $\exp[i(k_S R - \omega_I t)]$, is independent of $\mathbf{R}_o$. This exponential can be factored out of the integrand. The result is

$$\begin{aligned}\mathbf{E}_S(\mathbf{R}, t) = \nabla_{\mathbf{R}} \times \nabla_{\mathbf{R}} \\ \times \mathbf{E}_I^o \exp[i(k_S R - \omega_I t)]\Big(\frac{\varepsilon_o}{4\pi\varepsilon R}\Big) \int d^3 R_o \Delta\chi_e(\mathbf{R}_o) \\ \exp[i(\mathbf{k}_I - \mathbf{k}_S)\cdot\mathbf{R}_o]\end{aligned} \tag{8.11.11}$$

By comparison of Eq. (8.11.11) with Eq. (8.9.7), we conclude that

$$\begin{aligned}\mathbf{\Pi}_S(\mathbf{R}, t) = \exp[i(k_S R - \omega_I t)]\Big(\frac{\varepsilon_o}{4\pi\varepsilon R}\Big)\mathbf{E}_I^o \int d^3 R_o \Delta\chi_e(\mathbf{R}_o) \\ \exp[i(\mathbf{k}_I - \mathbf{k}_S)\cdot\mathbf{R}_o]\end{aligned} \tag{8.11.12}$$

Equation (8.11.12) is the solution to Eq. (8.9.9).

8.12 Calculation of the Electric Field of the Scattered Radiation

We follow Chu [1] by grouping together terms multiplying $\exp[i(k_S R - \omega_I t)]$ in Eq. (8.11.12), so that $\mathbf{\Pi}_S(\mathbf{r}, t)$ can be written in the form

$$\mathbf{\Pi}_S(\mathbf{R}, t) = \exp[i(k_S R - \omega_I t)]\Big(\frac{\mathbf{p}}{4\pi\varepsilon R}\Big) \tag{8.12.1a}$$

where $\mathbf{p}$ has the meaning,

$$\mathbf{p} = \varepsilon_o \mathbf{E}_I^o \int d^3 R_o \Delta\chi_e(\mathbf{R}_o) \exp[i(\mathbf{k}_I - \mathbf{k}_S)\cdot\mathbf{R}_o] \tag{8.12.1b}$$

The vector, $\mathbf{p}$, represents an electric dipole induced by the electric field, $\mathbf{E}_I$, of the light incident on the scattering center. Both $\mathbf{E}_I$ and $\mathbf{p}$ are parallel to $\mathbf{E}_I^o$. The latter makes an angle θ with the scattered light direction of propagation, $\mathbf{k}_S$. We can decompose $\mathbf{p}$ into a component parallel to $\mathbf{k}_S$ (or what is the same thing, $\mathbf{R}$) and a component perpendicular to $\mathbf{k}_S$. Whereas a positive value of θ is associated with $\mathbf{E}_I^o$ as the polar axis (i.e. z-axis), a negative value of θ is associated with $\mathbf{k}_S$ serving as the polar axis. Hence, the projection of $\mathbf{p}$ onto $\mathbf{k}_S$ is

$$p_R = p\cos(-\theta) = p\cos\theta \tag{8.12.2a}$$

while the component perpendicular to $\mathbf{k}_S$ is

$$p_\theta = p\sin(-\theta) = -p\sin\theta \tag{8.12.2b}$$

In Eqs. (8.12.2a) and (8.12.2b), p is the magnitude of the vector, $\mathbf{p}$. According to Eq. (8.12.1a), $\mathbf{\Pi}_S$ is parallel to $\mathbf{p}$. The corresponding components of $\mathbf{\Pi}_S$ are thus

$$\Pi_{S,R}(R,t) = \exp[i(k_S R - \omega_I t)]\left(\frac{p_R}{4\pi\varepsilon R}\right) \tag{8.12.3a}$$

and

$$\Pi_{S,\theta}(R,t) = \exp[i(k_S R - \omega_I t)]\left(\frac{p_\theta}{4\pi\varepsilon R}\right) \tag{8.12.3b}$$

respectively. Since $\mathbf{p}$ is parallel to $\mathbf{E}_I^o$, which is parallel to the z-axis, the azimuthal angle, ϕ, made by $\mathbf{p}$ with respect to the z-axis is zero. Hence, by virtue of Eq. (8.12.1a)

$$\Pi_{S,\phi} = 0 \tag{8.12.3c}$$

In order to calculate $\mathbf{E}_S$ using Eq. (8.9.7), we shall first need to compute the curl of $\mathbf{\Pi}_S$, which we denote by

$$\mathbf{C}_S = \nabla \times \mathbf{\Pi}_S \tag{8.12.4}$$

Equation (8.12.4) can be written in terms of spherical polar coordinates by invoking the determinant,

$$\begin{aligned} \mathbf{C}_S = \nabla \times \mathbf{\Pi}_S &= \left(\frac{1}{R^2\sin\theta}\right)\begin{bmatrix} \hat{\mathbf{R}} & R\hat{\theta} & R\sin\theta\hat{\varphi} \\ \frac{\partial}{\partial R} & \frac{\partial}{\partial\theta} & \frac{\partial}{\partial\phi} \\ \Pi_{S,R} & R\Pi_{S,\theta} & 0 \end{bmatrix} \\ &= C_{S,R}\hat{\mathbf{R}} + C_{S,\theta}\hat{\theta} + C_{S,\phi}\hat{\varphi} \end{aligned} \tag{8.12.5}$$

where $\hat{\theta}$ and $\hat{\varphi}$ are unit vectors in the θ and ϕ-directions, respectively. According to Eqs. (8.12.3a) and (8.12.3b), $\mathbf{\Pi}_S$ is independent of ϕ. Hence, when the determinant in Eq. (8.12.5) is expanded by minors of the first row, the radial component must be zero,

$$C_{S,R} = -\frac{1}{R^2\sin\theta}\frac{\partial\left(R\Pi_{S,\theta}\right)}{\partial\phi} = 0 \tag{8.12.6a}$$

and the polar component must be zero,

$$C_{S,\theta} = \frac{1}{R^2\sin\theta}\frac{\partial\Pi_{S,R}}{\partial\phi} = 0 \tag{8.12.6b}$$

Only the azimuthal component, $C_{s,\phi}$, is non-zero. It is

$$\begin{aligned} C_{S,\phi} &= \left(\frac{R\sin\theta}{R^2\sin\theta}\right)\left[\frac{\partial\left(R\Pi_{S,\theta}\right)}{\partial R} - \frac{\partial\Pi_{S,R}}{\partial\theta}\right] \\ &= \frac{1}{R}\left[\frac{\partial\left(R\Pi_{S,\theta}\right)}{\partial R} - \frac{\partial\Pi_{S,R}}{\partial\theta}\right] \end{aligned} \tag{8.12.6c}$$

Before we can evaluate the right hand side of Eq. (8.12.6c), we need explicit forms for $\Pi_{S,R}$ and $\Pi_{S,\theta}$ written in terms of R and θ. The first task is accomplished by substituting Eq. (8.12.2a) into Eq. (8.12.3a). The result is

$$\Pi_{S,R}(R,t) = \exp[i(k_S R - \omega_I t)]\left(\frac{p\cos\theta}{4\pi\varepsilon R}\right) \tag{8.12.7a}$$

The second is accomplished by substituting Eq. (8.12.2b) into Eq. (8.12.3b). That result is

$$\Pi_{S,\theta}(R,t) = \exp[i(k_S R - \omega_I t)]\left(\frac{-p\sin\theta}{4\pi\varepsilon R}\right) \tag{8.12.7b}$$

Now, when Eqs. (8.12.7) are substituted into Eq. (8.12.6c), one obtains

$$\begin{aligned} C_{S,\phi} &= \frac{1}{R}\left[\frac{\partial}{\partial R}\left((-p\sin\theta)\frac{\exp[i(k_S R - \omega_I t)]}{4\pi\varepsilon}\right) - \frac{\partial}{\partial\theta}\left((p\cos\theta)\frac{\exp[i(k_S R - \omega_I t)]}{4\pi\varepsilon R}\right)\right] \\ &= \frac{1}{R}\left[\frac{-ik_S\exp[i(k_S R - \omega_I t)]p\sin\theta}{4\pi\varepsilon} + \frac{\exp[i(k_S R - \omega_I t)]p\sin\theta}{4\pi\varepsilon R}\right] \end{aligned} \tag{8.12.8}$$

The two terms in the bracket on the right hand side of Eq. (8.12.8) can be grouped together to read,

$$C_{S,\phi} = \frac{[-ik_S + (1/R)]\exp[i(k_S R - \omega_I t)]p\sin\theta}{4\pi\varepsilon R} \tag{8.12.9}$$

We begin the process of obtaining a final form for $\mathbf{E}_S$ by combining Eqs. (8.9.7) and (8.12.4). The result is

$$\mathbf{E}_S = \nabla\times\mathbf{C}_S \tag{8.12.10}$$

The determinant representation of the curl in Eq. (8.12.10) is

$$\mathbf{E}_S = \nabla\times\mathbf{C}_S = \frac{1}{R^2\sin\theta}\begin{bmatrix} \hat{\mathbf{R}} & R\hat{\theta} & R\sin\theta\hat{\varphi} \\ \frac{\partial}{\partial R} & \frac{\partial}{\partial\theta} & \frac{\partial}{\partial\phi} \\ 0 & 0 & R\sin\theta C_{S,\phi} \end{bmatrix} \tag{8.12.11a}$$

where in the last row of the determinant in Eq. (8.12.11a), we have used Eqs. (8.12.6a) and (8.12.6b) in order to replace both $C_{S,R}$ and $C_{s,\phi}$ by zeroes. After expanding the determinant in Eq. (8.12.11a) about minors of the first row and noting that $C_{S,\phi}$ is independent of ϕ, we obtain the vector,

$$\begin{aligned}\mathbf{E}_S &= E_{S,R}\hat{\mathbf{R}} + E_{S,\theta}\hat{\theta} + E_{S,\phi}\hat{\varphi} \\ &= \frac{1}{R^2 \sin\theta}\left\{\hat{\mathbf{R}}\left[\frac{\partial}{\partial\theta}\left(R\sin\theta C_{S,\phi}\right)\right] - R\hat{\theta}\left[\frac{\partial}{\partial R}\left(R\sin\theta C_{S,\phi}\right)\right]\right\} \\ &+ 0\cdot\left(\frac{R\hat{\varphi}\sin\theta}{R^2\sin\theta}\right)\end{aligned} \tag{8.12.11b}$$

According to Eq. (8.12.11b), the radial component, $E_{S,R}$, of the scattered electric field is given by

$$E_{S,R} = \frac{1}{R^2\sin\theta}\left[\frac{\partial}{\partial\theta}\left(R\sin\theta C_{S,\phi}\right)\right] \tag{8.12.12a}$$

Substitute Eq. (8.12.9) for $C_{S,\phi}$ into Eq. (8.12.12a). The result is

$$E_{S,R} = \frac{1}{R^2\sin\theta}\left\{\frac{\partial}{\partial\theta}\left(\frac{R\sin\theta(-ik_S + (1/R))\exp[i(k_S R - \omega_I t]p\sin\theta}{4\pi\varepsilon R}\right)\right\} \tag{8.12.12b}$$

When the common factors of R are cancelled, Eq. (8.12.12b) simplifies to

$$E_{S,R} = \frac{1}{R\sin\theta}\frac{\partial}{\partial\theta}\left[p\sin^2\theta\frac{\exp[i(k_S R - \omega_I t)]}{4\pi\varepsilon R}\left(\frac{1}{R} - ik_S\right)\right] \tag{8.12.12c}$$

Calculation of the derivative on the right hand side of Eq. (8.12.12c) yields the result

$$E_{S,R} = \frac{1}{R}\left[\frac{2p(\cos\theta)\exp[i(k_S R - \omega_I t)]}{4\pi\varepsilon R}\left(\frac{1}{R} - ik_S\right)\right] \tag{8.12.12d}$$

Upon cancelling factors of two and grouping factors of R in Eq. (8.12.12d), the final result for $E_{S,R}$ can be written in the form

$$E_{S,R} = \frac{p(\cos\theta)\exp[i(k_S R - \omega_I t)]}{2\pi\varepsilon R^2}\left(\frac{1}{R} - ik_S\right) \tag{8.12.13}$$

We next return to Eq. (8.12.11b) where the polar angle component, $E_{S,\theta}$, of the scattered electric field is given by

$$E_{S,\theta} = -\frac{1}{R^2\sin\theta}\left[R\frac{\partial}{\partial R}\left(R\sin\theta C_{S,\phi}\right)\right] \tag{8.12.14a}$$

Upon substitution of Eq. (8.12.9) for $C_{S,\phi}$, Eq. (8.12.14a) reads,

$$E_{S,\theta} = -\frac{1}{R^2 \sin\theta}\left\{R\frac{\partial}{\partial R}\left(\frac{R(\sin\theta)[-ik_S + (1/R)]\exp[i(k_S R - \omega_I t]p\sin\theta}{4\pi\varepsilon R}\right)\right\} \tag{8.12.14b}$$

After grouping and cancelling common factors, Eq. (8.12.14b) takes the form,

$$E_{S,\theta} = -\frac{1}{R}\frac{\partial}{\partial R}\left[p(\sin\theta)\frac{\exp[i(k_S R - \omega_I t)]}{4\pi\varepsilon}\left(\frac{1}{R} - ik_S\right)\right] \tag{8.12.14c}$$

Calculation of the derivative with respect to R leads to

$$\begin{aligned} E_{S,\theta} = -&\left[\frac{p(\sin\theta)(ik_S)\exp[i(k_S R - \omega_I t)]}{4\pi\varepsilon R}\left(\frac{1}{R} - ik_S\right)\right. \\ &\left.+\frac{p(\sin\theta)\exp[i(k_S R - \omega_I t)]}{4\pi\varepsilon R}\left(-\frac{1}{R^2}\right)\right] \end{aligned} \tag{8.12.14d}$$

After grouping common factors in Eq. (8.12.14d), $E_{S,\theta}$ reads

$$E_{S,\theta} = \frac{p(\sin\theta)\exp[i(k_S R - \omega_I t)]}{4\pi\varepsilon R}\left(\frac{1}{R^2} - \frac{ik_S}{R} - k_S^2\right) \tag{8.12.15}$$

We turn finally to the azimuthal component, $E_{S,\phi}$, of the scattered electric field. On the basis of the the last term on the right of Eq. (8.12.11b), we have

$$E_{S,\phi} = 0 \tag{8.12.16}$$

Our calculation of the components of $\mathbf{E}_S$ can now be summarized by Eqs. (8.12.13), (8.12.15), and (8.12.16).

As we shall see, the terms in Eqs. (8.12.13) and (8.12.15), which decrease with distance faster than $1/R$, do not contribute to the radiation. These are terms proportional to $1/R^2$ and $1/R^3$. Such terms die out before they have propagated very far and as consequence do not contribute to the transmitted radiation. They are said to constitute the *near field*. By contrast, terms which decay in proportion to $1/R$ as a function of the distance contribute to the transmitted radiation. Such terms are said to constitute the *far field*. We will examine this distinction in more detail below. If in Eq. (8.12.13) we collect the various factors involving the reciprocal powers of R, we find only terms proportional to $1/R^2$ and $1/R^3$. Neither is a far field term. If the same procedure is carried out in Eq. (8.12.15), the sole far field term is that involving $\left(-k_S^2/R\right)$.

Equations (8.12.13) and (8.12.15) describe the electric field components associated with an oscillating electric dipole. The associated emission is said to be *dipole* radiation. As such, it is useful to compare these equations with those we found in

Chap. 1 for the electric field components associated with a *static* electric dipole. Specifically, if we write the fractions on the right hand side of Eq. (8.12.13) over a common denominator, the result is

$$E_{S,R} = \frac{p(\cos\theta)\exp[i(k_S R - \omega_I t)]}{2\pi\varepsilon R^3}(1 - ik_S R) \qquad (8.12.17a)$$

In the limit where $k_S R = (2\pi/\lambda)R \ll 1$, which corresponds to the static case (no propagation), $E_{S,R}$ as given by Eq. (8.12.17a) reads,

$$E_{S,R} = \frac{p(\cos\theta)\exp(-i\omega_I t)}{2\pi\varepsilon R^3} \qquad (8.12.17b)$$

The angular and radial dependence of $E_{S,R}$ in Eq. (8.12.17b) is the same as that in Eq. (1.7.22a), which describes the radial component of the electric field of a static dipole in vacuum. The significant difference is the time dependent factor in Eq. (8.12.17b).

If we factor $(1/R)^2$ out of Eq. (8.12.15), the result is

$$E_{S,\theta} = \frac{p(\sin\theta)\exp[i(k_S R - \omega_I t)]}{4\pi\varepsilon R^3}\left[1 - ik_S R - (k_S R)^2\right] \qquad (8.12.18a)$$

In the long wavelength limit where $k_S R = (2\pi/\lambda)R \ll 1$, Eq. (8.12.18a) becomes

$$E_{S,\theta} = \frac{p(\sin\theta)\exp(-i\omega_I t)}{4\pi\varepsilon R^3} \qquad (8.12.18b)$$

Except for the time dependent exponential factor, $E_{S,\theta}$ in Eq. (8.12.18b) has the same angular and radial dependence as in Eq. (1.7.22b), which represents the angular component of the electric field of a static electric dipole.

Neither Eq. (8.12.17b) nor Eq. (8.12.18b) represents a propagating wave, because in the long wave length limit, neither contains the traveling wave factor, $\exp[i(k_S R - \omega_I t)]$. Rather, in both cases, we find simply a localized electric dipole oscillating in time.

We can summarize as follows: When the detector is so close to the dipole that $k_S R = 0$, the dipole is seen to oscillate but not to radiate. This is called the *static field* region. By contrast, when the detector is located a bit further away so that $0 < k_S R = (2\pi/\lambda)R \ll 1$, the electric field propagates but decays rapidly with distance. This defines the *near field.* Finally, if the detector of the radiation is sufficiently distant from the source that $k_S R = (2\pi/\lambda)R \gg 1$, any signal which is detected necessarily constitutes propagating radiation. This defines the *far field*. In the case of visible light, [400 nm $< \lambda <$ 700 nm], the wave length is sufficiently short that any detector having macroscopic dimensions necessarily lies in the far field.

In the far field region ($(1/R) \ll k_s$), Eq. (8.12.13) can be approximated by

$$E_{S,R} = \frac{-ik_S p(\cos\theta)\exp[i(k_S R - \omega_I t)]}{2\pi\varepsilon R^2} \tag{8.12.19a}$$

whereas Eq. (8.12.15) can be approximated by

$$E_{S,\theta} = \frac{-k_S^2 p(\sin\theta)\exp[i(k_S R - \omega_I t)]}{4\pi\varepsilon R} \tag{8.12.19b}$$

According to Eq. (8.12.19a), as R increases, $E_{S,R}$ decreases as $1/R^2$. By comparison, in Eq. (8.12.19b), as R increases, $E_{S,\theta}$ decreases only as $1/R$. As a consequence, in the far field region, $E_{S,\theta}$ is much greater than $E_{S,R}$, so that the latter can be ignored.

Equation (8.12.1b) is a formula for the dipole vector,**p**. The magnitude of this vector is

$$p = \varepsilon_o E_I^o \int d^3R_o\{\Delta\chi_e(\mathbf{R}_o)\exp[i(\mathbf{k}_I - \mathbf{k}_S)\cdot\mathbf{R}_o]\} \tag{8.12.20a}$$

Use Eq. (8.12.20a) to replace p in Eq. (8.12.19b). The result is

$$E_{S,\theta} = -\frac{k_S^2\sin\theta}{4\pi R}\left(\frac{\varepsilon_o}{\varepsilon}\right)E_I^o\exp[i(k_S R - \omega_I t)] \int d^3R_o\{\Delta\chi_e(\mathbf{R}_o)\exp[i(\mathbf{k}_I - \mathbf{k}_S)\cdot\mathbf{R}_o]\} \tag{8.12.20b}$$

As $E_{S,\theta}$ is the sole substantial component of the scattered electric field vector, $\mathbf{E}_S$, Eq. (8.12.20b) can be rewritten in the vector form as

$$\mathbf{E}_S = -\hat{\theta}\frac{k_S^2\sin\theta}{4\pi R}\left(\frac{\varepsilon_o}{\varepsilon}\right)E_I^o\exp[i(k_S R - \omega_I t)] \int d^3R_o\{\Delta\chi_e(\mathbf{R}_o)\exp[i(\mathbf{k}_I - \mathbf{k}_S)\cdot\mathbf{R}_o]\} \tag{8.12.21}$$

In the case of Rayleigh scattering, the two important vector directions are the direction of polarization of the incident light,

$$\mathbf{E}_I^o = E_I^o\cos\theta\hat{\mathbf{R}} + E_I^o\sin\theta\hat{\theta} \tag{8.12.22}$$

and the direction of propagation of the scattered light, which is given by

$$\mathbf{k}_S = k_S\hat{\mathbf{R}} \tag{8.12.23}$$

It will prove worthwhile to express the electric field of the scattered light, $\mathbf{E}_S$, in terms of $\mathbf{E}_I^o$ and $\mathbf{k}_S$. The basis for this projection is to be found in the cross product relations among the unit vectors in the spherical polar coordinate system. These relations are

$$\hat{\mathbf{R}} \times \hat{\theta} = \hat{\varphi} \tag{8.12.24a}$$

$$\hat{\mathbf{R}} \times \hat{\varphi} = -\hat{\theta} \tag{8.12.24b}$$

$$\hat{\theta} \times \hat{\varphi} = \hat{\mathbf{R}} \tag{8.12.24c}$$

Equation (8.12.24a), supplemented with $\hat{\mathbf{R}} \times \hat{\mathbf{R}} = 0$ permits the calculation of the cross product of the vectors in Eqs. (8.12.23) and (8.12.22). The result is

$$\begin{aligned}\mathbf{k}_S \times \mathbf{E}_I^o &= k_S \hat{\mathbf{R}} \times E_I^o (\hat{\mathbf{R}} \cos\theta + \hat{\theta} \sin\theta) \\ &- +k_S E_I^o \sin\theta \hat{\varphi}\end{aligned} \tag{8.12.25a}$$

Next we compute

$$\begin{aligned}\mathbf{k}_S \times \left(\mathbf{k}_S \times \mathbf{E}_I^o\right) &= k_S \hat{\mathbf{R}} \times E_I^o (k_S E_I^o \sin\theta)(\hat{\varphi}) \\ &= -k_S^2 E_I^o \sin\theta \hat{\theta}\end{aligned} \tag{8.12.25b}$$

where we have used Eq. (8.12.24b). We can employ Eq. (8.12.25b) to replace the vector, $-k_S^2 E_I^o \sin\theta\hat{\theta}$, on the right hand side of Eq. (8.12.21). The result is

$$\begin{aligned}\mathbf{E}_S(R, t) = \mathbf{k}_S \times \left(\mathbf{k}_S \times \mathbf{E}_I^o\right)\left(\frac{\varepsilon_o}{\varepsilon}\right)\left(\frac{\exp[i(k_S R - \omega_I t)]}{4\pi R}\right) \\ \int d^3 R_o \{\Delta\chi_e(\mathbf{R}_o) \exp[i(\mathbf{k}_I - \mathbf{k}_S) \cdot \mathbf{R}_o]\}\end{aligned} \tag{8.12.26}$$

Equation (8.12.26) is included because it is often seen in the literature.

8.13 Calculation of the Magnetic Field of the Scattered Radiation

Combine Eqs. (8.9.5) and (8.12.4) to obtain

$$\mathbf{B}_S = \frac{1}{c_m^2} \frac{\partial \mathbf{C}_S}{\partial t} \tag{8.13.1}$$

By virtue of Eqs. (8.12.6a) through (8.12.6c), the sole non-zero component of $\mathbf{C}_S$ is $C_{S,\phi}$. Upon substitution of Eq. (8.12.9) into Eq. (8.13.1), the result is

$$B_{S,\phi} = \left(-\frac{i\omega_I}{c_m^2}\right) \frac{[-ik_S + (1/R)] \exp[i(k_S R - \omega_I t]p \sin\theta}{4\pi\varepsilon R} \tag{8.13.2a}$$

Substitute Eq. (8.13.2a) into Eq. (8.2.2c) and solve for $H_{s,\phi}$. One obtains,

$$H_{S,\phi} = \frac{B_{S,\phi}}{\mu} = \frac{1}{\mu}\left(\frac{1}{4\pi\varepsilon R}\right)\left(\frac{p\sin\theta}{c_m^2}\right)\left(\frac{1}{R} - ik_S\right)(-i\omega_I) \times \exp[i(k_S R - \omega_I t)] \tag{8.13.2b}$$

The far field component of $H_{S,\phi}$ can be selected by dropping the factor of $(1/R)$ as compared to $(-ik_S)$ in the third parenthetic factor in Eq. (8.13.2b). The result is

$$H_{S,\phi} = \frac{1}{\mu}\left(\frac{1}{4\pi\varepsilon R}\right)\left(\frac{p\sin\theta}{c_m^2}\right)(-ik_S)(-i\omega_I) \times \exp[i(k_S R - \omega_I t)] \tag{8.13.2c}$$

By multiplying factors of i and re-grouping terms in Eq. (8.13.2c), one obtains

$$H_{S,\phi} = -\frac{\omega_I}{\mu c_m^2}\frac{k_S p(\sin\theta)\exp[i(k_S R - \omega_I t)]}{4\pi\varepsilon R} \tag{8.13.2d}$$

Use $\omega_I = c_m k_S$ and $c_m = 1/\sqrt{\mu\varepsilon}$ to rewrite Eq. (8.13.2d) in the form,

$$H_{S,\phi} = -\frac{c_m k_S}{\mu c_m}\cdot\sqrt{\mu\varepsilon}\cdot\frac{k_S p(\sin\theta)\exp[i(k_S R - \omega_I t)]}{4\pi\varepsilon R} \tag{8.13.2e}$$

which simplifies to read

$$H_{S,\phi} = -\left(\frac{\varepsilon}{\mu}\right)^{1/2}\frac{k_S^2 p(\sin\theta)\exp[i(k_S R - \omega_I t)]}{4\pi\varepsilon R} \tag{8.13.2f}$$

Combine Eq. (8.13.2f) with Eq. (8.12.19b). The result is

$$H_{S,\phi} = \left(\frac{\varepsilon}{\mu}\right)^{1/2} E_{S,\theta} \tag{8.13.3}$$

Equation (8.13.3) will find use in computing the scattered energy flux.

8.14 Intensity of the Scattered Radiation

We assume that the detector is sensitive only to the time average of the intensity of the scattered radiation. To compute this average, we shall make use of $E_{S,\theta}$ and $H_{S,\phi}$, as these are the only non-zero components of the scattered electromagnetic field. Using Eq. (7.8.11) of Chap. 7, the time average of the Poynting vector for the scattered radiation is

$$\langle \mathbf{N}_S \rangle = \frac{1}{2}\mathrm{Re}\left(\mathbf{E}_S \times \mathbf{H}_S^*\right) = \frac{1}{2}\mathrm{Re}\begin{vmatrix} \hat{\mathbf{R}} & \hat{\theta} & \hat{\varphi} \\ 0 & E_{S,\theta} & 0 \\ 0 & 0 & H_{S,\phi}^* \end{vmatrix} \tag{8.14.1}$$

Expansion of the determinant about minors of the first row leads to

$$\langle \mathbf{N}_S \rangle = \frac{1}{2} E_{S,\theta} H_{S,\phi}^* \hat{\mathbf{R}} \tag{8.14.2}$$

The complex conjugate of Eq. (8.13.3) is

$$H_{S,\phi}^* = \left(\frac{\varepsilon}{\mu}\right)^{1/2} E_{S,\theta}^* \tag{8.14.3}$$

Upon substitution of Eq. (8.14.3) into Eq. (8.14.2), the magnitude of the time averaged Poynting vector is

$$\langle N_S \rangle = \frac{1}{2}\left(\frac{\varepsilon}{\mu}\right)^{1/2} \left|E_{S,\theta}\right|^2 \tag{8.14.4}$$

According to Eq. (8.12.21), the magnitude of the vector, $\mathbf{E}_s$, is

$$E_S = -\frac{k_S^2 \sin\theta}{4\pi R}\left(\frac{\varepsilon_o}{\varepsilon}\right) E_I^o \exp[i(k_S R - \omega_I t)] \int d^3 R_o \{\Delta\chi_e(\mathbf{R}_o) \exp[i(\mathbf{k}_I - \mathbf{k}_S)\cdot\mathbf{R}_o]\} \tag{8.14.5a}$$

According to Eq. (8.14.5a), E_s is a complex valued scalar. Its absolute value is defined as

$$\begin{aligned} |E_S|^2 = E_S E_S^* = &\left[-\frac{k_S^2 \sin\theta}{4\pi R}\left(\frac{\varepsilon_o}{\varepsilon}\right) E_I^o \exp[i(k_S R - \omega_I t)] \int d^3 r_1 \{\Delta\chi_e(\mathbf{r}_1) \exp[i(\mathbf{k}_I - \mathbf{k}_S)\cdot\mathbf{r}_1]\}\right] \\ &\times \left[-\frac{k_S^2 \sin\theta}{4\pi R}\left(\frac{\varepsilon_o}{\varepsilon}\right) E_I^o \exp[-i(k_S R - \omega_I t)] \int d^3 r_2 \{\Delta\chi_e(\mathbf{r}_2) \exp[-i(\mathbf{k}_I - \mathbf{k}_S)\cdot\mathbf{r}_2]\}\right] \end{aligned} \tag{8.14.5b}$$

When calculating $|E|^2$ according to Eq. (8.14.5b), we distinguish one integral from the other by replacing the integration variable, $\mathbf{R}_o$ in Eq. (8.14.5a) by $\mathbf{r}_1$ in the first integral and replacing it by $\mathbf{r}_2$ in the second. If in Eq. (8.14.5b) the real factors are grouped together and the complex factors are multiplied by their complex conjugates, the result is

$$
\begin{aligned}
|E_S|^2 = {} & \left(\frac{\varepsilon_o}{\varepsilon}\right)^2 (E_I^o)^2 \frac{k_S^4 \sin^2\theta}{(4\pi R)^2} \\
& \times \int d^3 r_1 \{\Delta\chi_e(r_1) \exp[i(\mathbf{k}_I - \mathbf{k}_S)\cdot \mathbf{r}_1]\} \\
& \times \int d^3 r_2 \{\Delta\chi_e(r_2) \exp[-i(\mathbf{k}_I - \mathbf{k}_S)\cdot \mathbf{r}_2]\}
\end{aligned} \tag{8.14.6a}
$$

Upon substitution of Eq. (8.14.6a) into Eq. (8.14.4), the result is

$$
\begin{aligned}
\langle N_S \rangle = {} & \frac{1}{2}\left(\frac{\varepsilon}{\mu}\right)^{1/2} \left(\frac{\varepsilon_o}{\varepsilon}\right)^2 (E_I^o)^2 \frac{k_S^4 \sin^2\theta}{(4\pi R)^2} \\
& \times \left[\int d^3 r_1 \{\Delta\chi_e(r_1) \exp[i(\mathbf{k}_I - \mathbf{k}_S)\cdot \mathbf{r}_1]\}\right. \\
& \left.\cdot \int d^3 r_2 \{\Delta\chi_e(r_2) \exp[-i(\mathbf{k}_I - \mathbf{k}_S)\cdot \mathbf{r}_2]\}\right]
\end{aligned} \tag{8.14.6b}
$$

Spatial fluctuations in the density cause fluctuations in the electric susceptibility, $\Delta\chi_e$. Light is scattered due to correlation between the fluctuation, $\Delta\chi_e$, at position $\mathbf{r}_1$ and the fluctuation in $\Delta\chi_e$ at a neighboring position, $\mathbf{r}_2$. In Eq. (8.14.6a), the product, $\Delta\chi_e(\mathbf{r}_1)\Delta\chi_e(\mathbf{r}_2)$, contains hidden random variables. After averaging over these random variables, the result is the statistical correlation, $\langle\Delta\chi_e(\mathbf{r}_1)\Delta\chi_e(\mathbf{r}_2)\rangle$. By contrast, all of the other factors in Eq. (8.14.6b) are deterministic. Hence, the product of integrals in Eq. (8.14.6b) can be written as the double integral,

$$
\begin{aligned}
\langle N_S \rangle = {} & \frac{1}{2}\left(\frac{\varepsilon}{\mu}\right)^{1/2} \left(\frac{\varepsilon_o}{\varepsilon}\right)^2 (E_I^o)^2 \frac{k_S^4 \sin^2\theta}{(4\pi R)^2} \\
& \times \int d^3 r_1 \int d^3 r_2 \{\langle\Delta\chi_e(\mathbf{r}_1)\Delta\chi_e(\mathbf{r}_2)\rangle \exp[i(\mathbf{k}_I - \mathbf{k}_S)\cdot (\mathbf{r}_1 - \mathbf{r}_2)]\}
\end{aligned} \tag{8.14.6c}
$$

If we follow Smoluchowski and Einstein and assume that $\langle\Delta\chi_e(\mathbf{r}_1)\Delta\chi_e(\mathbf{r}_2)\rangle$ is short ranged as a function of $\mathbf{r}_1 - \mathbf{r}_2$, we can replace it by a constant. If that constant is denoted by $\left\langle(\Delta\chi_e)^2\right\rangle$, it can be factored out of the integral. Equation (8.14.6c) then simplifies to read,

$$
\begin{aligned}
\langle N_S \rangle = {} & \frac{1}{2}\left(\frac{\varepsilon}{\mu}\right)^{1/2} \left(\frac{\varepsilon_o}{\varepsilon}\right)^2 (E_I^o)^2 \frac{k_S^4 \sin^2\theta}{(4\pi R)^2} \left\langle(\Delta\chi_e)^2\right\rangle \\
& \times \int d^3 r_1 \int d^3 r_2 \{\exp[i(\mathbf{k}_I - \mathbf{k}_S)\cdot (\mathbf{r}_1 - \mathbf{r}_2)]\}
\end{aligned} \tag{8.14.7}
$$

In the Smoluchowski – Einstein model, the incident light is scattered from small volume elements in the fluid. These elements, which have volume, v, are assumed to be so small that the distance between them, $|\mathbf{r}_1 - \mathbf{r}_2|$, is much smaller than the wavelength of the incident light. Since $k_I = k_S = 2\pi/\lambda$, this approximation allows us to replace $(\mathbf{k}_I - \mathbf{k}_S) \cdot (\mathbf{r}_1 - \mathbf{r}_2)$ by zero in the argument of the exponential in Eq. (8.14.7). Equation (8.14.7) then reads

$$\langle N_S \rangle = \frac{1}{2}\left(\frac{\varepsilon}{\mu}\right)^{1/2}\left(\frac{\varepsilon_o}{\varepsilon}\right)^2 (E_I^o)^2 \frac{k_S^4 \sin^2\theta}{(4\pi R)^2}\left\langle(\Delta\chi_e)^2\right\rangle \int d^3r_1 \int d^3r_2 \tag{8.14.8}$$

Given that scattering is assumed to be localized in v, defined by $v = \int d^3r$, both of the remaining integrals in Eq. (8.14.8) can be replaced by v. Hence, Eq. (8.14.8) becomes simply

$$\langle N_S \rangle = \frac{1}{2}\left(\frac{\varepsilon}{\mu}\right)^{1/2}\left(\frac{\varepsilon_o}{\varepsilon}\right)^2 (E_I^o)^2 \frac{k_S^4 \sin^2\theta}{(4\pi R)^2}\left\langle(\Delta\chi_e)^2\right\rangle v^2 \tag{8.14.9}$$

8.15 Incident Radiation Intensity

The intensity of the scattered light is proportional to the intensity of the incident light. To calculate the intensity of the incident light, we shall need an equation linking the incident electric field, $\mathbf{E}_I$, to the incident magnetic flux density, $\mathbf{B}_I$. The calculation begins with the application of Eq. (8.2.5) to the incident light,

$$\nabla \times \mathbf{E}_I = -\frac{\partial \mathbf{B}_I}{\partial t} \tag{8.15.1}$$

In the case of a plane wave traveling in the y-direction with electric field polarized in the z-direction, Eq. (8.11.8) assumes the form

$$\mathbf{E}_I(y, t) = E_I^o \hat{\mathbf{e}}_z \exp\left[i(k_I y - \omega_I t)\right] \tag{8.15.2}$$

The curl operation, which is required to the evaluate the left hand side of Eq. (8.15.1), can be calculated in Cartesian coordinates as

$$\nabla \times \mathbf{E}_I = \begin{vmatrix} \hat{\mathbf{e}}_x & \hat{\mathbf{e}}_y & \hat{\mathbf{e}}_z \\ \frac{\partial}{\partial x} & \frac{\partial}{\partial y} & \frac{\partial}{\partial z} \\ 0 & 0 & E_I^o e^{i(k_I y - \omega_I t)} \end{vmatrix}$$
$$= \hat{\mathbf{e}}_x \frac{\partial}{\partial y}\left[E_I^o e^{i(k_I y - \omega_I t)}\right] = i k_I E_I^o \exp\left[i(k_I y - \omega_I t)\right]\hat{\mathbf{e}}_x \tag{8.15.3}$$

The goal is to use Eq. (8.15.3) to calculate $\mathbf{H}_I$. Use Eq. (8.2.2c) to express $\mathbf{H}_I$ in terms of $\mathbf{B}_I$, then substitute Eq. (8.15.3) into the left hand side of Eq. (8.15.1) and integrate with respect time. The result is

$$\begin{aligned}\mathbf{H}_I &= \frac{\mathbf{B}_I}{\mu} = -\frac{1}{\mu}\int dt(\nabla\times\mathbf{E}_I) = -i\left(\frac{k_I}{\mu}\right)E_I^o e^{ik_I y}\hat{\mathbf{e}}_x\int dte^{-i\omega_I t} \\ &= \left(\frac{k_I}{\mu\omega_I}\right)E_I^o\exp\big[i(k_I y-\omega_I t)\big]\hat{\mathbf{e}}_x\end{aligned}\tag{8.15.4}$$

Applied to the case of the incident radiation, the time average of the Poynting vector as given by Eq. (7.8.11) of Chap. 7 reads,

$$\langle\mathbf{N}_I(y)\rangle = \frac{1}{2}\mathrm{Re}\big(\mathbf{E}_I(y,t)\times\mathbf{H}_I^*(y,t)\big)\tag{8.15.5}$$

Use of Eqs. (8.15.2) and (8.15.4) renders Eq. (8.15.5) in the form,

$$\begin{aligned}\langle\mathbf{N}_I\rangle &= \frac{1}{2}\mathrm{Re}\Big[E_I^o\hat{\mathbf{e}}_z\exp\big[i(k_I y-\omega_I t)\big] \\ &\quad\times\left(\frac{k_I}{\mu\omega_I}\right)E_I^o\exp\big[-i(k_I y-\omega_I t)\big]\hat{\mathbf{e}}_x\Big] \\ &= \frac{1}{2}\left(\frac{k_I}{\mu\omega_I}\right)\left(E_I^o\right)^2\hat{\mathbf{e}}_y\end{aligned}\tag{8.15.6a}$$

where we have used $\hat{\mathbf{e}}_z\times\hat{\mathbf{e}}_x=\hat{\mathbf{e}}_y$. Upon substitution of $\omega_I=c_m k_I$ and $c_m=1/\sqrt{\varepsilon\mu}$, into Eq. (8.15.6a), the magnitude of $\langle\mathbf{N}_I\rangle$ can be written as,

$$\langle N_I\rangle = \frac{1}{2}\left(\frac{k_I}{\mu c_m k_I}\right)\left(E_I^o\right)^2 = \frac{1}{2}\left(\frac{k_I\sqrt{\varepsilon\mu}}{\mu k_I}\right)\left(E_I^o\right)^2\tag{8.15.6b}$$

The final result is

$$\langle N_I\rangle = \frac{1}{2}\left(\frac{\varepsilon}{\mu}\right)^{1/2}\left(E_I^o\right)^2\tag{8.15.7}$$

where $\langle\mathbf{N}_I\rangle$, the incident radiation intensity, is a function of the absolute square of the amplitude of the electric field, $\left(E_I^o\right)^2$, of the incident light.

8.16 Rayleigh Ratio and the Turbidity

The Rayleigh ratio, Θ_R, is defined by

$$\Theta_R = \frac{R^2\langle N_S\rangle}{v\langle N_I\rangle}\tag{8.16.1}$$

Substitute Eqs. (8.14.9) and (8.15.7) into Eq. (8.16.1). The result can be simplified as follows:

$$\Theta_R(\theta) = \left(\frac{R^2}{\nu}\right)\left(\frac{1}{2}\right)\left(\frac{\varepsilon}{\mu}\right)^{1/2}\left(\frac{\varepsilon_o}{\varepsilon}\right)^2 (E_I^o)^2$$
$$\left(\frac{k_S^4 \sin^2\theta}{(4\pi R)^2}\langle(\Delta\chi_e)^2\rangle\nu^2\right)(2)\left(\frac{\mu}{\varepsilon}\right)^{1/2}(E_I^o)^{-2} \tag{8.16.2a}$$

After cancelation of common factors, Eq. (8.16.2a) reads

$$\Theta_R(\theta) = \frac{k_S^4 \sin^2\theta}{16\pi^2}\left(\frac{\varepsilon_o}{\varepsilon}\right)^2\langle(\Delta\chi_e)^2\rangle\nu \tag{8.16.2b}$$

In Eq. (8.16.2b), $\langle(\Delta\chi_e)^2\rangle$ is dimensionless, whereas ε_o and ε have the same units. The wave vector, k_S, has the units of m^{-1}, while ν has the units of m^3. Hence $\Theta_R(\theta)$ has the units of m^{-1}.

A second measure of the scattering, called the turbidity, τ^*, also with the units of m^{-1}, can be computed by integrating the Rayleigh ratio in Eq. (8.16.2b) over all angles. Recognizing that θ is the polar angle and ϕ is the azimuthal angle, the required integral is

$$\tau^* = \int_0^{2\pi} d\phi \int_0^{\pi} d\theta \sin\theta(\Theta_R(\theta))$$
$$= \frac{k_S^4}{16\pi^2}\left(\frac{\varepsilon_o}{\varepsilon}\right)^2\langle(\Delta\chi_e)^2\rangle\nu \int_0^{2\pi} d\phi \int_0^{\pi} d\theta \sin\theta \sin^2\theta \tag{8.16.3}$$

To evaluate Eq. (8.16.3), we need to compute,

$$K = \int_0^{2\pi} d\phi \int_0^{\pi} d\theta \sin\theta \sin^2\theta \tag{8.16.4}$$

Let

$$x = \cos\theta \tag{8.16.5a}$$

so that

$$dx = -\sin\theta d\theta \tag{8.16.5b}$$

and

$$\sin^2\theta = 1 - \cos^2\theta = 1 - x^2 \tag{8.16.5c}$$

When $\theta = 0$, x has the value, $x = 1$, and when $\theta = \pi$, it has the value, $x = -1$. Thus, the range of integration over dx is $x = -1$ to $x = +1$. With these substitutions, Eq. (8.16.4) becomes

$$K = \int_0^{2\pi} d\phi \int_{-1}^{+1} dx(1 - x^2) = 2\pi \left[x - \frac{x^3}{3} \right]_{-1}^{+1} = \frac{8\pi}{3} \tag{8.16.6a}$$

Upon substitution of Eq. (8.16.6a) into Eq. (8.16.3), the result is

$$\tau^* = \frac{k_S^4}{16\pi^2} \left(\frac{\varepsilon_o}{\varepsilon} \right)^2 \langle (\Delta \chi_e)^2 \rangle v \left(\frac{8\pi}{3} \right) \tag{8.16.6b}$$

After cancelation of common factors, Eq. (8.16.6b) reads,

$$\tau^* = \frac{k_S^4}{6\pi} \left(\frac{\varepsilon_o}{\varepsilon} \right)^2 \langle (\Delta \chi_e)^2 \rangle v \tag{8.16.7}$$

To obtain the dependence of τ^* upon the wavelength, we replace k_S in Eq. (8.16.7) by $k_S = 2\pi/\lambda$. The result is

$$\tau^* = \frac{8\pi^3}{3\lambda^4} \left(\frac{\varepsilon_o}{\varepsilon} \right)^2 \langle (\Delta \chi_e)^2 \rangle v \tag{8.16.8}$$

If we assume that the detector for the scattered radiation is located outside the medium responsible for the scattering, we can replace ε in the denominator of Eq. (8.16.8) by ε_o. This leads to the simplified result,

$$\tau^* = \frac{8\pi^3}{3\lambda^4} \langle (\Delta \chi_e)^2 \rangle v \tag{8.16.9}$$

As we shall see, Eq. (8.16.9) serves as the starting point for calculating the light scattering properties of molecules in gases and liquids. By virtue of the factor of $(1/\lambda)^4$, which appears in Eq. (8.16.9), it is apparent that short wavelengths are more strongly scattered than long wavelengths.

The turbidity can be used to determine the fraction of the incident light that passes unscattered through a unit length of the material. If $d\langle N_I \rangle$ is the change in the incident light intensity encountered upon passing through a differential thickness, dx, then the Beer–Lambert Law requires that

$$\frac{d\langle N_I \rangle}{dx} = -\tau^* \langle N_I \rangle \tag{8.16.10}$$

The integral of Eq. (8.16.10) is

$$\langle N_I(x)\rangle = \langle N_I(0)\rangle \exp\left(-\tau^* x\right) \qquad (8.16.11)$$

where $\langle N_I(0)\rangle$ is the intensity of the incident light at $x = 0$.

Visible light covers the wavelength region of the electromagnetic spectrum extending from about 400–700 nm. The light from the Sun contains a mixture of wavelengths in the visible region and appears white. On the basis of the dependence of τ^* on $1/\lambda^4$, Lord Rayleigh pointed out that blue light (the short wavelength region) is scattered more strongly than red light (the long wavelength region). Looking at the Sun at sunset, the Sun appears to be red or orange, due to the fact that the blue component of the sunlight has been scattered out of the line of sight. When looking away from the Sun during daylight, however, the sky looks blue. The color consists of the blue component of the sunlight, which has been scattered from dust particles and density fluctuations in the atmosphere. This remarkable discovery of why the sky is blue solved a mystery that had puzzled hominids from their first appearance on earth until Rayleigh provided a proper explanation in the nineteenth century!

8.17 Scattering from the Molecules of an Ideal Gas

In an ideal gas, the molecules respond individually to the application of an external electric field. Let the polarizability of a molecule be α. The electric field, $\mathbf{E}_I$, of light passing through the gas induces an electric dipole moment, $\mathbf{p}_{Mol}$, in each molecule, where

$$\mathbf{p}_{Mol} = \alpha \mathbf{E}_I \qquad (8.17.1)$$

If there are N such molecules in a volume, V, then the polarization is proportional to the product of the induced electric dipole moment, $\mathbf{p}_{Mol}$, on a molecule and the molecular number density, (N/V). We have

$$\mathbf{P}_{Gas} = (N/V)\mathbf{p}_{Mol} = (N/V)\alpha \mathbf{E}_I \qquad (8.17.2)$$

where $\mathbf{P}_{Gas}$ is the polarization in the gas. The space around the molecules is taken to be a vacuum. Because the electric susceptibility, χ_e, of the vacuum is zero, we can write

$$\mathbf{P}_{Vac} = 0 \qquad (8.17.3)$$

The fluctuation, $\Delta \mathbf{P}$, in the polarization due to the presence of the molecules is

$$\Delta \mathbf{P} = \mathbf{P}_{Gas} - \mathbf{P}_{Vac} \qquad (8.17.4)$$

According to Eq. (8.2.1a),

$$\Delta \mathbf{P} = \varepsilon_o \Delta \chi_e \mathrm{E}_I \tag{8.17.5}$$

By virtue of Eqs. (8.17.3) and (8.17.5),

$$\mathbf{P}_{Gas} = \varepsilon_o \Delta \chi_e \mathrm{E}_I \tag{8.17.6}$$

Upon comparing Eq. (8.17.2) with Eq. (8.17.6), one concludes that $\Delta \chi_e$ satisfies,

$$\Delta \chi_e = \frac{N\alpha}{\varepsilon_o V} \tag{8.17.7}$$

The volume associated with one of the N scattering centers is

$$v = \frac{V}{N} \tag{8.17.8}$$

Upon substitution of Eqs. (8.17.7) and (8.17.8) into Eq. (8.16.9), the turbidity is found to be

$$\tau^* = \frac{8\pi^3}{3\lambda^4}\left(\frac{N}{V}\right)^2\left(\frac{\alpha}{\varepsilon_o}\right)^2\left(\frac{V}{N}\right) \tag{8.17.9}$$

which can be simplified to read

$$\tau^* = \frac{8\pi^3}{3\lambda^4}\left(\frac{N}{V}\right)\left(\frac{\alpha}{\varepsilon_o}\right)^2 \tag{8.17.10}$$

The polarizability, α, is roughly proportional to the molecular volume. Hence, the turbidity increases with the square of the geometric size of the molecule.

8.18 Scattering from a Compressible Fluid

The term "fluid" encompasses both liquids and gases. The speed of light in a fluid medium is written as

$$c_m = \frac{1}{\sqrt{\varepsilon\mu}} \tag{8.18.1}$$

By comparison, the speed of light in vacuum is

$$c = \frac{1}{\sqrt{\varepsilon_o\mu_o}} \tag{8.18.2}$$

The ratio of the two speeds is referred to as the refractive index, n, where

$$n = \frac{c}{c_m} \tag{8.18.3}$$

If the fluid is diamagnetic or paramagnetic, its magnetic permeability $\mu \approx \mu_o$, in which case Eqs. (8.18.1) through (8.18.3) can be combined to express the refractive index of the material in the form,

$$n = \sqrt{\frac{\varepsilon}{\varepsilon_o}} \tag{8.18.4}$$

By virtue of Eq. (8.2.1d), Eq. (8.18.4) can be written as

$$n^2 = 1 + \chi_e \tag{8.18.5}$$

A small change, ΔN, in the number of molecules in a scattering volume, v, results in a change, $\Delta\chi_e$, in the electric permittivity given by

$$\Delta\chi_e = \left(\frac{\partial\chi_e}{\partial N}\right)\Delta N \tag{8.18.6}$$

According to Eq. (8.18.6), the ensemble average of the square of the change in the susceptibility, $\langle(\Delta\chi_e)^2\rangle$, is

$$\langle(\Delta\chi_e)^2\rangle = \left(\frac{\partial\chi_e}{\partial N}\right)^2\langle(\Delta N)^2\rangle \tag{8.18.7}$$

Use Eq. (8.18.5) to replace χ_e, so that Eq. (8.18.7) can be rewritten as

$$\langle(\Delta\chi_e)^2\rangle = \left(\frac{\partial n^2}{\partial N}\right)^2\langle(\Delta N)^2\rangle \tag{8.18.8}$$

Equation (8.18.8) can also be rendered in the form,

$$\langle(\Delta\chi_e)^2\rangle = 4n^2\left(\frac{\partial n}{\partial N}\right)^2\langle(\Delta N)^2\rangle \tag{8.18.9}$$

Equation (8.18.9) serves as the basis for the calculation of $\langle(\Delta\chi_e)^2\rangle$ in the case of a compressible fluid. To apply Eq. (8.18.9), it is useful to express $\langle(\Delta N)^2\rangle$ in terms of the isothermal compressibility, κ_T, which is a measurable quantity. The required relation is Eq. (2.13.13) in Chap. 2. We reprise Eq. (2.13.13) below:

$$\langle(\Delta N)^2\rangle = \frac{N^2 k_B T}{V}\kappa_T \tag{8.18.10}$$

Upon substitution of Eq. (8.18.10) into Eq. (8.18.9), the result is

$$\left\langle (\Delta\chi_e)^2 \right\rangle = 4n^2 \left(\frac{\partial n}{\partial N} \right)^2 \frac{N^2 k_B T}{V} \kappa_T \tag{8.18.11}$$

If we rewrite Eq. (8.16.9) with v replaced by V, then

$$\tau^* = \frac{8\pi^3}{3\lambda^4} \left\langle (\Delta\chi_e)^2 \right\rangle V \tag{8.18.12a}$$

Upon substitution of Eq. (8.18.11) into Eq. (8.18.12a), the result is

$$\tau^* = \frac{32\pi^3}{3\lambda^4} n^2 \left(\frac{\partial n}{\partial N} \right)^2 N^2 k_B T \kappa_T \tag{8.18.12b}$$

We denote the molecular number density by

$$\rho = \frac{N}{V} \tag{8.18.13}$$

With V fixed, $dN = V d\rho$. With a change in variable from N to ρ, the derivative of the refractive index with respect the N, reads

$$\left(\frac{\partial n}{\partial N} \right) = \frac{1}{V} \left(\frac{\partial n}{\partial \rho} \right) \tag{8.18.14a}$$

Upon substitution of (8.18.14a) into Eq. (8.18.12b), the turbidity is given by

$$\tau^* = \frac{32\pi^3}{3\lambda^4} n^2 \left(\frac{1}{V^2} \right) \left(\frac{\partial n}{\partial \rho} \right)^2 N^2 k_B T \kappa_T \tag{8.18.14b}$$

Substitute Eq. (8.18.13) into Eq. (8.18.14b) to obtain the final result for the turbidity,

$$\tau^* = \frac{32\pi^3}{3\lambda^4} n^2 \left(\frac{\partial n}{\partial \rho} \right)^2 \rho^2 k_B T \kappa_T \tag{8.18.15}$$

At the liquid–vapor critical point of a pure fluid, the isothermal compressibility diverges, namely,

$$\lim_{T \to T_c} \kappa_T = \lim_{T \to T_c} \left(\frac{-1}{V} \left(\frac{\partial V}{\partial P} \right)_T \right) \to \infty \tag{8.18.16}$$

This behavior has its origin in the fact that the slope, $(\partial P/\partial V)_T$, of the critical isotherm is zero at the critical point. See Fig. 8.2 for P versus V.

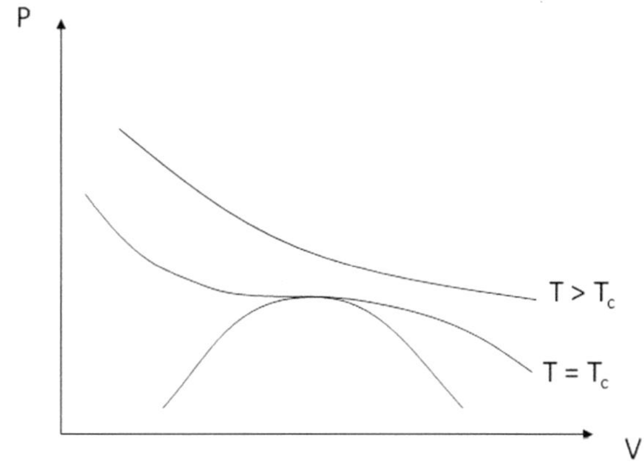

Fig. 8.2 Pressure, P, versus volume, V, for real gas near the critical temperature, T_c

Density fluctuations in fluids occur naturally, although ordinarily their effects are too small to be noticed. At the critical point, however, where the difference between the liquid and the vapor disappears, they grow and can extend over distances of the same order of magnitude as the wavelength of light. A density fluctuation of this size can scatter any light that is incident on the fluid. Indeed, the fluid may become so cloudy as to no longer be transparent. This strong light scattering effect is known as critical opalescence. Its theoretical basis is to be found in Eqs. (8.8.15) and (8.8.16) which when combined predict,

$$\lim_{T \to T_c} \tau^* \to \infty \tag{8.18.17}$$

According to Eq. (8.16.11), when Eq. (8.18.17) applies, it is the case that the transmitted light intensity is $\langle N_I(x) \rangle = 0$.

8.19 Scattering from a Macromolecular Solute

We shall consider a binary solution consisting of a solvent denoted as component (1) and a macromolecular solute denoted as component (2). Macromolecular solutions are known to scatter light. The scattering is dominated by the fluctuations in the solute concentration to the extent that the scattering from the solvent can be ignored. To calculate the fluctuations in the solute concentration, we refer to Eq. (2.14.4) of Chap. 2 which we reprise below:

$$\left\langle (\Delta N_2)^2 \right\rangle = N_A^2 V k_B T \left(\frac{\partial c_2}{\partial \mu_2} \right) \tag{8.19.1}$$

where c_2 is the molar concentration of the solute, and V is the volume of the solution. To specialize Eq. (8.18.9) so that it refers to the solute, we write it in the form

$$\left\langle (\Delta\chi_e)^2 \right\rangle = 4n^2 \left(\frac{\partial n}{\partial N_2} \right)^2 \left\langle (\Delta N_2)^2 \right\rangle \tag{8.19.2}$$

where N_2 is the number of molecules of the solute in the mixture. If the volume is constant, then $dN_2 = N_A V dc_2$, and Eq. (8.19.2) can be rewritten as

$$\left\langle (\Delta\chi_e)^2 \right\rangle = \frac{4n^2}{N_A^2 V^2} \left(\frac{\partial n}{\partial c_2} \right)^2 \left\langle (\Delta N_2)^2 \right\rangle \tag{8.19.3}$$

Combine Eqs. (8.19.1) and (8.19.3) to obtain

$$\left\langle (\Delta\chi_e)^2 \right\rangle = \frac{4n^2}{N_A^2 V^2} \left(\frac{\partial n}{\partial c_2} \right)^2 N_A^2 V k_B T \left(\frac{\partial c_2}{\partial \mu_2} \right) \tag{8.19.4a}$$

which upon cancelation of common factors reads

$$\left\langle (\Delta\chi_e)^2 \right\rangle = \frac{4n^2}{V} \left(\frac{\partial n}{\partial c_2} \right)^2 k_B T \left(\frac{\partial c_2}{\partial \mu_2} \right) \tag{8.19.4b}$$

Substitute Eq. (8.19.4b) into Eq. (8.18.12a) to obtain

$$\tau^* = \frac{8\pi^3}{3\lambda^4} \cdot \frac{4n^2}{V} \left(\frac{\partial n}{\partial c_2} \right)^2 k_B T \left(\frac{\partial c_2}{\partial \mu_2} \right) V \tag{8.19.5a}$$

which simplifies to read

$$\tau^* = \frac{32\pi^3 n^2 k_B T}{3\lambda^4} \left(\frac{\partial n}{\partial c_2} \right)^2 \left(\frac{\partial c_2}{\partial \mu_2} \right) \tag{8.19.5b}$$

For comparison, we can refer to Chap. 2, where Eq. (2.15.12) reads

$$\left(\frac{\partial \mu_2}{\partial c_2} \right) = \frac{1}{c_2} \left(\frac{\partial \Pi}{\partial c_2} \right) \tag{8.19.6a}$$

In the low concentration limit, Eq. (2.15.14) assumes the form,

$$\Pi = RTc_2 \tag{8.19.6b}$$

Differentiation of Eq. (8.19.6b) with respect to c_2 gives

$$\left(\frac{\partial \Pi}{\partial c_2} \right) = RT \tag{8.19.6c}$$

so that Eq. (8.19.6a) can be written as

$$\left(\frac{\partial \mu_2}{\partial c_2}\right) = \frac{RT}{c_2} \tag{8.19.6d}$$

Upon substitution of Eq. (8.19.6d), Eq. (8.19.5b) reads,

$$\tau^* = \frac{32\pi^3 n^2 k_B T}{3\lambda^4}\left(\frac{\partial n}{\partial c_2}\right)^2 \cdot \frac{c_2}{RT} \tag{8.19.7a}$$

We can use $R = N_A k_B$ to write Eq. (8.19.7a) in the form,

$$\tau^* = \frac{32\pi^3}{3N_A\lambda^4} n^2 \left(\frac{\partial n}{\partial c_2}\right)^2 c_2 \tag{8.19.7b}$$

Equation (8.19.7b) is the law governing the turbidity of a dilute ideal solution of a macromolecule.

Although we have specialized our treatment to a macromolecular solution, Eq. (8.19.1) is of interest in understanding the critical opalescence observed in a binary liquid mixture with a miscibility gap ending in a critical point of solution. A miscibility gap consists of a concentration range in the phase diagram where the two liquids have only partial miscibility. At a temperature and a concentration within this gap, the mixture exists in the form of two liquid layers having different compositions. See Fig. 8.3.

In the case of a binary liquid mixture, the composition fluctuates strongly at temperatures near T_c. The mixture exhibits pronounced light scattering known as

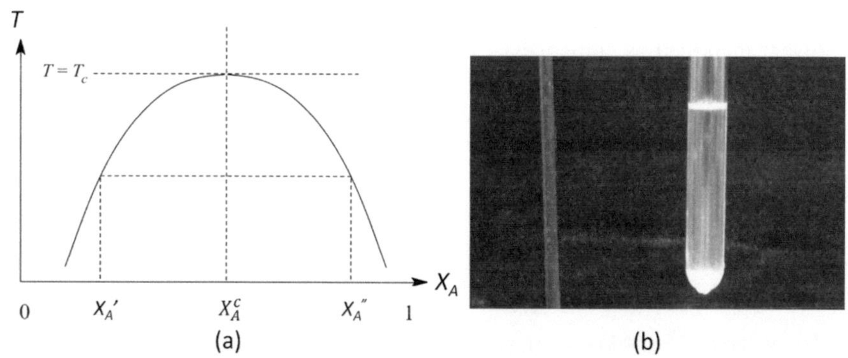

Fig. 8.3 **a** Phase diagram for a binary liquid mixture exhibiting a critical point of solution. The horizontal axis is the mole fraction, X_A, of component, A, in a mixture of liquid components, A and B. The critical composition corresponds to X_A^c. The vertical axis is the temperature, T. The critical temperature is T_c. On the convex side of the phase boundary, the mixture exists as a single liquid phase. On the concave side, the mixture splits into two immiscible liquid layers with compositions, X_A' and X_A''. **b** A binary liquid mixture exhibiting opalescence near its critical point of solution

critical opalescence. This is illustrated in Fig. 8.3b. More than 1000 such pairs are known. Examples include isobutyric acid + water and nitrobenzene + dodecane [4].

8.20 Turbidity in Terms of the Osmotic Compressibility

The derivative, $(\partial c_2/\partial \Pi)$, where Π is the osmotic pressure, is known as the osmotic compressibility. If we substitute Eq. (8.19.6a) into Eq. (8.19.5b), the result is

$$\tau^* = \frac{32\pi^3 n^2 k_B T}{3\lambda^4}\left(\frac{\partial n}{\partial c_2}\right)^2 \cdot c_2\left(\frac{\partial c_2}{\partial \Pi}\right) \tag{8.20.1}$$

As $(\partial c_2/\partial \Pi)$ increases, the two components in the solution mix more easily. The enhanced ability to mix accentuates the effect of the fluctuations in the concentration of the macromolecule and serves to increase the value of τ^*.

8.21 Determination of Polymer Number Average Molecular Weight by Light Scattering

Seldom can a polymer be prepared in monodisperse form. Rather, the typical synthesis scheme leads to a distribution of molecules having the same atomic stoichiometric ratios but different chain lengths and hence different molecular weights. As above, we shall consider the polymer to be component "2" in a solution with a solvent identified as component "1". The *i*-th molecular weight fraction of the polymer consists of N_2^i molecules with molecular weight, M_2^i. Neither N_2^i nor M_2^i is known a priori. Nonetheless, we will show that light scattering can be used to determine a composite of these two variables, termed the "number average" molecular weight, $\overline{M}_2$. The latter is defined as

$$\overline{M}_2 = \frac{\sum_{i=1} M_2^i N_2^i}{\sum_i N_2^i} \tag{8.21.1}$$

The value of $\overline{M}_2$, is known in statistical theory as the average or "first moment" of the underlying distribution, N_2^i, of molecular weights, M_2^i.

In analogy with the pressure of a real gas expressed as a power series in its density, the osmotic pressure of the polymer solution can be represented as a power series in the molar concentration, c_2, of the polymer. As indicated in Eq. (2.15.14) of Chap. 2, just two terms are required for an accurate representation of the data, namely

$$\Pi(c_2) = RT\left[c_2 + bc_2^2\right] \tag{8.21.2}$$

In Eq. (8.21.2), b is an empirical parameter known as the second virial coefficient. By contrast, the first virial coefficient is the coefficient of c_2, which is a unity. The concentration derivative of Eq. (8.21.2) is

$$\left(\frac{\partial \Pi}{\partial c_2}\right) = RT[1 + 2bc_2] \tag{8.21.3}$$

Upon substitution of Eq. (8.21.3) into Eq. (8.20.1), we obtain for the turbidity,

$$\tau^* = \frac{32\pi^3 n^2 k_B T}{3\lambda^4}\left(\frac{\partial n}{\partial c_2}\right)^2 \cdot \frac{c_2}{RT[1 + 2bc_2]} \tag{8.21.4a}$$

Using the relation, $R = N_A k_B$, Eq. (8.21.4a) reads

$$\tau^* = \frac{32\pi^3 n^2}{3\lambda^4 N_A}\left(\frac{\partial n}{\partial c_2}\right)^2 \cdot \frac{c_2}{[1 + 2bc_2]} \tag{8.21.4b}$$

Because $\overline{M}_2$ is not initially known, it is impossible to calculate the apparent average molar concentration, c_2, using the equation $c_2 = w_2/\overline{M}_2 V$ where w_2 is the weighed out mass of the polymer. One can calculate the mass density, $c = w_2/V$; however, where V is the volume of the solution. On the basis of these two definitions, c and c_2 are related to $\overline{M}_2$ by the equation,

$$c = \overline{M}_2 c_2 \tag{8.21.5}$$

According to Eq. (8.21.5), $dc_2 = dc/\overline{M}_2$. Upon substitution of this result and use of Eq. (8.21.5), Eq. (8.21.4b) can be transformed to read,

$$\tau^* = \frac{32\pi^3 n^2}{3\lambda^4 N_A}\left(\frac{\partial n}{\partial c}\right)^2 \cdot \overline{M}_2^2 \cdot \frac{c/\overline{M}_2}{[1 + 2Bc]} \tag{8.21.6a}$$

where $B = b/\overline{M}_2$ is a new empirical second virial coefficient, which is appropriate to the concentration measured in the units of c. Upon cancellation of common factors, Eq. (8.21.6a) reads,

$$\tau^* = \frac{32\pi^3 n^2}{3\lambda^4 N_A}\overline{M}_2\left(\frac{\partial n}{\partial c}\right)^2 \frac{c}{[1 + 2Bc]} \tag{8.21.6b}$$

For the purpose of fitting data, we can rewrite Eq. (8.21.6b) in the form,

$$\frac{32\pi^3 n^2}{3N_A\lambda^4}\left(\frac{\partial n}{\partial c}\right)^2 \frac{c}{\tau^*} = \frac{1}{\overline{M}_2}[1 + 2Bc] \tag{8.21.7a}$$

If we define the composite factor, H, by

$$H = \frac{32\pi^3 n^2 (\partial n/\partial c)^2}{3N_A \lambda^4} \tag{8.21.7b}$$

then Eq. (8.21.7a) can be written more simply as

$$\frac{Hc}{\tau^*} = \frac{1}{\overline{M}_2}[1 + 2Bc] \tag{8.21.8}$$

The turbidity, τ^*, is ordinarily measured as a function of the mass density, c. To evaluate H on the left hand side of Eq. (8.21.8) requires independent measurements of the refractive index, n, and its concentration derivative, $(\partial n/\partial c)$. Once this is done, the data in the form of Hc/τ^* versus c, should form a straight line with intercept $1/\overline{M}_2$ and slope $2B/\overline{M}_2$. The value of the slope serves to evaluate B, which provides information concerning the potential of mean force describing the interactions of the polymer molecules with one another.

References

1. B.J. Berne, R. Pecora, *Dynamic Light Scattering with Applications to Chemistry, Biology, and Physics*, chap. 3 (Dover, Mineola, NY, 2000)
2. B. Chu, *Laser Light Scattering*, chap. 1 (Academic Press, New York, 1974)
3. J.B. Lastovka, Light mixing spectroscopy and the spectrum of light scattered by thermal fluctuations in liquids. Ph.D. thesis, Massachusetts Institute of Technology, Cambridge, MA (1967)
4. X. Wang, C.M. Lyons, J.K. Baird, Quenching of critical opalescence in binary liquid mixtures. J. Mol. Liqs. **386**, 122326 (2023)

Recommended Additional Reading

1. A. Hinchliffe, R.W. Munn, *Molecular Electromagnetism* (Wiley, Chichester, W. Sussex, UK, 1985)
2. D.J. Griffiths, *Introduction to Electrodynamics*, 4th edn. (Pearson Education, India, 2013)
3. P. Norman, K. Ruud, T. Saue, *Principles and Practices of Molecular Properties (Theory and Simulations)* (Wiley, Hoboken, NJ, 2018)
4. L. Barron, *Molecular Light Scattering and Optical Activity*, 2nd edn. (Cambridge U. P., Cambridge, UK, 2004)
5. I.L. Fabelinskii, *Molecular Scattering of Light* (Plenum Press, NY, 1968)

Chapter 9
X-Ray Diffraction from Crystals

X-ray diffraction serves as the foundation for the science of crystallography. After the introduction of the unit cell concept in this chapter, the basis vectors in the space lattice are defined, followed by the Miller indices that are used to identify atomic planes in the space lattice that serve to diffract the X-rays. Building upon the scattering theory introduced in Chap. 8, the X-ray structure factor is introduced and used to draw the distinction between coherent and incoherent scattering. The reciprocal lattice, in which the scattering is analyzed, is then introduced, and its relation to the space lattice is demonstrated. The von Laue conditions for constructive interference of the scattered waves and Bragg's law governing the angular dependence of the scattering are derived. As examples of the application of the above concepts, the structure factor is evaluated explicitly for the simple cubic, the body-centered cubic, and the face-centered cubic unit cells.

9.1 Introduction to the Structure Factor

As pointed out in Chap. 8, the scattering of electromagnetic waves by a material has its origin in the quantity, $\langle \Delta\chi_e(\mathbf{r}_1)\Delta\chi_e(\mathbf{r}_2)\rangle$, which describes the spatial correlation of the fluctuations in the electric susceptibility. We found in Chap. 8 that the intensity of the scattered light was given by Eq. (8.14.6c), which we copy below in the form,

$$\langle N_S\rangle = \frac{1}{2}\left(\frac{\varepsilon}{\mu}\right)^{1/2}\left(\frac{\varepsilon_o}{\varepsilon}\right)^2 (E_I^o)^2 \frac{k_S^4 \sin^2\vartheta}{(4\pi R)^2} \times \int d^3r_1 \int d^3r_2 \langle \Delta\chi_e(\mathbf{r}_1)\Delta\chi_e(\mathbf{r}_2)\rangle \exp[i(\mathbf{k}_I - \mathbf{k}_S)\cdot(\mathbf{r}_1 - \mathbf{r}_2)] \quad (9.1.1)$$

J. K. Baird, *Electromagnetism for Chemistry*,
https://doi.org/10.1007/978-3-031-88508-2_9

Equation (9.1.1) is the same as Eq. (8.14.6c) except that script ϑ has replaced Greek θ denoting the angle of polarization. In what follows, script ϑ will denote the angle of polarization, while Greek θ will denote the scattering angle.

In Chap. 8, the incident light was considered to be in the visible range, which made the wavelength long when compared with the distance, $|\mathbf{r}_1 - \mathbf{r}_2|$, between any two scattering centers located at $\mathbf{r}_1$ and $\mathbf{r}_2$, respectively. By contrast, in the case of the scattering of X-rays, the wavelength is similar to the distance between the individual atoms. The atoms function as scattering centers located by the set of vectors, $\{\mathbf{r}_i\}$. Due to the fact that the scattering power can no longer be regarded as distributed in a continuous fashion as represented by the functions $\Delta\chi_e(\mathbf{r}_1)$ and $\Delta\chi_e(\mathbf{r}_2)$, Eq. (9.1.1) requires modification. Consider two atomic scattering centers located at $\mathbf{r}_j$ and $\mathbf{r}_k$ with scattering cross sections $\alpha_j = \alpha_j(\vartheta)$ and $\alpha_k = \alpha_k(\vartheta)$, respectively. To take into account the discrete nature of the atom scattering centers, we shall need to replace the double integral, $\int d^3r_1 \int d^3r_2 \langle \Delta\chi_e(\mathbf{r}_1)\Delta\chi_e(\mathbf{r}_2)\rangle$, in Eq. (9.1.1) by a double sum. Given two scattering centers located by the position vectors, $\mathbf{r}_j$ and $\mathbf{r}_k$, the susceptibilities, $\Delta\chi_e(\mathbf{r}_1)$ and $\Delta\chi_e(\mathbf{r}_2)$, will be replaced by scattering power densities, $\alpha_j = \alpha_j(\vartheta)\delta(\mathbf{r}' - \mathbf{r}_j)$ and $\alpha_k = \alpha_k(\vartheta)\delta(\mathbf{r}'' - \mathbf{r}_k)$, respectively. To make Eq. (9.1.1) consistent with the arbitrary locations of two scattering centers, we replace $\exp[i(\mathbf{k}_I - \mathbf{k}_S)\cdot(\mathbf{r}_1 - \mathbf{r}_2)]$ by $\exp\left[i(\mathbf{k}_I - \mathbf{k}_S)\cdot\left(\mathbf{r}_j - \mathbf{r}_k\right)\right]$. Delta functions have been introduced so that the factor, $\exp\left[i(\mathbf{k}_I - \mathbf{k}_S)\cdot\left(\mathbf{r}_j - \mathbf{r}_k\right)\right]$, can be represented by the double integral, $\int d^3r' \int d^3r'' \delta(\mathbf{r}' - \mathbf{r}_j)\delta(\mathbf{r}'' - \mathbf{r}_k)\exp\left[i(\mathbf{k}_I - \mathbf{k}_S)\cdot\left(\mathbf{r}' - \mathbf{r}''\right)\right]$, which has the same meaning. Incorporating these changes, the factor to the right of multiplication symbol, $\times$, in Eq. (9.1.1) now reads

$$
|\Sigma|^2 = \int d^3r' \int d^3r'' \sum_{j=1}\sum_{k=1} \alpha_j(\vartheta)\alpha_k(\vartheta)\delta(\mathbf{r}' - \mathbf{r}_j)\delta(\mathbf{r}'' - \mathbf{r}_k)
\times \exp\left[i(\mathbf{k}_I - \mathbf{k}_S)\cdot(\mathbf{r}' - \mathbf{r}'')\right] \tag{9.1.2}
$$

For simplicity, we shall assume that the solid contains only one type of atom so that $\alpha_j(\vartheta) = \alpha_k(\vartheta) = \alpha(\vartheta)$. Equation (9.1.2) then reads,

$$
|\Sigma|^2 = [\alpha(\vartheta)]^2 \int d^3r' \int d^3r'' \sum_{j=1}\sum_{k=1} \delta(\mathbf{r}' - \mathbf{r}_j)\delta(\mathbf{r}'' - \mathbf{r}_k)
\times \exp\left[i(\mathbf{k}_I - \mathbf{k}_S)\cdot(\mathbf{r}' - \mathbf{r}'')\right] \tag{9.1.3a}
$$

Upon performing the integrations over the delta functions, we obtain

$$
|\Sigma|^2 = [\alpha(\vartheta)]^2 \sum_j \sum_k \exp\left[i(\mathbf{k}_I - \mathbf{k}_S)\cdot(\mathbf{r}_j - \mathbf{r}_k)\right] \tag{9.1.3b}
$$

Equation (9.1.3b) can be factored to read

$$
\begin{aligned}
|\Sigma|^2 &= [\alpha(\vartheta)]^2 \left[\sum_{j=1} \exp\big[i(\mathbf{k}_I - \mathbf{k}_S) \cdot \mathbf{r}_j\big] \right] \left[\sum_{k=1} \exp[-i(\mathbf{k}_I - \mathbf{k}_S) \cdot \mathbf{r}_k] \right] \\
&= [\alpha(\vartheta)]^2 \left[\sum_{i=1} \exp[i(\mathbf{k}_I - \mathbf{k}_S) \cdot \mathbf{r}_i] \right] \left[\sum_{i=1} \exp[i(\mathbf{k}_I - \mathbf{k}_S) \cdot \mathbf{r}_i] \right]^* \qquad (9.1.3c)
\end{aligned}
$$

where the symbol, $*$, indicates the complex conjugate. Now introduce the structure factor defined by

$$S(\mathbf{q}) = \sum_{j=1} \exp\big[i(\mathbf{q} \cdot \mathbf{r}_j)\big] \qquad (9.1.4)$$

where

$$\mathbf{q} = \mathbf{k}_I - \mathbf{k}_S \qquad (9.1.5)$$

The structure factor is a function of the scattering angle, θ. We shall restrict the index, j, in the sum in Eq. (9.1.4) to the atoms in a single unit cell. If there are N such unit cells in the crystal then upon incorporation of Eq. (9.1.4), Eq. (9.1.3c) reads

$$|\Sigma(\theta)|^2 = N[\alpha(\vartheta)]^2 |S(\mathbf{q})|^2 \qquad (9.1.6)$$

Equation (9.1.1) for the scattering intensity becomes

$$\langle N_S \rangle = \frac{1}{2} \left(\frac{\varepsilon}{\mu} \right)^{1/2} \left(\frac{\varepsilon_o}{\varepsilon} \right)^2 (E_I^o)^2 \frac{k_S^4 \sin^2 \vartheta}{(4\pi R)^2} N[\alpha(\vartheta)]^2 |S(\mathbf{q})|^2 \qquad (9.1.7)$$

Figure 9.1 illustrates two vectors $\mathbf{r}_i$ and $\mathbf{r}_j$ in a simple cubic unit cell.

The structure factor, $S(\mathbf{q})$, determines the interference pattern, which is produced by waves scattered from neighboring atoms in the unit cell. The value of the exponential factor in Eq. (9.1.4) depends upon the interatomic structure of the unit cell

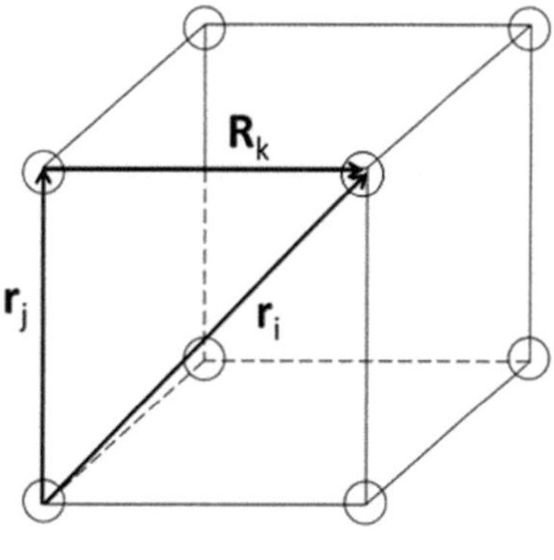

Fig. 9.1 Simple cubic unit cell with a lattice point (atomic location) at each corner of the cube where $\mathbf{r}_i$, $\mathbf{r}_j$, and $\mathbf{R}_k = \mathbf{r}_i - \mathbf{r}_j$ are lattice vectors

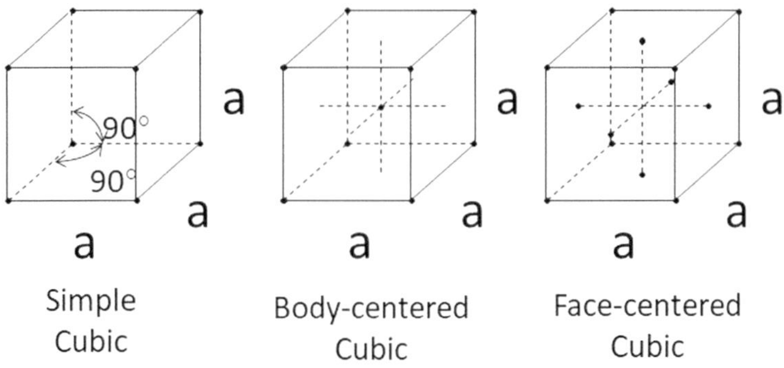

Fig. 9.2 The cubic system of unit cells. In the simple cubic system, nearest neighbor atoms are touching along the edges of the unit cell. In the body centered cubic, the nearest neighbors are touching along the body diagonal. In the face centered cubic, the nearest neighbors are touching along the face diagonal

and the magnitude and direction of the scattering vector, $\mathbf{q}$. Given that the atomic spacings in crystals are of the order of a few tenths of a nanometer, the magnitude of the vector $\mathbf{q}$ in Eq. (9.1.4) that will lead to constructive interference must be of the form $q = 2\pi/\lambda$, where λ is in the X-ray band.

Figure 9.2 illustrates the three unit cell types that are based on the cube. They are known as simple cubic (sc), face centered cubic (fcc), and body centered cubic (bcc) unit cells.

The bulk of the information, which we possess concerning the geometric arrangements of atoms in crystalline solids, has been determined by examining the dependence of $S(\mathbf{q})$ on the direction of $\mathbf{q}$. The behavior of $S(\mathbf{q})$ as a function of $\mathbf{q}$ is the subject of the remainder of this chapter.

A unit cell is said to be "primitive" if it contains within its volume the sum of pieces of atoms which taken together add up to a single atom. This concept can be understood as follows: In the three unit cells of the cubic system shown in Fig. 9.2, there is an atom located at each corner lattice point. A corner position is shared by 8 adjacent unit cells. Hence, one-eighth of an atom is contributed to each cell by that corner position. In sc, there are eight corners, which add up to one whole atom. This allows the sc unit cell to be classified as primitive. By contrast, the bcc lattice has one-eighth of an atom at each corner and an entire atom in the center for a total of two atoms per unit cell. As such, this cell cannot be said to be primitive. Finally, in fcc an atom is located on each face of the cube. A face is shared by two unit cells, making for one-half of atom in each unit cell. As a cube has 6 faces, an fcc unit cell contains 3 atoms contributed by the faces and 1 atom contributed by the corners for a total of 4 atoms. It also is not primitive.

9.2 Relation Between the Scattering Vector and the Scattering Angle

According to Eq. (9.1.5), the square of the scattering vector is

$$q^2 = (\mathbf{k}_I - \mathbf{k}_S) \cdot (\mathbf{k}_I - \mathbf{k}_S) = k_I^2 - 2(\mathbf{k}_I \cdot \mathbf{k}_S) + k_S^2 \tag{9.2.1}$$

We define the scattering angle, 2θ, as the angle between the incident X-ray propagation vector, $\mathbf{k}_I$, and the scattered X-ray propagation vector, $\mathbf{k}_S$. According to the definition of 2θ, the dot product involving the two propagation vectors is

$$\mathbf{k}_I \cdot \mathbf{k}_S = k_I k_S \cos(2\theta) \tag{9.2.2a}$$

Substitution of Eq. (9.2.2a) into Eq. (9.2.1) yields

$$q^2 = k_I^2 + k_S^2 - 2k_I k_S \cos(2\theta) \tag{9.2.2b}$$

If the scattering is elastic (no energy transfer from the radiation to the atom or vice versa), then $k_I = k_S$, which allows Eq. (9.2.2b) to be solved for q in the form,

$$q = k_I \sqrt{2(1 - \cos(2\theta))} \tag{9.2.2c}$$

If we invoke the trigonometric identity,

$$\sin^2\theta = \frac{1 - \cos(2\theta)}{2} \tag{9.2.2d}$$

then Eq. (9.2.2c) reads,

$$q = 2k_I \sin\theta \tag{9.2.3}$$

Equation (9.2.3) forms the basis for the famous Bragg's Law of X-ray scattering (William Lawrence Bragg, English, 1890–1971).

9.3 Space Lattice Defined in Terms of the Basis Vectors, (a_1, a_2, a_3)

Let the basis vectors defining the size and shape of the general unit cell in space be $\mathbf{a}_1, \mathbf{a}_2, \mathbf{a}_3$. See Fig. 9.3a.

The basis vectors, $\mathbf{a}_1$, $\mathbf{a}_2$, and $\mathbf{a}_3$, are parallel to the sides of the unit cell and have lengths equal to the lengths of the edges of the cell. Unlike the common case of unit vectors, these vectors need not be mutually perpendicular. No two should be parallel,

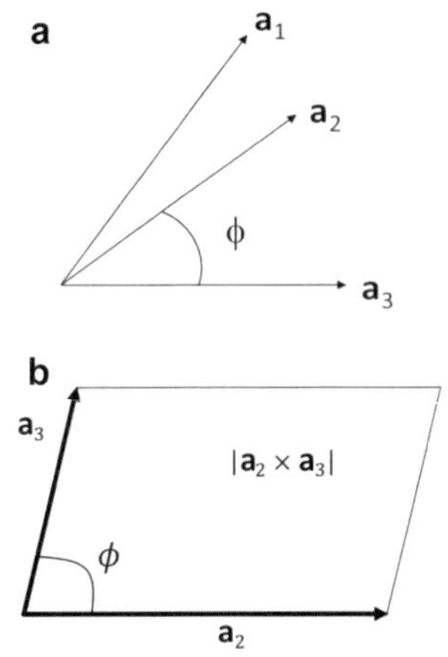

Fig. 9.3 **a** Basis vectors, $\mathbf{a}_1$, $\mathbf{a}_2$, and $\mathbf{a}_3$ defining the directions and lengths of the edges in a unit cell. The vectors $\mathbf{a}_2$ and $\mathbf{a}_3$ form a plane. The angle between $\mathbf{a}_2$ and $\mathbf{a}_3$ is ϕ. The vector $\mathbf{a}_1$ lies outside this plane. **b** Geometry of the cross product representing the area of a slanted rectangle

however. This triplet of vectors is said to define a unit cell in the *space lattice*. Let the base of the unit cell be spanned by the vectors, $\mathbf{a}_2$ and $\mathbf{a}_3$. See Fig. 9.3b. Since the area of the base is the width of the base times its height, the area can be expressed in vector terms as

$$\text{Area of base} = |\mathbf{a}_2||\mathbf{a}_3| \sin \phi = |\mathbf{a}_2 \times \mathbf{a}_3| \tag{9.3.1}$$

where ϕ is the angle between $\mathbf{a}_2$ and $\mathbf{a}_3$. Equation (9.3.1) makes sense, because the area is zero when $\phi = 0$ and $\mathbf{a}_2$ and $\mathbf{a}_3$ are parallel, whereas the area equals the product, $|\mathbf{a}_2||\mathbf{a}_3|$, when $\phi = \pi/2$ and $\mathbf{a}_2$ and $\mathbf{a}_3$ are perpendicular. The volume, Ω, of the unit cell is given by

$$\Omega = (\text{Area of base}) \cdot (\text{height perpendicular to the base}) \tag{9.3.2}$$

In vector terms, Eq. (9.3.2) is

$$\Omega = \mathbf{a}_1 \cdot (\mathbf{a}_2 \times \mathbf{a}_3) \tag{9.3.3}$$

9.4 Miller Indices Used to Identify Atomic Planes

The orientation of a plane in a lattice is determined by specifying its Miller indices (William Hallowes Miller, Welsh, 1801–1880), which we describe below. Let the lattice vectors spanning the unit cell be $\mathbf{a}_1$, $\mathbf{a}_2$, $\mathbf{a}_3$. As shown in Fig. 9.4, suppose that a plane cuts the axes specified by the vectors at the distances, x_1, x_2, and x_3, respectively, from the origin.

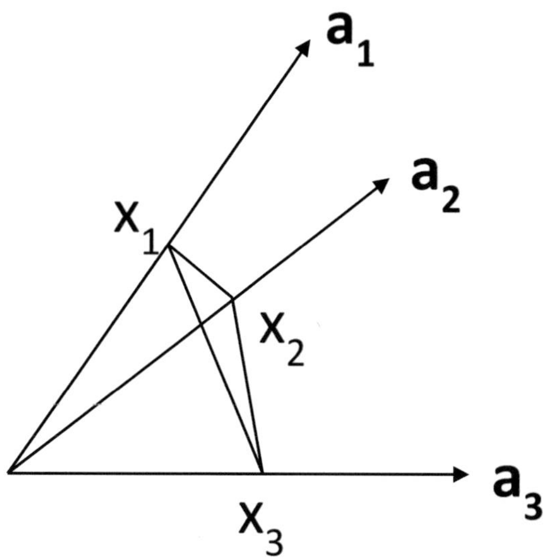

Fig. 9.4 The plane cuts the basis vectors, $\mathbf{a}_1$, $\mathbf{a}_2$, and $\mathbf{a}_3$ at x_1, x_2 and x_3, respectively

Ordinarily, x_1, x_2, and x_3 are rational multiples of the lengths, a_1, a_2, a_3, which are the magnitudes of $\mathbf{a}_1$, $\mathbf{a}_2$, $\mathbf{a}_3$. The following procedure is used to construct the Miller indices:

(1) Form the triplet of fractions, $\left(\frac{x_1}{a_1}, \frac{x_2}{a_2}, \frac{x_3}{a_3}\right)$.

(2) Invert each entry in the triplet to obtain $\left(\frac{a_1}{x_1}, \frac{a_2}{x_2}, \frac{a_3}{x_3}\right)$.

(3) Multiply the triplet $\left(\frac{a_1}{x_1}, \frac{a_2}{x_2}, \frac{a_3}{x_3}\right)$ through by the smallest integral factor that makes each fraction in the triplet an integer. After this has been achieved, one says that the plane has been "indexed".

Example Determine the Miller indices of a plane that cuts the axes at $x_1 = 2a_1$, $x_2 = \frac{3}{2}a_2$, $x_3 = a_3$, respectively.

Step (1): $\left(\frac{x_1}{a_1}, \frac{x_2}{a_2}, \frac{x_3}{a_3}\right) = \left(2, \frac{3}{2}, 1\right)$.
Step (2): $\left(\frac{a_1}{x_1}, \frac{a_2}{x_2}, \frac{a_3}{x_3}\right) = \left(\frac{1}{2}, \frac{1}{(3/2)}, \frac{1}{1}\right) = \left(\frac{1}{2}, \frac{2}{3}, \frac{1}{1}\right)$.
Step (3): Multiply the triplet through by 6: $6 \times \left(\frac{1}{2}, \frac{2}{3}, 1\right) = (3, 4, 6)$.

The standard Miller index designation for a plane is $(hk\ell)$. We have shown that the plane defined by $x_1 = 2a_1$, $x_2 = 3a_2/2$, $x_3 = a_3$, corresponds to $h = 3$, $k = 4$, and $\ell = 6$. Hence, it is referred to as the (346) plane.

Example Consider a unit cell in the cubic system, and determine the Miller indices of a plane that is parallel to the $\mathbf{a}_3$ axis and cuts the $\mathbf{a}_1$ axis at $x_1 = a$ and the $\mathbf{a}_2$ axis at $x_2 = a$. See Fig. 9.5.

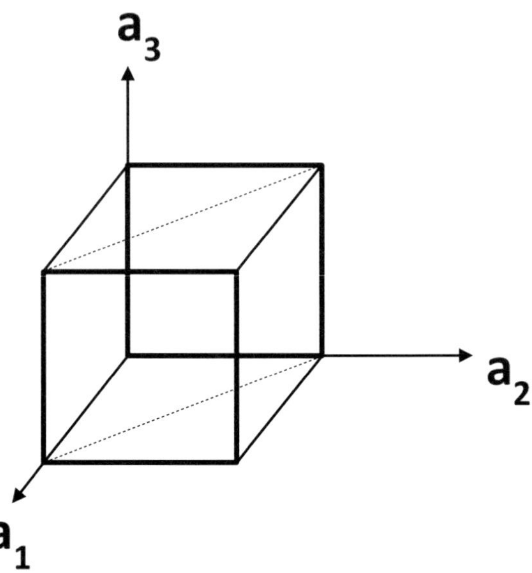

Fig. 9.5 Cubic cell of side, a, with $\mathbf{a}_3$ axis parallel to a plane which cuts the $\mathbf{a}_1$ axis at $x_1 = a$ and the $\mathbf{a}_2$ axis at $x_2 = a$

Being parallel to the $\mathbf{a}_3$ axis, the plane is regarded for the purposes of indexing as if it intersected the $\mathbf{a}_3$ axis at $x_3 = \infty$. With this convention, the intersections along to three axes can be identified as $x_1 = a$, $x_2 = a$, and $x_3 = \infty$.

Step (1): $\left(\frac{x_1}{a}, \frac{x_2}{a}, \frac{x_3}{a}\right) = \left(\frac{a}{a}, \frac{a}{a}, \frac{\infty}{a}\right) = (1, 1, \infty)$.
Step (2): $\left(\frac{1}{1}, \frac{1}{1}, \frac{1}{\infty}\right) = (1, 1, 0)$.
Step (3): The index triplet in Step (2) involves solely integers, so no multiplication by a common multiple is necessary. The plane has the identification, $(hk\ell) = (110)$.

The following are other Miller index conventions:

(1) A plane can intercept an axis on the negative side of the origin resulting in a Miller calculation involving a negative integer as in $(-h, k, \ell)$. The Miller index designation in this case is written as $(\bar{h}k\ell)$, where the minus sign appears as a superscript.
(2) An individual plane is denoted by its Miller symbol, $(hk\ell)$. If we want to indicate a set of planes that are parallel to $(hk\ell)$, we write $\{hk\ell\}$. These are said to be $\{hk\ell\}$ "type" planes.

In the remainder of this chapter, we shall limit ourselves to crystal lattices where the three basis vectors make right angles with one another. In so doing, the results we derive will be greatly simplified, but will be restricted in applicability to the cubic, the tetragonal, and the orthorhombic systems, examples of which are shown in Fig. 9.6.

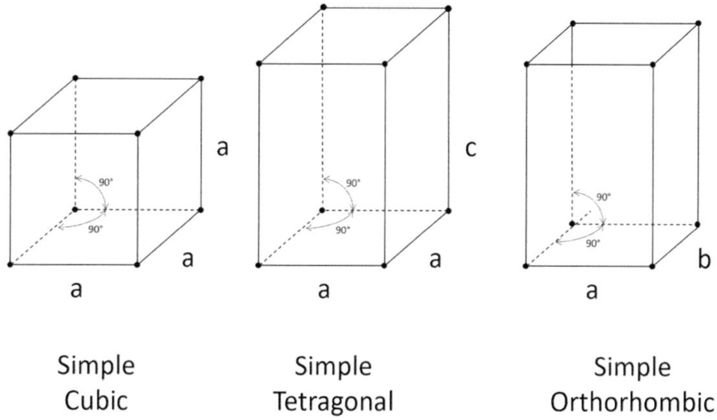

Fig. 9.6 Simple cubic, tetragonal, and orthorhombic unit cells

9.5 Spacing Between {*hkl*} Type Planes

If a plane cuts the *mutually perpendicular axes*, $\mathbf{a}_1, \mathbf{a}_2, \mathbf{a}_3$, at distances x_1, x_2, and x_3 respectively, we can define the direction cosines, $\cos\alpha$, $\cos\beta$, and $\cos\gamma$ so that the distance, d, from the origin to the plane is given by

$$x_1 \cos\alpha = x_2 \cos\beta = x_3 \cos\gamma = d \tag{9.5.1}$$

The angles involved are shown in Fig. 9.7. The sum of squares of the direction cosines is unity.

$$\cos^2\alpha + \cos^2\beta + \cos^2\gamma = 1 \tag{9.5.2}$$

[See the Mathematical Methods Appendix]. To construct a formula for d, substitute Eqs. (9.5.1) into Eq. (9.5.2). The result is

$$\left(\frac{d}{x_1}\right)^2 + \left(\frac{d}{x_2}\right)^2 + \left(\frac{d}{x_3}\right)^2 = 1 \tag{9.5.3}$$

Construction of the Miller indices describing this plane proceeds as follows:

Step 1: $\left(\frac{x_1}{a_1}, \frac{x_2}{a_2}, \frac{x_3}{a_3}\right)$.
Step 2: $\left(\frac{a_1}{x_1}, \frac{a_2}{x_2}, \frac{a_3}{x_3}\right)$.
Step 3:$(h, k, \ell) = n\left(\frac{a_1}{x_1}, \frac{a_2}{x_2}, \frac{a_3}{x_3}\right)$.

where n is an integer. Step 3 above relates each of h, k, and ℓ to one of x_1, x_2, and x_3, respectively. Specifically, one has

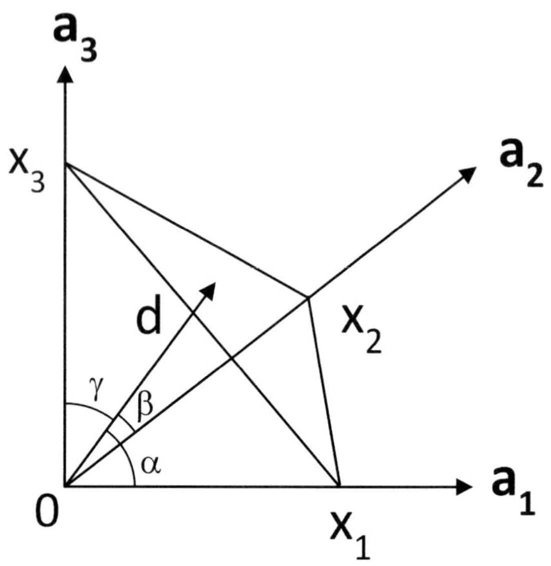

Fig. 9.7 A plane cutting basis vectors, $\mathbf{a}_1$, $\mathbf{a}_2$, and $\mathbf{a}_3$ at x_1, x_2 and x_3, respectively. The distance from the origin, O, in a direction normal to the plane is d. The angles defining the direction cosines are α, β, and γ

$$x_1 = \frac{na_1}{h} \tag{9.5.4a}$$

$$x_2 = \frac{na_2}{k} \tag{9.5.4b}$$

$$x_3 = \frac{na_3}{\ell} \tag{9.5.4c}$$

respectively. Now substitute Eqs. (9.5.4a) through (9.5.4c) into Eq. (9.5.3). With the designation, $d = d_{hk\ell}$, the results are

$$d_{hk\ell}^2 \left[\frac{h^2}{n^2 a_1^2} + \frac{k^2}{n^2 a_2^2} + \frac{\ell^2}{n^2 a_3^2} \right] = 1 \tag{9.5.5}$$

or upon taking the square root and solving,

$$d_{hk\ell} = \frac{n}{\sqrt{\frac{h^2}{a_1^2} + \frac{k^2}{a_2^2} + \frac{\ell^2}{a_3^2}}} \tag{9.5.6}$$

Equation (9.5.6) applies in the case of a cubic, tetragonal, or orthorhombic lattice. It expresses the distance between a set of parallel planes designated by the integer, n.

9.6 Reciprocal Lattice Defined by the Basis Vectors (b_1, b_2, b_3)

Given a crystal structure defined by the space lattice basis vectors, $\mathbf{a}_1$, $\mathbf{a}_2$, and $\mathbf{a}_3$, one can introduce a *reciprocal lattice* in order to specify the conditions that must be met for constructive interference of X-rays. In terms of $\mathbf{a}_1$, $\mathbf{a}_2$, and $\mathbf{a}_3$, the basis vectors for the reciprocal lattice are defined by the equations,

$$\mathbf{b}_1 = \frac{2\pi}{\Omega}(\mathbf{a}_2 \times \mathbf{a}_3) \tag{9.6.1a}$$

$$\mathbf{b}_2 = \frac{2\pi}{\Omega}(\mathbf{a}_3 \times \mathbf{a}_1) \tag{9.6.1b}$$

$$\mathbf{b}_3 = \frac{2\pi}{\Omega}(\mathbf{a}_1 \times \mathbf{a}_2) \tag{9.6.1c}$$

According to Eqs. (9.6.1a) to (9.6.1c), the reciprocal lattice basis vectors, $\mathbf{b}_1$, $\mathbf{b}_2$, and $\mathbf{b}_3$, point in the directions specified by the vector cross products, $(\mathbf{a}_2 \times \mathbf{a}_3)$,$(\mathbf{a}_3 \times \mathbf{a}_1)$, and $(\mathbf{a}_1 \times \mathbf{a}_2)$. These cross products define vectors which are normal to planes parallel to the facets of the unit cell in the space lattice. The reciprocal lattice vectors satisfy three interesting theorems.

Theorem 9.6.1 If $(\mathbf{a}_1, \mathbf{a}_2, \mathbf{a}_3)$ are mutually orthogonal, so also are $(\mathbf{b}_1, \mathbf{b}_2, \mathbf{b}_3)$.

Proof The orthogonality of $(\mathbf{a}_1, \mathbf{a}_2, \mathbf{a}_3)$ is summarized by the dot products,

$$\mathbf{a}_i \cdot \mathbf{a}_j = 0 \quad (i \neq j) \tag{9.6.2a}$$

Introduce the vector identity [See the Mathematical Methods Appendix],

$$(\mathbf{a}_2 \times \mathbf{a}_3) \cdot (\mathbf{a}_3 \times \mathbf{a}_1) = (\mathbf{a}_2 \cdot \mathbf{a}_3)(\mathbf{a}_3 \cdot \mathbf{a}_1) - (\mathbf{a}_2 \cdot \mathbf{a}_1)(\mathbf{a}_3 \cdot \mathbf{a}_3) \tag{9.6.2b}$$

As a test of the statement of the theorem, compute

$$\begin{aligned}\mathbf{b}_1 \cdot \mathbf{b}_2 &= \left(\frac{2\pi}{\Omega}\right)^2 (\mathbf{a}_2 \times \mathbf{a}_3) \cdot (\mathbf{a}_3 \times \mathbf{a}_1) \\ &= \left(\frac{2\pi}{\Omega}\right)^2 [(\mathbf{a}_2 \cdot \mathbf{a}_3)(\mathbf{a}_3 \cdot \mathbf{a}_1) - (\mathbf{a}_2 \cdot \mathbf{a}_1)(\mathbf{a}_3 \cdot \mathbf{a}_3)] = 0\end{aligned} \tag{9.6.2c}$$

where we have used Eqs. (9.6.2a) and (9.6.2b). The remaining two cases consisting of $\mathbf{b}_1 \cdot \mathbf{b}_3$ and $\mathbf{b}_2 \cdot \mathbf{b}_3$ can be calculated analogously. Hence, we can conclude

$$\mathbf{b}_i \cdot \mathbf{b}_j = 0\, i \neq j \quad \blacksquare \tag{9.6.3}$$

Theorem 9.6.2 $\mathbf{a}_i \cdot \mathbf{b}_j = 2\pi\delta_{ij}$, where $\delta_{ij} = 0$, if $i \neq j$ and $\delta_{ij} = 1$ if $i = j$.

Proof Case A: $i \neq j$.

For example, according to Eq. (9.6.1b),

$$\mathbf{a}_1 \cdot \mathbf{b}_2 = \frac{2\pi}{\Omega}\mathbf{a}_1 \cdot (\mathbf{a}_3 \times \mathbf{a}_1) = -\frac{2\pi}{\Omega}\mathbf{a}_3 \cdot (\mathbf{a}_1 \times \mathbf{a}_1) = 0 \tag{9.6.4a}$$

where we used the vector identity

$$\mathbf{A} \cdot (\mathbf{B} \times \mathbf{C}) = -\mathbf{B} \cdot (\mathbf{A} \times \mathbf{C}) \tag{9.6.4b}$$

plus $(\mathbf{a}_1 \times \mathbf{a}_1) = 0$. [See the Mathematical Methods Appendix].

Case B: $i = j$.

For example, according to Eqs. (9.6.1a) and (9.3.3),

$$\mathbf{a}_1 \cdot \mathbf{b}_1 = \frac{2\pi}{\Omega}\mathbf{a}_1 \cdot (\mathbf{a}_2 \times \mathbf{a}_3) = \left(\frac{2\pi}{\Omega}\right)\Omega = 2\pi \tag{9.6.4c}$$

Hence,

$$\mathbf{a}_i \cdot \mathbf{b}_j = 2\pi\delta_{ij} \quad \blacksquare \tag{9.6.5}$$

Theorem 9.6.3 The unit cell in the reciprocal lattice possesses the same rotational symmetry as the unit cell in the space lattice. For example, the simple cubic unit cell in the space lattice has a dual in the reciprocal lattice which is simple cubic.

Proof By virtue of the right-hand rule, we define the Cartesian unit vectors, $\hat{\mathbf{a}}_1$, $\hat{\mathbf{a}}_2$, $\hat{\mathbf{a}}_3$, by

$$\hat{\mathbf{a}}_1 \times \hat{\mathbf{a}}_2 = \hat{\mathbf{a}}_3 \tag{9.6.6a}$$

$$\hat{\mathbf{a}}_3 \times \hat{\mathbf{a}}_1 = \hat{\mathbf{a}}_2 \tag{9.6.6b}$$

$$\hat{\mathbf{a}}_2 \times \hat{\mathbf{a}}_3 = \hat{\mathbf{a}}_1 \tag{9.6.6c}$$

where

$$\hat{\mathbf{a}}_i \cdot \hat{\mathbf{a}}_i = \delta_{ij} \quad (i, j = 1, 2, 3) \tag{9.6.6d}$$

In terms of $\hat{\mathbf{a}}_1$, $\hat{\mathbf{a}}_2$, $\hat{\mathbf{a}}_3$, the basis vectors spanning the simple cubic unit cell in the space lattice are:

$$\mathbf{a}_1 = a\hat{\mathbf{a}}_1 \tag{9.6.7a}$$

$$\mathbf{a}_2 = a\hat{\mathbf{a}}_2 \tag{9.6.7b}$$

$$\mathbf{a}_3 = a\hat{\mathbf{a}}_3 \tag{9.6.7c}$$

The volume of the simple cubic unit cell in the space lattice is:

$$\Omega = a^3 \tag{9.6.8}$$

According to Eqs. (9.6.1a) through (9.6.1c), Eqs. (9.6.6a) through (9.6.6c), and Eqs. (9.6.7a) through (9.6.7c), the basis vectors spanning the unit cell in the reciprocal lattice are:

$$\mathbf{b}_1 = \frac{2\pi}{\Omega}(\mathbf{a}_2 \times \mathbf{a}_3) = \frac{2\pi}{a^3}(a\hat{\mathbf{a}}_2 \times a\hat{\mathbf{a}}_3) = \frac{2\pi a^2}{a^3}(\hat{\mathbf{a}}_2 \times \hat{\mathbf{a}}_3) = \frac{2\pi}{a}\hat{\mathbf{a}}_1 \tag{9.6.9a}$$

$$\mathbf{b}_2 = \frac{2\pi}{\Omega}(\mathbf{a}_3 \times \mathbf{a}_1) = \frac{2\pi}{a^3}(a\hat{\mathbf{a}}_3 \times a\hat{\mathbf{a}}_1) = \frac{2\pi a^2}{a^3}(\hat{\mathbf{a}}_3 \times \hat{\mathbf{a}}_1) = \frac{2\pi}{a}\hat{\mathbf{a}}_2 \tag{9.6.9b}$$

$$\mathbf{b}_3 = \frac{2\pi}{\Omega}(\mathbf{a}_1 \times \mathbf{a}_2) = \frac{2\pi}{a^3}(a\hat{\mathbf{a}}_1 \times a\hat{\mathbf{a}}_2) = \frac{2\pi a^2}{a^3}(\hat{\mathbf{a}}_1 \times \hat{\mathbf{a}}_2) = \frac{2\pi}{a}\hat{\mathbf{a}}_3 \tag{9.6.9c}$$

Equations (9.6.9a) to (9.6.9c) demonstrate that if the unit cell in the space lattice is a simple cube, the unit cell in the reciprocal lattice is a simple cube. The length of the side of the cubic unit cell in the reciprocal lattice is $2\pi/a$ ■

9.7 Reciprocal Lattice Vector and Miller Indices of a Plane in the Space Lattice

The equation,

$$\mathbf{G}_{hk\ell} = h\mathbf{b}_1 + k\mathbf{b}_2 + \ell\mathbf{b}_3 \tag{9.7.1}$$

defines a lattice vector in the reciprocal lattice. This definition satisfies certain simple theorems, which we prove below.

Theorem 9.7.1 $\mathbf{G}_{hk\ell} = h\mathbf{b}_1 + k\mathbf{b}_2 + \ell\mathbf{b}_3$ is normal to the $(hk\ell)$ plane in the space lattice.

Proof According to Eqs. (9.5.4a) through (9.5.4c),

$$\frac{hx_1}{a_1} = n \tag{9.7.2a}$$

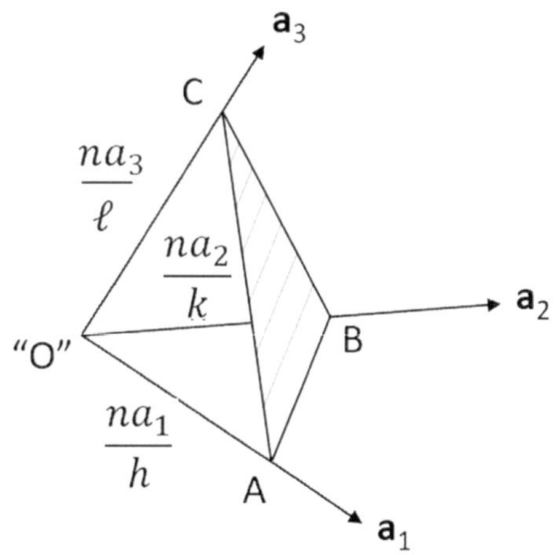

Fig. 9.8 A plane ABC which cuts the $\mathbf{a}_1$, $\mathbf{a}_2$, and $\mathbf{a}_3$ directions at the positions $x_1 = na_1/h$, $x_2 = na_2/k$, and $x_3 = na_3/\ell$, respectively. The lengths of the space lattice basis vectors, $\mathbf{a}_1$, $\mathbf{a}_2$, and $\mathbf{a}_3$ are a_1, a_2, and a_3. The plane has Miller indices, $(hk\ell)$. The origin is at "O"

$$\frac{kx_2}{a_2} = n \tag{9.7.2b}$$

$$\frac{\ell x_3}{a_3} = n \tag{9.7.2c}$$

where n is an integer. Figure 9.8 illustrates a plane in the space lattice.

Equations (9.6.7a) through (9.6.7c) can be rewritten as

$$\hat{\mathbf{a}}_1 = \left(\frac{\mathbf{a}_1}{a_1}\right) \tag{9.7.3a}$$

$$\hat{\mathbf{a}}_2 = \left(\frac{\mathbf{a}_2}{a_2}\right) \tag{9.7.3b}$$

$$\hat{\mathbf{a}}_3 = \left(\frac{\mathbf{a}_3}{a_3}\right) \tag{9.7.3c}$$

With reference to Fig. 9.8, use Eqs. (9.7.3a) through (9.7.3c) to define the vector, $\mathbf{u}$, as

$$\mathbf{u} \equiv \mathbf{BC} = -\mathbf{OB} + \mathbf{OC} = -x_2\hat{\mathbf{a}}_2 + x_3\hat{\mathbf{a}}_3 = -x_2\left(\frac{\mathbf{a}_2}{a_2}\right) + x_3\left(\frac{\mathbf{a}_3}{a_3}\right) \tag{9.7.4a}$$

The vector $\mathbf{u}$ lies in the $(hk\ell)$ plane. Another vector lying in the $(hk\ell)$ plane is $\mathbf{v}$, which is represented by the sum,

$$\mathbf{v} \equiv \mathbf{AB} = -\mathbf{OA} + \mathbf{OB} = -x_1\hat{\mathbf{a}}_1 + x_2\hat{\mathbf{a}}_2 = -x_1\left(\frac{\mathbf{a}_1}{a_1}\right) + x_2\left(\frac{\mathbf{a}_2}{a_2}\right) \tag{9.7.4b}$$

The two vectors, **u** and **v** are sufficient to define the plane. The projection of $\mathbf{G}_{hk\ell}$ onto **u** is

$$\begin{aligned}\mathbf{G}_{hk\ell}\cdot\mathbf{u} &= (h\mathbf{b}_1 + k\mathbf{b}_2 + \ell\mathbf{b}_3)\cdot\left[-x_2\left(\frac{\mathbf{a}_2}{a_2}\right) + x_3\left(\frac{\mathbf{a}_3}{a_3}\right)\right]\\ &= -\frac{kx_2}{a_2}(\mathbf{b}_2\cdot\mathbf{a}_2) + \frac{\ell x_3}{a_3}(\mathbf{b}_3\cdot\mathbf{a}_3)\\ &= -2\pi n + 2\pi n\\ &= 0\end{aligned} \tag{9.7.4c}$$

where we have used Eqs. (9.6.5), (9.7.2b), and (9.7.2c). The projection of $\mathbf{G}_{hk\ell}$ onto **v** is

$$\begin{aligned}\mathbf{G}_{hk\ell}\cdot\mathbf{v} &= (h\mathbf{b}_1 + k\mathbf{b}_2 + \ell\mathbf{b}_3)\cdot\left[-x_1\left(\frac{\mathbf{a}_1}{a_1}\right) + x_2\left(\frac{\mathbf{a}_2}{a_2}\right)\right]\\ &= -\frac{hx_1}{a_1}(\mathbf{b}_1\cdot\mathbf{a}_1) + \frac{kx_2}{a_2}(\mathbf{b}_2\cdot\mathbf{a}_2)\\ &= -2\pi n + 2\pi n\\ &= 0\end{aligned} \tag{9.7.4d}$$

where we have used Eqs. (9.6.5), (9.7.2a), and (9.7.2b). Since there is no projection of $\mathbf{G}_{hk\ell}$ onto either **u** or **v**, $\mathbf{G}_{hk\ell}$ is necessarily perpendicular to the $(hk\ell)$ plane.■

Theorem 9.7.2

$$d_{hk\ell} = \frac{2\pi n}{|\mathbf{G}_{hk\ell}|}$$

Proof Since $\mathbf{G}_{hk\ell} = h\mathbf{b}_1 + k\mathbf{b}_2 + \ell\mathbf{b}_3$ is perpendicular to the $(hk\ell)$ plane in the space lattice, the unit vector

$$\hat{\mathbf{e}} = \frac{\mathbf{G}_{hk\ell}}{|\mathbf{G}_{hk\ell}|} \tag{9.7.5a}$$

is also perpendicular to the $(hk\ell)$ plane. Two parallel $(hk\ell)$ planes are shown in Fig. 9.9.

According to the definition $\mathbf{G}_{hk\ell}$ in Eq. (9.7.1), the unit vector, $\hat{\mathbf{e}}$, which is perpendicular to the $(hk\ell)$ plane, can be written as

$$\hat{\mathbf{e}} = [h\mathbf{b}_1 + k\mathbf{b}_2 + \ell\mathbf{b}_3]\left(\frac{1}{|\mathbf{G}_{hk\ell}|}\right) \tag{9.7.5b}$$

With reference to Fig. 9.7, the projection of the vector, $\mathbf{x}_1 = x_1\hat{\mathbf{a}}_1$, onto $\hat{\mathbf{e}}$ equals $d_{hk\ell}$. This property is also shared by the projections of $\mathbf{x}_2 = x_2\hat{\mathbf{a}}_2$ and $\mathbf{x}_3 = x_3\hat{\mathbf{a}}_3$ onto $\hat{\mathbf{e}}$. In summary,

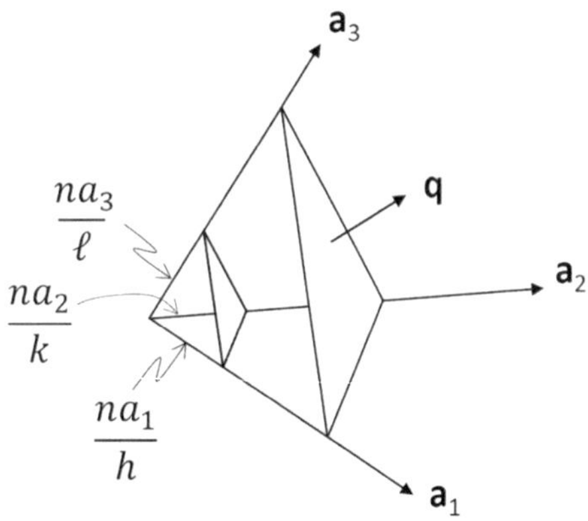

Fig. 9.9 Geometric relation between the scattering vector, $\mathbf{q}$, and planes of the $\{hk\ell\}$ type. The space lattice basis vectors are $\mathbf{a}_1$, $\mathbf{a}_2$, and $\mathbf{a}_3$. The inner plane cuts these vectors at $(n\mathbf{a}_1/h)$ $(n\mathbf{a}_2/k)$, and $(n\mathbf{a}_3/\ell)$, respectively, where n is an integer

$$d_{hk\ell} = \mathbf{x}_1 \cdot \hat{\mathbf{e}} = \mathbf{x}_2 \cdot \hat{\mathbf{e}} = \mathbf{x}_3 \cdot \hat{\mathbf{e}} \tag{9.7.6a}$$

The far left hand side of Eq. (9.7.6a) can be developed as follows:

$$\begin{aligned} d_{hk\ell} &= \mathbf{x}_1 \cdot \hat{\mathbf{e}} = \mathbf{x}_1 \cdot \frac{\mathbf{G}_{hk\ell}}{|\mathbf{G}_{hk\ell}|} = x_1 \hat{\mathbf{a}}_1 \cdot [h\mathbf{b}_1 + k\mathbf{b}_2 + \ell\mathbf{b}_3]\left(\frac{1}{|\mathbf{G}_{hk\ell}|}\right) \\ &= x_1\left(\frac{\mathbf{a}_1}{a_1}\right) \cdot [h\mathbf{b}_1 + k\mathbf{b}_2 + \ell\mathbf{b}_3]\left(\frac{1}{|\mathbf{G}_{hk\ell}|}\right) \\ &= \left(\frac{x_1 h}{a_1}\right)\frac{(\mathbf{a}_1 \cdot \mathbf{b}_1)}{|\mathbf{G}_{hk\ell}|} \\ &= \frac{n(2\pi)}{|\mathbf{G}_{hk\ell}|} \end{aligned} \tag{9.7.6b}$$

where we used Eqs. (9.6.5), (9.7.2a), and (9.7.3a). Equation (9.7.6b) can be rearranged to give the general result,

$$d_{hk\ell} = \frac{2\pi n}{|\mathbf{G}_{hk\ell}|} \quad \blacksquare \tag{9.7.7}$$

9.8 Lattice Vectors in the Space Lattice and the Condition for Constructive Interference

The space lattice basis vectors, $\mathbf{a}_1$, $\mathbf{a}_2$, and $\mathbf{a}_3$, can be used to define the vector distance between a point in the unit cell and any other equivalent point. A space lattice vector, $\mathbf{R}_c$, which connects two equivalent points has the representation,

$$\mathbf{R}_c = m_c\mathbf{a}_1 + n_c\mathbf{a}_2 + p_c\mathbf{a}_3 \tag{9.8.1}$$

where $[m_c, n_c, p_c]$ is a triplet of rational numbers. To achieve constructive interference, the sum on the right hand side of Eq. (9.1.4) must be non-zero. If in Eq. (9.1.4), we let the vector, $\mathbf{r}_j$, run over the space lattice vectors, $\mathbf{R}_c$, then the condition that a term in Eq. (9.1.4) contributes to constructive interference is

$$\exp\left[i(\mathbf{q} \cdot \mathbf{R}_c)\right] = \exp\left[i\mathbf{q} \cdot (m_c\mathbf{a}_1 + n_c\mathbf{a}_2 + p_c\mathbf{a}_3)\right] = 1 \tag{9.8.2}$$

Equation (9.8.2) is satisfied by the von Laue conditions (Max von Laue, German, 1879–1960). These are

$$\mathbf{a}_1 \cdot \mathbf{q} = 2\pi h \tag{9.8.3a}$$

$$\mathbf{a}_2 \cdot \mathbf{q} = 2\pi k \tag{9.8.3b}$$

$$\mathbf{a}_3 \cdot \mathbf{q} = 2\pi \ell \tag{9.8.3c}$$

where $(hk\ell)$ are the integers that the define the orientation of the scattering plane. Figure 9.9 illustrates the geometric relation between $\mathbf{q}$ and $\{hk\ell\}$ type planes. This relationship forms the basis for some interesting theorems satisfied by the scattering vector, $\mathbf{q}$.

Theorem 9.8.1 The scattering vector, $\mathbf{q} = \mathbf{k}_I - \mathbf{k}_S$, as defined in Eq. (9.1.5), can be shown to be perpendicular to the $(hk\ell)$ plane.

Proof According to Fig. 9.8, the vector **AB** has a component of length $x_1 = -\frac{na_1}{h}$ in the $\mathbf{a}_1$-direction, and a component of length, $x_2 = \frac{na_2}{k}$, in the $\mathbf{a}_2$-direction. Hence, upon use of Eqs. (9.5.4a) and (9.5.4b) plus and Eqs. (9.7.3a) and (9.7.3b), we have

$$\mathbf{AB} = -x_1\hat{\mathbf{a}}_1 + x_2\hat{\mathbf{a}}_2 = -\frac{na_1\hat{\mathbf{a}}_1}{h} + \frac{na_2\hat{\mathbf{a}}_2}{k} = \frac{-n\mathbf{a}_1}{h} + \frac{n\mathbf{a}_2}{k} \tag{9.8.4a}$$

According to Fig. 9.8, the vector **AC** has a component of length $x_1 = -\frac{na_1}{h}$ in the $\mathbf{a}_1$-direction, and a component of length $x_3 = \frac{na_3}{\ell}$ in the $\mathbf{a}_3$-direction. Hence,

$$\mathbf{AC} = -x_1\hat{\mathbf{a}}_1 + x_3\hat{\mathbf{a}}_3 = -\frac{na_1\hat{\mathbf{a}}_1}{h} + \frac{na_3\hat{\mathbf{a}}_3}{\ell} = \frac{-n\mathbf{a}_1}{h} + \frac{n\mathbf{a}_3}{\ell} \tag{9.8.4b}$$

where we have used Eqs. (9.5.4a) and (9.5.4c) plus Eqs. (9.7.3a) and (9.7.3c). The dot product of $\mathbf{q} \cdot (\mathbf{AB})$ is given by

$$\mathbf{q} \cdot (\mathbf{AB}) = -\frac{n(\mathbf{q} \cdot \mathbf{a}_1)}{h} + \frac{n(\mathbf{q} \cdot \mathbf{a}_2)}{k} = -2\pi n + 2\pi n = 0 \tag{9.8.4c}$$

where we have used Eqs. (9.8.3a) and (9.8.3b). The dot product, $\mathbf{q} \cdot (\mathbf{AC})$, is

$$\mathbf{q} \cdot (\mathbf{AC}) = -\frac{n(\mathbf{q} \cdot \mathbf{a}_1)}{h} + \frac{n(\mathbf{q} \cdot \mathbf{a}_3)}{\ell} = -2\pi n + 2\pi n = 0 \tag{9.8.4d}$$

where we used Eqs. (9.8.3a) and (9.8.3c). Because $\mathbf{q}$ is perpendicular to the two vectors, **AB** and **AC**, which span the $(hk\ell)$ plane, it is perpendicular to the plane. ■

Theorem 9.8.2 The scattering vector, $\mathbf{q} = \mathbf{k}_I - \mathbf{k}_S$, bisects the angle between $\mathbf{k}_I$ and $\mathbf{k}_S$.

Proof The relevant geometry is shown in Fig. 9.10. Use the definition of $\mathbf{q}$ in Eq. (9.1.5) to compute the dot product,

$$\mathbf{q} \cdot (-\mathbf{k}_I) = (\mathbf{k}_I - \mathbf{k}_S) \cdot (-\mathbf{k}_I) = -k_I^2 + \mathbf{k}_S \cdot \mathbf{k}_I = -\left(k_I^2 - k_I k_S \cos(2\theta)\right) \tag{9.8.5a}$$

where the cosine appears by virtue of Eq. (9.2.2a). The scattering is elastic, so $k_I = k_S$. This permits Eq. (9.8.5a) to be rewritten as

$$\mathbf{q} \cdot (-\mathbf{k}_I) = -k_I^2(1 - \cos(2\theta)) \tag{9.8.5b}$$

Next, we compute

$$\begin{aligned}\mathbf{q} \cdot (\mathbf{k}_S) &= (\mathbf{k}_I - \mathbf{k}_S) \cdot \mathbf{k}_S = \mathbf{k}_I \cdot \mathbf{k}_S - k_S^2 = -\left(k_S^2 - \mathbf{k}_I \cdot \mathbf{k}_S\right) \\ &= -\left(k_S^2 - k_I k_S \cos(2\theta)\right)\end{aligned} \tag{9.8.5c}$$

Again, $k_I = k_S$, so Eq. (9.8.5c) can be written as

$$\mathbf{q} \cdot (\mathbf{k}_S) = \quad k_I^2(1 - \cos(2\theta)) \tag{9.8.5d}$$

Since the projection of $\mathbf{q}$ on $\mathbf{k}_I$ is the same as the projection of $\mathbf{q}$ on $\mathbf{k}_S$, $\mathbf{q}$ bisects $\mathbf{k}_I$ and $\mathbf{k}_S$. ■

Theorem 9.8.3 Bragg's Law, $2d_{hk\ell} \sin\theta = n\lambda$.

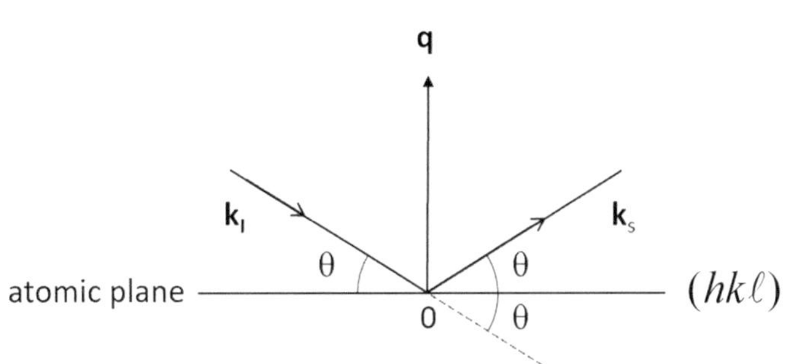

Fig. 9.10 Geometric relation between the propagation vector, $\mathbf{k}_I$, of the incident radiation, and the propagation vector, $\mathbf{k}_S$, of the scattered radiation. The scattering angle is 2θ. The scattering vector, $\mathbf{q} = \mathbf{k}_I - \mathbf{k}_S$, which is normal to the atomic $(hk\ell)$ plane bisects the angle, between $\mathbf{k}_I$ and $\mathbf{k}_S$

Proof Consider a plane that cuts the $\mathbf{a}_1$ axis at $\left(\frac{n\mathbf{a}_1}{h}\right)$ as shown in Fig. 9.8. By virtue of Theorem 9.8.1 of this section, the unit vector, $\left(\frac{\mathbf{q}}{q}\right)$, is perpendicular to the $(hk\ell)$ plane. With reference to Fig. 9.7, the projection of $\left(\frac{n\mathbf{a}_1}{h}\right)$ onto $\left(\frac{\mathbf{q}}{q}\right)$ equals the interplanar distance, $d_{hk\ell}$. Hence,

$$d_{hk\ell} = \frac{n\mathbf{a}_1}{h} \cdot \frac{\mathbf{q}}{q} = \left(\frac{\mathbf{a}_1 \cdot \mathbf{q}}{h}\right)\left(\frac{n}{q}\right) = \left(\frac{2\pi h}{h}\right)\left(\frac{n}{q}\right) = \frac{2\pi n}{q} = \frac{2\pi n}{2k_I \sin\theta} = \frac{\pi n}{k_I \sin\theta} \tag{9.8.6a}$$

where we used Eqs. (9.2.3) and (9.8.3a). Upon substitution of $k_I = 2\pi/\lambda$, Eq. (9.8.6a) becomes

$$d_{hk\ell} = \frac{\pi n}{(2\pi/\lambda)\sin\theta} \tag{9.8.6b}$$

which can be rewritten as,

$$2d_{hk\ell}\sin\theta = n\lambda \tag{9.8.7}$$

Equation (9.8.7) is known as Bragg's law. ■

Theorem 9.8.4 A statement equivalent to Bragg's law is $2\mathbf{k}_S \cdot \mathbf{G}_{hk\ell} = |\mathbf{G}_{hk\ell}|^2$.

Proof Recall from Theorem 9.8.1 of this section that $\mathbf{q}$ is perpendicular to the $(hk\ell)$ plane. Also recall from Theorem 9.8.2 of this section, that $\mathbf{q}$ bisects the angle between $\mathbf{k}_I$ and $\mathbf{k}_S$. Since by virtue of theorem 9.7.1, $\mathbf{G}_{hk\ell}$ is perpendicular to the $(hk\ell)$ plane, it is parallel to $\mathbf{q}$ and must necessarily also bisect the angle between $\mathbf{k}_I$ and $\mathbf{k}_S$. Thus, according to Fig. 9.11,

$$\mathbf{k}_S \cdot \mathbf{G}_{hk\ell} = |\mathbf{k}_S||\mathbf{G}_{hk\ell}|\cos\left(\frac{\pi}{2} - \theta\right) = |\mathbf{k}_S||\mathbf{G}_{hk\ell}|\sin\theta \tag{9.8.8a}$$

Solve Eq. (9.8.7), for

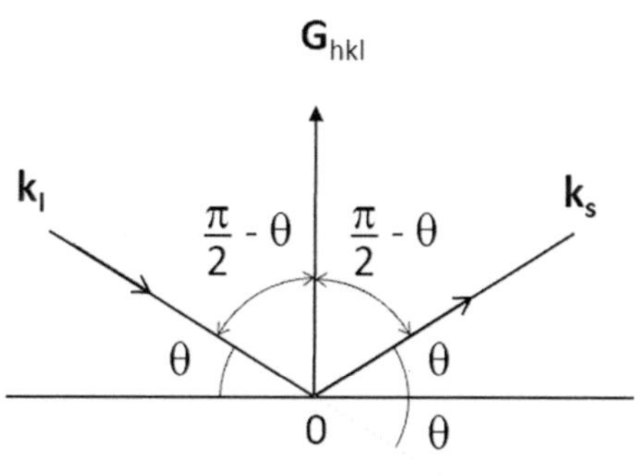

Fig. 9.11 Relation between the reciprocal lattice vector, $\mathbf{G}_{hk\ell}$, and the scattering plane in the space lattice. The wave vector for the incident radiation is $\mathbf{k}_I$. The wave vector for the scattered radiation is $\mathbf{k}_S$. The scattering angle is 2θ

$$\sin\theta = n\lambda/2d_{hk\ell} \tag{9.8.8b}$$

and substitute Eq. (9.8.8b) into Eq. (9.8.8a) to obtain

$$\mathbf{k}_S \cdot \mathbf{G}_{hk\ell} = |\mathbf{k}_S||\mathbf{G}_{hk\ell}|\left(\frac{n\lambda}{2d_{hk\ell}}\right) \tag{9.8.8c}$$

By using Eq. (9.7.7), Eq. (9.8.8c) reads

$$\mathbf{k}_S \cdot \mathbf{G}_{hk\ell} = |\mathbf{k}_S||\mathbf{G}_{hk\ell}|\left(\frac{n\lambda}{2}\right)\left(\frac{|\mathbf{G}_{hk\ell}|}{2\pi n}\right) \tag{9.8.8d}$$

If we introduce the definition $|\mathbf{k}_S| = 2\pi/\lambda$ into the right hand side of Eq. (9.8.8d), we obtain,

$$\mathbf{k}_S \cdot \mathbf{G}_{hk\ell} = |\mathbf{G}_{hk\ell}|\left(\frac{2\pi}{\lambda}\right)\left(\frac{n\lambda}{2}\right)\left(\frac{|\mathbf{G}_{hk\ell}|}{2\pi n}\right) = \frac{|\mathbf{G}_{hk\ell}|^2}{2} \tag{9.8.8e}$$

or what is the same thing,

$$2(\mathbf{k}_S \cdot \mathbf{G}_{hk\ell}) = |\mathbf{G}_{hk\ell}|^2 \quad \blacksquare \tag{9.8.9}$$

We conclude that the condition for constructive interference can be described either by identifying a plane in the space lattice [Eq. (9.8.7)] or by specifying a vector in the reciprocal lattice [Eq. (9.8.9)].

Example Construct the reciprocal lattice and determine the condition for constructive interference in the case of waves scattered from a square unit cell in a planar space lattice.

The basis vectors describing a square planar unit cell are

$$\mathbf{a}_1 = a\hat{\mathbf{a}}_1 \tag{9.8.10a}$$

and

$$\mathbf{a}_2 = a\hat{\mathbf{a}}_2 \tag{9.8.10b}$$

According to Eq. (9.6.5), which states that $\mathbf{a}_i \cdot \mathbf{b}_j = 2\pi\delta_{ij}$, the corresponding basis vectors for the unit cell in the reciprocal lattice are

$$\mathbf{b}_1 = (2\pi/a)\hat{\mathbf{a}}_1 \tag{9.8.11a}$$

and

$$\mathbf{b}_2 = (2\pi/a)\hat{\mathbf{a}}_2 \tag{9.8.11b}$$

Equations (9.8.11a) and (9.8.11b) describe a unit cell in the reciprocal lattice which is a square of side $(2\pi/a)$. By virtue of the same pair of equations, a lattice vector in this reciprocal lattice is given by

$$\mathbf{G}_{hk} = h\mathbf{b}_1 + k\mathbf{b}_2 = \left(\frac{2\pi h}{a}\right)\hat{\mathbf{a}}_1 + \left(\frac{2\pi k}{a}\right)\hat{\mathbf{a}}_2 \tag{9.8.12a}$$

Two such vectors are

$$\mathbf{G}_{10} = \left(\frac{2\pi}{a}\right)\hat{\mathbf{a}}_1 \tag{9.8.12b}$$

and

$$\mathbf{G}_{01} = \left(\frac{2\pi}{a}\right)\hat{\mathbf{a}}_2 \tag{9.8.12c}$$

We note that in the case of the three-dimensional space lattice, parallel planes of atoms are responsible for the scattering. By contrast, in the case of the two-dimensional space lattice, lines of atoms are responsible for the scattering. For constructive interference to occur from the $(hk) = (10)$ line, Eq. (9.8.9) implies that $\mathbf{k}_S$ should satisfy the equation,

$$2\mathbf{k}_S \cdot \mathbf{G}_{10} = |\mathbf{G}_{10}|^2 \tag{9.8.13a}$$

or upon substitution of Eq. (9.8.12b)

$$2\mathbf{k}_S \cdot \left(\frac{2\pi}{a}\right)\hat{\mathbf{a}}_1 = \left(\frac{2\pi}{a}\right)\hat{\mathbf{a}}_1 \cdot \left(\frac{2\pi}{a}\right)\hat{\mathbf{a}}_1 \tag{9.8.13b}$$

which simplifies to read

$$\mathbf{k}_S \cdot \hat{\mathbf{a}}_1 = \left(\frac{\pi}{a}\right) \tag{9.8.13c}$$

Hence, for there to be constructive interference in the case of wave scattered into the $[hk]$=[10] direction, the projection of the scattering vector, $\mathbf{k}_S$, onto the $\hat{\mathbf{a}}_1$-direction must be equal to π/a.

9.9 Evaluation of the Structure Factor for a Fixed Array of Scattering Centers

9.9.1 Incoherent Versus Coherent Scattering

We can show that the condition for constructive interference is

$$\mathbf{q} = \mathbf{G}_{hk\ell} \tag{9.9.1}$$

The vector, $\mathbf{r}_j$, in Eq. (9.1.4) runs over the space lattice vectors, $\mathbf{R}_c$. If we use Eq. (9.9.1), Eq. (9.1.4) reads

$$S(\mathbf{G}_{hk\ell}) = \sum_c \exp[i(\mathbf{G}_{hk\ell} \cdot \mathbf{R}_c)] \tag{9.9.2}$$

The reciprocal lattice vector, $\mathbf{G}_{hk\ell}$, in Eq. (9.7.1) is specified by the integers, h, k, and ℓ, whereas $\mathbf{R}_c$ in Eq. (9.8.1) is defined in terms of the rational numbers, m_c, n_c, and p_c. The typical term in Eq. (9.9.2) is of the form,

$$\begin{aligned}\exp[i(\mathbf{G}_{hk\ell} \cdot \mathbf{R}_c)] &= \exp\left[i(h\mathbf{b}_1 + k\mathbf{b}_2 + \ell\mathbf{b}_3\mathbf{R}_c) \cdot (m_c\mathbf{a}_1 + n_c\mathbf{a}_2 + p_c\mathbf{a}_3)\right] \\ &= \exp\left[2\pi i(hm_c + kn_c + \ell p_c)\right]\end{aligned} \tag{9.9.3}$$

where we have used Eq. (9.6.5). The origin of coordinates corresponds with $(m_c, n_c, p_c) = (0, 0, 0)$. According to Eq. (9.8.1), the atom at the origin is located by the lattice vector, $\mathbf{R}_c = 0$. Thus, when $(m_c, n_c, p_c) = (0, 0, 0)$, the radiation scattered from an atom constructively interferes with itself. This is known as *incoherent* scattering. By contrast, when one or more of (m_c, n_c, p_c) is non-zero, the constructive interference involves waves scattered from two different atoms. This type of scattering is said to be *coherent*. The identification in Eq. (9.9.1) makes Eq. (9.9.3) consistent with Eq. (9.8.2) and the von Laue conditions expressed by Eqs. (9.8.3a) through (9.8.3c).

9.9.2 Simple Cubic Unit Cell

The lattice vectors in the simple cubic space lattice are determined by linear combinations of the three mutually perpendicular unit vectors, $\hat{\mathbf{a}}_1$, $\hat{\mathbf{a}}_2$, and $\hat{\mathbf{a}}_3$. The basis vectors for the cube are given by Eqs. (9.6.7a) through (9.6.7c). Each atom in the unit cell has its center located at one of these corners of the cube. Measured from the origin of coordinates, which is ordinarily taken to be the lower, back, left hand corner of the cube, the position of any other atom in the unit cell is determined by the appropriate vector, $\mathbf{R}_c$, as given by Eq. (9.8.1).

In the case of the simple cubic lattice, the index, c, identifying the atomic positions in the unit cell runs over the triplets, $(m_c, n_c, p_c) = (0, 0, 0), (1, 0, 0), (0, 1, 0), (0, 0, 1), (1, 1, 0), (1, 0, 1), (0, 1, 1)$ and $(1, 1, 1)$, respectively. For example, the lattice vector corresponding to the triplet, $c = (1, 1, 1)$ is

$$\mathbf{R}_c = a\hat{\mathbf{a}}_1 + a\hat{\mathbf{a}}_2 + a\hat{\mathbf{a}}_3 \tag{9.9.4a}$$

The reciprocal lattice vector corresponding to the $(hk\ell)$ plane is

$$\mathbf{G}_{hk\ell} = h\mathbf{b}_1 + k\mathbf{b}_2 + \ell\mathbf{b}_3 = \frac{2\pi}{a}(h\hat{\mathbf{a}}_1 + k\hat{\mathbf{a}}_2 + \ell\hat{\mathbf{a}}_3) \tag{9.9.4b}$$

where we used Eq. (9.7.1) and Eqs. (9.6.9a) through (9.6.9c). Equations (9.9.4a) and (9.9.4b) imply that

$$\begin{aligned}\mathbf{G}_{hk\ell} \cdot \mathbf{R}_c &= \frac{2\pi}{a}(h\hat{\mathbf{a}}_1 + k\hat{\mathbf{a}}_2 + \ell\hat{\mathbf{a}}_3) \cdot (a\hat{\mathbf{a}}_1 + a\hat{\mathbf{a}}_2 + a\hat{\mathbf{a}}_3) \\ &= 2\pi(h + k + \ell)\end{aligned} \tag{9.9.4c}$$

where we used $(\hat{\mathbf{a}}_i \cdot \hat{\mathbf{a}}_j) = \delta_{ij}$. Eight unit cells meet at a corner of the cube in the simple cubic lattice, so that an atom at a corner is part of eight unit cells. In terms of atomic volume, each corner atom contributes one eighth of its volume and hence one eighth of its scattering power to each unit cell. The typical term in $S(\mathbf{G}_{hk\ell})$ is thus the exponential in Eq. (9.9.2) multiplied by (1/8). We write

$$\left(\frac{1}{8}\right)\exp[i\mathbf{G}_{hk\ell} \cdot \mathbf{R}_c] = \left(\frac{1}{8}\right)\exp[2\pi i(h + k + \ell)] \tag{9.9.4d}$$

where we have used Eq. (9.9.4c). To evaluate $S(\mathbf{q})$ in Eq. (9.9.2) with $\mathbf{q} = \mathbf{G}_{hk\ell}$, we must sum terms of the type in Eq. (9.9.4d) over all allowed values of c. The result is

$$\begin{aligned}S(\mathbf{G}_{hk\ell}) &= \sum_{c=1}^{8}\left(\frac{1}{8}\right)\exp(i\mathbf{G}_{hk\ell} \cdot \mathbf{R}_c) \\ &= \sum_{\{m_c,n_c,p_c\}}^{\{1,1,1\}}\left(\frac{1}{8}\right)\exp\big[2\pi i(hm_c + kn_c + \ell p_c)\big] \\ &= \frac{1}{8}e^{2\pi i0} + \frac{1}{8}e^{2\pi ih} + \frac{1}{8}e^{2\pi ik} + \frac{1}{8}e^{2\pi i\ell} + \frac{1}{8}e^{2\pi i(h+k)} \\ &\quad + \frac{1}{8}e^{2\pi i(h+\ell)} + \frac{1}{8}e^{2\pi i(k+\ell)} + \frac{1}{8}e^{2\pi i(h+k+\ell)} \\ &= \frac{1}{8} + \frac{1}{8}e^{2\pi ih} + \frac{1}{8}e^{2\pi ik} + \frac{1}{8}e^{2\pi i\ell} + \frac{1}{8}e^{2\pi i(h+k)}\end{aligned}$$

$$+\frac{1}{8}e^{2\pi i(h+\ell)}+\frac{1}{8}e^{2\pi i(k+\ell)}+\frac{1}{8}e^{2\pi i(h+k+\ell)} \tag{9.9.5a}$$

The Miller indices, h,k, and ℓ, of a plane are all whole numbers. Hence for any combination of h,k, and ℓ, every exponential in Eq. (9.9.5a) is unity. We conclude that

$$S(\mathbf{G}_{hk\ell})=8\left(\frac{1}{8}\right)=1 \tag{9.9.5b}$$

One should note that the same value of $S(\mathbf{G}_{hk\ell})$ is obtained if we replace the fractional atom at each corner by a full atom at the origin (0, 0, 0). This arrangement is said to be a simple cubic unit cell with a basis. Equation (9.9.5a) then reads,

$$S(\mathbf{G}_{hk\ell})=\exp(i\mathbf{G}_{hk\ell}\cdot 0)=1 \tag{9.9.6}$$

For a cubic unit cell, $a_1=a_2=a_3=a$; hence, according to Eq. (9.5.6) with $n=1$, the spacing between the scattering planes is

$$d_{hk\ell}=\frac{a}{\sqrt{h^2+k^2+\ell^2}} \tag{9.9.7}$$

In order to define a plane, at least one of the h, k, and ℓ indices must be unity. Thus taking into account the possible combinations of integers, including zero, which are capable of scattering, leads to the sum of squares: $h^2+k^2+\ell^2=1, 2, 3, 4, 5, 6, 8, 9, \ldots$. The value, $h^2+k^2+\ell^2=7$, however, is missing. The absence from the scattering pattern of a wavelength, $\lambda_{hk\ell}$, corresponding to $h^2+k^2+\ell^2=7$ is proof that the crystal is simple cubic. The $\{hk\ell\}$ scattering planes fall into three classes. In the first class are the planes separated by the distance d_{100}, in the second class are the planes separated by the distance d_{110}, and in the third class are those separated by the distance d_{111}. According to Eq. (9.9.7), these distances are in the ratio $\frac{d_{100}}{d_{100}}=1$, $\frac{d_{110}}{d_{100}}=\frac{1}{\sqrt{2}}$, $\frac{d_{111}}{d_{100}}=\frac{1}{\sqrt{3}}$. In the case of a fixed wavelength, Bragg's law, Eq. (9.8.7), predicts that the scattering angles are in the ratio $\frac{\sin\theta_{100}}{\sin\theta_{100}}=1$, $\frac{\sin\theta_{110}}{\sin\theta_{100}}=\sqrt{2}$, $\frac{\sin\theta_{111}}{\sin\theta_{100}}=\sqrt{3}$.

9.9.3 Body Centered Cubic Unit Cell

Like the simple cubic unit cell, the body centered cubic unit cell contains 1/8th of an atom located at each cube corner. The corners are located by the space lattice coordinates, $(m_c, n_c, p_c)=(0,0,0), (1,0,0), (0,1,0), (0,0,1), (1,1,0), (1,0,1), (0,1,1)$ and $(1,1,1)$. In contrast to the simple cubic unit cell, there is a full atom at the body center of the cube. The coordinates of the body center are $(m_c, n_c, p_c)=(1/2, 1/2, 1/2)$. As in the case of the cubic unit cell with a basis, the body centered arrangement can be replaced by a body centered cubic unit cell with a basis, the basis being a full atom at $(0, 0, 0)$ and a second full atom located at $(1/2, 1/2, 1/2)$. With the addition of a term associated with the presence of the central atom, the value of $S(\mathbf{G}_{hk\ell})$ can be calculated in the same way as for the simple cubic unit cell.The analog of Eq. (9.9.6) is

$$S(\mathbf{G}_{hk\ell}) = 1 + \exp\left[\frac{2\pi}{a}(h\hat{\mathbf{a}}_1 + k\hat{\mathbf{a}}_2 + \ell\hat{\mathbf{a}}_3)\cdot\left(\frac{a}{2}\hat{\mathbf{a}}_1 + \frac{a}{2}\hat{\mathbf{a}}_2 + \frac{a}{2}\hat{\mathbf{a}}_3\right)\right] \tag{9.9.8}$$

or

$$S(\mathbf{G}_{hk\ell}) = 1 + \exp[\pi i(h + k + \ell)] \tag{9.9.9}$$

The first term in Eq. (9.9.9) accounts for the atom located at the corner. The exponential takes into account the scattering from the atom at the body center.

According to Eq. (9.9.9), destructive interference occurs when $(h + k + \ell) =$ an odd integer in which case $S(\mathbf{G}_{hk\ell}) = 0$. Constructive interference occurs when $(h + k + \ell) =$ an even integer. In this case, $S(\mathbf{G}_{hk\ell}) = 2$. Constructive interference involves planes, $d_{hk\ell}$, for which $h^2 + k^2 + \ell^2 = 2, 4, 6, 8, 10, 12, 14, 16, \ldots$ This combination distinguishes the body centered cubic lattice from the simple cubic lattice.

According to Eq. (9.9.7), in the body centered cubic, three of the planes having the widest separation are characterized by the distances, d_{110}, d_{200}, and d_{211}. Compared to d_{110}, the distances are in the ratio $\frac{d_{110}}{d_{110}} = 1, \frac{d_{200}}{d_{110}} = \frac{1}{\sqrt{2}}, \frac{d_{211}}{d_{110}} = \frac{1}{\sqrt{3}}$.The corresponding scattering angles are in the ratio, $\frac{\sin\theta_{110}}{\sin\theta_{110}} = 1, \frac{\sin\theta_{200}}{\sin\theta_{100}} = \sqrt{2}, \frac{\sin\theta_{211}}{\sin\theta_{110}} = \sqrt{3}$.

9.9.4 Face Centered Cubic Unit Cell

Like the simple cubic unit cell, the face centered cubic unit cell has 1/8th of an atom at each corner of the cube at the space lattice coordinates, $(m_c, n_c, p_c) = (0, 0, 0), (1, 0, 0), (0, 1, 0), (0, 0, 1), (1, 1, 0), (1, 0, 1), (0, 1, 1)$ and $(1, 1, 1)$. In contrast to the simple cubic unit cell, however, in the face centered cubic unit cell there are atomic locations on the face centers. Half of each face centered atom lies within the unit cell. The other half lies within the neighboring unit cell. The atoms at the face centers are located by the coordinates, $(m_c, n_c, p_c) = (1/2, 1/2, 0), (1/2, 1/2, 1), (0, 1/2, 1/2), (1, 1/2, 1/2), (1/2, 0, 1/2)$, and $(1/2, 1, 1/2)$. The face centered cubic unit cell with a basis consists of a full atom located at $(0, 0, 0)$ and full atoms located on half of the six faces. These locations can be chosen to be $(1/2, 1/2, 0), (1/2, 0, 1/2)$, and $(0, 1/2, 1/2)$. Evaluated for atoms at these locations, $S(\mathbf{G}_{hk\ell})$ has the form,

$$S(\mathbf{G}_{hk\ell}) = 1 + \exp\left[2\pi i\left(\frac{h}{2} + \frac{k}{2}\right)\right] + \exp\left[2\pi i\left(\frac{h}{2} + \frac{\ell}{2}\right)\right] + \exp\left[2\pi i\left(\frac{k}{2} + \frac{\ell}{2}\right)\right] \tag{9.9.10}$$

Equation (9.9.10) simplifies to read,

$$S(\mathbf{G}_{hk\ell}) = 1 + \exp[i\pi(h+k)] + \exp[i\pi(k+\ell)] + \exp[i\pi(h+\ell)] \tag{9.9.11}$$

According to Eq. (9.9.11), constructive interference occurs when h, k, and ℓ are all even or all odd, for which $S(\mathbf{G}_{hk\ell}) = 4$. For the purposes of this calculation, zero is an even integer. Destructive interference occurs when h, k, and ℓ are mixed, for which $S(\mathbf{G}_{hk\ell}) = 0$. Constructive interference thus corresponds to $h^2 + k^2 + \ell^2 = 3$, 4, 8, 11, ..., which distinguishes face centered cubic from both the body centered cubic and the simple cubic space lattices.

According to Eq. (9.9.7), in the face centered cubic system, the most widely spaced scattering planes correspond to (111), (200) and (220) The corresponding separations d_{111}, d_{200}, and d_{220} are in the ratio, $\frac{d_{111}}{d_{111}} = 1, \frac{d_{200}}{d_{111}} = \frac{\sqrt{3}}{2}, \frac{d_{220}}{d_{111}} = \frac{\sqrt{3}}{2\sqrt{2}}$. The corresponding scattering angles are in the ratio $\frac{\sin\theta_{110}}{\sin\theta_{110}} = 1, \frac{\sin\theta_{200}}{\sin\theta_{111}} = \frac{2}{\sqrt{3}}, \frac{\sin\theta_{220}}{\sin\theta_{111}} = \frac{2\sqrt{2}}{\sqrt{3}}$.

Recommended Additional Reading

1. R.J. Naumann, *Introduction to the Physics and Chemistry of Materials* (CRC Press, Boca Raton, FL, 2009)
2. N.W. Ashcroft, N.D. Mermin, *Solid State Physics* (Saunders College, Philadelphia, PA, 1976)

Chapter 10
Classical Mechanics and the Electromagnetic Field

The goal of this chapter is to derive the formula for the electromagnetic Hamiltonian. This function describes the energy of interaction of a charged particle with the electromagnetic field. We start by reformulating Newton's second law of motion in terms of Hamilton's dynamical equations. The electric and magnetic fields are expressed in terms of the scalar potential and the vector potential, respectively, which were introduced in Chap. 8. Upon including these potentials in the equations of motion, we are led to the Lagrange function, which is the difference between the kinetic and potential energies. From Lagrange's function, we derive the Hamiltonian function which determines the total energy. This Hamiltonian will serve as the basis for our subsequent calculation of the quantum effects associated with the interaction of a charged particle with an electromagnetic field.

10.1 Introduction

Classical mechanics might seem to be a subject far afield from either electromagnetism or chemistry. Nonetheless, when a charged particle, such as an electron in a molecule, is acted on by an external electromagnetic field, mechanics comes into play. Fortunately, the whole of classical mechanics is not required, so this chapter is short. The goal of the chapter is to learn how the energy of a charged particle depends upon the space and time dependence of the electromagnetic field. The function which determines this energy is known as the Hamiltonian. It can be derived as a consequence of Newton's second law of motion.

J. K. Baird, *Electromagnetism for Chemistry*,
https://doi.org/10.1007/978-3-031-88508-2_10

10.2 Newton's Second Law

Newton's second law (Sir Isaac Newton, English, 1642–1727) governing motion in one dimension states that the force, F, required to produce an acceleration, a, of a particle of mass, m, is given by

$$F = ma \tag{10.2.1}$$

The particle's location is specified by the space coordinate, q, and its speed, v, is given by the derivative,

$$v = \frac{dq}{dt} \tag{10.2.2}$$

The linear momentum, p, is defined by

$$p = mv = m\frac{dq}{dt} \tag{10.2.3}$$

The acceleration is

$$a = \frac{dv}{dt} = \frac{d^2q}{dt^2} \tag{10.2.4}$$

With these definitions, Newton's law can be rendered in the familiar form,

$$m\frac{d^2q}{dt^2} = \frac{dp}{dt} = F \tag{10.2.5}$$

When many particles are interacting with one another and moving in three dimensions, Eq. (10.2.5) gets replaced by a system of differential equations determining the various momenta in terms of the various forces. Identification of these forces by examining the physical situation is often a challenge.

If the forces are conservative, however, the force identification problem can be greatly simplified. The task is replaced by constructing a potential energy function from which the forces can be derived by differentiation. For motion in one dimension, the potential energy is $V = V(q)$, and the force is

$$F = -\frac{\partial V}{\partial q} \tag{10.2.6}$$

In the case of Eq. (10.2.6), Newton's second law reads,

$$m\frac{d^2q}{dt^2} = \frac{dp}{dt} = -\frac{\partial V(q)}{\partial q} \tag{10.2.7}$$

Equation (10.2.7) is a second order ordinary differential equation to be solved for $q = q(t)$. By introducing a total energy function, called the Hamiltonian, we show in the next section that Eq. (10.2.7) can be replaced by two first order ordinary differential equations.

10.3 Hamilton's Equations

The kinetic energy of a particle of mass, m, is

$$KE = \frac{1}{2}mv^2 = \frac{p^2}{2m} \tag{10.3.1}$$

where we have used Eq. (10.2.3) for p. When the forces are derivable from a potential, as in Eq. (10.2.6), the sum of the kinetic energy, $p^2/2m$, and the potential energy, V, equals the total energy. In terms of the momentum and space coordinates, the total energy is

$$H(p, q) = \frac{p^2}{2m} + V(q) \tag{10.3.2}$$

where $H = H(p, q)$ is known as the Hamiltonian (Sir William Rowan Hamilton, Irish, 1805–1865).

As mentioned, Newton's law, Eq. (10.2.7), is a second order ordinary differential equation satisfied by $q = q(t)$. In Hamiltonian mechanics, Eq. (10.2.7) is replaced by two coupled first order ordinary differential equations involving derivatives of $H(p, q)$, where $p = p(t)$ and $q = q(t)$ are regarded as independent variables. To determine these differential equations, we first differentiate Eq. (10.3.2) with respect to q. Since $(\partial H/\partial q) = (\partial V/\partial q)$, Eq. (10.2.7) can be rewritten as

$$\frac{dp}{dt} = -\frac{\partial H}{\partial q} \tag{10.3.3}$$

Equation (10.3.3) is a differential equation determining the time dependence of p. To derive the differential equation specifying the time dependence of q, we differentiate Eq. (10.3.2) with respect to p. Since $(\partial H/\partial p) = p/m$, Eq. (10.2.3) becomes,

$$\frac{dq}{dt} = \frac{\partial H}{\partial p} \tag{10.3.4}$$

Equations (10.3.3) and (10.3.4) are known as Hamilton's equations of motion. If the initial conditions, $q = q(0)$ and $p = p(0)$, are known, Eqs. (10.3.3) and (10.3.4) can be integrated to obtain $q = q(t)$ and $p = p(t)$. The pair, $q = q(t)$ and $p = p(t)$, are said to be conjugate variables. One of the purposes of Eqs. (10.3.3) and (10.3.4) is to suppress the concept of force in favor of the concept of energy. The use of $H = H(p, q)$ also serves to place the momentum, p, and the space coordinate, q, on equal footing.

10.4 Lagrange's Equation

Lagrange discovered another way of avoiding direct reference to the force. This method begins with the definition of a function, L,

$$L = \frac{1}{2}m\left(\frac{dq}{dt}\right)^2 - V(q) \tag{10.4.1}$$

known as the Lagrangian. In the Lagrangian, the independent variables are considered to be the position coordinate, q, and the speed, $\dot{q}$, where $\dot{q} = dq/dt$. In this notation, Eq. (10.4.1) reads,

$$L(q, \dot{q}) = \frac{1}{2}m\dot{q}^2 - V(q) \tag{10.4.2}$$

The momentum can be calculated from Eq. (10.4.2) using the definition in Eq. (10.2.3) and the derivative,

$$p = \frac{\partial L}{\partial \dot{q}} \tag{10.4.3}$$

The corresponding force is computed in the form of the derivative,

$$\frac{\partial L}{\partial q} = -\frac{\partial V}{\partial q} = F \tag{10.4.4}$$

By substituting Eqs. (10.4.3) and (10.4.4), we can rewrite Eq. (10.2.7) for Newton's law in the form,

$$\frac{d}{dt}\left(\frac{\partial L}{\partial \dot{q}}\right) - \left(\frac{\partial L}{\partial q}\right) = 0 \tag{10.4.5}$$

Equation (10.4.5), is known as Lagrange's equation. Both q and $\dot{q}$ are considered to be functions of the time.

A useful relation between the Hamiltonian and the Lagrangian can be obtained by adding Eqs. (10.3.2) and (10.4.2). The result is

$$H + L = \frac{p^2}{2m} + V + \frac{1}{2}m\dot{q}^2 - V = \frac{pm\dot{q}}{2m} + \frac{p\dot{q}}{2} = p\dot{q} \tag{10.4.6}$$

where we have used Eq. (10.2.3) with $p = m\dot{q}$. Equation (10.4.6) simplifies to

$$H = \dot{q}p - L \tag{10.4.7}$$

10.5 The Electromagnetic Hamiltonian

We now turn to the case of particle motion in three dimensions. For a particle of mass, m, traveling with velocity, $\mathbf{v}$, the product, $m\mathbf{v}$, is said to be the kinetic momentum. If the particle has charge, e, and is acted on simultaneously by an electric field, $\mathbf{E}$, and a magnetic field, $\mathbf{H} = \mathbf{B}/\mu$, Newton's law takes the form,

$$m\frac{d\mathbf{v}}{dt} = e[\mathbf{E} + (\mathbf{v} \times \mathbf{B})] \tag{10.5.1}$$

where $e(\mathbf{v} \times \mathbf{B})$ is known as the Lorentz force. This force acts in a sense perpendicular to the direction of motion of the particle and perpendicular to the direction of the magnetic field. In terms of the vector potential, $\mathbf{A}$,

$$\mathbf{B} = \nabla \times \mathbf{A} \tag{10.5.2}$$

According to Chap. 8, Eq. (8.4.3), the electric field can be written in terms of the scalar potential, Φ, and the vector potential, $\mathbf{A}$, as

$$\mathbf{E} = -\nabla\Phi - \frac{\partial \mathbf{A}}{\partial t} \tag{10.5.3}$$

Upon substitution of Eqs. (10.5.2) and (10.5.3) into Eq. (10.5.1), the result is

$$m\frac{d\mathbf{v}}{dt} = e\left[-\nabla\Phi - \frac{\partial \mathbf{A}}{\partial t} + (\mathbf{v} \times (\nabla \times \mathbf{A}))\right] \tag{10.5.4}$$

The components of the velocity vector are $\mathbf{v} = (v_1, v_2, v_3)$. The components of the vector potential are $\mathbf{A} = (A_1, A_2, A_3)$. In order to write out the vector quantity, $\mathbf{v} \times (\nabla \times \mathbf{A})$, in terms of its Cartesian components, we use the determinant expansion of $\nabla \times \mathbf{A}$, which is

$$\begin{aligned}\nabla \times \mathbf{A} &= \begin{vmatrix} \hat{\mathbf{e}}_1 & \hat{\mathbf{e}}_2 & \hat{\mathbf{e}}_3 \\ \frac{\partial}{\partial x_1} & \frac{\partial}{\partial x_2} & \frac{\partial}{\partial x_3} \\ A_1 & A_2 & A_3 \end{vmatrix} \\ &= \hat{\mathbf{e}}_1\left(\frac{\partial A_3}{\partial x_2} - \frac{\partial A_2}{\partial x_3}\right) + \hat{\mathbf{e}}_2\left(\frac{\partial A_1}{\partial x_3} - \frac{\partial A_3}{\partial x_1}\right) + \hat{\mathbf{e}}_3\left(\frac{\partial A_2}{\partial x_1} - \frac{\partial A_1}{\partial x_2}\right)\end{aligned} \tag{10.5.5}$$

By employing the unit vector cross products,

$$\hat{\mathbf{e}}_1 \times \hat{\mathbf{e}}_2 = -\hat{\mathbf{e}}_2 \times \hat{\mathbf{e}}_1 = \hat{\mathbf{e}}_3 \tag{10.5.6a}$$

$$\hat{\mathbf{e}}_1 \times \hat{\mathbf{e}}_3 = -\hat{\mathbf{e}}_3 \times \hat{\mathbf{e}}_1 = -\hat{\mathbf{e}}_2 \tag{10.5.6b}$$

$$\hat{\mathbf{e}}_2 \times \hat{\mathbf{e}}_3 = -\hat{\mathbf{e}}_3 \times \hat{\mathbf{e}}_2 = \hat{\mathbf{e}}_1 \tag{10.5.6c}$$

the evaluation of $\mathbf{v} \times (\nabla \times \mathbf{A})$, proceeds as follows:

$$\begin{aligned} \mathbf{v} \times (\nabla \times \mathbf{A}) &= (v_1\hat{\mathbf{e}}_1 + v_2\hat{\mathbf{e}}_2 + v_3\hat{\mathbf{e}}_3) \\ &\quad \times \left[\left(\hat{\mathbf{e}}_1 \frac{\partial A_3}{\partial x_2} - \frac{\partial A_2}{\partial x_3}\right) + \hat{\mathbf{e}}_2\left(\frac{\partial A_1}{\partial x_3} - \frac{\partial A_3}{\partial x_1}\right) + \hat{\mathbf{e}}_3\left(\frac{\partial A_2}{\partial x_1} - \frac{\partial A_1}{\partial x_2}\right)\right] \end{aligned} \tag{10.5.7a}$$

The result is

$$\begin{aligned} \mathbf{v} \times (\nabla \times \mathbf{A}) = \hat{\mathbf{e}}_3 v_1 \left(\frac{\partial A_1}{\partial x_3} - \frac{\partial A_3}{\partial x_1}\right) + \hat{\mathbf{e}}_2 v_1 \left(\frac{\partial A_1}{\partial x_2} - \frac{\partial A_2}{\partial x_1}\right) \\ + \hat{\mathbf{e}}_3 v_2 \left(\frac{\partial A_2}{\partial x_3} - \frac{\partial A_3}{\partial x_2}\right) + \hat{\mathbf{e}}_1 v_2 \left(\frac{\partial A_2}{\partial x_1} - \frac{\partial A_1}{\partial x_2}\right) \\ + \hat{\mathbf{e}}_2 v_3 \left(\frac{\partial A_3}{\partial x_2} - \frac{\partial A_2}{\partial x_3}\right) + \hat{\mathbf{e}}_1 v_3 \left(\frac{\partial A_3}{\partial x_1} - \frac{\partial A_1}{\partial x_3}\right) \end{aligned} \tag{10.5.7b}$$

According to Eq. (10.5.7b), the individual components of $\mathbf{v} \times (\nabla \times \mathbf{A})$ are

$$[\mathbf{v} \times (\nabla \times \mathbf{A})]_1 = \hat{\mathbf{e}}_1 v_2 \left(\frac{\partial A_2}{\partial x_1} - \frac{\partial A_1}{\partial x_2}\right) + \hat{\mathbf{e}}_1 v_3 \left(\frac{\partial A_3}{\partial x_1} - \frac{\partial A_1}{\partial x_3}\right) \tag{10.5.8a}$$

$$[\mathbf{v} \times (\nabla \times \mathbf{A})]_2 = \hat{\mathbf{e}}_2 v_1 \left(\frac{\partial A_1}{\partial x_2} - \frac{\partial A_2}{\partial x_1}\right) + \hat{\mathbf{e}}_2 v_3 \left(\frac{\partial A_3}{\partial x_2} - \frac{\partial A_2}{\partial x_3}\right) \tag{10.5.8b}$$

$$[\mathbf{v} \times (\nabla \times \mathbf{A})]_3 = \hat{\mathbf{e}}_3 v_1 \left(\frac{\partial A_1}{\partial x_3} - \frac{\partial A_3}{\partial x_1}\right) + \hat{\mathbf{e}}_3 v_2 \left(\frac{\partial A_2}{\partial x_3} - \frac{\partial A_3}{\partial x_2}\right) \tag{10.5.8c}$$

By reference to Eq. (10.5.4), the first component of the velocity vector, v_1, satisfies the differential equation,

$$m\frac{dv_1}{dt} = e\left[-\frac{\partial \Phi}{\partial x_1} - \frac{\partial A_1}{\partial t} + [\mathbf{v} \times (\nabla \times \mathbf{A})]_1\right] \tag{10.5.9a}$$

Upon substitution of Eq. (10.5.8a) into Eq. (10.5.9a), the result is

$$m\frac{dv_1}{dt} = -e\frac{\partial\Phi}{\partial x_1} + e\left[-\frac{\partial A_1}{\partial t} + v_2\left(\frac{\partial A_2}{\partial x_1} - \frac{\partial A_1}{\partial x_2}\right)\right.$$
$$\left. + v_3\left(\frac{\partial A_3}{\partial x_1} - \frac{\partial A_1}{\partial x_3}\right) + v_1\left(\frac{\partial A_1}{\partial x_1} - \frac{\partial A_1}{\partial x_1}\right)\right] \tag{10.5.9b}$$

The term, $v_1\left(\frac{\partial A_1}{\partial x_1} - \frac{\partial A_1}{\partial x_1}\right)$, which is identically zero, has been added within the bracket on the right hand side of Eq. (10.5.9b) in order to produce a symmetrical result. Its presence permits the terms in Eq. (10.5.9b) to be expressed in terms of a sum involving all three dimensions. The result is

$$m\frac{dv_1}{dt} = -e\frac{\partial\Phi}{\partial x_1} + e\left[-\frac{\partial A_1}{\partial t} + \sum_{j=1}^{3} v_j\left(\frac{\partial A_j}{\partial x_1} - \frac{\partial A_1}{\partial x_j}\right)\right] \tag{10.5.10}$$

The vector potential, $\mathbf{A} = \mathbf{A}(x_1(t), x_2(t), x_3(t); t)$, can involve both implicit and explicit time dependence. In Eq. (10.5.10), the effect of the explicit time dependence is taken into account by the derivative, $(\partial A_1/\partial t)$. The total derivative of $A_1 = A_1(x_1(t), x_2(t), x_3(t); t)$ is

$$\frac{dA_1}{dt} = \frac{\partial A_1}{\partial t} + \left(\frac{\partial A_1}{\partial x_1}\right)\frac{dx_1}{dt} + \left(\frac{\partial A_1}{\partial x_2}\right)\frac{dx_2}{dt} + \left(\frac{\partial A_1}{\partial x_3}\right)\frac{dx_3}{dt} \tag{10.5.11a}$$

The components of the particle velocity are $v_1 = (dx_1/dt)$, $v_2 = (dx_2/dt)$, and $v_3 = (dx_3/dt)$. These identifications permit Eq. (10.5.11a) to be rewritten as

$$\frac{dA_1}{dt} = \frac{\partial A_1}{\partial t} + v_1\left(\frac{\partial A_1}{\partial x_1}\right) + v_2\left(\frac{\partial A_1}{\partial x_2}\right) + v_3\left(\frac{\partial A_1}{\partial x_3}\right) \tag{10.5.11b}$$

Solve Eq. (10.5.11b) for $\partial A_i/\partial t$ and substitute the result into Eq. (10.5.10). The result is

$$m\frac{dv_1}{dt} = -e\frac{\partial\Phi}{\partial x_1} - e\frac{dA_1}{dt} + e\left[v_1\left(\frac{\partial A_1}{\partial x_1}\right) + v_2\left(\frac{\partial A_1}{\partial x_2}\right) + v_3\left(\frac{\partial A_1}{\partial x_3}\right)\right]$$
$$+ e\left[v_1\left(\frac{\partial A_1}{\partial x_1} - \frac{\partial A_1}{\partial x_1}\right) + v_2\left(\frac{\partial A_2}{\partial x_1} - \frac{\partial A_1}{\partial x_2}\right) + v_3\left(\frac{\partial A_3}{\partial x_1} - \frac{\partial A_1}{\partial x_3}\right)\right] \tag{10.5.11c}$$

where the group of terms within the second square bracket are the explicit representation of the sum in Eq. (10.5.10). When terms of opposite sign are cancelled, and $-e(dA_1/dt)$ is transferred to the left, Eq. (10.5.11c) reads,

$$m\frac{dv_1}{dt} + e\frac{dA_1}{dt} = -e\frac{\partial\Phi}{\partial x_1} + e\left[v_1\left(\frac{\partial A_1}{\partial x_1}\right) + v_2\left(\frac{\partial A_2}{\partial x_1}\right) + v_3\left(\frac{\partial A_3}{\partial x_1}\right)\right] \tag{10.5.11d}$$

By introducing the "generalized" or "canonical" momentum,

$$p_i = mv_i + eA_i \quad (i = 1, 2, 3) \tag{10.5.11e}$$

Equation (10.5.11d) can be generalized to read,

$$\frac{dp_i}{dt} = -e\frac{\partial \Phi}{\partial x_i} + e\sum_{j=1}^{3} v_j\left(\frac{\partial A_j}{\partial x_i}\right) \quad (i = 1, 2, 3) \tag{10.5.12}$$

Equation (10.5.12) takes into account the possibility of motion in each of the three dimensions. As defined by Eqs. (10.5.11e), p_i contains a mechanical or kinetic part, mv_i, and an electromagnetic part, eA_i. Equations (10.5.12) constitute Newton's law of motion for a charged particle interacting with electric and magnetic fields as derived from the scalar potential, Φ, and the vector potential, $\mathbf{A}$.

We now turn to the task of deriving explicit forms for the electromagnetic Lagrangian and the electromagnetic Hamiltonian. The process begins with rewriting Eq. (10.4.3) in terms of the i-th component of the velocity, $v_i = \dot{q}_i$:

$$\left(\frac{\partial L}{\partial v_i}\right) = p_i \quad (i = 1, 2, 3) \tag{10.5.13a}$$

Upon substitution of Eqs. (10.5.11e) into Eqs. (10.5.13a), the result is

$$\left(\frac{\partial L}{\partial v_i}\right) = p_i = mv_i + eA_i \quad (i = 1, 2, 3) \tag{10.5.13b}$$

A formula for the Lagrangian, $L = L(x_1, x_2, x_3; v_1, v_2, v_3)$, is obtained by integration of each of Eqs. (10.5.13b) with respect to v_i in order to obtain,

$$L = \frac{m}{2}\sum_{i=1}^{3} v_i^2 + e\sum_{i=1}^{3} v_i A_i + f(x_1, x_2, x_3) \tag{10.5.14}$$

where $f(x_1, x_2, x_3)$ is an arbitrary function of the space coordinates. Because $(\partial f/\partial v_i) = 0, f(x_1, x_2, x_3)$ serves as a "constant" of integration.

In Eq. (10.4.5), introduce $\dot{q}_i = v_i$ and $q_i = x_i$. The result is

$$\frac{d}{dt}\left(\frac{\partial L}{\partial v_i}\right) = \left(\frac{\partial L}{\partial x_i}\right) \quad (i = 1, 2, 3) \tag{10.5.15a}$$

Upon substitution of Eqs. (10.5.13a), Eqs. (10.5.15a) read

$$\frac{dp_i}{dt} = \left(\frac{\partial L}{\partial x_i}\right) \quad (i = 1, 2, 3) \tag{10.5.15b}$$

Compare Eqs. (10.5.15b) with Eqs. (10.5.12). The conclusion is

$$\left(\frac{\partial L}{\partial x_i}\right) = -e\frac{\partial \Phi}{\partial x_i} + e\sum_{j=1}^{3} v_j\left(\frac{\partial A_j}{\partial x_i}\right) \quad (i = 1, 2, 3) \tag{10.5.15c}$$

The integral of Eqs. (10.5.15c) with respect to the x_i $(i = 1, 2, 3)$ is

$$L = -e\Phi + e\sum_{j=1}^{3} v_j A_j + g(v_1, v_2, v_3) \tag{10.5.16}$$

where $g(v_1, v_2, v_3)$ is an arbitrary function of the components of the velocity vector. Because $(\partial g/\partial x_i) = 0$, $g(v_1, v_2, v_3)$ serves as a "constant" of integration. Equations (10.5.14) and (10.5.16) will be consistent, if we make the identifications,

$$f(x_1, x_2, x_3) = -e\Phi(x_1, x_2, x_3) \tag{10.5.17a}$$

and

$$g(v_1, v_2, v_3) = \frac{m}{2}\sum_{i=1}^{3} v_i^2 \tag{10.5.17b}$$

With these identifications, the Lagrangian assumes the form,

$$L = \frac{m}{2}\sum_{i=1}^{3} v_i^2 + e\sum_{i=1}^{3} v_i A_i - e\Phi \tag{10.5.18}$$

In vector dot product notation, $\sum_{i=1}^{3} v_i A_i = \mathbf{v} \cdot \mathbf{A}$. The components of $\mathbf{v}$ satisfy $v^2 = v_1^2 + v_2^2 + v_3^2$, where v is the magnitude of $\mathbf{v}$. These identifications permit Eq. (10.5.18) to be rendered in vector notation as

$$L = \frac{1}{2}mv^2 + e\mathbf{v} \cdot \mathbf{A} - e\Phi \tag{10.5.19}$$

Taking into account all three components of each of the vectors, $\mathbf{p} = (p_1, p_2, p_2)$, $\mathbf{v} = (v_1, v_2, v_3)$, and $\mathbf{A} = (A_1, A_2, A_3)$, we can rewrite Eqs. (10.5.11e) as

$$\mathbf{p} = m\mathbf{v} + e\mathbf{A} \tag{10.5.20}$$

In Eq. (10.5.20), $m\mathbf{v}$ is the kinetic momentum, and $e\mathbf{A}$ is the momentum carried by the electromagnetic field. In vector notation, Eq. (10.4.7) assumes the form,

$$H = \mathbf{v} \cdot \mathbf{p} - L \tag{10.5.21}$$

Upon substitution of Eq. (10.5.19), Eq. (10.5.21) for H takes the form,

$$H = \mathbf{v} \cdot \mathbf{p} - \frac{1}{2} m v^2 - e\mathbf{v} \cdot \mathbf{A} + e\Phi \tag{10.5.22a}$$

Replace **p** in Eq. (10.5.22a) with its definition on Eq. (10.5.20). The result is

$$\begin{aligned} H &= \mathbf{v} \cdot [m\mathbf{v} + e\mathbf{A}] - \frac{1}{2} m v^2 - e\mathbf{v} \cdot \mathbf{A} + e\Phi \\ &= m v^2 + e\mathbf{v} \cdot \mathbf{A} - \frac{1}{2} m v^2 - e\mathbf{v} \cdot \mathbf{A} + e\Phi \end{aligned} \tag{10.5.22b}$$

After algebraic simplification, Eq. (10.5.22b) reads,

$$H = \frac{1}{2} m v^2 + e\Phi \tag{10.5.23}$$

Equation (10.5.23) is the total energy. Because the action of the magnetic induction, **B**, on a charge moving with a velocity, **v**, produces a force in the direction, $\mathbf{v} \times \mathbf{B}$, which is perpendicular to **v**, the magnetic induction does no work on the charge and makes no contribution to the total energy. That is why Eq. (10.5.23) does not depend on **A**.

Equation (10.5.20) can be solved for **v** to obtain

$$\mathbf{v} = \frac{\mathbf{p} - e\mathbf{A}}{m} \tag{10.5.24}$$

Upon substitution of Eq. (10.5.24) into Eq. (10.5.23), H assumes the final form,

$$H = \frac{(\mathbf{p} - e\mathbf{A})^2}{2m} + e\Phi \tag{10.5.25}$$

Although **A** appears explicitly in Eq. (10.5.25), its role is to cancel out the contribution of the electromagnetic momentum, which is part of **p**. Equation (10.5.25) serves as the basis for the quantum theory of the interaction of electromagnetic radiation with matter.

Chapter 11
Quantum Mechanics

In this chapter, we cover the basic concepts of quantum mechanics that are required to understand the interaction of electromagnetic radiation with matter. The chapter begins with a review of the Planck theory of temperature radiation and the De Broglie equation relating wavelength to particle momentum. The basis for the time-dependent Schrödinger equation is introduced, and its structure is motivated by comparing the dispersion relation for quantum waves with the dispersion relation associated with the classical wave equation. The space and time variables in the time-dependent Schrödinger equation are separated, and the resulting time-independent Schrödinger equation is solved for the case of a square well potential. The concepts of quantum mechanical operators and their eigenfunctions and eigenvalues are introduced. We show that quantum mechanical operators are Hermitian, and their eigenvalues are real. Perturbation methods are then introduced in order to obtain approximate solutions to the Schrödinger equation for arbitrary potentials, which can be both space and time-dependent. In the case of a time-dependent potential, the perturbation theory is used to obtain Fermi's Golden rule. The latter forms the basis for deriving formulae for the Einstein "A" and "B" coefficients that govern the spontaneous emission and absorption of electromagnetic radiation. These two coefficients find use in the analysis of the absorption and emission phenomena which we will encounter in subsequent chapters.

11.1 Introduction

As described in Sect. 2.10 of Chap. 2, Max Planck was searching for a theoretical scheme for analyzing the frequency dependence of the intensity of the light emitted by a heated object. We can think of a hot poker or the interior of a pottery kiln. The frequency spectrum of the light exhibits a broad maximum, which shifts toward shorter wavelengths as the temperature is increased.

J. K. Baird, *Electromagnetism for Chemistry*,
https://doi.org/10.1007/978-3-031-88508-2_11

In order to make sense of this experimental observation, Planck adopted a model in which the atoms in the material were regarded as oscillators having a spectrum of natural frequencies, ν. He suggested that the frequency of the light emitted or absorbed by an oscillator should coincide with the frequency, ν, or one of its overtones, $n\nu$, where n is an integer. He further postulated that at a fixed temperature, absorbed radiation and emitted radiation were in equilibrium with one another and with the walls of the enclosure. As we saw in Chap. 2, if the required partition function can be evaluated, this equilibrium can be described in terms of statistical mechanics. The evaluation of the partition function required a knowledge of the energies, ε, of the various oscillators. Since the light frequencies, ν, were fundamental in Planck's model, he needed to relate the frequencies to the energies is some way. At the time, however, there was nothing in classical physics to suggest what that relation should be. Making a guess that turned out to be prescient, Planck posited

$$\varepsilon = nh\nu \tag{11.1.1}$$

where he took h to be an adjustable parameter. We now know it as a constant of nature. As was shown in Chap. 2 in Eq. (2.10.22), Planck employed Eq. (11.1.1) and statistical mechanics to derive the formula,

$$\rho(\nu, T) = \frac{8\pi h\nu^3/c^3}{(e^{h\nu/k_B T} - 1)} \tag{11.1.2}$$

where $\rho(\nu, T)$ is the energy density per unit frequency of the light emitted or absorbed at frequency, ν, and temperature, T. Although in Eq. (11.1.1) the frequency serves as the theoretically appropriate variable, one ordinarily renders Eq. (11.1.2) in terms of wavelength when comparing with experiment. To make this change, $\rho(\nu, T)$ is replaced by $\overline{\rho}(\lambda, T)$, where

$$\overline{\rho}(\lambda, T)d\lambda = \rho(\nu, T)d\nu \tag{11.1.3a}$$

Upon division by $d\lambda$, Eq. (11.1.3a) reads

$$\overline{\rho}(\lambda, T) = \rho(\nu, T)\frac{d\nu}{d\lambda} \tag{11.1.3b}$$

Because $\nu = c/\lambda$, where c is the vacuum speed of light, the frequency increases as the wavelength decreases; hence, $(d\nu/d\lambda) = -c/\lambda^2$. When this result is substituted into Eq. (11.1.3b), $\overline{\rho}(\lambda, T)$ becomes negative. As it is only the magnitude of $\overline{\rho}(\lambda, T)$ which is of interest, however, we shall replace $(d\nu/d\lambda)$ by its absolute value. With this change, Eq. (11.1.3b) becomes

$$\overline{\rho}(\lambda, T) = \rho(\nu, T)\frac{c}{\lambda^2} \tag{11.1.3c}$$

Substitute $\nu = c/\lambda$ into Eq. (11.1.2) and then substitute that result into Eq. (11.1.3c), the following equation is the result:

$$\rho(\nu, T) = \frac{8\pi h(c/\lambda)^3}{c^3(e^{hc/\lambda k_B T} - 1)} \cdot \frac{c}{\lambda^2} \tag{11.1.4a}$$

Equation (11.1.4a) simplifies to read,

$$\overline{\rho}(\lambda, T) = \frac{8\pi hc}{\lambda^5\left(e^{hc/\lambda k_B T} - 1\right)} \tag{11.1.4b}$$

With the value of h appropriately adjusted, Planck found Eq. (11.1.4b) to be in good agreement with experiment.

Prior to the work of Planck, Lord Rayleigh and James Jeans (English, 1877–1946), combined the theory of electromagnetism with classical statistical mechanics to show that

$$\overline{\rho}(\lambda, T) = \frac{8\pi k_B T}{\lambda^4} \tag{11.1.4c}$$

Plots of Eqs. (11.1.4b) and (11.1.4c) can be seen in Fig. 11.1.

As illustrated in the figure, Eq. (11.1.4c) fails to agree with experiment in the short wavelength limit. This disagreement became known as the "ultraviolet catastrophe". Note that Eq. (11.1.4b) goes to zero as $\lambda \to 0$, thus avoiding the ultraviolet catastrophe. Nevertheless, the Rayleigh–Jeans law is not without applicability. In the limit, $\lambda \to \infty$, Eqs. (11.1.4b) and (11.1.4c) agree. As this has the same effect as letting $h \to 0$, the Rayleigh–Jeans law is said to be the classical limit of the Planck spectrum. As proof of this statement, consider the limiting process,

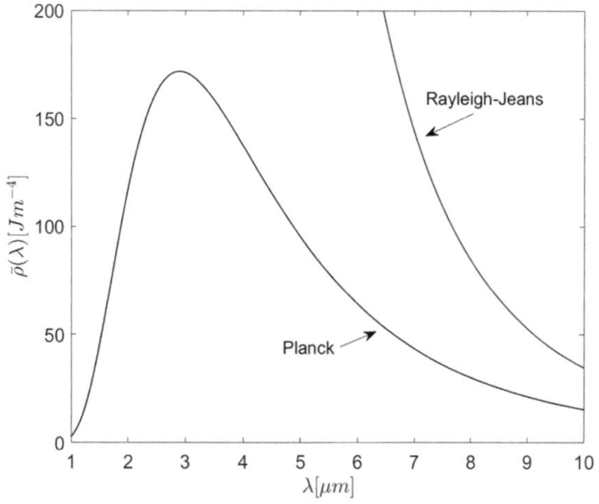

Fig. 11.1 Planck spectrum for thermal radiation at a temperature $T = 1000$ K. The curve with the maximum is a plot of Eq. (11.1.4b). The curve which decreases monotonically with increasing wavelength is the Rayleigh–Jeans law given Eq. (11.1.4c)

$$\lim_{h\to 0} \overline{\rho}(\lambda, T) = \lim_{h\to 0} \frac{8\pi hc/\lambda^5}{(e^{hc/\lambda k_B T} - 1)}$$
$$\approx \frac{8\pi hc/\lambda^5}{[1 - (hc/\lambda k_B T) - 1]} = \frac{8\pi hc}{\lambda^5 (hc/\lambda k_B T)} = \frac{8\pi k_B T}{\lambda^4} \tag{11.1.4d}$$

Early in the twentieth century, the photoelectric effect provided another challenge to classical theory. In the photoelectric effect, electrons are released from the surface of a metal in vacuum when the metal is illuminated by light with a sufficiently high frequency. In classical theory, the release of the electron was insensitive to the frequency of the light. Rather, classical theory predicted that if the binding energy of the electron was less than the Poynting energy absorbed by the metal, the electron would be released. For frequencies below a threshold value, which depended upon the chemical identity of the metal, experiment showed, however, that there was no photoelectric current regardless of the intensity of the light. For frequencies above threshold, photoelectrons appeared. Einstein plotted the kinetic energy of the electrons released as a function of the light frequency. See Fig. 11.2.

Einstein found that for frequencies above the threshold value given by, $\nu_o = W/h$, where W was the work function, the kinetic energy increased linearly with the frequency. The slope of the plot of electron kinetic energy versus frequency turned out to have a value that was identical to Planck's value for h. The number of electrons released per second depended upon the brightness of the light. For a light of frequency, $\nu > \nu_o$, the number of electrons was equal to the magnitude of Poynting vector divided by $h\nu$.

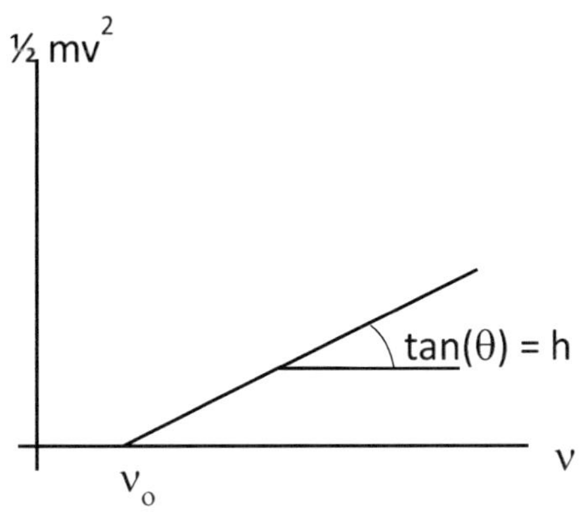

Fig. 11.2 Plot of the kinetic energy, $\frac{1}{2}m\text{v}^2$, of a photoelectron released versus the light frequency, ν. The velocity of the electron is v and its mass is m. The line is the graph of the function, $\frac{1}{2}m\text{v}^2 = h\nu - W$, where h is the Planck constant and $W = h\nu_o$ is the binding energy of the electron to the metal, also known as the "work function". The slope of the line, denoted by $\tan\theta$, proved to be equal to h

Still another challenge to classical physics at the time was the pattern of wavelengths of light absorbed by an isolated hydrogen atom. The various discrete emission wavelengths that were observed followed a pattern that could be reproduced by applying the empirical formula developed by J. R. Rydberg, (Swedish, 1854–1919),

$$\frac{1}{\lambda} = Ry\left(\frac{1}{n_1^2} - \frac{1}{n_2^2}\right) \tag{11.1.5}$$

where $n_1 < n_2$ are integers, and Ry is a multiplicative constant, known as the Rydberg constant. The value of Ry was determined by fitting Eq. (11.1.5) to the spectral data. The pattern of frequencies could be broken into groups depending upon the value assigned to n_1. With $n_1 = 1$, one group of wavelengths could be fitted by letting the value of n_2 run over $n_2 = 2, 3, 4$, etc. This series, which was in the ultraviolet, was dubbed the Lyman series (Theodore Lyman, American, 1874–1954). A second group of wavelengths in the visible could be fitted by setting $n_1 = 2$, and then letting n_2 run over $n_2 = 3, 4, 5$, etc. This series, which was in the visible, was known as the Balmer series (Johann Jakob Balmer, Swiss, 1825–1898). The value of Ry was the same for both series. The impressive agreement of a simple formula with experiment was not inconsequential.

As described in Sect. 1.15 of Chap. 1, Neils Bohr's theory of the hydrogen atom spectrum involved the Planck constant. The proton had been discovered in 1911 by Ernest Rutherford (New Zealand/British, 1871–1937). In 1913, Bohr suggested that a hydrogen atom consisted of an electron in a circular orbit about a proton. The centrifugal force associated with the orbital motion was balanced by the Coulomb force of attraction between the electron and the proton. Bohr focused his attention on the quantization of the "action" S, which is defined in Hamiltonian mechanics by the integral,

$$S = \oint p dq \tag{11.1.6}$$

If m and v are the mass and speed, respectively, of the electron, the electron linear momentum is $p = mv$. In the Bohr circular orbit model, the linear momentum is tangent to the circumference of the orbit. In the case of an orbit having radius r, the differential length of the orbit is $dq = rd\theta$, where θ is the angle between the initial direction of the radius and a subsequent value, and $d\theta$ is the differential change in this angle. See Fig. 11.3.

To describe a complete circle, the angle, θ, needs to cover the range, $0 \le \theta \le 2\pi$. Because in Eq. (11.1.6) a loop integral is involved, the total length of the path of integration is the circumference, $2\pi r$. To account for the appearance of integers in Rydberg's formula, which is Eq. (11.1.5), Bohr proposed that S should be proportional to an integer, n. Noting that pdq has the same units as Planck's constant, h, Bohr suggested that

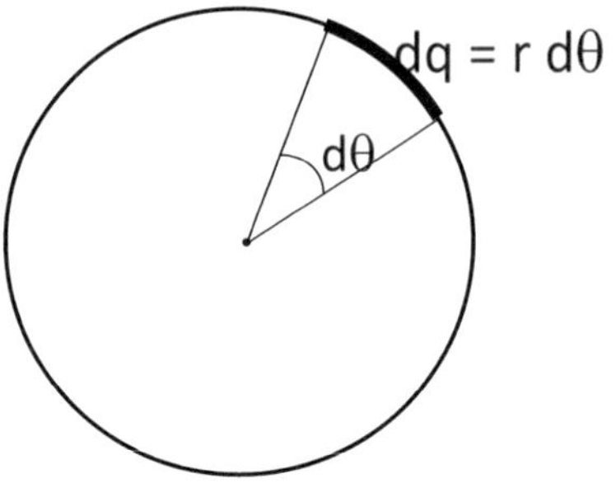

Fig. 11.3 Bohr orbit of an electron orbiting a proton in a hydrogen atom. The radius of the orbit is r. The differential element of length, dq, of the circular orbit subtends a differential angle $d\theta$

$$\oint pdq = nh \tag{11.1.7}$$

Since $p = mv$ is parallel to $dq = rd\theta$, Eq. (11.1.7) can be evaluated by writing it in the form,

$$nh = \int_0^{2\pi} mvrd\theta = mvr \int_0^{2\pi} d\theta = 2\pi mvr \tag{11.1.8}$$

When Bohr introduced Eq. (11.1.7) in 1913, it lacked a physical interpretation. A decade later, Louis De Broglie (French, 1892–1987) suggested that an electron in an H-atom could be represented as a self reinforcing wave traversing a circular orbit. For there to be constructive interference, the wave had to return to its point of origin on the circle with the same phase as when it started. A wave satisfying this restriction needed to have a wavelength, λ, satisfying the condition,

$$n\lambda = 2\pi r \tag{11.1.9}$$

where n was the same integer as in Eq. (11.1.7). Equations (11.1.8) and (11.1.9) could be consistent if the wavelength and the linear momentum satisfied the relation,

$$\lambda = \frac{h}{mv} = \frac{h}{p} \tag{11.1.10}$$

Equation (11.1.10) became known as the De Broglie relation.

The wave nature of the electron was subsequently confirmed experimentally by Clinton Joseph Davisson, (American, 1881–1958) and Lester Germer (American, 1896–1971). They showed that electrons scattered from the atoms in the surface of a crystal formed diffraction patterns governed by the structure factor discussed in Chap. 9.

In the case of light, h relates the frequency of the light wave to the energy it transfers when admitted or absorbed, whereas in the case of particle motion, h relates the momentum of the particle to its wavelength. If as De Broglie proposed, the electron could be replaced by a wave, then in analogy with Maxwell's theory of light, there had to be some wave equation that determined the motion of the electron. The solution to this problem was provided by Erwin Schrödinger (Austrian, 1887–1961) in 1926. In the section below, we begin our exploration of wave equations by deriving the wave equation obeyed by the propagation of a light.

11.2 The Wave Equation for Electromagnetic Radiation

In the absence of an electric current, $\mathbf{J} = 0$, the Maxwell equations read,

$$\nabla \times \mathbf{H} = \frac{\partial \mathbf{D}}{\partial t} \quad \text{(Ampere's Law)} \tag{11.2.1a}$$

$$\nabla \times \mathbf{E} = -\frac{\partial \mathbf{B}}{\partial t} \quad \text{(Faraday's Law)} \tag{11.2.1b}$$

If the medium is free space, the appropriate constitutive relations are

$$\mathbf{D} = \varepsilon_o \mathbf{E} \tag{11.2.2a}$$

$$\mathbf{B} = \mu_o \mathbf{H} \tag{11.2.2b}$$

Upon substitution of Eqs. (11.2.2a) and (11.2.2b), the Ampere and Faraday Laws read, respectively,

$$\nabla \times \mathbf{H} = \varepsilon_o \frac{\partial \mathbf{E}}{\partial t} \tag{11.2.3a}$$

$$\nabla \times \mathbf{E} = -\mu_o \frac{\partial \mathbf{H}}{\partial t} \tag{11.2.3b}$$

The curl of Eq. (11.2.3b) is

$$\nabla \times \nabla \times \mathbf{E} = -\mu_o \frac{\partial (\nabla \times \mathbf{H})}{\partial t} \tag{11.2.4a}$$

Upon substitution of Eq. (11.2.3a) into Eq. (11.2.4a), the result is

$$\nabla \times \nabla \times \mathbf{E} = -\mu_o \varepsilon_o \frac{\partial^2 \mathbf{E}}{\partial t^2} \tag{11.2.4b}$$

In the Mathematical Methods Appendix, one finds the vector identity,

$$\nabla \times \nabla \times \mathbf{E} = \nabla(\nabla \cdot \mathbf{E}) - \nabla^2 \mathbf{E} \tag{11.2.4c}$$

In free space where the charge density, $\rho = 0$, Gauss' Law reads,

$$\nabla \cdot \mathbf{D} = \varepsilon_o(\nabla \cdot \mathbf{E}) = 0 \tag{11.2.4d}$$

Substitute Eq. (11.2.4d) into Eq. (11.2.4c) to obtain

$$\nabla \times \nabla \times \mathbf{E} = -\nabla^2 \mathbf{E} \tag{11.2.4e}$$

Upon substitution of Eq. (11.2.4e) into Eq. (11.2.4b), one obtains the wave equation,

$$\nabla^2 \mathbf{E} = \frac{1}{c^2}\frac{\partial^2 \mathbf{E}}{\partial t^2} \tag{11.2.5}$$

which governs the propagation of an electromagnetic disturbance in vacuum. The wave speed is $c = 1/\sqrt{\varepsilon_o \mu_o}$, which is the speed of light in vacuum.

Consider a transverse wave traveling in the x-direction with electric field perpendicular to x-axis. In this case, the electric vector can be written in terms of the single component, $E(x, t)$. Equation (11.2.5) simplifies to read

$$\frac{\partial^2 E(x,t)}{\partial x^2} = \frac{1}{c^2}\frac{\partial^2 E(x,t)}{\partial t^2} \tag{11.2.6}$$

Equation (11.2.6) is a one dimensional wave equation. As such, it admits solutions for $E(x, t)$ that are periodic in both time and space. One of these solutions is

$$E(x,t) = E_o \sin\left[2\pi\left(\frac{x}{\lambda} - \nu t\right)\right] \tag{11.2.7}$$

where E_o is the amplitude of the electric field, λ is the wavelength, and ν is the circular frequency. One can show that Eq. (11.2.7) is periodic in both time and space by substituting $x + \lambda$ for x and $t + (1/\nu)$ for t. Specifically,

$$\begin{aligned} E(x+\lambda, t+1/\nu) &= E_o \sin\left[2\pi\left(\frac{(x+\lambda)}{\lambda} - \nu\left(t + \frac{1}{\nu}\right)\right)\right] \\ &= E_o \sin\left[2\pi\left(\frac{x}{\lambda} + 1 - \nu t - 1\right)\right] \\ &= E_o \sin\left[2\pi\left(\frac{x}{\lambda} - \nu t\right)\right] = E(x,t) \end{aligned} \tag{11.2.8}$$

Thus values of $E(x, t)$ repeat in space with a period, λ, and reoccur in time with a period, $T = 1/\nu$. To demonstrate that Eq. (11.2.7) satisfies Eq. (11.2.6), we shall need to compute,

$$\frac{\partial E(x,t)}{\partial x} = \left(\frac{2\pi}{\lambda}\right) E_o \cos\left[2\pi\left(\frac{x}{\lambda} - \nu t\right)\right] \tag{11.2.9a}$$

The second derivative is

$$\frac{\partial^2 E(x,t)}{\partial x^2} = -\left(\frac{2\pi}{\lambda}\right)^2 E_o \sin\left[2\pi\left(\frac{x}{\lambda} - \nu t\right)\right] \tag{11.2.9b}$$

We shall need the derivative with respect to time,

$$\frac{\partial E(x,t)}{\partial t} = -(2\pi\nu)E_o \cos\left[2\pi\left(\frac{x}{\lambda} - \nu t\right)\right] \tag{11.2.10a}$$

The second time derivative is

$$\frac{\partial^2 E(x,t)}{\partial t^2} = -(2\pi\nu)^2 E_o \sin\left[2\pi\left(\frac{x}{\lambda} - \nu t\right)\right] \tag{11.2.10b}$$

Upon substitution of Eqs. (11.2.9b) and (11.2.10b) into Eq. (11.2.6), the result is

$$-\left(\frac{2\pi}{\lambda}\right)^2 E_o \sin\left[2\pi\left(\frac{x}{\lambda} - \nu t\right)\right] = -\left(\frac{(2\pi\nu)^2}{c^2}\right) E_o \sin\left[2\pi\left(\frac{x}{\lambda} - \nu t\right)\right] \tag{11.2.11a}$$

The two sides of Eq. (11.2.11a) are equal if

$$\frac{1}{\lambda^2} = \frac{\nu^2}{c^2} \tag{11.2.11b}$$

The square root of Eq. (11.2.11b) is

$$c = \lambda\nu \tag{11.2.12}$$

If we define the wave number, k, by

$$k = \frac{2\pi}{\lambda} \tag{11.2.13}$$

and the angular frequency, ω, by

$$\omega = 2\pi\nu \tag{11.2.14}$$

then Eq. (11.2.12) becomes

$$\omega = ck \tag{11.2.15}$$

Equation (11.2.15) predicts that the angular frequency, ω, and wave number, k, are directly proportional. The coefficient of proportionality is the vacuum speed of light, c. Consider two waves traveling with speed, c, but with different values of ω and k. If these two waves are launched from the same point at the same time, the relative distance between the peak of one and the peak of the other, as well as the relative distance between the trough of one and the trough of the other, will remain constant with time. This correlation is due to the fact that they share the same wave speed. If that were not the case, the faster traveling wave would leave the slower wave behind. This happens in cases where the medium supporting the wave causes the wave speed $v(k)$ to be a function of k. Such a medium is said to be dispersive. Equation (11.2.15) is replaced by $\omega = v(k)k$. The latter equation is said to be a *dispersion relation.*

Equations (11.2.13) and (11.2.14) permit Eq. (11.2.7) to be written in the form,

$$E(x, t) = E_o \sin(kx - \omega t) \tag{11.2.16}$$

which is the standard representation of a classical wave traveling in one dimension.

11.3 The Quantum Wave Equation for a Particle

11.3.1 Free Particle

In classical mechanics, the total energy, ε, of a free particle of mass, m, moving in the absence of external forces through space with velocity, v, is

$$\varepsilon = \frac{1}{2}mv^2 \tag{11.3.1}$$

The linear momentum, p, is defined as

$$p = mv \tag{11.3.2}$$

If we eliminate v between Eqs. (11.3.1) and (11.3.2), the result is

$$\varepsilon = \frac{p^2}{2m} \tag{11.3.3}$$

According to Planck and Bohr, the energy is related to the frequency of the wave representing the particle by

$$\varepsilon = \hbar\omega \tag{11.3.4}$$

where ω is the angular frequency of the wave, and $\hbar = h/2\pi$ is pronounced "h bar".

Upon substitution of Eq. (11.2.13) into Eq. (11.1.10), the De Broglie relation can be rewritten in the form,

$$p = \hbar k \tag{11.3.5}$$

Next, substitute Eqs. (11.3.4) and (11.3.5) into Eq. (11.3.3). The result is

$$\hbar\omega = \frac{\hbar^2 k^2}{2m} \tag{11.3.6}$$

Equation (11.3.6) can be shown to be the quantum version of Eq. (11.2.15). If we write Eq. (11.3.6) with ω proportional to k, we obtain

$$\omega = \left(\frac{\hbar k}{2m}\right)k \tag{11.3.7a}$$

In Eq. (11.3.7a), the coefficient of proportionality, $(\hbar k/2m)$, serves as the apparent speed. It is not a constant but is instead a function of k. Quantum waves having different values of k will thus travel at different speeds. The result will be dispersion.

Our task is to find a wave equation consistent with Eq. (11.3.6). This task is just the reverse of the problem posed in Sect. 11.2, where the wave equation, Eq. (11.2.6) was given and the condition, Eq. (11.2.15) that made Eq. (11.2.16) a solution had to be found.

We will show that Eq. (11.3.6) prevents a quantum wave from having the mathematical form

$$\Psi(x, t) = \Psi_o \sin(kx - \omega t) \tag{11.3.7b}$$

which is the analog of Eq. (11.2.16), where $\Psi(x, t)$ replaces the electric field, $E(x, t)$, and the wave amplitude, Ψ_0, replaces the electric field amplitude E_o.

The first step in our proof involves multiplying Eq. (11.3.7b) by $-\hbar$ and differentiating the result with respect to time to get

$$-\hbar\frac{\partial\Psi(x, t)}{\partial t} = \hbar\omega\Psi_o \cos(kx - \omega t) \tag{11.3.8a}$$

We next multiply Eq. (11.3.7b) by $(-\hbar^2/2m)$ and differentiate twice with respect to space. The result is

$$-\frac{\hbar^2}{2m}\frac{\partial^2\Psi}{\partial x^2} = \frac{\hbar^2 k^2}{2m}\Psi_o \sin(kx - \omega t) \tag{11.3.8b}$$

In the hope of identifying

$$-\hbar\frac{\partial\Psi}{\partial t} = -\frac{\hbar^2}{2m}\frac{\partial^2\Psi}{\partial x^2} \tag{11.3.8c}$$

as the appropriate wave equation, we equate the right hand sides of Eqs. (11.3.8a) and (11.3.8b). Although we have $\hbar\omega$ on one side and $\frac{\hbar^2 k^2}{2m}$ on the other, the trigonometric functions, $\cos(kx - \omega t)$ and $\sin(kx - \omega t)$ fail to cancel. Hence, Eq. (11.3.7b) fails as a solution.

There is, however, a wave function, $\Psi(x, t) = \Psi(kx - \omega t)$, which is consistent with Eq. (11.3.6). That function takes the form of the complex valued exponential,

$$\Psi(x, t) = \Psi_o e^{i(kx-\omega t)} \tag{11.3.9}$$

which is periodic in both space and time. Using Eq. (11.3.9), it is possible to construct a wave equation consistent with Eq. (11.3.6). First use Eq. (11.3.9) to compute

$$\begin{aligned} i\hbar\frac{\partial\Psi(x,t)}{\partial t} &= i\hbar\frac{\partial\left(\Psi_o e^{i(kx-\omega t)}\right)}{\partial t} = i\hbar\Psi_o e^{ikx}\frac{\partial\left(e^{-i\omega t}\right)}{\partial t} \\ &= i\hbar\Psi_o(-i\omega)e^{ikx}\cdot e^{-i\omega t} = \hbar\omega\Psi_o e^{i(kx-\omega t)} = \hbar\omega\Psi(x,t) \end{aligned} \tag{11.3.10a}$$

In a similar fashion, compute

$$\begin{aligned} -\frac{\hbar^2}{2m}\frac{\partial^2\Psi(x,t)}{\partial x^2} &= -\frac{\hbar^2}{2m}\frac{\partial^2\left(\Psi_o e^{i(kx-\omega t)}\right)}{\partial x^2} = -\frac{\hbar^2}{2m}\Psi_o e^{-i\omega t}\frac{\partial^2\left(e^{ikx}\right)}{\partial x^2} \\ = -\frac{\hbar^2}{2m}\Psi_o(ik)^2 e^{-i\omega t}\cdot e^{ikx} &= \frac{\hbar^2k^2}{2m}\Psi_o e^{i(kx-\omega t)} = \frac{\hbar^2k^2}{2m}\Psi(x,t) \end{aligned} \tag{11.3.10b}$$

Equate the right hand sides of Eqs. (11.3.10a) and (11.3.10b) to obtain

$$\hbar\omega\Psi(x,t) = \frac{\hbar^2k^2}{2m}\Psi(x,t) \tag{11.3.10c}$$

After cancellation of $\Psi(x, t)$ from either side of Eq. (11.10.3c), the result can be seen to agree with Eq. (11.3.6).

We thus conclude that the quantum wave equation for a free particle is

$$i\hbar\frac{\partial\Psi}{\partial t} = -\frac{\hbar^2}{2m}\frac{\partial^2\Psi}{\partial x^2} \tag{11.3.11}$$

Equation (11.3.11) is known as the free particle, time-dependent Schrödinger equation. It differs from the classical wave equation, Eq. (11.2.6), in several important respects. Equation (11.3.11) involves a *first* time derivative, whereas Eq. (11.2.6) involves a *second* time derivative. More important, the coefficient of the time derivative in Eq. (11.3.11) involves the complex number, i, whereas the coefficient of the time derivative in Eq. (11.2.6) involves the real number, c.

Although complex numbers are useful for calculation, only their real and imaginary parts can be used to represent physically measurable quantities. Because neither $\cos(kx - \omega t)$ nor $\sin(kx - \omega t)$ by itself satisfies Eq. (11.3.11), it is not possible to choose either the real or the imaginary parts of $\Psi(x, t) = \Psi_o e^{i(kx-\omega t)}$ to represent the motion of the particle.

The Heisenberg uncertainty principle (Werner Heisenberg, German, 1901–1976) rules out the possibility of acquiring simultaneous knowledge of both the linear momentum and the position of a quantum particle. Equation (11.3.5) specifies the linear momentum as $\hbar k$. So, if the momentum is fixed at $\hbar k$, all that can be known about the particle's position, x, is that is somewhere in the range, $-\infty < x < +\infty$. Likewise, according to the uncertainty principle, if it is known that the particle is located at position, x, all that can be said about its momentum is that it lies somewhere in the range, $-\infty < p < +\infty$. This uncertainty property, combined with the fact that $\Psi(x, t)$ is complex valued, led Max Born to suggest that the real quantity, $|\Psi(x, t)|^2$, might have some physical meaning. Based upon the statistical interpretation of the electron motion implicit in the uncertainty principle, he defined the product, $|\Psi(x, t)|^2 dx$ as representing the probability of finding the particle at a point lying between x and $x + dx$. With this definition, $|\Psi(x, t)|^2$ can be interpreted as a probability density. Since the sum of all probabilities must equal one, there is unit probability that the particle will lie between $x = -\infty$ and $x = +\infty$. The Born interpretation thus requires that the probability, $|\Psi(x, t)|^2 dx$, should satisfy,

$$\int_{-\infty}^{+\infty} |\Psi(x, t)|^2 dx = 1 \tag{11.3.12}$$

A wave function, $\Psi(x, t)$, which satisfies Eq. (11.3.12) is said to be "normalized".

Another concept which is important in quantum mechanics is the idea that the value of any dynamical variable can be calculated by applying the appropriate operator to $\Psi(x, t)$. For example, the linear momentum p_x of a free particle moving in the x-direction can be calculated by applying the derivative operator

$$\mathrm{p}_x = -i\hbar\frac{\partial}{\partial x} \tag{11.3.13}$$

to the wave function of $\Psi(x, t)$. Application of p_x to $\Psi(x, t)$ has the meaning,

$$\mathrm{p}_x\Psi(x, t) = -i\hbar\frac{\partial\Psi(x, t)}{\partial x} \tag{11.3.14a}$$

If we substitute Eq. (11.3.9) into Eq. (11.3.14a), the calculation proceeds as follows:

$$\begin{aligned}\mathrm{p}_x\Psi(x, t) &= -i\hbar\frac{\partial\left(\Psi_o e^{i(kx-\omega t)}\right)}{\partial x} = -i\hbar\Psi_o e^{-i\omega t}(ik)e^{ikx} \\ &= \hbar k\Psi_o e^{i(kx-i\omega t)} = \hbar k\Psi(x, t)\end{aligned} \tag{11.3.14b}$$

We recognize the multiplier, $\hbar k$, as the particle momentum as defined in Eq. (11.3.5). According to Eq. (11.3.14b), the application of the operator, p_x, to $\Psi(x, t)$ has the effect of returning $\Psi(x, t)$ multiplied by the constant coefficient, $\hbar k$. The later corresponds to the momentum, p_x, of the particle. As such, $\Psi(x, t)$, as defined in Eq. (11.3.9), is said to be an *eigenfunction* of p_x. The coefficient, $\hbar k$, is said to be the *eigenvalue* of p_x associated with $\Psi(x, t)$. The factor of $-i$ in the definition in Eq. (11.3.13) guarantees that the eigenvalue is real.

According to Eq. (11.3.13),

$$\mathrm{p}_x^2 = -\hbar^2 \frac{\partial^2}{\partial x^2} \tag{11.3.15}$$

Application of $(\mathrm{p}_x^2/2m)$ to Eq. (11.3.9) yields

$$\begin{aligned} \frac{\mathrm{p}_x^2}{2m}\Psi(x, t) &= \frac{-\hbar^2}{2m}\frac{\partial^2\left(\Psi_o e^{i(kx-\omega t)}\right)}{\partial x^2} = -\frac{\hbar^2}{2m}\Psi_o e^{-i\omega t}(ik)^2 e^{ikx} \\ &= \frac{\hbar^2 k^2}{2m}\Psi_o e^{i(kx-\omega t)} = \frac{\hbar^2 k^2}{2m}\Psi(x, t) \end{aligned} \tag{11.3.16}$$

On the basis of Eqs. (11.3.3) and (11.3.5), the coefficient, $\hbar^2 k^2/2m$, multiplying $\Psi(x, t)$ on the far right hand side of Eq. (11.3.16) can be recognized as the value of the kinetic energy of a free particle. This identifies $\mathrm{p}_x^2/2m$ as the kinetic energy operator. The wave function, $\Psi(x, t)$, as defined by Eq. (11.3.9), is said to be an *eigenfunction* of the energy operator. The quantity, $\hbar^2 k^2/2m$, is said to be the kinetic energy *eigenvalue*.

11.3.2 Particle Acted upon by an External Potential

With x serving as the space coordinate, we found in Chap. 10 that the classical equations of motion of a particle moving in one dimension with potential energy, $V(x)$, could be derived from the Hamiltonian,

$$H(p, x) = \frac{p^2}{2m} + V(x) \tag{11.3.17a}$$

To construct a quantum mechanical operator based on Eq. (11.3.17a), we replace p^2 with the operator, $\mathrm{p}_x^2 = -\hbar^2 \frac{\partial^2}{\partial x^2}$. The result is

$$\mathrm{H} = -\frac{\hbar^2}{2m}\frac{\partial^2}{\partial x^2} + V(x) \tag{11.3.17b}$$

The analog of Eq. (11.3.11) is

$$i\hbar\frac{\partial\Psi(x, t)}{\partial t} = -\frac{\hbar^2}{2m}\frac{\partial^2\Psi(x, t)}{\partial x^2} + V(x)\Psi(x, t) \tag{11.3.18}$$

Equation (11.3.18) is known as the *time-dependent* Schrödinger equation.

By virtue of the method of separation of variables, any function of the form,

$$\Psi(x,t) = F(t)\psi(x) \tag{11.3.19a}$$

serves as a general solution to Eq. (11.3.18). This can be proved upon substitution of Eq. (11.3.19a) into Eq. (11.3.18). The result is

$$i\hbar\frac{\partial(F(t)\psi(x))}{\partial t} = -\frac{\hbar^2}{2m}\frac{\partial^2(F(t)\psi(x))}{\partial x^2} + V(x)(F(t)\psi(x)) \tag{11.3.19b}$$

or upon evaluation of the indicated derivatives,

$$i\hbar\psi(x)\frac{\partial F(t)}{\partial t} = -F(t)\frac{\hbar^2}{2m}\frac{\partial^2\psi(x)}{\partial x^2} + V(x)(F(t)\psi(x)) \tag{11.3.19c}$$

Now divide both sides of Eq. (11.3.19c) by the product, $F(t)\psi(x)$. The result is

$$i\hbar\frac{1}{F(t)}\frac{\partial F(t)}{\partial t} = -\frac{1}{\psi(x)}\frac{\hbar^2}{2m}\frac{\partial^2\psi(x)}{\partial x^2} + V(x) \tag{11.3.19d}$$

The left hand side of Eq. (11.3.19d) is a function of t alone, while the right hand side is a function of x alone. Suppose we took a snapshot of a quantum wave at some fixed time. Being the picture of a wave, the snapshot would show the wave varying in space. Because the time was fixed by the snapshot, the left hand side of Eq. (11.3.19d) would necessarily be a constant. We call this constant, E. As a result, Eq. (11.3.19d) can be separated into two equations. The first is

$$i\hbar\frac{1}{F(t)}\frac{\partial F(t)}{\partial t} = E \tag{11.3.19e}$$

The second is

$$-\frac{1}{\psi(x)}\frac{\hbar^2}{2m}\frac{\partial^2\psi(x)}{\partial x^2} + V(x) = E \tag{11.3.19f}$$

Equation (11.3.19f) can be rewritten as

$$-\frac{\hbar^2}{2m}\frac{\partial^2\psi(x)}{\partial x^2} + V(x)\psi(x) = E\psi(x) \tag{11.3.20}$$

Equation (11.3.20) is known as the *time-independent* Schrödinger equation. Equation (11.3.20) is often rewritten in the form,

$$\frac{d^2\psi(x)}{dx^2} + \frac{2m}{\hbar^2}(E - V(x))\psi(x) = 0 \tag{11.3.21}$$

The solution to Eq. (11.3.19e) is

$$F(t) = \exp(-iEt/\hbar) \tag{11.3.22}$$

When it is combined with Eq. (11.3.19a), we have

$$\Psi(x, t) = \psi(x)\exp(-iEt/\hbar) \tag{11.3.23}$$

The separation constant, E, can be shown to be equal to the energy. We return to the differential operator in Eq. (11.3.20), rendering it in classical form by replacing $-\frac{\hbar^2}{2m}\frac{\partial^2}{\partial x^2}$ by its classical value $p^2/2m$ and at the same time ignoring $\psi(x)$. The result is

$$\frac{p^2}{2m} + V(x) = E \tag{11.3.24}$$

The left-hand side is clearly the total energy.

On the basis of Eq. (11.3.17a), the left hand side of Eq. (11.3.24) can be recognized as the Hamiltonian, H. We can thus write H in its operator form,

$$\mathrm{H} = -\frac{\hbar^2}{2m}\frac{d}{dx^2} + V(x) \tag{11.3.25}$$

With this definition of H, Eq. (11.3.18) becomes

$$\mathrm{H}\Psi(x, t) = i\hbar\frac{\partial \Psi(x, t)}{\partial t} \tag{11.3.26}$$

Equation (11.3.26) is the standard abstract form of the time-dependent Schrödinger equation. With the definition in Eq. (11.3.25), Eq. (11.3.20) can be rewritten as

$$\mathrm{H}\psi(x) = E\psi(x) \tag{11.3.27}$$

Equation (11.3.27) is the standard abstract form of the time-independent Schrödinger equation.

When $V(x)$ is independent of the time, $|\Psi(x, t)|^2$ is independent of time. Given Eq. (11.3.23), the probability density interpretation for $\Psi(x, t)$ leads to the simple result,

$$|\Psi(x, t)|^2 = \Psi^*\Psi = e^{iEt/\hbar}\psi^*(x)e^{-iEt/\hbar}\psi(x) = |\psi(x)|^2 \tag{11.3.28}$$

When Eq. (11.3.28) prevails, the quantum mechanical system is said to be in a "stationary state".

11.4 Solution to the Time-Independent Schrödinger Equation for a Square Well Potential

Consider an idealized, time-independent potential, $V(x)$, which is zero over the open domain $0 < x < a$ and infinite otherwise. See Fig. 11.4.

In the limit that the sides of $V(x)$ are vertical, the force acting at $x = 0$ is $f = -(\partial V/\partial x) \to +\infty$. At $x = a$, the force is $f = -(\partial V/\partial x) \to -\infty$. These two forces confine the particle so that it has zero probability of being at either $x = 0$ or $x = a$, respectively. This combination of forces defines a "square well potential". The particle is said to be a "particle in a box." Over the domain $0 < x < a$, where $V(x) = 0$, the Hamiltonian in Eq. (11.3.25) is simply

$$\mathrm{H} = -\frac{\hbar^2}{2m}\frac{d^2}{dx^2} \tag{11.4.1a}$$

Equation (11.3.27) then reads

$$\frac{-\hbar^2}{2m}\frac{d^2\psi(x)}{dx^2} = E\psi(x) \tag{11.4.1b}$$

Equation (11.4.1b) can be recast in the form,

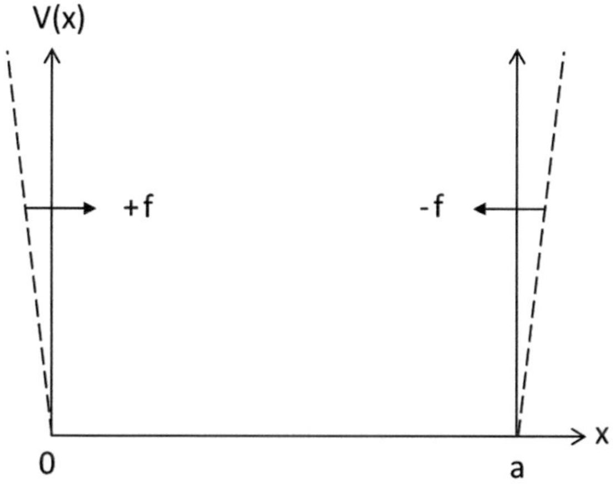

Fig. 11.4 One dimensional square well potential, $V(x)$, spanning the domain $0 \le x \le a$. The solid lines are the idealized sides of the potential function, which is the limit of a potential defined by the splayed sides represented by the dashed lines. The dashed lines are the source of a finite force, $f = -\partial V(x)/\partial x$, which acts on the particle when it reaches $x = 0$ and $x = a$, respectively. At $x = 0$, the force is $+f$ which acts in the positive x-direction. At $x = a$, the force is $-f$ which acts in the negative x-direction

$$\frac{d^2\psi(x)}{dx^2} + \frac{2mE}{\hbar^2}\psi(x) = 0 \tag{11.4.2}$$

Recall that in Eq. (11.3.28) that $|\psi(x)|^2$ serves as a probability density. Hence, because the particle cannot pass to the left of $x = 0$ or to the right of $x = a$, the boundary conditions that make $|\psi(x)|^2$ equal zero on the edges of the potential well are

$$\psi(0) = 0 \tag{11.4.3a}$$

and

$$\psi(a) = 0 \tag{11.4.3b}$$

respectively. If we make the identification

$$\alpha^2 = \frac{2mE}{\hbar^2} \tag{11.4.4}$$

Equation (11.4.2) can be rewritten in the form,

$$\frac{d^2\psi(x)}{dx^2} + \alpha^2\psi(x) = 0 \tag{11.4.5}$$

The general solution to Eq. (11.4.5) is

$$\psi(x) = A\sin(\alpha x) + B\cos(\alpha x) \tag{11.4.6}$$

where A and B are constants of integration. Equation (11.4.6) can be substituted into Eqs. (11.4.3a) and (11.4.3b) with the results

$$0 = A\sin(0) + B\cos(0) \tag{11.4.7a}$$

and

$$0 = A\sin(\alpha a) + B\cos(\alpha a) \tag{11.4.7b}$$

respectively. Since $\sin(0) = 0$ and $\cos(0) = 1$, Eq. (11.4.7a) is satisfied by

$$B = 0 \tag{11.4.7c}$$

Substitution of Eq. (11.4.7c) into Eq. (11.4.6) then gives

$$\psi(x) = A\sin(\alpha x) \tag{11.4.8a}$$

Equation (11.4.8a) must satisfy Eq. (11.4.7b) with $B = 0$. This implies that

$$0 = A\sin(\alpha a) \tag{11.4.8b}$$

Equation (11.4.8b) has more than one solution. The first, $A = 0$, is of no interest, because when combined with Eqs. (11.4.6) and (11.4.7c), it implies that $\psi(x) \equiv 0$, which can only mean that the particle was never in the potential box to start with. The other solutions to Eq. (11.4.8b) are determined by

$$\alpha = \frac{n\pi}{a} \quad (n = 1, 2, 3, ...) \tag{11.4.8c}$$

If we substitute Eq. (11.4.8c) into Eq. (11.4.4), we obtain a result which reads,

$$E_n = \frac{n^2\pi^2\hbar^2}{2ma^2} \quad (n = 1, 2, 3, ...) \tag{11.4.9}$$

where we have written $E = E_n$ in order to stress the fact that the energy of the particle takes on discrete values depending upon the value of n.

Equation (11.4.8c) can be substituted into Eq. (11.4.8a) to obtain,

$$\psi_n(x) = A\sin\left(\frac{n\pi x}{a}\right) \tag{11.4.10}$$

According to Eq. (11.3.23), the time-dependent wave function for a particle in a square well potential can be written as

$$\Psi_n(x, t) = e^{-iE_nt/\hbar}\psi_n(x) \tag{11.4.11}$$

where E_n is given by Eq. (11.4.9) and $\psi_n(x)$ by Eq. (11.4.10). The sole parameter left undetermined is the multiplicative factor A. To determine A, we call upon Eq. (11.3.12). Upon substitution of Eqs. (11.4.11) and (11.4.10) into Eq. (11.3.12). The result is

$$1 = \int_{-\infty}^{+\infty} |\Psi(x, t)|^2 dx = \int_{-\infty}^{+\infty} |\psi_n(x)|^2 dx = AA^* \int_0^a dx \sin^2\left(\frac{n\pi x}{a}\right) \tag{11.4.12a}$$

The integral on the right of Eq. (11.4.12a) can be evaluated by using the trigonometric identity,

$$\sin^2\theta = \left(\frac{1}{2}\right)(1 - \cos(2\theta)) \tag{11.4.12b}$$

Substitution of Eq. (11.4.12b) with $\theta = n\pi x/a$ into Eq. (11.4.12a) gives

$$1 = AA^* \int_0^a dx \sin^2\left(\frac{n\pi x}{a}\right) = AA^* \int_0^a dx\left(\frac{1}{2}\right)\left(1 - \cos\left(\frac{2n\pi x}{a}\right)\right)$$
$$= AA^* \left\{ \left[\frac{x}{2}\right]_0^a - \frac{1}{2}\left(\frac{a}{2n\pi}\right)\left[\sin\left(\frac{2n\pi x}{a}\right)\right]_0^a \right\} \tag{11.4.12c}$$

The second term in brackets on the right of Eq. (11.4.12c) is identically zero, so that Eq. (11.4.12c) simplifies to

$$1 = |A|^2\left(\frac{a}{2}\right) \tag{11.4.12d}$$

In Eq. (11.4.12d), $AA^* = |A|^2$. Unless the complex conjugate, A^*, is needed for some specific application, we can take A to be real. This permits Eq. (11.4.12d) to be solved with the result,

$$A = \sqrt{\frac{2}{a}} \tag{11.4.12e}$$

Hence, upon substitution of Eq. (11.4.12e) into Eq. (11.4.10), the solution to the time-independent Schrödinger equation, Eq. (11.4.2), is

$$\psi_n(x) = \sqrt{\frac{2}{a}} \sin\left(\frac{n\pi x}{a}\right) \; (n = 1, 2, 3, \ldots) \tag{11.4.13}$$

The energies, $\{E_n\}$, given by Eq. (11.4.9) and wave functions $\{\psi_n(x)\}$, given by Eq. (11.4.13) are said to be the eigenvalues and eigenfunctions, respectively, of the Schrödinger equation defined by Eq. (11.4.2). As such, they satisfy the eigenvalue equation, $\mathrm{H}\psi_n(x) = E_n\psi_n(x)$, where H is as defined by Eq. (11.4.1a).

Because the functions defined in Eq. (11.4.13) satisfy

$$\int_0^a dx \psi_n(x)\psi_n(x) = 1 \tag{11.4.14}$$

they are said to be *normalized*. We next inquire into the value of the integral in Eq. (11.4.14) in the case that one of $\psi_n(x)$ is replaced by $\psi_{n'}(x)$, where $n \neq n'$. The indicated integral is

$$\int_0^a dx \psi_n(x)\psi_{n'}(x) = \left(\sqrt{\frac{2}{a}}\right)^2 \int_0^a dx\left[\sin\left(\frac{n\pi x}{a}\right) \sin\left(\frac{n'\pi x}{a}\right)\right] \tag{11.4.15a}$$

In the case of Eq. (11.4.15a), the trigonometric identity

$$\sin(\gamma)\sin(\delta) = \left(\frac{1}{2}\right)\cos(\gamma - \delta) - \left(\frac{1}{2}\right)\cos(\gamma + \delta) \tag{11.4.15b}$$

will prove useful. [See the Mathematical Methods Appendix]. With $\gamma = (n\pi x/a)$ and $\delta = (n'\pi x/a)$. Equation (11.4.15b) permits Eq. (11.4.15a) to be rewritten in the form

$$\begin{aligned}
&\int_0^a dx\psi_n(x)\psi_{n'}(x) \\
&= \left(\sqrt{\frac{2}{a}}\right)^2 \left\{\left(\frac{1}{2}\right)\int_0^a dx\cos\left(\frac{(n-n')\pi x}{a}\right) - \left(\frac{1}{2}\right)\int_0^a dx\cos\left(\frac{(n+n')\pi x}{a}\right)\right\} \\
&= \left(\sqrt{\frac{2}{a}}\right)^2 \left\{\left(\frac{1}{2}\right)\left(\frac{a}{(n-n')\pi}\right)\left[\sin\left(\frac{(n-n')\pi x}{a}\right)\right]_0^a \right. \\
&\left. -\left(\frac{1}{2}\right)\left(\frac{a}{(n+n')\pi}\right)\left[\sin\left(\frac{(n+n')\pi x}{a}\right)\right]_0^a\right\}
\end{aligned} \tag{11.4.15c}$$

When evaluated at their lower and upper limits, both sine functions on the right hand side of Eq. (11.4.15c) are zero. Hence,

$$\int_0^a dx\psi_n(x)\psi_{n'}(x) = 0 \tag{11.4.16}$$

A set of eigenfunctions, such as $\{\psi_n(x)\}$, which obey Eq. (11.4.16) are said to be *orthogonal*. In summary, the eigenfunctions $\{\psi_n(x)\}$ of the operator, H, are both orthogonal and normal. They are said to be *orthonormal*.

In general, if the domain of a particle is extended to include the entire real line, $-\infty \le x \le +\infty$, the orthonormality condition is summarized by the statement,

$$\langle n \mid n' \rangle = \langle \psi_n(x) \mid \psi_{n'}(x) \rangle = \int_{-\infty}^{+\infty} dx\psi_n^*(x)\psi_{n'}(x) = \delta_{nn'} \tag{11.4.17}$$

where $\langle n \mid n' \rangle$ and $\langle \psi_n(x) \mid \psi_{n'}(x) \rangle$ are the Dirac notation (P. A. M. Dirac, English, 1902–1984) for the integral on the right, and $\delta_{nn'}$ is the Kronecker "delta", which by definition is unity when $n = n'$ and zero when $n \neq n'$.

Quantum mechanical operators possess several properties, which are worth proving. Recall that the linear momentum operator is $\mathrm{p}_x = -i\hbar\frac{\partial}{\partial x}$. If $\psi_n(x)$ is not an eigenfunction of p_x, the momentum in the state $\psi_n(x)$ is not fixed. Rather, upon measurement, one obtains only an average value, which in Dirac notation is defined by

$$\langle n|\mathrm{p}_x|n\rangle = \langle \psi_m(x)|\mathrm{p}_x|\psi_n(x)\rangle = \int_{-\infty}^{+\infty} dx\psi_n^*(x)\left[-i\hbar\frac{\partial}{\partial x}\right]\psi_n(x) \tag{11.4.18}$$

Given the infinite domain of the integration, the appropriate boundary conditions are

$$\psi_n(-\infty) = 0 \tag{11.4.19a}$$

and

$$\psi_n(+\infty) = 0 \tag{11.4.19b}$$

Equations (11.4.18), (11.4.19a) and (11.4.19b) permit us to prove a theorem.

Theorem 11.1 With the integers, m and n, distinguishing the wave functions $\psi_m(x)$ and $\psi_n(x)$ and p_x serving as an example of a dynamical operator, the collection of integrals, $\langle m|\mathrm{p}_x|n\rangle$, can be treated as elements of *a matrix*. [See the Mathematical Methods Appendix]. Using the definition,

$$\langle m|\mathrm{p}_x|n\rangle = \int_{-\infty}^{+\infty} dx\psi_m^*(x)\left[-i\hbar\frac{\partial}{\partial x}\right]\psi_n(x) \tag{11.4.20}$$

one may show that $\langle m|\mathrm{p}_x|n\rangle = \langle n|\mathrm{p}_x|m\rangle^*$.

Proof An integral of the type in Eq. (11.4.20) can be performed by parts. The calculation is

$$\begin{aligned}\langle m|\mathrm{p}_x|n\rangle &= \int_{-\infty}^{+\infty} dx\psi_m^*(x)\left[-i\hbar\frac{\partial}{\partial x}\right]\psi_n(x)\\ &= \left[\psi_m^*(x)(-i\hbar)\psi_n(x)\right]_{-\infty}^{+\infty} - \int_{-\infty}^{+\infty} dx\psi_n(x)\left[-i\hbar\frac{\partial}{\partial x}\right]\psi_m^*(x)\\ &= 0 - \int_{-\infty}^{+\infty} dx\psi_n(x)\left[-i\hbar\frac{\partial}{\partial x}\right]\psi_m^*(x) = \int_{-\infty}^{+\infty} dx\psi_n(x)\left[i\hbar\frac{\partial}{\partial x}\right]\psi_m^*(x)\end{aligned} \tag{11.4.21a}$$

where in obtaining the zero term in the last line in Eq. (11.4.21a), we used the boundary conditions in Eqs. (11.4.19a) and (11.4.19b). The remaining integral can be rewritten as

$$\langle m|\mathrm{p}_x|n\rangle = \int_{-\infty}^{+\infty} dx\psi_n(x)\left[-i\hbar\frac{\partial}{\partial x}\right]^* \psi_m^*(x)$$

$$= \left[\int_{-\infty}^{+\infty} dx\psi_n^*(x)\left[-i\hbar\frac{\partial}{\partial x}\right]\psi_m(x)\right]^* = \langle n|\mathrm{p}_x|m\rangle^* \tag{11.4.21b}$$

which proves the theorem. ■

An operator, such as p_x, which satisfies $\langle n|\mathrm{p}_x|n\rangle = \langle n|\mathrm{p}_x|n\rangle^*$, is said to be Hermitian (Charles Hermite, French, 1822–1901). *All quantum mechanical operators corresponding to dynamical variables are Hermitian.*

Another theorem of interest is:

Theorem 11.2 *The eigenvalues* of *a Hermitian operator* are real.

Proof As an example, we recall Eq. (11.3.27), which can be written for a specific eigenstate, $\psi_n(x)$, in the form,

$$\mathrm{H}\psi_n(x) = E_n\psi_n(x) \tag{11.4.22a}$$

If we multiply both sides by $\psi_n^*(x)$ and integrate, the result is

$$\int_{-\infty}^{+\infty} dx\psi_n^*(x)\mathrm{H}\psi_n(x) = E_n\int_{-\infty}^{+\infty} dx\psi_n^*(x)\psi_n(x) \tag{11.4.22b}$$

Written in Dirac notation, Eq. (11.4.22b) reads,

$$\langle n|\mathrm{H}|n\rangle = E_n\langle n \mid n\rangle \tag{11.4.22c}$$

The complex conjugate of Eq. (11.4.22c) is

$$\langle n|\mathrm{H}|n\rangle^* = E_n^*\langle n \mid n\rangle^* \tag{11.4.23a}$$

By virtue of Theorem 11.1,

$$\langle n|\mathrm{H}|n\rangle^* = \langle n|\mathrm{H}|n\rangle \tag{11.4.23b}$$

According to Eq. (11.4.17),

$$\langle n \mid n\rangle = \int_{-\infty}^{+\infty} dx\psi_n^*(x)\psi_n(x) = \int_{-\infty}^{+\infty} dx\psi_n(x)\psi_n^*(x) = \langle n \mid n\rangle^* \tag{11.4.24a}$$

Upon substitution of Eqs. (11.4.23b) and (11.4.24a) into Eq. (11.4.23a), we obtain

$$\langle n|\mathrm{H}|n\rangle = E_n^*\langle n \mid n\rangle \tag{11.4.24b}$$

If Eqs. (11.4.22c) and (11.4.24b) are not to be in conflict, we must have

$$E_n = E_n^* \tag{11.4.25}$$

which proves the theorem.

11.5 Time-Independent Perturbation Theory

We now turn to the problem of solving for the eigenfunctions, $\{\psi_n\}$, and associated eigenvalues, $\{E_n\}$, in the case of an arbitrary time-independent Hamiltonian operator, H. More often than not one finds that the Schrödinger equation,

$$\mathrm{H}\psi_n = E_n\psi_n \tag{11.5.1}$$

defies all attempts at an analytic solution. Nonetheless, it is also frequently the case that some portion of the potential energy has little effect on the state of the quantum particle. When this part of the potential is ignored, H often assumes the simpler form, denoted by $\mathrm{H}^{(\mathrm{o})}$, for which the Schrödinger equation

$$\mathrm{H}^{(o)}\psi_n^{(o)} = E_n^{(o)}\psi_n^{(o)} \tag{11.5.2}$$

has known eigenfunctions, $\left\{\psi_n^{(o)}\right\}$, and eigenvalues, $\left\{E_n^{(o)}\right\}$. We will show that Eq. (11.5.2) can be used as the basis for an accurate, approximate solution to Eq. (11.5.1).

To launch our procedure for finding approximations to the $\{\psi_n\}$ and $\{E_n\}$, we write H in the form,

$$\mathrm{H} = \mathrm{H}^{(o)} + \mathrm{H}' \tag{11.5.3}$$

where H' is the part of H which has little effect. The strategy for using Eq. (11.5.3) to solve Eq. (11.5.1) is known as Rayleigh–Schrödinger perturbation theory or alternatively time-independent perturbation theory. It is based upon the idea that whatever the form of ψ_n, it can be represented, at least approximately, as a weighted linear sum of the eigenfunctions, $\left\{\psi_n^{(o)}\right\}$, associated with $\mathrm{H}^{(o)}$. The concept is similar to the practice of expanding arbitrary functions in Fourier series. The weighted sum is said to constitute a linear superposition. For a fixed value of the index, n, it is proposed that ψ_n can be approximated by

$$\psi_n = \sum_k c_k \psi_k^{(o)} \tag{11.5.4}$$

where the set of numbers, $\{c_k\}$, are expansion coefficients, which serve as the "weights" in the weighted sum. Upon substitution of Eq. (11.5.4) into Eq. (11.5.1), the result is

$$\mathrm{H} \sum_k c_k \psi_k^{(o)} = E_n \sum_k c_k \psi_k^{(o)} \tag{11.5.5a}$$

Equation (11.5.5a) can be rewritten as

$$\sum_k c_k \mathrm{H} \psi_k^{(o)} = E_n \sum_k c_k \psi_k^{(o)} \tag{11.5.5b}$$

On the left hand side of Eq. (11.5.5b), substitute Eq. (11.5.3). The result is

$$\sum_k c_k \left(\mathrm{H}^{(o)} + \mathrm{H}'\right) \psi_k^{(o)} = E_n \sum_k c_k \psi_k^{(o)} \tag{11.5.5c}$$

After expanding the sum on the left, Eq. (11.5.5c) reads,

$$\sum_k c_k \mathrm{H}^{(o)} \psi_k^{(o)} + \sum_k c_k \mathrm{H}' \psi_k^{(o)} = E_n \sum_k c_k \psi_k^{(o)} \tag{11.5.5d}$$

Equation (11.5.2) can be used to transform the first sum on the left hand side of Eq. (11.5.5d) so that the result reads,

$$\sum_k c_k E_k^{(o)} \psi_k^{(o)} + \sum_k c_k \mathrm{H}' \psi_k^{(o)} = E_n \sum_k c_k \psi_k^{(o)} \tag{11.5.5e}$$

We can solve Eq. (11.5.5e) for the second sum on the left. The result is

$$\sum_k c_k \mathrm{H}' \psi_k^{(o)} = E_n \sum_k c_k \psi_k^{(o)} - \sum_k c_k E_k^{(o)} \psi_k^{(o)} \tag{11.5.5f}$$

After combining the two sums on the right of Eq. (11.5.5f), we obtain

$$\sum_k c_k \mathrm{H}' \psi_k^{(o)} = \sum_k c_k \left(E_n - E_k^{(o)}\right) \psi_k^{(o)} \tag{11.5.5g}$$

To solve for the expansion coefficients, $\{c_k\}$, we will need some integrals. Written in Dirac notation, the required integrals are

$$\left\langle \psi_m^{(o)} | \mathrm{H}' | \psi_k^{(o)} \right\rangle = \int dx \left[\psi_m^{(o)*}(x) \mathrm{H}' \psi_k^{(o)}(x)\right] \tag{11.5.6a}$$

and

$$\left\langle \psi_m^{(o)} \,\middle|\, \psi_k^{(o)} \right\rangle = \int dx \left[\psi_m^{(o)*}(x) \psi_k^{(o)}(x) \right] = \delta_{mk} \tag{11.5.6b}$$

where we have used Eq. (11.4.17). If we were to treat the multidimensional case, we would let the single coordinate, x, serve as a short hand notation for the totality of the spatial coordinates involved. The symbol, $\int dx$, would then denote integration over the appropriate ranges of those coordinates.

Multiply both sides of Eq. (11.5.5g) by $\psi_m^{(o)*} = \left\langle \psi_m^{(o)} \right|$, and integrate. [see Eq. (11.5.6a) as example]. Rendered in Dirac notation, the integration reads,

$$\sum_k c_k \left\langle \psi_m^{(o)} | \mathrm{H}' | \psi_k^{(o)} \right\rangle = \sum_k c_k \left(E_n - E_k^{(o)} \right) \left\langle \psi_m^{(o)} | \psi_k^{(o)} \right\rangle \tag{11.5.6c}$$

According to Eq. (11.5.6b), Eq. (11.5.6c) reads,

$$\sum_k c_k \left\langle \psi_m^{(o)} | \mathrm{H}' | \psi_k^{(o)} \right\rangle = \sum_k c_k \left(E_n - E_k^{(o)} \right) \delta_{mk} = c_m \left(E_n - E_m^{(o)} \right) \tag{11.5.6d}$$

Equation (11.5.6d) actually represents a system of equations, one for each value of the index, m. We can summarize these equations by solving them for $\left(E_n - E_m^{(o)} \right)$ and noting that the index, m, runs over $m = 1, 2, 3, \ldots$. The result is

$$\left(E_n - E_m^{(o)} \right) c_m = \sum_k c_k \left\langle \psi_m^{(o)} | \mathrm{H}' | \psi_k^{(o)} \right\rangle \quad (m = 1, 2, 3, \ldots) \tag{11.5.7}$$

Equation (11.5.7) consists of a mixture of quantities of different magnitudes.

To sort out the magnitude question, we can write each energy eigenvalue, E_n, as a sum of terms

$$E_n = E_n^{(o)} + E_n^{(1)} + E_n^{(2)} + \cdots \tag{11.5.8a}$$

where $E_n^{(o)} > E_n^{(1)} > E_n^{(2)} > \cdots$. Assume also that c_n can be analogously represented by the sum,

$$c_n = c_n^{(o)} + c_n^{(1)} + c_n^{(2)} + \cdots \tag{11.5.8b}$$

where $c_n^{(o)} > c_n^{(1)} > c_n^{(2)} > \cdots$. In Eqs. (11.5.8a) and (11.5.8b), the notation indicates that $c_n^{(o)}$ and $E_n^{(o)}$ are the same order of magnitude, while $c_n^{(1)}$ and $E_n^{(1)}$ are much smaller than $c_n^{(o)}$ and $E_n^{(o)}$, respectively, while $c_n^{(2)}$ and $E_n^{(2)}$ are much smaller than $c_n^{(1)}$ and $E_n^{(1)}$, respectively, etc. If H' is sufficiently weak, we can expect that ψ_n and $\psi_n^{(o)}$ will not differ by much. Thus, in the case, $m \neq n$, we can let $c_m^{(o)}$ be zero, whereas in the case, $m = n$, we can approximate $c_m^{(o)}$ by unity. The choice of these values can be expressed in terms of the Kronecker delta symbol by writing

$$c_m^{(o)} = \delta_{mn} \tag{11.5.8c}$$

By combining Eqs. (11.5.8b) and (11.5.8c), c_k can be expanded in the form,

$$c_k = \delta_{kn} + c_k^{(1)} + c_k^{(2)} + \cdots \tag{11.5.8d}$$

Next, we require ψ_n to be normalized. Using Eq. (11.5.4) and (11.5.6b), the normalization condition can be written as

$$\begin{aligned}\int dx\psi_n^*\psi_n &= \int dx\left[\sum_\ell c_\ell \psi_\ell^{(o)}\right]^*\left[\sum_k c_k\psi_k^{(o)}\right] = \sum_\ell\sum_k c_\ell^* c_k \int dx(\psi_\ell^{(o)})^*\psi_k^{(o)} \\ &= \sum_\ell\sum_k c_\ell^* c_k\left\langle\psi_\ell^{(o)}\;\middle|\;\psi_k^{(o)}\right\rangle = \sum_\ell\sum_k c_\ell^* c_k\delta_{\ell k} = \sum_k |c_k|^2 = 1\end{aligned} \tag{11.5.9}$$

Equation (11.5.8d) can be used to rewrite the second line of Eq. (11.5.9) in the form,

$$1 = \sum_k |c_k|^2 = \sum_k \left(\delta_{kn} + c_k^{(1)*} + c_k^{(2)*} + \cdots\right)\left(\delta_{kn} + c_k^{(1)} + c_k^{(2)} + \cdots\right) \tag{11.5.10a}$$

We ignore terms involving a product $c_k^{(m)*}\cdot c_k^{(n)}$ where $m+n\geq 2$. In comparison with terms reading $c_k^{(1)}$ or $c_k^{(1)*}$, which are said to be "first order" in magnitude, terms with $m+n\geq 2$ are an order of magnitude smaller and are said to be of "second order or higher". Keeping only the first order terms, Eq. (11.5.10a) reads,

$$1 = \sum_k \delta_{kn}\delta_{kn} + \sum_k \delta_{kn}\left(c_k^{(1)} + c_k^{(1)*}\right) = 1 + c_n^{(1)} + c_n^{(1)*} \tag{11.5.10b}$$

Hence, according to Eq. (11.5.10b), the sum of the first order coefficients, $c_n^{(1)}$ and $c_n^{(1)*}$, must satisfy,

$$c_n^{(1)} + c_n^{(1)*} = 0 \tag{11.5.11a}$$

With the recognition that $c_n^{(1)}$ can be a complex number, we write

$$c_n^{(1)} = \mathrm{Re}\left(c_n^{(1)}\right) + i\mathrm{Im}\left(c_n^{(1)}\right) \tag{11.5.11b}$$

and

$$c_n^{(1)*} = \mathrm{Re}\left(c_n^{(1)}\right) - i\mathrm{Im}\left(c_n^{(1)}\right) \tag{11.5.11c}$$

Upon substitution of Eqs. (11.5.11b) and (11.5.11c) into Eq. (11.5.11a), the conclusion is

$$\mathrm{Re}\left(c_n^{(1)}\right) + i\mathrm{Im}(c_n^{(1)}) + \mathrm{Re}\left(c_n^{(1)}\right) - i\mathrm{Im}(c_n^{(1)}) = 2\mathrm{Re}\left(c_n^{(1)}\right) = 0 \tag{11.5.11d}$$

Hence,

$$c_n^{(1)} = i\mathrm{Im}\left(c_n^{(1)}\right) \tag{11.5.12a}$$

For most applications, however, ψ_n and $\psi_k^{(o)}$ in Eq. (11.5.4) are real valued functions, which implies that the members of the set, $\{c_k\}$, are all real. Hence, if $c_n^{(1)}$ in Equation (11.5.11a) is to be real, Eq. (11.5.12a) implies that

$$c_n^{(1)} = i\mathrm{Im}\left(c_n^{(1)}\right) = 0 \tag{11.5.12b}$$

When the coefficients are real, then by use of Eqs. (11.5.8a) through (11.5.8d), Eq. (11.5.7) can be written in the form, correct to second order,

$$\begin{aligned}&\left(E_n^{(o)} + E_n^{(1)} + E_n^{(2)} - E_m^{(0)}\right)\left(\delta_{mn} + c_m^{(1)} + c_m^{(2)}\right)\\ &= \sum_k \left(\delta_{kn} + c_k^{(1)} + c_k^{(2)}\right)\left\langle \psi_m^{(o)} | \mathrm{H}' | \psi_k^{(o)} \right\rangle\end{aligned} \tag{11.5.13a}$$

Like $E_n^{(o)}$ and $E_m^{(o)}$ individually, the difference, $E_n^{(o)} - E_m^{(o)}$, is zero order. We assume that H′ is of first order in magnitude, so that the factor $\left\langle \psi_m^{(o)} | \mathrm{H}' | \psi_k^{(o)} \right\rangle$ on the right of Eq. (11.5.13a) is first order. We note that the product of a zeroth order term and a first order term is first order. The product of zeroth order term and a second order term is second order as is the product of two first order terms. These observations permit us to multiply out the terms on the left and right, grouping them together with respect order of magnitude. If we retain terms which are zeroth order, first order, and second order in magnitude, and drop the higher order terms, the result is

$$\begin{aligned}&\left(E_n^{(o)} - E_m^{(o)}\right)\delta_{mn} + \left(c_m^{(1)}\left(E_n^{(o)} - E_m^{(o)}\right) + E_n^{(1)}\delta_{mn}\right)\\ &+ \left(\delta_{mn}E_n^{(2)} + c_m^{(1)}E_n^{(1)} + c_m^{(2)}\left(E_n^{(o)} - E_m^{(o)}\right)\right) + \cdots\\ &= \sum_k \delta_{kn}\left\langle \psi_m^{(o)} | \mathrm{H}' | \psi_k^{(o)} \right\rangle + \sum_k c_k^{(1)}\left\langle \psi_m^{(o)} | \mathrm{H}' | \psi_k^{(o)} \right\rangle + \cdots\end{aligned} \tag{11.5.13b}$$

If on the opposite sides of Eq. (11.5.13b), we equate terms of like magnitude, the following system of linear equations results,

$$(E_n^{(o)} - E_m^{(o)})\delta_{mn} = 0 \tag{11.5.14a}$$

$$c_m^{(1)}\left(E_n^{(o)} - E_m^{(o)}\right) + E_n^{(1)}\delta_{mn} = \sum_k \delta_{kn}\left\langle\psi_m^{(o)}|\mathrm{H}'|\psi_k^{(o)}\right\rangle = \left\langle\psi_m^{(o)}|\mathrm{H}'|\psi_n^{(o)}\right\rangle \quad (11.5.14b)$$

$$\left(\delta_{mn}E_n^{(2)} + c_m^{(1)}E_n^{(1)} + c_m^{(2)}\left(E_n^{(o)} - E_m^{(o)}\right)\right) = \sum_k c_k^{(1)}\left\langle\psi_m^{(o)}|\mathrm{H}'|\psi_k^{(o)}\right\rangle \quad (11.5.14c)$$

Because δ_{mn} is zero when $m \neq n$ and unity when $m = n$, Eq. (11.5.14a) is an identity for all values of m and n. When $m = n$, Eq. (11.5.14b) reads

$$E_n^{(1)} = \left\langle\psi_n^{(o)}|\mathrm{H}'|\psi_n^{(o)}\right\rangle \quad (11.5.15a)$$

When $m \neq n$, Eq. (11.5.14b) can be solved for $c_m^{(1)}$. The result is

$$c_m^{(1)} = \frac{\left\langle\psi_m^{(o)}|\mathrm{H}'|\psi_n^{(o)}\right\rangle}{\left(E_n^{(o)} - E_m^{(o)}\right)} \quad (11.5.15b)$$

If we set $m = n$ in Eq. (11.5.14c), the result is

$$E_n^{(2)} + c_n^{(1)}E_n^{(1)} = \sum_k c_k^{(1)}\left\langle\psi_n^{(o)}|\mathrm{H}'|\psi_k^{(o)}\right\rangle \quad (11.5.16a)$$

Notice that according to Eq. (11.5.12b), $c_n^{(1)} = 0$. This recognition permits Eq. (11.5.16a) to be rewritten as

$$E_n^{(2)} = \sum_{k\neq n} c_k^{(1)}\left\langle\psi_n^{(o)}|\mathrm{H}'|\psi_k^{(o)}\right\rangle \quad (11.5.16b)$$

Next, let $m = k$ in Eq. (11.5.15b) so that it reads,

$$c_k^{(1)} = \frac{\left\langle\psi_k^{(o)}|\mathrm{H}'|\psi_n^{(o)}\right\rangle}{\left(E_n^{(o)} - E_k^{(o)}\right)} \quad (11.5.16c)$$

Upon substitution of Eq. (11.5.16c) into Eq. (11.5.16a), the result is

$$E_n^{(2)} = \sum_{k\neq n} \frac{\left\langle\psi_n^{(o)}|\mathrm{H}'|\psi_k^{(o)}\right\rangle\left\langle\psi_k^{(o)}|\mathrm{H}'|\psi_n^{(o)}\right\rangle}{\left(E_n^{(o)} - E_k^{(o)}\right)} \quad (11.5.16d)$$

By taking advantage of the Hermitian character of H′, we have

$$\left\langle\psi_k^{(o)}|\mathrm{H}'|\psi_n^{(o)}\right\rangle = \left\langle\psi_n^{(o)}|\mathrm{H}'|\psi_k^{(o)}\right\rangle^* \quad (11.5.16e)$$

Equation (11.5.16e) permits Eq. (11.5.16d) to be written in the form,

$$E_n^{(2)} = \sum_{k \neq n} \frac{\left|\left\langle \psi_k^{(o)} | \mathrm{H}' | \psi_n^{(o)} \right\rangle\right|^2}{\left(E_n^{(o)} - E_k^{(o)}\right)} \tag{11.5.16f}$$

Substitute Eqs. (11.5.15a) and (11.5.16f) into Eq. (11.5.8a). The final perturbation theory result estimating the energy, E_n, is

$$E_n = E_n^{(o)} + \left\langle \psi_n^{(o)} | \mathrm{H}' | \psi_n^{(o)} \right\rangle + \sum_{k \neq n} \frac{\left|\left\langle \psi_k^{(o)} | \mathrm{H}' | \psi_n^{(o)} \right\rangle\right|^2}{\left(E_n^{(o)} - E_k^{(o)}\right)} \tag{11.5.17}$$

Equation (11.5.17) is correct to second order.

11.6 Time-Dependent Perturbation Theory

Time-dependent perturbation theory is invoked when it becomes necessary to solve the time-dependent Schrödinger equation

$$i\hbar \frac{\partial \Psi(x,t)}{\partial t} = \mathrm{H(t)}\Psi(x,t) \tag{11.6.1}$$

In time-dependent perturbation theory, the Hamiltonian

$$\mathrm{H(t)} = \mathrm{H}^{(o)} + \mathrm{H}'(t) \tag{11.6.2}$$

has a time-dependent part, $\mathrm{H}'(t)$, which is small as compared with the time-independent part, $\mathrm{H}^{(o)}$. In time-dependent perturbation theory, it is assumed that we know the solutions to

$$i\hbar \frac{\partial \Psi_m^{(o)}(x,t)}{\partial t} = \mathrm{H}^{(o)}\Psi_m^{(o)}(x,t) \tag{11.6.3}$$

which according to Eq. (11.3.23) are of the form,

$$\Psi_m^{(o)}(x,t) = \psi_m^{(o)}(x)e^{-iE_m^{(o)}t/\hbar} \tag{11.6.4}$$

In Eq. (11.6.4), the wave functions, $\left\{\psi_m^{(o)}(x)\right\}$, and their associated eigenvalues, $\left\{E_m^{(o)}\right\}$, satisfy the time-independent Schrödinger equation,

$$\mathrm{H}^{(o)}\psi_m^{(o)}(x) = E_m^{(o)}\psi_m^{(o)}(x) \tag{11.6.5}$$

We propose to write $\Psi(x, t)$, which solves Eq. (11.6.1), as a linear combination of the wave functions, $\left\{\Psi_k^{(o)}(x, t)\right\}$. By linear combination, we mean the sum

$$\Psi(x, t) = \sum_k c_k(t)\Psi_k^{(o)}(x, t) \tag{11.6.6}$$

where the expansion coefficients, $\{c_k(t)\}$, are functions of the time, t. If we substitute Eq. (11.6.2) and our proposed solution, Eq. (11.6.6), into Eq. (11.6.1), the result is

$$i\hbar\frac{d}{dt}\left[\sum_k c_k\Psi_k^{(o)}\right] = \left(\mathrm{H}^{(o)} + \mathrm{H}'\right)\sum_k c_k\Psi_k^{(o)} \tag{11.6.7a}$$

Let the operators in Eq. (11.6.7a) operate on the wave functions in the sums. The resulting equation reads,

$$i\hbar\left[\sum_k \frac{dc_k}{dt}\Psi_k^{(o)} + \sum_k c_k\frac{d\Psi_k^{(o)}}{dt}\right] = \sum_k c_k\mathrm{H}^{(o)}\Psi_k^{(o)} + \sum_k c_k\mathrm{H}'\Psi_k^{(o)} \tag{11.6.7b}$$

In Eq. (11.6.7b), transfer $\sum_k c_k\mathrm{H}^{(o)}\Psi_k^{(o)}$ to the left with the result,

$$i\hbar\sum_k \frac{dc_k}{dt}\Psi_k^{(o)} + \sum_k c_k\left[i\hbar\frac{d\Psi_k^{(o)}}{dt} - \mathrm{H}^{(o)}\Psi_k^{(o)}\right] = \sum_k c_k\mathrm{H}'\Psi_k^{(o)} \tag{11.6.7c}$$

By virtue of Eq. (11.6.3), the quantity within the square bracket is zero. The remaining terms satisfy

$$i\hbar\sum_k \frac{dc_k}{dt}\Psi_k^{(o)} = \sum_k c_k\mathrm{H}'\Psi_k^{(o)} \tag{11.6.7d}$$

Apply the Dirac integral operation $\int dx\Psi_m^{(o)*} = \left\langle\Psi_m^{(o)}\right|$ to both sides of Eq. (11.6.7d). This operation produces the result,

$$i\hbar\sum_k \frac{dc_k}{dt}\left\langle\Psi_m^{(o)} \mid \Psi_k^{(o)}\right\rangle = \sum_k c_k\left\langle\Psi_m^{(o)}|\mathrm{H}'|\Psi_k^{(o)}\right\rangle \tag{11.6.7e}$$

Because the set of eigenfunctions, $\left\{\Psi_m^{(o)}\right\}$, are orthonormal, the left hand side simplifies so that Eq. (11.6.7e) reads,

$$i\hbar\sum_k \frac{dc_k}{dt}\delta_{mk} = \sum_k c_k\left\langle\Psi_m^{(o)}|\mathrm{H}'|\Psi_k^{(o)}\right\rangle \tag{11.6.7f}$$

After taking into account the Kronecker delta, the final result is

$$i\hbar\frac{dc_m}{dt} = \sum_k c_k\left\langle \Psi_m^{(o)}|H'|\Psi_k^{(o)}\right\rangle \quad (m = 1, 2, 3, \ldots) \tag{11.6.8}$$

Equation (11.6.8) is a system of first order differential equations to be solved for the set of coefficients $\{c_m(t)\}$. The system is assumed to be initially in the state $\psi_n^{(o)}(x)$. This initial condition is summarized by the equation,

$$c_m(0) = \delta_{mn} \tag{11.6.9}$$

On the right hand side of Eq. (11.6.8), the term, $c_k\left\langle \Psi_m^{(o)}|\mathrm{H}'|\Psi_k^{(o)}\right\rangle$, with $k = n$ is first order. All other $c_k\left\langle \Psi_m^{(o)}|\mathrm{H}'|\Psi_k^{(o)}\right\rangle$ terms with $k \neq n$ are of second order. If $\mathrm{H}'(\mathrm{t})$ is a weak function of the time, we can expect $c_n(t)$ to be slowly varying. This permits $c_n(t)$ in the sum on the right hand side of Eq. (11.6.8) to be replaced by its initial value, $c_n(0) = 1$. All of other terms involving c_k with $k \neq n$ will be of higher order in magnitude and can be dropped. After making these simplifications, Eq. (11.6.8) reads,

$$i\hbar\frac{dc_m}{dt} = \left\langle \Psi_m^{(o)}|\mathrm{H}'|\Psi_n^{(o)}\right\rangle \quad (m = 1, 2, 3, \ldots) \tag{11.6.10}$$

The next task is to solve Eq. (11.6.10) for the coefficients, $\{c_m\}$, subject to the initial conditions in Eq. (11.6.9). The result is known as Fermi's Golden Rule.

11.7 Fermi's Golden Rule

We are interested in a system initially in the state, $\psi_n^{(o)}$, which is then subjected to a perturbation, $\mathrm{H}'(\mathrm{t})$, which varies sinusoidally in time. The solution to Eq. (11.6.10) can then be used to evaluate the probability, $|c_m(t)|^2$, that the system is transferred to the state m, at a later time, t. The resulting expression for $|c_m(t)|^2$ can be used to calculate the rate of transfer from state n to state m. This formula is known as Fermi's Golden Rule (Enrico Fermi, Italian and American, 1901–1954). When teaching classes in quantum theory, Fermi dubbed it the "Golden Rule", but the formula, itself, is due to P. A. M. Dirac (English, 1902–1984). To derive the Golden Rule, we begin by representing $H'(x, t)$ in the form,

$$\mathrm{H}'(\mathrm{x}, \mathrm{t}) = 2V(x)\cos(\omega t) = 2V(x)\left(\frac{e^{i\omega t} + e^{-i\omega t}}{2}\right) \tag{11.7.1}$$

where $V(x)$ is a potential, which accounts for the possible spatial dependence of H′. As we shall see, only the term in $e^{-i\omega t}$ will prove to be resonant with any of the natural frequencies of the quantum system acted on by H′. This allows us to drop the positive exponential in Eq. (11.7.1) and write simply,

$$\mathrm{H}'(\mathrm{x},\mathrm{t}) = V(x)e^{-i\omega t} \tag{11.7.2a}$$

In preparation for the integration of Eq. (11.6.10), we can use Eqs. (11.6.4) and (11.7.2a) to compute

$$\begin{aligned}
&\left\langle \Psi_m^{(o)}(x,t)|\mathrm{H}'(x,t)|\Psi_n^{(o)}(x,t)\right\rangle = \left\langle e^{-iE_m^{(o)}t}\psi_m^{(o)}(x)|V(x)e^{-iot|}|e^{-iE_n^{(o)}t}\psi_n^{(o)}(x)\right\rangle \\
&= \int dx\left(e^{-iE_m^{(o)}t}\psi_m^{(o)}(x)\right)^*\left(V(x)e^{-iot}\right)\left(e^{-iE_n^{(o)}t}\psi_n^{(o)}(x)\right) \\
&= e^{i\left(E_m^{(o)}-E_n^{(o)}-\hbar\omega\right)t/\hbar}\int dx\left[\psi_m^{(o)*}(x)V(x)\psi_n^{(o)}(x)\right] \\
&= \left\langle \psi_m^{(o)}|V|\psi_n^{(o)}\right\rangle e^{i\left(E_m^{(o)}-E_n^{(o)}-\hbar\omega\right)t/\hbar} \\
&= \langle m|V|n\rangle e^{i\left(E_m^{(o)}-E_n^{(o)}-\hbar\omega\right)t/\hbar}
\end{aligned} \tag{11.7.2b}$$

Substitute Eq. (11.7.2b) into the right hand side of Eq. (11.6.10). The result is

$$i\hbar\frac{dc_m}{dt} = \langle m|V|n\rangle e^{i\left(E_m^{(o)}-E_n^{(o)}-\hbar\omega\right)t/\hbar} \tag{11.7.3}$$

The formal integral of Eq. (11.7.3) is

$$c_m(t) = \frac{\langle m|V|n\rangle}{i\hbar}\int_0^t dt\left[e^{i\left(E_m^{(o)}-E_n^{(o)}-\hbar\omega\right)t/\hbar}\right] \tag{11.7.4a}$$

where we have made use of the initial condition as stated in Eq. (11.6.9). Evaluation of the integral on the right of Eq. (11.7.4a) leads to the result,

$$c_m(t) = \frac{\langle m|V|n\rangle}{i\hbar}\left[\frac{\exp\left(i(E_m^{(o)}-E_n^{(o)}-\hbar\omega)t/\hbar\right)-1}{i(E_m^{(o)}-E_n^{(o)}-\hbar\omega)/\hbar}\right] \tag{11.7.4b}$$

Factor the square root of the exponential out the bracket. Equation (11.7.4b) then reads,

$$c_m(t) = \frac{\langle m|V|n\rangle}{i\hbar}e^{i\left(E_m^{(o)}-E_n^{(o)}-\hbar\omega\right)t/2\hbar}\left[\frac{e^{i\left(E_m^{(o)}-E_n^{(o)}-\hbar\omega\right)t/2\hbar}-e^{-i\left(E_m^{(o)}-E_n^{(o)}-\hbar\omega\right)t/2\hbar}}{i\left(E_m^{(o)}-E_n^{(o)}-\hbar\omega\right)/\hbar}\right] \tag{11.7.4c}$$

Multiply the right hand side of Eq. (11.7.4c) by two and divide it also by two.

Rewrite the result in the form,

$$c_m(t) = \frac{2\langle m|V|n\rangle}{i\hbar} e^{i\left(E_m^{(o)} - E_n^{(o)} - \hbar\omega\right)t/2\hbar} \frac{1}{\left(E_m^{(o)} - E_n^{(o)} - \hbar\omega\right)/\hbar} \times \left[\frac{e^{i\left(E_m^{(o)} - E_n^{(o)} - \hbar\omega\right)t/2\hbar} - e^{-i\left(E_m^{(o)} - E_n^{(o)} - \hbar\omega\right)t/2\hbar}}{2i}\right] \tag{11.7.4d}$$

The expression within the square bracket in Eq. (11.7.4d) can be replaced by a sine function. The result can be written in the form,

$$c_m(t) = \frac{\langle m|V|n\rangle}{i\hbar} e^{i\left(E_m^{(o)} - E_n^{(o)} - \hbar\omega\right)t/2\hbar} \left\{\frac{\sin\left[\left(E_m^{(o)} - E_n^{(o)} - \hbar\omega\right)t/2\hbar\right]}{\left(E_m^{(o)} - E_n^{(o)} - \hbar\omega\right)/2\hbar}\right\} \tag{11.7.4e}$$

The absolute square of $c_m(t)$ is

$$|c_m(t)|^2 = \frac{|\langle m|V|n\rangle|^2}{\hbar^2} \left\{\frac{\sin^2\left[\left(E_m^{(o)} - E_n^{(o)} - \hbar\omega\right)t/2\hbar\right]}{\left[\left(E_m^{(o)} - E_n^{(o)} - \hbar\omega\right)/2\hbar\right]^2}\right\} \tag{11.7.5}$$

The quantity, $|c_m(t)|^2$, can be interpreted as the probability at time, t, that a system initially with energy, $E_n^{(o)}$, has reached the state with energy, $E_m^{(o)}$. According to Eq. (11.7.5), $|c_m(t)|^2$ has a maximum when $\hbar\omega = E_m^{(o)} - E_n^{(o)}$. This phenomenon is termed "resonance".

To simplify the proof of Fermi's Golden Rule, we let

$$E = E_m^{(o)} - E_n^{(o)} - \hbar\omega \tag{11.7.6}$$

so that Eq. (11.7.5) reads

$$|c_m(t)|^2 = \frac{|\langle m|V|n\rangle|^2}{\hbar^2} \frac{\sin^2(Et/2\hbar)}{(E/2\hbar)^2} \tag{11.7.7}$$

The system may support multiple excited states, all having in common the energy, $E_m^{(o)}$. Quantum states having the same energy are said to be *degenerate*. Thus, the single value of E is capable of describing the transition from the ground state, $E_n^{(o)}$, to any one of these degenerate excited states. To enumerate the possible transitions with the same E, we introduce the *density of states* function, $D(E)$. The value of this function equals the number of allowed transitions per unit energy, which connect $E_n^{(o)}$ to $E_m^{(o)}$. The integral, $\int D(E)dE$, which is dimensionless, enumerates the total possibilities. To take into account the degeneracy phenomenon, we multiply Eq. (11.7.7) by $D(E)$. The result is

$$|c_m(t)|^2 = \frac{|\langle m|V|n\rangle|^2}{\hbar^2} \frac{\sin^2(Et/2\hbar)}{(E/2\hbar)^2} D(E) \tag{11.7.8}$$

The total probability, P, that the system has transitioned from its initial state, $E_n^{(o)}$, to any of the degenerate excited states is given by the integral,

$$P = \int \frac{|\langle m|V|n\rangle|^2}{\hbar^2} \frac{\sin^2(Et/2\hbar)}{(E/2\hbar)^2} D(E)dE \tag{11.7.9}$$

The required integral can be evaluated by making the change in variable,

$$x = Et/2\hbar \tag{11.7.10}$$

where $dE = (2\hbar/t)dx$. With this strategy, Eq. (11.7.9) reads,

$$\begin{aligned} P &= \int \frac{|\langle m|V|n\rangle|^2}{\hbar^2} t^2 \frac{\sin^2(Et/2\hbar)}{(Et/2\hbar)^2} D(E)dE \\ &= \frac{|\langle m|V|n\rangle|^2}{\hbar^2} \cdot D(E)t^2 \left(\frac{2\hbar}{t}\right) \cdot \int_{-\infty}^{+\infty} dx \frac{\sin^2(x)}{x^2} \end{aligned} \tag{11.7.11}$$

where we have factored out of the integrand constants and the density of states function, $D(E)$, because it varies slowly with E. This latter factorization is further justified because the function, $\frac{\sin^2(x)}{x^2}$, is strongly peaked around $E = 0$, which corresponds to $x = 0$. As found in tables of integrals [See the Mathematical Methods Appendix], the integral in Eq. (11.7.11) has the value

$$\int_{-\infty}^{+\infty} \frac{\sin^2(x)}{x^2} = \pi \tag{11.7.12}$$

Upon substitution of Eq. (11.7.12) into Eq. (11.7.11), the latter reads

$$P = \frac{2\pi}{\hbar} |\langle m|V|n\rangle|^2 D(E)t \tag{11.7.13}$$

Since P is the dimensionless probability of transition, its time derivative, $W = dP/dt$, is the rate of transition. Upon differentiating Eq. (11.7.13) with respect to time, we get

$$W = \frac{2\pi}{\hbar} |\langle m|V|n\rangle|^2 D(E) \tag{11.7.14}$$

As mentioned, Eq. (11.7.14) was heavily used and publicized by Enrico Fermi. It has become known as Fermi's Golden Rule.

11.8 Einstein's A and B Coefficients

11.8.1 Construction of the Hamiltonian Operator

The Einstein A and B coefficients describe the absorption, stimulated emission, and spontaneous emission of electromagnetic radiation by matter on the atomic and molecular scale. We begin with the Chap. 10 picture of a classical particle interacting with the electromagnetic field. We then introduce the quantum mechanical additions to this theory which are required to make it apply to the resonance behavior exhibited by atomic and molecular systems. A completely rigorous theory of the absorption of radiation by matter requires quantization of both the matter and the radiation. Fortunately, for the purposes of chemistry, the quantization of matter is all that is required. Limited in this way, the theory is said to be "semi-classical".

In classical mechanics, the Hamiltonian describing the motion of a particle of mass, m, and charge, q, interacting with applied electric and magnetic fields is given by Eq. (10.5.25), which we copy below:

$$H = \frac{(\mathbf{p} - q\mathbf{A})^2}{2m} + q\Phi \tag{11.8.1}$$

To convert Eq. (11.8.1) into a quantum mechanical operator, we replace the momentum vector, $\mathbf{p}$, by the differential operator, $\mathbf{p} = -i\hbar\nabla$, which serves as the three dimensional generalization of Eq. (11.3.13). The result is

$$\mathrm{H} = \frac{(-i\hbar\nabla - q\mathbf{A})^2}{2m} + q\Phi \tag{11.8.2}$$

Consider H operating on a wave function, ψ:

$$\mathrm{H}\psi = \frac{1}{2m}(-i\hbar\nabla - q\mathbf{A})^2\psi + q\Phi\psi \tag{11.8.3}$$

By virtue of the fact that it is squared, the operator, $-i\hbar\nabla - q\mathbf{A}$, in Eq. (11.8.3) gets applied twice. This procedure can best be understood if we restrict our attention to the motion of a particle confined to one dimension. In one dimension, the operator assumes the simplified form, $-i\hbar(d/dx) - qA(x)$. Its square then proceeds as two sequential applications of $-i\hbar(d/dx) - qA(x)$ to the wave function, $\psi(x)$. The calculation is shown below:

$$\left(-i\hbar\frac{d}{dx} - qA(x)\right)^2\psi(x)$$
$$= \left(-i\hbar\frac{d}{dx} - qA(x)\right)\left(-i\hbar\frac{d}{dx} - qA(x)\right)\psi(x)$$

$$= \left(-i\hbar\frac{d}{dx} - qA(x)\right)\left(-i\hbar\frac{d\psi(x)}{dx} - qA(x)\psi(x)\right)$$
$$= -\hbar^2\frac{d^2\psi(x)}{dx^2} + i\hbar q\frac{d}{dx}(A(x)\psi(x)) + i\hbar qA(x)\frac{d\psi(x)}{dx} + q^2(A(x))^2\psi(x) \tag{11.8.4a}$$

With this illustration, we can now cope with Eq. (11.8.3). As in Eq. (11.8.4a), the calculation proceeds in steps.

$$\begin{aligned} \mathrm{H}\psi &= \frac{1}{2m}(-i\hbar\nabla - q\mathbf{A})^2\psi + q\Phi\psi \\ &= \frac{1}{2m}(-i\hbar\nabla - q\mathbf{A})\cdot(-i\hbar\nabla - q\mathbf{A})\psi + q\Phi\psi \\ &= \frac{1}{2m}(-i\hbar\nabla - q\mathbf{A})\cdot(-i\hbar\nabla\psi - q\mathbf{A}\psi) + q\Phi\psi \end{aligned} \tag{11.8.4b}$$

Calculation of the remaining dot product in Eq. (11.18.8.4b) proceeds with the results:

$$\mathrm{H}\psi = \frac{1}{2m}\left(-\hbar^2\nabla^2\psi + i\hbar q\nabla\cdot(\mathbf{A}\psi) + i\hbar q\mathbf{A}\cdot\nabla\psi + q^2A^2\psi\right) + q\Phi\psi \tag{11.8.4c}$$

where we have noted that $\mathbf{A}\cdot\mathbf{A} = A^2$. Equation (11.8.4c) is the generalization of Eq. (11.8.4a) to three dimensions. To proceed further, we expand $\nabla\cdot(\mathbf{A}\psi)$ in Eq. (11.8.4c) using the vector operator identity [see the Mathematical Methods Appendix],

$$\nabla\cdot(\mathbf{A}\psi) = \psi\nabla\cdot\mathbf{A} + \mathbf{A}\cdot\nabla\psi \tag{11.8.4d}$$

Upon substitution of Eq. (11.8.4d) into Eq. (11.8.4c), the result is

$$\mathrm{H}\psi = \frac{1}{2m}\left(-\hbar^2\nabla^2\psi + i\hbar q\psi\nabla\cdot\mathbf{A} + i\hbar q\mathbf{A}\cdot\nabla\psi + i\hbar q\mathbf{A}\cdot\nabla\psi + q^2A^2\psi\right) + q\Phi\psi \tag{11.8.4e}$$

It would simplify things if we could rid ourselves of $\nabla\cdot\mathbf{A}$ in Eq. (11.8.4e). We know that $\mathbf{B} = \nabla\times\mathbf{A}$. Since $\mathbf{B}$ is entirely determined by $\nabla\times\mathbf{A}$, it is independent of $\nabla\cdot\mathbf{A}$. It thus may be possible to adjust $\mathbf{A}$ in a way that preserves $\mathbf{B} = \nabla\times\mathbf{A}$ but makes $\nabla\cdot\mathbf{A} = 0$. The gauge theory comes to mind. The argument proceeds as follows: If χ is an arbitrary scalar valued function, the curl of its gradient satisfies the identity $\nabla\times\nabla\chi = 0$ [See the Mathematical Methods Appendix]. Thus, the vector field $\mathbf{A}$ and the vector field defined by $\mathbf{A}' = \mathbf{A} + \nabla\chi$ give the same result for the magnetic flux density, because $\mathbf{B} = \nabla\times\mathbf{A} = \nabla\times\mathbf{A}'$. This permits us to replace $\mathbf{A}$ as needed by a new vector potential, $\mathbf{A}'$, that satisfies

$$\nabla\cdot\mathbf{A}' = \nabla\cdot\mathbf{A} + \nabla^2\chi = 0 \tag{11.8.5a}$$

Equation (11.8.5a) can be rewritten as $\nabla^2\chi = -\nabla\cdot\mathbf{A}$. This is Poisson's equation. It is structurally analogous to $\nabla^2\Phi = -\rho/\varepsilon_r\varepsilon_o$, which determines the Coulomb's law potential, Φ, associated with the charge density, ρ. For this reason, the corresponding gauge function, χ, is sometimes referred to as "Coulomb gauge".

Example Construction of a gauge function.

The vector potential

$$\mathbf{A} = Bx\hat{\mathbf{e}}_y + Bz\hat{\mathbf{e}}_z \tag{11.8.5b}$$

suffices to generate a magnetic induction, B, in the z-direction. The calculation proceeds as follows:

$$\mathbf{B} = \nabla\times\mathbf{A} = \begin{bmatrix} \hat{\mathbf{e}}_x & \hat{\mathbf{e}}_y & \hat{\mathbf{e}}_z \\ \frac{\partial}{\partial x} & \frac{\partial}{\partial y} & \frac{\partial}{\partial z} \\ 0 & Bx & Bz \end{bmatrix} = B\hat{\mathbf{e}}_z \tag{11.8.5c}$$

Below, we show that the divergence of $\mathbf{A}$ is not zero:

$$\nabla\cdot\mathbf{A} = \left(\hat{\mathbf{e}}_x\frac{\partial}{\partial x} + \hat{\mathbf{e}}_y\frac{\partial}{\partial y} + \hat{\mathbf{e}}_z\frac{\partial}{\partial z}\right)\cdot\left(Bx\hat{\mathbf{e}}_y + Bz\hat{\mathbf{e}}_z\right) = (\hat{\mathbf{e}}_z\cdot\hat{\mathbf{e}}_z)\frac{\partial(Bz)}{\partial z} = B \tag{11.8.5d}$$

If we want to replace $\mathbf{A}$ with a vector potential, $\mathbf{A}'$, for which the divergence is zero, we can let

$$\mathbf{A}' = \mathbf{A} + \nabla\chi = \mathbf{A} - Bz\hat{\mathbf{e}}_z \tag{11.8.5e}$$

According to Eq. (11.8.5e),

$$\nabla\chi = -Bz\hat{\mathbf{e}}_z \tag{11.8.5f}$$

Equation (11.8.5f) can be solved for χ to obtain

$$\chi = -B\frac{z^2}{2} \tag{11.8.5g}$$

Notice that

$$\nabla^2\chi = -\nabla\cdot\mathbf{A} \tag{11.8.5h}$$

which qualifies Eq. (11.8.5g) as a Coulomb gauge function.

Now that we understand that it is always possible to replace $\mathbf{A}$ by a vector potential, $\mathbf{A}'$, with the property, $\nabla\cdot\mathbf{A}' = 0$, we will assume that this has already been done in the case of Eq. (11.8.4e). With this assumption, Eq. (11.8.4e) reads,

$$\mathrm{H}\psi = \frac{1}{2m}\left(-\hbar^2\nabla^2\psi + 2i\hbar q\mathbf{A}\cdot\nabla\psi + q^2A^2\psi\right) + q\Phi\psi \tag{11.8.6}$$

Equation (11.8.6) can be rearranged to read,

$$\mathrm{H}\psi = \frac{-\hbar^2}{2m}\nabla^2\psi + q\Phi\psi + \frac{i\hbar q\mathbf{A}\cdot\nabla\psi}{m} + \frac{q^2A^2\psi}{2m} \tag{11.8.7}$$

In the case that there is no magnetic field, we can set $\mathbf{A} = 0$. With $\mathbf{A}$ set to zero, the first two terms on the right hand side of Eq. (11.8.7) define a time-independent Hamiltonian for a particle with charge, q, and mass, m, acted on by an electrostatic potential, Φ. The third and fourth terms in Eq. (11.8.7) come into play when the particle is interacting with a magnetic field. The third term is ordinarily smaller than the first two, whereas the fourth term is smaller still. Including only the first three terms, Eq. (11.8.7) reads

$$\mathrm{H}\psi = \frac{-\hbar^2}{2m}\nabla^2\psi + q\Phi\psi + \frac{i\hbar q\mathbf{A}\cdot\nabla\psi}{m} \tag{11.8.8}$$

Since the momentum operator is $\mathbf{p} = -i\hbar\nabla$, the last term in Eq. (11.8.8) can be rewritten so that

$$\mathrm{H}\psi = \frac{-\hbar^2}{2m}\nabla^2\psi + q\Phi\psi - \frac{q\mathbf{A}}{m}\cdot\mathbf{p}\psi \tag{11.8.9}$$

For the purposes of incorporating Eq. (11.8.9) into the time-dependent perturbation theory, we can identify the time independent, zeroth order Hamiltonian and a time-dependent first order perturbation. The time-independent Hamiltonian is

$$\mathrm{H}^{(o)} = \frac{-\hbar^2}{2m}\nabla^2 + q\Phi \tag{11.8.10a}$$

where Φ contains all of the electrostatic interactions between the nuclei and electrons in the molecule. As such, it is assigned the role of the potential energy function, $U(x)$, that determines the molecular electronic structure. In Eq. (11.8.9), the time dependent, first order perturbation is

$$\mathrm{H}'(t) = -\frac{q\mathbf{A}}{m}\cdot\mathbf{p} \tag{11.8.10b}$$

We have now identified the leading terms in the Hamiltonian operator. The next task is to calculate the matrix elements of $\mathrm{H}'(t)$.

11.8.2 Matrix Elements of the Perturbation

If we assume that the applied field is a wave that varies sinusoidally in both space and time, then **A** takes the form,

$$\mathbf{A}(\mathbf{r}, t) = \mathbf{A}_o \cos(\mathbf{k} \cdot \mathbf{r} - \omega t) \tag{11.8.11}$$

where $\mathbf{A}_o$ is the vector amplitude. Upon substitution of Eq. (11.8.11) into Eq. (11.8.10b), $\mathrm{H}'(t)$ can be written as,

$$\mathrm{H}'(t) = -\frac{q}{m}\mathbf{A}_o \cdot \mathbf{p} \cos(\mathbf{k} \cdot \mathbf{r} - \omega t) \tag{11.8.12}$$

In the case of light with wavelength, λ, in the infrared, visible, or ultraviolet regions of the spectrum, λ is much longer than the size of the molecule. Applied to the wave number, $k = 2\pi/\lambda$, this observation indicates that $\mathbf{k} \cdot \mathbf{r} \ll 1$. In this approximation, Eq. (11.8.12) simplifies to read,

$$\mathrm{H}'(t) = -\frac{q}{m}\mathbf{A}_o \cos(\omega t) \cdot \mathbf{p} \tag{11.8.13}$$

Equation (11.8.13) can be further simplified by replacing **p** by an operator which involves **r** and H. The required relation is best understood by proving it for the one-dimensional case.

Theorem 11.3 $[x, \mathrm{H}] = \frac{i\hbar}{m}\mathrm{p}_x$, where $[x, \mathrm{H}] \equiv x\mathrm{H} - \mathrm{H}x$.

Proof The operator, $[x, \mathrm{H}] = x\mathrm{H} - \mathrm{H}x$, is an example of a "commutator", which when applied to ψ, reads

$$[x, \mathrm{H}]\psi = x(\mathrm{H}\psi) - \mathrm{H}(x\psi) \tag{11.8.14}$$

In one dimension, the Hamiltonian assumes the form

$$\mathrm{H} = \frac{-\hbar^2}{2m}\frac{d^2}{dx^2} + U(x) \tag{11.8.15a}$$

where $U(x)$ is the potential energy function. Using the definition in Eq. (11.18.15a), the first term on the right hand side of Eq. (11.8.14) reads,

$$x(\mathrm{H}\psi) = x\left[\frac{-\hbar^2}{2m}\frac{d^2}{dx^2} + U(x)\right]\psi(x) = -x\frac{\hbar^2}{2m}\frac{d^2\psi(x)}{dx^2} + xU(x)\psi(x) \tag{11.8.15b}$$

The second term on the right hand side of Eq. (11.8.14) can be expanded as follows

$$\mathrm{H}(x\psi) = \left[\frac{-\hbar^2}{2m}\frac{d^2}{dx^2} + U(x)\right](x\psi(x)) = \frac{-\hbar^2}{2m}\frac{d^2(x\psi(x))}{dx^2} + U(x)x\psi(x)$$

$$= \frac{-\hbar^2}{2m}\frac{d}{dx}\left[x\frac{d\psi}{dx} + \psi\right] + U(x)x\psi$$

$$= \frac{-\hbar^2}{2m}\left[x\frac{d^2\psi}{dx^2} + \frac{d\psi}{dx} + \frac{d\psi}{dx}\right] + U(x)x\psi$$

$$= \frac{-\hbar^2}{2m}\left[x\frac{d^2\psi}{dx^2} + 2\frac{d\psi}{dx}\right] + U(x)x\psi \tag{11.8.15c}$$

When Eqs. (11.8.15b) and (11.8.15c) are substituted into Eq. (11.8.14), the result is

$$\begin{aligned}[x, \mathrm{H}]\psi &= x(\mathrm{H}\psi) - \mathrm{H}(x\psi)\\ &= -x\frac{\hbar^2}{2m}\frac{d^2\psi(x)}{dx^2} + xU(x)\psi(x) + x\frac{\hbar^2}{2m}\frac{d^2\psi}{dx^2} + 2\frac{\hbar^2}{2m}\frac{d\psi}{dx} - U(x)x\psi\\ &= \frac{\hbar^2}{m}\frac{d\psi(x)}{dx}\end{aligned} \tag{11.8.15d}$$

Equation (11.8.15d) can be rewritten as

$$[x, \mathrm{H}]\psi = \frac{i\hbar}{m}\left(-i\hbar\frac{d}{dx}\right)\psi \tag{11.8.15e}$$

By recognizing the definition of the linear momentum operator, $\mathrm{p}_x = -i\hbar(d/dx)$, Eq. (11.8.15e) can be rendered in symbolic terms as

$$[x, \mathrm{H}] = \frac{i\hbar}{m}\mathrm{p}_x \blacksquare \tag{11.8.16}$$

The generalization of Eq. (11.8.16) to three dimensions then follows:

$$[\mathbf{r}, \mathrm{H}] = \frac{i\hbar}{m}\mathbf{p} \tag{11.8.17}$$

We can solve Eq. (11.8.17) for **p** to get

$$\mathbf{p} = \frac{m}{i\hbar}[\mathbf{r}, \mathrm{H}] \tag{11.8.18a}$$

Upon substitution of Eq. (11.8.18a) into Eq. (11.8.13), the result is

$$\mathrm{H}'(t) = -\frac{q}{m}\mathbf{A}_o\cos(\omega t)\cdot\frac{m}{i\hbar}[\mathbf{r}, \mathrm{H}] \tag{11.8.18b}$$

If we use the exponential representation of the cosine, Eq. (11.8.18b) becomes

$$\mathrm{H}'(t) = -\frac{q}{i\hbar}\mathbf{A}_o\left[\frac{e^{i\omega t} + e^{-i\omega t}}{2}\right]\cdot[\mathbf{r}, \mathrm{H}] \tag{11.8.18c}$$

As we shall see, only $e^{-i\omega t}$ is capable in participating in a resonance, so we can ignore $e^{i\omega t}$ and approximate Eq. (11.8.18c) as

$$\mathrm{H}'(t) = -\frac{q}{2i\hbar}\mathbf{A}_o\cdot[\mathbf{r}, \mathrm{H}]e^{-i\omega t} \tag{11.8.19}$$

Substitute Eq. (11.6.2) into Eq. (11.8.19) to obtain,

$$\mathrm{H}'(t) = -\frac{q}{2i\hbar}\mathbf{A}_o\cdot\left[\mathbf{r}, \left(\mathrm{H}^{(o)} + \mathrm{H}'(t)\right)\right]e^{-i\omega t} \tag{11.8.20a}$$

Since $\mathrm{H}'(t) \ll \mathrm{H}^{(o)}$, Eq. (11.8.20a) can be simplified to read,

$$\mathrm{H}'(t) = -\frac{q}{2i\hbar}\mathbf{A}_o\cdot\left[\mathbf{r}, \mathrm{H}^{(o)}\right]e^{-i\omega t} \tag{11.8.20b}$$

which is correct to lowest order in $\mathbf{A}_o$.

By virtue of Eq. (11.8.20b), the right hand side of Eq. (11.6.10) is

$$\left\langle\Psi_m^{(o)}\middle|\mathrm{H}'\middle|\Psi_n^{(o)}\right\rangle = -\frac{q}{2i\hbar}\mathbf{A}_o\cdot\left\langle\Psi_m^{(o)}\middle|\left[\mathbf{r}, \mathrm{H}^{(o)}\right]\middle|\Psi_n^{(o)}\right\rangle e^{-i\omega t} \tag{11.8.21}$$

We can expand the matrix element on the right hand side of Eq. (11.8.21) as follows:

$$\begin{aligned}\left\langle\Psi_m^{(o)}\middle|\left[\mathbf{r}, \mathrm{H}^{(o)}\right]\middle|\Psi_n^{(o)}\right\rangle &= \left\langle\Psi_m^{(o)}\middle|\mathbf{r}\mathrm{H}^{(o)} - \mathrm{H}^{(o)}\mathbf{r}\middle|\Psi_n^{(o)}\right\rangle \\ &= \left\langle\Psi_m^{(o)}\middle|\mathbf{r}\mathrm{H}^{(o)}\middle|\Psi_n^{(o)}\right\rangle - \left\langle\Psi_m^{(o)}\middle|\mathrm{H}^{(o)}\mathbf{r}\middle|\Psi_n^{(o)}\right\rangle\end{aligned} \tag{11.8.22}$$

The matrix element, $\left\langle\Psi_m^{(o)}\middle|\mathbf{r}\mathrm{H}^{(o)}\middle|\Psi_n^{(o)}\right\rangle$, in Eq. (11.8.22) can be simplified by substituting the wave function given by Eq. (11.6.4). Using d^3r to represent the differential element of space in three dimensions, the effect of the substitution is

$$\begin{aligned}\left\langle\Psi_m^{(o)}\middle|\mathbf{r}\mathrm{H}^{(o)}\middle|\Psi_n^{(o)}\right\rangle &= \int d^3r\left(\psi_m^{(o)}(\mathbf{r})e^{-iE_m^{(o)}t/\hbar}\right)^*\left(\mathbf{r}\mathrm{H}^{(o)}\right)\left(\psi_n^{(o)}(\mathbf{r})e^{-iE_n^{(o)}t/\hbar}\right) \\ &= e^{iE_m^{(o)}t/\hbar}\left[\int d^3r\left(\psi_m^{(o)}(\mathbf{r})\right)^*\left(\mathbf{r}\mathrm{H}^{(o)}\right)\psi_n^{(o)}(\mathbf{r})\right]e^{-iE_n^{(o)}t/\hbar} \\ &= \left\langle\psi_m^{(o)}\middle|\mathbf{r}\mathrm{H}^{(o)}\middle|\psi_n^{(o)}\right\rangle e^{i(E_m^{(o)} - E_n^{(o)})t/\hbar}\end{aligned} \tag{11.8.23a}$$

where we have used the Dirac integral definition,

$$\left\langle\psi_m^{(o)}\middle|\mathbf{r}\mathrm{H}^{(o)}\middle|\psi_n^{(o)}\right\rangle = \int d^3r\left(\psi_m^{(o)}(\mathbf{r})\right)^*\left(\mathbf{r}\mathrm{H}^{(o)}\right)\psi_n^{(o)}(\mathbf{r}) \tag{11.8.23b}$$

By following the same procedure, the second term on the right of Eq. (11.8.22) is

$$\left\langle\Psi_m^{(o)}\middle|\mathrm{H}^{(o)}\mathbf{r}\middle|\Psi_n^{(o)}\right\rangle=\left\langle\psi_m^{(o)}\middle|\mathrm{H}^{(o)}\mathbf{r}\middle|\psi_n^{(o)}\right\rangle e^{i(E_m^{(o)}-E_n^{(o)})t/\hbar} \tag{11.8.24a}$$

where

$$\left\langle\psi_m^{(o)}\middle|\mathrm{H}^{(o)}\mathbf{r}\middle|\psi_n^{(o)}\right\rangle=\int d^3r\left(\psi_m^{(o)}(\mathbf{r})\right)^*\left(\mathrm{H}^{(o)}\mathbf{r}\right)\psi_n^{(o)}(\mathbf{r}) \tag{11.8.24b}$$

Substitute Eqs. (11.8.23a) and (11.8.24a) into Eq. (11.8.22). The result is

$$\begin{aligned}\left\langle\Psi_m^{(o)}\middle|\left[\mathbf{r},\mathrm{H}^{(o)}\right]\middle|\Psi_n^{(o)}\right\rangle&=\left\langle\psi_m^{(o)}\middle|\mathbf{r}\mathrm{H}^{(o)}\middle|\psi_n^{(o)}\right\rangle e^{i\left(E_m^{(o)}-E_n^{(o)}\right)t/\hbar}-\left\langle\psi_m^{(o)}\middle|\mathrm{H}^{(o)}\mathbf{r}\middle|\psi_n^{(o)}\right\rangle e^{i\left(E_m^{(o)}-E_n^{(o)}\right)t/\hbar}\\&=\left\langle\psi_m^{(o)}\middle|\left[\mathbf{r},\mathrm{H}^{(o)}\right]\middle|\psi_n^{(o)}\right\rangle e^{i\left(E_m^{(o)}-E_n^{(o)}\right)t/\hbar}\end{aligned} \tag{11.8.24c}$$

When Eq. (11.8.24c) is substituted into Eq. (11.8.21), we have

$$\left\langle\Psi_m^{(o)}\middle|\mathrm{H}'\middle|\Psi_n^{(o)}\right\rangle=-\frac{q}{2i\hbar}\mathbf{A}_o\cdot\left\langle\psi_m^{(o)}\middle|\left[\mathbf{r},\mathrm{H}^{(o)}\right]\middle|\psi_n^{(o)}\right\rangle e^{i(E_m^{(o)}-E_n^{(o)}-\hbar\omega)t/\hbar} \tag{11.8.25}$$

In Eq. (11.8.25), we are assuming $E_m^{(o)}-E_n^{(o)}>0$. Because resonance corresponds to $\hbar\omega=E_m^{(o)}-E_n^{(o)}$, this reflects our neglect of $e^{i\omega t}$ in going from Eq. (11.8.18c) to Eq. (11.8.19). In the simplified Dirac notation, the matrix element of the commutator, $\left[\mathbf{r},H^{(o)}\right]$, can be written in the form,

$$\left\langle\psi_m^{(o)}\middle|\left[\mathbf{r},\mathrm{H}^{(o)}\right]\middle|\psi_n^{(o)}\right\rangle=\left\langle m\middle|\left[\mathbf{r},\mathrm{H}^{(o)}\right]\middle|n\right\rangle \tag{11.8.26}$$

The following operator identity will prove useful.

Theorem 11.4 $\left\langle m\middle|\left[\mathbf{r},\mathrm{H}^{(o)}\right]\middle|n\right\rangle=\left(E_n^{(o)}-E_m^{(o)}\right)\langle m|\mathbf{r}|n\rangle$

Proof Consider the commutator expansion,

$$\left\langle m\middle|\left[\mathbf{r},H^{(o)}\right]\middle|n\right\rangle=\langle m|\mathbf{r}\mathrm{H}^{(o)}-\mathrm{H}^{(o)}\mathbf{r}|n\rangle=\langle m|\mathbf{r}\mathrm{H}^{(o)}|n\rangle-\langle m|\mathrm{H}^{(o)}\mathbf{r}|n\rangle \tag{11.8.27a}$$

Since $\langle m|\mathbf{r}\mathrm{H}^{(o)}|n\rangle$ and $\langle m|\mathrm{H}^{(o)}\mathbf{r}|n\rangle$ are identified by the two indices, m and n, they can be treated as if they were elements of the matrices representing the operators, $\mathbf{r}\mathrm{H}^{(o)}$ and $\mathrm{H}^{(o)}\mathbf{r}$, respectively. Pursuing this analogy further, the "operators", $\mathbf{r}$ and $\mathrm{H}^{(o)}$, have their own matrix representations. For example, the generic element of the matrix representation of $\mathbf{r}$ can be written as $\langle m|\mathbf{r}|k\rangle$, while the generic element of the matrix representation of $\mathrm{H}^{(o)}$ can be written as $\langle k|\mathrm{H}^{(o)}|n\rangle$. In both of these representations, k is an integer. According to the rules of matrix algebra, the matrix of $\mathbf{r}\mathrm{H}^{(o)}$ can be represented as the product of the matrix representations of $\mathbf{r}$ and $\mathrm{H}^{(o)}$. The element, $\langle m|\mathbf{r}\mathrm{H}^{(o)}|n\rangle$, of the matrix representation of the product, $\mathbf{r}\mathrm{H}^{(o)}$, is computed by evaluating the sum,

$$\langle m|\mathbf{r}\mathrm{H}^{(o)}|n\rangle = \sum_k \langle m|\mathbf{r}|k\rangle\langle k|\mathrm{H}^{(o)}|n\rangle \tag{11.8.27b}$$

By analogy, the matrix element, $\langle m|\mathrm{H}^{(o)}\mathbf{r}|n\rangle$, can be computed as the sum,

$$\langle m|\mathrm{H}^{(o)}\mathbf{r}|n\rangle = \sum_k \langle m|H^{(o)}|k\rangle\langle k|\mathbf{r}|n\rangle \tag{11.8.27c}$$

Equations (11.8.27a) through (11.8.27c) permit us to write in Dirac notation,

$$\left\langle m\middle|\left[\mathbf{r}, H^{(o)}\right]\middle|n\right\rangle = \sum_k \langle m|\mathbf{r}|k\rangle\langle k|\mathrm{H}^{(o)}|n\rangle - \sum_k \langle m|\mathrm{H}^{(o)}|k\rangle\langle k|\mathbf{r}|n\rangle \tag{11.8.28}$$

The matrix element, $\langle k|\mathrm{H}^{(o)}|n\rangle$, of the zeroth order Hamiltonian, $\mathrm{H}^{(o)}$, has the explicit integral representation,

$$\begin{aligned}\langle k|\mathrm{H}^{(o)}|n\rangle &= \int d^3r\,\psi_k^{(o)*}(\mathbf{r})\mathrm{H}^{(o)}\psi_n^{(o)}(\mathbf{r}) = \int d^3r\,\psi_k^{(o)*}(\mathbf{r})\left(\mathrm{H}^{(o)}\psi_n^{(o)}(\mathbf{r})\right)\\ &= \int d^3r\,\psi_k^{(o)*}(\mathbf{r})E_n^{(o)}\psi_n^{(o)}(\mathbf{r}) = E_n^{(o)}\int d^3r\,\psi_k^{(o)*}(\mathbf{r})\psi_n^{(o)}(\mathbf{r})\end{aligned} \tag{11.8.29a}$$

When generalized to three dimensions, Eq. (11.4.17) reads

$$\int d^3r\,\psi_k^{(o)*}(\mathbf{r})\psi_n^{(o)}(\mathbf{r}) = \left\langle \psi_k^{(o)} \middle| \psi_n^{(o)}\right\rangle = \langle k \mid n\rangle = \delta_{kn} \tag{11.8.29b}$$

When Eq. (11.8.29b) is substituted into Eq. (11.8.29a), the result is

$$\langle k|\mathrm{H}^{(o)}|n\rangle = E_n^{(o)}\delta_{kn} \tag{11.8.29c}$$

A similar calculation leads to the equation,

$$\langle m|\mathrm{H}^{(o)}|k\rangle = E_k^{(o)}\int d^3r\,\psi_m^{(o)*}(\mathbf{r})\psi_k^{(o)}(\mathbf{r}) = E_k^{(o)}\delta_{mk} \tag{11.8.29d}$$

The results of Eqs. (11.8.29c) and (11.8.29d) can be combined to evaluate Eq. (11.8.28) as follows:

$$\begin{aligned}\left\langle m\middle|\left[\mathbf{r}, H^{(o)}\right]\middle|n\right\rangle &= \langle m|\mathbf{r}\mathrm{H}^{(o)}|n\rangle - \langle m|\mathrm{H}^{(o)}\mathbf{r}|n\rangle\\ &= \sum_k \langle m|\mathbf{r}|k\rangle\langle k|\mathrm{H}^{(o)}|n\rangle - \sum_k \langle m|\mathrm{H}^{(o)}|k\rangle\langle k|\mathbf{r}|n\rangle\\ &= \sum_k \langle m|\mathbf{r}|k\rangle E_n^{(o)}\langle k \mid n\rangle - \sum_k E_k^{(o)}\langle m \mid k\rangle\langle k|\mathbf{r}|n\rangle\end{aligned}$$

$$
\begin{aligned}
&= E_n^{(o)} \sum_k \langle m|\mathbf{r}|k\rangle \delta_{kn} - \sum_k E_k^{(o)} \delta_{mk} \langle k|\mathbf{r}|n\rangle \\
&= E_n^{(o)} \langle m|\mathbf{r}|n\rangle - \langle m|\mathbf{r}|n\rangle E_m^{(o)} \\
&= \left(E_n^{(o)} - E_m^{(o)}\right) \langle m|\mathbf{r}|n\rangle \blacksquare
\end{aligned} \tag{11.8.30}
$$

Expressed in Dirac notation, and on the basis of the calculation in Eq. (11.7.2b), Eq. (11.8.25) assumes the form,

$$\left\langle \Psi_m^{(o)} |\mathrm{H}'| \Psi_n^{(o)} \right\rangle = -\frac{q}{2i\hbar} \mathbf{A}_o \cdot \langle m|\left[\mathbf{r}, \mathrm{H}^{(o)}\right]|n\rangle e^{i\left(E_m^{(o)} - E_n^{(o)} - \hbar\omega\right)t/\hbar} \tag{11.8.31a}$$

Upon substitution of Eq. (11.8.30), Eq. (11.8.31a) becomes,

$$\left\langle \Psi_m^{(o)} |\mathrm{H}'| \Psi_n^{(o)} \right\rangle = -\frac{q}{2i\hbar} \mathbf{A}_0 \cdot \left(E_n^{(o)} - E_m^{(o)}\right) \langle m|\mathbf{r}|n\rangle e^{i\left(E_m^{(o)} - E_n^{(o)} - \hbar\omega\right)t/\hbar} \tag{11.8.31b}$$

Based upon (11.8.31b), the role of $\langle m|V|n\rangle$ in Eq. (11.7.2b) is played by

$$\langle m|V|n\rangle = -\frac{q}{2i\hbar} \mathbf{A}_o \cdot \left(E_n^{(o)} - E_m^{(o)}\right) \langle m|\mathbf{r}|n\rangle \tag{11.8.31c}$$

11.8.3 The Transition Probability in the Case of Unpolarized Radiation

Equation (11.8.31c) allows us to write Eq. (11.7.3) in the form,

$$i\hbar \frac{dc_m}{dt} = -\frac{q}{2i\hbar} \left(E_n^{(o)} - E_m^{(o)}\right) \left(\mathbf{A}_o \cdot \langle m|\mathbf{r}|n\rangle\right) e^{i\left(E_m^{(o)} - E_n^{(o)} - \hbar\omega\right)t/\hbar} \tag{11.8.32}$$

According to Eqs. (11.7.4e), the integral of Eq. (11.8.32) is

$$
\begin{aligned}
c_m(t) = \left(\frac{1}{i\hbar}\right) &\left[-\frac{q}{2i\hbar} \mathbf{A}_o \cdot \left(E_n^{(o)} - E_m^{(o)}\right) \langle m|\mathbf{r}|n\rangle\right] e^{i\left(E_m^{(o)} - E_n^{(o)} - \hbar\omega\right)t/2\hbar} \\
&\times \frac{\sin\left[\left(E_m^{(o)} - E_n^{(o)} - \hbar\omega\right)t/2\hbar\right]}{\left(E_m^{(o)} - E_n^{(o)} - \hbar\omega\right)/2\hbar}
\end{aligned} \tag{11.8.33}
$$

By comparison with Eq. (11.7.5), the absolute square of Eq. (11.8.33) is

$$|c_m(t)|^2 = \frac{1}{4\hbar^4} q^2 \left(E_n^{(o)} - E_m^{(o)}\right)^2 |\mathbf{A}_o \cdot \langle m|\mathbf{r}|n\rangle|^2 \frac{\sin^2\left[\left(E_m^{(o)} - E_n^{(o)} - \hbar\omega\right)t/2\hbar\right]}{\left[\left(E_m^{(o)} - E_n^{(o)} - \hbar\omega\right)/2\hbar\right]^2} \tag{11.8.34}$$

To obtain the total probability of transition, we shall integrate Eq. (11.8.34) with respect to the circular frequency, $\nu = \omega/2\pi$, where $\hbar\omega = h\nu$. The range of integration will extend from $\nu = -\infty$ to $\nu = +\infty$. The integral has the form,

$$\int_{-\infty}^{+\infty} d\nu |c_m(t)|^2 = \frac{1}{4\hbar^4} q^2 \left(E_n^{(o)} - E_m^{(o)}\right)^2 |\mathbf{A}_o \cdot \langle m|\mathbf{r}|n\rangle|^2 \times \int_{-\infty}^{+\infty} d\nu \frac{\sin^2\left[(E_m^{(o)} - E_n^{(o)} - h\nu)t/2\hbar\right]}{\left[(E_m^{(o)} - E_n^{(o)} - h\nu)/2\hbar\right]^2} \tag{11.8.35a}$$

Let $x = (h\nu - (E_m^{(o)} - E_n^{(o)}))/2\hbar$, so that $d\nu = 2\hbar dx/h = 2h dx/2\pi h = dx/\pi$. Equation (11.8.35a) then reads,

$$\int_{-\infty}^{+\infty} d\nu |c_m(t)|^2 = \frac{1}{4\hbar^4} q^2 \left(E_n^{(o)} - E_m^{(o)}\right)^2 |\mathbf{A}_o \cdot \langle m|\mathbf{r}|n\rangle|^2 \frac{1}{\pi} \int_{-\infty}^{+\infty} dx \frac{\sin^2(tx)}{x^2} \tag{11.8.35b}$$

Previously in Eq. (11.7.12) we have encountered an integral similar to that in Eq. (11.8.35b). As it appears in Eq. (11.8.35b), the integral has the value [See the Mathematical Methods Appendix],

$$\int_{-\infty}^{+\infty} dx \frac{\sin^2(tx)}{x^2} = \pi t \tag{11.8.35c}$$

Upon substitution of Eq. (11.8.35c) into Eq. (11.8.35b), the result is

$$\int_{-\infty}^{+\infty} d\nu |c_m(t)|^2 = \frac{1}{4\hbar^4} q^2 \left(E_n^{(o)} - E_m^{(o)}\right)^2 |\mathbf{A}_o \cdot \langle m|\mathbf{r}|n\rangle|^2 t \tag{11.8.35d}$$

With the substitutions, $\hbar = h/2\pi$ and $h\nu_{mn} = E_m^{(o)} - E_n^{(o)}$, Eq. (11.8.35d) simplifies to read,

$$\int_{-\infty}^{+\infty} d\nu |c_m(t)|^2 = \frac{\pi^2 q^2 \nu_{mn}^2}{\hbar^2} |\mathbf{A}_o \cdot \langle m|\mathbf{r}|n\rangle|^2 t \tag{11.8.36}$$

The position coordinate locating the charge is

$$\mathbf{r} = x_1 \hat{\boldsymbol{e}}_1 + x_2 \hat{\boldsymbol{e}}_2 + x_3 \hat{\boldsymbol{e}}_3 \tag{11.8.37a}$$

Assuming that the electromagnetic wave is not polarized, the most general representation of the amplitude of the vector potential is

$$\mathbf{A}_o = A_{o,1}\hat{\boldsymbol{e}}_1 + A_{o,2}\hat{\boldsymbol{e}}_2 + A_{o,3}\hat{\boldsymbol{e}}_3 \tag{11.8.37b}$$

The absolute square of the dot product, $|\mathbf{A}_o \cdot \langle m|\mathbf{r}|n\rangle|^2$, contains contributions from all three components of $\mathbf{A}_o$. With the dot product expanded in terms of Cartesian components, Eq. (11.8.36) becomes,

$$\int_{-\infty}^{+\infty} d\nu |c_m(t)|^2 = \frac{\pi^2 q^2 \nu_{mn}^2}{\hbar^2}\left[A_{o,1}^2|\langle m|x_1|n\rangle|^2 + A_{o,2}^2|\langle m|x_2|n\rangle|^2 + A_{o,3}^2|\langle m|x_3|n\rangle|^2\right]t$$

$$+\frac{2\pi^2 q^2 \nu_{mn}^2}{\hbar^2}\left[A_{o,1}A_{o,2}\langle m|x_1|n\rangle\langle m|x_2|n\rangle + A_{o,1}A_{o,3}\langle m|x_1|n\rangle\langle m|x_3|n\rangle\right.$$
$$\left.+A_{o,2}A_{o,3}\langle m|x_2|n\rangle\langle m|x_3|n\rangle\right]t \tag{11.8.37c}$$

where the first line of Eq. (11.8.37c) contains diagonal terms, while the second line contains the off-diagonal terms. The square of the length of $\mathbf{A}_o$ is

$$A_o^2 = A_{o,1}^2 + A_{o,2}^2 + A_{o,3}^2 \tag{11.8.37d}$$

The contribution of each component of $\mathbf{A}_o$ to Eq. (11.8.37c) is determined by the angles which each component makes with respect to the vector, $\mathbf{r}$. In the case of unpolarized radiation, no direction is preferred over any other. This observation permits us to replace the individual contributions by their angle averages. The calculation of the average involves integrating over the differential solid angle, $d\Omega$. In spherical polar coordinates, the differential solid angle [See the Mathematical Methods Appendix] is,

$$d\Omega = \sin\theta d\theta d\phi \tag{11.8.38a}$$

The integral of $d\Omega$ over the extremes of θ and ϕ gives for the total solid angle,

$$\Omega = \int d\Omega = \int_0^{\pi} d\theta \sin\theta \int_0^{2\pi} d\phi = [-\cos\theta]_0^{\pi}(2\pi) = [-(-1)+1](2\pi) = 4\pi \tag{11.8.38b}$$

We turn first to the averages of the diagonal components in Eq. (11.18.37c). In spherical polar coordinates, the component of $\mathbf{A}_o$ in the x_1-direction is

$$A_{o,1} = A_o \sin\theta\cos\phi \tag{11.8.39a}$$

Its square is

$$A_{o,1}^2 = A_o^2 \sin^2\theta \cos^2\phi \tag{11.8.39b}$$

The angle average of $A_{o,1}^2$ is then

$$\left\langle A_{o,1}^2 \right\rangle = A_o^2 \left(\frac{1}{\Omega}\right) \int d\Omega \sin^2\theta \cos^2\phi \tag{11.8.39c}$$

In Eq. (11.8.39c), it is helpful to replace $\cos^2\phi$ with the trigonometric identity,

$$\cos^2\phi = \left(\frac{1+\cos 2\phi}{2}\right) \tag{11.8.39d}$$

[See the Mathematical Methods Appendix]. The integral over θ in Eq. (11.8.39c) can be cast in algebraic form by making the substitution $x = \cos\theta$, $dx = -\sin\theta d\theta$. The original limits of integration, $\theta = 0$ and $\theta = \pi$, then correspond to $x = +1$ and $x = -1$, respectively. With these substitutions, Eq. (11.8.39c) reads,

$$\begin{aligned}
\left\langle A_{o,1}^2 \right\rangle &= A_o^2\left(\frac{1}{\Omega}\right)\int_0^{\pi} d\theta(\sin\theta\sin^2\theta)\int_0^{2\pi} d\phi\left(\frac{1+\cos 2\phi}{2}\right) \\
&= A_o^2\left(\frac{1}{\Omega}\right)\int_{-1}^{+1} dx(1-x^2)\int_0^{2\pi} d\phi\left(\frac{1+\cos 2\phi}{2}\right) \\
&= \frac{A_o^2}{4\pi}\left[x-\frac{x^3}{3}\right]_{-1}^{+1}\left[\frac{\phi}{2}+\frac{\sin 2\phi}{4}\right]_0^{2\pi} = \left(\frac{A_o^2}{4\pi}\right)\left[\left(1-\frac{1}{3}\right)-\left(-1+\frac{1}{3}\right)\right]\left(\frac{2\pi}{2}\right) \\
&= \left(\frac{A_o^2}{4\pi}\right)(2)\left[1-\frac{1}{3}\right]\left(\frac{2\pi}{2}\right) = \frac{4\pi A_o^2}{(4\pi)(2)}\left(\frac{2}{3}\right) = \frac{1}{3}A_o^2
\end{aligned} \tag{11.8.39e}$$

The component of $\mathbf{A}_o$ in the x_2-direction is

$$A_{o,2} = A_o \sin\theta \sin\phi \tag{11.8.40a}$$

Its square is

$$A_{o,2}^2 = A_o^2 \sin^2\theta \sin^2\phi \tag{11.8.40b}$$

The angle average is

$$\left\langle A_{o,2}^2 \right\rangle = A_o^2 \left(\frac{1}{\Omega}\right) \int d\Omega \sin^2\theta \sin^2\phi \tag{11.8.40c}$$

With the substitution

$$\sin^2\phi = \left(\frac{1-\cos 2\phi}{2}\right) \tag{11.8.40d}$$

See trigonometric identities in the Mathematical Methods Appendix. A change of integration variable from θ to $x = \cos\theta$, renders Eq. (11.8.40c) in the form,

$$\begin{aligned}
\langle A_{o,2}^2\rangle &= A_o^2\left(\frac{1}{\Omega}\right)\int_0^{\pi} d\theta\left(\sin\theta\sin^2\theta\right)\int_0^{2\pi} d\phi\left(\sin^2\phi\right) \\
&= \int_{-1}^{+1} dx(1-x^2)\int_0^{2\pi} d\phi\left(\frac{1-\cos 2\phi}{2}\right) \\
&= \frac{A_o^2}{4\pi}\left[x-\frac{x^3}{3}\right]_{-1}^{+1}\left[\frac{\phi}{2}-\frac{\sin 2\phi}{4}\right]_0^{2\pi} = \left(\frac{A_o^2}{4\pi}\right)(2)\left[1-\frac{1}{3}\right]\left(\frac{2\pi}{2}\right) \\
&= \frac{4\pi A_o^2}{(4\pi)(3)} = \frac{1}{3}A_o^2
\end{aligned} \tag{11.8.40e}$$

The component of $\mathbf{A}_o$ in the z-direction is

$$A_{o,2} = A_o\cos\theta \tag{11.8.41a}$$

Its square is

$$A_{o,3}^2 = A_o^2\cos^2\theta \tag{11.8.41b}$$

The angle average is

$$\langle A_{o,3}^2\rangle = A_o^2\left(\frac{1}{\Omega}\right)\int d\Omega\cos^2\theta \tag{11.8.41c}$$

With the change in variables from θ to x, Eq. (11.8.41c) reads

$$\begin{aligned}
\langle A_{o,3}^2\rangle &= A_o^2\left(\frac{1}{\Omega}\right)\int_0^{\pi} d\theta\left(\sin\theta\cos^2\theta\right)\int_0^{2\pi} d\phi \\
&= A_o^2\left(\frac{1}{\Omega}\right)\int_{-1}^{+1} dx x^2\int_0^{2\pi} d\phi = \left[\frac{x^3}{3}\right]_{-1}^{+1}(2\pi) \\
&= A_o^2\left[\frac{1}{3}+\frac{1}{3}\right]\left(\frac{2\pi}{4\pi}\right) = \frac{1}{3}A_o^2
\end{aligned} \tag{11.8.41d}$$

There are three cross terms. The angle average of the first is

$$\begin{aligned}\langle A_{o,1}A_{o,2}\rangle &= A_o^2\left(\frac{1}{\Omega}\right)\int d\Omega \sin^2\theta\cos\phi\sin\phi\\ &= \int_0^{\pi} d\theta\left(\sin\theta\sin^2\theta\right)\int_0^{2\pi} d\phi(\cos\phi\sin\phi)\\ &= \int_0^{\pi} d\theta\left(\sin\theta\sin^2\theta\right)\left[\frac{\sin^2\phi}{2}\right]_0^{2\pi} = 0\end{aligned} \tag{11.8.42a}$$

The angle average of the second cross term is

$$\begin{aligned}\langle A_{o,1}A_{o,3}\rangle &= A_o^2\left(\frac{1}{\Omega}\right)\int d\Omega \sin\theta\cos\phi\cos\theta\\ &= A_o^2\left(\frac{1}{\Omega}\right)\int_0^{2\pi} d\theta\left(\sin^2\theta\cos\theta\right)\int_0^{2\pi} d\phi(\cos\phi)\\ &= A_o^2\left(\frac{1}{\Omega}\right)\int_0^{2\pi} d\theta\left(\sin^2\theta\cos\theta\right)[\sin\phi]_0^{2\pi} = 0\end{aligned} \tag{11.8.42b}$$

The angle average of the third cross term is

$$\begin{aligned}\langle A_{o,2}A_{o,3}\rangle &= A_o^2\left(\frac{1}{\Omega}\right)\int d\Omega \sin\theta\sin\phi\cos\theta\\ &= A_o^2\left(\frac{1}{\Omega}\right)\int_0^{\pi} d\theta\left(\sin^2\theta\cos\theta\right)\int_0^{2\pi} d\phi(\sin\phi)\\ &= \int_0^{\pi} d\theta\left(\sin^2\theta\cos\theta\right)[-\cos\phi]_0^{2\pi} = 0\end{aligned} \tag{11.8.42c}$$

On the basis of Eqs. (11.8.39e), (11.8.40e), and (11.8.41d), we conclude that in the case of unpolarized radiation that

$$\langle A_{o,1}^2\rangle = \langle A_{o,2}^2\rangle = \langle A_{o,3}^2\rangle = \frac{1}{3}A_o^2 \tag{11.8.43}$$

By contrast, according to Eqs. (11.8.42a), (11.8.42b), and (11.8.42c), the cross correlations are all zero. Using these results and Eq. (11.8.43), we obtain for the angle average of Eq. (11.8.37c),

$$\int_{-\infty}^{+\infty} d\nu |c_m(t)|^2 = \frac{\pi^2 q^2 \nu_{mn}^2}{3\hbar^2}\left[|\langle m|x_1|n\rangle|^2 + |\langle m|x_2|n\rangle|^2 + |\langle m|x_3|n\rangle|^2\right] A_o^2 t \quad (11.8.44)$$

11.8.4 Time Averaged Electromagnetic Energy Density

The final task is to express A_o^2 in terms of the time average electromagnetic energy density, $\langle \rho \rangle$, where ρ is given by

$$\rho = \frac{1}{2}(\mathbf{D}\cdot\mathbf{E} + \mathbf{B}\cdot\mathbf{H}) \quad (11.8.45)$$

Equation (11.8.45) is the same as Eq. (7.3.7) in Chap. 7. If the electromagnetic fields are acting on a molecule in vacuum, we can set $\mathbf{D} = \varepsilon_o \mathbf{E}$ and $\mathbf{B} = \mu_o \mathbf{H}$, which upon substitution into Eq. (11.8.45) gives

$$\rho = \frac{1}{2}\left(\varepsilon_o E^2 + \mu_o H^2\right) \quad (11.8.46)$$

In order to express Eq. (11.8.46) in terms of the vector potential, we begin with the calculation of

$$\nabla \times \mathbf{A} = \begin{bmatrix} \hat{\boldsymbol{e}}_1 & \hat{\boldsymbol{e}}_2 & \hat{\boldsymbol{e}}_3 \\ \frac{\partial}{\partial x_1} & \frac{\partial}{\partial x_2} & \frac{\partial}{\partial x_3} \\ A_1 & A_2 & A_3 \end{bmatrix} = \hat{\boldsymbol{e}}_1\left(\frac{\partial A_3}{\partial x_2} - \frac{\partial A_2}{\partial x_3}\right) + \hat{\boldsymbol{e}}_2\left(\frac{\partial A_1}{\partial x_3} - \frac{\partial A_3}{\partial x_1}\right) + \hat{\boldsymbol{e}}_3\left(\frac{\partial A_2}{\partial x_1} - \frac{\partial A_1}{\partial x_2}\right) \quad (11.8.47a)$$

We now assume that the vector potential is given by Eq. (11.8.11). With the argument of the cosine written in terms of components, Eq. (11.8.11) reads,

$$\mathbf{A}(\mathbf{r}, t) = \mathbf{A}_o \cos(k_1 x_1 + k_2 x_2 + k_3 x_3 - \omega t) \quad (11.8.47b)$$

The component of $\mathbf{A}(\mathbf{r}, t)$ in the x_3-direction is

$$A_3 = A_{0,3} \cos(k_1 x_1 + k_2 x_2 + k_3 x_3 - \omega t) \quad (11.8.47c)$$

To illustrate the calculations required to evaluate the right hand side of Eq. (11.8.47a), we compute the example,

$$\frac{\partial A_3}{\partial x_2} = -k_2 A_{o3} \sin(k_1 x_1 + k_2 x_2 + k_3 x_3 - \omega t) \quad (11.8.47d)$$

With Eq. (11.8.47d) as a model, the right hand side of Eq. (11.8.47a) becomes

$$\nabla \times \mathbf{A} = -\left[\hat{\boldsymbol{e}}_1\left(k_2A_{o,3} - k_3A_{o,2}\right) + \hat{\boldsymbol{e}}_2\left(k_3A_{o,1} - k_1A_{o,3}\right) + \hat{\boldsymbol{e}}_3\left(k_1A_{o,2} - k_2A_{o,1}\right)\right] \cdot [\sin(k_1x_1 + k_2x_2 + k_3x_3 - \omega t)] \tag{11.8.47e}$$

Using Eq. (11.8.11), and the definition of the cross product, Eq. (11.8.47e) can rewritten as

$$\nabla \times \mathbf{A} = -(\mathbf{k} \times \mathbf{A}_o) \sin(\mathbf{k} \cdot \mathbf{r} - \omega t) \tag{11.8.47f}$$

The magnetic induction is given by

$$\mathbf{B} = \nabla \times \mathbf{A} \tag{11.8.47g}$$

The combination of Eqs. (11.8.47f) and (11.8.47g) gives

$$\mathbf{B} = -(\mathbf{k} \times \mathbf{A}_o) \sin(\mathbf{k} \cdot \mathbf{r} - \omega t) \tag{11.8.47h}$$

Since $\mathbf{B} = \mu_o\mathbf{H}$, the magnetic field is

$$\mathbf{H} = -\frac{(\mathbf{k} \times \mathbf{A}_o)}{\mu_o} \sin(\mathbf{k} \cdot \mathbf{r} - \omega t) \tag{11.8.48}$$

With reference to Eq. (8.4.3), the electric field is given by

$$\mathbf{E} = -\frac{\partial \mathbf{A}}{\partial t} - \nabla\Phi \tag{11.8.49a}$$

According to Eq. (11.8.10a), Φ has already been included as part of the zeroth order Hamiltonian, so Eq. (11.8.49a) simplifies to read

$$\mathbf{E} = -\frac{\partial \mathbf{A}}{\partial t} \tag{11.8.49b}$$

Upon substitution of Eq. (11.8.11), Eq. (11.8.49b) becomes

$$\mathbf{E} = -\frac{\partial \mathbf{A}}{\partial t} = -\mathbf{A}_o\frac{\partial}{\partial t}[\cos(\mathbf{k} \cdot \mathbf{r} - \omega t)] = \mathbf{A}_o\omega[\sin(\mathbf{k} \cdot \mathbf{r} - \omega t)] \tag{11.8.50}$$

With the representations of **H** in Eq. (11.8.48) and **E** in Eq. (11.8.50), Eq. (11.8.46) can be evaluated with the result,

$$\rho = \frac{1}{2}\left(\varepsilon_o\omega^2A_o^2 \sin^2(\mathbf{k} \cdot \mathbf{r} - \omega t) + \mu_o\frac{(\mathbf{k} \times \mathbf{A}_o)^2}{\mu_o^2} \sin^2(\mathbf{k} \cdot \mathbf{r} - \omega t)\right) \tag{11.8.51}$$

We can simplify $(\mathbf{k} \times \mathbf{A}_o)^2$ by invoking the vector identity [See the Mathematical Methods Appendix],

$$(\mathbf{A} \times \mathbf{B}) \cdot (\mathbf{C} \times \mathbf{D}) = (\mathbf{A} \cdot \mathbf{C})(\mathbf{B} \cdot \mathbf{D}) - (\mathbf{A} \cdot \mathbf{D})(\mathbf{B} \cdot \mathbf{C}) \tag{11.8.52a}$$

With the identifications, $\mathbf{A} = \mathbf{C} = \mathbf{k}$ and $\mathbf{B} = \mathbf{D} = \mathbf{A}_o$, the coefficient of the second term on the right of Eq. (11.8.52a) becomes,

$$(\mathbf{k} \times \mathbf{A}_o)^2 = (\mathbf{k} \times \mathbf{A}_o) \cdot (\mathbf{k} \times \mathbf{A}_o) = (\mathbf{k} \cdot \mathbf{k})(\mathbf{A}_o \cdot \mathbf{A}_o) - (\mathbf{k} \cdot \mathbf{A}_o)(\mathbf{A}_o \cdot \mathbf{k}) \tag{11.8.52b}$$

According to Eq. (11.8.50), **E** and **A** are parallel. Since for electromagnetic radiation, **E** is necessarily perpendicular to **k**, we can conclude that **A** is perpendicular to **k**. Hence, $(\mathbf{A}_o \cdot \mathbf{k}) = (\mathbf{k} \cdot \mathbf{A}_o) = 0$, and Eq. (11.8.52b) simplifies to

$$(\mathbf{k} \times \mathbf{A}_o)^2 = k^2 A_o^2 \tag{11.8.52c}$$

Upon substitution of Eq. (11.8.52c) into Eq. (11.8.51), one obtains,

$$\rho = \frac{1}{2}\left(\varepsilon_o \omega^2 + \frac{k^2}{\mu_o}\right) A_o^2 \sin^2(\mathbf{k} \cdot \mathbf{r} - \omega t) \tag{11.8.52d}$$

The next task is to compute the time average of the energy density, $\langle \rho \rangle$. The vacuum speed of light is $c = 1/\sqrt{\varepsilon_o \mu_o}$. With reference to Eq. (11.2.15), we can use this relation to link the angular frequency, ω, and the wave number, k, according to

$$\omega^2 = c^2 k^2 = \frac{k^2}{\varepsilon_o \mu_o} \tag{11.8.52e}$$

Equation (11.8.52e) implies that $k^2 = \varepsilon_o \mu_o \omega^2$. When this result is substituted into Eq. (11.8.52d), the result is

$$\rho = \frac{1}{2}\left(\varepsilon_o \omega^2 + \frac{\varepsilon_o \mu_o \omega^2}{\mu_o}\right) A_o^2 \sin^2(\mathbf{k} \cdot \mathbf{r} - \omega t) \tag{11.8.52f}$$

Equation (11.8.52f) simplifies to read,

$$\rho = \varepsilon_o \omega^2 A_o^2 \sin^2(\mathbf{k} \cdot \mathbf{r} - \omega t) \tag{11.8.52g}$$

Since we are considering radiation with wavelengths long as compared to the molecular dimension, we are reminded that $\mathbf{k} \cdot \mathbf{r} \ll 1$ holds. This simplification permits Eq. (11.8.52g) to be rewritten as

$$\rho = \varepsilon_o \omega^2 A_o^2 \sin^2(\omega t) \tag{11.8.53}$$

Since the period of the wave is $1/\nu = 2\pi/\omega$, the time average of $\sin^2(\omega t)$ is

$$\left\langle \sin^2(\omega t)\right\rangle = \left[\int_0^{2\pi/\omega} dt\right]^{-1}\left[\int_o^{2\pi/\omega} dt\,\sin^2(\omega t)\right] = \frac{\omega}{2\pi}\int_0^{2\pi/\omega} dt\left(\frac{1}{2}\right)[1-\cos(2\omega t)]$$
$$= \frac{\omega}{2\pi}\left(\frac{1}{2}\right)\left\{[t]_o^{2\pi/\omega} - \left[\frac{\sin(2\omega t)}{2\omega}\right]_0^{2\pi/\omega}\right\} = \frac{1}{2} \tag{11.8.54a}$$

Using Eq. (11.8.54a), the time average of Eq. (11.8.53) is

$$\langle\rho\rangle = \frac{1}{2}\varepsilon_o\omega^2 A_o^2 \tag{11.8.54b}$$

We can now specialize Eq. (11.8.54b) to a transition where $\omega = \omega_{mn} = 2\pi\nu_{mn}$. With this identification, Eq. (11.8.54b) can be solved for A_o^2 with the result,

$$A_o^2 = \frac{\langle\rho\rangle}{2\pi^2\varepsilon_o\nu_{mn}^2} \tag{11.8.55}$$

It is useful to define,

$$R_{mn}^2 = \left[|\langle m|x_1|n\rangle|^2 + |\langle m|x_2|n\rangle|^2 + |\langle m|x_3|n\rangle|^2\right] \tag{11.8.56a}$$

so that Eq. (11.8.44) reads,

$$\int_{-\infty}^{+\infty} d\nu|c_m(t)|^2 = \frac{\pi^2 q^2\nu_{mn}^2 R_{mn}^2}{3\hbar^2}A_o^2 t \tag{11.8.56b}$$

Upon substitution of Eq. (11.8.55) into Eq. (11.8.56b), the result is

$$\int_{-\infty}^{+\infty} d\nu|c_m(t)|^2 = \frac{\pi^2 q^2\nu_{mn}^2}{3\hbar^2}R_{mn}^2\left(\frac{\langle\rho\rangle}{2\pi^2\varepsilon_o\nu_{mn}^2}\right)t \tag{11.8.56c}$$

After cancellation of common factors, Eq. (11.8.56c) reads

$$\int_{-\infty}^{+\infty} d\nu|c_m(t)|^2 = \frac{q^2}{6\hbar^2\varepsilon_o}R_{mn}^2\langle\rho\rangle t \tag{11.8.57}$$

If $\int_{-\infty}^{+\infty} d\nu|c_m(t)|^2$ is the probability of an $n \to m$ transition, then the transition rate is the time derivative,

$$\frac{d}{dt}\int_{-\infty}^{+\infty} d\nu |c_m(t)|^2 = \frac{q^2}{6\hbar^2\varepsilon_o} R_{mn}^2 \langle\rho\rangle \tag{11.8.58}$$

which we calculated using Eq. (11.8.57).

11.8.5 Rate Constants for Absorption and Stimulated Emission of Radiation

In Eq. (11.8.58), the coefficient of $\langle\rho\rangle$, is denoted by

$$B_{n\to m} = \frac{q^2}{6\hbar^2\varepsilon_o} R_{mn}^2 \tag{11.8.59}$$

Defined as in Eq. (11.8.59), $B_{n\to m}$ plays the role of a second order kinetic rate coefficient, which makes its units worthy of some consideration. The units of q are Coulombs (C), the units of R_{mn} are meters (m), the units of $\hbar$ are Joules times seconds (J s), while the units of ε_o are (C^2/J m). Hence, as defined by Eq. (11.8.59), the units of $B_{n\to m}$ are (m^3/J s^2). When $B_{n\to m}$ is multiplied by $\rho(\nu, T)$, where $\rho(\nu, T)$ is the Planck energy density with units Js/m^3, as given by Eq. (11.1.2), the product has the units of s^{-1}, and thus represents the rate at which photons of frequency, ν, disappear by absorption.

We now introduce the terminology, *Einstein "B" coefficient*. The first B-coefficient, coincides with $B_{n\to m}$, which serves as the rate constant governing the rate at which a system absorbs a photon and makes a transition from $E_n^{(o)}$ to $E_m^{(o)}$. A second B-coefficient, denoted by $B_{m\to n}$, is the rate constant that governs the rate at which a system makes a transition from $E_m^{(o)}$ to $E_n^{(o)}$ by emission of a photon.

Einstein suggested the novel idea that the emission of radiation from $E_m^{(o)}$ to $E_n^{(o)}$, as well as the absorption radiation from $E_n^{(o)}$ to $E_m^{(o)}$, were both dependent upon the presence of photons. Emission in the presence of photons is termed *stimulated emission*. He identified the rate of stimulated emission with $B_{m\to n}\rho(\nu, T)$. The basis for Einstein's argument was his assertion that if photons with density $\rho(\nu, T)$ were to be in equilibrium with matter during both absorption and emission, then the observed rate of absorption from the ground state $E_n^{(o)}$ to the excited state $E_m^{(o)}$ had to be equal to the observed rate of emission from the excited state $E_m^{(o)}$ to the ground state $E_n^{(o)}$. This equilibrium criterion is known as the principle of *detailed balance*. If P_n is the probability that a molecule has energy, $E_n^{(o)}$, and P_m is the probability that it has energy, $E_m^{(o)}$, then if there is no other coupling between these two states, the principle of detailed balance is satisfied by

$$B_{n\to m}\rho(\nu, T)P_n = B_{m\to n}\rho(\nu, T)P_m \tag{11.8.60}$$

This criterion is identical to the dynamic equilibrium concept in chemical kinetics. Equation (11.8.60) implies that

$$\frac{B_{n\to m}}{B_{m\to n}} = \frac{P_m}{P_n} \tag{11.8.61}$$

According to Eq. (2.9.22) of Chap. 2, the respective probabilities of occupation are given by the Boltzmann factors,

$$P_n = \frac{e^{-E_n^{(o)}/k_B T}}{Z} \tag{11.8.62}$$

and

$$P_m = \frac{e^{-E_m^{(o)}/k_B T}}{Z} \tag{11.8.63}$$

The frequency, ν_{mn}, of the radiation linking the two states satisfies, $h\nu_{mn} = E_m^{(o)} - E_n^{(o)}$. Using this relation, we can conclude on the basis of Eqs. (11.8.61), (11.8.62), and (11.8.63) that

$$\frac{B_{n\to m}}{B_{m\to n}} = \frac{P_m}{P_n} = \exp\left[-(E_m^{(o)} - E_n^{(o)})/k_B T\right] = \exp[-h\nu_{mn}/k_B T] \tag{11.8.64}$$

According to Eq. (11.8.64), the ratio, $B_{n\to m}/B_{m\to n}$, is a function of temperature. Yet this conclusion is belied by Eq. (11.8.59). Thus, it is obvious that the picture of emission and absorption described so far is incomplete.

11.8.6 Detailed Balance and the Rate Constant for Spontaneous Emission of Radiation

Einstein proposed that the detailed balance principle had to include the fact that molecules in excited states could emit radiation in the *absence* of stimulation. This process is known as *spontaneous emission.* When both stimulated and spontaneous emission are taken into account, the principle of detailed balance assumes the form,

$$B_{n\to m}\rho(\nu, T)P_n = B_{m\to n}\rho(\nu, T)P_m + A_{m\to n}P_m \tag{11.8.66a}$$

where $A_{m\to n}$ is the rate constant for spontaneous emission. Equation (11.8.66a) can be solved for $\rho(\nu, T)$ to obtain

$$\rho(\nu, T) = \frac{A_{m\to n}P_m}{B_{n\to m}P_n - B_{m\to n}P_m} \tag{11.8.66b}$$

Upon dividing numerator and denominator by P_n, Eq. (11.8.66b) reads,

$$\rho(\nu, T) = \frac{A_{m\to n}(P_m/P_n)}{B_{n\to m} - B_{m\to m}(P_m/P_n)} \quad (11.8.66c)$$

According to Eq. (11.8.64),

$$\frac{P_m}{P_n} = \frac{e^{-E_m^{(o)}/k_BT}}{e^{-E_n^{(o)}/k_BT}} = e^{-(E_m^o - E_n^o)/k_BT} = e^{-h\nu_{mn}/k_BT} \quad (11.8.67a)$$

Substitute Eq. (11.8.67a) into Eq. (11.8.66c). The result is

$$\rho(\nu_{mn}, T) = \frac{A_{m\to n}e^{-h\nu_{mn}/k_BT}}{B_{n\to m} - B_{m\to m}e^{-h\nu_{mn}/k_BT}} \quad (11.8.67b)$$

Equation (11.8.67b) can be rearranged to read,

$$\rho(\nu_{mn}, T) = \frac{(A_{m\to n}/B_{n\to m})}{e^{h\nu_{mn}/k_BT} - (B_{m\to m}/B_{n\to m})} \quad (11.8.68a)$$

If Eqs. (11.1.2) and (11.8.68a) are to agree, one must conclude that

$$\frac{8\pi h\nu_{mn}^3/c^3}{e^{h\nu_{mn}/k_BT} - 1} = \frac{(A_{m\to n}/B_{n\to m})}{e^{h\nu_{mn}/k_BT} - (B_{m\to n}/B_{n\to m})} \quad (11.8.68b)$$

The left and right hand sides of Eq. (11.8.68b) can be made equal if

$$B_{n\to m} = B_{m\to n} \quad (11.8.69)$$

and

$$A_{m\to n} = \frac{8\pi h\nu_{mn}^3}{c^3}B_{n\to m} \quad (11.8.70)$$

Equation (11.8.69) shows that the rate coefficients for absorption and stimulated emission are equal, while Eq. (11.8.70) shows that the rate coefficient for spontaneous emission is proportional to the rate coefficient for absorption. In Eq. (11.8.70), $A_{m\to n}$ has the units, s^{-1}. According to Eq. (11.8.70), $A_{m\to n}$ increases strongly with increasing frequency, ν_{mn}, of the transition. Whereas spontaneous emission is weak in the microwave and infrared regions of the spectrum, where the frequency is low and the wavelength is long, it plays an important role in the visible and ultraviolet regions.

11.9 Basic Results Governing Spectroscopy

By combining Eqs. (11.8.59) and (11.8.70), we find that the Einstein A-coefficient is given by

$$\begin{aligned} A_{m\to n} &= \left(\frac{8\pi h\nu_{m\to n}^3}{c^3}\right)\left(\frac{q^2}{6\hbar^2\varepsilon_o}R_{mn}^2\right) = \left(\frac{8\pi(2\pi\hbar)\nu_{m\to n}^3}{c^3}\right)\left(\frac{q^2}{6\hbar^2\varepsilon_o}R_{mn}^2\right) \\ &= \left(\frac{16\pi^2\hbar\nu_{m\to n}^3}{c^3}\right)\left(\frac{q^2}{6\hbar^2\varepsilon_o}R_{mn}^2\right) \end{aligned} \tag{11.9.1}$$

or upon cancellation of common factors,

$$A_{m\to n} = \left(\frac{8\pi^2\nu_{m\to n}^3}{3\hbar\varepsilon_o c^3}\right)q^2R_{mn}^2 \tag{11.9.2}$$

According to Eqs. (11.8.59) and (11.8.69), the corresponding B-coefficients are

$$B_{n\to m} = B_{m\to n} = \frac{q^2}{6\hbar^2\varepsilon_o}R_{mn}^2 \tag{11.9.3}$$

Equations (11.9.2) and (11.9.3) constitute the fundamental equations governing the interaction of electromagnetic radiation with matter. We will rely upon them when it comes to understanding the phenomenon of optical rotation as well as when discussing the electromagnetic basis for the London force of intermolecular attraction.

Recommended Additional Reading

1. H. Eyring, J. Walter, G.E. Kimball, *Quantum Chemistry* (Wiley, New York, 1944)
2. G.C. Schatz, M.A. Ratner, *Quantum Mechanics in Chemistry* (Dover Publications, Mineola, NY, 2002)
3. A. Hinchliffe, R.W. Munn, *Molecular Electromagnetism* (Wiley, Chichester, W. Sussex, UK, 1985)
4. P. Norman, K. Ruud, T. Saue, *Principles and Practices of Molecular Properties (Theory and Simulations)* (Wiley, Hoboken, NJ, 2018)

Chapter 12
Semiclassical Theory of the Interaction of Radiation with Matter

This chapter introduces the semi-classical theory of radiation. The theory is semi-classical in the sense that, the radiation is described by Maxwell equations, while the matter which is absorbing and emitting the radiation, is treated quantum mechanically. The theory provides an explanation for the vast majority of spectroscopic and electromagnetic phenomena that are encountered in chemistry. This chapter builds on the time-dependent perturbation theory previously encountered in Chap. 11. The density matrix is introduced and used to calculate the real and imaginary parts of the molecular polarizability, a function which was previously encountered in its classical form in Chap. 7. Afterwards, the Kramers–Kronig relations are derived. This pair of equations serves to link the real part of the polarizability to the imaginary part and vice versa.

12.1 Introduction

Following the lead of Gordon [1], we shall in this chapter analyze in detail the frequency dependence of the spectroscopic transitions alluded to in Chap. 11. We begin with the time-dependent Schrödinger equation,

$$i\hbar\frac{\partial\Psi(\mathbf{r},t)}{\partial t} = \left(\mathrm{H}^{(o)} + \mathrm{H}'(\mathrm{t})\right)\Psi(\mathbf{r},t) \tag{12.1.1}$$

As before, the energy levels of the molecule will be assumed to be quantized, whereas we will continue to treat the electromagnetic field classically. This point of view is known as the *semi-classical* theory of the interaction of radiation with matter. The Hamiltonian, $\mathrm{H}^{(o)}$, in Eq. (12.1.1) determines the energy levels, $\{E_n^{(o)}\}$, of the molecule in its stationary states. The time-dependent perturbation, $\mathrm{H}'(\mathrm{t})$, represents the interaction of this molecule with the electromagnetic field and determines the

J. K. Baird, *Electromagnetism for Chemistry*,
https://doi.org/10.1007/978-3-031-88508-2_12

probability that a transition between quantum levels will occur. Our objective in solving Eq. (12.1.1) is to develop a quantum mechanical model for the complex valued polarizability, $\tilde{\alpha}(\omega)$, which was introduced in its classical form in Chap. 7.

As in Chap. 11, we denote the set of stationary state eigenfunctions of $\mathrm{H}^{(o)}$ simply by $\left\{\psi_n^{(o)}(\mathbf{r})\right\}$. The time-independent Schrödinger equation reads,

$$\mathrm{H}^{(o)}\psi_n^{(o)}(\mathbf{r}) = E_n^{(o)}\psi_n^{(o)}(\mathbf{r}) \tag{12.1.2}$$

The set of wave functions, $\left\{\psi_n^{(o)}(\mathbf{r})\right\}$, satisfies the orthonormality condition,

$$\langle k \mid n\rangle = \langle\psi_k^{(o)}(\mathbf{r}) \mid \psi_n^{(o)}(\mathbf{r})\rangle = \int d^3r\,\psi_k^{(o)*}(\mathbf{r})\psi_n^{(o)}(\mathbf{r}) = \delta_{kn} \tag{12.1.3}$$

where δ_{kn} is the Kronecker delta symbol [See the Mathematical Methods Appendix].

12.2 The Matrix Element of the Perturbation

In the case of a time varying electric field interacting with a molecule, the operator, $\mathrm{H}'(t)$, given in Eq. (12.1.1), can be replaced by the operator, $\mathrm{V}(t) = -\mathbf{E}(t)\cdot\boldsymbol{\mu}$ where $\mathbf{E}(t)$ is the electric field vector of the electromagnetic wave, and $\boldsymbol{\mu}$ is the electric dipole moment induced in the molecule.

To show this, we follow Chap. 11 and exploit the vector potential representation,

$$\mathbf{A}(t) = \mathbf{A}_o\exp(-i\omega t) \tag{12.2.1}$$

According to Eq. (11.8.10b), the time-dependent perturbation assumes the form,

$$\mathrm{H}'(t) = -\frac{q}{m}\mathbf{A}_o e^{-i\omega t}\cdot\mathbf{p} \tag{12.2.2}$$

On the basis of Theorem 11.3 of Chap. 11,

$$\mathbf{p} = \frac{m}{i\hbar}\left[\mathbf{r}, \mathrm{H}^{(o)}\right] \tag{12.2.3}$$

Upon substitution of Eq. (12.2.3) into Eq. (12.2.2), the result in Eq. (11.8.20b) follows:

$$\mathrm{H}'(t) = -\frac{q}{i\hbar}\mathbf{A}_o\cdot\left[\mathbf{r}, \mathrm{H}^{(o)}\right]e^{-i\omega t} \tag{12.2.4}$$

In this chapter, we are interested in the matrix element of $\mathrm{H}'(t)$ between the time-independent states, $\psi_k^{(o)}(\mathbf{r})$ and $\psi_\ell^{(o)}(\mathbf{r})$. According to Eq. (12.2.4), this matrix element is

$$\langle k|\mathrm{H}'(\mathrm{t})|\ell\rangle = \langle \psi_k^{(o)}|\mathrm{H}'(\mathrm{t})|\psi_\ell^{(o)}\rangle = -\frac{q}{i\hbar}\mathbf{A}_o \cdot \langle k|\left[\mathbf{r}, \mathrm{H}^{(\mathrm{o})}\right]|\ell\rangle e^{-i\omega t} \tag{12.2.5}$$

where k and ℓ are integers. By virtue of Theorem 11.4 of Chap. 11, an element of the matrix representation of the commutator, $\left[\mathbf{r}, H^{(o)}\right]$ is

$$\langle k|\left[\mathbf{r}, \mathrm{H}^{(\mathrm{o})}\right]|\ell\rangle = \left(E_\ell^{(o)} - E_k^{(o)}\right)\langle k|\mathbf{r}|\ell\rangle \tag{12.2.6}$$

Equation (12.2.6) permits the general element of the matrix of $\mathrm{H}'(t)$ in Eq. (12.2.5) to be written as

$$\langle k|\mathrm{H}'(\mathrm{t})|\ell\rangle = -\frac{q}{i\hbar}\mathbf{A}_o e^{-i\omega t} \cdot \left(E_\ell^{(o)} - E_k^{(o)}\right)\langle k|\mathbf{r}|\ell\rangle \tag{12.2.7}$$

With the charge, q, included, the matrix element on the right of Eq. (12.2.7) can be written as $\langle k|q\mathbf{r}|\ell\rangle$. This combination allows us to introduce the electric dipole moment operator,

$$\boldsymbol{\mu} = q\mathbf{r} \tag{12.2.8a}$$

Accordingly, with the definition

$$\langle k|q\mathbf{r}|\ell\rangle = \langle k|\boldsymbol{\mu}|\ell\rangle \tag{12.2.8b}$$

Equation (12.2.7) takes the form,

$$\langle k|\mathrm{H}'(\mathrm{t})|\ell\rangle = -\frac{1}{i\hbar}\mathbf{A}_o e^{-i\omega t} \cdot \left(E_\ell^{(o)} - E_k^{(o)}\right)\langle k|\boldsymbol{\mu}|\ell\rangle \tag{12.2.9}$$

For simplicity of notation, we replace the Dirac bracket notation by a subscript notation. For example, we write $\langle k|\boldsymbol{\mu}|\ell\rangle$ in Eq. (12.2.9) as

$$\boldsymbol{\mu}_{k\ell} = \langle k|\boldsymbol{\mu}|\ell\rangle = \int d^3r \psi_k^{(o)*}(\mathbf{r})\boldsymbol{\mu}\psi_\ell^{(o)}(\mathbf{r}) \tag{12.2.10}$$

Generally, the electromagnetic field depends upon the vector potential, **A**, and the scalar potential, Φ. The scalar potential, Φ, can be assumed to be absorbed into $\mathrm{H}^{(\mathrm{o})}$. Thus, the electric field of the light wave is derivable from the vector potential alone. If we use Eq. (12.2.1), then

$$\mathbf{E}(t) = -\frac{\partial \mathbf{A}}{\partial t} = -\frac{\partial}{\partial t}\mathbf{A}_o e^{-i\omega t} = i\omega \mathbf{A}_o e^{-i\omega t} \tag{12.2.11}$$

We can combine Eq. (12.2.10) with Eq. (12.2.11) to rewrite Eq. (12.2.9) in the form,

$$\langle k|\mathrm{H}'(\mathrm{t})|\ell\rangle = \left(\frac{-1}{i\hbar}\right)\frac{\mathbf{E}(t)}{i\omega}\cdot\left(E_\ell^{(o)} - E_k^{(o)}\right)\boldsymbol{\mu}_{k\ell} = \frac{\mathbf{E}(t)}{\hbar\omega}\cdot\left(E_\ell^{(o)} - E_k^{(o)}\right)\boldsymbol{\mu}_{k\ell} \tag{12.2.13}$$

If $E_\ell^{(o)}$ is the ground state energy, and $E_k^{(o)}$ is the energy of the molecule in the excited state, then when the system is exposed to electromagnetic radiation with frequency ω, which is near the resonance, it is the case that $\hbar\omega \approx \left(E_k^{(o)} - E_\ell^{(o)}\right)$. This permits Eq. (12.2.13) to be simplified to read,

$$\langle k|\mathrm{H}'(t)|\ell\rangle \approx -\mathbf{E}(t)\cdot\boldsymbol{\mu}_{k\ell} \tag{12.2.14}$$

To convert to standard notation, we define $\mathrm{V}(\mathbf{r}, t) = -\mathbf{E}(t)\cdot\boldsymbol{\mu}$, which permits $\langle k|\mathrm{H}'(t)|\ell\rangle$ in Eq. (12.2.14) to be replaced by

$$\mathrm{V}_{k\ell}(t) = -\mathbf{E}(t)\cdot\boldsymbol{\mu}_{k\ell} \tag{12.2.15}$$

12.3 Time-Dependent Schrödinger Equation and the Density Matrix

We shall construct a solution to Eq. (12.1.1), by representing $\Psi(\mathbf{r}, t)$ as a sum of eigenfunctions of the time-independent Hamiltonian, $\mathrm{H}^{(o)}$. This expansion reads,

$$\Psi(\mathbf{r}, t) = \sum_k a_k(t)\psi_k^{(o)}(\mathbf{r}) \tag{12.3.1a}$$

where the expansion coefficients, $\{a_k(t)\}$, are assumed to be functions of the time. The complex conjugate of Eq. (12.3.1a) is

$$\Psi^*(\mathbf{r}, t) = \sum_j a_j^*(t)\psi_j^{(o)*}(\mathbf{r}) \tag{12.3.1b}$$

We now use $\boldsymbol{\mu}(t)$ to represent the expectation value of the time dependence of the electric dipole, $\boldsymbol{\mu}$, where

$$\boldsymbol{\mu}(t) = \int d^3r\Psi^*(\mathbf{r}, t)\boldsymbol{\mu}\Psi(\mathbf{r}, t) \tag{12.3.2}$$

Eq. (12.3.2) will form the basis for our calculation of the frequency dependence of the polarizability.

Upon substitution of Eqs. (12.3.1a) and (12.3.1b), Eq. (12.3.2) becomes

$$\boldsymbol{\mu}(t) = \sum_j \sum_k a_j^* a_k \int d^3 r \psi_j^{(o)*}(\mathbf{r}) \boldsymbol{\mu} \psi_k^{(o)}(\mathbf{r}) \tag{12.3.3a}$$

In the notation of Eq. (12.2.10), Eq. (12.3.3a) reads

$$\boldsymbol{\mu}(t) = \sum_j \sum_k a_j^* a_k \boldsymbol{\mu}_{jk} = \sum_j \sum_k a_k a_j^* \boldsymbol{\mu}_{jk} \tag{12.3.3b}$$

We can show that the double sum on the right hand side of Eq. (12.3.3b) involves the product of two matrices.

First, we identify $a_k a_j^*$ on the far right hand side of Eq. (12.3.3b) as an element of what is known as the "density matrix", $\rho(t)$. Specifically,

$$\rho_{kj}(t) = a_k(t) a_j^*(t) \tag{12.3.3c}$$

Using the density matrix identification in Eq. (12.3.3c) permits Eq. (12.3.3b) to be written in the form,

$$\boldsymbol{\mu}(t) = \sum_k \sum_j \rho_{kj}(t) \boldsymbol{\mu}_{jk} \tag{12.3.3d}$$

For the sake of clarity, we shall for the moment restrict ourselves to the coupling between just two quantum states. With this limitation, the "density" matrix ρ assumes the 2×2 form,

$$\rho(t) = \begin{bmatrix} \rho_{11} & \rho_{12} \\ \rho_{21} & \rho_{22} \end{bmatrix} \tag{12.3.4a}$$

One should note that this use of ρ as a matrix operator is not to be confused with the use of ρ in Chap. 11, where it served as a scalar representing the energy density.

The second required matrix, is the matrix of the dipole moment, $\boldsymbol{\mu}$. This matrix has the definition,

$$\boldsymbol{\mu} = \begin{bmatrix} \boldsymbol{\mu}_{11} & \boldsymbol{\mu}_{12} \\ \boldsymbol{\mu}_{21} & \boldsymbol{\mu}_{22} \end{bmatrix} \tag{12.3.4b}$$

According to the rule of matrix multiplication, left multiplication of $\boldsymbol{\mu}$ by ρ can be rendered in the form,

$$\rho\boldsymbol{\mu} = \begin{bmatrix} \rho_{11} & \rho_{12} \\ \rho_{21} & \rho_{22} \end{bmatrix} \begin{bmatrix} \mu_{11} & \mu_{12} \\ \mu_{21} & \mu_{22} \end{bmatrix} = \begin{bmatrix} \rho_{11}\mu_{11} + \rho_{12}\mu_{21} & \rho_{11}\mu_{12} + \rho_{12}\mu_{22} \\ \rho_{21}\mu_{11} + \rho_{22}\mu_{21} & \rho_{21}\mu_{12} + \rho_{22}\mu_{22} \end{bmatrix} \quad (12.3.4c)$$

The trace of a matrix is defined as the sum of its diagonal elements. In the case of $\rho\boldsymbol{\mu}$ in Eq. (12.3.4c), the trace is given by

$$Tr(\rho\boldsymbol{\mu}) = \rho_{11}\mu_{11} + \rho_{12}\mu_{21} + \rho_{21}\mu_{12} + \rho_{22}\mu_{22} \quad (12.3.4d)$$

We can show that Eq. (12.3.3d) is the same as Eq. (12.3.4d). To see this equivalence, we fix the upper limits on the indices j and k in Eq. (12.3.3d) at $j = 2$ and $k = 2$, respectively, and evaluate the two sums in sequence. The calculation proceeds as follows:

$$\boldsymbol{\mu}(t) = \sum_{k=1}^{2}\sum_{j=1}^{2} \rho_{kj}(t)\mu_{jk} = \sum_{k=1}^{2} (\rho_{k1}\mu_{1k} + \rho_{k2}\mu_{2k}) = \rho_{11}\mu_{11} + \rho_{12}\mu_{21} + \rho_{21}\mu_{12} + \rho_{22}\mu_{22} \quad (12.3.4e)$$

Notice that the right hand sides of Eqs. (12.3.4d) and (12.3.4e) are identical. Thus Eq. (12.3.3d) can be written symbolically as

$$\boldsymbol{\mu}(t) = Tr(\rho\boldsymbol{\mu}) \quad (12.3.5)$$

Returning now to the general case of multiple quantum states, substitution of Eq. (12.3.1a) into Eq. (12.1.1) with $\mathrm{H}'(t) = \mathrm{V}(\mathbf{r}, t)$ yields

$$i\hbar \sum_n \frac{da_n}{dt}\psi_n^{(o)} = \sum_n a_n \mathrm{H}^{(o)}\psi_n^{(o)} + \sum_n a_n \mathrm{V}(\mathbf{r}, t)\psi_n^{(o)} = \sum_n a_n E_n^{(o)}\psi_n^{(o)} + \sum_n a_n \mathrm{V}(\mathbf{r}, t)\psi_n^{(o)} \quad (12.3.6a)$$

To simplify the first sum on the right of Eq. (12.3.6a), we have used Eq. (12.1.2). Next, apply the integral operation, $\int d^3r\psi_k^{(o)*}$, to the left side and also to the far right hand side of Eq. (12.3.6a). The result is

$$i\hbar \sum_n \frac{da_n}{dt} \int d^3r\psi_k^{(o)*}(\mathbf{r})\psi_n^{(o)}(\mathbf{r}) = \sum_n a_n E_n^{(o)} \int d^3r\psi_k^{(o)*}(\mathbf{r})\psi_n^{(o)}(\mathbf{r}) + \sum_n a_n \int d^3r\psi_k^{(o)*}(\mathbf{r})\mathrm{V}(\mathbf{r}, t)\psi_n^{(o)}(\mathbf{r}) \quad (12.3.6b)$$

Upon use of Eq. (12.1.3) to evaluate the integral, $\int d^3r\psi_k^{(o)*}(\mathbf{r})\psi_n^{(o)}(\mathbf{r})$, Eq. (12.3.6b) reads,

$$i\hbar \sum_n \frac{da_n}{dt}\delta_{kn} = \sum_n a_n E_n^{(o)}\delta_{kn} + \sum_n a_n \mathrm{V}_{kn}(t) \tag{12.3.6c}$$

where

$$\mathrm{V}_{kn}(t) = \int d^3 r \psi_k^{(o)*}(\mathbf{r})\mathrm{V}(\mathbf{r}, t)\psi_n^{(o)}(\mathbf{r}) \tag{12.3.6d}$$

After computing the sums involving the Kronecker delta, Eq. (12.3.6c) takes the form,

$$i\hbar \frac{da_k}{dt} = E_k^{(o)} a_k + \sum_n a_n \mathrm{V}_{kn}(t) \tag{12.3.7a}$$

If we replace each energy eigenvalue, $E_k^{(o)}$, by an equivalent frequency

$$\omega_k = \frac{E_k^{(o)}}{\hbar} \tag{12.3.7b}$$

Equation (12.3.7a) takes the form,

$$\frac{da_k}{dt} = -i\omega_k a_k + \left(\frac{1}{i\hbar}\right) \sum_n a_n \mathrm{V}_{kn}(t) \tag{12.3.7c}$$

Equation (12.3.7c) can be generalized to include the effects of energy dissipation. An excited molecule can lose energy in both spontaneous emission of photons and non-radiative transitions, such as those involving the energy transfer taking place in collisions of the excited molecule with molecules of the solvent. Both constitute a form of heat transfer, and the rates of both are proportional to $a_k(t)$. To take such mechanisms into account, we introduce on the right hand side of Eq. (12.3.7c), a semi-empirical term, $-(1/2)\gamma_k a_k$, where $\gamma_k/2$ is termed the composite energy loss rate coefficient. This is the analogue of the damping term, $m\gamma(dx/dt)$, that appears in Eq. (3.11.1) in the classical harmonic oscillator model of the bound electron. With addition of the energy loss term, Eq. (12.3.7c) becomes

$$\frac{da_k}{dt} = -i\omega_k a_k - \frac{1}{2}\gamma_k a_k + \left(\frac{1}{i\hbar}\right) \sum_n a_n \mathrm{V}_{kn}(t) \tag{12.3.8a}$$

After substituting the subscript ℓ for k, and forming the complex conjugate, Eq. (12.3.8a) reads

$$\frac{da_\ell^*}{dt} = i\omega a_\ell^* - \frac{1}{2}\gamma_\ell a_\ell^* - \left(\frac{1}{i\hbar}\right) \sum_n a_n^* \mathrm{V}_{\ell n}^*(t) \tag{12.3.8b}$$

Multiply both sides of Eq. (12.3.8a) by a_ℓ^*. Multiply both sides of Eq. (12.3.8b) by a_k. Add the two resulting equations to obtain

$$\frac{da_k}{dt}a_\ell^* + \frac{da_\ell^*}{dt}a_k = -i\omega_k a_k a_\ell^* + i\omega_\ell a_k a_\ell^* - \frac{1}{2}(\gamma_k + \gamma_\ell)a_k a_\ell^* + \left(\frac{1}{i\hbar}\right)\sum_n \left(a_\ell^* a_n \mathrm{V}_{kn}(t) - a_k a_n^* \mathrm{V}_{\ell n}^*(t)\right) \tag{12.3.8c}$$

The left hand side of Eq. (12.3.8c) is the time derivative of the product, $a_k a_\ell^*$. The right hand side of this equation can be simplified by grouping the terms proportional to the frequencies, ω_k and ω_ℓ, together and by multiplying the numerator and denominator of the sum on the far right by $i = \sqrt{-1}$. With these observations taken into account, Eq. (12.3.8c) reads,

$$\frac{d(a_k a_\ell^*)}{dt} = -i(\omega_k - \omega_\ell)a_k a_\ell^* - \frac{1}{2}(\gamma_k + \gamma_\ell)a_k a_\ell^* + \left(\frac{i}{\hbar}\right)\sum_n \left(a_k a_n^* \mathrm{V}_{\ell n}^*(t) - a_\ell^* a_n \mathrm{V}_{kn}(t)\right) \tag{12.3.8d}$$

We can use the definition in Eq. (12.3.3c) to identify the elements of the density matrix in Eq. (12.3.8d). In addition, it will be helpful to exploit the Hermitian property,

$$\mathrm{V}_{\ell n}^*(t) = \mathrm{V}_{n\ell}(t) \tag{12.3.9}$$

With these changes, Eq. (12.3.8d) reads,

$$\frac{d\rho_{k\ell}}{dt} = -i(\omega_k - \omega_\ell)\rho_{k\ell} - \frac{1}{2}(\gamma_k + \gamma_\ell)\rho_{k\ell} + \left(\frac{i}{\hbar}\right)\sum_n \left(\rho_{kn}\mathrm{V}_{n\ell}(t) - \mathrm{V}_{kn}(t)\rho_{n\ell}\right) \tag{12.3.10}$$

Equation (12.3.10) is an ordinary differential equation, which determines the time dependence of $\rho_{k\ell}(t)$. Because the matrix elements $\mathrm{V}_{n\ell}(t)$ and $\mathrm{V}_{kn}(t)$ are functions of the time, Eq. (12.3.10) is not a member of the class of easily integrated constant coefficient differential equations. Nonetheless, by making reference to the physical conditions which apply in most cases, it is possible to convert Eq. (12.3.10) into a constant coefficient differential equation.

At $t = 0$, the molecule is assumed to be in the quantum state $\psi_\ell^{(o)}(\boldsymbol{r})$ with energy, $E_\ell^{(o)}$. The corresponding initial conditions that apply to the wave functions in Eqs. (12.3.1a) and (12.3.1b) are $a_k(0) = a_k^*(0) = 1$ when $k = \ell$, and $a_k(0) = a_k^*(0) = 0$ when $k \neq \ell$. The initial conditions satisfied by the density matrix are then

$$\rho_{\ell\ell}(0) = 1 \tag{12.3.11a}$$

$$\rho_{mn}(0) = 0, \text{ if } m \text{ and/or } n \text{ is not equal to } \ell \tag{12.3.11b}$$

In order to construct a constant coefficient differential equation satisfied by $\rho_{k\ell}$, we substitute Eqs. (12.3.11a) and (12.3.11b) into the sums on the right hand side of Eq. (12.3.10). The first sum does not include the term involving $\rho_{\ell\ell}$, so the sum can be replaced by zero. All the terms in the second sum satisfy Eq. (12.3.11b) except for one which reads $-\mathrm{V}_{k\ell}(t)\rho_{\ell\ell}$. According to Eq. (12.3.11a) we can replace this term by $-\mathrm{V}_{k\ell}(t)$. With these substitutions, Eq. (12.3.10) simplifies to read

$$\frac{d\rho_{k\ell}(t)}{dt} = -i(\omega_k - \omega_\ell)\rho_{k\ell}(t) - \frac{1}{2}(\gamma_k + \gamma_\ell)\rho_{k\ell}(t) - \left(\frac{i}{\hbar}\right)\mathrm{V}_{k\ell}(t) \tag{12.3.12}$$

Using Eq. (12.2.15), Eq. (12.3.12) now becomes,

$$\frac{d\rho_{k\ell}(t)}{dt} = -i(\omega_k - \omega_\ell)\rho_{k\ell}(t) - \frac{1}{2}(\gamma_k + \gamma_\ell)\rho_{k\ell}(t) + \left(\frac{i}{\hbar}\right)\boldsymbol{\mu}_{k\ell}\cdot\mathbf{E}(t) \tag{12.3.13}$$

Equation (12.3.13) contains all terms of lowest order in $\rho_{k\ell}(t)$ and $\mathrm{V}_{k\ell}(t)$. Move the first two terms on the right hand side to the left, so that Eq. (12.3.13) reads

$$\frac{d\rho_{k\ell}(t)}{dt} + i(\omega_k - \omega_\ell)\rho_{k\ell}(t) + \frac{1}{2}(\gamma_k + \gamma_\ell)\rho_{k\ell}(t) = \left(\frac{i}{\hbar}\right)\boldsymbol{\mu}_{k\ell}\cdot\mathbf{E}(t) \tag{12.3.14}$$

Multiply Eq. (12.3.14) on both sides by the integrating factor, $\exp\{[i(\omega_k - \omega_\ell) + (1/2)(\gamma_k + \gamma_\ell)]t\}$. By noting that $e^{(ia+b)t}\frac{d\rho}{dt} + (ia+b)e^{(ia+b)t}\rho = \frac{d}{dt}\left[e^{(ia+b)t}\rho\right]$, where $ia + b = i(\omega_k - \omega_\ell) + (1/2)(\gamma_k + \gamma_\ell)$, Eq. (12.3.14) can be written as

$$\begin{aligned} &\frac{d}{dt}\left(\rho_{k\ell}\exp\{[i(\omega_k - \omega_\ell) + (1/2)(\gamma_k + \gamma_\ell)]t\}\right) \\ &\quad = \left(\frac{i}{\hbar}\right)\boldsymbol{\mu}_{k\ell}\cdot\mathbf{E}(t)\exp\{[i(\omega_k - \omega_\ell) + (1/2)(\gamma_k + \gamma_\ell)]t\} \end{aligned} \tag{12.3.15a}$$

Integration of Eq. (12.3.15a) from $t = 0$ to $t = t$ produces the result,

$$\begin{aligned} &\rho_{k\ell}(t)\exp\{[i(\omega_k - \omega_\ell) + (1/2)(\gamma_k + \gamma_\ell)]t\} \\ &\quad - \rho_{k\ell}(0)\exp\{[i(\omega_k - \omega_\ell) + (1/2)(\gamma_k + \gamma_\ell)]0\} \end{aligned} \tag{12.3.15b}$$

$$= \left(\frac{i}{\hbar}\right)\boldsymbol{\mu}_{k\ell}\cdot\int_0^t d\tau\,\mathbf{E}(\tau)\exp\{[i(\omega_k - \omega_\ell) + (1/2)(\gamma_k + \gamma_\ell)]\tau\}$$

If we incorporate the initial condition $\rho_{k\ell}(0) = 0$, Eq. (12.3.15b) assumes the form,

$$\rho_{k\ell}(t)\exp\{[i(\omega_k-\omega_\ell)+(1/2)(\gamma_k+\gamma_\ell)]t\}$$
$$=\left(\frac{i}{\hbar}\right)\boldsymbol{\mu}_{k\ell}\cdot\int_0^t d\tau\mathbf{E}(\tau)\exp\{[i(\omega_k-\omega_\ell)+(1/2)(\gamma_k+\gamma_\ell)]\tau\} \quad (12.3.15c)$$

In our short hand notation, where $a=\omega_k-\omega_\ell$ and $b=(1/2)(\gamma_k+\gamma_\ell)$, Eq. (12.3.15c) reads,

$$\rho(t)e^{(ia+b)t}=\left(\frac{i}{\hbar}\right)\mu\cdot\int_0^t d\tau E(\tau)e^{(ia+b)\tau} \quad (12.3.16a)$$

When both sides of Eq. (12.3.16a) are multiplied by $e^{-(ia+b)t}$, the result is

$$\rho(t)=\left(\frac{i}{\hbar}\right)\mu\cdot\int_0^t d\tau E(\tau)e^{-[(ia+b)(t-\tau)]} \quad (12.3.16b)$$

On the basis of Eqs. (12.3.16a) and (12.3.16b), Eq. (12.3.15c) can be rewritten as

$$\rho_{k\ell}(t)=\left(\frac{i}{\hbar}\right)\boldsymbol{\mu}_{k\ell}\cdot\int_0^t d\tau\mathbf{E}(\tau)\exp\{[-i(\omega_k-\omega_\ell)-(1/2)(\gamma_k+\gamma_\ell)](t-\tau)\} \quad (12.3.17)$$

Equation (12.3.17) is the solution for $\rho_{k\ell}(t)$ and serves as the basis for subsequent calculations in this chapter.

12.4 Derivation of the Response Function

Assume for the moment that there are three quantum states and that the system is initially in the ground state $\psi_1^{(o)}$ with energy $E_1^{(o)}$. We replace the system dipole moment $\boldsymbol{\mu}(t)$ by $\mathbf{p}_1(t)$ to indicate that the system is initially in the ground state. With reference to Eqs. (12.3.5) and (12.3.3d), we can write for the dipole moment, $\mathbf{p}_1(t)$, as

$$\mathbf{p}_1(t)=Tr(\rho\boldsymbol{\mu})=\sum_{k=1}^{3}\sum_{j=1}^{3}\rho_{kj}\boldsymbol{\mu}_{jk} \quad (12.4.1)$$

The double sum in Eq. (12.4.1) can be written out explicitly in the form,

$$\mathbf{p}_1(t) = Tr(\rho\boldsymbol{\mu}) = \rho_{11}\boldsymbol{\mu}_{11} + \rho_{12}\boldsymbol{\mu}_{21} + \rho_{13}\boldsymbol{\mu}_{31} + \rho_{21}\boldsymbol{\mu}_{12} + \rho_{22}\boldsymbol{\mu}_{22} + \rho_{23}\boldsymbol{\mu}_{32} + \rho_{31}\boldsymbol{\mu}_{13} + \rho_{32}\boldsymbol{\mu}_{23} + \rho_{33}\boldsymbol{\mu}_{33} \tag{12.4.2}$$

For simplicity, we assume that the expectation value of the induced dipole moment in the ground state and in each excited state is zero, that is $\boldsymbol{\mu}_{11} = \boldsymbol{\mu}_{22} = \boldsymbol{\mu}_{33} = 0$. This is possible, because in the case of an atom, each quantum state has definite parity; namely, $\psi_n^{(o)}(-\mathbf{r}) = \pm\psi_n^{(o)}(\mathbf{r})$, where the upper sign corresponds to even parity while the lower sign corresponds to odd parity. In either case, if we replace $\mathbf{r}$ by $-\mathbf{r}$, the product, $\psi_n^{(o)*}(\mathbf{r})\psi_n^{(o)}(\mathbf{r})$, will be even, while the product, $\psi_n^{(o)*}(\mathbf{r})\mathbf{r}\psi_n^{(o)}(\mathbf{r})$, will be odd. The integral over all space of an odd integrand is zero. Hence, we have

$$\boldsymbol{\mu}_{nn} = \langle n|\boldsymbol{\mu}|n\rangle = \int d^3r\left[\psi_n^{(o)*}(\mathbf{r})\mathbf{r}\psi_n^{(o)}(\mathbf{r})\right] = 0 \quad n = 1, 2, 3 \tag{12.4.3a}$$

By virtue of Eq. (12.4.3a), Eq. (12.4.2) reads,

$$\mathbf{p}_1(t) = Tr(\rho\boldsymbol{\mu}) = \rho_{12}\boldsymbol{\mu}_{21} + \rho_{13}\boldsymbol{\mu}_{31} + \rho_{21}\boldsymbol{\mu}_{12} + \rho_{23}\boldsymbol{\mu}_{32} + \rho_{31}\boldsymbol{\mu}_{13} + \rho_{32}\boldsymbol{\mu}_{23} \tag{12.4.3b}$$

We shall assume that the system is initially in the ground state with quantum number $\ell = 1$.

Depending upon the frequency of the light, it can make transitions to the states with $k = 2$ and $k = 3$, respectively. Either of these states can transition back to the ground state, $\ell = 1$. The population of excited molecules in the states with $k = 2$ and $k = 3$, are low which makes the product, $a_2(t)a_3^*(t)$, of second order. Thus, a transition, which couples these two states, has a negligible chance of occurrence, so we can set, $\rho_{23} = 0 = \rho_{32}$. With this assumption, Eq. (12.4.3b) further simplifies to read,

$$\mathbf{p}_1(t) = Tr(\rho\boldsymbol{\mu}) = \rho_{12}\boldsymbol{\mu}_{21} + \rho_{13}\boldsymbol{\mu}_{31} + \rho_{21}\boldsymbol{\mu}_{12} + \rho_{31}\boldsymbol{\mu}_{13} \tag{12.4.3c}$$

The right hand side of Eq. (12.4.3c) can be rewritten in summation notation as

$$\mathbf{p}_1(t) = \rho_{12}\boldsymbol{\mu}_{21} + \rho_{13}\boldsymbol{\mu}_{31} + \rho_{21}\boldsymbol{\mu}_{12} + \rho_{31}\boldsymbol{\mu}_{13} = \sum_{k=2}^{3}(\rho_{1k}\boldsymbol{\mu}_{k1} + \rho_{k1}\boldsymbol{\mu}_{1k}) \tag{12.4.3d}$$

If in Eq. (12.4.3d), we replace the subscript, "1" by "ℓ", we obtain the general result,

$$\mathbf{p}_\ell(t) = \sum_{\substack{k=1\\k\neq\ell}}\left(\rho_{\ell k}\boldsymbol{\mu}_{k\ell} + \rho_{k\ell}\boldsymbol{\mu}_{\ell k}\right) \tag{12.4.4}$$

By reference to Eq. (12.3.3c), we have

$$\rho_{\ell k} = \rho_{k\ell}^* \tag{12.4.5a}$$

As the elements of $\boldsymbol{\mu}_{\ell k}$ are Hermitian, the following is the result:

$$\boldsymbol{\mu}_{\ell k} = \boldsymbol{\mu}_{k\ell}^* \tag{12.4.5b}$$

Use Eqs. (12.4.5a) and (12.4.5b) to write Eq. (12.4.4) in the form,

$$\mathbf{p}_\ell(t) = \sum_{\substack{k=1 \\ k\neq\ell}} \left(\rho_{k\ell}^* \boldsymbol{\mu}_{k\ell} + \rho_{k\ell}\boldsymbol{\mu}_{k\ell}^*\right) \tag{12.4.6}$$

The complex conjugate of Eq. (12.3.17) is

$$\rho_{k\ell}^*(t) = \left(\frac{-i}{\hbar}\right)\boldsymbol{\mu}_{k\ell}^* \cdot \int_0^t d\tau \mathbf{E}(\tau) \exp\{[i(\omega_k - \omega_\ell) - (1/2)(\gamma_k + \gamma_\ell)](t-\tau)\} \tag{12.4.7}$$

Substitution of Eqs. (12.3.17) and (12.4.7) into Eq. (12.4.6) leads to the result,

$$\begin{aligned}
\mathbf{p}_\ell(t) &= \left(\frac{-i}{\hbar}\right)\sum_{k\neq\ell} \boldsymbol{\mu}_{k\ell}\boldsymbol{\mu}_{k\ell}^* \cdot \int_0^t d\tau \mathbf{E}(\tau)\exp\{[i(\omega_k - \omega_\ell) - (1/2)(\gamma_k + \gamma_\ell)](t-\tau)\} \\
&\quad + \left(\frac{i}{\hbar}\right)\sum_{k\neq\ell} \boldsymbol{\mu}_{k\ell}^*\boldsymbol{\mu}_{k\ell} \cdot \int_0^t d\tau \mathbf{E}(\tau)\exp\{[-i(\omega_k - \omega_\ell) - (1/2)(\gamma_k + \gamma_\ell)](t-\tau)\} \\
&= \left(\frac{-i}{\hbar}\right)\sum_{k\neq\ell} \boldsymbol{\mu}_{k\ell}^*\boldsymbol{\mu}_{k\ell} \\
&\quad \cdot \int_0^t d\tau \mathbf{E}(\tau)\left\{e^{[i(\omega_k-\omega_\ell)-(1/2)(\gamma_k+\gamma_\ell)](t-\tau)} - e^{[-i(\omega_k-\omega_\ell)-(1/2)(\gamma_k+\gamma_\ell)](t-\tau)}\right\}
\end{aligned} \tag{12.4.8}$$

To simplify the notation in Eq. (12.4.8), it is worthwhile to define the response tensor, $\overleftrightarrow{\Phi}_\ell(t)$, which we write in dyadic form as

$$\overleftrightarrow{\Phi}_\ell(t) = \left(\frac{-i}{\hbar}\right)\sum_{k\neq\ell} \boldsymbol{\mu}_{k\ell}^*\boldsymbol{\mu}_{k\ell}\left\{e^{[i(\omega_k-\omega_\ell)-(1/2)(\gamma_k+\gamma_\ell)]t} - e^{[-i(\omega_k-\omega_\ell)-(1/2)(\gamma_k+\gamma_\ell)]t}\right\} \tag{12.4.9a}$$

$\overleftrightarrow{\Phi}_\ell(t)$ is a tensor by virtue of its dependence on the dyad product, $\boldsymbol{\mu}_{k\ell}^*\boldsymbol{\mu}_{k\ell}$. Using Eq. (12.4.9a), we can condense Eq. (12.4.8) to read,

$$\mathbf{p}_\ell(t) = \int_0^t d\tau \overset{\leftrightarrow}{\Phi}_\ell(t-\tau)\cdot\mathbf{E}(\tau) \tag{12.4.9b}$$

When acting on a vector, a tensor produces a new vector. Whereas the dot product, $\boldsymbol{\mu}_{k\ell}\cdot\mathbf{E}(t)$, is a scalar, the dyadic product $\boldsymbol{\mu}_{k\ell}^*\boldsymbol{\mu}_{k\ell}\cdot\mathbf{E}(t) = \left(\boldsymbol{\mu}_{k\ell}^*\right)\left(\boldsymbol{\mu}_{k\ell}\cdot\mathbf{E}(t)\right)$ is a vector. For another example, see Sect. 4.1 of Chap. 4.

After removing the factor, $\exp\left[-(1/2)(\gamma_k+\gamma_\ell)t\right]$, from the braces in Eq. (12.4.9a), and multiplication and division of the resulting sum by a factor of $2i$, the resulting equation reads,

$$\overset{\leftrightarrow}{\Phi}_\ell(t) = \left(\frac{2}{\hbar}\right)\sum_k \boldsymbol{\mu}_{k\ell}^*\boldsymbol{\mu}_{k\ell}e^{-\frac{1}{2}(\gamma_k+\gamma_\ell)t}\left\{\frac{e^{i(\omega_k-\omega_\ell)t}-e^{-i(\omega_k-\omega_\ell)t}}{2i}\right\} \tag{12.4.10a}$$

which is equivalent to

$$\overset{\leftrightarrow}{\Phi}_\ell(t) = \left(\frac{2}{\hbar}\right)\sum_k \boldsymbol{\mu}_{k\ell}^*\boldsymbol{\mu}_{k\ell}e^{-\frac{1}{2}(\gamma_k+\gamma_\ell)t}\sin[(\omega_k-\omega_\ell)t] \tag{12.4.10b}$$

When the material under consideration is electrically isotropic, such as a gas or a liquid, the vectors, $\mathbf{p}_\ell(t)$ and $\mathbf{E}(t)$, will be parallel to one another. When this is the case, the non-zero elements of $\overset{\leftrightarrow}{\Phi}_\ell(t)$ lie solely along the diagonal, and are all equal to one another. As a result, $\overset{\leftrightarrow}{\Phi}_\ell(t)$ becomes a scalar, which can be written as $\Phi_\ell(t)$. The latter is known as the response function. With this simplification, Eq. (12.4.9b) reduces to

$$\mathbf{p}_\ell(t) = \int_0^t d\tau\,\Phi_\ell(t-\tau)\mathbf{E}(\tau) \tag{12.4.11}$$

12.5 Real and Imaginary Parts of the Polarizability

Define the three Fourier transforms,

$$\tilde{\mathbf{p}}_\ell(\omega) = \int_{-\infty}^{+\infty} dt\, e^{i\omega t}\mathbf{p}_\ell(t) \tag{12.5.1}$$

$$\tilde{\Phi}_\ell(\omega) = \int_{-\infty}^{+\infty} dt\, e^{i\omega t}\Phi_\ell(t) \tag{12.5.2}$$

$$\tilde{\mathbf{E}}(\omega) = \int_{-\infty}^{+\infty} dt\, e^{i\omega t} \mathbf{E}(t) \tag{12.5.3}$$

By virtue of Eq. (12.4.11) and the convolution theorem [see the Mathematical Methods Appendix], the three are related by

$$\tilde{\mathbf{p}}_\ell(\omega) = \tilde{\Phi}_\ell(\omega)\tilde{\mathbf{E}}(\omega) \tag{12.5.4}$$

The complex valued molecular polarizability, $\tilde{\alpha}(\omega)$, is defined by the relation,

$$\tilde{\mathbf{p}}_\ell(\omega) = \tilde{\alpha}(\omega)\tilde{\mathbf{E}}(\omega) \tag{12.5.5}$$

Hence,

$$\tilde{\alpha}(\omega) = \tilde{\Phi}_\ell(\omega) \tag{12.5.6}$$

We assume that the system is unpolarized at $t = 0$. This can be assured if $\Phi_\ell(t)$ is zero for $t < 0$. As a result, the Fourier transform of $\Phi_\ell(t)$ can be adjusted to read,

$$\tilde{\Phi}_\ell(\omega) = \int_{0}^{+\infty} dt\, e^{i\omega t} \Phi_\ell(t) \tag{12.5.7}$$

Since $\overleftrightarrow{\Phi}_\ell(t)$ is now considered to be replaced by $\Phi_\ell(t)$, Eq. (12.4.9a) takes the form,

$$\Phi_\ell(t) = \left(\frac{-i}{\hbar}\right) \sum_{k \neq \ell} \left|\boldsymbol{\mu}_{k\ell}\right|^2 \left\{ e^{[i(\omega_k - \omega_\ell) - (1/2)(\gamma_k + \gamma_\ell)]t} - e^{[-i(\omega_k - \omega_\ell) - (1/2)(\gamma_k + \gamma_\ell)]t} \right\} \tag{12.5.8a}$$

Substitution of Eq. (12.5.8a) into Eq. (12.5.7) reads,

$$\begin{aligned} \tilde{\Phi}_\ell(\omega) = \left(\frac{-i}{\hbar}\right) \sum_{k} \left|\boldsymbol{\mu}_{k\ell}\right|^2 \int_{0}^{\infty} dt\, e^{i\omega t} \\ \times \left\{ e^{[i(\omega_k - \omega_\ell) - (1/2)(\gamma_k + \gamma_\ell)]t} - e^{[-i(\omega_k - \omega_\ell) - (1/2)(\gamma_k + \gamma_\ell)]t} \right\} \end{aligned} \tag{12.5.8b}$$

Define

$$\omega_{k\ell} = \omega_k - \omega_\ell \tag{12.5.9a}$$

and

$$\gamma_{k\ell} = \gamma_k + \gamma_\ell \tag{12.5.9b}$$

Write Eq. (12.5.8b) in the form,

$$\tilde{\Phi}_\ell(\omega) = \left(\frac{-i}{\hbar}\right)\sum_k |\mu_{k\ell}|^2 \int_0^\infty dt\, e^{i\omega t}\left\{e^{[i\omega_{k\ell}-(1/2)\gamma_{k\ell}]t} - e^{[-i\omega_{k\ell}-(1/2)\gamma_{k\ell}]t}\right\} \tag{12.5.9c}$$

Finally, multiply the terms within braces in Eq. (12.5.9c) by $e^{i\omega t}$ as indicated so that Eq. (12.5.9c) takes the form,

$$\tilde{\Phi}_\ell(\omega) = \left(\frac{-i}{\hbar}\right)\sum_k |\mu_{k\ell}|^2 \int_0^\infty dt\left\{e^{[i(\omega+\omega_{k\ell})-(1/2)\gamma_{k\ell}]t} - e^{[i(\omega-\omega_{k\ell})-(1/2)\gamma_{k\ell}]t}\right\} \tag{12.5.10}$$

Eq. (12.5.10) can be rewritten as

$$\tilde{\Phi}_\ell(\omega) = \left(\frac{-i}{\hbar}\right)\sum_k |\mu_{k\ell}|^2\left(I^+_{k\ell}(\omega) - I^-_{k\ell}(\omega)\right) \tag{12.5.11}$$

where $I^+_{k\ell}(\omega)$ and $I^-_{k\ell}(\omega)$ are defined respectively by the following equation,

$$I^\pm_{k\ell}(\omega) = \int_0^\infty dt\, \exp\left\{\left[i(\omega \pm \omega_{k\ell}) - (1/2)\gamma_{k\ell}\right]t\right\} \tag{12.5.12a}$$

Evaluation of the integral on the right hand side of Eq. (12.5.12a) leads to the result,

$$I^\pm_{k\ell}(\omega) = \left[\frac{\exp\left\{\left[i(\omega \pm \omega_{k\ell}) - (1/2)\gamma_{k\ell}\right]t\right\}}{i(\omega \pm \omega_{k\ell}) - (1/2)\gamma_{k\ell}}\right]_0^\infty = \frac{-1}{i(\omega \pm \omega_{k\ell}) - (1/2)\gamma_{k\ell}} \tag{12.5.12b}$$

In Eq. (12.5.12b), the $t = \infty$ evaluation of the term in the bracket has the result, $\exp\left\{\left[i(\omega \pm \omega_{k\ell}) - (1/2)\gamma_{k\ell}\right](\infty)\right\}$, which is zero because of the negative term, $-(1/2)\gamma_{k\ell}$, in the argument of the exponential. Using the results for $I^+_{k\ell}(\omega)$ and $I^-_{k\ell}(\omega)$, which are implicit in Eq. (12.5.12b), it is possible to write Eq. (12.5.11) as

$$\tilde{\Phi}_\ell(\omega) = \left(-\frac{i}{\hbar}\right)\sum_k |\mu_{k\ell}|^2\left[\frac{-1}{i(\omega + \omega_{k\ell}) - (1/2)\gamma_{k\ell}} - \frac{-1}{i(\omega - \omega_{k\ell}) - (1/2)\gamma_{k\ell}}\right]$$

$$= \left(\frac{i}{\hbar}\right) \sum_k |\boldsymbol{\mu}_{k\ell}|^2 \left[\frac{1}{i(\omega + \omega_{k\ell}) - (1/2)\gamma_{k\ell}} - \frac{1}{i(\omega - \omega_{k\ell}) - (1/2)\gamma_{k\ell}} \right] \tag{12.5.13}$$

The first fraction on the right hand side of the second line of Eq. (12.5.13) can be rationalized to read,

$$\frac{1}{i(\omega + \omega_{k\ell}) - (1/2)\gamma_{k\ell}} = \frac{-i(\omega + \omega_{k\ell}) - (1/2)\gamma_{k\ell}}{\left[i(\omega + \omega_{k\ell}) - (1/2)\gamma_{k\ell}\right]\left[-i(\omega + \omega_{k\ell}) - (1/2)\gamma_{k\ell}\right]}$$
$$= \frac{-i(\omega + \omega_{k\ell}) - (1/2)\gamma_{k\ell}}{(\omega + \omega_{k\ell})^2 + (1/4)\gamma_{k\ell}^2} \tag{12.5.14a}$$

The second fraction on the right hand side of Eq. (12.5.13) can be rationalized to read,

$$\frac{1}{i(\omega - \omega_{k\ell}) - (1/2)\gamma_{k\ell}} = \frac{-i(\omega - \omega_{k\ell}) - (1/2)\gamma_{k\ell}}{[i(\omega - \omega_{k\ell}) - (1/2)\gamma_{k\ell}][-i(\omega - \omega_{k\ell}) - (1/2)\gamma_{k\ell}]}$$
$$= \frac{-i(\omega - \omega_{k\ell}) - (1/2)\gamma_{k\ell}}{(\omega - \omega_{k\ell})^2 + (1/4)\gamma_{k\ell}^2} \tag{12.5.14b}$$

If Eqs. (12.5.14a) and (12.5.14b) are substituted into Eq. (12.5.13), the result is

$$\tilde{\Phi}_\ell(\omega) = \left(\frac{i}{\hbar}\right) \sum_k |\boldsymbol{\mu}_{k\ell}|^2 \left[\frac{-i(\omega + \omega_{k\ell}) - (1/2)\gamma_{k\ell}}{(\omega + \omega_{k\ell})^2 + (1/4)\gamma_{k\ell}^2} - \frac{-i(\omega - \omega_{k\ell}) - (1/2)\gamma_{k\ell}}{(\omega - \omega_{k\ell})^2 + (1/4)\gamma_{k\ell}^2} \right] \tag{12.5.15a}$$

Upon multiplying each term in Eq. (12.5.15a) by the factor, i, that appears outside the sum, one obtains,

$$\tilde{\Phi}_\ell(\omega) = \left(\frac{1}{\hbar}\right) \sum_k |\boldsymbol{\mu}_{k\ell}|^2 \left[\frac{(\omega + \omega_{k\ell}) - i(1/2)\gamma_{k\ell}}{(\omega + \omega_{k\ell})^2 + (1/4)\gamma_{k\ell}^2} - \frac{(\omega - \omega_{kl}) - i(1/2)\gamma_{k\ell}}{(\omega - \omega_{k\ell})^2 + (1/4)\gamma_{k\ell}^2} \right] \tag{12.5.15b}$$

The minus sign in the second fraction can be incorporated into the numerator, so that Eq. (12.5.15b) then reads,

$$\tilde{\Phi}_\ell(\omega) = \left(\frac{1}{\hbar}\right) \sum_k |\boldsymbol{\mu}_{k\ell}|^2 \left[\frac{(\omega + \omega_{k\ell}) - i(1/2)\gamma_{k\ell}}{(\omega + \omega_{k\ell})^2 + (1/4)\gamma_{k\ell}^2} + \frac{(\omega_{kl} - \omega) + i(1/2)\gamma_{k\ell}}{(\omega - \omega_{k\ell})^2 + (1/4)\gamma_{k\ell}^2} \right] \tag{12.5.15c}$$

The real and imaginary parts in Eq. (12.5.15c) can be separated to obtain the result,

$$\tilde{\Phi}_\ell(\omega) = \left(\frac{1}{\hbar}\right)\sum_k |\boldsymbol{\mu}_{k\ell}|^2\left[\frac{(\omega_{k\ell}-\omega)}{(\omega-\omega_{k\ell})^2+(1/4)\gamma_{k\ell}^2}+\frac{(\omega+\omega_{k\ell})}{(\omega+\omega_{k\ell})^2+(1/4)\gamma_{k\ell}^2}\right]$$
$$+\left(\frac{i}{2\hbar}\right)\sum_k |\boldsymbol{\mu}_{k\ell}|^2\left[\frac{\gamma_{k\ell}}{(\omega-\omega_{k\ell})^2+(1/4)\gamma_{k\ell}^2}-\frac{\gamma_{k\ell}}{(\omega+\omega_{k\ell})^2+(1/4)\gamma_{k\ell}^2}\right] \tag{12.5.16}$$

The polarizability, $\tilde{\alpha}(\omega)$, being a complex valued function, can be written in terms of its real and imaginary parts as

$$\tilde{\alpha}(\omega) = \alpha'(\omega) + i\alpha''(\omega) \tag{12.5.17}$$

By virtue of Eq. (12.5.6), we can now use Eq. (12.5.17) to identify the real and imaginary parts of Eq. (12.5.16) with the real and imaginary parts, respectively, of the polarizability. Upon equating the real parts, one has,

$$\alpha'(\omega) = \left(\frac{1}{\hbar}\right)\sum_k |\boldsymbol{\mu}_{k\ell}|^2\left[\frac{(\omega_{k\ell}-\omega)}{(\omega-\omega_{k\ell})^2+(1/4)\gamma_{k\ell}^2}+\frac{(\omega+\omega_{k\ell})}{(\omega+\omega_{k\ell})^2+(1/4)\gamma_{k\ell}^2}\right] \tag{12.5.18}$$

Near resonance, where $\omega \approx \omega_{k\ell}$, the first term within the bracket in Eq. (12.5.18) varies rapidly, whereas the second term varies slowly. Taking into account this observation, Eq. (12.5.18) can for all practical purposes be replaced by

$$\alpha'(\omega) = \left(\frac{1}{\hbar}\right)\sum_k |\boldsymbol{\mu}_{k\ell}|^2\frac{(\omega_{k\ell}-\omega)}{(\omega-\omega_{k\ell})^2+(1/4)\gamma_{k\ell}^2} \tag{12.5.19}$$

According to Eq. (12.5.19), $\alpha'(\omega)$ changes sign and exhibits a point of inflection as ω passes through resonance.

In those cases where $\gamma_{k\ell}$ is negligible, Eq. (12.5.18) takes the form,

$$\alpha'(\omega) = \left(\frac{1}{\hbar}\right)\sum_k |\boldsymbol{\mu}_{k\ell}|^2\left[\frac{(\omega_{kl}-\omega)}{(\omega_{kl}-\omega)^2}+\frac{(\omega+\omega_{k\ell})}{(\omega+\omega_{k\ell})^2}\right] \tag{12.5.20a}$$

The fractions in Eq. (12.5.20a) can be simplified to read,

$$\alpha'(\omega) = \left(\frac{1}{\hbar}\right)\sum_k |\boldsymbol{\mu}_{k\ell}|^2\left[\frac{1}{(\omega_{kl}-\omega)}+\frac{1}{(\omega+\omega_{k\ell})}\right]$$
$$= \left(\frac{1}{\hbar}\right)\sum_k |\boldsymbol{\mu}_{k\ell}|^2\left[\frac{(\omega+\omega_{k\ell})+(\omega_{kl}-\omega)}{(\omega_{k\ell}^2-\omega^2)}\right] \tag{12.5.20b}$$

After making the obvious cancellations, Eq. (12.5.20b) becomes,

$$\alpha'(\omega) = \left(\frac{2}{\hbar}\right) \sum_k \frac{\omega_{k\ell} \left|\boldsymbol{\mu}_{k\ell}\right|^2}{(\omega_{k\ell}^2 - \omega^2)} \tag{12.5.21}$$

Equation (12.5.21) is a good representation of the real part of the polarizability when the damping is sufficiently weak that $\gamma_{k\ell} \ll \omega_{k\ell}$, which is a case that often occurs.

We now turn to the imaginary part of the polarizability. Referring to Eq. (12.5.16), we conclude that

$$\alpha''(\omega) = \left(\frac{1}{2\hbar}\right) \sum_k \left|\boldsymbol{\mu}_{k\ell}\right|^2 \left[\frac{\gamma_{k\ell}}{(\omega - \omega_{k\ell})^2 + (1/4)\gamma_{k\ell}^2} - \frac{\gamma_{k\ell}}{(\omega + \omega_{k\ell})^2 + (1/4)\gamma_{k\ell}^2}\right] \tag{12.5.22}$$

Close to a resonance where $\omega \approx \omega_{k\ell}$, the first fraction within the brackets in Eq. (12.5.22) is much larger than the second. This permits the second to be dropped in favor the first. The result is

$$\alpha''(\omega) = \left(\frac{1}{2\hbar}\right) \sum_k \frac{\gamma_{k\ell} \left|\boldsymbol{\mu}_{k\ell}\right|^2}{(\omega - \omega_{k\ell})^2 + (1/4)\gamma_{k\ell}^2} \tag{12.5.23}$$

According to Eq. (12.5.23), $\alpha''(\omega)$ exhibits a maximum at $\omega = \omega_{k\ell}$. See Fig. 12.1. This contrasts with Eq. (12.5.19) for $\alpha'(\omega)$, which exhibits a point of inflection at $\omega = \omega_{k\ell}$. Eq. (12.5.21) applies in the limit that $\gamma_{k\ell} \to 0$. In this limit, Eq. (12.5.23) indicates that $\alpha''(\omega)$ vanishes.

12.6 Kramers–Kronig Relations

In Sect. 12.5, the functional forms assumed by the real and imaginary parts, $\alpha'(\omega)$ and $\alpha''(\omega)$, of the complex valued polarizability, $\tilde{\alpha}(\omega)$, were derived on the basis of quantum mechanical perturbation theory. H. A. Kramers (Dutch, 1894–1952) and R. Kronig (German, 1904–1995) showed that regardless of the model, the two polarizabilities, $\alpha'(\omega)$ and $\alpha''(\omega)$, were related to one another by a pair of integral equations. This pair of equations has since become known as the Kramers–Kronig relations. Specifically, if $\alpha'(\omega)$ is known, $\alpha''(\omega)$ can be calculated by evaluating a certain integral involving $\alpha'(\omega)$, whereas if $\alpha''(\omega)$ is known, $\alpha'(\omega)$ can be calculated by evaluating an analogous integral involving $\alpha''(\omega)$. In deriving the Kramers–Kronig relations, we will follow the methods introduced by Harbecke [2] and by Norman et al. [3].

We begin by considering the Fourier transform, $\tilde{G}(\omega)$, of an arbitrary function of time,$G(t)$. The transform can be written as

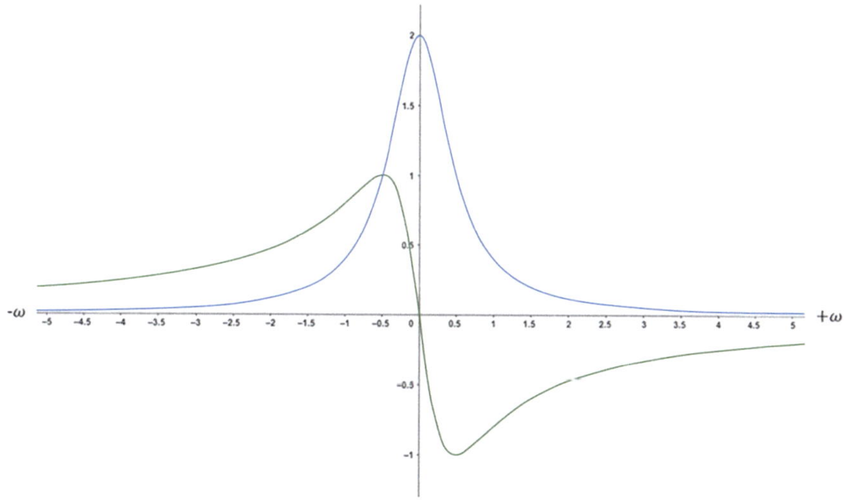

Fig. 12.1 Graphs of Eqs. (12.5.19) and (12.5.23) as functions of ω. For ease in plotting, the summations in Eqs. (12.5.19) and (12.5.23) are ignored, and the parameters in the summand have been given the convenient values, $\omega_{k\ell} = 0$, $\gamma_{k\ell} = 1$, and $\left|\mu_{k\ell}\right|^2/\hbar = 1$. The condition, $\omega_{k\ell} = 0$, serves to place the center of the graphs at the origin, $\omega = 0$. The vertical axis is dimensionless. The symmetric curve, which corresponds to $\alpha''(\omega)$, exhibits a peak at $\omega = \omega_{k\ell}$, and is known as a Lorentzian. In contrast, the asymmetric curve, which corresponds to $\alpha'(\omega)$, exhibits an inflection point at $\omega = \omega_{k\ell}$

$$\tilde{G}(\omega) = \int_{-\infty}^{+\infty} dt\, e^{i\omega t} G(t) = \int_{-\infty}^{+\infty} dt \cos(\omega t) G(t) + i \int_{-\infty}^{+\infty} dt \sin(\omega t) G(t) \tag{12.6.1}$$

In expanding the exponential in the integrand, we have used the identity

$$e^{\pm i\omega t} = \cos(\omega t) \pm i \sin(\omega t) \tag{12.6.2}$$

with the positive sign. According to Eq. (12.6.1), $\tilde{G}(\omega)$ has real and imaginary parts. Since $\tilde{G}(\omega)$ is arbitrary, we can identify $\tilde{G}(\omega)$ with $\tilde{\alpha}(\omega)$. With reference to Eq. (12.5.17), we then have,

$$\alpha'(\omega) = \int_{-\infty}^{+\infty} dt \cos(\omega t) G(t) \tag{12.6.3}$$

and

$$\alpha''(\omega) = \int_{-\infty}^{+\infty} dt \sin(\omega t) G(t) \tag{12.6.4}$$

It will now prove useful to express $G(t)$ as the sum of its even and odd parts. A short theorem establishes the fact that this is always possible.

Theorem: An arbitrary function, $G(t)$ can always be written as the sum

$$G(t) = G_{even}(t) + G_{odd}(t) \tag{12.6.5}$$

where $G_{even}(t)$ is defined by

$$G_{even}(-t) = G_{even}(t) \tag{12.6.6a}$$

while $G_{odd}(t)$ is defined by

$$G_{odd}(-t) = -G_{odd}(t) \tag{12.6.6b}$$

Proof As a means of constructing Eq. (12.6.5), we introduce the following proposed equations:

$$G_{even}(t) = \frac{1}{2}(G(t) + G(-t)) \tag{12.6.7a}$$

and

$$G_{odd}(t) = \frac{1}{2}(G(t) - G(-t)) \tag{12.6.7b}$$

Note that the sum of Eqs. (12.6.7a) and (12.6.7b) is $G(t)$. If we replace t by $-t$ everywhere in Equations (12.6.7a) and (12.6.7b), it can be seen that

$$G_{even}(-t) = G_{even}(t) \tag{12.6.8a}$$

and

$$G_{odd}(-t) = -G_{odd}(t) \tag{12.6.8b}$$

which establishes the meaning of Eqs. (12.6.7a) and (12.6.7b) and proves the theorem. ■

To proceed further, some integrals involving the products of $\sin(\omega t)$ and $\cos(\omega t)$ with $G_{even}(t)$ and $G_{odd}(t)$ will be prove useful. As an aid to evaluating these integrals, we note that $\cos(-\omega t) = \cos(\omega t)$ and $\sin(-\omega t) = -\sin(\omega t)$. By reference to Eqs. (12.6.8a) and (12.6.8b) we can write respectively,

$$\sin(-\omega t)G_{even}(-t) = -\sin(\omega t)G_{even}(t) \tag{12.6.9a}$$

$$\cos(-\omega t)G_{odd}(-t) = -\cos(\omega t)G_{odd}(t) \tag{12.6.9b}$$

Consider the integral,

$$\int_{-\infty}^{+\infty} dt\,\sin(\omega\, t) G_{even}(t) = \int_{-\infty}^{0} dt\,\sin(\omega\, t) G_{even}(t) + \int_{0}^{+\infty} dt\,\sin(\omega\, t) G_{even}(t) \tag{12.6.10a}$$

where we have broken up the range of integration into positive and negative segments. In the first integral on the right of Eq. (12.6.10a) replace t by $-t$. The result is

$$\begin{aligned}\int_{-\infty}^{+\infty} dt\,\sin(\omega t) G_{even}(t) &= \int_{+\infty}^{0} -dt\,\sin(-\omega t) G_{even}(-t) + \int_{0}^{+\infty} dt\,\sin(\omega t) G_{even}(t)\\ &= -\int_{0}^{+\infty} dt\,\sin(\omega t) G_{even}(t) + \int_{0}^{+\infty} dt\,\sin(\omega t) G_{even}(t) = 0\end{aligned} \tag{12.6.10b}$$

where Eq. (12.6.9a) has been used to obtain the final result.

We now turn to the integral,

$$\int_{-\infty}^{+\infty} dt\,\cos(\omega t) G_{odd}(t) = \int_{-\infty}^{0} dt\,\cos(\omega t) G_{odd}(t) + \int_{0}^{+\infty} dt\,\cos(\omega t) G_{odd}(t) \tag{12.6.10c}$$

where we have broken up the range of integration into positive and negative segments. If in the first integral on the right of Eq. (12.6.10c), we replace t by $-t$, the result is

$$\begin{aligned}\int_{-\infty}^{+\infty} dt\,\cos(\omega t) G_{odd}(t) &= \int_{+\infty}^{0} -dt\,\cos(-\omega t) G_{odd}(-t) + \int_{0}^{+\infty} dt\,\cos(\omega t) G_{odd}(t)\\ &= -\int_{0}^{+\infty} dt\,\cos(\omega t) G_{odd}(t) + \int_{0}^{+\infty} dt\,\cos(\omega t) G_{odd}(t) = 0\end{aligned} \tag{12.6.10d}$$

where we have used Eq. (12.6.9b) to obtain the final result.

Now substitute Eq. (12.6.5) into Eq. (12.6.3) to obtain,

$$\alpha'(\omega) = \int_{-\infty}^{+\infty} dt\,\cos(\omega t) G_{even}(t) + \int_{-\infty}^{+\infty} dt\,\cos(\omega t) G_{odd}(t) \tag{12.6.11a}$$

According to Eq. (12.6.10d), the second integral on the right of Eq. (12.6.11a) is zero.

Hence, Eq. (12.6.11a) becomes simply,

$$\alpha'(\omega) = \int_{-\infty}^{+\infty} dt \cos(\omega t) G_{even}(t) \tag{12.6.11b}$$

Now substitute Eq. (12.6.5) into Eq. (12.6.4) to obtain

$$\alpha''(\omega) = \int_{-\infty}^{+\infty} dt \sin(\omega t) G_{even}(t) + \int_{-\infty}^{+\infty} dt \sin(\omega t) G_{odd}(t) \tag{12.6.12a}$$

According to Eq. (12.6.10b), the first integral on the right of Eq. (12.6.12a) is zero; thus, the net result is

$$\alpha''(\omega) = \int_{-\infty}^{+\infty} dt \sin(\omega t) G_{odd}(t) \tag{12.6.12b}$$

The Fourier transform of $G_{even}(t)$ is

$$\tilde{G}_{even}(\omega) = \int_{-\infty}^{+\infty} dt e^{i\omega t} G_{even}(t) \tag{12.6.13a}$$

Substitute Eq. (12.6.2) with the upper sign into Eq. (12.6.13a). The result is

$$\begin{aligned} \tilde{G}_{even}(\omega) &= \int_{-\infty}^{+\infty} dt \cos(\omega t) G_{even}(t) + i \int_{-\infty}^{+\infty} dt \sin(\omega t) G_{even}(t) \\ &= \int_{-\infty}^{+\infty} dt \cos(\omega t) G_{even}(t) \end{aligned} \tag{12.6.13b}$$

where we have used Eq. (12.6.10b). Upon comparison of Eq. (12.6.13b) with Eq. (12.6.11b), we conclude that

$$\tilde{G}_{even}(\omega) = \alpha'(\omega) \tag{12.6.14}$$

The Fourier transform of $G_{odd}(t)$ is

$$\tilde{G}_{odd}(\omega) = \int_{-\infty}^{+\infty} dt e^{i\omega t} G_{odd}(t) \tag{12.6.15a}$$

Substitution of Eq. (12.6.2) with the upper sign yields the result,

$$\begin{aligned}\tilde{G}_{odd}(\omega) &= \int_{-\infty}^{+\infty} dt\, e^{i\omega t} G_{odd}(t) = \int_{-\infty}^{+\infty} dt \cos(\omega\, t) G_{odd}(t) + i \int_{-\infty}^{+\infty} dt \sin(\omega\, t) G_{odd}(t) \\ &= i \int_{-\infty}^{+\infty} dt \sin(\omega t) G_{odd}(t)\end{aligned} \tag{12.6.15b}$$

where we have used Eq. (12.6.10d). Upon comparison of Eq. (12.6.15b) with Eq. (12.6.12b), the conclusion is

$$\alpha''(\omega) = \int_{-\infty}^{+\infty} dt \sin(\omega t) G_{odd}(t) = -i\tilde{G}_{odd}(\omega) \tag{12.6.15c}$$

Eq. (12.6.15c) can be rewritten as

$$\tilde{G}_{odd}(\omega) = i\alpha''(\omega) \tag{12.6.16}$$

We now introduce the *sign* function, $s(t)$, which has the definition,

$$\mathrm{s}(t) = \begin{cases} 1, & t > 0 \\ 0, & t = 0 \\ -1, & t < 0 \end{cases} \tag{12.6.17}$$

A plot of Eq. (12.6.17) is shown in Fig. 12.2.
Note that

$$s(-t) = -s(t) \tag{12.6.18}$$

In terms of $s(t)$, we propose to write,

$$G_{even}(t) = \mathrm{s}(t) G_{odd}(t) \tag{12.6.19a}$$

One can confirm that Eq. (12.6.19a) represents an even function. If we substitute $-t$ for t, and make reference to Eq. (12.6.18), the proof follows:

$$\begin{aligned}G_{even}(-t) &= \mathrm{s}(-t) G_{odd}(-t) = (-1)s(t) \cdot (-1) G_{odd}(t) \\ &= s(t) G_{odd}(t) = G_{even}(t)\end{aligned} \tag{12.6.19b}$$

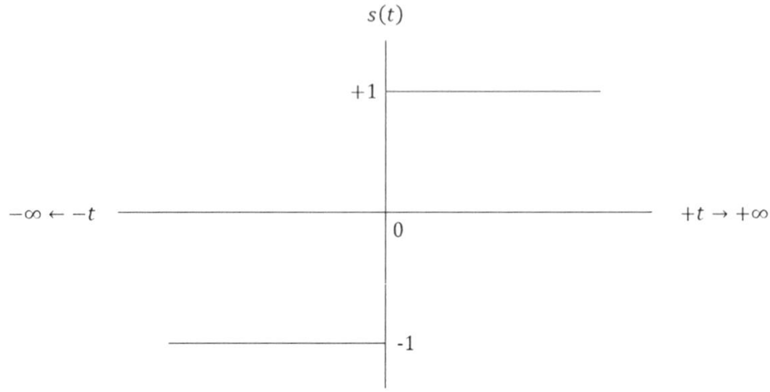

Fig. 12.2 Plot of the sign function, $s(t)$ as defined by Eq. (12.6.17)

Likewise, we propose,

$$G_{odd}(t) = \mathrm{s}(t)G_{even}(t) \tag{12.6.20a}$$

If we substitute $-t$ for t, and make reference to Eq. (12.6.18), we can show that Eq. (12.6.20a) represents an odd function, The proof follows:

$$G_{odd}(-t) = \mathrm{s}(-t)G_{even}(-t) = (-1)s(t) \cdot G_{even}(t) = -G_{odd}(t) \tag{12.6.20b}$$

The Fourier transforms of $s(t)$, $G_{even}(t)$, and $G_{odd}(t)$ are respectively,

$$\tilde{s}(\omega) = \int_{-\infty}^{+\infty} e^{i\omega t} s(t)dt \tag{12.6.21a}$$

$$\tilde{G}_{even}(\omega) = \int_{-\infty}^{+\infty} e^{i\omega t} G_{even}(t)dt \tag{12.6.21b}$$

$$\tilde{G}_{odd}(\omega) = \int_{-\infty}^{+\infty} e^{i\omega t} G_{odd}(t)dt \tag{12.6.21c}$$

By virtue of Eq. (12.6.19a) and the inverse convolution theorem [See the Mathematical Methods Appendix], the Fourier transform, $\tilde{G}_{even}(\omega)$, satisfies

$$\tilde{G}_{even}(\omega) = \frac{1}{2\pi} \int_{-\infty}^{+\infty} d\omega_o \tilde{s}(\omega - \omega_o)\tilde{G}_{odd}(\omega_o) \tag{12.6.22}$$

where $\tilde{s}(\omega-\omega_o)$ is the Fourier transform of $s(t)$ evaluated at $\omega-\omega_o$, and $\tilde{G}_{odd}(\omega_o)$ is the Fourier transform of $G_{odd}(t)$ evaluated at frequency, ω_o, where ω_o is a variable.

In Sect. A.17 of the Mathematical Methods Appendix, it is shown that

$$\tilde{s}(\omega) = \frac{2i}{\omega} \tag{12.6.23a}$$

If in Eq. (12.6.23a), we replace ω by $\omega - \omega_o$, the result is

$$\tilde{s}(\omega - \omega_o) = \frac{2i}{\omega - \omega_o} \tag{12.6.23b}$$

Upon substitution of Eq. (12.6.23b) into Eq. (12.6.22), the result is

$$\tilde{G}_{even}(\omega) = \frac{1}{2\pi}\int_{-\infty}^{+\infty} d\omega_o\left(\frac{2i}{\omega-\omega_o}\right)\tilde{G}_{odd}(\omega_o) \tag{12.6.24a}$$

Upon substitution of Eqs. (12.6.14) and (12.6.16), Eq. (12.6.24a) reads,

$$\alpha'(\omega) = \frac{1}{2\pi}\int_{-\infty}^{+\infty} d\omega_o\left(\frac{2i}{\omega-\omega_o}\right)\left(i\alpha''(\omega_o)\right) \tag{12.6.24b}$$

Upon using $i^2 = -1$, we obtain the final result,

$$\alpha'(\omega) = -\frac{1}{\pi}\int_{-\infty}^{+\infty} d\omega_o\frac{\alpha''(\omega_o)}{(\omega-\omega_o)} \tag{12.6.25}$$

Equation (12.6.25) is the first Kramers–Kronig relation.

With reference to Eq. (12.6.20a), the analog of Eq. (12.6.22) is

$$\tilde{G}_{odd}(\omega) = \frac{1}{2\pi}\int_{-\infty}^{+\infty} d\omega_o\tilde{s}(\omega-\omega_o)\tilde{G}_{even}(\omega_o) \tag{12.6.26}$$

Upon substitution of Eq. (12.6.23b), we obtain

$$\tilde{G}_{odd}(\omega) = \frac{1}{2\pi}\int_{-\infty}^{+\infty} d\omega_o\left(\frac{2i}{\omega-\omega_o}\right)\tilde{G}_{even}(\omega_o) \tag{12.6.27a}$$

Substitute Eqs. (12.6.14) and (12.6.16) into Eq. (12.6.27a). The result is

$$i\alpha''(\omega) = \frac{1}{2\pi}\int_{-\infty}^{+\infty} d\omega_o\left(\frac{2i}{\omega - \omega_o}\right)\alpha'(\omega_o) \tag{12.6.27b}$$

After cancellation of the factor, i, from either side of Eq. (12.6.27b), our final result is

$$\alpha''(\omega) = \frac{1}{\pi}\int_{-\infty}^{+\infty} d\omega_o \frac{\alpha'(\omega_o)}{(\omega - \omega_o)} \tag{12.6.28}$$

Equation (12.6.28) is the second Kramers–Kronig relation.

Equations (12.6.25) and (12.6.28) are of substantial theoretical significance. No matter what model is adopted to calculate, $\tilde{\alpha}(\omega)$, be it classical or quantum, this pair of equations shows that the real and imaginary parts of $\tilde{\alpha}(\omega)$ are not arbitrary, but rather involve integrals of one another. In the case of an optically absorbing material, the equations serve to link the refractive index, which depends upon $\alpha'(\omega)$, to the absorptivity, which depends upon $\alpha''(\omega)$.

12.7 Thermal Averages

For convenience, we reprise Eq. (12.5.18), renumbering it below as Eq. (12.7.1):

$$\alpha'(\omega) = \left(\frac{1}{\hbar}\right)\sum_k |\boldsymbol{\mu}_{k\ell}|^2\left[\frac{(\omega_{k\ell} - \omega)}{(\omega - \omega_{k\ell})^2 + (1/4)\gamma_{k\ell}^2} + \frac{(\omega + \omega_{k\ell})}{(\omega + \omega_{k\ell})^2 + (1/4)\gamma_{k\ell}^2}\right] \tag{12.7.1}$$

Equation (12.7.1) represents the contribution to $\alpha'(\omega)$ made by transitions from quantum state $\psi_\ell^{(o)}(\mathbf{r})$, having energy, $E_\ell^{(o)}$. In many cases $\psi_\ell^{(o)}(\mathbf{r})$ represents a manifold of states having energies nearly equal to $E_\ell^{(o)}$. According to Eq. (2.9.22) of Chap. 2, if the molecule is in thermal equilibrium with its surroundings, the probability, P_ℓ, that it has energy $E_\ell^{(o)}$ is given by the Boltzmann factor,

$$P_\ell = \exp(-\beta E_\ell^{(o)})/Z \tag{12.7.2a}$$

where $\beta = 1/k_B T$, and Z is the canonical partition function. This observation allows us to calculate the electromagnetic response of a set of absorbing states whose quantum populations are determined by the temperature [1].

To construct the average value of $\alpha'(\omega)$ characterized by this manifold of states, we begin by multiplying each term in the sum in Eq. (12.7.1) by P_ℓ and summing the result over ℓ. One obtains,

$$\alpha'(\omega) = \left(\frac{1}{\hbar}\right)\sum_{k,\ell}\left|\boldsymbol{\mu}_{k\ell}\right|^2\left[\frac{P_\ell(\omega_{k\ell}-\omega)}{(\omega-\omega_{k\ell})^2+(1/4)\gamma_{k\ell}^2}+\frac{P_\ell(\omega+\omega_{k\ell})}{(\omega+\omega_{k\ell})^2+(1/4)\gamma_{k\ell}^2}\right] \tag{12.7.2b}$$

Next write the sum in Eq. (12.7.2b) as two sums:

$$\begin{aligned}\alpha'(\omega) &= \left(\frac{1}{\hbar}\right)\sum_{k,\ell}\left|\boldsymbol{\mu}_{k\ell}\right|^2\left[\frac{P_\ell(\omega_{k\ell}-\omega)}{(\omega-\omega_{k\ell})^2+(1/4)\gamma_{k\ell}^2}\right]\\ &+\left(\frac{1}{\hbar}\right)\sum_{k,\ell}\left|\boldsymbol{\mu}_{k\ell}\right|^2\left[\frac{P_\ell(\omega+\omega_{k\ell})}{(\omega+\omega_{k\ell})^2+(1/4)\gamma_{k\ell}^2}\right]\end{aligned} \tag{12.7.2c}$$

In Eq. (12.7.2c), interchange the indices, k and ℓ, in the second sum, so that it reads,

$$\begin{aligned}\alpha'(\omega) &= \left(\frac{1}{\hbar}\right)\sum_{k,\ell}\left|\boldsymbol{\mu}_{k\ell}\right|^2\left[\frac{P_\ell(\omega_{k\ell}-\omega)}{(\omega-\omega_{k\ell})^2+(1/4)\gamma_{k\ell}^2}\right]\\ &+\left(\frac{1}{\hbar}\right)\sum_{\ell,k}\left|\boldsymbol{\mu}_{\ell k}\right|^2\left[\frac{P_k(\omega+\omega_{\ell k})}{(\omega+\omega_{\ell k})^2+(1/4)\gamma_{\ell k}^2}\right]\end{aligned} \tag{12.7.2d}$$

In the second sum in Eq. (12.7.2d), we can make use of the Hermitian property to write $\left|\boldsymbol{\mu}_{\ell k}\right|^2 = \left|\boldsymbol{\mu}_{k\ell}\right|^2$; moreover, since $\omega_{k\ell} = \omega_k - \omega_\ell$, we have $\omega_{\ell k} = -\omega_{k\ell}$. We can assume that $\gamma_{k\ell} = \gamma_{\ell k}$. Upon making these substitutions in the second sum in Eq. (12.7.2d), that equation reads,

$$\begin{aligned}\alpha'(\omega) &= \left(\frac{1}{\hbar}\right)\sum_{k,\ell}\left|\boldsymbol{\mu}_{k\ell}\right|^2\left[\frac{P_\ell(\omega_{k\ell}-\omega)}{(\omega-\omega_{k\ell})^2+(1/4)\gamma_{k\ell}^2}\right]\\ &+\left(\frac{1}{\hbar}\right)\sum_{k,\ell}\left|\boldsymbol{\mu}_{k\ell}\right|^2\left[\frac{P_k(\omega-\omega_{k\ell})}{(\omega-\omega_{k\ell})^2+(1/4)\gamma_{k\ell}^2}\right]\end{aligned} \tag{12.7.3a}$$

The sums in Eq. (12.7.3a) can now be combined to read,

$$\alpha'(\omega) = \left(\frac{1}{\hbar}\right)\sum_{k,\ell}\left|\boldsymbol{\mu}_{k\ell}\right|^2\left[\frac{(P_\ell-P_k)(\omega_{k\ell}-\omega)}{(\omega-\omega_{k\ell})^2+(1/4)\gamma_{k\ell}^2}\right] \tag{12.7.3b}$$

In Eq. (12.7.3b), evaluate P_ℓ using Eq. (12.7.2a), and evaluate P_k using Eq. (12.7.2a) with the change, $\ell = k$. One then obtains,

$$\alpha'(\omega) = \left(\frac{1}{\hbar Z}\right)\sum_{k,\ell}\left|\boldsymbol{\mu}_{k\ell}\right|^2\left[\frac{(e^{-\beta E_\ell^{(o)}}-e^{-\beta E_k^{(o)}})(\omega_{k\ell}-\omega)}{(\omega-\omega_{k\ell})^2+(1/4)\gamma_{k\ell}^2}\right] \tag{12.7.3c}$$

We can remove a factor of $\exp(-\beta E_\ell^{(o)})$ from the term in brackets with the result

$$\alpha'(\omega) = \left(\frac{1}{\hbar Z}\right) \sum_{k,\ell} |\mathbf{\mu}_{k\ell}|^2 \left[\frac{e^{-\beta E_\ell^{(o)}} \left(1 - e^{-\beta(E_k^{(o)} - E_\ell^{(o)})}\right)(\omega_{k\ell} - \omega)}{(\omega - \omega_{k\ell})^2 + (1/4)\gamma_{k\ell}^2} \right] \tag{12.7.3d}$$

Upon substitution of $\hbar\omega_{k\ell} = (E_k^{(o)} - E_\ell^{(o)})$, Eq. (12.7.3d) reads

$$\alpha'(\omega) = \left(\frac{1}{\hbar Z}\right) \sum_{k,\ell} |\mathbf{\mu}_{k\ell}|^2 \left[\frac{e^{-\beta E_\ell^{(o)}} \left(1 - e^{-\beta\hbar\omega_{k\ell}}\right)(\omega_{k\ell} - \omega)}{(\omega - \omega_{k\ell})^2 + (1/4)\gamma_{k\ell}^2} \right] \tag{12.7.4}$$

Equation (12.7.4) is our final result for $\alpha'(\omega)$. We note that Eq. (12.7.4) can be combined with the Lorenz–Lorentz equation,

$$\frac{n^2(\omega) - 1}{n^2(\omega) + 2} = \frac{N\alpha'(\omega)}{3\varepsilon_o} \tag{12.7.5}$$

to obtain a formula capable of determining the temperature and frequency dependence of the refractive index.

Below, we copy Eq. (12.5.22), renumbering it as Eq. (12.7.6):

$$\alpha''(\omega) = \left(\frac{1}{2\hbar}\right) \sum_{k} |\mathbf{\mu}_{k\ell}|^2 \left[\frac{\gamma_{k\ell}}{(\omega - \omega_{k\ell})^2 + (1/4)\gamma_{k\ell}^2} - \frac{\gamma_{k\ell}}{(\omega + \omega_{k\ell})^2 + (1/4)\gamma_{k\ell}^2} \right] \tag{12.7.6}$$

To convert Eq. (12.7.6) to a thermal average, multiply each term in the bracket by P_ℓ and sum over ℓ. The result is

$$\alpha''(\omega) = \left(\frac{1}{2\hbar}\right) \sum_{k,\ell} |\mathbf{\mu}_{k\ell}|^2 \left[\frac{P_\ell\gamma_{k\ell}}{(\omega - \omega_{k\ell})^2 + (1/4)\gamma_{k\ell}^2} - \frac{P_\ell\gamma_{k\ell}}{(\omega + \omega_{k\ell})^2 + (1/4)\gamma_{k\ell}^2} \right] \tag{12.7.7a}$$

Next, write the two sums in Eq. (12.7.7a) separately so that,

$$\begin{aligned} \alpha''(\omega) = {} & \left(\frac{1}{2\hbar}\right) \sum_{k,\ell} |\mathbf{\mu}_{k\ell}|^2 \left[\frac{P_\ell\gamma_{k\ell}}{(\omega - \omega_{k\ell})^2 + (1/4)\gamma_{k\ell}^2} \right] \\ & - \left(\frac{1}{2\hbar}\right) \sum_{k,\ell} |\mathbf{\mu}_{k\ell}|^2 \left[\frac{P_\ell\gamma_{k\ell}}{(\omega + \omega_{k\ell})^2 + (1/4)\gamma_{k\ell}^2} \right] \end{aligned} \tag{12.7.7b}$$

Interchange k and ℓ in the second sum. Eq. (12.7.7b) then reads,

$$\alpha''(\omega) = \left(\frac{1}{2\hbar}\right)\sum_{k,\ell}\frac{|\boldsymbol{\mu}_{k\ell}|^2 P_\ell \gamma_{k\ell}}{(\omega-\omega_{k\ell})^2+(1/4)\gamma_{k\ell}^2} - \left(\frac{1}{2\hbar}\right)\sum_{\ell,k}\frac{|\boldsymbol{\mu}_{\ell k}|^2 P_k \gamma_{\ell k}}{(\omega+\omega_{\ell k})^2+(1/4)\gamma_{\ell k}^2} \tag{12.7.7c}$$

As before, make the substitutions, $|\boldsymbol{\mu}_{\ell k}|^2 = |\boldsymbol{\mu}_{k\ell}|^2$, $\omega_{\ell k} = -\omega_{k\ell}$, and $\gamma_{k\ell} = \gamma_{\ell k}$ in the second sum. The result is

$$\alpha''(\omega) = \left(\frac{1}{2\hbar}\right)\sum_{k,\ell}\frac{|\boldsymbol{\mu}_{k\ell}|^2 P_\ell \gamma_{k\ell}}{(\omega-\omega_{k\ell})^2+(1/4)\gamma_{k\ell}^2} - \left(\frac{1}{2\hbar}\right)\sum_{\ell,k}\frac{|\boldsymbol{\mu}_{k\ell}|^2 P_k \gamma_{k\ell}}{(\omega-\omega_{k\ell})^2+(1/4)\gamma_{k\ell}^2} \tag{12.7.8a}$$

The two sums in Eq. (12.7.8a) can be combined to read

$$\alpha''(\omega) = \left(\frac{1}{2\hbar}\right)\sum_{k,\ell}\frac{|\boldsymbol{\mu}_{k\ell}|^2 (P_\ell - P_k)\gamma_{k\ell}}{(\omega-\omega_{k\ell})^2+(1/4)\gamma_{k\ell}^2} \tag{12.7.8b}$$

Upon replacement of P_ℓ and P_k by their respective Boltzmann factors, Eq. (12.7.8b) takes the form,

$$\alpha''(\omega) = \left(\frac{1}{2Z\hbar}\right)\sum_{k,\ell}|\boldsymbol{\mu}_{k\ell}|^2\left[\frac{\left(e^{-\beta E_\ell^{(o)}} - e^{-\beta E_k^{(o)}}\right)\gamma_{k\ell}}{(\omega-\omega_{k\ell})^2+(1/4)\gamma_{k\ell}^2}\right] \tag{12.7.8c}$$

Factor out $\exp(-\beta E_\ell^{(o)})$ in Eq. (12.7.8c). The result is

$$\alpha''(\omega) = \left(\frac{1}{2Z\hbar}\right)\sum_{k,\ell}|\boldsymbol{\mu}_{k\ell}|^2\left[\frac{e^{-\beta E_\ell^{(o)}}\left(1 - e^{-\beta(E_k^{(o)}-E_\ell^{(o)})}\right)\gamma_{k\ell}}{(\omega-\omega_{k\ell})^2+(1/4)\gamma_{k\ell}^2}\right] \tag{12.7.8d}$$

Finally, use $\hbar\omega_{k\ell} = (E_k^{(o)} - E_\ell^{(o)})$, so that Eq. (12.7.8d) becomes,

$$\alpha''(\omega) = \left(\frac{1}{2Z\hbar}\right)\sum_{k,\ell}|\boldsymbol{\mu}_{k\ell}|^2\left[\frac{e^{-\beta E_\ell^{(o)}}\left(1 - e^{-\beta\hbar\omega_{k\ell}}\right)\gamma_{k\ell}}{(\omega-\omega_{k\ell})^2+(1/4)\gamma_{k\ell}^2}\right] \tag{12.7.9}$$

Equations (12.7.4) and (12.7.9) can be combined with the Clausius–Mossotti equation,

$$\frac{\tilde{\varepsilon}_r(\omega)-1}{\tilde{\varepsilon}_r(\omega)+2} = \frac{N}{3\varepsilon_o}\left(\alpha'(\omega) + i\alpha''(\omega)\right) \tag{12.7.10}$$

to obtain a formula for predicting the temperature dependence and frequency dependence of the complex valued relative permittivity, $\tilde{\varepsilon}_r(\omega)$.

12.8 Discussion

If we set $\omega = 0$ in Eq. (12.5.18), the result is

$$\alpha'(0) = \left(\frac{1}{\hbar}\right)\sum_k |\boldsymbol{\mu}_{k\ell}|^2\left[\frac{\omega_{k\ell}}{\omega_{k\ell}^2 + (1/4)\gamma_{k\ell}^2} + \frac{\omega_{k\ell}}{\omega_{k\ell}^2 + (1/4)\gamma_{k\ell}^2}\right] \tag{12.8.1a}$$

or

$$\alpha'(0) = \left(\frac{2}{\hbar}\right)\sum_k \left[\frac{|\boldsymbol{\mu}_{k\ell}|^2\omega_{k\ell}}{\omega_{k\ell}^2 + (1/4)\gamma_{k\ell}^2}\right] \tag{12.8.1b}$$

Equation (12.8.1b) gives us a formula for the static polarizability of a molecule in its ground state as identified by the quantum number, ℓ.

The frequency dependence of the polarizabilities, $\alpha'(\omega)$ and $\alpha''(\omega)$, as represented by Eqs. (12.5.19) and (12.5.23), respectively, do not match the frequency dependence shown in the corresponding formulae, Eqs. (7.9.9a) and (7.9.9b) of Chap. 7. This is because the former were calculated quantum mechanically, whereas the latter were calculated using a classical harmonic oscillator model. In classical mechanics, the dynamical equation describing the state of the electron is Newton's law, which is Eq. (7.9.1). The equivalent equation in quantum mechanics is Eq. (12.1.1).

The Kramers–Kronig relations in Eqs. (12.6.25) and (12.6.28) are not restricted to the evaluation of the polarizability. Rather, they can be used to link the real and imaginary parts of any complex valued function. The Kramers–Kronig relations are a special case of what in mathematics are known as the Hilbert transforms (David Hilbert, German, 1862–1943). The Hilbert transforms and the Fourier transforms introduced in Chap. 7 are members of a still more general class of equations known as integral transforms.

References

1. R.G. Gordon, Correlation functions for molecular motion, in *Advances in Magnetic Resonance*, ed. by J.S. Waugh, vol. 3 (Academic, New York, 1968), pp. 1–41
2. B. Harbecke, Application of Fourier's allied integrals to the Kramers–Kronig transformation of reflectance data. Appl. Phys. A **40**, 1151–1156 (1986)
3. P. Norman, K. Ruud, T. Saue, *Principles and Practices of Molecular Properties: Theory, Modeling, and Simulations* (Oxford, UK, 2018)

Recommended Additional Reading

1. A. Hinchliffe, R.W. Munn, *Molecular Electromagnetism* (Wiley, Chichester, W. Sussex, UK, 1985)
2. A. Zangwill, *Modern Electrodynamics* (Cambridge University Press, Cambridge, UK, 2013)

Chapter 13
Optical Rotation

When a molecule cannot be superimposed on its mirror image, it is said to be chiral. Beams of left and right circularly polarized light travel at different speeds through a chiral medium. Because a linearly polarized beam can be decomposed into two such circularly polarized beams, the phase of one circularly polarized beam lags behind the other, which has the effect of rotating the direction of linear polarization about the axis of propagation. To understand this phenomenon, we must take into account the fact that the light induces both electric and magnetic dipoles in molecules. The magnetic interaction has not been encountered in the previous chapters, so the radiation theory of Chap. 12 is extended in this chapter to include it. Taking both the electric and magnetic interactions into account, the time dependent perturbation theory is used to calculate the induced molecular electric and magnetic dipole moments. The electric and magnetic constitutive relations and the Lorenz–Lorentz equation are then invoked to obtain a formula for the specific angle of rotation of the light passing through the medium.

13.1 Introduction

The mirror image of a right hand is a left hand, but when the palm of the left hand is placed over the back of a right hand, the five fingers of the left fail to superimpose on the five fingers of the right. Similarly, if a molecule is not superimposable on its mirror image in the same sense that a left hand is not superimposable on a right, the molecule is said to be *chiral*. A chiral molecule is capable of rotating the direction of the time dependent electric field of a linearly polarized beam of light.

J. K. Baird, *Electromagnetism for Chemistry*,
https://doi.org/10.1007/978-3-031-88508-2_13

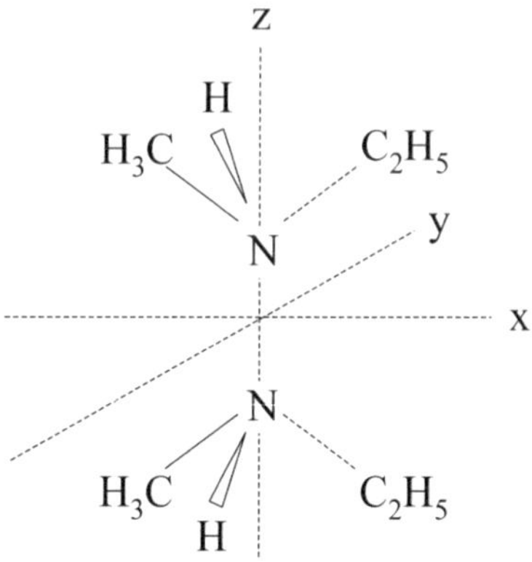

Fig. 13.1 Reflection of a molecule of methylethylamine in a mirror lying in the *x*, *y*-plane

A pair of chiral molecular forms of opposite handedness are said to be optical isomers or *enantiomers*. An example of a chiral molecular pair is shown in Fig. 13.1.

In general, the necessary and sufficient condition that a molecule be chiral is that its molecular geometry should *not* contain a reflection plane or a center of symmetry.

In the case of a plane electromagnetic wave, the fields **E** and **B** are at right angles. Their magnitudes oscillate in the plane normal to the propagation vector, **k**. When a light wave propagates with the directions of **E** and **B** fixed, the wave is said to be *linearly polarized*. A linearly polarized light wave can be represented mathematically as the sum of a right circularly polarized wave and a left circularly polarized wave. See Fig. 13.2.

A wave is said to be circularly polarized when the plane containing **E** and **B** rotates about the direction of **k**. If the observer is looking into the source of the light, and the sense of rotation is to the right, then the rotation is said to be clockwise. If the rotation is to the left, it is said to be counterclockwise. This is the convention that has been adopted in chemistry and in optics. In electrical engineering and physics, however, the convention has the observer looking along the direction in which the light is traveling. In what follows, we shall use the convention common to chemistry and optics.

In an achiral medium, left and right circularly polarized beams travel at the same speed. As shown in Fig. 13.2, the magnitudes of their angles of rotation are the same. By contrast, in a chiral medium, the refractive indices sensed by the right and left circularly polarized components have different values. This causes any two waves having opposite directions of circular polarization to travel at different speeds in the medium. The difference in speeds leads to different times of passage through the material. The difference in passage times causes the two polarizations to rotate for different lengths of time when traversing the material. As a result, the respective angles of rotation of the two polarizations are different, and their vector sum no longer coincides with the *x*-axis.

As shown in Fig. 13.3, the source sends a mixture of light waves with various polarizations into the polarizer. Only light polarized in the *x*-direction, which is represented by the vertical slit in the polarizer, can exit the polarizer. This light

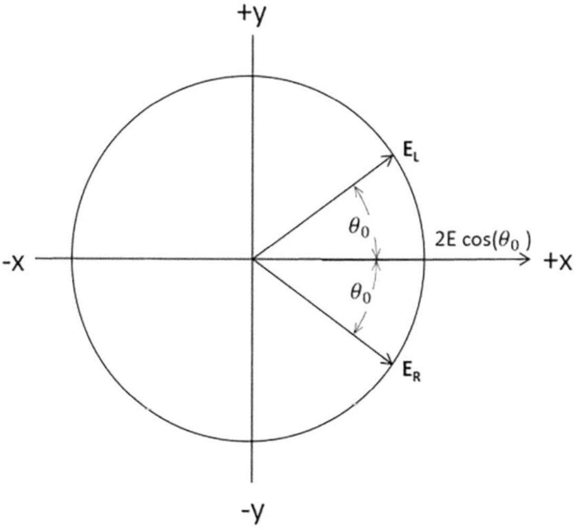

Fig. 13.2 Superposition of circular polarized light beams passing through an achiral medium. The electric field rotating in the clockwise direction (i.e. right circular polarization) is $\mathbf{E}_R$, while the electric field rotating in the counter clockwise direction (i.e. left circular polarization) is $\mathbf{E}_L$. If the medium in achiral, both waves travel with the same speed. The instantaneous value of the angle of rotation is θ_o. Since the two electric fields have the same magnitude, E, the sum of the projections of $\mathbf{E}_R$ and $\mathbf{E}_L$ on the horizontal direction (linear polarization) is $2E \cos \theta_o$

beam then enters the sample. While passing through the sample, the left and right circular polarizations making up the linear polarization travel with different speeds. Upon exiting the sample, the light enters the analyzer. The analyzer consists of the same material as the polarizer. Like the polarizer, it passes only linearly polarized light. The analyzer is initially oriented so that it passes light polarized along the x-direction. It can be rotated as needed, however, so that it can pass light polarized along any x'-direction that makes an angle with the original x-direction. In the figure, the orientation of the x'-direction is indicated by the diagonal double arrow and the direction of rotation of the analyzer is indicated by the rotated line.

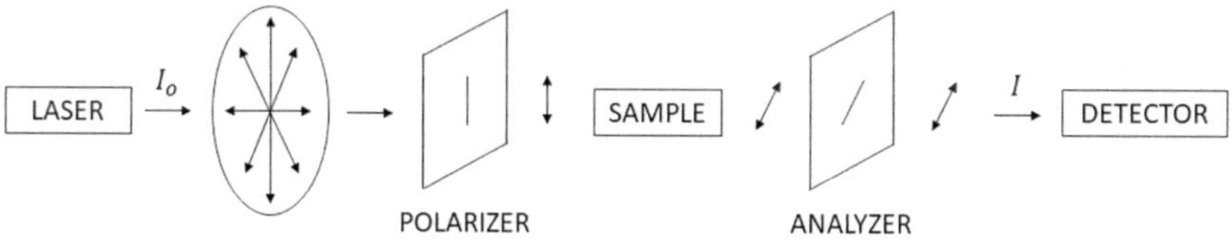

Fig. 13.3 Laboratory arrangement for determining the direction of linear polarization of a light wave. The intensity of the light incident on the polarizer is I_o, while the intensity of the light wave incident on the detector is I

We define a reference angle, θ_o, known as the central angle, whose magnitude accounts for a rotations of left and right circularly polarized light that are just sufficient to keep the net linear polarization oriented along the x-axis. In Fig. 13.3, a linearly polarized beam has departed the polarizer with its electric field oriented in the x-direction. This beam of intensity, I_o, is made up of an equal mixture of clockwise rotating circular polarization (right circular polarization) and counter clockwise rotating circular polarization (left circular polarization). With reference to the central angle, the angle of rotation of the right circularly polarized beam is $\psi_R = \theta_o + \delta$. Decomposed into its x- and y-components, the electric field of the right circular polarized beam is

$$\mathbf{E}_R = E_o\left(\hat{\mathbf{e}}_x \cos\psi_R - \hat{\mathbf{e}}_y \sin\psi_R\right) \tag{13.1.1}$$

Note in Eq. (13.1.1), as the value of ψ_R increases, the component of $\mathbf{E}_R$ in the $-\hat{\mathbf{e}}_y$ direction increases. This makes $\mathbf{E}_R$ rotate in the clockwise direction.

With reference to the central angle, θ_o, the angle of rotation of the left circularly polarized beam is $\psi_L = \theta_o - \delta$. Decomposed into its x- and y-components, the electric field of the left circular polarized component is

$$\mathbf{E}_L = E_o\left(\hat{\mathbf{e}}_x \cos\psi_L + \hat{\mathbf{e}}_y \sin\psi_L\right) \tag{13.1.2}$$

Note in Eq. (13.1.2), as the value of ψ_L increases, the component of $\mathbf{E}_L$ in the $\hat{\mathbf{e}}_y$ direction increases. This makes $\mathbf{E}_R$ rotate in the counter clockwise direction.

If the right circularly polarized component travels more slowly than the left circularly polarized component, then $\delta > 0$, and the right circularly polarized component will spend a longer time passing through the material. Upon exiting the material, its angle of orientation will be

$$\psi_R = \theta_o + \delta \tag{13.1.3a}$$

and its electric field will be

$$\mathbf{E}_R = E_o\left(\hat{\mathbf{e}}_x \cos(\theta_o + \delta) - \hat{\mathbf{e}}_y \sin(\theta_o + \delta)\right) \tag{13.1.3b}$$

If the left circularly polarized component of the beam travels faster than the right circularly polarized component, the left circularly polarized component will spend less time passing through the medium. Its angle of rotation will be only

$$\psi_L = \theta_o - \delta \tag{13.1.4a}$$

Its electric field is given by

$$\mathbf{E}_L = E_o\left(\hat{\mathbf{e}}_x \cos(\theta_o - \delta) + \hat{\mathbf{e}}_y \sin(\theta_o - \delta)\right) \tag{13.1.4b}$$

If we add Eqs. (13.1.3b) and (13.1.4b), the total electric field is

$$\begin{aligned}\mathbf{E} &= \mathbf{E}_R + \mathbf{E}_L \\ &= E_o\big(\hat{\mathbf{e}}_x \cos(\theta_o + \delta) - \hat{\mathbf{e}}_y \sin(\theta_o + \delta)\big) + E_o\big(\hat{\mathbf{e}}_x \cos(\theta_o - \delta) + \hat{\mathbf{e}}_y \sin(\theta_o - \delta)\big)\end{aligned} \tag{13.1.5a}$$

By use of the trigonometric identities (see the Mathematical Methods Appendix),

$$\cos(\theta_o \pm \delta) = \cos\theta_o \cos\delta \mp \sin\theta_o \sin\delta \tag{13.1.5b}$$

and

$$\sin(\theta_o \pm \delta) = \sin\theta_o \cos\delta \pm \cos\theta_o \sin\delta \tag{13.1.5c}$$

Equation (13.1.5a) can be rewritten as

$$\begin{aligned}\mathbf{E} = {} & E_o\hat{\mathbf{e}}_x(\cos\theta_o \cos\delta - \sin\theta_o \sin\delta) + E_o\hat{\mathbf{e}}_y(-\sin\theta_o \cos\delta - \cos\theta_o \sin\delta) \\ & + E_o\hat{\mathbf{e}}_x(\cos\theta_o \cos\delta + \sin\theta_o \sin\delta) + E_o\hat{\mathbf{e}}_y(\sin\theta_o \cos\delta - \cos\theta_o \sin\delta)\end{aligned} \tag{13.1.6a}$$

Upon cancellation of like terms, Eq. (13.1.6a) reads,

$$\mathbf{E} = 2E_o\hat{\mathbf{e}}_x(\cos\theta_o \cos\delta) - 2E_o\hat{\mathbf{e}}_y(\cos\theta_o \sin\delta) \tag{13.1.6b}$$

Upon factoring out $\cos\theta_o$, Eq. (13.1.6b) reads,

$$\mathbf{E} = 2E_o \cos\theta_o\big(\hat{\mathbf{e}}_x \cos\delta - \hat{\mathbf{e}}_y \sin\delta\big) \tag{13.1.7}$$

The interpretation of Eq. (13.1.7), requires us to consider three possibilities: (1) If $\delta = 0$, the electric field remains parallel to the x-axis. This corresponds to no optical rotation. (2) If as we have assumed $\delta > 0$, the electric field is rotated toward the $-y$-axis. This rotation is said to be *dextrorotatory*. (3) If, by contrast, we had assumed $\delta < 0$, the electric field would be rotated toward the $+\,y$-axis. This rotation is said to be *levorotatory*.

The value of δ also depends upon of the wavelength of the radiation. A medium whose optical properties are a function of wavelength is said to be dispersive. For that reason, the rotation of linearly polarized light, which occurs as the light passes through a chiral medium, is often referred to as *optical rotatory dispersion* (ORD). The value of δ also depends upon the thickness of the sample. This is a property which ORD shares with the optical absorptivity. In a chiral medium, clockwise polarized light is absorbed differently from counter clockwise polarized light. This effect is termed *circular dichroism* (CD). Taken together, ORD and CD are said to constitute optical activity.

The original theory of ORD is due to Leon Rosenfeld (Belgian, 1904–1974) and applies in the case where the wavelength of the light is long compared with the size of the molecular unit responsible for the rotation of the polarization. The Rosenfeld theory relies upon the time dependent perturbation theory developed in Chap. 11. According to Eq. (11.6.6), the wave function, $\Psi(t)$, describing a system interacting with a time dependent perturbation can be written as

$$\Psi(x,t) = \sum_k c_k(t)\Psi_k^{(o)}(x,t) \tag{13.1.8}$$

where the set, $\left\{\Psi_k^{(o)}(x,t)\right\}$, are the wave functions that determine the states with energy $E_k^{(o)}$ when they are not acted on by the perturbation. According to Eq. (11.6.8) in Chap. 11, the set of coefficients, $\{c_m(t)\}$, denoted by the index, m, should satisfy the family of differential equations,

$$i\hbar\frac{dc_m}{dt} = \sum_k c_k\left\langle\Psi_m^{(o)}\middle|\mathrm{H}'\middle|\Psi_k^{(o)}\right\rangle \qquad (m = 1, 2, 3, \ldots) \tag{13.1.9}$$

In Eq. (13.1.9), H′(t) is the time dependent perturbation connecting the state, "k", to the excited states, "m". Without loss of generality, we can specialize Eq. (13.1.9) to the situation where just two quantum states play a role. Denoting the ground state by "a" and the excited state by "b", Eq. (13.1.9) simplifies to read,

$$i\hbar\frac{dc_b}{dt} = \left\langle\Psi_b^{(o)}\middle|\mathrm{H}'(t)\middle|\Psi_a^{(o)}\right\rangle \tag{13.1.10}$$

The integral of Eq. (13.1.10) can be written formally as

$$c_b(t) = \frac{1}{i\hbar}\int_0^t \left\langle\Psi_b^{(o)}\middle|\mathrm{H}'(t)\middle|\Psi_a^{(o)}\right\rangle dt + c_b(0) \tag{13.1.11}$$

Since "b" is an excited state, we can take its initial value to be $c_b(0) = 0$, in which case Eq. (13.1.11) simplifies to read,

$$c_b(t) = \frac{1}{i\hbar}\int_0^t \left\langle\Psi_b^{(o)}\middle|\mathrm{H}'(t)\middle|\Psi_a^{(o)}\right\rangle dt \tag{13.1.12}$$

We shall use the method of Eyring, Walter, and Kimball [1] to evaluate the integral in Eq. (13.1.12). Unlike all of the optical phenomena encountered in previous chapters, the evaluation of Eq. (13.1.12) will require both electric and magnetic effects to be taken into account.

13.2 The Perturbation

13.2.1 A Taylor Series for the Vector Potential

In Chap. 11, the effect of a time varying electromagnetic field on a charge q was found to have the form (see Eq. 11.8.10b),

$$\mathrm{H}'(\mathbf{r},t) = -\frac{q}{m}\mathbf{A}(\mathbf{r},t)\cdot\mathbf{p} \tag{13.2.1}$$

where m is the mass of the particle, and $\mathbf{p}$ is its linear momentum. The vector potential, $\mathbf{A}(\mathbf{r},t)$, is a function of the space coordinate, $\mathbf{r}$, and the time, t. For convenience in determining the optical rotation, we shall assume that the electromagnetic field is associated with a standing wave, which can be represented by two plane waves moving in opposite directions. The vector potential then assumes the form,

$$\mathbf{A}(\mathbf{r},t) = \frac{1}{2}\mathbf{A}_o\left[e^{i(\mathbf{k}\cdot\mathbf{r}-\omega t)} + e^{i(\mathbf{k}\cdot\mathbf{r}+\omega t)}\right] = \mathbf{A}_o e^{i\mathbf{k}\cdot\mathbf{r}}\cos(\omega t) \tag{13.2.2}$$

where $\mathbf{A}_o$ is a constant vector amplitude. Note that according to Eq. (13.2.2), $\mathbf{A}(\mathbf{r},t)$ is complex valued. When substituted into Eq. (13.2.1), the result is

$$\mathrm{H}'(\mathbf{r},t) = -\frac{q}{m}\mathbf{A}_o e^{i\mathbf{k}\cdot\mathbf{r}}\cos(\omega t)\cdot\mathbf{p} \tag{13.2.3}$$

which makes $\mathrm{H}'(\mathbf{r},t)$ complex-valued. This will not prove to be a problem, because at the end of the calculation, we shall be able to assign separate interpretations to the real and imaginary parts.

Next expand $\mathbf{A}(\mathbf{r},t)$ in a Taylor series about the center of the molecule, which is assumed to be located at $\mathbf{r} = 0$. The first two terms of the Taylor series are

$$\mathbf{A}(\mathbf{r},t) = \mathbf{A}(0,t) + \mathbf{r}\cdot[\nabla\mathbf{A}]_{\mathbf{r}=0} \tag{13.2.4a}$$

where

$$\mathbf{r}\cdot[\nabla\mathbf{A}]_{\mathbf{r}=0} = \left[x_1\left(\frac{\partial\mathbf{A}}{\partial x_1}\right)_{\mathbf{r}=0} + \left(x_2\frac{\partial\mathbf{A}}{\partial x_2}\right)_{\mathbf{r}=0} + \left(x_3\frac{\partial\mathbf{A}}{\partial x_3}\right)_{\mathbf{r}=0}\right] \tag{13.2.4b}$$

Employing Eq. (13.2.4a), the dot product of the vector potential with the momentum is

$$\mathbf{A}(\mathbf{r},t)\cdot\mathbf{p} = \mathbf{A}(0,t)\cdot\mathbf{p} + \mathbf{r}\cdot[\nabla\mathbf{A}]_{\mathbf{r}=0}\cdot\mathbf{p} \tag{13.2.5}$$

Notice in Eq. (13.2.5) that $[\nabla \mathbf{A}]_{\mathbf{r}=0}$ functions as a tensor, whereas $[\nabla \mathbf{A}]_{\mathbf{r}=0} \cdot \mathbf{p}$ is a vector; hence, $\mathbf{r} \cdot [\nabla \mathbf{A}]_{\mathbf{r}=0} \cdot \mathbf{p}$ is a scalar. If $\mathbf{r} = (x_1, x_2, x_3)$ and $\mathbf{k} = (k_1, k_2, k_3)$, then using Eq. (13.2.2), the dot product of $\mathbf{r}$ with $\nabla \mathbf{A}(\mathbf{r}, t)$ is

$$\begin{aligned}\mathbf{r} \cdot \nabla \mathbf{A} &= \left(x_1 \frac{\partial}{\partial x_1} + x_2 \frac{\partial}{\partial x_2} + x_3 \frac{\partial}{\partial x_3} \right) \exp[i(k_1 x_1 + k_2 x_2 + k_3 x_3)] \mathbf{A}_o \cos(\omega t) \\ &= i(k_1 x_1 + k_2 x_2 + k_3 x_3) \exp[i(k_1 x_1 + k_2 x_2 + k_3 x_3)] \mathbf{A}_o \cos(\omega t) \\ &= i(\mathbf{k} \cdot \mathbf{r}) e^{i\mathbf{k} \cdot \mathbf{r}} \mathbf{A}_o \cos(\omega t) \\ &= i(\mathbf{k} \cdot \mathbf{r}) \mathbf{A}(\mathbf{r}, t) \end{aligned} \tag{13.2.6}$$

Upon substitution of Eq. (13.2.6), we can write Eq. (13.2.5) in the form,

$$\begin{aligned}\mathbf{A}(\mathbf{r}, t) \cdot \mathbf{p} &= \mathbf{A}(0, t) \cdot \mathbf{p} + \mathbf{r} \cdot [\nabla \mathbf{A}]_{\mathbf{r}=0} \cdot \mathbf{p} \\ &= \mathbf{A}(0, t) \cdot \mathbf{p} + i(\mathbf{k} \cdot \mathbf{r})[\mathbf{A}(\mathbf{r}, t)]_{\mathbf{r}=0} \cdot \mathbf{p} \\ &= \mathbf{A}(0, t) \cdot \mathbf{p} + i(\mathbf{k} \cdot \mathbf{r})(\mathbf{A}(0, t) \cdot \mathbf{p}) \end{aligned} \tag{13.2.7}$$

The combination $(\mathbf{k} \cdot \mathbf{r})(\mathbf{A}(0, t) \cdot \mathbf{p})$ can be expanded into a more recognizable form, where we subtract $(1/2)(\mathbf{A} \cdot \mathbf{r})(\mathbf{k} \cdot \mathbf{p})$ from $(1/2)(\mathbf{k} \cdot \mathbf{r})(\mathbf{A} \cdot \mathbf{p})$ and add $(1/2)(\mathbf{A} \cdot \mathbf{r})(\mathbf{k} \cdot \mathbf{p})$ to $(1/2)(\mathbf{k} \cdot \mathbf{r})(\mathbf{A} \cdot \mathbf{p})$. The result is

$$\begin{aligned}(\mathbf{k} \cdot \mathbf{r})(\mathbf{A} \cdot \mathbf{p}) &= \frac{1}{2}\left[(\mathbf{k} \cdot \mathbf{r})(\mathbf{A} \cdot \mathbf{p}) - (\mathbf{A} \cdot \mathbf{r})(\mathbf{k} \cdot \mathbf{p})\right] \\ &\quad + \frac{1}{2}\left[(\mathbf{k} \cdot \mathbf{r})(\mathbf{A} \cdot \mathbf{p}) + (\mathbf{A} \cdot \mathbf{r})(\mathbf{k} \cdot \mathbf{p})\right] \end{aligned} \tag{13.2.8}$$

The second bracket, $(1/2)\left[(\mathbf{k} \cdot \mathbf{r})(\mathbf{A} \cdot \mathbf{p}) + (\mathbf{A} \cdot \mathbf{r})(\mathbf{k} \cdot \mathbf{p})\right]$, involves the interaction of the radiation with the molecular electric quadrupole moment [2]. The quadrupole interaction is an order of magnitude weaker than the electric dipole interaction.

Consequently, the second bracket in Eq. (13.2.8) can be dropped in favor of the first bracket. Equation (13.2.8) can then be written approximately as

$$(\mathbf{k} \cdot \mathbf{r})(\mathbf{A} \cdot \mathbf{p}) \approx \frac{1}{2}\left[(\mathbf{k} \cdot \mathbf{r})(\mathbf{A} \cdot \mathbf{p}) - (\mathbf{A} \cdot \mathbf{r})(\mathbf{k} \cdot \mathbf{p})\right] \tag{13.2.9}$$

We can now exploit the vector identity (see the Mathematical Methods Appendix),

$$(\mathbf{a} \cdot \mathbf{c})(\mathbf{b} \cdot \mathbf{d}) - (\mathbf{b} \cdot \mathbf{c})(\mathbf{a} \cdot \mathbf{d}) = (\mathbf{a} \times \mathbf{b}) \cdot (\mathbf{c} \times \mathbf{d}) \tag{13.2.10}$$

to condense the terms in the bracket on the right hand side of Eq. (13.2.9). With $\mathbf{a} = \mathbf{k}$, $\mathbf{c} = \mathbf{r}$, $\mathbf{b} = \mathbf{A}$, and $\mathbf{d} = \mathbf{p}$, Eq. (13.2.10) reads,

$$(\mathbf{k} \cdot \mathbf{r})(\mathbf{A} \cdot \mathbf{p}) - (\mathbf{A} \cdot \mathbf{r})(\mathbf{k} \cdot \mathbf{p}) = (\mathbf{k} \times \mathbf{A}) \cdot (\mathbf{r} \times \mathbf{p}) \tag{13.2.11}$$

Use Eq. (13.2.11) to rewrite Eq. (13.2.9) in the form,

$$(\mathbf{k}\cdot\mathbf{r})(\mathbf{A}\cdot\mathbf{p}) = \frac{1}{2}(\mathbf{k}\times\mathbf{A})\cdot(\mathbf{r}\times\mathbf{p}) \tag{13.2.12}$$

Equations (13.2.6) and (13.2.12) can be combined to evaluate $\mathbf{r}\cdot[\nabla\mathbf{A}]_{\mathbf{r}=0}\cdot\mathbf{p}$ in the form,

$$\mathbf{r}\cdot[\nabla\mathbf{A}]_{\mathbf{r}=0}\cdot\mathbf{p} = (i\mathbf{k}\cdot\mathbf{r})(\mathbf{A}(0,t)\cdot\mathbf{p}) = \frac{1}{2}[i\mathbf{k}\times\mathbf{A}(0,t)]\cdot(\mathbf{r}\times\mathbf{p}) \tag{13.2.13}$$

We next show that the right hand side of Eq. (13.2.13) can be expressed in terms of the curl of $\mathbf{A}(\mathbf{r},t)$. According to Eq. (13.2.2), the curl is

$$\nabla\times\mathbf{A}(\mathbf{r},t) = \cos(\omega t)\nabla\times\left(\mathbf{A}_o e^{i\mathbf{k}\cdot\mathbf{r}}\right) \tag{13.2.14}$$

As an aid in computing the far right hand side of Eq. (13.2.14), we introduce the vector valued function,

$$\mathbf{V}(\mathbf{r}) = V_1(\mathbf{r})\hat{\mathbf{e}}_1 + V_2(\mathbf{r})\hat{\mathbf{e}}_2 + V_3(\mathbf{r})\hat{\mathbf{e}}_3 \tag{13.2.15a}$$

The curl of $\mathbf{V}(\mathbf{r})$ is

$$\begin{aligned}\nabla\times\mathbf{V}(\mathbf{r}) = \begin{vmatrix}\hat{\mathbf{e}}_1 & \hat{\mathbf{e}}_2 & \hat{\mathbf{e}}_3\\ \frac{\partial}{\partial x_1} & \frac{\partial}{\partial x_2} & \frac{\partial}{\partial x_3}\\ V_1 & V_2 & V_3\end{vmatrix} &= \left(\frac{\partial V_3}{\partial x_2}-\frac{\partial V_2}{\partial x_3}\right)\hat{\mathbf{e}}_1 + \left(\frac{\partial V_1}{\partial x_3}-\frac{\partial V_3}{\partial x_1}\right)\hat{\mathbf{e}}_2\\ &+\left(\frac{\partial V_2}{\partial x_1}-\frac{\partial V_1}{\partial x_2}\right)\hat{\mathbf{e}}_3\end{aligned} \tag{13.2.15b}$$

Let

$$V_1(\mathbf{r}) = V_1^o e^{i\mathbf{k}\cdot\mathbf{r}} = V_1^o \exp[i(k_1x_1 + k_2x_2 + k_3x_3)] \tag{13.2.15c}$$

where V_1^o is the amplitude of $V_1(\mathbf{r})$. As an example of one of the derivatives required to evaluate $\nabla\times\mathbf{V}(\mathbf{r})$, consider the calculation of $\frac{\partial V_1(\mathbf{r})}{\partial x_2}$ which appears in Eq. (13.2.15b):

$$\begin{aligned}\frac{\partial V_1(\mathbf{r})}{\partial x_2} &= V_1^o\frac{\partial}{\partial x_2}\exp[i(k_1x_1 + k_2x_2 + k_3x_3)]\\ &= V_1^o(ik_2)\exp[i(k_1x_1 + k_2x_2 + k_3x_3)] = ik_2V_1(\mathbf{r})\end{aligned} \tag{13.2.15d}$$

Substitute Eq. (13.2.15d) and its generalizations into the right hand side of Eq. (13.2.15b). The result is

$$\nabla \times \mathbf{V}(\mathbf{r}) = (ik_2V_3 - ik_3V_2)\hat{\mathbf{e}}_1 + (ik_3V_1 - ik_1V_3)\hat{\mathbf{e}}_2 + (ik_1V_2 - ik_2V_1)\hat{\mathbf{e}}_3 \tag{13.2.15e}$$

Note that the right hand side of Eq. (13.2.15e) is just $i\mathbf{k} \times \mathbf{V}(\mathbf{r})$. Hence, Eq. (13.2.15b) can be rewritten as

$$\nabla \times \mathbf{V}(\mathbf{r}) = i\mathbf{k} \times \mathbf{V}(\mathbf{r}) \tag{13.2.15f}$$

With the identification $\mathbf{V}(\mathbf{r}) = \mathbf{A}_o e^{i\mathbf{k}\cdot\mathbf{r}}$, Eq. (13.2.15f) permits $\nabla \times \mathbf{A}(\mathbf{r}, t)$ in Eq. (13.2.14) to be written as

$$\nabla \times \mathbf{A}(\mathbf{r}, t) = \cos(\omega t)\nabla \times \left(\mathbf{A}_o e^{i\mathbf{k}\cdot\mathbf{r}}\right) = i\mathbf{k} \times \left(\mathbf{A}_o e^{i\mathbf{k}\cdot\mathbf{r}}\right)\cos(\omega t) = i\mathbf{k} \times \mathbf{A}(\mathbf{r}, t) \tag{13.2.16}$$

Evaluate Eq. (13.2.16) at $\mathbf{r} = 0$. The result is

$$[\nabla \times \mathbf{A}(\mathbf{r}, t)]_{\mathbf{r}=0} = i\mathbf{k} \times \mathbf{A}(0, t) \tag{13.2.17}$$

Equation (13.2.17) permits Eq. (13.2.13) to be rewritten as

$$\mathbf{r} \cdot [\nabla\mathbf{A}]_{\mathbf{r}=0} \cdot \mathbf{p} = \frac{1}{2}[\nabla \times \mathbf{A}(\mathbf{r}, t)]_{\mathbf{r}=0} \cdot (\mathbf{r} \times \mathbf{p}) \tag{13.2.18}$$

Finally, using Eq. (13.2.18), Eq. (13.2.5) becomes

$$\mathbf{A}(\mathbf{r}, t) \cdot \mathbf{p} = \mathbf{A}(0, t) \cdot \mathbf{p} + \frac{1}{2}([\nabla \times \mathbf{A}]_{\mathbf{r}=0}) \cdot (\mathbf{r} \times \mathbf{p}) \tag{13.2.19}$$

We will find that the term $\mathbf{A}(0, t) \cdot \mathbf{p}$ is the origin of the electric component of ORD, whereas the term involving $([\nabla \times \mathbf{A}]_{\mathbf{r}=0}) \cdot (\mathbf{r} \times \mathbf{p}) = \mathbf{B} \cdot (\mathbf{r} \times \mathbf{p})$, where $\mathbf{B}$ is the magnetic induction, is the origin of the magnetic component.

13.2.2 Electric and Magnetic Components of the Interaction Hamiltonian

When Eq. (13.2.19) is substituted into Eq. (13.2.1), the perturbation reads,

$$\mathrm{H}'(\mathbf{r}, t) = -\frac{q}{m}\mathbf{A}(\mathbf{r}, t) \cdot \mathbf{p} = \left[-\frac{q}{m}\mathbf{A}(0, t) \cdot \mathbf{p} - ([\nabla \times \mathbf{A}]_{\mathbf{r}=0}) \cdot \frac{q}{2m}(\mathbf{r} \times \mathbf{p})\right] \tag{13.2.20}$$

In mechanics, the orbital angular momentum, $\mathbf{L}$, of a particle moving with linear momentum, $\mathbf{p}$, is given by

$$\mathbf{L} = \mathbf{r} \times \mathbf{p} \tag{13.2.21}$$

The magnetic dipole moment, **M**, associated with an orbiting particle having charge, q, is defined by

$$\mathbf{M} = \gamma \mathbf{L} = \gamma(\mathbf{r} \times \mathbf{p}) \tag{13.2.22}$$

where

$$\gamma = \frac{q}{2m} \tag{13.2.23}$$

Equation (13.2.23)introduccs thc gyromagnctic ratio, γ. In Eq. (13.2.23), m is the particle mass. Equations (13.2.21)–(13.2.23) permit the vector cross product, $\frac{q}{2m}(\mathbf{r} \times \mathbf{p})$, in Eq. (13.2.20) to be replaced by **M** with the result,

$$\mathrm{H}'(t) = \left[-\frac{q}{m}\mathbf{A}(0, t) \cdot \mathbf{p} - \mathbf{M} \cdot ([\nabla \times \mathbf{A}]_{\mathbf{r}=0})\right] \tag{13.2.24}$$

The coefficients, $-\frac{q}{m}\mathbf{A}(0, t)$, and $([\nabla \times \mathbf{A}]_{\mathbf{r}=0})$, in Eq. (13.2.24) have been evaluated at the origin and are consequently constants with respect to space. The remaining dependence of Eq. (13.2.24) on the space coordinate appears in **p** and **M**, which involve **r** implicitly.

The right hand side of Eq. (13.1.12) requires evaluation of the matrix element, $\left\langle \Psi_b^{(o)} \middle| \mathrm{H}'(t) \middle| \Psi_a^{(o)} \right\rangle$, connecting states $\Psi_a^{(o)}$ and $\Psi_b^{(o)}$. According to Eq. (13.2.24), the required matrix element has the form,

$$\left\langle \Psi_b^{(o)} \middle| \mathrm{H}' \middle| \Psi_a^{(o)} \right\rangle = \left[-\frac{q}{m}\mathbf{A}(0, t) \cdot \left\langle \Psi_b^{(o)} |\mathbf{p}| \Psi_a^{(o)} \right\rangle - \left\langle \Psi_b^{(o)} |\mathbf{M}| \Psi_a^{(o)} \right\rangle \cdot ([\nabla \times \mathbf{A}]_{\mathbf{r}=0})\right] \tag{13.2.25}$$

The matrix element of $\mathbf{p} = m(d\mathbf{r}/dt)$ can be calculated as,

$$\left\langle \Psi_b^{(o)} |\mathbf{p}| \Psi_a^{(o)} \right\rangle = m\frac{d}{dt}\left\langle \Psi_b^{(o)} |\mathbf{r}| \Psi_a^{(o)} \right\rangle \tag{13.2.26}$$

Recall that we can write $\Psi_a^{(o)}(\mathbf{r}, t)$ in the form,

$$\Psi_a^{(o)}(\mathbf{r}, t) = \psi_a^{(o)}(\mathbf{r})e^{-iE_a^{(o)}t/\hbar} \tag{13.2.27a}$$

and $\Psi_b^{(o)*}(\mathbf{r}, t)$ in the form,

$$\Psi_b^{(o)*}(\mathbf{r}, t) = \psi_b^{(o)*}(\mathbf{r})e^{iE_b^{(o)}t/\hbar} \tag{13.2.27b}$$

By use of Eqs. (13.2.27a) and (13.2.27b), the calculation of the matrix element of **r** in Eq. (13.2.26) proceeds as follows:

$$\begin{aligned}\left\langle \Psi_b^{(o)} |\mathbf{r}| \Psi_a^{(o)} \right\rangle &= \int d^3 r e^{iE_b^{(o)} t/\hbar} \psi_b^{(o)*}(\mathbf{r})\, \mathbf{r} \psi_a^{(o)}(\mathbf{r}) e^{-iE_a^{(o)} t/\hbar} \\ &= \left[\int d^3 r \psi_b^{(o)*}(\mathbf{r})\, \mathbf{r} \psi_a^{(o)}(\mathbf{r}) \right] e^{i[(E_b^{(o)} - E_a^{(o)})t/\hbar]} \\ &= \left\langle \psi_b^{(o)} |\mathbf{r}| \psi_a^{(o)} \right\rangle e^{i[(E_b^{(o)} - E_a^{(o)})t/\hbar]} \end{aligned} \tag{13.2.28a}$$

Define

$$E_{ba}^{(o)} = E_b^{(o)} - E_a^{(o)} \tag{13.2.28b}$$

then Eq. (13.2.28a) reads

$$\left\langle \Psi_b^{(o)} |\mathbf{r}| \Psi_a^{(o)} \right\rangle = \left\langle \psi_b^{(o)} |\mathbf{r}| \psi_a^{(o)} \right\rangle e^{iE_{ba}^{(o)} t/\hbar} \tag{13.2.28c}$$

Upon substitution of Eq. (13.2.28c), Eq. (3.2.26) assumes the form

$$\begin{aligned}\left\langle \Psi_b^{(o)} |\mathbf{p}| \Psi_a^{(o)} \right\rangle &= m \frac{d}{dt} \left\langle \Psi_b^{(o)} |\mathbf{r}| \Psi_a^{(o)} \right\rangle = \left\langle \psi_b^{(o)} |\mathbf{r}| \psi_a^{(o)} \right\rangle \frac{d}{dt} e^{iE_{ba}^{(o)} t/\hbar} \\ &= m \left\langle \psi_b^{(o)} |\mathbf{r}| \psi_a^{(o)} \right\rangle \left(\frac{iE_{ba}^{(o)}}{\hbar} \right) e^{iE_{ba}^{(o)} t/\hbar} \end{aligned} \tag{13.2.29}$$

By analogy with Eq. (13.2.28a),

$$\left\langle \Psi_b^{(o)} |\mathbf{M}| \Psi_a^{(o)} \right\rangle = \left\langle \psi_b^{(o)} |\mathbf{M}| \psi_a^{(o)} \right\rangle e^{i[(E_b^{(o)} - E_a^{(o)})t/\hbar]} = \left\langle \psi_b^{(o)} |\mathbf{M}| \psi_a^{(o)} \right\rangle e^{iE_{ba}^{(o)} t/\hbar} \tag{13.2.30}$$

Upon substitution of Eqs. (13.2.29) and (13.2.30) into Eq. (13.2.25), we obtain

$$\begin{aligned}\left\langle \Psi_b^{(o)} |\mathrm{H}'| \Psi_a^{(o)} \right\rangle = \Bigg[&-\frac{1}{m} \mathbf{A}(0, t) \cdot m \left(\frac{iE_{ba}^{(o)}}{\hbar} \right) q \left\langle \psi_b^{(o)} |\mathbf{r}| \psi_a^{(o)} \right\rangle e^{iE_{ba}^{(o)} t/\hbar} \\ &- \left\langle \psi_b^{(o)} |\mathbf{M}| \psi_a^{(o)} \right\rangle e^{iE_{ba}^{(o)} t/\hbar} \cdot ([\nabla \times \mathbf{A}]_{\mathbf{r}=0}) \Bigg] \end{aligned} \tag{13.2.31}$$

Define the instantaneous electric dipole moment by

$$\mathbf{R} = q\mathbf{r} \tag{13.2.32}$$

Upon substitution of Eq. (13.2.32) into Eq. (13.2.31) followed by factoring out the complex valued exponential, the result is

$$\left\langle\psi_b^{(o)}|\mathrm{H}'|\psi_a^{(o)}\right\rangle = \left[-\left(i\mathbf{E}_{ba}^{(o)}/\hbar\right)\left\langle\psi_b^{(o)}|\mathbf{R}|\psi_a^{(o)}\right\rangle\cdot\mathbf{A}(0,t) - \left\langle\psi_b^{(o)}|\mathbf{M}|\psi_a^{(o)}\right\rangle\cdot[\nabla\times\mathbf{A}]_{\mathbf{r}=0}\right]e^{i\mathbf{E}_{ba}^{(o)}t/\hbar} \tag{13.2.33}$$

In Eq. (13.2.33), the term involving $\left\langle\psi_b^{(o)}|\mathbf{R}|\psi_a^{(o)}\right\rangle$ accounts for the electric dipole moment induced by the electromagnetic field. The term involving $\left\langle\psi_b^{(o)}|\mathbf{M}|\psi_a^{(o)}\right\rangle$ accounts for the magnetic dipole moment induced by the electromagnetic field.

13.3 Evaluation of $c_b(t)$

Return to Eq. (13.1.12) and substitute Eq. (13.2.33). The result is

$$c_b(t) = \frac{1}{i\hbar}\int_0^t dt\left[-\left(i\mathbf{E}_{ba}^{(o)}/\hbar\right)\left\langle\psi_b^{(o)}|\mathbf{R}|\psi_a^{(o)}\right\rangle\cdot\mathbf{A}(0,t) - \left\langle\psi_b^{(o)}|\mathbf{M}|\psi_a^{(o)}\right\rangle\cdot[\nabla\times\mathbf{A}]_{\mathbf{r}=0}\right]e^{i\mathbf{E}_{ba}^{(o)}t/\hbar} \tag{13.3.1a}$$

Upon factoring (-1) from the integrand, Eq. (13.3.1a) reads,

$$c_b(t) = \frac{i}{\hbar}\int_0^t dt\left[\left(i\mathbf{E}_{ba}^{(o)}/\hbar\right)\left\langle\psi_b^{(o)}|\mathbf{R}|\psi_a^{(o)}\right\rangle\cdot\mathbf{A}(0,t) + \left\langle\psi_b^{(o)}|\mathbf{M}|\psi_a^{(o)}\right\rangle\cdot[\nabla\times\mathbf{A}]_{\mathbf{r}=0}\right]e^{i\mathbf{E}_{ba}^{(o)}t/\hbar} \tag{13.3.1b}$$

In order to evaluate the first term in Eq. (13.3.1b), we return to Eq. (13.2.2) and set $\mathbf{r} = 0$. The result is

$$\mathbf{A}(0,t) = \mathbf{A}_o\cos(\omega t) \tag{13.3.2}$$

To evaluate the second term within the bracket in Eq. (13.3.1b), we shall need $[\nabla\times\mathbf{A}]_{\mathbf{r}=0}$. This evaluation is aided if we set $t = 0$ in Eq. (13.2.2) to obtain

$$\mathbf{A}(\mathbf{r},0) = \mathbf{A}_o e^{i\mathbf{k}\cdot\mathbf{r}} \tag{13.3.3a}$$

Equation (13.3.3a) permits Eq. (13.2.2) to be rewritten as

$$\mathbf{A}(\mathbf{r},t) = \mathbf{A}(\mathbf{r},0)\cos(\omega t) \tag{13.3.3b}$$

The curl of Eq. (13.3.3b) is

$$\nabla\times\mathbf{A}(\mathbf{r},t) = [\nabla\times\mathbf{A}(\mathbf{r},0)]\cos(\omega t) \tag{13.3.3c}$$

Equation (13.3.3c) can be evaluated at $\mathbf{r} = 0$ by writing it symbolically as

$$[\nabla\times\mathbf{A}(\mathbf{r},t)]_{\mathbf{r}=0} = [\nabla\times\mathbf{A}(\mathbf{r},0)]_{\mathbf{r}=0}\cos(\omega t) \tag{13.3.3d}$$

Upon substitution of Eq. (13.3.2) into the first term in the bracket in Eq. (13.3.1b) and Eq. (13.3.3d) into the second term, the result is

$$c_b = \frac{i}{\hbar}\left[\left(\frac{iE_{ba}^{(o)}}{\hbar}\right)\left\langle\psi_b^{(o)}|\mathbf{R}|\psi_a^{(o)}\right\rangle\right]\cdot\mathbf{A}_o\int_0^t e^{iE_{ba}^{(o)}\tau/\hbar}\cos(\omega\tau)d\tau$$
$$+\frac{i}{\hbar}\left[\left\langle\psi_b^{(o)}|\mathbf{M}|\psi_a^{(o)}\right\rangle\cdot\left([\nabla\times\mathbf{A}(\mathbf{r},0)]_{\mathbf{r}=0}\right)\right]\int_0^t e^{iE_{ba}^{(o)}\tau/\hbar}\cos(\omega\tau)d\tau \quad (13.3.4)$$

In Eq. (13.3.4), let $s = \hbar\omega$. The result is

$$c_b = \frac{i}{\hbar}\left(\frac{iE_{ba}^{(o)}}{\hbar}\right)\left\langle\psi_b^{(o)}|\mathbf{R}|\psi_a^{(o)}\right\rangle\cdot\mathbf{A}_o\int_0^t e^{iE_{ba}^{(o)}\tau/\hbar}\cos(s\tau/\hbar)d\tau$$
$$+\frac{i}{\hbar}\left\langle\psi_b^{(o)}|\mathbf{M}|\psi_a^{(o)}\right\rangle\cdot\left([\nabla\times\mathbf{A}(\mathbf{r},0)]_{\mathbf{r}=0}\right)\int_0^t e^{iE_{ba}^{(o)}\tau/\hbar}\cos(s\tau/\hbar)d\tau \quad (13.3.5a)$$

With this substitution, the indicated integration over $d\tau$ can be performed as follows:

$$\begin{aligned}\int_0^t e^{iE_{ba}^{(o)}\tau/\hbar}\cos(s\tau/\hbar)d\tau &= \frac{1}{2}\int_0^t d\tau e^{iE_{ba}^{(o)}\tau/\hbar}\left(e^{is\tau/\hbar}+e^{-is\tau/\hbar}\right)\\ &= \frac{1}{2}\int_0^t d\tau e^{i(E_{ba}^{(o)}+s)\tau/\hbar}+\frac{1}{2}\int_0^t d\tau e^{i(E_{ba}^{(o)}-s)\tau/\hbar}\\ &= \frac{1}{2}\frac{\left[e^{i(E_{ba}^{(o)}+s)\tau/\hbar}\right]_0^t}{i(E_{ba}^{(o)}+s)/\hbar}+\frac{1}{2}\frac{\left[e^{i(E_{ba}^{(o)}-s)\tau/\hbar}\right]_0^t}{i(E_{ba}^{(o)}-s)/\hbar}\end{aligned} \quad (13.3.5b)$$

The bracket limits in Eq. (13.3.5b) can be simplified if we evaluate each bracket at the upper limit and combine the results, then evaluate each bracket at the lower limit and combine those results. The result then reads,

$$\int_0^t e^{iE_{ba}^{(o)}\tau/\hbar}\cos(s\tau/\hbar)d\tau = \left[\frac{1}{2}\frac{e^{i(E_{ba}^{(o)}+s)\tau/\hbar}}{i(E_{ba}^{(o)}+s)/\hbar}+\frac{1}{2}\frac{e^{i(E_{ba}^{(o)}-s)\tau/\hbar}}{i(E_{ba}^{(o)}-s)/\hbar}\right]$$
$$-\left[\frac{1/2}{i(E_{ba}^{(o)}+s)/\hbar}+\frac{1/2}{i(E_{ba}^{(o)}-s)/\hbar}\right] \quad (13.3.5c)$$

The sum in the second bracket in Eq. (13.3.5c) serves as a constant of integration, which we denote by C, where

$$C = -\left[\frac{1/2}{i(E_{ba}^{(o)} + s)/\hbar} + \frac{1/2}{i(E_{ba}^{(o)} - s)/\hbar}\right] \tag{13.3.5d}$$

Equation (13.3.5d) then permits Eq. (13.3.5c) to be rewritten as,

$$\int_0^t e^{iE_{ba}^{(o)}\tau/\hbar}\cos(s\tau/\hbar)d\tau = \frac{1}{2}\frac{e^{i(E_{ba}^{(o)}+s)t/\hbar}}{i(E_{ba}^{(o)} + s)/\hbar} + \frac{1}{2}\frac{e^{i(E_{ba}^{(o)}-s)t/\hbar}}{i(E_{ba}^{(o)} - s)/\hbar} + C \tag{13.3.5e}$$

Upon substitution of Eq. (13.3.5e) into Eq. (13.3.5a), the result is

$$\begin{aligned}
c_b = {} & \frac{i}{\hbar}\left(\frac{iE_{ba}^{(o)}}{\hbar}\right)\left\langle\psi_b^{(o)}|\mathbf{R}|\psi_a^{(o)}\right\rangle\cdot\mathbf{A}_o\left[\frac{1}{2}\frac{e^{i(E_{ba}^{(o)}+s)t/\hbar}}{i(E_{ba}^{(o)} + \varepsilon)/\hbar} + \frac{1}{2}\frac{e^{i(E_{ba}^{(o)}-s)t/\hbar}}{i(E_{ba}^{(o)} - \varepsilon)/\hbar}\right] \\
& + \frac{i}{\hbar}\left\langle\psi_b^{(o)}|\mathbf{M}|\psi_a^{(o)}\right\rangle\cdot([\nabla\times\mathbf{A}(\mathbf{r},0)]_{\mathbf{r}=0})\left[\frac{1}{2}\frac{e^{i(E_{ba}^{(o)}+s)t/\hbar}}{i(E_{ba}^{(o)} + \varepsilon)/\hbar} + \frac{1}{2}\frac{e^{i(E_{ba}^{(o)}-s)t/\hbar}}{i(E_{ba}^{(o)} - \varepsilon)/\hbar}\right] \\
& + C\left[\frac{i}{\hbar}\left(\frac{iE_{ba}^{(o)}}{\hbar}\right)\left\langle\psi_b^{(o)}|\mathbf{R}|\psi_a^{(o)}\right\rangle\cdot\mathbf{A}_o + \frac{i}{\hbar}\left\langle\psi_b^{(o)}|\mathbf{M}|\psi_a^{(o)}\right\rangle\cdot([\nabla\times\mathbf{A}(\mathbf{r},0)]_{\mathbf{r}=0})\right]
\end{aligned} \tag{13.3.5f}$$

The time dependent terms involving $e^{i(E_{ba}^{(o)}+s)t/\hbar}$ and $e^{i(E_{ba}^{(o)}-s)t/\hbar}$ in Eq. (13.3.5f) depend upon the light frequency. These time dependent terms will determine the resonance. As a consequence, the terms multiplying C can be dropped. After factoring, Eq. (13.3.5f) then assumes the simplified form,

$$\begin{aligned}
c_b = {} & \left[\frac{i}{\hbar}\left(\frac{iE_{ba}^{(o)}}{\hbar}\right)\left\langle\psi_b^{(o)}|\mathbf{R}|\psi_a^{(o)}\right\rangle\cdot\mathbf{A}_o + \frac{i}{\hbar}\left\langle\psi_b^{(o)}|\mathbf{M}|\psi_a^{(o)}\right\rangle\cdot([\nabla\times\mathbf{A}(\mathbf{r},0)]_{\mathbf{r}=0})\right] \\
& \times\left[\frac{1}{2}\frac{e^{i(E_{ba}^{(o)}+s)t/\hbar}}{i(E_{ba}^{(o)} + s)/\hbar} + \frac{1}{2}\frac{e^{i(E_{ba}^{(o)}-s)t/\hbar}}{i(E_{ba}^{(o)} - s)/\hbar}\right]
\end{aligned} \tag{13.3.6}$$

13.4 Calculation of the Induced Electric Dipole Moment

13.4.1 A Formula for the Expectation Value

For ease in calculation in the case of a quantum state with index, a, we choose the real part of the expectation value of the instantaneous electric dipole moment. In the absence of a perturbing electromagnetic field, this is

$$\mathbf{R}_a = \mathrm{Re}\left\langle \Psi_a^{(o)} | \mathbf{R} | \Psi_a^{(o)} \right\rangle \tag{13.4.1}$$

In the presence of the time varying electromagnetic field, the expectation value of the instantaneous electric dipole moment at time, t, is

$$\mathbf{R}_a = \mathrm{Re}[\langle \Psi_a(t) | \mathbf{R} | \Psi_a(t) \rangle] \tag{13.4.2}$$

Because the system that we are considering is initially in the state, "a", $\Psi_a^{(o)}(t)$ dominates in the sum in Eq. (13.1.8), so to a good approximation, we can set $c_k = 1$ when $k = a$ and separate $\Psi_a^{(o)}(t)$ out of the sum. For convenience, we replace the index, k, in the remaining sum by b. With suppression of explicit reference to the variable, x, $\Psi(x, t)$ on the left hand side of Eq. (13.1.8) becomes $\Psi_a(t)$. With the same procedure employed in the sum on the right, Eq. (13.1.8) reads,

$$\Psi_a(t) = \Psi_a^{(o)}(t) + \sum_{b \neq a} c_b(t) \Psi_b^{(o)}(t) \tag{13.4.3a}$$

Upon substitution of Eq. (13.4.3a) into Eq. (13.4.2), the result is

$$\begin{aligned}
\mathbf{R}_a &= \mathrm{Re}\left[\int d^3 r \left(\Psi_a^{(o)*}(t) + \sum_{b \neq a} c_b^*(t) \Psi_b^{(o)*}(t) \right) \mathbf{R} \left(\Psi_a^{(o)}(t) + \sum_{b \neq a} c_b(t) \Psi_b^{(o)}(t) \right) \right] \\
&\approx \mathrm{Re}\left[\langle \Psi_a^{(o)} | \mathbf{R} | \Psi_a^{(o)} \rangle + \sum_{b \neq a} c_b(t) \left\langle \Psi_a^{(o)} | \mathbf{R} | \Psi_b^{(o)} \right\rangle + \sum_{b \neq a} c_b^*(t) \left\langle \Psi_b^{(o)} | \mathbf{R} | \Psi_a^{(o)} \right\rangle \right]
\end{aligned} \tag{13.4.3b}$$

Because each of the $\{c_b\}$ coefficients is small, terms involving the products, $c_b^* c_b$, are all of second order and have been dropped in passing from the first line of Eq. (13.4.3b) to the second line. The Hermitian property of quantum mechanical operators can be applied in the form, $\left\langle \Psi_b^{(o)} | \mathbf{R} | \Psi_a^{(o)} \right\rangle = \left\langle \Psi_a^{(o)} | \mathbf{R} | \Psi_b^{(o)} \right\rangle^*$, in order to modify the third term in the bracket of Eq. (13.4.3b). The result is

$$\mathbf{R}_a = \text{Re}\left[\left\langle\Psi_a^{(o)}|\mathbf{R}|\Psi_a^{(o)}\right\rangle + \sum_b c_b\left\langle\Psi_a^{(o)}|\mathbf{R}|\Psi_b^{(o)}\right\rangle + \sum_b c_b^*\left\langle\Psi_a^{(o)}|\mathbf{R}|\Psi_b^{(o)}\right\rangle^*\right] \tag{13.4.3c}$$

Now consider the complex number $z = x+iy$ with complex conjugate, $z^* = x-iy$. The real part of the sum of z and z^* is

$$\text{Re}(z + z^*) = 2\text{Re}(z) \tag{13.4.3d}$$

The two sums on the right of Eq. (13.4.3c) are complex conjugates of one another. By comparing Eq. (13.4.3c) with Eq. (13.4.3d), the former can be rewritten as

$$\mathbf{R}_a = \text{Re}\left\langle\Psi_a^{(o)}|\mathbf{R}|\Psi_a^{(o)}\right\rangle + 2\text{Re}\sum_b c_b\left\langle\Psi_a^{(o)}|\mathbf{R}|\Psi_b^{(o)}\right\rangle \tag{13.4.4}$$

By referencing Eqs. (13.2.27a) and (13.2.27b), we have

$$\Psi_a^{(o)}(\mathbf{r}, t) = \psi_a^{(o)}(\mathbf{r})e^{-iE_a^{(o)}t/\hbar} \tag{13.4.5a}$$

$$\Psi_a^{(o)*}(\mathbf{r}, t) = \psi_a^{(o)*}(\mathbf{r})e^{iE_a^{(o)}t/\hbar} \tag{13.4.5b}$$

$$\Psi_b^{(o)}(\mathbf{r}, t) = \psi_b^{(o)}(\mathbf{r})e^{-iE_b^{(o)}t/\hbar} \tag{13.4.5c}$$

Upon substitution of Eqs. (13.4.5a) through (13.4.5c) into Eq. (13.4.4), the result reads,

$$\begin{aligned}\mathbf{R}_a &= \text{Re}\langle a|\mathbf{R}|a\rangle e^{i(E_a^{(o)}-E_a^{(o)})t/\hbar} + 2\text{Re}\sum_b c_b\langle a|\mathbf{R}|b\rangle e^{i(E_a^{(o)}-E_b^{(o)})t/\hbar} \\ &= \text{Re}\langle a|\mathbf{R}|a\rangle + 2\text{Re}\sum_b c_b\langle a|\mathbf{R}|b\rangle e^{i(E_a^{(o)}-E_b^{(o)})t/\hbar}\end{aligned} \tag{13.4.5d}$$

where

$$\langle a|\mathbf{R}|a\rangle = \int d^3r\, \psi_a^{(o)*}(\mathbf{r})\mathbf{R}\psi_a^{(o)}(\mathbf{r}) \tag{13.4.5e}$$

and

$$\langle a|\mathbf{R}|b\rangle = \int d^3r\, \psi_a^{(o)*}(\mathbf{r})\mathbf{R}\psi_b^{(o)}(\mathbf{r}) \tag{13.4.5f}$$

In Eq. (13.4.5e), $\langle a|\mathbf{R}|a\rangle$ is the expectation value representing the permanent electric dipole moment of a molecule in the state a. We assume for convenience that the permanent electric dipole moment is zero. By denoting the left hand side of

Eq. (13.4.5d) by $\mathbf{R}'_a$, this restriction has the result,

$$\mathbf{R}'_a = 2\text{Re}\sum_b c_b \langle a|\mathbf{R}|b\rangle e^{i(E_a^{(o)} - E_b^{(o)})t/\hbar} = 2\text{Re}\sum_b c_b \langle a|\mathbf{R}|b\rangle e^{-iE_{ba}^{(o)}t/\hbar} \qquad (13.4.5g)$$

where in Eq. (13.4.5g), we have used Eq. (13.2.28b) to write $(E_a^{(o)} - E_b^{(o)}) = -E_{ba}^{(o)}$.

13.4.2 Induced Electric Dipole Moment Expressed in Terms of the Vector Potential

Next, substitute Eq. (13.3.6) for c_b into Eq. (13.4.5g). The result is

$$\begin{aligned} \mathbf{R}'_a = 2\text{Re}\Bigg\{ \sum_b \langle a|\mathbf{R}|b\rangle e^{-iE_{ba}^{(o)}t/\hbar} \Bigg[\left(\frac{i}{\hbar}\right)\left(\frac{iE_{ba}^{(o)}}{\hbar}\right) \langle b|\mathbf{R}|a\rangle \cdot \mathbf{A}_o \\ + \left(\frac{i}{\hbar}\right) \langle b|\mathbf{M}|a\rangle \cdot ([\nabla \times \mathbf{A}(\mathbf{r},0)]_{\mathbf{r}=0}) \Bigg] \\ \times \left[\frac{1}{2}\frac{e^{iE_{ba}^{(o)}t/\hbar}e^{ist/\hbar}}{(i/\hbar)(E_{ba}^{(o)} + s)} + \frac{1}{2}\frac{e^{iE_{ba}^{(o)}t/\hbar}e^{-ist/\hbar}}{(i/\hbar)(E_{ba}^{(o)} - s)} \right] \Bigg\} \end{aligned} \qquad (13.4.6a)$$

A factor of $(i/\hbar)$ in both terms in the first bracket of Eq. (13.4.6a) can be cancelled against the factors of $(i/\hbar)$ in the denominators of the terms in the second bracket. Moreover, the factor of $e^{-iE_{ba}^{(o)}t/\hbar}$ to the left of the first bracket cancels against the factors of $e^{iE_{ba}^{(o)}t/\hbar}$ within the second bracket. As a result of these cancellations, Eq. (13.4.6a) reads

$$\begin{aligned} \mathbf{R}'_a = \text{Re}\Bigg\{ \sum_b \Bigg[\left(\frac{iE_{ba}^{(o)}}{\hbar}\right) \langle a|\mathbf{R}|b\rangle \langle b|\mathbf{R}|a\rangle \cdot \mathbf{A}_o + \langle a|\mathbf{R}|b\rangle \langle b|\mathbf{M}|a\rangle \\ \cdot ([\nabla \times \mathbf{A}(\mathbf{r},0)]_{\mathbf{r}=0}) \Bigg] \times \left[\frac{e^{ist/\hbar}}{(E_{ba}^{(o)} + s)} + \frac{e^{-ist/\hbar}}{(E_{ba}^{(o)} - s)} \right] \Bigg\} \end{aligned} \qquad (13.4.6b)$$

The fractions in the second bracket in Eq. (13.4.6b) can be simplified by constructing a common denominator. The calculation proceeds as follows:

$$\begin{aligned} \frac{e^{ist/\hbar}}{(E_{ba}^{(o)} + s)} + \frac{e^{-ist/\hbar}}{(E_{ba}^{(o)} - s)} &= \frac{(E_{ba}^{(o)} - s)e^{ist/\hbar} + (E_{ba}^{(o)} + s)e^{-ist/\hbar}}{(E_{ba}^{(o)} + s)(E_{ba}^{(o)} - s)} \\ &= \frac{(E_{ba}^{(o)} - s)e^{ist/\hbar} + (E_{ba}^{(o)} + s)e^{-ist/\hbar}}{\left((E_{ba}^{(o)})^2 - s^2\right)} \end{aligned}$$

$$= \frac{E_{ba}^{(o)}(e^{ist/\hbar} + e^{-ist/\hbar})}{\left((E_{ba}^{(o)})^2 - s^2\right)} - \frac{s(e^{ist/\hbar} - e^{-ist/\hbar})}{\left((E_{ba}^{(o)})^2 - s^2\right)} \tag{13.4.6c}$$

Next insert Eq. (13.4.6c) into Eq. (13.4.6b) and write the result as two separate sums, so that

$$\begin{aligned}
\mathbf{R}_a' = {} & \mathrm{Re} \sum_b \left(\frac{iE_{ba}^{(o)}}{\hbar}\right) \langle a|\mathbf{R}|b\rangle \langle b|\mathbf{R}|a\rangle \\
& \cdot \mathbf{A}_o \left[\frac{E_{ba}^{(o)}\left(e^{ist/\hbar} + e^{-ist/\hbar}\right)}{\left((E_{ba}^{(o)})^2 - s^2\right)} - \frac{s\left(e^{ist/\hbar} - e^{-ist/\hbar}\right)}{\left((E_{ba}^{(o)})^2 - s^2\right)}\right] \\
& + \mathrm{Re} \sum_b \langle a|\mathbf{R}|b\rangle \langle b|\mathbf{M}|a\rangle \cdot ([\nabla \times \mathbf{A}(\mathbf{r}, 0)]_{\mathbf{r}=0}) \\
& \times \left[\frac{E_{ba}^{(o)}\left(e^{ist/\hbar} + e^{-ist/\hbar}\right)}{\left((E_{ba}^{(o)})^2 - s^2\right)} - \frac{s\left(e^{ist/\hbar} - e^{-ist/\hbar}\right)}{\left((E_{ba}^{(o)})^2 - s^2\right)}\right]
\end{aligned} \tag{13.4.6d}$$

Each of the brackets in Eq. (13.6.4d) gives rise to two sums. Writing out all sums explicitly, the resulting expression is

$$\begin{aligned}
\mathbf{R}_a' = {} & \mathrm{Re} \sum_b \langle a|\mathbf{R}|b\rangle \langle b|\mathbf{R}|a\rangle \cdot \mathbf{A}_o\left(e^{ist/\hbar} + e^{-ist/\hbar}\right) \left[\frac{i\left(E_{ba}^{(o)}\right)^2}{\hbar\left((E_{ba}^{(o)})^2 - s^2\right)}\right] \\
& + \mathrm{Re} \sum_b \langle a|\mathbf{R}|b\rangle \langle b|\mathbf{R}|a\rangle \cdot \mathbf{A}_o\left(e^{ist/\hbar} - e^{-ist/\hbar}\right) \left[\frac{-iE_{ba}^{(o)} s}{\hbar\left((E_{ba}^{(o)})^2 - s^2\right)}\right] \\
& + \mathrm{Re} \sum_b \langle a|\mathbf{R}|b\rangle \langle b|\mathbf{M}|a\rangle \cdot ([\nabla \times \mathbf{A}(\mathbf{r}, 0)]_{\mathbf{r}=0})\left(e^{ist/\hbar} + e^{-ist/\hbar}\right) \\
& \times \left[\frac{E_{ba}^{(o)}}{\left((E_{ba}^{(o)})^2 - s^2\right)}\right] + \mathrm{Re} \sum_b \langle a|\mathbf{R}|b\rangle \langle b|\mathbf{M}|a\rangle \\
& \cdot ([\nabla \times \mathbf{A}(\mathbf{r}, 0)]_{\mathbf{r}=0})\left(e^{ist/\hbar} - e^{-ist/\hbar}\right) \left[-\frac{s}{\left((E_{ba}^{(o)})^2 - s^2\right)}\right]
\end{aligned} \tag{13.4.6e}$$

According to Eq. (13.4.6e), the induced electric dipole moment, $\mathbf{R}_a'$, depends upon both $\mathbf{A}_o$ and $[\nabla \times \mathbf{A}(\mathbf{r}, 0)]_{\mathbf{r}=0}$. Because the vector potential is ordinarily not measured, we shall recast Eq. (13.4.6e) in terms of the electric and magnetic fields of the incident electromagnetic wave.

13.4.3 Separating the Electric and Magnetic Effects

By using $s = \hbar\omega$ to replace ω in Eq. (13.2.2), the vector potential evaluated at the origin of coordinates is

$$\mathbf{A}(0, t) = \frac{1}{2}\mathbf{A}_o\left(e^{ist/\hbar} + e^{-ist/\hbar}\right) \tag{13.4.7a}$$

Due to the fact that we have included the scalar potential, Φ, as a part of the time independent Hamilton $\mathrm{H}^{(o)}$, the time varying electric field at the origin is calculated simply by differentiating the vector potential with respect to time according to Eq. (8.4.3) of Chap. 8; hence,

$$\mathbf{E}(0, t) = -\frac{\partial \mathbf{A}(0, t)}{\partial t} = -\frac{is}{2\hbar}\mathbf{A}_o\left(e^{ist/\hbar} - e^{-ist/\hbar}\right) \tag{13.4.7b}$$

The time derivative of the electric field, $\mathbf{E}(0, t)$, will also prove to be useful. According to Eq. (13.4.7b), it is

$$\frac{\partial \mathbf{E}(0, t)}{\partial t} = -\frac{1}{2}\left(\frac{is}{\hbar}\right)^2 \mathbf{A}_o\left(e^{ist/\hbar} + e^{-ist/\hbar}\right) \tag{13.4.7c}$$

With reference to Eq. (8.4.1) of Chap. 8, the magnetic induction is

$$\mathbf{B}(\mathbf{r}, t) = \nabla \times \mathbf{A}(\mathbf{r}, t) \tag{13.4.8a}$$

The vector potential is given by Eq. (13.3.3b). Equations (13.3.3b) and (13.4.8a) can be combined to evaluate $\mathbf{B}(0, t)$ to obtain

$$\mathbf{B}(0, t) = [\nabla \times \mathbf{A}(\mathbf{r}, 0)]_{\mathbf{r}=0} \cos(\omega t) \tag{13.4.8b}$$

The cosine in Eq. (13.4.8b) can be expanded in terms of exponentials. With $s = \hbar\omega$, the result is

$$\mathbf{B}(0, t) = \frac{1}{2}[\nabla \times \mathbf{A}(\mathbf{r}, 0)]_{\mathbf{r}=0}\left(e^{ist/\hbar} + e^{-ist/\hbar}\right) \tag{13.4.8c}$$

Use Eq. (13.4.8c) to calculate the time derivative of the magnetic flux density. That result is

$$\frac{\partial \mathbf{B}(0, t)}{\partial t} = \left(\frac{is}{2\hbar}\right)[\nabla \times \mathbf{A}(\mathbf{r}, 0)]_{\mathbf{r}=0}\left(e^{ist/\hbar} - e^{-ist/\hbar}\right) \tag{13.4.8d}$$

The above results permit Eq. (13.4.6e) to be expressed in terms of $\mathbf{E}(0, t)$ and $\mathbf{B}(0, t)$ and their time derivatives. Equation (13.4.7c) can be used with the first sum on the right of Eq. (13.4.6e) to replace $\mathbf{A}_o\left(e^{ist/\hbar} + e^{-ist/\hbar}\right)$ with

$(2\hbar^2/s^2)(\partial\mathbf{E}(0,t)/\partial t))$. Equation (13.4.7b) can be used in the second sum to replace $\mathbf{A}_o\left(e^{ist/\hbar}-e^{-ist/\hbar}\right)$ with $(-2\hbar/is)\mathbf{E}(0,t)$. Equation (13.4.8c) can be used to replace $[\nabla\times\mathbf{A}(\mathbf{r},0)]_{\mathbf{r}=0}\left(e^{ist/\hbar}+e^{-ist/\hbar}\right)$ with $2\mathbf{B}(0,t)$, while Eq. (13.4.8d) can be used to replace $[\nabla\times\mathbf{A}(\mathbf{r},0)]_{\mathbf{r}=0}\left(e^{ist/\hbar}-e^{-ist/\hbar}\right)$ with $(2\hbar/is)(\partial\mathbf{B}(0,t)/\partial t)$. The result which follows from these substitutions is

$$
\begin{aligned}
\mathbf{R}'_a = {} & 2\mathrm{Re}\sum_b \langle a|\mathbf{R}|b\rangle\langle b|\mathbf{R}|a\rangle\cdot\left(\frac{\hbar}{s}\right)^2\left[\frac{i\left(E_{ba}^{(o)}\right)^2}{\hbar\left((E_{ba}^{(o)})^2-s^2\right)}\right]\frac{\partial\mathbf{E}(0,t)}{\partial t}\\
& + 2\mathrm{Re}\sum_b \langle a|\mathbf{R}|b\rangle\langle b|\mathbf{R}|a\rangle\cdot\left(\frac{\hbar}{-is}\right)\left[\frac{-iE_{ba}^{(o)}s}{\hbar\left((E_{ba}^{(o)})^2-s^2\right)}\right]\mathbf{E}(0,t)\\
& + 2\mathrm{Re}\sum_b \langle a|\mathbf{R}|b\rangle\langle b|\mathbf{M}|a\rangle\cdot\left[\frac{E_{ba}^{(o)}}{\left((E_{ba}^{(o)})^2-s^2\right)}\right]\mathbf{B}(0,t)\\
& + 2\mathrm{Re}\sum_b \langle a|\mathbf{R}|b\rangle\langle b|\mathbf{M}|a\rangle\cdot\left(\frac{\hbar}{is}\right)\left[\frac{-s}{\left((E_{ba}^{(o)})^2-s^2\right)}\right]\frac{\partial\mathbf{B}(0,t)}{\partial t}
\end{aligned}
\tag{13.4.9a}
$$

After algebraic cancellation of common factors and interchange of the first two sums, Eq. (13.4.9a) reads,

$$
\begin{aligned}
\mathbf{R}'_a = {} & 2\mathrm{Re}\sum_b \langle a|\mathbf{R}|b\rangle\langle b|\mathbf{R}|a\rangle\cdot\left[\frac{E_{ba}^{(o)}}{\left((E_{ba}^{(o)})^2-s^2\right)}\right]\mathbf{E}(0,t)\\
& + 2\mathrm{Re}\sum_b \langle a|\mathbf{R}|b\rangle\langle b|\mathbf{R}|a\rangle\cdot\left[\frac{i\hbar\left(E_{ba}^{(o)}\right)^2}{s^2\left((E_{ba}^{(o)})^2-s^2\right)}\right]\frac{\partial\mathbf{E}(0,t)}{\partial t}\\
& + 2\mathrm{Re}\sum_b \langle a|\mathbf{R}|b\rangle\langle b|\mathbf{M}|a\rangle\cdot\left[\frac{E_{ba}^{(o)}}{\left((E_{ba}^{(o)})^2-s^2\right)}\right]\mathbf{B}(0,t)\\
& + 2\mathrm{Re}\sum_b \langle a|\mathbf{R}|b\rangle\langle b|\mathbf{M}|a\rangle\cdot\left[\frac{i\hbar}{\left((E_{ba}^{(o)})^2-s^2\right)}\right]\frac{\partial\mathbf{B}(0,t)}{\partial t}
\end{aligned}
\tag{13.4.9b}
$$

In Eq. (13.4.9b), the effects of the electric and magnetic fields of the light wave are clearly separated. Equation (13.4.9b) applies in the case of a molecule whose orientation is fixed in space.

13.4.4 The Angle Average Over Molecular Orientations

A molecule in the gas phase experiences free rotation. If it is in solution, it experiences a restricted rotation known as libration. In either case, optical measurements on the macroscopic scale are sensitive only to the angle averages of the tensor products, $\langle a|\mathbf{R}|b\rangle\langle b|\mathbf{R}|a\rangle \cdot \mathbf{E}(0,t)$ and $\langle a|\mathbf{R}|b\rangle\langle b|\mathbf{R}|a\rangle \cdot (\partial\mathbf{E}(0,t)/\partial t)$, which appear in the first and second sums of Eq. (13.4.9b) and the angle averages of the tensor products, $\langle a|\mathbf{R}|b\rangle\langle b|\mathbf{M}|a\rangle \cdot \mathbf{B}(0,t)$ and $\langle a|\mathbf{R}|b\rangle\langle b|\mathbf{M}|a\rangle \cdot (\partial\mathbf{B}(0,t)/\partial t)$, which appear in the second and fourth sums in Eq. (13.4.9b).

In the case of $\langle a|\mathbf{R}|b\rangle\langle b|\mathbf{M}|a\rangle \cdot \mathbf{B}(0,t)$, we are required to calculate the angle average of the vector $\langle a|\mathbf{R}|b\rangle$ as weighted by the scalar, $\langle b|\mathbf{M}|a\rangle \cdot \mathbf{B}(0,t)$. The method used to compute the angle average of $\langle a|\mathbf{R}|b\rangle\langle b|\mathbf{M}|a\rangle \cdot \mathbf{B}(0,t)$ is general and can be adopted with only minor changes to compute the angle average of $\langle a|\mathbf{R}|b\rangle\langle b|\mathbf{R}|a\rangle \cdot \mathbf{E}(0,t)$, for example. The other products in Eq. (13.4.9b) can be treated by analogy with these calculations. Hence, we shall turn our attention to $\langle a|\mathbf{R}|b\rangle\langle b|\mathbf{M}|a\rangle \cdot \mathbf{B}(0,t)$.

Let $\mathbf{B}(0,t)$ lie along the z-axis, and write $\mathbf{B}(0,t) = \mathbf{B} = B\hat{\mathbf{e}}_z$, where $\hat{\mathbf{e}}_z$ is a unit vector in the z-direction. Since the angle averages of the matrix elements, $\langle a|\mathbf{R}|b\rangle$ and $\langle b|\mathbf{M}|a\rangle$ are proportional to the angle averages of $\mathbf{R}$ and $\mathbf{M}$, we can replace these quantum operators by their classical equivalents, $\mathbf{R} = (R_x, R_y, R_z)$ and $\mathbf{M} = (M_x, M_y, M_z)$, respectively. In this classical context, our task is to compute the average of $\mathbf{RM} \cdot \mathbf{B}$, where $\mathbf{RM}$ is a dyad. Written in dyadic form,

$$\begin{aligned}\mathbf{RM} = {} & R_xM_x\hat{\mathbf{e}}_x\hat{\mathbf{e}}_x + R_xM_y\hat{\mathbf{e}}_x\hat{\mathbf{e}}_y + R_xM_z\hat{\mathbf{e}}_x\hat{\mathbf{e}}_z \\ & + R_yM_x\hat{\mathbf{e}}_y\hat{\mathbf{e}}_x + R_yM_y\hat{\mathbf{e}}_y\hat{\mathbf{e}}_y + R_yM_z\hat{\mathbf{e}}_y\hat{\mathbf{e}}_z \\ & + R_zM_x\hat{\mathbf{e}}_z\hat{\mathbf{e}}_x + R_zM_y\hat{\mathbf{e}}_z\hat{\mathbf{e}}_y + R_zM_z\hat{\mathbf{e}}_z\hat{\mathbf{e}}_z\end{aligned} \tag{13.4.10a}$$

The dot product of $\mathbf{RM}$ with $\mathbf{B} = B\hat{\mathbf{e}}_z$, the magnetic induction in the z-direction, can be written as

$$\begin{aligned}\mathbf{RM}\cdot\mathbf{B} &= \mathbf{RM}\cdot B\hat{\mathbf{e}}_z \\ &= (R_xM_x\hat{\mathbf{e}}_x\hat{\mathbf{e}}_x + R_xM_y\hat{\mathbf{e}}_x\hat{\mathbf{e}}_y + R_xM_z\hat{\mathbf{e}}_x\hat{\mathbf{e}}_z)\cdot B\hat{\mathbf{e}}_z \\ &\quad + (R_yM_x\hat{\mathbf{e}}_y\hat{\mathbf{e}}_x + R_yM_y\hat{\mathbf{e}}_y\hat{\mathbf{e}}_y + R_yM_z\hat{\mathbf{e}}_y\hat{\mathbf{e}}_z)\cdot B\hat{\mathbf{e}}_z \\ &\quad + (R_zM_x\hat{\mathbf{e}}_z\hat{\mathbf{e}}_x + R_zM_y\hat{\mathbf{e}}_z\hat{\mathbf{e}}_y + R_zM_z\hat{\mathbf{e}}_z\hat{\mathbf{e}}_z)\cdot B\hat{\mathbf{e}}_z \\ &= \left(R_xM_z\hat{\mathbf{e}}_x + R_yM_z\hat{\mathbf{e}}_y + R_zM_z\hat{\mathbf{e}}_z\right)B\end{aligned} \tag{13.4.10b}$$

According to the last line of Eq. (13.4.10b), the angle average, $\langle\mathbf{RM}\cdot\mathbf{B}\rangle$, is

$$\langle\mathbf{RM}\cdot\mathbf{B}\rangle = \left(\langle R_xM_z\rangle\hat{\mathbf{e}}_x + \langle R_yM_z\rangle\hat{\mathbf{e}}_y + \langle R_zM_z\rangle\hat{\mathbf{e}}_z\right)B \tag{13.4.11}$$

To calculate the last angle average specified by Eq. (13.4.11), we shall exploit an indirect method by which we first evaluate separately the angle averages, $\langle\mathbf{R}\cdot\mathbf{M}\rangle$ and $\langle R_xM_x\rangle + \langle R_yM_y\rangle$, and use the results to determine $\langle R_zM_z\rangle$. We will also show that $\langle R_xM_z\rangle = \langle R_yM_z\rangle = 0$.

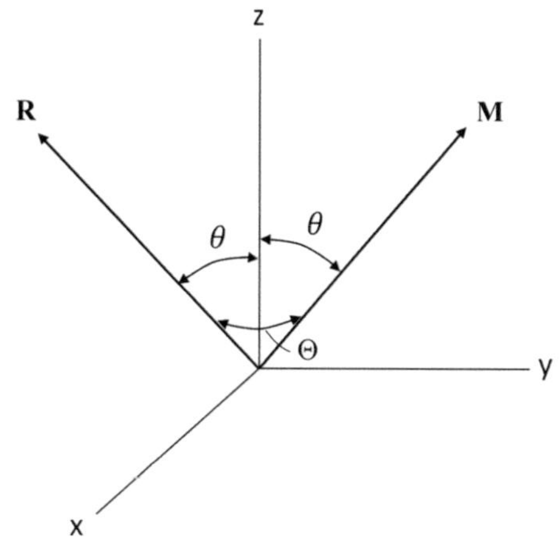

Fig. 13.4 Vectors **R** and **M** are separated by the angle, Θ. Each of **R** and **M** makes an angle, θ, with respect to z-axis

By virtue of the directionality of the chemical bonds, the vectors **R** and **M** are fixed in the molecular frame. The value of the angle, Θ, bounded by **R** and **M** is fixed. A rotation of the molecule as a whole does not change the value of Θ. The two vectors, **R** and **M**, define a plane whose surface is canted at an angle θ with respect to the z-axis. See Fig. 13.4.

The components of the projections of **R** and **M** onto the x, y, z-axes are thus

$$R_x = R\sin(\theta)\cos(\phi_R) \tag{13.4.12a}$$

$$R_y = R\sin(\theta)\sin(\phi_R) \tag{13.4.12b}$$

$$R_z = R\cos(\theta) \tag{13.4.12c}$$

$$M_x = R\sin(\theta)\cos(\phi_M) \tag{13.4.13a}$$

$$M_y = R\sin(\theta)\sin(\phi_M) \tag{13.4.13b}$$

$$M_z = M\,\cos(\theta) \tag{13.4.13c}$$

where θ, as shown in Fig. 13.4, is the polar angle made by their common plane. The angle ϕ_R, not shown in Fig. 13.4, is the angle between the x-axis and the projection of **R** onto the (x,y)-plane. The angle ϕ_M is the angle between the x-axis and the projection of **M** onto the (x,y)- plane. Written out in components, the dot product of **R** and **M**, $RM\cos\Theta$, is

$$\mathbf{R}\cdot\mathbf{M} = RM\cos\Theta = (R_x\hat{\mathbf{e}}_x + R_y\hat{\mathbf{e}}_y + R_z\hat{\mathbf{e}}_z)\cdot(M_x\hat{\mathbf{e}}_x + M_y\hat{\mathbf{e}}_y + M_z\hat{\mathbf{e}}_z)$$

$$= R_xM_x + R_yM_y + R_zM_z \tag{13.4.14}$$

According to Eqs. (13.4.12a)–(13.4.12c) and (13.4.13a)–(13.4.13c), the product, $R_xM_x + R_yM_y$, which involves the vector projections of **R** and **M** onto the (x, y)-plane, can be written as,

$$R_xM_x + R_yM_y = RM\ \sin^2(\theta)\cos(\phi_R)\cos(\phi_M) + RM\ \sin^2(\theta)\sin(\phi_R)\sin(\phi_M) \tag{13.4.15a}$$

Use of the trigonometric identity [See the Mathematical Methods Appendix],

$$\cos(\phi_R - \phi_M) = \cos(\phi_R)\cos(\phi_M) + \sin(\phi_R)\sin(\phi_M) \tag{13.4.15b}$$

allows Eq. (13.4.15a) to be rewritten as

$$R_xM_x + R_yM_y = RM\ \sin^2(\theta)\cos(\phi_R - \phi_M) \tag{13.4.15c}$$

Because both **R** and **M** make an angle θ, with respect to the z-axis, the angle between their projections onto the (x, y)-plane is the angle, Θ. In terms of the angles ϕ_R and ϕ_M, which these projections make with the x-axis, we thus have $\Theta = \phi_R - \phi_M$; hence, Eq. (13.4.15c) can be rewritten as

$$R_xM_x + R_yM_y = RM\ \sin^2(\theta)\cos(\Theta) \tag{13.4.15d}$$

According to Eq. (13.4.14), the angle average of the dot product of **R** with **M**, takes the form,

$$\langle \mathbf{R}\cdot\mathbf{M}\rangle = RM\ \cos\Theta = \langle R_xM_x\rangle + \langle R_yM_y\rangle + \langle R_zM_z\rangle \tag{13.4.16}$$

The normal to the plane formed by **R** and **M** is $\mathbf{N} = (\mathbf{R}\times\mathbf{M})$. Because **R** and **M** each makes an angle of θ with the z-axis, **N** makes an angle $(\pi/2) - \theta$ with the z-axis. Over the angle domain, $0 \le \theta \le \pi$, **N** can assume all possible orientations with respect to the z-axis. As **N** takes on all possible orientations, the product, $R_xM_x + R_yM_y$, traces out its entire range of values.

The total solid angle covered by these rotations is

$$\Omega = \int_0^{2\pi} d\phi \int_0^{\pi} d\theta\ \sin\theta = 4\pi \tag{13.4.17a}$$

Using Eq. (13.4.15d), the solid angle average of the projection of the inner product of **R** and **M** onto the (x, y)-plane is

$$\langle R_xM_x\rangle + \langle R_yM_y\rangle = RM\ \cos\Theta\langle\sin^2\theta\rangle$$

$$= RM \cos\Theta\left(\frac{1}{\Omega}\int_0^{2\pi} d\phi \int_0^{\pi} d\theta \sin\theta \sin^2\theta\right)$$

$$= RM \cos\Theta\left(\frac{1}{\Omega}\int_0^{2\pi} d\phi \int_0^{\pi} d\theta \sin\theta\left(1-\cos^2\theta\right)\right)$$

$$= RM \cos\Theta\left(\frac{2\pi}{\Omega}\right)\left[-\cos\theta+\frac{\cos^3\theta}{3}\right]_0^{\pi}$$

$$= RM \cos\Theta\left(\frac{2\pi}{\Omega}\right)\left[\left(1-\frac{1}{3}\right)-\left(-1+\frac{1}{3}\right)\right]$$

$$= RM \cos\Theta\left(\frac{2\pi}{4\pi}\right)(2)\left(\frac{2}{3}\right) = \frac{2}{3}RM \cos\Theta \tag{13.4.17b}$$

Now that $\langle R_x M_x\rangle + \langle R_y M_y\rangle$ has been calculated, the angle average of $\langle R_z M_z\rangle$ which is required to evaluate Eq. (13.4.11) can be calculated by difference using Eq. (13.4.16). On the basis of Eq. (13.4.16) and (13.4.17b), $\langle R_z M_z\rangle$, is

$$\begin{aligned}\langle R_z M_z\rangle &= RM \cos\Theta - \langle R_x M_x\rangle - \langle R_y M_y\rangle \\ &= RM \cos\Theta - \frac{2}{3}RM \cos\Theta \\ &= \frac{1}{3}RM \cos\Theta = \frac{1}{3}(\mathbf{R}\cdot\mathbf{M})\end{aligned} \tag{13.4.18a}$$

We now turn to computation of the individual averages, $\langle R_x M_z\rangle$ and $\langle R_y M_z\rangle$. By integration over the polar angle, we can show that $\langle R_x M_z\rangle = \langle R_y M_z\rangle = 0$. According Eqs. (13.4.12a), (13.4.13c), and (13.4.12b), we have

$$\langle R_y M_z\rangle \propto \langle R_x M_z\rangle \propto \int_0^{\pi} d\theta \sin\theta \sin\theta \cos\theta = \int_0^{\pi} d(\sin\theta)\sin^2\theta$$

$$= \left[\frac{\sin^3\theta}{3}\right]_0^{\pi} = \frac{1}{3}[0-0] = 0 \tag{13.4.18b}$$

Using Eqs. (13.4.18a) and (13.4.18b), the evaluation of Eq. (13.4.11) proceeds as follows:

$$\begin{aligned}\langle \mathbf{RM}\cdot\mathbf{B}\rangle &= \left(\langle R_x M_z\rangle\hat{\mathbf{e}}_x + \langle R_y M_z\rangle\hat{\mathbf{e}}_y + \langle R_z M_z\rangle\hat{\mathbf{e}}_z\right)B \\ &= (0+0+\langle R_z M_z\rangle\hat{\mathbf{e}}_z)B = \langle R_z M_z\rangle B\hat{\mathbf{e}}_z \\ &= \frac{1}{3}(\mathbf{R}\cdot\mathbf{M})B\hat{\mathbf{e}}_z = \frac{1}{3}(\mathbf{R}\cdot\mathbf{M})\mathbf{B}\end{aligned} \tag{13.4.19}$$

We now refer back to the original task of evaluating the angle average of $\langle\langle a|\mathbf{R}|b\rangle\langle b|\mathbf{M}|a\rangle\cdot\mathbf{B}(0,t)\rangle$. On the basis of Eq. (13.4.19), we conclude that

$$\langle\langle a|\mathbf{R}|b\rangle\langle b|\mathbf{M}|a\rangle\cdot\mathbf{B}(0,t)\rangle=\frac{1}{3}\langle a|\mathbf{R}|b\rangle\cdot\langle b|\mathbf{M}|a\rangle\mathbf{B}(0,t) \tag{13.4.20a}$$

With the replacement of $\mathbf{B}(0,t)$ by $(\partial\mathbf{B}(0,t)/\partial t)$, Eq. (13.4.20a) can be used to evaluate the angle average of $\langle a|\mathbf{R}|b\rangle\langle b|\mathbf{M}|a\rangle\cdot(\partial\mathbf{B}(0,t)/\partial t)$. The result is

$$\left\langle\langle a|\mathbf{R}|b\rangle\langle b|\mathbf{M}|a\rangle\cdot\frac{\partial\mathbf{B}(0,t)}{\partial t}\right\rangle=\frac{1}{3}\langle a|\mathbf{R}|b\rangle\cdot\langle b|\mathbf{M}|a\rangle\frac{\partial\mathbf{B}(0,t)}{\partial t} \tag{13.4.20b}$$

We now turn our attention to the angle average of $\langle a|\mathbf{R}|b\rangle\langle b|\mathbf{R}|a\rangle\cdot\mathbf{E}(0,t)$. Start with the first line of Eq. (13.4.10b), and replace $\mathbf{M}$ by $\mathbf{R}$ and $\mathbf{B}$ by $\mathbf{E}$. The result is $\mathbf{RR}\cdot\mathbf{E}=\mathbf{RR}\cdot E\hat{\mathbf{e}}_z$. By comparison with Eq. (3.4.14), the angle between $\mathbf{R}$ and $\mathbf{R}$ is obviously, $\Theta=0$. With only this change, the steps leading from Eq. (13.4.10b) to Eq. (13.4.20a) can be repeated with the following conclusion:

$$\langle\langle a|\mathbf{R}|b\rangle\cdot\langle b|\mathbf{R}|a\rangle\mathbf{E}(0,t)\rangle=\frac{1}{3}\langle a|\mathbf{R}|b\rangle\cdot\langle b|\mathbf{R}|a\rangle\mathbf{E}(0,t) \tag{13.4.21a}$$

With the replacement of $\mathbf{E}(0,t)$ by $(\partial\mathbf{E}(0,t)/\partial t)$, Eq. (13.4.21a) reads

$$\left\langle\langle a|\mathbf{R}|b\rangle\cdot\langle b|\mathbf{R}|a\rangle\frac{\partial\mathbf{E}(0,t)}{\partial t}\right\rangle=\frac{1}{3}\langle a|\mathbf{R}|b\rangle\cdot\langle b|\mathbf{R}|a\rangle\frac{\partial\mathbf{E}(0,t)}{\partial t} \tag{13.4.21b}$$

Equations (13.4.20a), (13.4.20b) and (13.4.21a) and (13.4.21b) summarize the angle averages associated with of the interaction of the molecule with the electromagnetic field.

13.4.5 The Real Part of the Expectation Value of the Electric Dipole Moment

Upon incorporating the angle averages summarized by Eqs. (13.4.20a) through (13.4.21b) into Eq. (13.4.9b) the result is

$$\begin{aligned}\mathbf{R}'_a&=\frac{2}{3}\mathrm{Re}\sum_b\langle a|\mathbf{R}|b\rangle\cdot\langle b|\mathbf{R}|a\rangle\left[\frac{E^{(o)}_{ba}}{\left((E^{(o)}_{ba})^2-s^2\right)}\right]\mathbf{E}(0,t)\\&+\frac{2}{3}\mathrm{Re}\sum_b\langle a|\mathbf{R}|b\rangle\cdot\langle b|\mathbf{R}|a\rangle\left[\frac{i\hbar\left(E^{(o)}_{ba}\right)^2}{s^2\left((E^{(o)}_{ba})^2-s^2\right)}\right]\frac{\partial\mathbf{E}(0,t)}{\partial t}\end{aligned}$$

$$+\frac{2}{3}\text{Re}\sum_b \langle a|\mathbf{R}|b\rangle \cdot \langle b|\mathbf{M}|a\rangle \left[\frac{E_{ba}^{(o)}}{\left((E_{ba}^{(o)})^2 - s^2\right)}\right]\mathbf{B}(0,t)$$

$$+\frac{2}{3}\text{Re}\sum_b \langle a|\mathbf{R}|b\rangle \cdot \langle b|\mathbf{M}|a\rangle \left[\frac{i\hbar}{\left((E_{ba}^{(o)})^2 - s^2\right)}\right]\frac{\partial \mathbf{B}(0,t)}{\partial t} \tag{13.4.22}$$

To simplify the remaining analysis, we summarize Eq. (13.4.22) in the form,

$$\mathbf{R}_a' = R_E\mathbf{E}(0,t) + R_{\partial E/\partial t}\frac{\partial \mathbf{E}(0,t)}{\partial t} + R_B\mathbf{B}(0,t) + R_{\partial B/\partial t}\frac{\partial \mathbf{B}(0,t)}{\partial t} \tag{13.4.23}$$

where the meanings of the coefficients, R_E,$R_{\partial E/\partial t}$, R_B, and $R_{\partial B/\partial t}$, follow from a term by term comparison of Eq. (13.4.22) with Eq. (13.4.23). Below, we simplify each coefficient in the order in which it appears in Eq. (13.4.23).

Beginning with the coefficient R_E,

$$R_E = \frac{2}{3}\text{Re}\sum_b \langle a|\mathbf{R}|b\rangle \cdot \langle b|\mathbf{R}|a\rangle \left[\frac{E_{ba}^{(o)}}{\left((E_{ba}^{(o)})^2 - s^2\right)}\right] \tag{13.4.24a}$$

we note that $\langle a|\mathbf{R}|b\rangle$ is Hermitian. This permits the matrix element product to be written as

$$\langle a|\mathbf{R}|b\rangle \cdot \langle b|\mathbf{R}|a\rangle = \langle a|\mathbf{R}|b\rangle \cdot \langle a|\mathbf{R}|b\rangle^* = |\langle a|\mathbf{R}|b\rangle|^2 \tag{13.4.24b}$$

Using Eq. (13.4.24b), Eq. (13.4.24a) reads,

$$R_E = \frac{2}{3}\text{Re}\sum_b |\langle a|\mathbf{R}|b\rangle|^2 \left[\frac{E_{ba}^{(o)}}{\left((E_{ba}^{(o)})^2 - s^2\right)}\right] \tag{13.4.24c}$$

Now use $h\nu_{ba} = E_{ba}^{(o)}$, and set $\omega = 2\pi\nu$ and $h = 2\pi\hbar$, so that $s = \hbar\omega = h\nu$. With these changes, Eq. (13.4.24c) becomes,

$$R_E = \frac{2}{3}\text{Re}\sum_b |\langle a|\mathbf{R}|b\rangle|^2 \left[\frac{h\nu_{ba}}{\left(h^2\nu_{ba}^2 - h^2\nu^2\right)}\right] \tag{13.4.24d}$$

After cancellation of the factors of h from the numerator and denominator of the fraction in Eq. (13.4.24d), the result is

$$R_E = \frac{2}{3h}\text{Re}\sum_b |\langle a|\mathbf{R}|b\rangle|^2 \left[\frac{\nu_{ba}}{\left(\nu_{ba}^2 - \nu^2\right)}\right] \tag{13.4.24e}$$

As every term in the sum on the right hand side of Eq. (13.4.24e) is real, we can drop the designation "Re" and simply write Eq. (13.4.24e) as

$$R_E = \frac{2}{3h}\sum_b\left[\frac{\nu_{ba}|\langle a|\mathbf{R}|b\rangle|^2}{\left(\nu_{ba}^2 - \nu^2\right)}\right] \tag{13.4.25}$$

We turn next to $R_{\partial E/\partial t}$, which reads,

$$R_{\partial E/\partial t} = \frac{2}{3}\mathrm{Re}\sum_b \langle a|\mathbf{R}|b\rangle\cdot\langle b|\mathbf{R}|a\rangle\left[\frac{i\hbar\left(E_{ba}^{(o)}\right)^2}{s^2\left((E_{ba}^{(o)})^2 - s^2\right)}\right] \tag{13.4.26a}$$

We use the Hermitian property of $\langle a|\mathbf{R}|b\rangle$ to write $R_{\partial E/\partial t}$ in the form,

$$R_{\partial E/\partial t} = \frac{2}{3}\mathrm{Re}\sum_b |\langle a|\mathbf{R}|b\rangle|^2\left[\frac{i\hbar\left(E_{ba}^{(o)}\right)^2}{s^2\left((E_{ba}^{(o)})^2 - s^2\right)}\right] \tag{13.4.26b}$$

The right hand side of Eq. (13.4.26b) is seen to be purely imaginary. We conclude that

$$R_{\partial E/\partial t} = 0 \tag{13.4.27}$$

Now consider R_B, which reads

$$R_B = \frac{2}{3}\mathrm{Re}\sum_b \langle a|\mathbf{R}|b\rangle\cdot\langle b|\mathbf{M}|a\rangle\left[\frac{E_{ba}^{(o)}}{\left((E_{ba}^{(o)})^2 - s^2\right)}\right] \tag{13.4.28}$$

Upon use of $s = h\nu$ and $h\nu_{ba} = E_{ba}^{(o)}$, Eq. (13.4.28) becomes,

$$R_B = \frac{2}{3h}\sum_b \mathrm{Re}\{\langle a|\mathbf{R}|b\rangle\cdot\langle b|\mathbf{M}|a\rangle\}\left[\frac{\nu_{ba}}{\left(\nu_{ba}^2 - \nu^2\right)}\right] \tag{13.4.29}$$

Upon comparison of Eq. (13.4.22) and Eq. (13.4.23), the last coefficient to be evaluated is

$$R_{\partial B/\partial t} = \frac{2}{3}\mathrm{Re}\sum_b \langle a|\mathbf{R}|b\rangle\cdot\langle b|\mathbf{M}|a\rangle\left[\frac{i\hbar}{\left((E_{ba}^{(o)})^2 - s^2\right)}\right] \tag{13.4.30a}$$

With the substitution of $\hbar = h/2\pi$, $s = h\nu$ and $h\nu_{ba} = E_{ba}^{(o)}$, Eq. (13.4.30a) can be rewritten as

$$R_{\partial B/\partial t} = \frac{2}{3}\mathrm{Re}\sum_b \langle a|\mathbf{R}|b\rangle \cdot \langle b|\mathbf{M}|a\rangle \left[\frac{ih/2\pi}{\left(h^2\nu_{ba}^2 - h^2\nu^2\right)}\right]$$
$$= \frac{1}{3\pi h}\mathrm{Re}\sum_b \frac{i\langle a|\mathbf{R}|b\rangle \cdot \langle b|\mathbf{M}|a\rangle}{\left(\nu_{ba}^2 - \nu^2\right)} \tag{13.4.30b}$$

If we let $z = x + iy$ be a complex number, then $iz = i(x + iy) = ix - y$, from which we conclude $\mathrm{Re}(iz) = -\mathrm{Im}(z)$. If we let $z = \langle a|\mathbf{R}|b\rangle \cdot \langle b|\mathbf{M}|a\rangle$, this identity permits us write,

$$\mathrm{Re}\{i[\langle a|\mathbf{R}|b\rangle \cdot \langle b|\mathbf{M}|a\rangle]\} = -\mathrm{Im}\{[\langle a|\mathbf{R}|b\rangle \cdot \langle b|\mathbf{M}|a\rangle]\} \tag{13.4.30c}$$

Upon substitution of Eq. (13.4.30c), Eq. (13.4.30b) reads,

$$R_{\partial B/\partial t} = -\frac{1}{3\pi h}\mathrm{Re}\sum_b \frac{\mathrm{Im}\{\langle a|\mathbf{R}|b\rangle \cdot \langle b|\mathbf{M}|a\rangle\}}{\left(\nu_{ba}^2 - \nu^2\right)} \tag{13.4.31}$$

Among the results in Eqs. (13.4.25), (13.4.27), (13.4.29), and (13.4.31), only R_E, R_B, and $R_{\partial B/\partial t}$ were found to be non-zero. We summarize below:

$$R_E = \frac{2}{3h}\sum_b \left[\frac{\nu_{ba}|\langle a|\mathbf{R}|b\rangle|^2}{\left(\nu_{ba}^2 - \nu^2\right)}\right] \tag{13.4.32a}$$

$$R_B = \frac{2}{3h}\sum_b \left[\frac{\nu_{ba}\mathrm{Re}\{\langle a|\mathbf{R}|b\rangle \cdot \langle b|\mathbf{M}|a\rangle\}}{\left(\nu_{ba}^2 - \nu^2\right)}\right] \tag{13.4.32b}$$

$$R_{\partial B/\partial t} = -\frac{1}{3\pi h}\sum_b \left[\frac{\mathrm{Im}\{\langle a|\mathbf{R}|b\rangle \cdot \langle b|\mathbf{M}|a\rangle\}}{\left(\nu_{ba}^2 - \nu^2\right)}\right] \tag{13.4.32c}$$

Upon substitution of Eqs. (13.4.32a) through (13.4.32c) into Eq. (13.4.23), our final result for $\mathbf{R}'_a$ is

$$\mathbf{R}'_a = \frac{2}{3h}\sum_b \left[\frac{\nu_{ba}|\langle a|\mathbf{R}|b\rangle|^2}{\left(\nu_{ba}^2 - \nu^2\right)}\right]\mathbf{E}(0,t) + \frac{2}{3h}\sum_b \left[\frac{\nu_{ba}\mathrm{Re}\{\langle a|\mathbf{R}|b\rangle \cdot \langle b|\mathbf{M}|a\rangle\}}{\left(\nu_{ba}^2 - \nu^2\right)}\right]\mathbf{B}(0,t)$$
$$-\frac{1}{3\pi h}\sum_b \left[\frac{\mathrm{Im}\{\langle a|\mathbf{R}|b\rangle \cdot \langle b|\mathbf{M}|a\rangle\}}{\left(\nu_{ba}^2 - \nu^2\right)}\right]\frac{\partial \mathbf{B}(0,t)}{\partial t} \tag{13.4.33}$$

In terms of our simplified notation, Eq. (13.4.33) reads,

$$\mathbf{R}'_a = R_E\mathbf{E}(0,t) + R_B\mathbf{B}(0,t) + R_{\partial B/\partial t}\frac{\partial \mathbf{B}(0,t)}{\partial t} \tag{13.4.34}$$

where we have reprised Eq. (13.4.23). Equations (13.4.33) and (13.4.34) complete the calculation of the angle averaged value of the induced electric dipole moment.

13.5 Calculation of the Induced Magnetic Dipole Moment

13.5.1 Analogy with the Induced Electric Dipole Moment

If we substitute **M** for **R** in Eq. (13.4.1) and follow the subsequent derivation, it is clear that the left-most appearance of $\langle a|\mathbf{R}|b\rangle$ in each of the sums in Eq. (13.4.9b) gets replaced by $\langle a|\mathbf{M}|b\rangle$. Thus, the magnetic dipole moment $\mathbf{M}'_a$ can be represented formally by replacing $\mathbf{R}'_a$ on the left hand side of Eq. (13.4.22) by $\mathbf{M}'_a$, and on the right hand side, by replacing $\langle a|\mathbf{R}|b\rangle \cdot \langle b|\mathbf{R}|a\rangle$ by $\langle a|\mathbf{M}|b\rangle \cdot \langle b|\mathbf{R}|a\rangle$, and replacing $\langle a|\mathbf{R}|b\rangle \cdot \langle b|\mathbf{M}|a\rangle$ by $\langle a|\mathbf{M}|b\rangle \cdot \langle b|\mathbf{M}|a\rangle$. The angle average of $\langle a|\mathbf{M}|b\rangle \cdot \langle b|\mathbf{R}|a\rangle$ is the same as the angle average of $\langle a|\mathbf{R}|b\rangle \cdot \langle b|\mathbf{M}|a\rangle$, while the angle average of $\langle a|\mathbf{R}|b\rangle \cdot \langle b|\mathbf{R}|a\rangle$ is the same as the angle average of $\langle a|\mathbf{M}|b\rangle \cdot \langle b|\mathbf{M}|a\rangle$. With these substitutions, Eq. (13.4.22) is transformed to read,

$$\begin{aligned}
\mathbf{M}'_a = {} & \frac{2}{3}\mathrm{Re}\sum_b \langle a|\mathbf{M}|b\rangle \cdot \langle b|\mathbf{R}|a\rangle \left[\frac{E_{ba}^{(o)}}{\left((E_{ba}^{(o)})^2 - s^2\right)}\right]\mathbf{E}(0,t) \\
& + \frac{2}{3}\mathrm{Re}\sum_b \langle a|\mathbf{M}|b\rangle \cdot \langle b|\mathbf{R}|a\rangle \left[\frac{i\hbar\left(E_{ba}^{(o)}\right)^2}{s^2\left((E_{ba}^{(o)})^2 - s^2\right)}\right]\frac{\partial \mathbf{E}(0,t)}{\partial t} \\
& + \frac{2}{3}\mathrm{Re}\sum_b \langle a|\mathbf{M}|b\rangle \cdot \langle b|\mathbf{M}|a\rangle \left[\frac{E_{ba}^{(o)}}{\left((E_{ba}^{(o)})^2 - s^2\right)}\right]\mathbf{B}(0,t) \\
& + \frac{2}{3}\mathrm{Re}\sum_b \langle a|\mathbf{M}|b\rangle \cdot \langle b|\mathbf{M}|a\rangle \left[\frac{i\hbar}{\left((E_{ba}^{(o)})^2 - s^2\right)}\right]\frac{\partial \mathbf{B}(0,t)}{\partial t}
\end{aligned} \tag{13.5.1}$$

According to Eq. (13.5.1), the magnetic analog of Eq. (13.4.23) is

$$\mathbf{M}'_a = M_E\mathbf{E}(0,t) + M_{\partial E/\partial t}\frac{\partial \mathbf{E}(0,t)}{\partial t} + M_B\mathbf{B}(0,t) + M_{\partial B/\partial t}\frac{\partial \mathbf{B}(0,t)}{\partial t} \tag{13.5.2}$$

The meaning of the coefficients, M_E,$M_{\partial E/\partial t}$, M_B, and $M_{\partial B/\partial t}$ can be found by a comparing Eqs. (13.5.1) and (13.5.2) term by term.

13.5.2 The Real Part of the Expectation Value of the Magnetic Dipole Moment

We begin with M_E, which can be identified as

$$M_E = \frac{2}{3}\mathrm{Re}\sum_b \langle a|\mathbf{M}|b\rangle \cdot \langle b|\mathbf{R}|a\rangle \left[\frac{E_{ba}^{(o)}}{\left((E_{ba}^{(o)})^2 - s^2\right)}\right] \tag{13.5.3a}$$

Upon substitution of $h\nu_{ba} = E_{ba}$ and $h\nu = s$, Eq. (13.5.3a) reads

$$M_E = \frac{2}{3h}\sum_b \left[\frac{\nu_{ba}\mathrm{Re}\{\langle a|\mathbf{M}|b\rangle \cdot \langle b|\mathbf{R}|a\rangle\}}{\left(\nu_{ba}^2 - \nu^2\right)}\right] \tag{13.5.3b}$$

By virtue of the Hermitian property of the operators, we have

$$\langle a|\mathbf{M}|b\rangle \cdot \langle b|\mathbf{R}|a\rangle = \langle b|\mathbf{M}|a\rangle^* \cdot \langle a|\mathbf{R}|b\rangle^* = [\langle b|\mathbf{M}|a\rangle \cdot \langle a|\mathbf{R}|b\rangle]^* \tag{13.5.3c}$$

According to of Eq. (13.5.3c),

$$\mathrm{Re}\{\langle a|\mathbf{M}|b\rangle \cdot \langle b|\mathbf{R}|a\rangle\} = \mathrm{Re}\left\{[\langle b|\mathbf{M}|a\rangle \cdot \langle a|\mathbf{R}|b\rangle]^*\right\} \tag{13.5.3d}$$

Given a complex number, z, it is the case that $\mathrm{Re}(z) = \mathrm{Re}(z^*)$, then using Eq. (13.5.3d) and commuting the vectors in a dot product, we obtain the results,

$$\mathrm{Re}\left\{[\langle b|\mathbf{M}|a\rangle \cdot \langle a|\mathbf{R}|b\rangle]^*\right\} = \mathrm{Re}\{\langle b|\mathbf{M}|a\rangle \cdot \langle a|\mathbf{R}|b\rangle\} = \mathrm{Re}\{\langle a|\mathbf{R}|b\rangle \cdot \langle b|\mathbf{M}|a\rangle\} \tag{13.5.3e}$$

Upon substitution of Eq. (13.5.3e) into Eq. (13.5.3b), we obtain

$$M_E = \frac{2}{3h}\sum_b \left[\frac{\nu_{ba}\mathrm{Re}\{\langle a|\mathbf{R}|b\rangle \cdot \langle b|\mathbf{M}|a\rangle\}}{\left(\nu_{ba}^2 - \nu^2\right)}\right] \tag{13.5.4}$$

Likewise, by comparison of Eqs. (13.5.1) and (13.5.2), the coefficient, $M_{\partial E/\partial t}$, is

$$M_{\partial E/\partial t} = \frac{2}{3}\mathrm{Re}\sum_b i\langle a|\mathbf{M}|b\rangle \cdot \langle b|\mathbf{R}|a\rangle\hbar\left(E_{ba}^{(o)}\right)^2\left[\frac{1}{s^2\left((E_{ba}^{(o)})^2 - s^2\right)}\right] \tag{13.5.5a}$$

Expand the fraction in the parentheses in Eq. (13.5.5a) by the method of partial fractions. The procedure begins with

$$\frac{1}{s^2\left((E_{ba}^{(o)})^2 - s^2\right)} = \frac{\widehat{A}}{s^2} + \frac{\widehat{B}}{\left((E_{ba}^{(o)})^2 - s^2\right)} \tag{13.5.5b}$$

Multiply both sides of Eq. (13.5.5b) through by $s^2\left((E_{ba}^{(o)})^2 - s^2\right)$ to obtain

$$1 = \widehat{A}\left((E_{ba}^{(o)})^2 - s^2\right) + \widehat{B}s^2 \tag{13.5.5c}$$

Equation (13.5.5c) holds for all values of s. Let $s = 0$. The result is

$$1 = \widehat{A}\left(E_{ba}^{(o)}\right)^2 \tag{13.5.5d}$$

from which the value of $\widehat{A}$ follows:

$$\widehat{A} = \frac{1}{(E_{ba}^{(o)})^2} \tag{13.5.5e}$$

Let $s = E_{ba}^{(o)}$ in Eq. (13.5.5c). The result is

$$1 = \widehat{B}(E_{ba}^{(o)})^2 \tag{13.5.5f}$$

from which it follows that

$$\widehat{B} = \frac{1}{(E_{ba}^{(o)})^2} \tag{13.5.5g}$$

Upon substitution of Eqs. (13.5.5e) and (13.5.5g) for $\widehat{A}$ and $\widehat{B}$, respectively, into Eq. (13.5.5b), we have as our final result,

$$\frac{1}{s^2\left((E_{ba}^{(o)})^2 - s^2\right)} = \frac{1}{\left(E_{ba}^{(o)}\right)^2 s^2} + \frac{1}{\left(E_{ba}^{(o)}\right)^2\left((E_{ba}^{(o)})^2 - s^2\right)} \tag{13.5.5h}$$

Substitute the partial fraction expansion in Eq. (13.5.5h) into Eq. (13.5.5a). The result is

$$M_{\partial E/\partial t} = \frac{2}{3}\mathrm{Re}\sum_b i\langle a|\mathbf{M}|b\rangle$$
$$\cdot\ \langle b|\mathbf{R}|a\rangle\hbar\left(E_{ba}^{(o)}\right)^2\left[\frac{1}{\left(E_{ba}^{(o)}\right)^2 s^2} + \frac{1}{\left(E_{ba}^{(o)}\right)^2\left((E_{ba}^{(o)})^2 - s^2\right)}\right] \tag{13.5.6a}$$

After cancellation of factors of $\left(E_{ba}^{(o)}\right)^2$, we can rewrite Eq. (13.5.6a) as two sums:

$$M_{\partial E/\partial t} = \frac{2}{3}\mathrm{Re}\left(\frac{i\hbar}{s^2}\sum_b \langle a|\mathbf{M}|b\rangle \cdot \langle b|\mathbf{R}|a\rangle\right) + \frac{2}{3}\left[\hbar\sum_b \frac{\mathrm{Re}\{i\langle a|\mathbf{M}|b\rangle \cdot \langle b|\mathbf{R}|a\rangle\}}{\left((E_{ba}^{(o)})^2 - s^2\right)}\right] \tag{13.5.6b}$$

By the law of matrix multiplication, the first sum in Eq. (13.5.6b) can be recognized as the diagonal element, $\langle a|\mathbf{M}\cdot\mathbf{R}|a\rangle$, of the matrix, $\mathbf{M}\cdot\mathbf{R}$. That is,

$$\langle a|\mathbf{M}\cdot\mathbf{R}|a\rangle = \sum_b \langle a|\mathbf{M}|b\rangle \cdot \langle b|\mathbf{R}|a\rangle \tag{13.5.6c}$$

In the second sum in Eq. (13.5.6b), we can use the Hermitian property to cast the product of matrix elements into the form,

$$\begin{aligned}\langle a|\mathbf{M}|b\rangle \cdot \langle b|\mathbf{R}|a\rangle &= \langle b|\mathbf{M}|a\rangle^* \cdot \langle a|\mathbf{R}|b\rangle^* \\ &= \langle a|\mathbf{R}|b\rangle^* \cdot \langle b|\mathbf{M}|a\rangle^* = [\langle a|\mathbf{R}|b\rangle \cdot \langle b|\mathbf{M}|a\rangle]^*\end{aligned} \tag{13.5.6d}$$

Using Eq. (13.5.6d), we can write

$$\mathrm{Re}\{i\langle a|\mathbf{M}|b\rangle \cdot \langle b|\mathbf{R}|a\rangle\} = \mathrm{Re}\left\{i[\langle a|\mathbf{R}|b\rangle \cdot \langle b|\mathbf{M}|a\rangle]^*\right\} \tag{13.5.6e}$$

If we note that because $\mathrm{Re}(iz^*) = \mathrm{Im}(z)$, (13.5.6e) reads,

$$\mathrm{Re}\{i\langle a|\mathbf{M}|b\rangle \cdot \langle b|\mathbf{R}|a\rangle\} = \mathrm{Re}\left\{i[\langle a|\mathbf{R}|b\rangle \cdot \langle b|\mathbf{M}|a\rangle]^*\right\} = \mathrm{Im}\{\langle a|\mathbf{R}|b\rangle \cdot \langle b|\mathbf{M}|a\rangle\} \tag{13.5.6f}$$

Now put Eqs. (13.5.6c) and (13.5.6f) into Eq. (13.5.6b), with the result,

$$M_{\partial E/\partial t} = \frac{2}{3}\mathrm{Re}\left(\frac{i\hbar}{s^2}\langle a|\mathbf{M}\cdot\mathbf{R}|a\rangle\right) + \frac{2}{3}\left(\hbar\sum_b \frac{\mathrm{Im}\{\langle a|\mathbf{R}|b\rangle \cdot \langle b|\mathbf{M}|a\rangle\}}{\left((E_{ba}^{(o)})^2 - s^2\right)}\right) \tag{13.5.7a}$$

As $\langle a|\mathbf{M}\cdot\mathbf{R}|a\rangle$ is physically observable, it is real; hence, the first term on the right hand side of Eq. (13.5.7a) is purely imaginary, so its real part is zero. Upon substitution of $\hbar = h/2\pi$, $h\nu_{ba} = E_{ba}^{(o)}$ and $h\nu = s$ into the second term on the right of Eq. (13.5.7a), we obtain

$$M_{\partial E/\partial t} = \frac{2}{3}\left(\frac{h}{2\pi}\sum_b \frac{\mathrm{Im}\{\langle a|\mathbf{R}|b\rangle \cdot \langle b|\mathbf{M}|a\rangle\}}{\left(h^2\nu_{ba}^2 - h^2\nu^2\right)}\right) \tag{13.5.7b}$$

After cancellation of common factors of h, the final result for $M_{\partial E/\partial t}$ is

$$M_{\partial E/\partial t} = \frac{1}{3\pi h} \sum_b \frac{\mathrm{Im}\{\langle a|\mathbf{R}|b\rangle \cdot \langle b|\mathbf{M}|a\rangle\}}{\left(\nu_{ba}^2 - \nu^2\right)} \tag{13.5.8}$$

Upon comparing Eqs. (13.5.1) and (13.5.2), the coefficient, M_B, can be identified as

$$M_B = \frac{2}{3}\mathrm{Re}\sum_b \langle a|\mathbf{M}|b\rangle \cdot \langle b|\mathbf{M}|a\rangle \left[\frac{E_{ba}^{(o)}}{\left((E_{ba}^{(o)})^2 - s^2\right)}\right] \tag{13.5.9a}$$

The Hermitian property can be used to write,

$$\langle a|\mathbf{M}|b\rangle \cdot \langle b|\mathbf{M}|a\rangle = \langle a|\mathbf{M}|b\rangle \cdot \langle a|\mathbf{M}|b\rangle^* = |\langle a|\mathbf{M}|b\rangle|^2 \tag{13.5.9b}$$

Substitute $h\nu_{ba} = E_{ba}^{(o)}$ and $h\nu = s$, and Eq. (13.5.9b) into Eq. (13.5.9a). By virtue of Eq. (13.5.9b), every term in Eq. (13.5.9a) can be seen to be real. As a result, Eq. (13.5.9a) simplifies to read,

$$M_B = \frac{2}{3h}\sum_b \left[\frac{\nu_{ba}|\langle a|\mathbf{M}|b\rangle|^2}{\left(\nu_{ba}^2 - \nu^2\right)}\right] \tag{13.5.10}$$

According to a comparison of Eqs. (13.5.1) and (13.5.2), the fourth and last term in Eq. (13.5.2) is given by

$$M_{\partial B/\partial t} = \frac{2}{3}\mathrm{Re}\sum_b \langle a|\mathbf{M}|b\rangle \cdot \langle b|\mathbf{M}|a\rangle \left[\frac{i\hbar}{\left((E_{ba}^{(o)})^2 - s^2\right)}\right] \tag{13.5.11a}$$

Upon substitution of Eq. (13.5.9b), Eq. (13.5.11a) reads

$$M_{\partial B/\partial t} = \frac{2}{3}\mathrm{Re}\sum_b \left[\frac{i\hbar|\langle a|\mathbf{M}|b\rangle|^2}{\left((E_{ba}^{(o)})^2 - s^2\right)}\right] \tag{13.5.11b}$$

Because each term in the sum is purely imaginary, we conclude that

$$M_{\partial B/\partial t} = 0 \tag{13.5.12}$$

Equations (13.5.4), (13.5.8), and (13.5.10) are formulae for the non-zero coefficients, M_E,$M_{\partial E/\partial t}$, and M_B, respectively, which appear in Eq. (13.5.2). Summarized, these equations read,

$$M_E = \frac{2}{3h}\sum_b \left[\frac{\nu_{ba}\mathrm{Re}\{\langle a|\mathbf{R}|b\rangle \cdot \langle b|\mathbf{M}|a\rangle\}}{\left(\nu_{ba}^2 - \nu^2\right)}\right] \tag{13.5.13a}$$

$$M_{\partial E/\partial t} = \frac{1}{3\pi h}\sum_b \left[\frac{\mathrm{Im}\{\langle a|\mathbf{R}|b\rangle \cdot \langle b|\mathbf{M}|a\rangle\}}{\left(\nu_{ba}^2 - \nu^2\right)}\right] \tag{13.5.13b}$$

$$M_B = \frac{2}{3h}\sum_b \left[\frac{\nu_{ba}|\langle a|\mathbf{M}|b\rangle|^2}{\left(\nu_{ba}^2 - \nu^2\right)}\right] \tag{13.5.13c}$$

Equations (13.5.13a) through (13.5.13c) permit Eq. (13.5.1) to be rewritten in the form,

$$\begin{aligned}\mathbf{M}_a' &= \frac{2}{3h}\sum_b \left[\frac{\nu_{ba}\mathrm{Re}\{\langle a|\mathbf{R}|b\rangle \cdot \langle b|\mathbf{M}|a\rangle\}}{\left(\nu_{ba}^2 - \nu^2\right)}\right]\mathbf{E}(0,t)\\ &+ \frac{1}{3\pi h}\sum_b \left[\frac{\mathrm{Im}\{\langle a|\mathbf{R}|b\rangle \cdot \langle b|\mathbf{M}|a\rangle\}}{\left(\nu_{ba}^2 - \nu^2\right)}\right]\frac{\partial \mathbf{E}(0,t)}{\partial t}\\ &+ \frac{2}{3h}\sum_b \left[\frac{\nu_{ba}|\langle a|\mathbf{M}|b\rangle|^2}{\left(\nu_{ba}^2 - \nu^2\right)}\right]\mathbf{B}(0,t)\end{aligned} \tag{13.5.14}$$

With the identifications in Eqs. (13.5.13a) through (13.5.13c), Eq. (13.5.14) can be summarized as

$$\mathbf{M}_a' = M_E\mathbf{E}(0,t) + M_{\partial E/\partial t}\frac{\partial \mathbf{E}(0,t)}{\partial t} + M_B\mathbf{B}(0,t) \tag{13.5.15}$$

Equations (13.5.14) and (13.5.15) complete the calculation of the value of the angle averaged induced magnetic dipole moment.

13.6 Summary of Results

Equation (13.4.33) expresses $\mathbf{R}_a'$ in terms of three sums, while Eq. (13.5.14) expresses $\mathbf{M}_a'$ in terms of three sums. A perusal of the six sums reveals that the coefficient of $\mathbf{B}(0, t)$ in Eq. (13.4.33) is the same as the coefficient of $\mathbf{E}(0, t)$ in Eq. (13.5.14). Likewise, except for sign, the coefficient of $\partial\mathbf{B}(0, t)/\partial t$ in Eq. (13.4.33) is the same as the coefficient of $\partial\mathbf{E}(0, t)/\partial t$ in Eq. (13.5.14). Thus, there are only four independent sums. Following Eyring et al. [1, 2], we identify them as,

$$\alpha_a = \frac{2}{3h}\sum_b \left[\frac{\nu_{ba}|\langle a|\mathbf{R}|b\rangle|^2}{\left(\nu_{ba}^2 - \nu^2\right)}\right] \tag{13.6.1a}$$

$$\kappa_a = \frac{2}{3h}\sum_b \left[\frac{\nu_{ba}|\langle a|\mathbf{M}|b\rangle|^2}{\left(\nu_{ba}^2 - \nu^2\right)}\right] \tag{13.6.1b}$$

$$\gamma_a = \frac{2}{3h}\sum_b \left[\frac{\nu_{ba}\mathrm{Re}\{\langle a|\mathbf{R}|b\rangle \cdot \langle b|\mathbf{M}|a\rangle\}}{\left(\nu_{ba}^2 - \nu^2\right)}\right] \tag{13.6.1c}$$

$$\beta_a = \frac{1}{3\pi h}\left[\sum_b \frac{\mathrm{Im}\{\langle a|\mathbf{R}|b\rangle \cdot \langle b|\mathbf{M}|a\rangle\}}{\left(\nu_{ba}^2 - \nu^2\right)}\right] \tag{13.6.1d}$$

With these definitions, Eq. (13.4.33) for $\mathbf{R}_a'$ can be rewritten as,

$$\mathbf{R}_a' = \alpha_a\mathbf{E} + \gamma_a\mathbf{B} - \beta_a\frac{\partial \mathbf{B}}{\partial t} \tag{13.6.2}$$

and Eq. (13.5.14) for $\mathbf{M}_a'$ can be rewritten as,

$$\mathbf{M}_a' = \kappa_a\mathbf{B} + \gamma_a\mathbf{E} + \beta_a\frac{\partial \mathbf{E}}{\partial t} \tag{13.6.3}$$

13.7 Evaluation of the Electromagnetic Fields

Experiment shows that the value of γ_a is negligible, so that Eqs. (13.6.2) and (13.6.3) can be further simplified to read,

$$\mathbf{R}_a' = \alpha_a\mathbf{E} - \beta_a\frac{\partial \mathbf{B}}{\partial t} \tag{13.7.1}$$

and

$$\mathbf{M}_a' = \kappa_a\mathbf{B} + \beta_a\frac{\partial \mathbf{E}}{\partial t} \tag{13.7.2}$$

respectively. In Eq. (13.7.1), α_a functions as an effective molecular polarizability. In Eq. (13.7.2), κ_a functions as an effective magnetic susceptibility. Of the two, α_a has a greater effect than κ_a. The coefficient, β_a, defined in Eq. (13.6.1d), represents the coupling between the electric and magnetic effects. By virtue of its appearance in Eq. (13.7.1), β_a indicates that part of the induced electric dipole moment, $\mathbf{R}_a'$, has its origin in the time derivative of the magnetic induction, while its appearance in Eq. (13.7.2) indicates that part of the induced magnetic dipole moment, $\mathbf{M}_a'$, has its origin in the time derivative of the electric field. Although the presence of β_a is essential to the coupling between the electric and magnetic effects, it is ordinarily smaller in magnitude than either α_a and κ_a. This recognition permits the final results to be greatly simplified.

13.7.1 Calculation of D

In order to use $\mathbf{R}_a'$ to analyze the optical rotation of molecules in condensed matter, we need to replace $\mathbf{E}$ with the Lorentz local field, $\mathbf{F}$. With this change, Eq. (13.7.1) reads,

$$\mathbf{R}_a' = \alpha_a \mathbf{F} - \beta_a \frac{\partial \mathbf{B}}{\partial t} \tag{13.7.3}$$

We recall from Eq. (3.5.4) of Chap. 3 that in the Lorenz–Lorentz model

$$\mathbf{F} = \mathbf{E} + \frac{\mathbf{P}}{3\varepsilon_o} \tag{13.7.4}$$

Substitute Eq. (13.7.4) into Eq. (13.7.3), to obtain

$$\mathbf{R}_a' = \alpha_a \left(\mathbf{E} + \frac{\mathbf{P}}{3\varepsilon_o} \right) - \beta_a \frac{\partial \mathbf{B}}{\partial t} \tag{13.7.5}$$

Since $\mathbf{R}_a'$ is the electric dipole moment induced in an individual molecule, the macroscopic polarization is

$$\mathbf{P} = N\mathbf{R}_a' \tag{13.7.6}$$

where N is the molecular number density (molecules per unit volume). Substitution of Eq. (13.7.6) into Eq. (13.7.5) has the result,

$$\mathbf{R}_a' = \alpha_a \left(\mathbf{E} + \frac{N\mathbf{R}_a'}{3\varepsilon_o} \right) - \beta_a \frac{\partial \mathbf{B}}{\partial t} \tag{13.7.7}$$

Equation (13.7.7) can be solved for $\mathbf{R}_a'$ using the following algebra:

$$\mathbf{R}_a' \left(1 - \frac{N\alpha_a}{3\varepsilon_o} \right) = \alpha_a \mathbf{E} - \beta_a \frac{\partial \mathbf{B}}{\partial t} \tag{13.7.8a}$$

The desired result is

$$\mathbf{R}_a' = \left[\frac{\alpha_a}{1 - (N\alpha_a/3\varepsilon_o)} \right] \mathbf{E} - \left[\frac{\beta_a}{1 - (N\alpha_a/3\varepsilon_o)} \right] \frac{\partial \mathbf{B}}{\partial t} \tag{13.7.8b}$$

As α_a serves as the molecular polarizability, the Clausius–Mossotti equation, which is Eq. (3.5.11) of Chap. 3, predicts for the relative permittivity, ε_r,

$$\frac{\varepsilon_r - 1}{\varepsilon_r + 2} = \frac{N\alpha_a}{3\varepsilon_o} \tag{13.7.9}$$

Equation (13.7.9) can be rearranged to read,

$$1 - \frac{N\alpha_a}{3\varepsilon_o} = 1 - \frac{\varepsilon_r - 1}{\varepsilon_r + 2} = \frac{3}{\varepsilon_r + 2} \tag{13.7.10}$$

Upon substitution of Eq. (13.7.10) into Eq. (13.7.8b), we obtain

$$\mathbf{R}'_a = \alpha_a\left(\frac{\varepsilon_r + 2}{3}\right)\mathbf{E} - \beta_a\left(\frac{\varepsilon_r + 2}{3}\right)\frac{\partial \mathbf{B}}{\partial t} \tag{13.7.11}$$

According to Eq. (7.2.13) of Chap. 7, the electric displacement is given by

$$\mathbf{D} = \varepsilon_o\mathbf{E} + \mathbf{P} \tag{13.7.12}$$

Upon substitution of Eq. (13.7.6) into Eq. (13.7.12), the result is

$$\mathbf{D} = \varepsilon_o\mathbf{E} + N\mathbf{R}'_a \tag{13.7.13a}$$

Substitute Eq. (13.7.11) into Eq. (13.7.13a) to obtain,

$$\mathbf{D} = \varepsilon_o\mathbf{E} + N\alpha_a\left(\frac{\varepsilon_r + 2}{3}\right)\mathbf{E} - N\beta_a\left(\frac{\varepsilon_r + 2}{3}\right)\frac{\partial \mathbf{B}}{\partial t} \tag{13.7.13b}$$

Use Eq. (13.7.9) to replace $N\alpha_a$ in the second term on the right in Eq. (13.7.13b). The result is

$$\mathbf{D} = \varepsilon_o\mathbf{E} + 3\varepsilon_o\left(\frac{\varepsilon_r - 1}{\varepsilon_r + 2}\right)\left(\frac{\varepsilon_r + 2}{3}\right)\mathbf{E} - N\beta_a\left(\frac{\varepsilon_r + 2}{3}\right)\frac{\partial \mathbf{B}}{\partial t} \tag{13.7.13c}$$

After cancellation of like terms, Eq. (13.7.13c) reads,

$$\mathbf{D} = \varepsilon_o\mathbf{E} + \varepsilon_o(\varepsilon_r - 1)\mathbf{E} - N\beta_a\left(\frac{\varepsilon_r + 2}{3}\right)\frac{\partial \mathbf{B}}{\partial t} \tag{13.7.13d}$$

Upon combination of the first two terms in Eq. (13.7.13d), the result is

$$\mathbf{D} = \varepsilon_r\varepsilon_o\mathbf{E} - N\beta_a\left(\frac{\varepsilon_r + 2}{3}\right)\frac{\partial \mathbf{B}}{\partial t} \tag{13.7.13e}$$

Now substitute $\varepsilon = \varepsilon_r \varepsilon_o$ into Eq. (13.7.13e) to obtain the final result, for **D** which is

$$\mathbf{D} = \varepsilon \mathbf{E} - N\beta_a \left(\frac{\varepsilon_r + 2}{3} \right) \frac{\partial \mathbf{B}}{\partial t} \tag{13.7.14}$$

We next turn to deriving the equivalent expression for **B**.

13.7.2 Calculation of B

In order to apply Eq. (13.7.2) to molecules in condensed matter, we replace **E** with the Lorentz field, **F**. The result is

$$\mathbf{M}_a' = \kappa_a \mathbf{B} + \beta_a \frac{\partial \mathbf{F}}{\partial t} \tag{13.7.15}$$

With the aid of Eq. (13.7.4), Eq. (13.7.15) can be rewritten as

$$\mathbf{M}_a' = \kappa_a \mathbf{B} + \beta_a \frac{\partial}{\partial t} \left(\mathbf{E} + \frac{\mathbf{P}}{3\varepsilon_o} \right) \tag{13.7.16a}$$

Next, use Eq. (13.7.6) to substitute for the polarization. The result is

$$\mathbf{M}_a' = \kappa_a \mathbf{B} + \beta_a \frac{\partial}{\partial t} \left(\mathbf{E} + \frac{N\mathbf{R}_a'}{3\varepsilon_o} \right) \tag{13.7.16b}$$

Since $\mathbf{M}_a'$ is the induced magnetic dipole moment per molecule, the magnetization is given by

$$\mathbf{M} = N\mathbf{M}_a' \tag{13.7.17}$$

According to Eq. (7.2.20) of Chap. 7, the flux density, **B**, and the magnetization, **M**, are related by

$$\mathbf{B} = \mu_o(\mathbf{H} + \mathbf{M}) \tag{13.7.18}$$

Equations (13.7.17) and (13.7.18) can be combined to read

$$\mathbf{B} = \mu_o \mathbf{H} + \mu_o N \mathbf{M}_a' \tag{13.7.19a}$$

Substitute Eq. (13.7.16b) into Eq. (13.7.19a). The result is

$$\mathbf{B} = \mu_o\mathbf{H} + \mu_o N\kappa_a\mathbf{B} + \mu_o N\beta_a\frac{\partial}{\partial t}\left(\mathbf{E} + \frac{N\mathbf{R}'_a}{3\varepsilon_o}\right) \tag{13.7.19b}$$

Equation (13.7.19b) can be solved for **B** by first writing it in the form

$$(1 - \mu_o N\kappa_a)\mathbf{B} = \mu_o\mathbf{H} + \mu_o N\beta_a\frac{\partial}{\partial t}\left(\mathbf{E} + \frac{N\mathbf{R}'_a}{3\varepsilon_o}\right) \tag{13.7.19c}$$

The result for **B** is then

$$\mathbf{B} = \left[\frac{\mu_o}{1 - \mu_o N\kappa_a}\right]\mathbf{H} + \left[\frac{\mu_o N\beta_a}{1 - \mu_o N\kappa_a}\right]\frac{\partial}{\partial t}\left(\mathbf{E} + \frac{N\mathbf{R}'_a}{3\varepsilon_o}\right) \tag{13.7.19d}$$

The second term in Eq. (13.7.19d) can be expanded with the result,

$$\mathbf{B} = \left[\frac{\mu_o}{1 - \mu_o N\kappa_a}\right]\mathbf{H} + \left[\frac{\mu_o N\beta_a}{1 - \mu_o N\kappa_a}\right]\frac{\partial\mathbf{E}}{\partial t} + \left[\frac{\mu_o N\beta_a}{1 - \mu_o N\kappa_a}\right]\left(\frac{N}{3\varepsilon_o}\right)\frac{\partial\mathbf{R}'_a}{\partial t} \tag{13.7.19e}$$

Use Eq. (13.7.11) to differentiate $\mathbf{R}'_a$ with respect to time and substitute the result into Eq. (13.7.19e). One finds,

$$\begin{aligned}\mathbf{B} = {} & \left[\frac{\mu_o}{1 - \mu_o N\kappa_a}\right]\mathbf{H} + \left[\frac{\mu_o N\beta_a}{1 - \mu_o N\kappa_a}\right]\frac{\partial\mathbf{E}}{\partial t} \\ & + \left[\frac{\mu_o N}{1 - \mu_o N\kappa_a}\right]\left(\frac{N}{3\varepsilon_o}\right)\left[\alpha_a\beta_a\left(\frac{\varepsilon_r + 2}{3}\right)\frac{\partial\mathbf{E}}{\partial t} - \beta_a^2\left(\frac{\varepsilon_r + 2}{3}\right)\frac{\partial^2\mathbf{B}}{\partial t^2}\right]\end{aligned} \tag{13.7.19f}$$

The coefficient of the term involving $\partial^2\mathbf{B}/\partial t^2$ is proportional the second power of, β_a. Since β_a is a small quantity this term can be dropped. After this truncation, Eq. (13.7.19f) reads,

$$\mathbf{B} = \left[\frac{\mu_o}{1 - \mu_o N\kappa_a}\right]\mathbf{H} + \left[\frac{\mu_o N\beta_a}{1 - \mu_o N\kappa_a}\right]\frac{\partial\mathbf{E}}{\partial t} + \left[\frac{\mu_o N\beta_a}{1 - \mu_o N\kappa_a}\right]\left(\frac{N\alpha_a}{3\varepsilon_o}\right)\left(\frac{\varepsilon_r + 2}{3}\right)\frac{\partial\mathbf{E}}{\partial t} \tag{13.7.19g}$$

In Eq. (13.7.19g), the terms proportional to $\partial\mathbf{E}/\partial t$ can be combined. The result is

$$\mathbf{B} = \left[\frac{\mu_o}{1 - \mu_o N\kappa_a}\right]\mathbf{H} + \left[\frac{\mu_o N\beta_a}{1 - \mu_o N\kappa_a}\right]\left[1 + \left(\frac{N\alpha_a}{3\varepsilon_o}\right)\left(\frac{\varepsilon_r + 2}{3}\right)\right]\frac{\partial\mathbf{E}}{\partial t} \tag{13.7.19h}$$

Upon substitution of Eq. (13.7.9), Eq. (13.7.19h) reads,

$$\mathbf{B} = \left[\frac{\mu_o}{1 - \mu_o N\kappa_a}\right]\mathbf{H} + \left[\frac{\mu_o N\beta_a}{1 - \mu_o N\kappa_a}\right]\left[1 + \left(\frac{\varepsilon_r - 1}{\varepsilon_r + 2}\right)\left(\frac{\varepsilon_r + 2}{3}\right)\right]\frac{\partial\mathbf{E}}{\partial t} \tag{13.7.19i}$$

Equation (13.7.19i) can be simplified by combining the terms within the third bracket to obtain our final result for **B**, which is

$$\mathbf{B} = \left[\frac{\mu_o}{1 - \mu_o N \kappa_a}\right]\mathbf{H} + \left[\frac{\mu_o N}{1 - \mu_o N \kappa_a}\right]\beta_a\left(\frac{\varepsilon_r + 2}{3}\right)\frac{\partial \mathbf{E}}{\partial t} \tag{13.7.20}$$

13.7.3 Approximate Forms for B and D

Ordinarily κ_a is sufficiently small, that $\mu_o N \kappa_a \ll 1$. Equation (13.7.20) for **B** can then be written in the simplified form,

$$\mathbf{B} = \mu_o \mathbf{H} + \mu_o N \beta_a \left(\frac{\varepsilon_r + 2}{3}\right)\frac{\partial \mathbf{E}}{\partial t} \tag{13.7.21}$$

Substitute Eq. (13.7.21) into Eq. (13.7.14) to obtain **D**:

$$\mathbf{D} = \varepsilon \mathbf{E} - N\beta_a\left(\frac{\varepsilon_r + 2}{3}\right)\frac{\partial}{\partial t}\left[\mu_o \mathbf{H} + \mu_o N \beta_a \left(\frac{\varepsilon_r + 2}{3}\right)\frac{\partial \mathbf{E}}{\partial t}\right] \tag{13.7.22a}$$

Upon computation of the indicated time derivatives, Eq. (13.7.22a) reads

$$\mathbf{D} = \varepsilon \mathbf{E} - \mu_o N \beta_a \left(\frac{\varepsilon_r + 2}{3}\right)\frac{\partial \mathbf{H}}{\partial t} - \mu_o N^2 \beta_a^2 \left(\frac{\varepsilon_r + 2}{3}\right)^2 \frac{\partial^2 \mathbf{E}}{\partial t^2} \tag{13.7.22b}$$

The last term on the right of Eq. (13.7.22b) involves the square of the small quantity, β_a, and can be dropped. Equation (13.7.22b) then assumes the simplified form,

$$\mathbf{D} = \varepsilon \mathbf{E} - \mu_o N \beta_a \left(\frac{\varepsilon_r + 2}{3}\right)\frac{\partial \mathbf{H}}{\partial t} \tag{13.7.23}$$

13.8 Summary of Equations Governing Optical Rotation

Equations (13.7.21) and (13.7.23) include all of the principal couplings that determine optical rotation. They can be summarized as follows:

$$\mathbf{B} = \mu_o \mathbf{H} + g\frac{\partial \mathbf{E}}{\partial t} \tag{13.8.1}$$

and

$$\mathbf{D} = \varepsilon \mathbf{E} - g \frac{\partial \mathbf{H}}{\partial t} \tag{13.8.2}$$

where g is defined by

$$g = \mu_o N \beta_a \left(\frac{\varepsilon_r + 2}{3} \right) \tag{13.8.3}$$

Because the fields, **B** and **D**, must also satisfy Maxwell's equations, they are not independent. In the absence of currents and free charges,

$$\nabla \times \mathbf{E} = -\frac{\partial \mathbf{B}}{\partial t} \tag{13.8.4}$$

and

$$\nabla \times \mathbf{H} = \frac{\partial \mathbf{D}}{\partial t} \tag{13.8.5}$$

If it is known that the electric field, **E**, has linear or circular polarization, then Eqs. (13.8.1) through (13.8.5) suffice to evaluate the fields, **D**, **B**, and **H**, as well as the coupling constant, g. That calculation follows in the next section.

13.9 Angle of Rotation of Polarized Light

In the case of a circularly polarized wave that has traveled through a distance, z, for a time, t, the central angle, θ_o, described in Sect. 13.1 is given by

$$\theta_o = \omega \left(t - \frac{n z}{c} \right) \tag{13.9.1}$$

where ω is the angular frequency, c is the vacuum speed of light, and the refractive index is

$$n = \sqrt{\varepsilon_r} \tag{13.9.2}$$

As mentioned in Sect. 13.1, θ_o is the angle of rotation that would prevail if the material were an equimolar mixture of components of opposite chirality. The phase velocity is c/n, where c is the vacuum speed of light.

Defined in terms of the refractive index, n_R, for right circularly polarized light, the angle of rotation is

$$\psi_R = \omega\left(t - \frac{n_R z}{c}\right) \tag{13.9.3}$$

In Sect. 13.1, we introduced the angle, δ, defined by

$$\psi_R = \theta_o + \delta \tag{13.9.4}$$

Solve Eq. (13.9.4) for δ and substitute Eqs. (13.9.1) and (13.9.3). The result is

$$\delta = \psi_R - \theta_o = \omega\left(t - \frac{n_R z}{c}\right) - \omega\left(t - \frac{nz}{c}\right) = \frac{\omega\,(n - n_R)z}{c} \tag{13.9.5}$$

In the case of left circularly polarized light, the formula equivalent to Eq. (13.9.3) is

$$\psi_L = \omega\left(t - \frac{n_L z}{c}\right) \tag{13.9.6}$$

In terms of ψ_L and θ_o, the equivalent of Eq. (13.9.4) is

$$\psi_L = \theta_o - \delta \tag{13.9.7}$$

If we solve Eq. (13.9.7) for δ and substitute Eqs. (13.9.1) and (13.9.6), the result is

$$\delta = \theta_o - \psi_L = \omega\left(t - \frac{nz}{c}\right) - \omega\left(t - \frac{n_L z}{c}\right) = -\frac{\omega\,(n - n_L)z}{c} \tag{13.9.8}$$

On the basis of Eqs. (13.9.5) and (13.9.8), it is clear that the consistency of Eqs. (13.9.4) and (13.9.7) demands that $(n - n_R) = -(n - n_L)$. The calculations in the next section below confirm this requirement.

13.10 The Fields, B and H, for Right Circular Polarization

According to Eq. (13.8.4), the differential equation satisfied by the flux density, **B**, depends upon a calculation of $\nabla \times \mathbf{E}$. Using Eq. (13.1.1), the latter is evaluated as follows:

$$\begin{aligned}\nabla \times \mathbf{E} &= \begin{vmatrix} \hat{\mathbf{e}}_x & \hat{\mathbf{e}}_y & \hat{\mathbf{e}}_z \\ \frac{\partial}{\partial x} & \frac{\partial}{\partial y} & \frac{\partial}{\partial z} \\ E_o \cos\psi_R & -E_o \sin\psi_R & 0 \end{vmatrix} \\ &= \hat{\mathbf{e}}_x E_o\left(\frac{\partial \sin\psi_R}{\partial z}\right) - \hat{\mathbf{e}}_y E_o\left(-\frac{\partial \cos\psi_R}{\partial z}\right)\end{aligned}$$

$$+\hat{\mathbf{e}}_z E_o\left[-\left(\frac{\partial \sin\psi_R}{\partial x}\right)-\left(\frac{\partial \cos\psi_R}{\partial y}\right)\right]$$
$$= E_o\left(\hat{\mathbf{e}}_x \cos\psi_R - \hat{\mathbf{e}}_y \sin\psi_R\right)\left(\frac{\partial \psi_R}{\partial z}\right) \tag{13.10.1a}$$

where by virtue of Eq. (13.9.3),

$$\left(\frac{\partial \sin\psi_R}{\partial x}\right)=\left(\frac{\partial \cos\psi_R}{\partial y}\right)=0 \tag{13.10.1b}$$

Also according to Eq. (13.9.3)

$$\frac{\partial \psi_R}{\partial z} = -\frac{\omega n_R}{c} \tag{13.10.1c}$$

Upon substitution of Eq. (13.10.1c) into Eq. (13.10.1a), the result is

$$\nabla\times\mathbf{E} = \frac{\omega n_R}{c} E_o\left(-\hat{\mathbf{e}}_x \cos\psi_R + \hat{\mathbf{e}}_y \sin\psi_R\right) \tag{13.10.2}$$

Substitute Eq. (13.10.2) into Eq. (13.8.4) to obtain the differential equation,

$$\frac{\partial \mathbf{B}}{\partial t} = -\frac{\omega n_R}{c} E_o\left(-\hat{\mathbf{e}}_x \cos\psi_R + \hat{\mathbf{e}}_y \sin\psi_R\right) = \frac{\omega n_R}{c}\mathbf{E}_o\left(\hat{\mathbf{e}}_x \cos\psi_R - \hat{\mathbf{e}}_y \sin\psi_R\right) \tag{13.10.3a}$$

If in Eq. (13.10.3a), we replace the time, t, by the dummy variable, τ, the indefinite integral of Eq. (13.10.3a) with respect to time is

$$\mathbf{B} = \frac{\omega n_R}{c} E_o\left(\hat{\mathbf{e}}_x \int^t d\tau \cos\psi_R(\tau) - \hat{\mathbf{e}}_y \int^t d\tau \sin\psi_R(\tau)\right) \tag{13.10.3b}$$

The integrals in Eq. (13.10.3b) are absent constants of integration, because we will later have to differentiate **B**. Upon substitution of $\psi(\tau)$ as per Eq. (13.9.3), the first indefinite integral can evaluated as follows:

$$\int^t d\tau \cos\psi_R(\tau) = \int^t d\tau \cos\left(\omega\tau - \frac{\omega n_R z}{c}\right) = \frac{1}{\omega}\int^{\omega t} d(\omega\tau)\cos\left(\omega\tau - \frac{\omega n_R z}{c}\right)$$
$$= \frac{1}{\omega}\sin\left(\omega t - \frac{\omega n_R z}{c}\right) = \frac{1}{\omega}\sin\left[\omega\left(t - \frac{n_R z}{c}\right)\right] = \frac{1}{\omega}\sin\psi_R(t,z) \tag{13.10.3c}$$

The second integral is

$$\int^{t} d\tau \sin \psi_R(\tau) = \int^{t} d\tau \sin\left(\omega\tau - \frac{\omega_R n z}{c}\right) = \frac{1}{\omega}\int^{\omega t} d(\omega\tau) \sin\left(\omega\tau - \frac{\omega n_R z}{c}\right)$$
$$= -\frac{1}{\omega}\cos\left(\omega t - \frac{\omega n_R z}{c}\right) = -\frac{1}{\omega}\cos\left[\omega\left(t - \frac{\omega_R n z}{c}\right)\right]$$
$$= -\frac{1}{\omega}\cos \psi_R(t, z) \tag{13.10.3d}$$

Upon substitution of Eqs. (13.10.3c) and (13.10.3d) into Eq. (13.10.3b), the result for the magnetic induction is

$$\mathbf{B} = \frac{n_R}{c} E_o\big(\hat{\mathbf{e}}_x \sin \psi_R(t, z) + \hat{\mathbf{e}}_y \cos \psi_R(t, z)\big) \tag{13.10.4}$$

Next, we use Eq. (13.1.1) to compute time derivative of the electric field:

$$\frac{\partial \mathbf{E}(z, t)}{\partial t} = \frac{\partial}{\partial t} E_o\big(\hat{\mathbf{e}}_x \cos \psi_R - \hat{\mathbf{e}}_y \sin \psi_R\big) = E_o\big(-\hat{\mathbf{e}}_x \sin \psi_R - \hat{\mathbf{e}}_y \cos \psi_R\big)\left(\frac{\partial \psi_R}{\partial t}\right) \tag{13.10.5a}$$

Differentiation of Eq. (13.9.3) gives

$$\frac{\partial \psi_R}{\partial t} = \omega \tag{13.10.5b}$$

Substitute Eq. (13.10.5b) into Eq. (13.10.5a) to obtain

$$\frac{\partial \mathbf{E}(z, t)}{\partial t} = -\omega E_o\big(\hat{\mathbf{e}}_x \sin \psi_R(t, z) + \hat{\mathbf{e}}_y \cos \psi_R(t, z)\big) \tag{13.10.6}$$

A comparison of Eqs. (13.10.4) and (13.10.6) allows us to write,

$$\frac{\partial \mathbf{E}}{\partial t} = \frac{-\omega\, c}{n_R}\mathbf{B} \tag{13.10.7}$$

Equation (13.10.7) provides a useful expression for **B** in terms of the time derivative of **E**.

To solve for **H**, put Eq. (13.10.7) into Eq. (13.8.1) to obtain

$$\mathbf{B} = \mu_o \mathbf{H} - g\frac{\omega\, c}{n_R}\mathbf{B} \tag{13.10.8}$$

Upon rearranging Eq. (13.10.8), we obtain for the magnetic field,

$$\mathbf{H} = \frac{1}{\mu_o}\left(1 + g\frac{\omega\, c}{n_R}\right)\mathbf{B} \tag{13.10.9}$$

13.11 Two Equations for D in Right Circular Polarization

To obtain the first equation for **D**, we exploit Eq. (13.8.5). Evaluation of Eq. (13.8.5) requires the calculation of the curl of **H**. Upon substitution of Eq. (13.10.4) into Eq. (13.10.9), **H** can be written in the form,

$$\mathbf{H} = \frac{1}{\mu_o}\left(1 + g\frac{\omega c}{n_R}\right)\left(\frac{n_R}{c}\right)E_o\big(\hat{\mathbf{e}}_x \sin\psi_R + \hat{\mathbf{e}}_y \cos\psi_R\big) \tag{13.11.1}$$

Using Eq. (13.11.1), the calculation of $\nabla \times \mathbf{H}$ proceeds as follows:

$$\begin{aligned}
\nabla \times \mathbf{H} &= \frac{1}{\mu_o}\left(1 + g\frac{\omega\, c}{n_R}\right)\left(\frac{n_R}{c}\right)\nabla \times \big[E_o\big(\hat{\mathbf{e}}_x \sin\psi_R + \hat{\mathbf{e}}_y \cos\psi_R\big)\big] \\
&= \frac{1}{\mu_o}\left(1 + g\frac{\omega\, c}{n_R}\right)\left(\frac{n_R}{c}\right)\begin{vmatrix} \hat{\mathbf{e}}_x & \hat{\mathbf{e}}_y & \hat{\mathbf{e}}_z \\ \frac{\partial}{\partial x} & \frac{\partial}{\partial y} & \frac{\partial}{\partial z} \\ E_o \sin\psi_R & E_o \cos\psi_R & 0 \end{vmatrix} \\
&= \frac{1}{\mu_o}\left(1 + g\frac{\omega\, c}{n_R}\right)\left(\frac{n_R}{c}\right)E_o\left[-\hat{\mathbf{e}}_x\left(\frac{\partial \cos\psi_R}{\partial z}\right) - \hat{\mathbf{e}}_y\left(-\frac{\partial \sin\psi_R}{\partial z}\right)\right] \\
&= \frac{1}{\mu_o}\left(1 + g\frac{\omega\, c}{n_R}\right)\left(\frac{n_R}{c}\right)E_o\big(\hat{\mathbf{e}}_x \sin\psi_R + \hat{\mathbf{e}}_y \cos\psi_R\big)\left(\frac{\partial\psi_R}{\partial z}\right)
\end{aligned} \tag{13.11.2a}$$

Upon substitution of Eq. (13.10.1c), Eq. (13.11.2a) becomes

$$\nabla \times \mathbf{H} = \frac{1}{\mu_o}\left(1 + g\frac{\omega c}{n_R}\right)\left(\frac{n_R}{c}\right)\left(\frac{-\omega n_R}{c}\right)E_o\big(\hat{\mathbf{e}}_x \sin\psi_R + \hat{\mathbf{e}}_y \cos\psi_R\big) \tag{13.11.2b}$$

After performing the indicated multiplications, Eq. (13.11.2b) reads

$$\nabla \times \mathbf{H} = -\frac{\omega n_R}{\mu_o c^2}(n_R + g\omega\, c)E_o\big(\hat{\mathbf{e}}_x \sin\psi_R + \hat{\mathbf{e}}_y \cos\psi_R\big) \tag{13.11.3}$$

Substitute Eq. (13.11.3) into Eq. (13.8.5) to obtain the differential equation

$$\frac{\partial \mathbf{D}}{\partial t} = -\frac{\omega n_R}{\mu_o c^2}(n_R + g\omega c)E_o\big(\hat{\mathbf{e}}_x \sin\psi_R + \hat{\mathbf{e}}_y \cos\psi_R\big) \tag{13.11.4a}$$

The time integral of Eq. (13.11.4a) is

$$\mathbf{D} = -\frac{\omega n_R}{\mu_o c^2}(n_R + g\omega c)E_o\left(\hat{\mathbf{e}}_x \int^t d\tau \sin\psi_R(\tau) + \hat{\mathbf{e}}_y \int^t d\tau \cos\psi_R(\tau)\right) \tag{13.11.4b}$$

Upon reference to the integrals in Eqs. (13.10.3c) and (13.10.3d), we can rewrite Eq. (13.11.4b) in the form,

$$\mathbf{D} = -\frac{\omega\, n_R}{\mu_o c^2}(n_R + g\omega\, c)E_o\left[\hat{\mathbf{e}}_x\left(\frac{-1}{\omega}\right)\cos\psi_R + \hat{\mathbf{e}}_y\left(\frac{1}{\omega}\right)\sin\psi_R\right] \tag{13.11.4c}$$

Equation (13.11.4c) simplifies to

$$\mathbf{D} = \frac{n_R}{\mu_o c^2}(n_R + g\omega\, c)E_o\big(\hat{\mathbf{e}}_x\cos\psi_R - \hat{\mathbf{e}}_y\sin\psi_R\big) \tag{13.11.4d}$$

On the right hand side of Eq. (13.11.4d), we recognize the vector corresponding to **E** as given by Eq. (13.1.1). With this substitution, Eq. (13.11.4d) becomes,

$$\mathbf{D} = \frac{n_R}{\mu_o c^2}(n_R + g\omega\, c)\mathbf{E} \tag{13.11.5}$$

Equation (13.11.5) is the first of the two equations linking **D** and **E**.

A second equation linking **D** and **E** can be found by using Eq. (13.11.1) to compute the time derivative of **H**. With the use of Eq. (13.10.5b), the calculation proceeds as follows:

$$\begin{aligned}
\frac{\partial\mathbf{H}}{\partial t} &= \frac{1}{\mu_o}\frac{\partial}{\partial t}\left(1 + g\frac{\omega\, c}{n_R}\right)\left(\frac{n_R}{c}\right)\mathbf{E}_o\big(\hat{\mathbf{e}}_x\sin\psi_R + \hat{\mathbf{e}}_y\cos\psi_R\big)\\
&= \frac{1}{\mu_o}\left(1 + g\frac{\omega\, c}{n_R}\right)\left(\frac{n_R}{c}\right)\mathbf{E}_o\left(\hat{\mathbf{e}}_x\frac{\partial\sin\psi_R}{\partial t} + \hat{\mathbf{e}}_y\frac{\partial\cos\psi_R}{\partial t}\right)\\
&= \frac{1}{\mu_o}\left(1 + g\frac{\omega c}{n_R}\right)\left(\frac{n_R}{c}\right)\mathbf{E}_o\big(\hat{\mathbf{e}}_x\cos\psi_R - \hat{\mathbf{e}}_y\sin\psi_R\big)\left(\frac{\partial\psi_R}{\partial t}\right)\\
&= \frac{1}{\mu_o}\left(1 + g\frac{\omega c}{n_R}\right)\left(\frac{n_R}{c}\right)\omega\mathbf{E}_o\big(\hat{\mathbf{e}}_x\cos\psi_R - \hat{\mathbf{e}}_y\sin\psi_R\big)
\end{aligned} \tag{13.11.6a}$$

On the basis of Eqs. (13.1.1), Eq. (13.11.6a) can be rewritten as

$$\frac{\partial\mathbf{H}}{\partial t} = \frac{1}{\mu_o}\left(\frac{n_R + g\omega c}{n_R}\right)\left(\frac{n_R}{c}\right)\omega\mathbf{E} \tag{13.11.6b}$$

After cancellation of common factors, Eq. (13.11.6b) reads,

$$\frac{\partial\mathbf{H}}{\partial t} = \left(\frac{n_R + g\omega c}{\mu_o c}\right)\omega\mathbf{E} \tag{13.11.7}$$

When Eq. (13.11.7) is solved for the electric field, **E**, the result is

$$\mathbf{E} = \frac{\partial \mathbf{H}/\partial t}{\left[\frac{\omega(n_R+g\omega c)}{\mu_o c}\right]} \tag{13.11.8a}$$

Equation (13.11.5) can be solved for **E** to obtain

$$\mathbf{E} = \frac{\mathbf{D}}{\left[\frac{n_R(n_R+g\omega c)}{\mu_o c^2}\right]} \tag{13.11.8b}$$

When **E** is eliminated between Eqs. (13.11.8a) and (13.11.8b), the result is

$$\frac{\mathbf{D}}{\left[\frac{n_R(n_R+g\omega c)}{\mu_o c^2}\right]} = \frac{\partial \mathbf{H}/\partial t}{\left[\frac{\omega(n_R+g\omega c)}{\mu_o c}\right]} \tag{13.11.8c}$$

After cancellation of common factors, Eq. (13.11.8c) reads

$$\frac{\partial \mathbf{H}}{\partial t} = \left(\frac{c\omega}{n_R}\right)\mathbf{D} \tag{13.11.8d}$$

Substitute Eq. (13.11.8d) into Eq. (13.8.2) to obtain

$$\mathbf{D} = \varepsilon\mathbf{E} - g\left(\frac{c\omega}{n_R}\right)\mathbf{D} \tag{13.11.8e}$$

When Eq. (13.11.8e) is solved for **D**, the result is

$$\mathbf{D} + g\left(\frac{c\omega}{n_R}\right)\mathbf{D} = \varepsilon\mathbf{E} \tag{13.11.9a}$$

$$\left[1 + g\left(\frac{c\omega}{n_R}\right)\right]\mathbf{D} = \varepsilon\mathbf{E} \tag{13.11.9b}$$

Upon forming a common denominator, Eq. (13.11.9b) reads,

$$\left(\frac{n_R + gc\omega}{n_R}\right)\mathbf{D} = \varepsilon\mathbf{E} \tag{13.11.9c}$$

which finally leads to

$$\mathbf{D} = \left(\frac{\varepsilon n_R}{n_R + g\omega c}\right)\mathbf{E} \tag{13.11.10}$$

Equation (13.11.9) is the second equation linking **D** and **E**.

13.12 The Refractive Indices for Circularly Polarized Light

Equations (13.11.5) and (13.11.10) must be consistent. This implies for right circularly polarized light,

$$\frac{n_R}{\mu_o c^2}(n_R + g\omega c) = \frac{\varepsilon n_R}{(n_R + g\omega c)} \tag{13.12.1}$$

Solve Eq. (13.12.1) for $(n_R + g\omega c)$. The first step is

$$(n_R + g\omega\, c)^2 = \mu_o c^2 \varepsilon \tag{13.12.2a}$$

In Eq. (13.12.2a), we can substitute $c^2 = 1/\varepsilon_o\mu_o$, plus $\varepsilon = \varepsilon_r\varepsilon_o$, so that $\mu_o c^2\varepsilon = \mu_o c^2\varepsilon_r\varepsilon_o = \varepsilon_r$. After substituting this result into Eq. (13.12.2a) and taking the square root of both sides, we obtain

$$(n_R + g\omega\, c) = \sqrt{\varepsilon_r} \tag{13.12.2b}$$

Solve Eq. (13.12.2b) to obtain

$$n_R = \sqrt{\varepsilon_r} - g\omega c \tag{13.12.3}$$

Upon substitution of $n = \sqrt{\varepsilon_r}$ in Eq. (13.12.3), the result is

$$n_R = n - g\omega\, c \tag{13.12.4}$$

If we compare Eqs. (13.1.1) with (13.1.2), we see that right circular polarization can be converted to left circular polarization simply by replacing ψ_R by $-\psi_L$. By virtue of Eq. (13.9.3), this is equivalent to replacing ω by $-\omega$. With this change plus the replacement of n_R by n_L, Eq. (13.12.4) reads,

$$n_L = n + g\omega\, c \tag{13.12.5}$$

With Eqs. (13.12.4) and (13.12.5), we are in a position to provide a theoretical interpretation of the angle of rotation as observed in the laboratory.

13.13 The Angle of Rotation

The angle of rotation depends upon the thickness of the sample. This is a property which ORD shares with the optical absorbtivity discussed in Sect. 7.10 of Chap. 7. Customarily, the distance along the z-axis is measured in centimeters, so with reference to Eqs. (13.9.5), (13.9.8), (13.12.4), and (13.12.5), it is possible to define a specific angle of rotation, ϕ', measured in radians/cm by

$$\phi' = \frac{\delta}{z} = g\omega^2 \tag{13.13.1}$$

Substitution of Eq. (13.8.3) for g and introduction of the circular frequency, $\nu = \omega/2\pi$, transforms Eq. (13.13.1) to read,

$$\phi' = 4\pi^2\nu^2\left(\frac{\varepsilon_r + 2}{3}\right)N\mu_o\beta_a \tag{13.13.2}$$

If in Eq. (13.13.2), we can replace β_a by Eq. (13.6.1d), the result is

$$\phi' = 4\pi^2\nu^2\left(\frac{\varepsilon_r + 2}{3}\right)N\mu_o\left(\frac{1}{3\pi h}\right)\left(\sum_b \frac{\mathrm{Im}[\langle a|\mathbf{R}|b\rangle \cdot \langle b|\mathbf{M}|a\rangle]}{(\nu_{ba}^2 - \nu^2)}\right) \tag{13.13.3}$$

After substitution of $\varepsilon_r = n^2$ and algebraic rearrangement, Eq. (13.13.3) simplifies to

$$\phi' = \frac{4\pi N\mu_o}{9h}(n^2 + 2)\left(\sum_b \frac{\nu^2\mathrm{Im}[\langle a|\mathbf{R}|b\rangle \cdot \langle b|\mathbf{M}|a\rangle]}{(\nu_{ba}^2 - \nu^2)}\right) \tag{13.13.4}$$

Define

$$R_{ba} = \mathrm{Im}[\langle a|\mathbf{R}|b\rangle \cdot \langle b|\mathbf{M}|a\rangle] \tag{13.13.5}$$

then Eq. (13.13.4) reads,

$$\phi' = \frac{4\pi N\mu_o}{9h}(n^2 + 2)\left[\sum_b \frac{\nu^2 R_{ba}}{(\nu_{ba}^2 - \nu^2)}\right] \tag{13.13.6}$$

which expresses the frequency dependence of the specific angle of rotation.

13.14 Discussion

According to Maxwell's equations, the interaction between the electric and magnetic fields is responsible for the propagation of light. In the cases which we have analyzed in previous chapters involving the interaction of light with matter, the material responded principally to the electric field. In the case of optical rotation, however, the response of the material to both the electric and magnetic fields plays a role.

Equation (13.13.6) indicates that the sign and strength of the rotation is determined by R_{ba}, which depends upon the product of the matrix elements of $\langle a|\mathbf{R}|b\rangle$ and $\langle b|\mathbf{M}|a\rangle$. In understanding optical rotation from a physical point of view, an examination of the symmetry property of the corresponding classical quantities is instructive. The classical electric dipole moment is

$$\mathbf{R} = q\mathbf{r} \tag{13.14.1}$$

According to Eq. (13.2.22) the classical magnetic dipole moment is

$$\mathbf{M} = \gamma\mathbf{L} \tag{13.14.2}$$

The angular momentum is given by Eq. (13.2.21) and can be written as,

$$\mathbf{L} = \mathbf{r} \times \mathbf{p} = m(\mathbf{r} \times \mathbf{v}) \tag{13.14.3}$$

where $\mathbf{v} = (d\,\mathbf{r}/dt)$ is the velocity of the particle. Combine Eqs. (13.14.2) and (13.14.3) to obtain

$$\mathbf{M} = \gamma m(\mathbf{r} \times \mathbf{v}) \tag{13.14.4}$$

The dot product of **R** with **M** is

$$\mathbf{R} \cdot \mathbf{M} = q\gamma m\mathbf{r} \cdot (\mathbf{r} \times \mathbf{v}) \tag{13.14.5}$$

In Cartesian coordinates

$$\mathbf{r} = x\hat{\mathbf{e}}_x + y\hat{\mathbf{e}}_y + z\hat{\mathbf{e}}_z \tag{13.14.6a}$$

while

$$\mathbf{v} = v_x\hat{\mathbf{e}}_x + v_y\hat{\mathbf{e}}_y + v_z\hat{\mathbf{e}}_z \tag{13.14.6b}$$

The cross product is

$$\mathbf{r} \times \mathbf{v} = \begin{vmatrix} \hat{\mathbf{e}}_x & \hat{\mathbf{e}}_y & \hat{\mathbf{e}}_z \\ x & y & z \\ v_x & v_y & v_z \end{vmatrix} = \hat{\mathbf{e}}_x\left(yv_z - zv_y\right) + \hat{\mathbf{e}}_y(zv_x - xv_z) + \hat{\mathbf{e}}_z\left(xv_y - yv_x\right) \tag{13.14.6c}$$

According to Eqs. (13.14.6a) and (13.14.6c), the dot product required in Eq. (13.14.5) is

$$\mathbf{r} \cdot (\mathbf{r} \times \mathbf{v}) = x\left(yv_z - zv_y\right) + y(zv_x - xv_z) + z\left(xv_y - yv_x\right) \tag{13.14.7a}$$

Consider a mirror lying in the x, y-plane. The normal to the plane of the mirror is the z-axis. Reflection in the mirror corresponds to replacing z by $-z$. Because $v_z = (dz/dt)$, reflection of v_z in the mirror corresponds to replacing v_z by $-v_z$. If we make these substitutions into the right hand side of Eq. (13.14.7a), the result is

$$\begin{aligned} &x\left(-yv_z + zv_y\right) + y(-zv_x + xv_z) - z\left(xv_y - yv_x\right) \\ &\quad = -x\left(yv_z - zv_y\right) - y(zv_x - xv_z) - z\left(xv_y - yv_x\right) = -\mathbf{r}\cdot(\mathbf{r}\times\mathbf{v}) \end{aligned} \tag{13.14.7b}$$

According to Eq. (13.14.7b), the reflection causes $(\mathbf{R}\cdot\mathbf{M})$ to be replaced by $-(\mathbf{R}\cdot\mathbf{M})$. This calculation implies that a chiral molecule and its mirror image have opposite values of $(\mathbf{R}\cdot\mathbf{M})$. By contrast, because an achiral molecule is superimposable on its mirror image, this is tolerable only if $(\mathbf{R}\cdot\mathbf{M}) = 0$. Hence, for an achiral molecule, Eq. (13.13.5) predicts that the rotation strength $R_{ab} = 0$. Thus achiral molecules are optically inactive.

When light is absorbed by a chiral molecule, we need to consider a complex-valued angle of rotation, $\tilde{\phi}(\nu)$, which can be expressed in terms of its real and imaginary parts as

$$\tilde{\phi}(\nu) = \phi'(\nu) + i\phi''(\nu) \tag{13.14.8}$$

The real part, $\phi'(\nu)$, goes through zero at resonance and determines the optical rotatory dispersion (ORD). The imaginary part, which exhibits a maximum at resonance, determines the circular dichroism (CD) [3].

References

1. H. Eyring, J. Walter, G.E. Kimball, *Quantum Chemistry* (Wiley, New York, 1944)
2. D.J. Caldwell, H. Eyring, *The Theory of Optical Rotation* (Wiley, New York, 1971)
3. E. Charney, *The Molecular Basis for Optical Activity* (Wiley, New York, 1979)

Recommended Additional Reading

1. L.D. Barron, *Molecular Light Scattering and Optical Activity*, 2nd edn. (Cambridge University Press, Cambridge, UK, 2004)
2. C. Djerassi, *Optical Rotatory Dispersion* (McGraw-Hill, New York, 1960)

Chapter 14
Dispersion Forces

Ionic charges and permanent electric dipoles are the sources of molecular forces which are long range and electrostatic in nature. These forces can be either repulsive or attractive. At sufficiently short range, however, every atom, molecule or ion exerts an additional attractive force on any other atom, molecule or ion in its vicinity. The existence of such a force was suggested on empirical grounds by van der Waals in the late nineteenth century. It was given a quantum mechanical interpretation by F. London in the early twentieth century. London's theory has found wide application in chemistry. Because the London force depends upon the frequencies of motion of the electrons in the molecule or atom, the force is said to be dispersive. In the first section of this chapter, we reprise London's original derivation. In the subsequent sections, we analyze two applications of the London theory. The first of these is Th. Förster's 1948 theory of the rate of radiationless energy transfer between the excited state of a donor molecule and the ground state of an acceptor molecule, a phenomenon which has since become known as Förster resonance energy transfer. Resonance energy transfer forms the basis for photosynthesis. By including all of the mathematical details of the proof, we reveal the origin of the famous multiplicative factor of 9000 which appeared in Förster's original paper. In the second application of the London theory, we describe Hamaker's 1937 calculation of the attractive force between two macroscopic objects, such as two colloidal particles. Hamaker's original proof lacks key mathematical details. To retrieve the situation, we include all of the missing steps in Hamaker's derivation.

J. K. Baird, *Electromagnetism for Chemistry*,
https://doi.org/10.1007/978-3-031-88508-2_14

14.1 London Theory of the van der Waals' Attractive Force Between Molecules

14.1.1 van der Waals' Equation of State

In the year following his doctoral dissertation of 1872, J. D. van der Waals (Dutch, 1837–1923) proposed an equation of state designed to reproduce the P-V-T behavior of real gases. His equation corrected the ideal gas law in a way that accounted empirically for the effects of the intermolecular forces. As in the ideal gas equation, P is the pressure, V is the volume, T is the Kelvin temperature, R is the gas law constant, and n is the number of moles. The van der Waals equation extends this list to include two empirical parameters, "a" and "b". The equation reads,

$$P = \frac{nRT}{(V - nb)} - a\left(\frac{n}{V}\right)^2 \tag{14.1.1}$$

The parameters, a and b, have values that are specific to each individual gas and can be determined by fitting Eq. (14.1.1) to (P, V, T) data. Although originally mostly empirical in nature, the van der Waals a and b, as they are now named, have since been given theoretical interpretations.

The parameter, b, serves as a measure of the short range repulsive intermolecular force. Because two molecules cannot occupy the same space at the same time, the product, nb, serves as an estimate of the volume that n moles of these molecules would occupy if they were in a condensed phase. According to van der Waals, molecules in the gas have access not to the total volume, V, but only to the "free" volume, $(V - nb)$, which is the difference between V, and the volume of exclusion, nb. Because the factor $(V - nb)$ appears in the denominator of the first term in Eq. (14.1.1), it has the effect of increasing the pressure.

The intermolecular force also has an attractive component, which is represented by the term, $-a(n/V)^2$ in Eq. (14.1.1). This attractive component is proportional to the probability that two molecules are close enough to interact with one another. The argument is as follows: The probability that a single molecule is at a certain point in space is uniform and proportional to the density, (n/V). The probability that two molecules are simultaneously close to this point corresponds to the coincidence of two independent random events, each with probability, (n/V). The probability of the coincidence of independent events is given by the product of the independent probabilities. Thus, the joint probability is proportional to $(n/V)^2$. The attractive force serves to confine the molecules to the free volume and reduces the likelihood that either molecule will hit the wall of the container and contribute to the pressure. In Eq. (14.1.1), this reduction in pressure is taken into account by multiplying the product $(n/V)^2$ by the empirical coefficient, $a > 0$, and subtracting the result from the right hand side of the equation.

14.1.2 Electrostatic Interaction of Two Hydrogen Atoms

Fritz London (German-American, 1900–1954) showed that the attractive component of the van der Waals intermolecular force was electrical in origin. Although we will not arrive at a formula for the parameter, *a*, we will, nonetheless, derive the London equation which serves as the basis for such a calculation. As an aid in the derivation of London's result, we shall rely on a conflation of the presentations of Kauzmann [1], and Eyring, Walter, and Kimball [2].

For simplicity, consider the intermolecular force acting between two hydrogen atoms positioned a distance R apart. The atoms are assumed to be sufficiently separated that covalence is not a factor. See Fig. 14.1.

The protons in the respective hydrogen atoms will be labeled "*a*" and "*b*", not to be confused with the van der Waals parameters, "*a*" and "*b*". When the atoms are infinitely far apart, electron "1" is confined to the vicinity of proton *a*, while electron "2" is confined to the vicinity of proton *b*. In a Cartesian coordinate system with origin on proton *a*, the coordinates of electron "1" are (x_1, y_1, z_1), while in a Cartesian coordinate system centered on proton *b*, the coordinates of electron "2" are (x_2, y_2, z_2). The origins of these coordinate frames are separated by the distance, R. The corresponding z-axes are collinear and lie along the R direction. In both cases, the positive z-direction points from left to right. In addition, the x_1-axis is parallel to the x_2-axis. The y_1-axis is parallel to the y_2-axis.

We denote the magnitude of the electron charge by e. If $r_{1a}, r_{1b}, r_{2a}, r_{2b}$, and r_{12} are the distances separating the charges designated by the corresponding subscripts, then the total electrostatic potential energy, V, of the charge configuration is

$$V = -\frac{e^2}{4\pi\varepsilon_o}\left[\frac{1}{r_{1a}} + \frac{1}{r_{2b}}\right] + \frac{e^2}{4\pi\varepsilon_o}\left[\frac{1}{r_{12}} - \frac{1}{r_{1b}} - \frac{1}{r_{2a}} + \frac{1}{R}\right] \tag{14.1.2a}$$

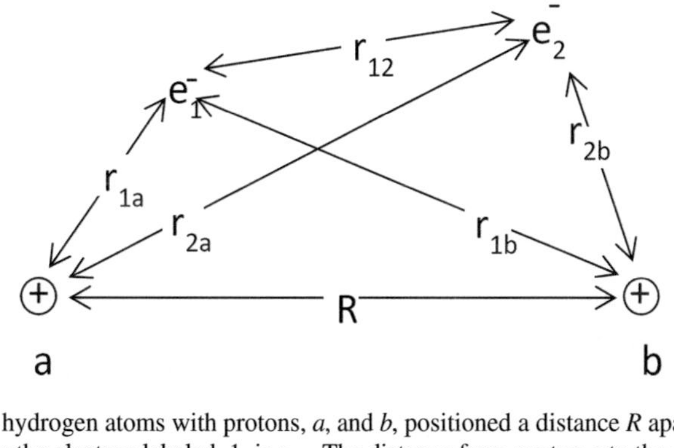

Fig. 14.1 Two hydrogen atoms with protons, *a*, and *b*, positioned a distance *R* apart. The distance from proton *a* to the electron labeled, 1, is r_{1a}. The distance from proton *a* to the electron labeled, 2, is r_{2a}. The corresponding distances with respect to proton, *b*, are r_{1b} and r_{2b}, respectively. The distance separating the two electrons is r_{12}

The terms in the first bracket account for the electrostatic potential energy of interaction of each electron with its parent nucleus. The terms within the second bracket are inter-atomic. The first term in the second bracket accounts for the interaction between the two electrons. The second and third terms represent the interaction of each electron with the nucleus of the atom which is not its parent. The fourth term accounts for the electrostatic repulsion acting between the nuclei.

As the terms within the second bracket account for all of the possible inter-atomic electrostatic interactions between the two hydrogen atoms, we shall group them together so as to define,

$$V' = \frac{e^2}{4\pi\varepsilon_o}\left[\frac{1}{R} + \frac{1}{r_{12}} - \frac{1}{r_{1b}} - \frac{1}{r_{2a}}\right] \tag{14.1.2b}$$

As mentioned, the origin of coordinates (x_1, y_1, z_1) of electron "1" is located on nucleus a. The origin of the coordinates (x_2, y_2, z_2) of electron "2" is located on nucleus b. Expressed in terms of these coordinates, the projection of r_{1b} on the z-axis has length, $(R - z_1)$. The projection of r_{2a} on the z-axis has length $(R + z_2)$. Note that for the electron configuration in Fig. 14.1, z_2 is negative. The distances, r_{1b}, r_{2a}, and r_{12} in Eq. (14.1.2b) have the following definitions:

$$r_{1b} = \left[x_1^2 + y_1^2 + (R - z_1)^2\right]^{1/2} \tag{14.1.3a}$$

$$r_{2a} = \left[x_2^2 + y_2^2 + (R + z_2)^2\right]^{1/2} \tag{14.1.3b}$$

$$r_{12} = \left[(x_1 - x_2)^2 + (y_1 - y_2)^2 + (R + z_2 - z_1)^2\right]^{1/2} \tag{14.1.3c}$$

According to Eqs. (14.1.3a) through (14.1.3c), when R is large, r_{1b}, r_{2a}, and r_{12} are all of the same order of magnitude as R. As a consequence, as $R \to \infty$, V' in Eq. (14.1.2b) approaches zero. We are interested in the intermediate distances, where r_{1a} and r_{2b} are finite and have values less than R.

To begin the task of identifying the leading terms in Eq. (14.1.2b), which determine V' at distances where $(x_1, y_1, z_1) < R$ and $(x_2, y_2, z_2) < R$, we shall remove a factor of R from Eq. (14.1.3a). The result is

$$r_{1b} = R\left[\frac{x_1^2 + y_1^2}{R^2} + \frac{(R - z_1)^2}{R^2}\right]^{1/2} \tag{14.1.4a}$$

Upon expanding the binomial square, $(R - z_1)^2$, the reciprocal of r_{1b} can be written and rearranged to read,

$$\frac{1}{r_{1b}} = \frac{1}{R}\left[\frac{x_1^2 + y_1^2}{R^2} + \frac{R^2 - 2z_1R + z_1^2}{R^2}\right]^{-1/2} = \frac{1}{R}\left[1 - \frac{2z_1}{R} + \frac{x_1^2 + y_1^2 + z_1^2}{R^2}\right]^{-1/2} \tag{14.1.4b}$$

The following approximations have the goal of preserving the terms in Eq. (14.1.4a) involving the first power of z_1 (referred to as being first order) and the squares x_1^2, y_1^2, and z_1^2, respectively (referred to as second order). All higher order terms can be dropped.

At distances, where $(x_1, y_1, z_1) < R$, we can simplify Eq. (14.1.4b) by writing it in the form,

$$\frac{1}{r_{1b}} = \frac{1}{R}(1+Z)^{-1/2} \tag{14.1.5}$$

where Z is defined by

$$Z = \left(-\frac{2z_1}{R} + \frac{x_1^2 + y_1^2 + z_1^2}{R^2}\right) \tag{14.1.6a}$$

For $Z < 1$, Eq. (14.1.5) can be expanded in a binomial series to read,

$$\frac{1}{r_{1b}} = \frac{1}{R}(1+Z)^{-1/2} = \frac{1}{R}\left[1 - \frac{1}{2}Z + \frac{(-1/2)(-3/2)}{2}Z^2 + \cdots\right] \tag{14.1.6b}$$

According to Eq. (14.1.6a),

$$Z^2 = \left(-\frac{2z_1}{R} + \frac{x_1^2 + y_1^2 + z_1^2}{R^2}\right)^2 = \left(-\frac{2z_1}{R}\right)^2 + O\left[\frac{z_1(x_1^2 + y_1^2 + z_1^2)}{R^3}\right] + \ldots \tag{14.1.6c}$$

where "O" should be read "order of". For large values of R, the term, $(-2z_1/R)^2$, dominates terms of higher order, e.g. $z_1x_1^2/R^3$, in the expansion in Eq. (14.1.6c). Hence, Z^2 can be approximated as

$$Z^2 \approx \left(-\frac{2z_1}{R}\right)^2 \tag{14.1.6d}$$

Upon substitution of Eqs. (14.1.6a) and (14.1.6d) into Eq. (14.1.6b), the result is

$$\frac{1}{r_{1b}} = \frac{1}{R}\left[1 - \frac{1}{2}\left(-\frac{2z_1}{R} + \frac{x_1^2 + y_1^2 + z_1^2}{R^2}\right) + \frac{(-1/2)(-3/2)}{2}\left(-\frac{2z_1}{R}\right)^2 + \cdots\right] \tag{14.1.6e}$$

After simplification, Eq. (14.1.6e) reads,

$$\frac{1}{r_{1b}} \approx \frac{1}{R}\left[1 + \frac{z_1}{R} - \frac{1}{2}\left(\frac{x_1^2 + y_1^2 + z_1^2}{R^2}\right) + \frac{3}{2}\left(\frac{z_1}{R}\right)^2\right] \tag{14.1.7}$$

Equation (14.1.7) is our final result for $1/r_{1b}$.

We now turn to Eq. (14.1.3b) and employ the same strategy that was applied in order to simplify Eq. (14.1.3a). Let $x_2, y_2, z_2 < R$. To take advantage of this inequality, we factor out R in Eq. (14.1.3b), so that it reads,

$$r_{2a} = R\left[\frac{x_2^2 + y_2^2}{R^2} + \frac{(R + z_2)^2}{R^2}\right]^{1/2} \tag{14.1.8a}$$

After expanding the binomial square, $(R + z_2)^2$, and simplifying the result, the reciprocal of r_{2a} reads,

$$\begin{aligned}\frac{1}{r_{2a}} &= \frac{1}{R}\left[\frac{x_2^2 + y_2^2}{R^2} + \frac{R^2 + 2z_2R + z_2^2}{R^2}\right]^{-1/2} \\ &= \frac{1}{R}\left[1 + \frac{2z_2}{R} + \frac{x_2^2 + y_2^2 + z_2^2}{R^2}\right]^{-1/2}\end{aligned} \tag{14.1.8b}$$

Let

$$\frac{1}{r_{2a}} = \frac{1}{R}(1 + Z)^{-1/2} \tag{14.1.9}$$

We identify

$$Z = \frac{2z_2}{R} + \frac{x_2^2 + y_2^2 + z_2^2}{R^2} \tag{14.1.10a}$$

Following our previous reasoning, we can approximate Z^2 by

$$Z^2 \approx \left(\frac{2z_2}{R}\right)^2 \tag{14.1.10b}$$

We next invoke the binomial expansion,

$$\frac{1}{r_{2a}} = \frac{1}{R}(1 + Z)^{-1/2} = \frac{1}{R}\left[1 - \frac{1}{2}Z + \frac{(-1/2)(-3/2)}{2}Z^2 + ...\right] \tag{14.1.10c}$$

Upon substitution of Eqs. (14.1.10a) and (14.10b) into Eq. (14.1.10c), we obtain

$$\frac{1}{r_{2a}} = \frac{1}{R}\left[1 - \frac{1}{2}\left(\frac{2z_2}{R} + \frac{x_2^2 + y_2^2 + z_2^2}{R^2}\right) + \frac{(-1/2)(-3/2)}{2}\left(\frac{2z_2}{R}\right)^2 + \cdots\right] \tag{14.1.10d}$$

which upon simplification leads to

$$\frac{1}{r_{2a}} \approx \frac{1}{R}\left[1 - \frac{z_2}{R} - \frac{1}{2}\left(\frac{x_2^2 + y_2^2 + z_2^2}{R^2}\right) + \frac{3}{2}\left(\frac{z_2}{R}\right)^2\right] \tag{14.1.11}$$

Equation (14.1.11) is our final result for $1/r_{2a}$.

We now turn to r_{12}. Factor out R^2 from the third term in the bracket in Eq. (14.1.3c). The result is

$$r_{12} = \left[(x_1 - x_2)^2 + (y_1 - y_2)^2 + R^2\left(1 + \frac{z_2 - z_1}{R}\right)^2\right]^{1/2} \tag{14.1.12a}$$

Compute the binomial squares, $(x_1 - x_2)^2$ and $(y_1 - y_2)^2$. Combine them and multiply and divide them by R^2. Finally, expand the binomial, $\left(1 + \frac{z_2 - z_1}{R}\right)^2$, so that Eq. (14.1.12a) reads,

$$\begin{aligned} r_{12} = \Bigg[& R^2\left(\frac{x_1^2 - 2x_1x_2 + x_2^2 + y_1^2 - 2y_1y_2 + y_2^2}{R^2}\right) \\ & + R^2\left(1 + 2\frac{(z_2 - z_1)}{R} + \frac{(z_2 - z_1)^2}{R^2}\right)\Bigg]^{1/2} \end{aligned} \tag{14.1.12b}$$

Factor out R^2 from each term in Eq. (14.1.12b). The result is

$$\begin{aligned} r_{12} = R\Bigg[& \left(\frac{x_1^2 - 2x_1x_2 + x_2^2 + y_1^2 - 2y_1y_2 + y_2^2}{R^2}\right) \\ & + \left(1 + 2\frac{(z_2 - z_1)}{R} + \frac{(z_2 - z_1)^2}{R^2}\right)\Bigg]^{1/2} \end{aligned} \tag{14.1.13a}$$

Upon rearranging terms, the reciprocal of r_{12} can be written in the form,

$$\begin{aligned} \frac{1}{r_{12}} = \frac{1}{R}\Bigg[& 1 + \frac{2(z_2 - z_1)}{R} + \frac{(z_2 - z_1)^2}{R^2} \\ & + \left(\frac{x_1^2 + y_1^2 + x_2^2 + y_2^2 - 2x_1x_2 - 2y_1y_2}{R^2}\right)\Bigg]^{-1/2} \end{aligned} \tag{14.1.13b}$$

Use $(z_2 - z_1)^2 = z_2^2 - 2z_1z_2 + z_1^2$, and combine the resulting quadratic terms with the other quadratic terms inside the parentheses involving x_1^2, x_2^2, etc. The result is

$$\begin{aligned} \frac{1}{r_{12}} = \frac{1}{R}\Bigg[& 1 + \frac{2(z_2 - z_1)}{R} \\ & + \left(\frac{x_1^2 + y_1^2 + z_1^2 + x_2^2 + y_2^2 + z_2^2 - 2x_1x_2 - 2y_1y_2 - 2z_1z_2}{R^2}\right)\Bigg]^{-1/2} \end{aligned} \tag{14.1.13c}$$

Next write,

$$\frac{1}{r_{12}} = \frac{1}{R}(1+Z)^{-1/2} \tag{14.1.14}$$

Identify Z in Eq. (14.1.14) with

$$Z = \frac{2(z_2 - z_1)}{R} + \left(\frac{x_1^2 + y_1^2 + z_1^2 + x_2^2 + y_2^2 + z_2^2}{R^2}\right) - \left(\frac{2x_1x_2 + 2y_1y_2 + 2z_1z_2}{R^2}\right) \tag{14.1.15a}$$

If Z^2 is computed using Eq. (14.1.15a), the leading term, which is proportional to $(z_2 - z_1)^2$, is of second order. The other terms in this evaluation of Z^2 are of third order (e.g. $x_1^2(z_2 - z_1)$) and fourth order (e.g. x_1^4) and can be ignored. Hence to capture the term, which is second order, we approximate,

$$Z^2 \approx \left(\frac{2(z_2 - z_1)}{R}\right)^2 \tag{14.1.15b}$$

Expand Eq. (14.1.14) to read

$$\frac{1}{r_{12}} = \frac{1}{R}(1+Z)^{-1/2} = \frac{1}{R}\left[1 - \frac{1}{2}Z + \frac{(-1/2)(-3/2)}{2}Z^2 + \cdots\right] \tag{14.1.16a}$$

After substituting Eqs. (14.1.15a) and (14.1.15b) into Eq. (14.1.16a), one obtains

$$\begin{aligned}\frac{1}{r_{12}} = \frac{1}{R}\Bigg\{1 - \left(\frac{1}{2}\right)\Bigg[\frac{2(z_2 - z_1)}{R} + \left(\frac{x_1^2 + y_1^2 + z_1^2 + x_2^2 + y_2^2 + z_2^2}{R^2}\right) \\ -\left(\frac{2x_1x_2}{R^2} + \frac{2y_1y_2}{R^2} + \frac{2z_1z_2}{R^2}\right)\Bigg]\Bigg\} + \frac{1}{R}\left[\frac{(-1/2)(-3/2)}{2} \times \frac{(2)^2(z_2 - z_1)^2}{R^2}\right]\end{aligned} \tag{14.1.16b}$$

Cancelling coefficients, rearranging terms, and squaring the remaining binomial in Eq. (14.1.16b) leads to

$$\begin{aligned}\frac{1}{r_{12}} \approx \frac{1}{R}\left[1 - \frac{(z_2 - z_1)}{R} - \frac{1}{2}\left(\frac{x_1^2 + y_1^2 + z_1^2 + x_2^2 + y_2^2 + z_2^2}{R^2}\right)\right] \\ + \frac{1}{R}\left[\frac{x_1x_2}{R^2} + \frac{y_1y_2}{R^2} + \frac{z_1z_2}{R^2} + \frac{3}{2}\frac{\left(z_2^2 - 2z_1z_2 + z_1^2\right)}{R^2}\right]\end{aligned} \tag{14.1.17}$$

Equation (14.1.17) is our final result for $1/r_{12}$.

To get a final result for V', substitute Eqs. (14.1.7), (14.1.11) and (14.1.17) into Eq. (14.1.2b). The result is

$$
\begin{aligned}
V' &= \frac{e^2}{4\pi\varepsilon_o}\left[\frac{1}{R}+\frac{1}{r_{12}}-\frac{1}{r_{1b}}-\frac{1}{r_{2a}}\right] \\
&= \frac{e^2}{4\pi\varepsilon_o R}+\frac{e^2}{4\pi\varepsilon_o R}\left[1-\frac{(z_2-z_1)}{R}-\frac{1}{2}\left(\frac{x_1^2+y_1^2+z_1^2+x_2^2+y_2^2+z_2^2}{R^2}\right)\right] \\
&\quad+\frac{e^2}{4\pi\varepsilon_o R}\left[\frac{x_1x_2}{R^2}+\frac{y_1y_2}{R^2}+\frac{z_1z_2}{R^2}+\frac{3}{2}\frac{\left(z_2^2-2z_1z_2+z_1^2\right)}{R^2}\right] \\
&\quad-\frac{e^2}{4\pi\varepsilon_o R}\left[1+\frac{z_1}{R}-\frac{1}{2}\left(\frac{x_1^2+y_1^2+z_1^2}{R^2}\right)+\frac{3}{2}\left(\frac{z_1}{R}\right)^2\right] \\
&\quad-\frac{e^2}{4\pi\varepsilon_o R}\left[1-\frac{z_2}{R}-\frac{1}{2}\left(\frac{x_2^2+y_2^2+z_2^2}{R^2}\right)+\frac{3}{2}\left(\frac{z_2}{R}\right)^2\right]
\end{aligned}
\tag{14.1.18}
$$

A careful perusal of Eq. (14.1.18) reveals that all of the terms involving $(1/R)$ alone cancel, as do all of the first order terms involving (z_1/R) and (z_2/R). Many, but not all, of the second order terms also cancel. After making the indicated cancellations, the remaining terms in Eq. (14.1.18) read

$$
V' = \frac{e^2}{4\pi\varepsilon_o R^3}(x_1x_2+y_1y_2-2z_1z_2) \tag{14.1.19}
$$

Equation (14.1.19) represents the energy of interaction of two hydrogen atoms calculated in classical electrostatics [1].

14.1.3 Quantum Mechanical Evaluation of the Interaction Hamiltonian

According to the Bohr theory, the electronic energy of a hydrogen atom is quantized. See Chap. 1, Sect. 1.15. In the case of a pair of hydrogen atoms, the long-range, non-covalent energy of interaction between them can be calculated by identifying V' in Eq. (14.1.19) with the Hamiltonian operator,

$$
\mathrm{H}' = \frac{e^2}{4\pi\varepsilon_o R^3}(x_1x_2+y_1y_2-2z_1z_2) \tag{14.1.20}
$$

We shall exploit Eq. (14.1.20) to evaluate the energy of interaction of two hydrogen atoms in their ground states. A ground state 1s wave function is spherically symmetrical. In the case of two widely separated hydrogen atoms, the two-electron ground state in Dirac notation, $|1s(1), 1s(2)\rangle$, is the product, $|1s(1)\rangle|1s(2)\rangle$, where $|1s(1)\rangle$ is the state function for the first electron, while $|1s(2)\rangle$ is the state function for the second electron. To calculate the effect of H' in first order perturbation theory, we will need to evaluate

$$
\begin{aligned}
&\langle 1s(1), 1s(2)|\mathrm{H}'|1s(1), 1s(2)\rangle \\
&= \frac{e^2}{4\pi\varepsilon_o R^3}[\langle 1s(1), 1s(2)|x_1x_2|1s(1), 1s(2)\rangle + \langle 1s(1), 1s(2)|y_1y_2|1s(1), 1s(2)\rangle] \\
&\quad - 2\cdot\frac{e^2}{4\pi\varepsilon_o R^3}\cdot\langle 1s(1), 1s(2)|z_1z_2|1s(1), 1s(2)\rangle
\end{aligned}
\tag{14.1.21a}
$$

Matrix elements of the form

$$
\langle 1s(1), 1s(2)|x_1x_2|1s(1), 1s(2)\rangle = \langle 1s(1)|x_1|1s(1)\rangle\langle 1s(2)|x_2|1s(2)\rangle \tag{14.1.21b}
$$

can be written out explicitly using the real valued wave functions, where $|1s(1)\rangle = \langle 1s(1)| = \psi_1^{(o)}(r_1)$ and $|1s(2)\rangle = \langle 1s(2)| = \psi_2^{(o)}(r_2)$. Equation (14.1.21b) reads,

$$
\begin{aligned}
&\int d^3r_1 \int d^3r_2\, \psi_1^{(o)}(x_1, y_1, z_1)\psi_2^{(o)}(x_2, y_2, z_2)x_1x_2\psi_1^{(o)}(x_1, y_1, z_1)\psi_2^{(o)}(x_2, y_2, z_2) \\
&= \int d^3r_1\, \psi_1^{(o)}x_1\psi_1^{(o)} \cdot \int d^3r_2\, \psi_2^{(o)}x_2\psi_2^{(o)}
\end{aligned}
\tag{14.1.21c}
$$

Being spherically symmetric, the $|1s(1)\rangle$ state is an even function of the coordinates (x_1, y_1, z_1) of electron "1", while the $|1s(2)\rangle$ state is an even function of the coordinates (x_2, y_2, z_2) of electron "2". By contrast, the integrands involving, x_1 and x_2, in $\langle 1s(1)|x_1|1s(1)\rangle = \int d^3r_1\, \psi_1^{(o)}x_1\psi_1^{(o)}$ and $\langle 1s(2)|x_2|1s(2)\rangle = \int d^3r_2\, \psi_2^{(o)}x_2\psi_2^{(o)}$, respectively, are odd functions. As the integral of an odd function over the entire extent of its variable is zero, we can conclude that $\langle 1s(1)|x_1|1s(1)\rangle = \langle 1s(2)|x_2|1s(2)\rangle = 0$, etc. Thus, the perturbation, H$'$, has no effect in first order. As a consequence, we must consider it in second order perturbation theory.

In second order perturbation theory, the correction to the ground state energy is given by Eq. (11.5.16f) of Chap. 11. To evaluate Eq. (11.5.16f), a representation of the operator, H'^2, is required. According to Eq. (14.1.20),

$$
\mathrm{H}'^2 = \left(\frac{e^2}{4\pi\varepsilon_o R^3}\right)^2 \left[x_1x_2 + y_1y_2 - 2z_1z_2\right]^2 \tag{14.1.22a}
$$

Upon expanding the square of the trinomial in Eq. (14.1.22a), the result is

$$
\mathrm{H}'^2 = \left(\frac{e^2}{4\pi\varepsilon_o R^3}\right)^2 \left[x_1^2x_2^2 + y_1^2y_2^2 + 4z_1^2z_2^2 + 2x_1x_2y_1y_2 - 4x_1x_2z_1z_2 - 4y_1y_2z_1z_2\right] \tag{14.1.22b}
$$

In the Cartesian coordinate systems specified by x_1, y_1, z_1 and x_2, y_2, z_2, the vectors $\mathbf{r}_{1b}$, $\mathbf{r}_{2a}$ and $\mathbf{r}_{12}$ make lots of angles with each other [2]. Since the 1s orbital of a hydrogen atom is spherically symmetric, and Eq. (14.1.2b) depends only upon the lengths of $\mathbf{r}_{1b}$, $\mathbf{r}_{2a}$ and $\mathbf{r}_{12}$, we propose to replace each factor in Eq. (14.1.22b) by its spherical polar average. Using the symbol, $\langle\rangle$, to represent angle average, we can rewrite Eq. (14.1.22b) as

$$
\begin{aligned}
H'^2 &= \left(\frac{e^2}{4\pi\varepsilon_o R^3}\right)^2 \left[\langle x_1^2\rangle\langle x_2^2\rangle + \langle y_1^2\rangle\langle y_2^2\rangle + 4\langle z_1^2\rangle\langle z_2^2\rangle\right] \\
&+ \left(\frac{e^2}{4\pi\varepsilon_o R^3}\right)^2 \left[2\langle x_1\rangle\langle x_2\rangle\langle y_1\rangle\langle y_2\rangle - 4\langle x_1\rangle\langle x_2\rangle\langle z_1\rangle\langle z_2\rangle - 4\langle y_1\rangle\langle y_2\rangle\langle z_1\rangle\langle z_2\rangle\right]
\end{aligned}
\tag{14.1.23}
$$

The angle averages, $\langle x_1\rangle$ and $\langle x_2\rangle$, can be computed by expressing them in spherical polar coordinate notation as

$$x = r\sin\theta\cos\phi \tag{14.1.24a}$$

$$y = r\sin\theta\sin\phi \tag{14.1.24b}$$

$$z = r\cos\theta \tag{14.1.24c}$$

As differential solid angel is, $d\Omega = \sin\theta\, d\theta\, d\phi$, and the total solid angle is $\Omega = 4\pi$, the solid angle average of x is computed as

$$\langle x\rangle = \left(\frac{1}{\Omega}\right) r \int_0^{\pi} d\theta \sin^2\theta \int_0^{2\pi} d\phi \cos\phi = \left(\frac{1}{\Omega}\right) r \int_0^{\pi} d\theta \sin^2\theta [\sin\phi]_0^{2\pi} = 0 \tag{14.1.25a}$$

The angle average of y is

$$\langle y\rangle = \left(\frac{1}{\Omega}\right) r \int_0^{\pi} d\theta \sin^2\theta \int_0^{2\pi} d\phi \sin\phi = \left(\frac{1}{\Omega}\right) r \int_0^{\pi} d\theta \sin^2\theta [-\cos\phi]_0^{2\pi} = 0 \tag{14.1.25b}$$

The angle averages, $\langle z_1\rangle$ and $\langle z_2\rangle$, require

$$\langle z\rangle = \left(\frac{1}{\Omega}\right) r \int_0^{2\pi} d\phi \int_0^{\pi} d\theta \sin\theta\cos\theta = -2\pi\left(\frac{1}{\Omega}\right) r \left[\frac{\cos^2\theta}{2}\right]_0^{\pi} = 0 \tag{14.1.25c}$$

On the basis of Eqs. (14.1.25a) through (14.1.25c), we conclude that

$$\langle x_i\rangle = \langle y_i\rangle = \langle z_i\rangle = 0 \qquad \text{for } i = 1, 2 \tag{14.1.26}$$

According to Eq. (14.1.26), each of the terms in the second bracket in Eq. (14.1.23) is zero.

The averages in the first bracket of Eq. (14.1.23), involve the squares of the Cartesian coordinates. By virtue of Eq. (14.1.24a),

$$\langle x^2 \rangle = \left(\frac{1}{\Omega}\right) r^2 \int_0^{2\pi} d\phi \cos^2\phi \int_0^{\pi} d\theta \sin\theta \sin^2\theta \tag{14.1.27a}$$

where $\Omega = 4\pi$. To compute the integral over ϕ in Eq. (14.1.27a), use

$$\cos^2\phi = \left(\frac{1}{2}\right)(1 + \cos 2\phi) \tag{14.1.27b}$$

To compute the integral over θ, let $w = \cos\theta$, $\sin^2\theta = 1 - \cos^2\theta = 1 - w^2$, and $dw = -\sin\theta d\theta$. The integration limit, $\theta = 0$, corresponds to $w = +1$, while the other limit, $\theta = \pi$ corresponds to $w = -1$. With these substitutions, the integrals in Eq. (14.1.27a) become

$$\begin{aligned} \langle x^2 \rangle &= \left(\frac{1}{\Omega}\right) r^2 \int_0^{2\pi} d\phi \left(\frac{1}{2}\right)(1 + \cos 2\phi) \int_{-1}^{+1} dw\left(1 - w^2\right) \\ &= \left(\frac{1}{\Omega}\right) r^2 \left(\frac{1}{2}\right)\left[\phi + \frac{1}{2}\sin 2\phi\right]_0^{2\pi} \left[w - \frac{w^3}{3}\right]_{-1}^{+1} \\ &= \left(\frac{1}{4\pi}\right) r^2 \left(\frac{1}{2}\right)(2\pi)(2)\left(\frac{2}{3}\right) = \frac{1}{3} r^2 \end{aligned} \tag{14.1.27c}$$

With reference to Eq. (14.1.24b),

$$\langle y^2 \rangle = \left(\frac{1}{\Omega}\right) r^2 \int_0^{2\pi} d\phi \sin^2\phi \int_0^{\pi} d\theta \sin\theta \sin^2\theta \tag{14.1.28a}$$

To compute the integral over ϕ in Eq. (14.1.28a), use

$$\sin^2\phi = \left(\frac{1}{2}\right)(1 - \cos 2\phi) \tag{14.1.28b}$$

As before, to compute the integral over θ, let $w = \cos\theta$. The result of these two angle substitutions converts Eq. (14.1.28a) to the form,

$$\begin{aligned}\langle y^2\rangle &= \left(\frac{1}{\Omega}\right)r^2\int_0^{2\pi}d\phi\left(\frac{1}{2}\right)(1-\cos 2\phi)\int_{-1}^{+1}dw\left(1-w^2\right)\\ &= \left(\frac{1}{\Omega}\right)r^2\left(\frac{1}{2}\right)\left[\phi-\frac{1}{2}\sin 2\phi\right]_0^{2\pi}\left[w-\frac{w^3}{3}\right]_{-1}^{+1}\\ &= \left(\frac{1}{4\pi}\right)r^2\left(\frac{1}{2}\right)(2\pi)(2)\left(\frac{2}{3}\right)=\frac{1}{3}r^2\end{aligned}\tag{14.1.28c}$$

Finally, we use Eq. (14.1.24c) to compute $\langle z^2\rangle$. The calculation proceeds as follows:

$$\begin{aligned}\langle z^2\rangle &= \left(\frac{1}{\Omega}\right)r^2\int_0^{2\pi}d\phi\int_0^{\pi}d\theta\sin\theta\cos^2\theta\\ &= \left(\frac{1}{4\pi}\right)r^2(2\pi)\left[-\frac{\cos^3\theta}{3}\right]_0^{\pi}\\ &= \left(\frac{1}{4\pi}\right)r^2(2\pi)\left(\frac{2}{3}\right)=\frac{1}{3}r^2\end{aligned}\tag{14.1.29}$$

In terms of the lengths, $r_{1a}^2 = x_1^2 + y_1^2 + z_1^2$ and $r_{2b}^2 = x_2^2 + y_2^2 + z_2^2$, the averages required to evaluate Eq. (14.1.23) are thus

$$\langle x_1^2\rangle = \langle y_1^2\rangle = \langle z_1^2\rangle = \frac{1}{3}r_{1a}^2\tag{14.1.30a}$$

and

$$\langle x_2^2\rangle = \langle y_2^2\rangle = \langle z_2^2\rangle = \frac{1}{3}r_{2b}^2\tag{14.1.30b}$$

respectively. Upon substitution of Eqs. (14.1.26), (14.1.30a), and (14.1.30b) into Eq. (14.1.23), the result is

$$\mathrm{H}'^2 = \left(\frac{e^2}{4\pi\varepsilon_o R^3}\right)^2\left[\left(\frac{1}{3}\right)^2+\left(\frac{1}{3}\right)^2+4\left(\frac{1}{3}\right)^2\right]r_{1a}^2r_{2b}^2 = \left(\frac{2}{3}\right)\left(\frac{e^2}{4\pi\varepsilon_o R^3}\right)^2 r_{1a}^2r_{2b}^2\tag{14.1.31}$$

14.1.4 Application of Second Order Perturbation Theory

Equation (11.5.16f) of Chap. 11 can be written in the form,

$$E_n^{(2)} = \sum_{k \neq n} \frac{\left|\left\langle \psi_k^{(o)} \middle| \mathrm{H}' \middle| \psi_n^{(o)} \right\rangle\right|^2}{\left(E_n^{(o)} - E_k^{(o)}\right)} \tag{14.1.32a}$$

To simplify the evaluation of Eq. (14.1.32a), we introduce the Dirac notation, where $\left\langle \psi_k^{(o)} \middle| \mathrm{H}' \middle| \psi_n^{(o)} \right\rangle = \langle k | \mathrm{H}' | n \rangle$. Equation (14.1.32a) then reads,

$$E_n^{(2)} = \sum_{k \neq n} \frac{\left|\langle k | \mathrm{H}' | n \rangle\right|^2}{\left(E_n^{(o)} - E_k^{(o)}\right)} \tag{14.1.32b}$$

We now turn to the evaluation of the matrix element in Eq. (14.1.32b). Consider two hydrogen atoms with electrons having principal quantum numbers, n_1 and n_2, respectively. Recall that the energy of an electron in a hydrogen atom depends only upon the principal quantum number. If we identify $|n\rangle$ as a two-particle state, then $|n\rangle = |1\rangle|1\rangle$ is the two-particle ground state corresponding to $n_1 = n_2 = 1$. By contrast, $|k\rangle = |n_1\rangle|n_2\rangle$ is the two-particle excited state in which at least one of n_1 and n_2 is not unity. The complex conjugate of $|k\rangle$ is $\langle k| = \langle n_1|\langle n_2|$. Since a single subscript on r is sufficient to specify the separation of an electron from its parent nucleus, we shall condense our notation by writing $r_1 = r_{1a}$ and $r_2 = r_{2b}$. In this notation, the matrix element of Eq. (14.1.31) can be written as

$$\left|\langle k | H' | n \rangle\right|^2 = \left|\langle n_1, n_2 | \mathrm{H}' | 1, 1 \rangle\right|^2 = \left(\frac{2}{3}\right)\left(\frac{e^2}{4\pi\varepsilon_o R^3}\right)^2 |\langle n_1 | r_1 | 1 \rangle|^2 |\langle n_2 | r_2 | 1 \rangle|^2 \tag{14.1.33}$$

The total energy in the ground state is

$$E_n^{(o)} = E_1^{(o)} + E_1^{(o)} \tag{14.1.34a}$$

The total energy of an excited state consisting of one atom with principal quantum number n_1 and the other with principal quantum number n_2 is

$$E_k^{(o)} = E_{n_1}^{(o)} + E_{n_2}^{(o)} \tag{14.1.34b}$$

Substitute Eq. (14.1.33) and Eqs. (14.1.34a) and (14.1.34b) into Eq. (14.1.32b). The result is

$$E_{1,1}^{(2)} = \left(\frac{2}{3}\right)\left(\frac{e^2}{4\pi\varepsilon_o}\right)^2 \frac{1}{R^6} \sum_{n_1,n_2} \frac{|\langle n_1|r_1|1\rangle|^2 |\langle n_2|r_2|1\rangle|^2}{\left(E_1^{(o)} + E_1^{(o)}\right) - \left(E_{n_1}^{(o)} + E_{n_2}^{(o)}\right)} \tag{14.1.35a}$$

In Eq. (14.1.35a), $E_{1,1}^{(2)}$ is the energy of interaction of two ground state hydrogen atoms located a distance, R, apart. It is understood that the term in the sum with $n_1 = n_2 = 1$ is omitted, else the denominator would be zero. Upon factoring out a minus sign and rearranging the terms in the denominator of the fraction, Eq. (14.1.35a) reads

$$E_{1,1,}^{(2)} = -\left(\frac{2}{3}\right)\left(\frac{e^2}{4\pi\varepsilon_o}\right)^2 \frac{1}{R^6} \sum_{n_1,n_2} \frac{|\langle n_1|r_1|1\rangle|^2 |\langle n_2|r_2|1\rangle|^2}{\left(E_{n_1}^{(o)} - E_1^{(o)}\right) + \left(E_{n_2}^{(o)} \quad E_1^{(o)}\right)} \tag{14.1.35b}$$

14.1.5 Interaction Energy in Terms of the Atomic Polarizability and the Ionization Potential

Part of the task posed by the evaluation of Eq. (14.1.35b) is to find simple representations for the matrix elements, $|\langle n_1|r_1|1\rangle|^2$ and $|\langle n_2|r_2|1\rangle|^2$. With this task in mind, we turn to the real part of the polarizability as given by Eq. (12.5.18) of Chap. 12. We reprise Eq. (12.5.18) below,

$$\alpha'(\omega) = \left(\frac{1}{\hbar}\right) \sum_k |\mu_{k\ell}|^2 \left[\frac{(\omega_{k\ell} - \omega)}{(\omega - \omega_{k\ell})^2 + (1/4)\gamma_{k\ell}^2} + \frac{(\omega + \omega_{k\ell})}{(\omega + \omega_{k\ell})^2 + (1/4)\gamma_{k\ell}^2}\right] \tag{14.1.36a}$$

To specialize Eq. (14.1.36a) to a single hydrogen atom, we will identify the index, $\ell = 1$, with the single particle state $|1\rangle$. The index, "k", will be identified with the single particle state, $|n\rangle$, with $n \neq 1$; thus $k = n$. Since the net interatomic electric attraction is parallel to the z-direction, the electric dipole matrix element, $|\mu_{k\ell}|^2$, will be replaced by $e^2|z_{n1}|^2 = e^2|\langle n|z|1\rangle|^2$. With these alterations, Eq. (14.1.36a) reads

$$\alpha'(\omega) = \left(\frac{e^2}{\hbar}\right) \sum_{n \neq 1} |z_{n1}|^2 \left[\frac{(\omega_{n1} - \omega)}{(\omega - \omega_{n1})^2 + (1/4)\gamma_{n1}^2} + \frac{(\omega + \omega_{n1})}{(\omega + \omega_{n1})^2 + (1/4)\gamma_{n1}^2}\right] \tag{14.1.36b}$$

The real part of the polarizability evaluated at $\omega = 0$ is ordinarily denoted as $\alpha'(0) = \alpha$. According to Eq. (14.1.36b), the polarizability evaluated at $\omega = 0$ (the so called static polarizability), can be written as

$$\alpha = \left(\frac{e^2}{\hbar}\right)\sum_{n\neq 1}|z_{n1}|^2\left[\frac{\omega_{n1}}{(\omega_{n1})^2+(1/4)\gamma_{n1}^2}+\frac{\omega_{n1}}{(\omega_{n1})^2+(1/4)\gamma_{n1}^2}\right]$$
$$= \left(\frac{2e^2}{\hbar}\right)\sum_{k\neq 1}|z_{n1}|^2\frac{\omega_{n1}}{(\omega_{n1})^2+(1/4)\gamma_{n1}^2} \qquad (14.1.36c)$$

If as is usually the case, $\gamma_{n1} \ll \omega_{n1}$, then Eq. (14.1.36c) simplifies to the form,

$$\alpha = \left(\frac{2e^2}{\hbar}\right)\sum_{n\neq 1}|z_{n1}|^2\frac{\omega_{n1}}{(\omega_{n1})^2} = \left(\frac{2e^2}{\hbar}\right)\sum_{n\neq 1}\frac{|z_{n1}|^2}{\omega_{n1}} \qquad (14.1.36d)$$

The energy levels, $E_n^{(o)}$ and $E_1^{(o)}$, are connected by the frequency, ω_{n1}, where

$$\hbar\omega_{n1} = E_n^{(o)} - E_1^{(o)} \qquad (14.1.37a)$$

With this substitution, Eq. (14.1.36d) reads,

$$\alpha = 2e^2\sum_{n\neq 1}\frac{|z_{n1}|^2}{\left(E_n^{(o)} - E_1^{(o)}\right)} \qquad (14.1.37b)$$

In Eq. (14.1.37b), the absolute square matrix element will behave in the same fashion as the classical averages given by Eqs. (14.1.30a) and (14.1.30b); namely,

$$\left|x_{n1}^2\right| = \left|y_{n1}^2\right| = \left|z_{n1}^2\right| = \frac{1}{3}|\langle n|r|1\rangle|^2 \qquad (14.1.37c)$$

Upon substitution of Eq. (14.1.37c), Eq. (14.13.137b) reads,

$$\alpha = \left(\frac{2e^2}{3}\right)\sum_{n\neq 1}\frac{|\langle n|r|1\rangle|^2}{\left(E_n^{(o)} - E_1^{(o)}\right)} \qquad (14.1.38)$$

To evaluate the denominator in Eq. (14.1.38), we return to Sect. 1.15 of Chap. 1, where the Bohr theory for the hydrogen atom was introduced. For the electron orbit to be stable, we found that the centrifugal force had to be balanced by the Coulomb force, which when rendered in MKSQ units yields the equation,

$$\frac{mv^2}{r} = \frac{e^2}{4\pi\varepsilon_o r^2} \qquad (14.1.39a)$$

Quantization was introduced by writing

$$mv = \frac{n\hbar}{r} \qquad (14.1.39b)$$

By combining Eqs. (14.1.39a) and (14.1.39b), we can write

$$\frac{e^2}{4\pi\varepsilon_o r^2} = \frac{mv^2}{r} = \frac{m^2v^2}{mr} = \frac{n^2\hbar^2}{r^2}\cdot\frac{1}{mr} \tag{14.1.39c}$$

which when solved for r reads,

$$r_n = \frac{4\pi\varepsilon_o n^2\hbar^2}{e^2 m} \tag{14.1.39d}$$

When $n = 1$, Eq. (14.1.39d) predicts, $r_1 = a_o$, the radius of the first Bohr orbit,

$$a_o = \frac{4\pi\varepsilon_o h^2}{e^2 m} \tag{14.1.39e}$$

In terms of a_o defined in Eq. (14.1.39e), Eq. (14.1.39d) reads

$$r_n = n^2 a_o \tag{14.1.39f}$$

In analogy with the derivation in Chap. 1, Sect. 1.15, the energy of an electron in a Bohr orbit in MKS units is

$$E_n = -\frac{1}{2}\frac{e^2}{4\pi\varepsilon_o r_n} \tag{14.1.39g}$$

Upon substitution of Eq. (14.1.39f), Eq. (14.1.39g) reads,

$$E_n^{(o)} = -\left(\frac{1}{2}\right)\left(\frac{e^2}{4\pi\varepsilon_o a_o}\right)\frac{1}{n^2} \tag{14.1.40a}$$

According to Eq. (14.1.40a), the difference between the excited state energy, $E_n^{(o)}$, and the ground state energy, $E_1^{(o)}$, is

$$E_n^{(o)} - E_1^{(o)} = \left(\frac{1}{2}\right)\left(\frac{e^2}{4\pi\varepsilon_o a_o}\right)\left(1 - \frac{1}{n^2}\right) \tag{14.1.40b}$$

The ionization energy, I, of the atom is calculated by setting $n = \infty$, in Eq. (14.1.40b). The result reads

$$I = \left(\frac{1}{2}\right)\left(\frac{e^2}{4\pi\varepsilon_o a_o}\right) \tag{14.1.40c}$$

According to Eq. (14.1.40b), the value of $E_n^{(o)} - E_1^{(o)}$ ranges between $(3/4)I$, when $n = 2$, and the maximum which is I, when $n = \infty$.

As a way of evaluating the denominator in Eq. (14.1.38), London suggested the approximation,

$$I \approx E_n^{(o)} - E_1^{(o)} \tag{14.1.41a}$$

In this approximation, the error in the energy difference as a function of n can never exceed 25%. With this approximation, Eq. (14.1.38) reads,

$$\alpha = \left(\frac{2e^2}{3I}\right) \sum_{n \neq 1} |\langle n|r|1\rangle|^2 \tag{14.1.41b}$$

Equation (14.1.41b) can be used to solve for the matrix element sum, $\sum_{n \neq 1} |\langle n|r|1\rangle|^2$. Upon application of Eq. (14.1.41b) to electron "1", we get for the sum,

$$\sum_{n_1 \neq 1} |\langle n_1|r_1|1\rangle|^2 = \left(\frac{3I_1\alpha_1}{2e^2}\right) \tag{14.1.42a}$$

Upon application of Eq. (14.1.41b) to electron "2", we get for the sum,

$$\sum_{n_2 \neq 1} |\langle n_2|r_1|1\rangle|^2 = \left(\frac{3I_2\alpha_2}{2e^2}\right) \tag{14.1.42b}$$

In the case that we are considering, the atoms are identical. Hence, the values of I_1 and I_2 are numerically the same. Nonetheless, we shall retain the distinction, $I_1 \not\to I_2$ so that Eq. (14.1.35b) can be later applied to the case where the atoms differ.

With the use of the substitutions, $I_1 \approx E_{n_1}^{(o)} - E_1^{(o)}$ and $I_2 \approx E_{n_2}^{(o)} - E_1^{(o)}$, Eq. (14.1.35b) can be rewritten as

$$E_{1,1}^{(2)} \approx -\left(\frac{2}{3}\right)\left(\frac{e^2}{4\pi\varepsilon_o}\right)^2 \frac{1}{R^6} \sum_{n_1,n_2} \frac{|\langle n_1|r_1|1\rangle|^2 |\langle n_2|r_2|1\rangle|^2}{I_1 + I_2} \tag{14.1.43a}$$

In Eq. (14.1.43a), the double sum can be factored into the product of two sums, so that

$$E_{1,1}^{(2)} = -\left(\frac{2}{3}\right)\left(\frac{e^2}{4\pi\varepsilon_o}\right)^2 \frac{1}{R^6} \frac{\left(\sum_{n_1 \neq 1} |\langle n_1|r_1|1\rangle|^2\right)\left(\sum_{n_2 \neq 1} |\langle n_2|r_2|1\rangle|^2\right)}{(I_1 + I_2)} \tag{14.1.43b}$$

Now substitute Eqs. (14.1.42a) and (14.1.42b), so that Eq. (14.1.43b) reads,

$$E_{1,1}^{(2)} = -\left(\frac{2}{3}\right)\left(\frac{e^2}{4\pi\varepsilon_o}\right)^2 \left(\frac{1}{R^6}\right) \frac{\left(\frac{3I_1\alpha_1}{2e^2}\right)\left(\frac{3I_2\alpha_2}{2e^2}\right)}{(I_1 + I_2)} \tag{14.1.43c}$$

After algebraic simplification, Eq. (14.1.43c) leads us to London's formula, which is

$$E_{1,1}^{(2)} = -\left(\frac{3}{2}\right)\left(\frac{1}{4\pi\varepsilon_o}\right)^2 \frac{I_1 I_2}{I_1 + I_2} \frac{\alpha_1\alpha_2}{R^6} \tag{14.1.44}$$

As written, Eq. (14.1.44) can be used when the two atoms are different. Since all atoms and molecules are polarizable, the London interaction is universal. The London force is obtained by differentiating $E_{1,1}^{(2)}$ with respect to R.

14.1.6 Discussion

According to London, a neutral atom becomes subject to an electric field when it finds itself in the vicinity of a neighboring atom. The resulting force acting between the two atoms has its origin in the atomic polarizability, which is frequency dependent. As electromagnetic waves having different frequencies travel at different speeds in such a medium, any starting wave pattern formed by superposition of waves of different frequencies will suffer dispersion in both time and distance. As a consequence, the London force is often referred to as a "dispersive" force.

At separations characteristic of ordinary intermolecular distances, the London force is weaker than most other intermolecular forces. Nonetheless, it acts between all atoms and molecules, whether they are charged or uncharged, polar or non-polar. In the case of the rare gases, for example, the London force accounts for the entire interatomic attraction. Indeed, the rare gases would not liquefy without it. In addition to the liquefaction of rare gases, the London force plays an essential role in determining the bulk cohesive energies in the case of molecular solids, as well as the attractive energies involved in surface phenomena, such as coagulation, adhesion, surface tension, and adsorption. In the next sections of this chapter, we shall exploit the London theory to derive Förster's formula for the non-radiative rate of energy transfer between molecules and then Hamaker's formula for the energy of interaction of two colloidal particles.

It can be shown that Eq. (14.1.19), which forms the basis for derivation of the London force, is identical in form to the dipole–dipole interaction formula derived in Sect. 1.9 of Chap. 1. The proof goes as follows: Rewrite Eq. (14.1.19) as,

$$V' = \frac{1}{4\pi\varepsilon_o}\left[\frac{e^2(x_1x_2 + y_1y_2 + z_1z_2)}{R^3} - 3\frac{e^2 z_1 z_2}{R^3}\right] \tag{14.1.45}$$

Define the instantaneous electric dipole moments,

$$\mathbf{p}_1 = e(x_1\hat{\mathbf{e}}_x + y_1\hat{\mathbf{e}}_y + z_1\hat{\mathbf{e}}_z) \tag{14.1.46a}$$

and

$$\mathbf{p}_2 = e(x_2\hat{\mathbf{e}}_x + y_2\hat{\mathbf{e}}_y + z_2\hat{\mathbf{e}}_z) \tag{14.1.46b}$$

of the two interacting atoms. The inner product of these two dipole moments is

$$\begin{aligned} \mathbf{p}_1 \cdot \mathbf{p}_2 &= e(x_1\hat{\mathbf{e}}_x + y_1\hat{\mathbf{e}}_y + z_1\hat{\mathbf{e}}_z) \cdot e(x_2\hat{\mathbf{e}}_x + y_2\hat{\mathbf{e}}_y + z_2\hat{\mathbf{e}}_z) \\ &= e^2(x_1x_2 + y_1y_2 + z_1z_2) \end{aligned} \tag{14.1.47}$$

The separation, R, is measured in the positive direction along the z-axis. Written in vector form, this separation distance is

$$\mathbf{R} = R\hat{\mathbf{e}}_z \tag{14.1.48a}$$

Equation (14.1.48a) can be solved to obtain

$$\hat{\mathbf{e}}_z = \frac{\mathbf{R}}{R} \tag{14.1.48b}$$

Next use Eqs. (14.1.46a) and (14.1.46b) with Eq. (14.1.48b) to compute

$$\mathbf{p}_1 \cdot \frac{\mathbf{R}}{R} = e(x_1\hat{\mathbf{e}}_x + y_1\hat{\mathbf{e}}_y + z_1\hat{\mathbf{e}}_z) \cdot \hat{\mathbf{e}}_z = ez_1 \tag{14.1.49a}$$

and

$$\mathbf{p}_2 \cdot \frac{\mathbf{R}}{R} = e(x_2\hat{\mathbf{e}}_x + y_2\hat{\mathbf{e}}_y + z_2\hat{\mathbf{e}}_z) \cdot \hat{\mathbf{e}}_z = ez_2 \tag{14.1.49b}$$

Finally, substitute Eqs. (14.1.47), (14.1.49a), and (14.1.1.49b) into Eq. (14.1.45) so that it reads,

$$V' = \frac{1}{4\pi\varepsilon_o}\left[\frac{\mathbf{p}_1 \cdot \mathbf{p}_2}{R^3} - 3\frac{(\mathbf{p}_1 \cdot \mathbf{R})(\mathbf{p}_2 \cdot \mathbf{R})}{R^5}\right] \tag{14.1.50}$$

Equation (14.1.50) has exactly the same structure as Eq. (1.9.12) of Chap. 1. According to Eq. (14.1.50), the basis for the London force is the instantaneous interaction of mutually induced atomic dipoles. As such, it is often referred to as the induced dipole-induced dipole force.

14.2 Förster Resonance Energy Transfer

14.2.1 Phenomenology

Electronic excitation can be transferred from a donor molecule to an acceptor molecule without photon emission by the donor followed by photon absorption by the acceptor molecule. This non-radiative exchange of energy is termed resonance energy transfer. In the resonance energy transfer process, the excited donor molecule makes a transition from an excited electronic state to its ground electronic state, and the energy released reappears instantaneously as electronic excitation in the neighboring acceptor molecule. In electrical terms, a transition dipole in the donor is annihilated, while a transition dipole in the acceptor is created. The perturbation energy at work is the London induced dipole-induced dipole force. Starting with the London model, Th. Förster (German, 1910–1974) used Fermi's Golden Rule to derive a formula for the rate of energy transfer between a donor and an acceptor. The phenomenon has since become known as Forster resonance energy transfer or FRET.

In chemical terms, non-radiative energy transfer can be represented by the stoichiometric equation,

$$D' + A \rightarrow D + A' \tag{14.2.1}$$

where D and D' are the ground state and excited state of the donor, respectively, while A an A' are the ground state and excited state of the acceptor.

14.2.2 Application of Fermi's Golden Rule

According to Eq. (14.2.1), the wave function corresponding to the initial quantum state of the system, made up of the donor in its excited state and the acceptor in its ground state, is $|I\rangle = |D'\rangle|A\rangle$, while the wave function corresponding to the final quantum state consisting of the donor in the ground state and the acceptor in the excited state is $|F\rangle = |D\rangle|A'\rangle$. We denote the induced dipole in the donor by

$$\mu_D = er_1 \tag{14.2.2a}$$

where r_1 is the position coordinate of the excited electron in the donor. The acceptor induced dipole is denoted by

$$\mu_A = er_2 \tag{14.2.2b}$$

where r_2 is the position coordinate of the electron in the acceptor. The matrix element connecting the electronic excited state of the donor, $|n_1\rangle$, with quantum number n_1

to the ground electronic state of the donor, $|1\rangle$, with quantum number $n_1 = 1$ is

$$\left|\langle D'|\mu_D|D\rangle\right| = |\langle n_1|er_1|1\rangle| \tag{14.2.2c}$$

The matrix element controlling the electronic transition in the acceptor from state $|1\rangle$ where $n_2 = 1$ to the state $|n_2\rangle$ with quantum number, n_2 is

$$\left|\langle A|\mu_A|A'\rangle\right| = |\langle 1|er_2|n_2\rangle| \tag{14.2.2d}$$

If the initial state is $|I\rangle = |D'\rangle|A\rangle$, and the final state is $|F\rangle = |D\rangle|A'\rangle$, the dipole–dipole perturbation matrix element, which replaces Eq. (14.1.33) is

$$\left|\langle F|\mathrm{H}'|I\rangle\right|^2 = \langle\kappa^2\rangle\frac{\left|\langle D'|\mu_D|D\rangle\right|^2\left|\langle A|\mu_A|A'\rangle\right|^2}{\left(4\pi\varepsilon_o R^3\right)^2} \tag{14.2.3}$$

where the generic symbol $\langle\kappa^2\rangle$ substitutes for the factor of $2/3$, which appears in Eq. (14.1.33). As shown in the previous section, $\langle\kappa^2\rangle = 2/3$ applies when the possible orientations of the transition dipole in the donor are not correlated with the orientations of the transition dipole in the acceptor. In the case of resonance energy transfer involving donor and acceptor molecules with directional covalent bonds, however, the assumption of uncorrelated orientation is not always justified. Random orientations are obviously ruled out in the case of molecules that have fixed positions as in a crystalline solid. To cope with this ambiguity, it is customary in the Förster theory to substitute the generic $\langle\kappa^2\rangle$ for the factor of (2/3) that appears Eq. (14.1.33).

When applied to the energy transfer process as defined by Eq. (14.2.1), the Fermi Golden rule in Eq. (11.7.14) of Chap. 11 reads,

$$W_{D\to A} = \frac{2\pi}{\hbar}\left|\langle F|H'|I\rangle\right|^2 D(E_F) \tag{14.2.4}$$

In Eq. (14.2.4), the absolute square matrix element $\left|\langle F|H'|I\rangle\right|^2$ is as defined by Eq. (14.2.3). The total energy in the initial state is E_I, while the total energy in the final state is E_F. The density of final states is $D(E_F)$. We can substitute Eq. (14.2.3) into Eq. (14.2.4) to obtain

$$W_{D\to A} = \frac{2\pi}{\hbar}\left[\frac{\langle\kappa^2\rangle}{\left(4\pi\varepsilon_o R^3\right)^2}\right]\left|\langle D'|\mu_D|D\rangle\right|^2\left|\langle A|\mu_A|A'\rangle\right|^2 D(E_F) \tag{14.2.5}$$

Starting with Eq. (14.2.5), the main task will be the evaluation of the matrix elements, $\left|\langle D|\mu_D|D'\rangle\right|^2$ and $\left|\langle A'|\mu_A|A\rangle\right|^2$[3]. This must take into account the conservation of energy.

14.2.3 Conservation of Energy

As an aid to the evaluation of the matrix elements for an arbitrary donor and acceptor pair, we refer to the energy level schemes diagrammed in Fig. 14.2.

In Fig. 14.2, we note the following symbols:

$$w_D = \text{energy of the donor ground state in } D \tag{14.2.6a}$$

$$w_{D'} = \text{energy of the donor excited state in } D' \tag{14.2.6b}$$

$$w_A = \text{energy of the acceptor ground state in A} \tag{14.2.6c}$$

$$w_{A'} = \text{energy of the acceptor excited state in } A' \tag{14.2.6d}$$

In terms of Eqs. (14.2.6b) and (14.2.6c), the initial energy of the system consisting of D' and A is

$$E_I = w_A + w_{D'} \tag{14.2.7}$$

According to Eqs. (14.2.6a) and (14.2.6d), the final energy of the system consisting of D and A' is

$$E_F = w_{A'} + w_D \tag{14.2.8}$$

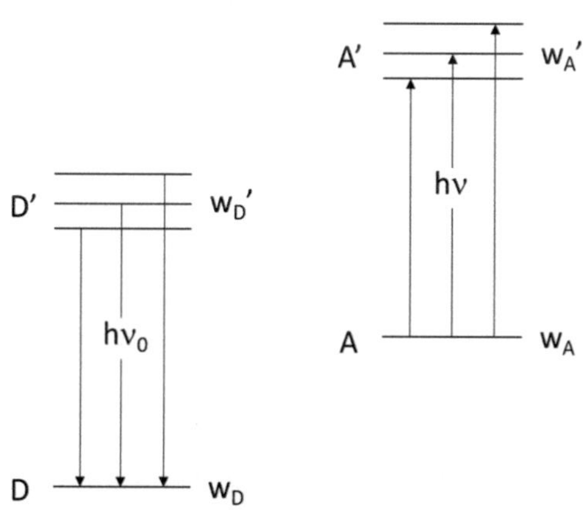

Fig. 14.2 Donor energy levels are on the left, and acceptor energy levels are on the right. The ground state of the donor is D. The excited state of the donor is D'. An energy level in D is w_D. An energy level in D' is $w_{D'}$ The quantum of energy separating $w_{D'}$ from w_D is $h\nu_o$. The corresponding states in the acceptor are A and A'. The corresponding energy levels are w_A and $w_{A'}$. The quantum of energy separating $w_{A'}$ from w_A is $h\nu$

The difference in energies is

$$E_F - E_I = (w_{A'} + w_D) - (w_A + w_{D'}) \tag{14.2.9a}$$

or upon rearranging

$$E_F - E_I = (w_{A'} - w_A) - (w_{D'} - w_D) \tag{14.2.9b}$$

Define

$$h\nu = w_{A'} - w_A \tag{14.2.10a}$$

and

$$h\nu_o = w_{D'} - w_D \tag{14.2.10b}$$

Substitute Eqs. (14.2.10a) and (14.2.10b) into Eq. (14.2.9b) to obtain

$$E_F - E_I = h\nu - h\nu_o \tag{14.2.11}$$

where as usual, $h = 2\pi\hbar$, is the Planck constant. In proceeding further, it will be helpful to replace the density of states per unit energy, $D(E)$, by the density of states per unit frequency, $n(\nu)$, evaluated at the acceptor resonance frequency, ν, where

$$D(E) = \frac{n(\nu)}{h} \tag{14.2.12}$$

The units of $n(\nu)$ are seconds. With this substitution, Eq. (14.2.5) reads,

$$W_{D\to A} = \frac{2\pi}{\hbar}\left[\frac{\langle\kappa^2\rangle}{\left(4\pi\varepsilon_o R^3\right)^2}\right]\left|\langle D'|\mu_D|D\rangle\right|^2\left|\langle A|\mu_A|A'\rangle\right|^2\frac{n(\nu)}{h} \tag{14.2.13}$$

To make a comparison of Eq. (14.2.13) with experiment, we search for representations of the matrix elements, which correspond to measurements in the laboratory. It will turn out that the matrix elements have spectroscopic interpretations.

14.2.4 Absorber Response in Terms of the Einstein B-Coefficient

The density of states function, $n(\nu)$, counts the number of A molecules per unit frequency. These molecules are capable of absorbing a photon in the frequency range, ν and $\nu + d\nu$. The energy absorbed per unit frequency by $n(\nu)$ ground states is $h\nu n(\nu)$.

Let $\rho(\nu)$ be the photon energy density in the units of J/m^3. In spectroscopic absorption, the photon energy flux passing through the material is diminished as a result of the bimolecular reactions between photons with energy density, $\rho(\nu)$, and absorbers with density, $n(\nu)$. The energy captured by the absorbers is $h\nu n(\nu)$. According to Eq. (11.8.59) of Chap. 11, the effective rate constant for energy transfer is the Einstein coefficient,B_A, which is given by

$$B_A = \frac{\left|\langle A'|\mu_A|A\rangle\right|^2}{6\varepsilon_o\hbar^2} \tag{14.2.14}$$

If S is the specific rate in the units of J/s at which the energy is absorbed by the acceptor molecules, then

$$S = B_A\rho(\nu)h\nu n(\nu) \tag{14.2.15a}$$

Alternatively, one can look at the absorption process in terms of the effective absorption cross section,$\sigma_A(\nu)$, which has the units of m^2. Here, we are referring to the absorption cross section model alluded to in the context of the Beer–Lambert law, which we derived by exclusively electromagnetic means in Chap. 7 (See Sect. 7.11). The speed of light in the material is $\mathrm{v} = (c/n)$, where n is the refractive index of the material. Hence, the photon absorption rate in the units of J/s is

$$S = \sigma_A(\nu)\rho(\nu)(c/n) \tag{14.2.15b}$$

The use of n to represent the refractive index should not be confused with $n(\nu)$, which represents a density of states at frequency, ν. If Eqs. (14.2.15a) and (14.2.15b) are set equal, the result is

$$B_A\rho(\nu)h\nu n(\nu) = \sigma_A(\nu)\rho(\nu)(c/n) \tag{14.2.16}$$

from which we conclude that B_A and $\sigma_A(\nu)$ are linked by the relation,

$$B_A = \frac{\sigma_A(\nu)c}{n(\nu)h\nu n} \tag{14.2.17}$$

When B_A is eliminated between Eqs. (14.2.14) and (14.2.17), one finds,

$$\frac{\left|\langle A'|\mu_A|A\rangle\right|^2}{6\varepsilon_o\hbar^2} = \frac{\sigma_A(\nu)c}{n(\nu)h\nu n} \tag{14.2.18}$$

Equation (14.2.18) can be solved for $\left|\langle A'|\mu_A|A\rangle\right|^2$. The result is,

$$\left|\langle A'|\mu_A|A\rangle\right|^2 = \frac{\sigma_A(\nu)c}{n(\nu)h\nu n}\left(6\varepsilon_o\hbar^2\right) \tag{14.2.19}$$

Equation (14.2.19) provides an expression for $\left|\langle A'|\mu_A|A\rangle\right|^2$ in terms of physical constants and the material parameters, $\sigma_A(\nu)$ and $n(\nu)$, which are potentially measurable.

14.2.5 Donor Action in Terms of the Einstein A-Coefficient

As explained in Chap. 11, the Einstein "A" coefficient (units are s^{-1}) determines the rate of spontaneous emission of a photon by an excited state. Applied to photon emission from the donor excited state, D', Eq. (11.9.2) reads,

$$A_{D'} = \frac{8\pi^2\nu^3\left|\langle D|\mu_D|D'\rangle\right|^2}{3\hbar\varepsilon_o(c/n)^3} \tag{14.2.20}$$

where we have replaced c in Eq. (11.9.2) by c/n to take into account the passage of light through matter of refractive index, n. The rate of spontaneous emission from D', and the excited state characteristic lifetime, $\tau_{D'}$, are related by

$$A_{D'} = 1/\tau_{D'} \tag{14.2.21}$$

Equations (14.2.20) and (14.2.21) can be solved for the electric dipole matrix element with the result,

$$\left|\langle D|\mu_D|D'\rangle\right|^2 = \left(\frac{3\hbar}{8\pi^2}\right)\frac{\varepsilon_o(c/n)^3}{\nu^3\tau_{D'}} \tag{14.2.22}$$

Equation (14.2.22) provides an expression for $\left|\langle D|\mu_D|D'\rangle\right|^2$ in terms of physical constants and the measureable excited state lifetime, $\tau_{D'}$.

14.2.6 The Transfer Rate

Now substitute Eqs. (14.2.19) and (14.2.22) into Eq. (14.2.13) with the result,

$$W_{D'\to A} = \frac{2\pi}{\hbar}\left(\frac{\langle\kappa^2\rangle}{(4\pi\varepsilon_o)^2R^6}\right)\left(\frac{3\hbar\varepsilon_o c^3}{8\pi^2\nu^3n^3}\right)\left(\frac{1}{\tau_{D'}}\right)\left(\frac{\sigma_A(\nu)c}{n(\nu)h\nu n}\right)\left(6\varepsilon_o\hbar^2\right)\left(\frac{n(\nu)}{h}\right) \tag{14.2.23}$$

As is often the case, the algebra required to simplify a complicated expression, such as Eq. (14.2.23), stands in the way of our having complete confidence in the final result. In this regard, we hope that the reader agrees that the work of reproducing the algebra in this case is worthwhile.

If we replace h by $2\pi\hbar$ where it occurs in Eq. (14.2.23), the result is

$$W_{D'\to A} = \frac{2\pi}{\hbar}\left(\frac{\langle\kappa^2\rangle}{(4\pi\varepsilon_o)^2R^6}\right)\left(\frac{3\hbar\varepsilon_o(c/n)^3}{8\pi^2\nu^3\tau_{D'}}\right)\left(\frac{\sigma_A(\nu)(c/n)\cdot 6\hbar^2\varepsilon_o}{2\pi\hbar\nu\cdot n(\nu)}\right)\frac{n(\nu)}{2\pi\hbar} \tag{14.2.24a}$$

We can factor Eq. (14.2.24a) so that like terms appear together. The result is

$$W_{D'\to A} = \left[\frac{2\pi\cdot 3\cdot 6}{16\pi^2\cdot 8\pi^2\cdot 2\pi\cdot 2\pi}\right]\cdot\left[\frac{\hbar\cdot\hbar^2}{\hbar\cdot\hbar\cdot\hbar}\right]\cdot\left[\frac{\varepsilon_o\cdot\varepsilon_o}{\varepsilon_o^2}\right]\cdot\left[\frac{(c/n)^3\cdot(c/n)}{\nu^3\cdot\nu}\right]\cdot\left[\frac{n(\nu)}{n(\nu)}\right]\cdot\frac{\langle\kappa^2\rangle\sigma_A(\nu)}{\tau_{D'}R^6} \tag{14.2.24b}$$

In Eq. (14.2.24b), the bracket involving the numerical constants and powers of π can be rewritten using factorization and rearrangement so that only constants and powers of 2π appear. The common algebraic factors found in the numerators and denominators of the remaining fractions can be cancelled where possible and the results grouped together. One finds,

$$W_{D'\to A} = \left[\frac{(2\pi)\cdot 3\cdot 3\cdot 2}{2^2\cdot(2\pi)^2\cdot 2\cdot(2\pi)^2\cdot(2\pi)^2}\right]\left[\frac{(c/n)^4}{\nu^4}\right]\cdot\frac{\langle\kappa^2\rangle\sigma_A(\nu)}{\tau_{D'}R^6} \tag{14.2.25a}$$

Replace ν with $\omega/2\pi$ in Eq. (14.2.25a) so that it reads,

$$W_{D'\to A} = \left[\frac{(2\pi)\cdot 3\cdot 3\cdot 2}{2^2\cdot(2\pi)^2\cdot 2\cdot(2\pi)^2\cdot(2\pi)^2}\right]\left[\frac{(2\pi)^4}{\omega^4}\right]\cdot\frac{(c/n)^4\langle\kappa^2\rangle\sigma_A(\nu)}{\tau_{D'}R^6} \tag{14.2.25b}$$

Upon making the indicated cancellations in Eq. (14.2.25b), we obtain the final result,

$$W_{D'\to A} = \left(\frac{9\langle\kappa^2\rangle}{8\pi}\right)\left(\frac{c}{\omega}\right)^4\left(\frac{\sigma_A(\nu)}{\tau_{D'}n^4}\right)\cdot\frac{1}{R^6} \tag{14.2.26}$$

The Golden Rule transition rate, $W_{D'\to A}$ defines a first order rate coefficient, $k_{D'A}$ (units are s^{-1}) for transfer of excitation from D' to A. As $k_{D'A}$ is equal to $W_{D'\to A}$, Eq. (14.2.26) provides the final result,

$$k_{D'A} = \frac{9\langle\kappa^2\rangle}{8\pi}\left(\frac{c}{\omega}\right)^4\left(\frac{\sigma_A(\nu)}{\tau_{D'}n^4}\right)\frac{1}{R^6} \tag{14.2.27}$$

14.2.7 The Transfer Rate Constant in Practical Units

We now express the cross section, $\sigma_A(\nu)$ (units are cm^2) in terms of the more familiar chemical spectroscopic absorbtivity coefficient, ε_A (units are L/mol cm). To do this, we introduce the photon flux, I (units are cm^{-2} s^{-1}), and the number of absorbers per unit volume, $\overline{N}_A$ (cm^{-3}). The thickness of the sample is ℓ (units are cm). The governing equation is

$$\frac{dI}{d\ell} = -\sigma_A(\nu)\overline{N}_A I \tag{14.2.28a}$$

Integrate both sides to obtain,

$$I = I_o \exp(-\sigma_A(\nu)\overline{N}_A \ell) \tag{14.2.28b}$$

where I_o is the photon flux incident on the material sample. Equation (14.2.28b) is the Beer–Lambert Law in exponential form. It is tradition to express the Beer–Lambert Law in decadic form as

$$I = I_o \cdot 10^{-\varepsilon_A [A]\ell} \tag{14.2.29}$$

where $[A]$ (units are mol/L) is the molar concentration of the absorbers. We can obtain a relation between $\sigma_A(\nu)$ and ε_A by equating the natural logarithms of Eqs. (14.2.28b) and (14.2.29) with the result

$$\ln I_o - \sigma_A(\nu)\overline{N}_A \ell = \ln I_o - \varepsilon_A [A]\ell \ln(10) \tag{14.2.30}$$

On the basis of Eq. (14.2.30), one can conclude that

$$\sigma_A(\nu) = \frac{\varepsilon_A [A] \ln(10)}{\overline{N}_A} \tag{14.2.31}$$

The number density, $\overline{N}_A$, can be expressed in terms of the molar concentration, $[A]$, by writing

$$\overline{N}_A(\text{cm}^{-3}) = [A](\text{mol/L}) \times \frac{\text{L}}{10^3\text{cm}^3} \times \tilde{N}(\text{mol}^{-1}) \tag{14.2.32a}$$

where $\tilde{N} = 6.022 \times 10^{23}$ is the Avogadro number. Equation (14.2.32a) can be solved for $[A]/\overline{N}_A$ with the result,

$$\frac{[A](\text{mol/L})}{\overline{N}_A(\text{cm}^{-3})} = \frac{10^3\ \text{cm}^3/\text{L}}{\tilde{N}(\text{mol}^{-1})} \tag{14.2.32b}$$

Substitution of Eq. (14.2.32b) into Eq. (14.2.31) leads to the result,

$$\sigma_A(\text{cm}^2) = \varepsilon_A(\text{L/mol cm}) \times \ln(10) \times \left(\frac{10^3\text{cm}^3/\text{L}}{\tilde{N}(\text{mol}^{-1})}\right) \tag{14.2.32c}$$

With the units suppressed in the conversion factor, Eq. (14.2.32c) gives for $\sigma_A(\nu)$ in the units of cm^2,

$$\sigma_A(\nu) = \frac{1000\varepsilon_A \ln(10)}{\tilde{N}} \tag{14.2.33}$$

Substitute Eq. (14.2.33) into Eq. (14.2.27). The result is

$$k_{D'A} = \frac{9\langle\kappa^2\rangle}{8\pi}\left(\frac{c}{\omega}\right)^4\left(\frac{1000\varepsilon_A \ln(10)}{\tau_{D'}n^4\tilde{N}}\right)\frac{1}{R^6} \tag{14.2.34}$$

After combining the numerical constants, one obtains finally,

$$k_{D'A} = \frac{9000\langle\kappa^2\rangle}{8\pi}\left(\frac{c}{\omega}\right)^4\left(\frac{\varepsilon_A \ln(10)}{\tau_{D'}n^4\tilde{N}}\right)\frac{1}{R^6} \tag{14.2.35}$$

Equation (14.2.35) exhibits the characteristic dependence of $k_{D'A}$ upon $1/R^6$ and also serves to reveal the origin of the famous factor of 9000 that appeared in Förster's original formula.

Because of the presence of closely spaced substates in A' and D, there may be a number of donor and the acceptor states within each manifold, that are able to satisfy the condition $h\nu_o = w_{D'} - w_D = w_{A'} - w_A = h\nu$. The fluorescence spectrum of the donor is the origin of the $D' \to D$ electronic transition, whereas the absorption spectrum of the acceptor is the origin of the $A \to A'$ transition. See Fig. 14.3.

As shown in Fig. 14.3, the absorption spectrum of the acceptor overlaps the fluorescence spectrum of the donor. To take this into account, we replace ε_A in Eq. (14.2.35) by the absorption spectrum of the acceptor, $\varepsilon_A(\omega)$, and multiply by the fluorescence spectrum, $f_D(\omega)$, of the donor. Equation (14.2.35) is then integrated over the frequency to obtain,

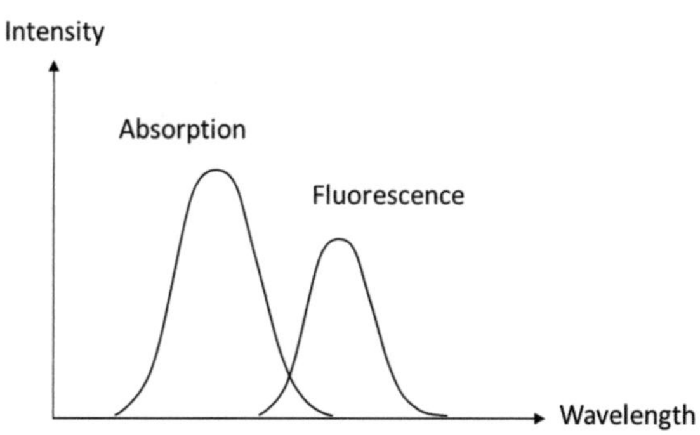

Fig. 14.3 Fluorescence spectrum of D' and absorption spectrum of A plotted in the form of intensity versus wavelength. The absorption peak is a measure of the density of A states while the fluorescence peak is a measure of the density of D' states

$$k_{D'A} = \frac{9000}{8\pi}\left(\frac{c}{n}\right)^4 \frac{\ln(10)}{\tilde{N}}\left(\frac{\langle\kappa^2\rangle}{\tau_{D'}}\right)\left(\int \frac{\varepsilon_A(\omega) f_D(\omega) d\omega}{\omega^4}\right)\frac{1}{R^6} \tag{14.2.36}$$

which shows how the rate, $k_{D'A}$, of the non-radiative transition can be expressed in terms of the spectra associated with the underlying radiative transitions.

14.2.8 Discussion

The analysis above shows that the physical basis for Förster resonant energy transfer is the collapse of the donor transition electric dipole followed by the excitation of the acceptor transition electric dipole. Because the mechanism involves the interaction between induced dipoles, the theory of FRET finds its basis in the London theory.

FRET, however, is not the only mechanism of decay of the excited state, D'. There is also fluorescence of D', which occurs at the specific rate, $(1/\tau_{D'})$. In addition, there are various non-radiative collision mechanisms. These non-radiative transition rates can be lumped together into a composite rate constant denoted by k_{NR}. Of interest in normalizing the FRET rate is the total rate of disappearance of donor excitation. This is given by the FRET rate plus the fluorescence rate plus the non-radiative rate or $k_{D'A} + (1/\tau_{D'}) + k_{NR}$. The FRET quantum yield, Φ_{FRET}, is defined as the ratio,

$$\Phi_{FRET} = \frac{k_{D'A}}{k_{D'A} + (1/\tau_{D'}) + k_{NR}} \tag{14.2.37a}$$

Equation (14.2.37a) can rearranged to read,

$$\Phi_{FRET} = \frac{1}{1 + \left(\frac{(1/\tau_{D'}) + k_{NR}}{k_{D'A}}\right)} \tag{14.2.37b}$$

For the purposes of comparison with experimental data, we can define an effective radius, R_o, by writing the numerator of the fraction in Eq. (14.2.37b) in the form,

$$\left(\frac{1}{\tau_{D'}} + k_{NR}\right) = \frac{1}{R_o^6}\left[\frac{9000}{8\pi}\left(\frac{c}{n}\right)^4 \frac{\ln(10)}{\tilde{N}}\left(\frac{\langle\kappa^2\rangle}{\tau_{D'}}\right)\left(\int \frac{\varepsilon_A(\omega) f_D(\omega) d\omega}{\omega^4}\right)\right] \tag{14.2.37c}$$

Upon using Eq. (14.2.36) to replace $k_{D'A}$ and substituting Eq. (14.2.37c), we can rewrite Eq. (14.2.37b) as

$$\Phi_{FRET} = \frac{1}{1 + (R/R_o)^6} \tag{14.2.38}$$

The parameter, R_o, known as the Förster radius, is the value of the radius that would make the FRET effect equal to the sum of the competing effects. The value of R_o ordinarily lies in the range of 3–6 nm. According to Eq. (14.2.38), when $R = R_o$, the quantum yield is $\Phi_{FRET} = 0.5$.

Equation (14.2.38) serves as the basis for measuring the distance between two separate molecules or between two chromophores on the same molecule. The equation has found use in biophysics and biochemistry as a yard stick for determining the molecular distances associated with protein conformational changes, protein–protein interactions, and protein–DNA interactions. When examining plant photosynthesis, for example, Förster found that after a chlorophyll molecule had absorbed a photon, it was possible for a sequence of resonant energy transfers to cause the excitation to migrate across 10,000 chlorophyll molecules before ultimately dying out due to competition from fluorescence and non-radiative processes. In inorganic chemistry, FRET has been used to measure the distance between chromophores on nano-particles. To obtain best results in the application of Eq. (14.2.38), the donor and acceptor should be separated by about 10 nm.

14.3 Hamaker Theory of the Dispersion Interaction Between Spherical Particles

14.3.1 Energy of Interaction of a Molecule with a Sphere

H. C. Hamaker (Dutch, 1905–1993) used London's theory to account for the dipolar induction forces acting between objects larger than molecules [4]. Below, we fill in the steps in the derivations of his main results.

Figure 14.4 shows a molecule located at point, P, interacting by way of the London force with the atoms on the surface of a spherical object, whose center is located at point O.

According to Eq. (14.1.44), the energy of interaction of the molecule at P with a molecule in the sphere located at a distance, r, from P is

$$E(r) = -\left(\frac{3}{2}\right)\left(\frac{1}{4\pi\varepsilon_o}\right)^2 \frac{I_1 I_2}{I_1 + I_2} \frac{\alpha_1 \alpha_2}{r^6} \tag{14.3.1a}$$

The total energy, E_P, of the molecule at P interacting with all of the molecules in the sphere can then be calculated by integration of $E(r)$ over the volume, V, of the sphere. The required integral is

$$E_P = -\lambda \int_V \left(\frac{1}{r^6}\right) Q dV \tag{14.3.1b}$$

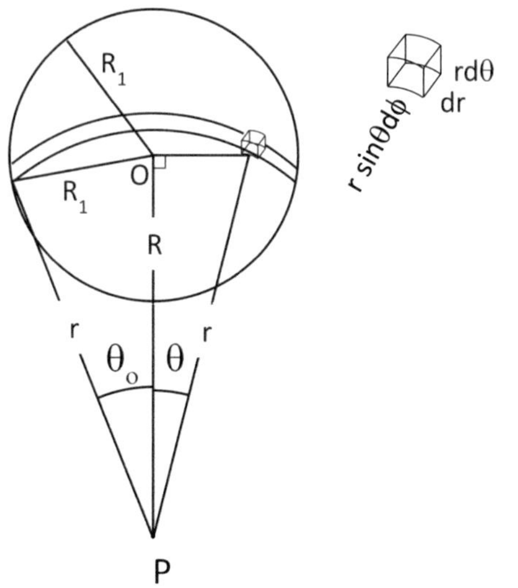

Fig. 14.4 A molecule at point P located a distance, R, from a sphere of radius, R_1, with center at O. Within the sphere, there is a strip of spherical shell having width, $rd\theta$, and circumference, $2\pi r \sin\theta$, where r is the distance from P to the strip. A rectangular volume having sides $rd\theta$, $r\sin\theta d\phi$, and dr is located on the shell. The angle θ_o corresponds to r being tangent to the sphere. The range of θ is $0 \le \theta \le \theta_o$

where

$$\lambda = \frac{3\alpha_1\alpha_2}{32\pi^2\varepsilon_o^2}\frac{I_1 I_2}{I_1 + I_2} \tag{14.3.1c}$$

In Eq. (14.3.1b), Q is the number density of molecules in the sphere, which makes QdV equal to the number of molecules in the differential volume element, dV. The integral runs over the total volume, V, of the sphere. The number density, Q, is a constant, independent of r.

As shown in Fig. 14.4, the distance between P and the center of the sphere at O is R. The radius of the sphere is R_1. The cubical differential volume element shown interior to the sphere contains atoms that are a distance r away from the molecule at point P. The radius that defines the band where the cube is located is $r\sin\theta$, where θ is the angle between r and R. In spherical polar coordinates, the width of the cube is $r\sin\theta d\phi$. The height is $rd\theta$, and the depth is dr. The circumference of the band containing the cube is thus $2\pi r\sin\theta$. The height of the band being, $rd\theta$, the area of the band is the product, $2\pi r\sin\theta \cdot rd\theta$, or $2\pi r^2\sin\theta d\theta$. The total surface area up to angle θ is given by the integral,

$$\text{Area} = 2\pi r^2 \int_0^\theta \sin\theta' d\theta' = 2\pi r^2\left[-\cos\theta'\right]_0^\theta = 2\pi r^2(1-\cos\theta) \tag{14.3.2a}$$

According to the figure, the three distances, r, R_1, and R form a triangle. Taking into account their respective directions, we can write them as the vectors, $\mathbf{r}$, $\mathbf{R_1}$, and $\mathbf{R}$, respectively. With the origin at P, the vectors satisfy the triangle rule $\mathbf{r} = \mathbf{R} + \mathbf{R_1}$ or alternatively,

$$\mathbf{R_1} = \mathbf{r} - \mathbf{R} \tag{14.3.2b}$$

To determine the angle between $\mathbf{r}$ and $\mathbf{R}$, we use Eq. (14.3.2b) to compute the inner product,

$$\mathbf{R_1} \cdot \mathbf{R_1} = (\mathbf{r} - \mathbf{R}) \cdot (\mathbf{r} - \mathbf{R}) = \mathbf{R} \cdot \mathbf{R} + \mathbf{r} \cdot \mathbf{r} - 2\mathbf{r} \cdot \mathbf{R} \tag{14.3.2c}$$

In terms of the lengths of the vectors, Eq. (14.3.2c) reads

$$R_1^2 = R^2 + r^2 - 2Rr\cos\theta \tag{14.3.2d}$$

As R and R_1 are fixed, θ varies as a function of r. According to Eq. (14.3.2d)

$$\cos\theta = \frac{R^2 + r^2 - R_1^2}{2Rr} \tag{14.3.2e}$$

Upon substitution of Eq. (14.3.2e) into Eq. (14.3.2a), one obtains for the area of the ring,

$$\text{Area} = 2\pi r^2\left(1 - \frac{R^2 + r^2 - R_1^2}{2Rr}\right) = 2\pi r^2\left(\frac{2Rr - R^2 - r^2 + R_1^2}{2Rr}\right) \tag{14.3.2f}$$

Taking note of the perfect square in the numerator, Eq. (14.3.2f) can be rewritten as

$$\text{Area} = \pi r\left[\frac{R_1^2 - (R - r)^2}{R}\right] \tag{14.3.3}$$

In terms of the area, the differential volume element is, $dV = (\text{Area})\cdot dr$. It then follows that the number of molecules at a distance r from P within dV is $QdV = Q(\text{Area})dr$. Hence, the total energy as given by Eq. (14.3.1b) takes the form,

$$E_P(R) = -\lambda Q \int dr \cdot \text{Area} \cdot \left(\frac{1}{r^6}\right) \tag{14.3.4a}$$

By writing, $E_P = E_P(R)$ in Eq. (14.3.4a), we stress the fact that R is the key adjustable parameter determining E_P.

With reference to the center at O, a point on the sphere is located by the vector, $\mathbf{R}_1$. The extreme values of r coincide with $\mathbf{R}_1$ pointing toward P and $\mathbf{R}_1$ pointing away from P. In the former case, $r = R - R_1$, and in the latter, $r = R + R_1$. These two equations establish the limits of integration in Eq. (14.3.4a).

The point, P, which is exterior to the sphere, serves as the origin of coordinates. [As a test of the feasibility of using this peculiar exterior point as a center of coordinates, we show in the Mathematical Methods Appendix that the integral of $dV = (Area)dr$ equals the volume of the sphere, $4\pi R_1^3/3$, as expected.]

After substitution of Eq. (14.3.3) into Eq. (14.3.4a) and taking into account the limits of integration, the latter becomes,

$$E_P(R) = -\lambda Q \int_{R-R_1}^{R+R_1} dr\pi r\left[\frac{R_1^2 - (R-r)^2}{R}\right]\left(\frac{1}{r^6}\right) \tag{14.3.4b}$$

As an aid to further calculation, we write Eq. (14.3.4b) in the form,

$$E_P(R) = \frac{-\pi\lambda Q}{R} I(R) \tag{14.3.4c}$$

where the integral, $I(R)$, is defined by

$$I(R) = \int_{R-R_1}^{R+R_1} dr \frac{R_1^2 - (r-R)^2}{r^5} \tag{14.3.5}$$

14.3.2 The Molecule–Sphere Energy Formula

The task now is to evaluate $I(R)$ in Eq. (14.3.5). Calculation is aided if we make the substitution

$$r = x + R \tag{14.3.6a}$$

or equivalently,

$$x = r - R \tag{14.3.6b}$$

With R held constant, Eq. (14.3.6a) implies $dr = dx$. The upper limit of integration, $r = R + R_1$, corresponds to $x = +R_1$, whereas the lower limit of integration, $r = R - R_1$, corresponds to $x = -R_1$. With the introduction of x as the variable of integration, $I(R)$ in Eq. (14.3.5) can be written as the sum of two integrals,

$$I(R) = R_1^2 \int_{-R_1}^{+R_1} \frac{dx}{(x+R)^5} - \int_{-R_1}^{+R_1} dx \frac{x^2}{(x+R)^5} \tag{14.3.7}$$

The first integral in Eq. (14.3.7) is elementary. The second integral can be reduced to elementary integrals by the method of partial fractions. Write the integrand in the second integral in the form,

$$\frac{-x^2}{(x+R)^5} = \frac{A_1}{(x+R)} + \frac{A_2}{(x+R)^2} + \frac{A_3}{(x+R)^3} + \frac{A_4}{(x+R)^4} + \frac{A_5}{(x+R)^5} \tag{14.3.8a}$$

where the values of the arbitrary coefficients, $(A_1, A_2, A_3, A_4, A_5)$, are yet to be determined. To determine them, multiply both sides of Eq. (14.3.8a) by $(x+R)^5$. The result is

$$\begin{aligned} -x^2 &= A_1(x+R)^4 + A_2(x+R)^3 + A_3(x+R)^2 + A_4(x+R) + A_5 \\ &= A_1\left(x^4 + 4x^3R + 6x^2R^2 + 4xR^3 + R^4\right) \\ &\quad + A_2\left(x^3 + 3x^2R + 3xR^2 + R^3\right) + A_3\left(x^2 + 2xR + R^2\right) + A_4(x+R) + A_5 \\ &= A_1x^4 + (4A_1R + A_2)x^3 + \left(6A_1R^2 + 3A_2R + A_3\right)x^2 \\ &\quad + \left(4A_1R^3 + 3A_2R^2 + 2A_3R + A_4\right)x \\ &\quad + \left(A_1R^4 + A_2R^3 + A_3R^2 + A_4R + A_5\right) \end{aligned} \tag{14.3.8b}$$

After equating coefficients of like powers of x on either side of the equals sign, the undetermined coefficients are found to satisfy the following system of linear equations,

$$A_1 = 0 \tag{14.3.9a}$$

$$4A_1R + A_2 = 0 \tag{14.3.9b}$$

$$6A_1R^2 + 3A_2R + A_3 = -1 \tag{14.3.9c}$$

$$4A_1R^3 + 3A_2R^2 + 2A_3R + A_4 = 0 \tag{14.3.9d}$$

$$A_1R^4 + A_2R^3 + A_3R^2 + A_4R + A_5 = 0 \tag{14.3.9e}$$

The solutions to Eq. (14.3.9) are

$$A_1 = A_2 = 0 \tag{14.3.10a}$$

$$A_3 = -1 \tag{14.3.10b}$$

$$A_4 = -2A_3R = 2R \tag{14.3.10c}$$

$$A_5 = -A_4R - A_3R^2 = -(2R)R - (-1)R^2 = -R^2 \tag{14.3.10d}$$

Upon substitution of Eqs. (14.3.10a) through (14.3.10d) into Eq. (14.3.8a), the result is

$$-\frac{x^2}{(x+R)^5} = -\frac{1}{(x+R)^3} + \frac{2R}{(x+R)^4} - \frac{R^2}{(x+R)^5}. \tag{14.3.11}$$

Now substitute Eq. (14.3.11) into Eq. (14.3.7) to obtain

$$\begin{aligned} I(R) &= R_1^2 \int_{-R_1}^{+R_1} \frac{dx}{(x+R)^5} - \int_{-R_1}^{+R_1} \frac{dx}{(x+R)^3} + 2R \int_{-R_1}^{+R_1} \frac{dx}{(x+R)^4} - R^2 \int_{-R_1}^{+R_1} \frac{dx}{(x+R)^5} \\ &= \left[\frac{-R_1^2}{4(x+R)^4} + \frac{1}{2(x+R)^2} - \frac{2R}{3}\frac{1}{(x+R)^3} + \frac{R^2}{4}\frac{1}{(x+R)^4} \right]_{-R_1}^{+R_1} \end{aligned} \tag{14.3.12a}$$

Upon collecting terms involving $(x+R)^4$, Eq. (14.3.12a) reads,

$$I(R) = \left[\frac{\left(R^2 - R_1^2\right)}{4(x+R)^4} + \frac{1}{2(x+R)^2} - \frac{2R}{3}\frac{1}{(x+R)^3} \right]_{-R_1}^{+R_1} \tag{14.3.12b}$$

Next factor $(x+R)^2$from the denominators of the fractions, so that

$$I(R) = \left[\frac{1}{(x+R)^2} \left(\frac{\left(R^2 - R_1^2\right)}{4(x+R)^2} + \frac{1}{2} - \frac{2R}{3}\frac{1}{(x+R)} \right) \right]_{-R_1}^{+R_1} \tag{14.3.12c}$$

Write the remaining fractions over a common denominator, so that Eq. (14.3.12c) reads,

$$\begin{aligned} I(R) &= \left[\frac{1}{(x+R)^2} \left(\frac{3\left(R^2 - R_1^2\right) + 6(x+R)^2 - 8R(x+R)}{12(x+R)^2} \right) \right]_{-R_1}^{+R_1} \\ &= \left[\frac{1}{12(x+R)^4} \left(3\left(R^2 - R_1^2\right) + 6(x+R)^2 - 8R(x+R) \right) \right]_{-R_1}^{+R_1} \end{aligned} \tag{14.3.12d}$$

Evaluation of the bracket in Eq. (14.3.12d) at the upper and lower limits of integration leads to the calculation,

$$\begin{aligned} I(R) &= \left\{ \frac{1}{12(R_1+R)^4} \left[3\left(R^2 - R_1^2\right) + 6(R_1+R)^2 - 8R(R_1+R) \right] \right\} \\ &\quad - \left\{ \frac{1}{12(R-R_1)^4} \left[3\left(R^2 - R_1^2\right) + 6(R-R_1)^2 - 8R(R-R_1) \right] \right\} \end{aligned} \tag{14.3.13a}$$

Next factor out $(R_1 + R)$ from the first bracket and $(R - R_1)$ from the second and cancel like terms from the numerators and the denominators. The result is

$$\begin{aligned} I(R) &= \left\{ \frac{(R+R_1)}{12(R_1+R)^4}[3(R-R_1)+6(R_1+R)-8R] \right\} \\ &\quad - \left\{ \frac{(R-R_1)}{12(R-R_1)^4}[3(R+R_1)+6(R-R_1)-8R] \right\} \\ &= \left[\frac{1}{12(R_1+R)^3}(3R-3R_1+6R_1+6R-8R) \right] \\ &\quad - \left[\frac{1}{12(R-R_1)^3}(3R+3R_1+6R-6R_1-8R) \right]^+ \end{aligned} \tag{14.3.13b}$$

After simplification of the numerators in the two brackets in Eq. (14.3.13b), the result is

$$I(R) = \left[\frac{(R+3R_1)}{12(R_1+R)^3} \right] - \left[\frac{(R-3R_1)}{12(R-R_1)^3} \right] \tag{14.3.13c}$$

The fractions in the first and second brackets in Eq. (14.3.13c) can be partitioned to read,

$$I(R) = \left[\frac{(R+R_1)}{12(R_1+R)^3} + \frac{2R_1}{12(R_1+R)^3} \right] - \left[\frac{(R-R_1)}{12(R-R_1)^3} - \frac{2R_1}{12(R-R_1)^3} \right] \tag{14.3.14}$$

Rearrange the fractions in Eq. (14.3.14), and then in cases where the numerators and denominators contain like powers, perform the indicated cancellations. The result is

$$I(R) = \frac{2R_1}{12(R_1+R)^3} + \frac{2R_1}{12(R-R_1)^3} + \frac{1}{12(R_1+R)^2} - \frac{1}{12(R-R_1)^2} \tag{14.3.15}$$

To obtain a formula for the energy of interaction of a molecule with the whole sphere, substitute Eq. (14.3.15) into Eq. (14.3.4c). The result is

$$E_P(R) = \frac{-\pi\lambda Q}{12R} \left[\frac{2R_1}{(R_1+R)^3} + \frac{2R_1}{(R-R_1)^3} + \frac{1}{(R_1+R)^2} - \frac{1}{(R-R_1)^2} \right] \tag{14.3.16}$$

Equation (14.3.16) is the first of Hamaker's main results. It expresses the energy of interaction between a molecule and a sphere. Equation (14.3.16) will appear again when we calculate the energy of interaction of two spheres.

14.3.3 Energy of Interaction of Two Spheres

In Fig. 14.4, the point, P, served as the source point, and the point in the sphere centered on O served as the field point. In the calculation leading to Eq. (14.3.16), P was fixed, and the position of the field point in the sphere was allowed to vary. We now turn to Fig. 14.5. which involves two spheres, one with radius, R_1, centered on O_1 and the other with radius, R_2, centered on O_2. As the energy of interaction between a point and a sphere can be associated with either, we can associate the energy given by $E_P(R)$ with the point, O_1, which in Fig. 14.5 is the center of the sphere of radius, R_1. The field point, which lies in the sphere of radius, R_2, can then be allowed to vary.

Let the shortest distance between O_1 and O_2 be C. At their closest approach, the contours of the two spheres are a distance, d, apart. Hence,

$$C = R_1 + R_2 + d. \tag{14.3.17a}$$

In the calculation that follows, we keep the point at O_1 fixed and integrate over the sphere centered on O_2 by varying, R. In evaluating the area of the sphere centered on O_2, we need to translate from the geometry of Fig. 14.4 to the geometry of Fig. 14.5. The following changes are required: (1) Replace R_1 in Fig. 14.4 by R_2 as defined in Fig. 14.5. (2) Replace r as defined in Fig. 14.4 by R where it appears in Fig. 14.5. (3) Replace R as it is defined in Fig. 14.4 by C. With these changes the area formula applicable to the sphere centered on O_2 in Fig. 14.5 reads,

$$\text{Area}(R) = \pi R\left[\frac{R_2^2 - (C - R)^2}{C}\right] \tag{14.3.17b}$$

The number of molecules per unit volume of the sphere centered on O_2 is Q. Thus, the sum of the energies of interaction of the molecules in the O_2 sphere with those in the O_1nsphere is given by the integral, E, where

$$E = \int dR(\text{Area}(R) \cdot Q \cdot E_P(R)) \tag{14.3.18a}$$

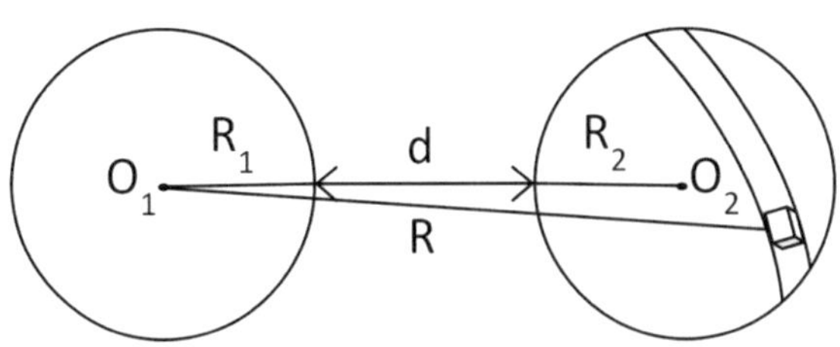

Fig. 14.5 Sphere of radius R_1 with center at O_1 and a second sphere of radius R_2 with center at O_2. At points on the circumference corresponding to the narrowest separation, the spheres are a distance d apart. The distance from O_1 to a differential volume element in the second sphere is R

and $E_P(R)$ is as defined as in Eq. (14.3.16). According to Fig. 14.5, the extreme values of R occur along the line connecting O_1 and O_2. The minimum value of this distance is $C - R_2$, while the maximum value is $C + R_2$. Upon substituting Eq. (14.3.17b) into Eq. (14.3.18a) followed by taking into account these limits, the energy of interaction of the two spheres is given by the integral,

$$E = \int_{C-R_2}^{C+R_2} dR\pi R\left[\frac{R_2^2 - (C-R)^2}{C}\right]QE_P(R) \tag{14.3.18b}$$

After replacing $E_P(R)$ in Eq. (14.3.18b) with Eq. (14.3.16), the integral takes the explicit form

$$E = \int_{C-R_2}^{C+R_2} dR\pi R\left[\frac{R_2^2 - (C-R)^2}{C}\right]Q\left(\frac{-\pi\lambda Q}{12R}\right) \times\left[\frac{2R_1}{(R_1+R)^3} + \frac{2R_1}{(R-R_1)^3} + \frac{1}{(R_1+R)^2} - \frac{1}{(R-R_1)^2}\right] \tag{14.3.18c}$$

The integrand in Eq. (14.3.18c) can be simplified by factoring out the constant terms and canceling a factor of R in the numerator against a factor of R in the denominator. The result is

$$E = -\frac{\pi^2 Q^2 \lambda}{12C}\int_{C-R_2}^{C+R_2} dR\left[R_2^2 - (C-R)^2\right] \times\left[\frac{2R_1}{(R_1+R)^3} + \frac{2R_1}{(R-R_1)^3} + \frac{1}{(R_1+R)^2} - \frac{1}{(R-R_1)^2}\right]. \tag{14.3.18d}$$

As an aid to further calculation, we write Eq. (14.3.18d) in the form,

$$E = -\frac{\pi^2 Q^2 \lambda}{12C} J(C) \tag{14.3.19}$$

where $J(C)$ is the integral defined by

$$J(C) = \int_{C-R_2}^{C+R_2} dR\left[R_2^2 - (C-R)^2\right] \times\left[\frac{2R_1}{(R+R_1)^3} + \frac{2R_1}{(R-R_1)^3} + \frac{1}{(R+R_1)^2} - \frac{1}{(R-R_1)^2}\right] \tag{14.3.20}$$

By identifying the integral as $J(C)$, we emphasize that the key adjustable parameter is C. Evaluation of $J(C)$ in Eq. (14.3.20) is simplified by replacing R as the variable of integration by x, where x is defined by

$$R = x + C \tag{14.3.21a}$$

or equivalently

$$x = R - C \tag{14.3.21b}$$

On the basis of Eq. (14.3.21a), $dR = dx$. With this change in variables, reference to Eq. (14.3.21b) indicates that the upper limit, $R = C + R_2$, in Eq. (14.3.20), can be replaced by $x = +R_2$, while the lower limit, $R = C - R_2$, can be replaced by $x = -R_2$. Further simplification can be achieved by defining the constants,

$$\alpha = C + R_1 \tag{14.3.22a}$$

$$\beta = C - R_1 \tag{14.3.22b}$$

$$\gamma = R_2 \tag{14.3.22c}$$

Equations (14.3.22a) and (14.3.22b) can be solved for

$$2R_1 = (\alpha - \beta) \tag{14.3.23a}$$

$$2C = (\alpha + \beta) \tag{14.3.23b}$$

Equations (14.3.21a) and (14.3.22a) and (14.3.22b) permit us to write

$$R + R_1 = x + C + R_1 = x + \alpha \tag{14.3.24a}$$

$$R - R_1 = x + C - R_1 = x + \beta \tag{14.3.24b}$$

In Eq. (14.3.20), we use Eq. (14.3.21b) to replace $(R - C)$ by x. We use Eq. (14.3.22c) to replace R_2 by γ and Eq. (14.3.23a) to replace by $2R_1$ by $(\alpha - \beta)$. We use Eq. (14.3.24a) to replace $(R + R_1)$ by $(x + \alpha)$ and Eq. (14.3.24b) to replace $(R - R_1)$ by $(x + \beta)$. Finally we let $dR = dx$. After these changes, Eq. (14.13.3.20) reads,

$$J(C) = \int_{-R_2}^{+R_2} dx\left(\gamma^2 - x^2\right)\left[\frac{(\alpha-\beta)}{(x+\alpha)^3} + \frac{(\alpha-\beta)}{(x+\beta)^3} + \frac{1}{(x+\alpha)^2} - \frac{1}{(x+\beta)^2}\right] \tag{14.3.25}$$

Evaluation of Eq. (14.3.25) requires the computation of four integrals. The method of partial fractions will prove useful.

14.3.4 Evaluation of Integrals

We first turn to the integral in Eq. (14.3.25) which involves $(x+\alpha)^3$ in the denominator. Use the method of partial fractions to expand $\left(\gamma^2 - x^2\right)/(x+\alpha)^3$ in the form,

$$\frac{\left(\gamma^2 - x^2\right)}{(x+\alpha)^3} = \frac{A_1}{(x+\alpha)} + \frac{A_2}{(x+\alpha)^2} + \frac{A_3}{(x+\alpha)^3}. \tag{14.3.26a}$$

Multiply through by $(x+\alpha)^3$. The result is

$$\gamma^2 - x^2 = A_1(x+\alpha)^2 + A_2(x+\alpha) + A_3 \tag{14.3.26b}$$

Upon expanding $(x+\alpha)^2$ on the right, Eq. (14.3.26b) reads

$$\begin{aligned}\gamma^2 - x^2 &= A_1\left(x^2 + 2\alpha x + \alpha^2\right) + A_2(x+\alpha) + A_3 \\ &= A_1 x^2 + (2\alpha A_1 + A_2)x + \left(A_1\alpha^2 + A_2\alpha + A_3\right)\end{aligned} \tag{14.3.26c}$$

After equating the two sides of Eq. (14.3.26c) term by term, the partial fraction coefficients (A_1, A_2, A_3) are found to satisfy the system of equations:

$$A_1 = -1 \tag{14.3.27a}$$

$$0 = 2\alpha A_1 + A_2 \tag{14.2.27b}$$

$$\gamma^2 = \alpha^2 A_1 + \alpha A_2 + A_3. \tag{14.3.27c}$$

By taking note of Eq. (14.3.27a), the solutions to Eqs. (14.3.27b) and (14.3.27c) are

$$A_2 = -2\alpha A_1 = 2\alpha. \tag{14.3.28a}$$

$$A_3 = \gamma^2 - \alpha^2 A_1 - \alpha A_2 = \gamma^2 + \alpha^2 - \alpha(2\alpha) = \gamma^2 - \alpha^2 \tag{14.3.28b}$$

Upon substitution of Eqs. (14.3.27a), (14.3.28a), and (14.3.28b) into Eq. (14.3.26a), the partial fraction expansion of the integrand involving $(\gamma^2 - x^2)/(x+\alpha)^3$ in Eq. (14.3.25) is

$$\frac{\gamma^2 - x^2}{(x+\alpha)^3} = \frac{-1}{(x+\alpha)} + \frac{2\alpha}{(x+\alpha)^2} + \frac{\gamma^2 - \alpha^2}{(x+\alpha)^3} \tag{14.3.29}$$

Next, we examine the partial fraction expansion of the integrand involving $(\gamma^2 - x^2)/(x+\beta)^3$ in Eq. (14.3.25). The result is obtained simply by substituting β in the place of α in Eq. (14.3.29); hence, the result is

$$\frac{\gamma^2 - x^2}{(x+\beta)^3} = \frac{-1}{(x+\beta)} + \frac{2\beta}{(x+\beta)^2} + \frac{\gamma^2 - \beta^2}{(x+\beta)^3} \tag{14.3.30}$$

We now turn to the integrand in Eq. (14.3.25) involving $(\gamma^2 - x^2)/(x+\alpha)^2$. Its partial fraction expansion can be obtained by multiplying Eq. (14.2.29) through by $(x+\alpha)$. The result is

$$\frac{\gamma^2 - x^2}{(x+\alpha)^2} = -1 + \frac{2\alpha}{(x+\alpha)} + \frac{\gamma^2 - \alpha^2}{(x+\alpha)^2} \tag{14.3.31}$$

Finally, upon substituting β for α in Eq. (14.3.31), the integrand in Eq. (14.3.25) involving $(\gamma^2 - x^2)/(x+\beta)^2$ is

$$\frac{\gamma^2 - x^2}{(x+\beta)^2} = -1 + \frac{2\beta}{(x+\beta)} + \frac{\gamma^2 - \beta^2}{(x+\beta)^2} \tag{14.3.32}$$

We are now ready to substitute the partial fraction expansions in Eqs. (14.3.29), (14.3.30), (14.3.31), and (14.3.32) into Eq. (14.3.25). The latter reads,

$$\begin{aligned} J(C) = {} & \left\{ (\alpha - \beta) \left[-\int_{-\gamma}^{+\gamma} \frac{dx}{(x+\alpha)} + 2\alpha \int_{-\gamma}^{+\gamma} \frac{dx}{(x+\alpha)^2} \right.\right. \\ & \left.\left. + (\gamma^2 - \alpha^2) \int_{-\gamma}^{+\gamma} \frac{dx}{(x+\alpha)^3} \right] \right\} \\ & + \left\{ (\alpha - \beta) \left[-\int_{-\gamma}^{+\gamma} \frac{dx}{(x+\beta)} + 2\beta \int_{-\gamma}^{+\gamma} \frac{dx}{(x+\beta)^2} \right.\right. \end{aligned}$$

$$+(\gamma^2-\beta^2)\int_{-\gamma}^{+\gamma}\frac{dx}{(x+\beta)^3}\Bigg]\Bigg\}$$

$$+\left\{-\int_{-\gamma}^{+\gamma}dx+2\alpha\int_{-\gamma}^{+\gamma}\frac{dx}{(x+\alpha)}+\left(\gamma^2-\alpha^2\right)\int_{-\gamma}^{+\gamma}\frac{dx}{(x+\alpha)^2}\right\}$$

$$+\left\{\int_{-\gamma}^{+\gamma}dx-2\beta\int_{-\gamma}^{+\gamma}\frac{dx}{(x+\beta)}-\left(\gamma^2-\beta^2\right)\int_{-\gamma}^{+\gamma}\frac{dx}{(x+\beta)^2}\right\} \quad (14.3.33)$$

where in the limits of integration, we have substituted $R_2=\gamma$. Although Eq. (14.3.33) involves twelve integrals, each is elementary. We can partition $J(C)$ into four groups, H_0, H_1, H_2, H_3, by writing

$$J(C)=H_0+H_1+H_2+H_3 \quad (14.3.34)$$

The first group, H_0, consists of the two integrals,

$$H_0=-\int_{-\gamma}^{+\gamma}dx+\int_{-\gamma}^{+\gamma}dx=0 \quad (14.3.35)$$

which add to zero.

The second group, H_1, consists of the four integrals involving the reciprocals of the first powers of the binomials involving x. They are

$$H_1=-(\alpha-\beta)\int_{-\gamma}^{+\gamma}\frac{dx}{(x+\alpha)}-(\alpha-\beta)\int_{-\gamma}^{+\gamma}\frac{dx}{(x+\beta)}$$

$$+2\alpha\int_{-\gamma}^{+\gamma}\frac{dx}{(x+\alpha)}-2\beta\int_{-\gamma}^{+\gamma}\frac{dx}{(x+\beta)}. \quad (14.3.36a)$$

which can be combined and evaluated as follows:

$$H_1=(-\alpha+\beta+2\alpha)\int_{-\gamma}^{+\gamma}\frac{dx}{(x+\alpha)}+(-\alpha+\beta-2\beta)\int_{-\gamma}^{+\gamma}\frac{dx}{(x+\beta)}$$

$$=(\alpha+\beta)\int_{-\gamma}^{+\gamma}\frac{dx}{(x+\alpha)}-(\alpha+\beta)\int_{-\gamma}^{+\gamma}\frac{dx}{(x+\beta)}$$

$$= (\alpha + \beta)[\ln(\alpha + \gamma) - \ln(\alpha - \gamma)] - (\alpha + \beta)[\ln(\beta + \gamma) - \ln(\beta - \gamma)] \tag{14.3.36b}$$

Upon combination of the logarithms, Eq. (14.3.36b) reads,

$$H_1 = (\alpha + \beta)\ln\left[\frac{(\alpha + \gamma)(\beta - \gamma)}{(\alpha - \gamma)(\beta + \gamma)}\right] \tag{14.3.37}$$

The third group, H_2, consists of the four integrals involving the second power binomials involving x. They are

$$\begin{aligned} H_2 = {} & 2\alpha(\alpha - \beta)\int_{-\gamma}^{+\gamma}\frac{dx}{(x + \alpha)^2} + 2\beta(\alpha - \beta)\int_{-\gamma}^{+\gamma}\frac{dx}{(x + \beta)^2} \\ & + \left(\gamma^2 - \alpha^2\right)\int_{-\gamma}^{+\gamma}\frac{dx}{(x + \alpha)^2} - \left(\gamma^2 - \beta^2\right)\int_{-\gamma}^{+\gamma}\frac{dx}{(x + \beta)^2} \end{aligned} \tag{14.3.38a}$$

Evaluation of the integrals in Eq. (14.3.38a) produces a sum of eight fractions. The result is

$$\begin{aligned} H_2 = {} & -\frac{2\alpha(\alpha - \beta)}{(\alpha + \gamma)} + \frac{2\alpha(\alpha - \beta)}{(\alpha - \gamma)} - \frac{2\beta(\alpha - \beta)}{(\beta + \gamma)} + \frac{2\beta(\alpha - \beta)}{(\beta - \gamma)} \\ & - \frac{\left(\gamma^2 - \alpha^2\right)}{(\alpha + \gamma)} + \frac{\left(\gamma^2 - \alpha^2\right)}{(\alpha - \gamma)} + \frac{\left(\gamma^2 - \beta^2\right)}{(\beta + \gamma)} - \frac{\left(\gamma^2 - \beta^2\right)}{(\beta - \gamma)} \end{aligned} \tag{14.3.38b}$$

The first four fractions in Eq. (14.3.38b) can be regrouped. In the case of the last four fractions, it is possible to factor each numerator into two binomials. With these operations completed, Eq. (14.3.38b) reads,

$$\begin{aligned} H_2 = {} & -\frac{2\alpha(\alpha - \beta)}{(\alpha + \gamma)} + \frac{2\beta(\alpha - \beta)}{(\beta - \gamma)} + \frac{2\alpha(\alpha - \beta)}{(\alpha - \gamma)} - \frac{2\beta(\alpha - \beta)}{(\beta + \gamma)} \\ & + \frac{(\alpha - \gamma)(\alpha + \gamma)}{(\alpha + \gamma)} - \frac{(\alpha - \gamma)(\alpha + \gamma)}{(\alpha - \gamma)} - \frac{(\beta - \gamma)(\beta + \gamma)}{(\beta + \gamma)} \\ & + \frac{(\beta - \gamma)(\beta + \gamma)}{(\beta - \gamma)} \end{aligned} \tag{14.3.38c}$$

The next step is to simplify Eq. (14.3.38c) by factoring and combining the first two fractions over a common denominator and then factoring the second two fractions and combining them over a separate common denominator. In the last four fractions, the common factors in the numerators and denominators can be cancelled. The net result of these operations is:

$$H_2 = 2(\alpha-\beta)\left[\frac{-\alpha(\beta-\gamma)+\beta(\alpha+\gamma)}{(\alpha+\gamma)(\beta-\gamma)}\right] + 2(\alpha-\beta)\left[\frac{\alpha(\beta+\gamma)-\beta(\alpha-\gamma)}{(\alpha-\gamma)(\beta+\gamma)}\right]$$
$$+ (\alpha-\gamma) - (\alpha+\gamma) - (\beta-\gamma) + (\beta+\gamma) \tag{14.3.38d}$$

Notice that the four binomial terms in the second line of Eq. (14.3.38d) sum to zero. That leaves the fractions in the first two brackets, which can be simplified. After cancellations of terms of opposite sign in the numerators of these fractions, H_2 reads finally,

$$H_2 = 2(\alpha-\beta)\left[\frac{(\alpha+\beta)\gamma}{(\alpha+\gamma)(\beta-\gamma)}\right] + 2(\alpha-\beta)\left[\frac{(\alpha+\beta)\gamma}{(\alpha-\gamma)(\beta+\gamma)}\right] \tag{14.3.39}$$

The final group, H_3, involves the third powers of binomials in x. This group consists of the two integrals,

$$H_3 = (\alpha-\beta)\left[\left(\gamma^2-\alpha^2\right)\int_{-\gamma}^{+\gamma}\frac{dx}{(x+\alpha)^3} + \left(\gamma^2-\beta^2\right)\int_{-\gamma}^{+\gamma}\frac{dx}{(x+\beta)^3}\right] \tag{14.3.40a}$$

After integration and evaluation at the upper and lower limits, H_3 reads,

$$H_3 = (\alpha-\beta)\left\{\left(\gamma^2-\alpha^2\right)\left[-\frac{1}{2(\alpha+\gamma)^2}+\frac{1}{2(\alpha-\gamma)^2}\right]\right.$$
$$\left.+\left(\gamma^2-\beta^2\right)\left[-\frac{1}{2(\beta+\gamma)^2}+\frac{1}{2(\beta-\gamma)^2}\right]\right\} \tag{14.3.40b}$$

Take out a factor of ½ everywhere, and then use $\left(\gamma^2-\alpha^2\right) = -\left(\alpha^2-\gamma^2\right)$ and $\left(\gamma^2-\beta^2\right) = -\left(\beta^2-\gamma^2\right)$ to change the signs before each bracket. The result is

$$H_3 = \frac{(\alpha-\beta)}{2}\left\{-\left(\alpha^2-\gamma^2\right)\left[-\frac{1}{(\alpha+\gamma)^2}+\frac{1}{(\alpha-\gamma)^2}\right]\right.$$
$$\left.-\left(\beta^2-\gamma^2\right)\left[-\frac{1}{(\beta+\gamma)^2}+\frac{1}{(\beta-\gamma)^2}\right]\right\} \tag{14.3.40c}$$

Multiply each bracket in Eq. (14.3.40c) by the coefficient preceding it and cancel common factors from the numerators and respective denominators of the resulting fractions. The result is

$$H_3 = \frac{(\alpha-\beta)}{2}\left\{\left[\frac{(\alpha-\gamma)}{(\alpha+\gamma)}-\frac{(\alpha+\gamma)}{(\alpha-\gamma)}\right]+\left[\frac{(\beta-\gamma)}{(\beta+\gamma)}-\frac{(\beta+\gamma)}{(\beta-\gamma)}\right]\right\} \tag{14.3.40d}$$

Rearrange the four fractions in Eq. (14.3.40d) into two groups to obtain

$$H_3 = \frac{(\alpha-\beta)}{2}\left\{\left[\frac{(\alpha-\gamma)}{(\alpha+\gamma)} - \frac{(\beta+\gamma)}{(\beta-\gamma)}\right] + \left[\frac{(\beta-\gamma)}{(\beta+\gamma)} - \frac{(\alpha+\gamma)}{(\alpha-\gamma)}\right]\right\}. \tag{14.3.40e}$$

Next, write the pairs of fractions over the appropriate common denominators, multiply binomials in the numerators, and cancel the common factors that result as shown below:

$$\begin{aligned} H_3 &= \frac{(\alpha-\beta)}{2}\left\{\left[\frac{(\alpha-\gamma)(\beta-\gamma)-(\alpha+\gamma)(\beta+\gamma)}{(\alpha+\gamma)(\beta-\gamma)}\right]\right. \\ &\quad \left. + \left[\frac{(\beta-\gamma)(\alpha-\gamma)-(\beta+\gamma)(\alpha+\gamma)}{(\beta+\gamma)(\alpha-\gamma)}\right]\right\} \\ &= \frac{(\alpha-\beta)}{2}\left\{\left[\frac{\alpha\beta-(\alpha+\beta)\gamma+\gamma^2-\alpha\beta-(\alpha+\beta)\gamma-\gamma^2}{(\alpha+\gamma)(\beta-\gamma)}\right]\right. \\ &\quad \left. + \left[\frac{\alpha\beta-(\alpha+\beta)\gamma+\gamma^2-\alpha\beta-(\alpha+\beta)\gamma-\gamma^2}{(\beta+\gamma)(\alpha-\gamma)}\right]\right\} \\ &= \frac{(\alpha-\beta)}{2}\left\{\left[\frac{-2(\alpha+\beta)\gamma}{(\alpha+\gamma)(\beta-\gamma)}\right] + \left[\frac{-2(\alpha+\beta)\gamma}{(\beta+\gamma)(\alpha-\gamma)}\right]\right\} \end{aligned} \tag{14.3.40f}$$

Equation (14.3.40f) can be simplified to read,

$$H_3 = -(\alpha-\beta)\left[\frac{(\alpha+\beta)\gamma}{(\alpha+\gamma)(\beta-\gamma)}\right] - (\alpha-\beta)\left[\frac{(\alpha+\beta)\gamma}{(\beta+\gamma)(\alpha-\gamma)}\right] \tag{14.3.41}$$

14.3.5 The Two Sphere Energy Formula

Our evaluations of H_0, H_1, H_2 and H_3 are now complete. The results are summarized by Eqs. (14.3.35), (14.3.37), (14.3.39), and (14.3.41), respectively. Upon substitution of the results into Eq. (14.3.34), we obtain

$$\begin{aligned} J(C) &= H_0 + H_1 + H_2 + H_3 \\ &= (\alpha+\beta)\ln\left[\frac{(\alpha+\gamma)(\beta-\gamma)}{(\alpha-\gamma)(\beta+\gamma)}\right] + 2(\alpha-\beta)\left[\frac{(\alpha+\beta)\gamma}{(\alpha+\gamma)(\beta-\gamma)}\right] \\ &\quad + 2(\alpha-\beta)\left[\frac{(\alpha+\beta)\gamma}{(\alpha-\gamma)(\beta+\gamma)}\right] - (\alpha-\beta)\left[\frac{(\alpha+\beta)\gamma}{(\alpha+\gamma)(\beta-\gamma)}\right] \\ &\quad - (\alpha-\beta)\left[\frac{(\alpha+\beta)\gamma}{(\beta+\gamma)(\alpha-\gamma)}\right] \\ &= (\alpha+\beta)\ln\left[\frac{(\alpha+\gamma)(\beta-\gamma)}{(\alpha-\gamma)(\beta+\gamma)}\right] + \left[\frac{(\alpha-\beta)(\alpha+\beta)\gamma}{(\alpha+\gamma)(\beta-\gamma)}\right] + \left[\frac{(\alpha-\beta)(\alpha+\beta)\gamma}{(\alpha-\gamma)(\beta+\gamma)}\right] \end{aligned} \tag{14.3.42}$$

To express Eq. (14.3.42) in terms of the original variables, we use Eqs. (14.3.22a) through (14.3.22c) to construct the identities,

$$(\alpha+\gamma)(\beta-\gamma) = (C+R_1+R_2)(C-R_1-R_2) = C^2-(R_1+R_2)^2 \quad (14.3.43a)$$

$$(\alpha-\gamma)(\beta+\gamma) = (C+R_1-R_2)(C-R_1+R_2) = C^2-(R_1-R_2)^2. \quad (14.3.43b)$$

On the basis of Eqs. (14.3.22a) and (14.3.22b), we have

$$\alpha+\beta = 2C \quad (14.3.43c)$$

and

$$\alpha-\beta = 2R_1 \quad (14.3.43d)$$

After substitution of Eqs. (14.3.43a) through (14.3.43d) into Eq. (14.3.42), the result is

$$J(C) = 2C\ln\left[\frac{C^2-(R_1+R_2)^2}{C^2-(R_1-R_2)^2}\right] + \left[\frac{4R_1CR_2}{C^2-(R_1+R_2)^2}\right] + \left[\frac{4R_1CR_2}{C^2-(R_1-R_2)^2}\right] \quad (14.3.44)$$

Upon substitution of Eq. (14.3.44) into Eq. (14.3.19), one then obtains

$$E = -\frac{\pi^2Q^2\lambda}{12C}\left\{2C\ln\left[\frac{C^2-(R_1+R_2)^2}{C^2-(R_1-R_2)^2}\right] + \left[\frac{4R_1CR_2}{C^2-(R_1+R_2)^2}\right] + \left[\frac{4R_1CR_2}{C^2-(R_1-R_2)^2}\right]\right\} \quad (14.3.45)$$

After cancellation of common factors, Eq. (14.3.45) can be rearranged to read

$$E = -\frac{\pi^2Q^2\lambda}{6}\left\{\frac{2R_1R_2}{C^2-(R_1+R_2)^2} + \frac{2R_1R_2}{C^2-(R_1-R_2)^2} + \ln\left[\frac{C^2-(R_1+R_2)^2}{C^2-(R_1-R_2)^2}\right]\right\} \quad (14.3.46)$$

Equation (14.3.46) is the second of Hamaker's main results. Equation (14.3.46) neatly expresses the energy of interaction of the two spheres in terms of the product of factors. The leading factors, Q and λ, take into account the material properties. The factor within the braces depends exclusively on the geometry. Of the three factors, the material parameter, λ, has attracted the most attention, being the subject of many revisions as our understanding of the atomic and molecular properties of condensed matter has evolved.

14.3.6 The Dispersion Energy for Special Cases

Although algebraically elaborate, Eq. (14.3.46) has two approximate forms which have proved useful in many practical applications. The first of these occurs in the case of the interaction of two spheres of equal radii. The second involves a sphere interacting with a flat plate.

To derive these two approximate forms, we will preserve the original notation of Hamaker [4] by redefining x as the dimensionless variable,

$$x = \frac{d}{2R_1} \tag{14.3.47a}$$

We introduce a second dimensionless variable, y, defined by

$$y = \frac{R_2}{R_1} \tag{14.3.47b}$$

The geometry continues to be that shown in Fig. 14.5. On the basis of Eqs. (14.3.47a) and (14.3.47b), we can solve for the following geometric quantities,

$$R_1 = \frac{d}{2x} \tag{14.3.48a}$$

$$R_2 = R_1 y = \left(\frac{d}{2x}\right)y \tag{14.3.48b}$$

$$R_1 + R_2 = \left(\frac{d}{2x}\right)(1+y) \tag{14.3.48c}$$

$$R_1 - R_2 = \left(\frac{d}{2x}\right)(1-y) \tag{14.3.48d}$$

By referring to Eq. (14.3.17a), we can then write,

$$C = \left(\frac{d}{2x}\right)(1+y) + d \tag{14.3.48e}$$

To evaluate Eq. (14.3.46) in terms of x and y, we shall need to combine Eqs. (14.3.48c) and (14.3.48e) to obtain

$$\begin{aligned} C^2 - (R_1+R_2)^2 &= \left[\left(\frac{d}{2x}\right)(1+y) + d\right]^2 - \left(\frac{d}{2x}\right)^2(1+y)^2 \\ &= \left(\frac{d}{2x}\right)^2(1+y)^2 + 2\left(\frac{d}{2x}\right)(1+y)d + d^2 \end{aligned}$$

$$
\begin{aligned}
&-\left(\frac{d}{2x}\right)^2(1+y)^2\\
&=2\left(\frac{d^2}{2x}\right)(1+y)+d^2\\
&=d^2\left[1+\left(\frac{1}{x}\right)(1+y)\right]\\
&=\left(d^2/x^2\right)\left[x^2+x(1+y)\right] \qquad (14.3.49a)
\end{aligned}
$$

and finally,

$$
C^2-(R_1+R_2)^2=\left(d^2/x^2\right)\left(x^2+x+xy\right) \qquad (14.3.49b)
$$

Next, Eqs. (14.3.48a), (14.3.48b), and (14.3.49b) can be combined to compute

$$
\frac{2R_1R_2}{C^2-(R_1+R_2)^2}=\frac{2(d/2x)^2y}{\left(d^2/x^2\right)\left(x^2+x+xy\right)}=\frac{y}{2\left(x^2+x+xy\right)} \qquad (14.3.49c)
$$

Equations (14.3.48d) and (14.3.48e) can be used to calculate the combination

$$
\begin{aligned}
C^2-(R_1-R_2)^2&=\left[\left(\frac{d}{2x}\right)(1+y)+d\right]^2-\left(\frac{d}{2x}\right)^2(1-y)^2\\
&=\left(\frac{d}{2x}\right)^2(1+y)^2+2\left(\frac{d^2}{2x}\right)(1+y)+d^2-\left(\frac{d}{2x}\right)^2(1-y)^2\\
&=\left(\frac{d}{2x}\right)^2\left(1+2y+y^2\right)+\left(\frac{d^2}{x}\right)(1+y)+d^2\\
&\quad-\left(\frac{d}{2x}\right)^2\left(1-2y+y^2\right)\\
&=d^2+\left(\frac{d^2}{x}\right)(1+y)+\left(\frac{d^2}{4x^2}\right)(4y)\\
&=d^2\left[1+\left(\frac{1}{x}\right)(1+y)+\frac{y}{x^2}\right]\\
&=\left(d^2/x^2\right)\left[x^2+x(1+y)+y\right] \qquad (14.3.50a)
\end{aligned}
$$

or

$$
C^2-(R_1-R_2)^2=\left(d^2/x^2\right)\left(x^2+x+xy+y\right) \qquad (14.3.50b)
$$

Next use Eqs. (14.3.48a), (14.3.48b) and (14.3.50b) to compute

$$\frac{2R_1R_2}{C^2-(R_1-R_2)^2}=\frac{2(d/2x)^2y}{(d^2/x^2)(x^2+x+xy+y)}=\frac{y}{2(x^2+x+xy+y)} \tag{14.3.50c}$$

The last required quantity is the ratio of Eq. (14.3.49b) to Eq. (14.3.50b). The result is

$$\frac{C^2-(R_1+R_2)^2}{C^2-(R_1-R_2)^2}=\frac{x^2+x+xy}{x^2+x+xy+y} \tag{14.3.51}$$

Upon substitution of Eqs. (14.3.49c), (14.3.50c) and (14.3.51) into Eq. (14.3.46), we have

$$E=-\frac{\pi^2Q^2\lambda}{6}\left\{\frac{y}{2(x^2+x+xy)}+\frac{y}{2(x^2+x+xy+y)}+\ln\left(\frac{x^2+x+xy}{x^2+x+xy+y}\right)\right\} \tag{14.3.52a}$$

Define

$$A=\pi^2Q^2\lambda \tag{14.3.52b}$$

Upon substitution of Eq. (14.3.52b) into Eq. (14.3.52a), we have finally

$$E=-\frac{A}{12}\left\{\frac{y}{(x^2+x+xy)}+\frac{y}{(x^2+x+xy+y)}+2\ln\left(\frac{x^2+x+xy}{x^2+x+xy+y}\right)\right\}. \tag{14.3.53}$$

Equation (14.3.53) serves as the basis for deriving approximate forms, which apply in special cases.

In deriving the *first* approximate form, we consider two closely spaced spheres of equal radii. In this case, d is small, and according to Eq. (14.3.47a), $x \ll 1$. In this limit, the argument of the logarithm in Eq. (14.3.53) is small. The appropriate expansion of $\ln z$ for small values of its argument is

$$\ln(z)\approx(z-1) \quad \text{for} \quad 0<z<2 \tag{14.3.54}$$

With

$$z=\frac{x^2+x+xy}{x^2+x+xy+y} \tag{14.3.55a}$$

we can exploit Eq. (14.3.54) for $\ln(z)$ to write

$$\ln\left(\frac{x^2+x+xy}{x^2+x+xy+y}\right) \approx \left(\frac{x^2+x+xy}{x^2+x+xy+y}\right) - 1$$
$$= \frac{x^2+x+xy-x^2-x-xy-y}{\left(x^2+x+xy+y\right)} = \frac{-y}{\left(x^2+x+xy+y\right)} \tag{14.3.55b}$$

Substitute Eq. (14.3.55b) into (14.3.53). The result is

$$E \approx -\frac{A}{12}\left\{\frac{y}{\left(x^2+x+xy\right)} + \frac{y}{\left(x^2+x+xy+y\right)} - \frac{2y}{\left(x^2+x+xy+y\right)}\right\} \tag{14.3.56a}$$

After performing the indicated subtraction, Eq. (14.3.56a) reads

$$E = -\frac{A}{12}\left\{\frac{y}{\left(x^2+x+xy\right)} - \frac{y}{\left(x^2+x+xy+y\right)}\right\} \tag{14.3.56b}$$

Combine the fractions in Eq. (14.3.56b) over a common denominator. The result is

$$E = -\frac{Ay}{12}\left\{\frac{x^2+x+xy+y-x^2-x-xy}{\left(x^2+x+xy\right)\left(x^2+x+xy+y\right)}\right\} \tag{14.3.56c}$$

which simplifies to

$$E = -\frac{A}{12}\left[\frac{y^2}{x(x+1+y)\left(x^2+x+xy+y\right)}\right] \tag{14.3.56d}$$

In the limit $x \ll 1$, the three terms in the first set of parentheses in the denominator of Eq. (14.3.56d) simplify to $(1+y)$. In this same limit, the four terms in the second set of parentheses simplify to y. With these approximations, Eq. (14.3.56d) reduces to

$$E = -\frac{A}{12}\left(\frac{y^2}{x(1+y)y}\right) \tag{14.3.56e}$$

or simply

$$E = -\frac{A}{12}\left(\frac{y}{x(1+y)}\right) \tag{14.3.57}$$

According to Eq. (14.3.47b), the condition, $y = 1$, describes two spheres of *equal radii*, in which case Eq. (14.3.57) reads,

$$E = -\frac{A}{24x} \tag{14.3.58}$$

By use of Eq. (14.3.47a), Eq. (14.3.58) can be written in terms of dimensioned parameters as

$$E = -\frac{AR_1}{12d} \tag{14.3.59}$$

Equation (14.3.59) applies in the case of two spheres of equal radii which are close together.

In deriving the *second* approximate form, we consider one sphere to be much larger than the other. According to Eq. (14.3.47b), if one sphere is much larger than the other, say $R_2 \gg R_1$, then $y \rightarrow \infty$. This extreme case corresponds to a sphere with radius R_1 interacting with a plane. In the limit, $y \rightarrow \infty$, Eq. (14.3.57) reads,

$$E = -\frac{A}{12x} \tag{14.3.60}$$

Using Eq. (14.3.47a), we can rewrite E in Eq. (14.3.60) in terms of dimensioned parameters as

$$E = -\frac{AR_1}{6d} \tag{14.3.61}$$

Equation (14.3.47a) has found many applications.

14.3.7 Discussion

The expression in braces in Eq. (14.3.53) represents a geometric factor, which is determined by the radii of the spheres and the distance between them. By contrast, the multiplying factor, A, which is defined in Eq. (14.3.52b) and is known as the Hamaker constant, is a function of the materials making up the spheres. According to Eq. (14.3.1c), A depends on the ionization potentials and static electric polarizabilities of the molecules in the spheres. Since London's original work, the parameterization of A has been greatly elaborated [5, 6]. For example, consideration of the finite speed of the electrons orbiting the nuclei in the molecules reintroduces the frequency dependence of the polarizabilities, thereby resurrecting the descriptive term *dispersion force*. In the case of two metal spheres, the polarizabilities disappear altogether and are replaced by an exchange interaction involving the electrons in their respective Fermi seas. In the case a colloidal suspension, the effect of the solvent in screening the force acting between spheres has to be taken into account. For these various cases, the estimates of the value of A have ranged between 10^{-19} J and 10^{-12} J [6].

References

1. W. Kauzmann, *Quantum Chemistry* (Academic, New York, 1957)
2. H. Eyring, J. Walter, G.E. Kimball, *Quantum Chemistry* (Wiley, New York, 1944)
3. D.L. Dexter, Theory of sensitized luminescence in solids. J. Chem. Phys. **21**, 836–850 (1953)
4. H.C. Hamaker, The London—van der Waals attraction between spherical particles. Physica **4**, 1058–1072 (1937)
5. J.N. Israelachvili, *Intermolecular and Surface Forces* (Academic, Boston, 1991)
6. K.S. Schmitz, *Macroions in Solution and Colloidal Suspensions* (VCH Publishers, New York, 1993)

Recommended Additional Reading

1. A. Hinchcliffe, R.W. Munn, *Molecular Electromagnetism* (Wiley, New York, 1985)
2. W.S. Struve, Theory of electronic energy transfer, in *Anoxygenic Photosynthetic Bacteria*, ed. by R.E. Blankenship, M.T. Madigan, C.E. Bauer (Kluwer Academic Publishers, Amsterdam, Netherlands, 1995), pp. 297–313
3. R.J. Hunter, *Foundations of Colloid Science* (Clarendon Press, Oxford, UK, 1993)

Appendix
Mathematical Methods

A.1 Introduction

Although many of the equations and theorems used in the previous chapters were proved at the point where they were first introduced, it is not the case for all. The purpose of this appendix is to provide a general compendium of formulae and theorems useful in electromagnetism including those which were used in the chapters but not proved there.

A.2 Trigonometric Functions and a Hyperbolic Function Identity

$$\sin^2\theta + \cos^2\theta = 1 \tag{A.2.1}$$

$$\sin^2\theta = \frac{1-\cos(2\theta)}{2} \tag{A.2.2}$$

$$\cos^2\theta = \frac{1+\cos(2\theta)}{2} \tag{A.2.3}$$

$$\sec^2\theta = 1 + \tan^2\theta \tag{A.2.4}$$

$$\sin(\alpha \pm \beta) = \sin\alpha\cos\beta \pm \cos\alpha\sin\beta \tag{A.2.5}$$

$$\cos(\alpha \pm \beta) = \cos\alpha\cos\beta \mp \sin\alpha\sin\beta \tag{A.2.6}$$

$$\sin\alpha\sin\beta = \frac{1}{2}\cos(\alpha-\beta) - \frac{1}{2}\cos(\alpha+\beta) \tag{A.2.7}$$

J. K. Baird, *Electromagnetism for Chemistry*,
https://doi.org/10.1007/978-3-031-88508-2

$$\tanh^{-1}(x) = \frac{1}{2}\ln\left(\frac{1+x}{1-x}\right) \tag{A.2.8}$$

A.3 Series Expansions of Some Functions

$$(1+x)^n = 1 + nx + \frac{n(n-1)}{2!}x^2 + \frac{n(n-1)(n-2)}{3!}x^3 + \cdots \quad \text{for } -1 < x < 1 \tag{A.3.1}$$

$$\ln\left(\frac{1+x}{1-x}\right) \approx 2x \cdots \quad \text{for } -1 < x < 1 \tag{A.3.2}$$

$$e^x = 1 + x + \frac{x^2}{2!} + \frac{x^3}{3!} + \cdots \tag{A.3.3}$$

$$\ln x \approx (x-1) \quad \text{for } 0 < \text{x} < 2 \tag{A.3.4}$$

$$\ln(1+x) = x - \frac{x^2}{2} + \frac{x^3}{3} - \frac{x^4}{4} + \cdots \tag{A.3.5}$$

$$\cos\theta = 1 - \frac{1}{2!}\theta^2 + \frac{1}{4!}\theta^4 - \cdots \tag{A.3.6}$$

$$\sin\theta - \theta - \frac{1}{3!}\theta^3 + \frac{1}{5!}\theta^5 - \ldots \tag{A.3.7}$$

$$f(x) = f(a) + \left(\frac{df}{dx}\right)_a (x-a) + \frac{1}{2!}\left(\frac{d^2f}{dx^2}\right)_a (x-a)^2 + \frac{1}{3!}\left(\frac{d^3f}{dx^3}\right)_a (x-a)^3 + \cdots \tag{A.3.8}$$

A.4 Some Derivatives and Integrals

$$\frac{d\tan\theta}{d\theta} = \sec^2\theta \tag{A.4.1}$$

$$\int \sin\theta d\theta = -\cos\theta \tag{A.4.2}$$

$$\int \cos\theta d\theta = \sin\theta \tag{A.4.3}$$

$$\int \operatorname{csch}(\theta)d\theta = \ln\left[\tanh\left(\frac{\theta}{2}\right)\right] \tag{A.4.4}$$

$$\int \frac{dx}{x} = \ln x \tag{A.4.5}$$

$$\int \frac{dx}{x\sqrt{x^2+1}} = -\operatorname{csch}^{-1}(x) \tag{A.4.6}$$

$$\int_{-\infty}^{+\infty} \exp\left(-a^2x^2\right)dx = 2\int_0^{\infty} \exp\left(-a^2x^2\right)dx = \frac{\sqrt{\pi}}{a} \tag{A.4.7}$$

$$\frac{1}{\pi}\int_0^{+\infty} d\omega \frac{\sin(\omega t)}{\omega} = \frac{1}{2}s(t) \tag{A.4.8}$$

$$\int_{-\infty}^{+\infty} dx \frac{\sin^2(ax)}{x^2} = \pi a \tag{A.4.9}$$

In Eq. (A.4.8), $s(t)$ is the sign function.

A.5 Stirling's Approximation

By integration by parts, it can be shown that the integral, $\int_0^{\infty} x^3e^{-x}dx$, is a representation of 3! The calculation proceeds as follows,

$$\begin{aligned}
\int_0^{\infty} x^3e^{-x}dx &= -\left[x^3e^{-x}\right]_0^{\infty} + \int_0^{\infty} 3x^2e^{-x}dx \\
&= -\left[x^3e^{-x}\right]_0^{\infty} - \left[3x^2e^{-x}\right]_0^{\infty} + \int_0^{\infty} (3)(2)xe^{-x}dx \\
&= -\left[x^3e^{-x}\right]_0^{\infty} - \left[3x^2e^{-x}\right]_0^{\infty} - \left[(3)(2)xe^{-x}\right]_0^{\infty} + \int_0^{\infty} (3)(2)e^{-x}dx \\
&= -\left[x^3e^{-x}\right]_0^{\infty} - \left[3x^2e^{-x}\right]_0^{\infty} - \left[(3)(2)xe^{-x}\right]_0^{\infty} - \left[(3)(2)e^{-x}\right]_0^{\infty}
\end{aligned}$$

$$\int_0^\infty x^3 e^{-x} dx = (3)(2)(1) = 3! \tag{A.5.1}$$

To extend the result in Eq. (A.5.1) to include, $n!$, we consider the integral, I, defined by

$$I = \int_0^\infty x^n e^{-x} dx \tag{A.5.2}$$

Make the substitution, $x^n = e^{n \ln x}$, so that Eq. (A.5.2) reads,

$$I = \int_0^\infty e^{n \ln x - x} dx \tag{A.5.3}$$

Next, define

$$f(x) = n \ln x - x \tag{A.5.4}$$

The first derivative of $f(x)$ is

$$f'(x) = \frac{n}{x} - 1 \tag{A.5.5}$$

Note that when $x = n$,

$$f'(n) = 0 \tag{A.5.6}$$

Use Eq. (A.5.5) to compute,

$$f''(x) = -\frac{n}{x^2} \tag{A.5.7}$$

Note that when, $x = n$,

$$f''(n) = -\frac{1}{n} \tag{A.5.8}$$

The Taylor series representation of $f(x)$ is

$$f(x) = f(n) + f'(n)(x-n) + \frac{1}{2!} f''(n)(x-n)^2 + \cdots \tag{A.5.9}$$

Upon substitution of Eqs. (A.5.6) and (A.5.8), Eq. (A.5.9) reads,

$$f(x) = f(n) - \frac{(x-n)^2}{2n} \tag{A.5.10}$$

Next substitute Eq. (A.5.4) into Eq. (A.5.3) to obtain

$$I = \int_0^\infty \exp(f(x))dx \tag{A.5.11}$$

Substitution of Eq. (A.5.10) into Eq. (A.5.11) leads to the result,

$$I = \exp(f(n)) \int_0^\infty \exp\left[-\frac{(x-n)^2}{2n}\right]dx \tag{A.5.12}$$

In the integrand of the integral in Eq. (A.5.12), substitute, $y = x - n$, and $dx = dy$, where $y = -n$ when $x = 0$. Eq. (A.5.12) then reads,

$$I = \exp(f(n)) \int_{-n}^\infty \exp\left[-\frac{y^2}{2n}\right]dy \tag{A.5.13}$$

For sufficiently large value of n, we can replace the lower limit in Eq. (A.5.13) by $-\infty$ with the result,

$$I = \exp(f(n)) \int_{-\infty}^\infty \exp\left[-\frac{y^2}{2n}\right]dy \tag{A.5.14}$$

Eq. (A.5.4) reads $f(n) = n \ln n - n$. The integral Eq. (A.5.14) can be evaluated by invoking Eq. (A.4.7) with $a = \sqrt{\frac{1}{2n}}$. With these substitutions, Eq. (A.5.14) reads,

$$I = \exp(f(n))\left(\frac{\sqrt{\pi}}{\sqrt{1/2n}}\right) = \exp(n \ln n - n) \cdot \sqrt{2\pi n} = \sqrt{2\pi n}\exp\left(\ln n^n\right) \cdot \exp(-n) \tag{A.5.15}$$

Upon simplification, Eq. (A.5.15) becomes,

$$n! \approx \sqrt{2\pi n} \cdot n^n e^{-n} \tag{A.5.16}$$

Equation (A.5.16) is known as Stirling's approximation (James Stirling, English, 1698–1770).

A.6 Rectangular, Cylindrical Polar, and Spherical Polar Coordinates

A.6.1 Rectangular Coordinates

A point in space is represented in rectangular coordinates by the three real numbers (x, y, z). A differential element of volume associated with the point is denoted by $d^3r = dxdydz$. The unit vectors in rectangular coordinates are $\hat{\mathbf{e}}_x$, $\hat{\mathbf{e}}_y$, and $\hat{\mathbf{e}}_z$, respectively. In rectangular coordinates, the gradient operator is

$$\nabla = \hat{\mathbf{e}}_x \frac{\partial}{\partial x} + \hat{\mathbf{e}}_y \frac{\partial}{\partial y} + \hat{\mathbf{e}}_z \frac{\partial}{\partial z} \tag{A.6.1a}$$

In the case of a scalar function, $F = F(x, y, z)$,

$$\nabla F = \hat{\mathbf{e}}_x \frac{\partial F}{\partial x} + \hat{\mathbf{e}}_y \frac{\partial F}{\partial y} + \hat{\mathbf{e}}_z \frac{\partial F}{\partial z}$$

The Laplacian in rectangular coordinates is

$$\nabla \cdot \nabla F = \left(\hat{\mathbf{e}}_x \frac{\partial}{\partial x} + \hat{\mathbf{e}}_y \frac{\partial}{\partial y} + \hat{\mathbf{e}}_z \frac{\partial}{\partial z} \right) \cdot \left(\hat{\mathbf{e}}_x \frac{\partial F}{\partial x} + \hat{\mathbf{e}}_y \frac{\partial F}{\partial y} + \hat{\mathbf{e}}_z \frac{\partial F}{\partial z} \right) \tag{A.6.1b}$$

Upon exploiting the orthornormality of $(\hat{\mathbf{e}}_x, \hat{\mathbf{e}}_y, \hat{\mathbf{e}}_z)$, the final result is

$$\nabla^2 F = \frac{\partial^2 F}{\partial x^2} + \frac{\partial^2 F}{\partial y^2} + \frac{\partial^2 F}{\partial z^2} \tag{A.6.1c}$$

A.6.2 Cylindrical Polar Coordinates

A given point in space is identified by the coordinates (ρ, ϕ, z), where ϕ is the angle of rotation about the z-axis, and ρ is the radial distance measured perpendicular to the z-axis. The equations linking the cylindrical coordinates to the rectangular coordinates are

$$x = \rho \cos \phi \tag{A.6.2a}$$

$$y = \rho \sin \phi \tag{A.6.2b}$$

$$z = z \tag{A.6.2c}$$

The differential area of the surface of the cylinder is $dA = \rho d\phi dz$. The differential of the volume is $d^3\rho = dA(d\rho) = \rho\, d\phi d\rho dz$.

The orthonormal unit vectors in cylindrical polar coordinates, $\hat{\mathbf{e}}_\rho$, $\hat{\mathbf{e}}_\phi$, and $\hat{\mathbf{e}}_\phi$ can be expressed in terms of the unit vectors, $\hat{\mathbf{e}}_x$, $\hat{\mathbf{e}}_y$, and $\hat{\mathbf{e}}_z$ in rectangular coordinates according to the following relations:

$$\hat{\mathbf{e}}_\rho = \hat{\mathbf{e}}_x \cos \phi + \hat{\mathbf{e}}_y \sin \phi \tag{A.6.2d}$$

$$\hat{\mathbf{e}}_\phi = -\hat{\mathbf{e}}_x \sin \phi + \hat{\mathbf{e}}_y \cos \phi \tag{A.6.2e}$$

The gradient operator in cylindrical polar coordinates is

$$\nabla = \hat{\mathbf{e}}_\rho \frac{\partial}{\partial \rho} + \hat{\mathbf{e}}_\phi \frac{\partial}{\rho \partial \phi} + \hat{\mathbf{e}}_z \frac{\partial}{\partial z} \tag{A.6.2f}$$

Operating on a function, $F(\rho, \phi, z)$, the gradient is

$$\nabla F = \hat{\mathbf{e}}_\rho \frac{\partial F}{\partial \rho} + \hat{\mathbf{e}}_\phi \frac{\partial F}{\rho \partial \phi} + \hat{\mathbf{e}}_z \frac{\partial F}{\partial z} \tag{A.6.2g}$$

The Laplacian in cylindrical polar coordinates is

$$\begin{aligned}
\nabla^2 F &= \left(\hat{\mathbf{e}}_\rho \frac{\partial}{\partial \rho} + \hat{\mathbf{e}}_\phi \frac{\partial}{\rho \partial \phi} + \hat{\mathbf{e}}_z \frac{\partial}{\partial z}\right) \cdot \left(\hat{\mathbf{e}}_\rho \frac{\partial F}{\partial \rho} + \hat{\mathbf{e}}_\phi \frac{\partial F}{\rho \partial \phi} + \hat{\mathbf{e}}_z \frac{\partial F}{\partial z}\right) \\
&= \left(\frac{\partial^2 F}{\partial \rho^2} + \frac{1}{\rho^2}\frac{\partial^2 F}{\partial \phi^2} + \frac{\partial^2 F}{\partial z^2}\right) + \left(\hat{\mathbf{e}}_\rho \cdot \frac{\partial F}{\partial \rho}\frac{\partial \hat{\mathbf{e}}_\rho}{\partial \rho} + \hat{\mathbf{e}}_\rho \cdot \frac{\partial F}{\rho \partial \phi}\frac{\partial \hat{\mathbf{e}}_\phi}{\partial \rho} + \hat{\mathbf{e}}_\rho \cdot \frac{\partial F}{\partial z}\frac{\partial \hat{\mathbf{e}}_z}{\partial \rho}\right) \\
&\quad + \left(\hat{\mathbf{e}}_\phi \cdot \frac{\partial F}{\partial \rho}\frac{\partial \hat{\mathbf{e}}_\rho}{\rho \partial \phi} + \hat{\mathbf{e}}_\phi \cdot \frac{\partial F}{\rho \partial \phi}\frac{\partial \hat{\mathbf{e}}_\phi}{\rho \partial \phi} + \hat{\mathbf{e}}_\phi \cdot \frac{\partial F}{\partial z}\frac{\partial \hat{\mathbf{e}}_z}{\rho \partial \phi}\right) \\
&\quad + \left(\hat{\mathbf{e}}_z \cdot \frac{\partial F}{\partial \rho}\frac{\partial \hat{\mathbf{e}}_\rho}{\partial z} + \hat{\mathbf{e}}_z \cdot \frac{\partial F}{\rho \partial \phi}\frac{\partial \hat{\mathbf{e}}_\phi}{\partial z} + \hat{\mathbf{e}}_z \cdot \frac{\partial F}{\partial z}\frac{\partial \hat{\mathbf{e}}_z}{\partial z}\right)
\end{aligned} \tag{A.6.2h}$$

According to Eq. (A.6.2d), the derivative of $\hat{\mathbf{e}}_\rho$ with respect to ϕ is

$$\frac{\partial \hat{\mathbf{e}}_\rho}{\partial \phi} = -\hat{\mathbf{e}}_x \sin \phi + \hat{\mathbf{e}}_y \cos \phi = \hat{\mathbf{e}}_\phi \tag{A.6.2i}$$

According to Eq. (A.6.2e), the derivative of $\hat{\mathbf{e}}_\phi$ with respect to ϕ is

$$\frac{\partial \hat{\mathbf{e}}_\phi}{\partial \phi} = -\hat{\mathbf{e}}_x \cos\phi - \hat{\mathbf{e}}_y \sin\phi = -\hat{\mathbf{e}}_\rho \tag{A.6.2j}$$

The remaining derivatives required to evaluate Eq. (A.6.2h) are all zero. Namely,

$$\frac{\partial \hat{\mathbf{e}}_\rho}{\partial \rho} = \frac{\partial \hat{\mathbf{e}}_\phi}{\partial \rho} = \frac{\partial \hat{\mathbf{e}}_z}{\partial \rho} = \frac{\partial \hat{\mathbf{e}}_z}{\partial \phi} = \frac{\partial \hat{\mathbf{e}}_\rho}{\partial z} = \frac{\partial \hat{\mathbf{e}}_\phi}{\partial z} = \frac{\partial \hat{\mathbf{e}}_z}{\partial z} = 0 \tag{A.6.2k}$$

When Eqs. (A.6.2i) through (A.6.2k) are substituted into Eq. (A.6.2h), the result is

$$\nabla^2 F = \left(\frac{\partial^2 F}{\partial \rho^2} + \frac{1}{\rho^2}\frac{\partial^2 F}{\partial \phi^2} + \frac{\partial^2 F}{\partial z^2}\right) + \hat{\mathbf{e}}_\phi \cdot \frac{\partial F}{\rho \partial \rho}\hat{\mathbf{e}}_\phi + \hat{\mathbf{e}}_\phi \cdot \frac{\partial F}{\rho^2 \partial \phi}\left(-\hat{\mathbf{e}}_\rho\right) \tag{A.6.2l}$$

By invoking the orthonormality conditions satisfied by the unit vectors, we have $\hat{\mathbf{e}}_\phi \cdot \hat{\mathbf{e}}_\phi = 1$ and $\hat{\mathbf{e}}_\phi \cdot \hat{\mathbf{e}}_\rho = 0$. After substitution of these results, Eq. (A.6.2l) reads,

$$\nabla^2 F = \frac{\partial^2 F}{\partial \rho^2} + \frac{\partial F}{\rho \partial \rho} + \frac{1}{\rho^2}\frac{\partial^2 F}{\partial \phi^2} + \frac{\partial^2 F}{\partial z^2} \tag{A.6.2m}$$

After combining the first two derivatives in Eq. (A.6.2m), the final result is

$$\nabla^2 F = \frac{1}{\rho}\frac{\partial}{\partial \rho}\left(\rho \frac{\partial F}{\partial \rho}\right) + \frac{1}{\rho^2}\frac{\partial^2 F}{\partial \phi^2} + \frac{\partial^2 F}{\partial z^2} \tag{A.6.2n}$$

A.6.3 Spherical Polar Coordinates

A given point in space is identified by spherical polar coordinates, (r, θ, ϕ), where θ is the polar angle, ϕ is the azimuthal angle, and r is the radial distance to the point. The equations relating the rectangular coordinates to spherical polar coordinates are

$$x = r \sin\theta \cos\phi \tag{A.6.3a}$$

$$y = r \sin\theta \sin\phi \tag{A.6.3b}$$

$$z = r \cos\theta \tag{A.6.3c}$$

In spherical polar coordinates, a differential element of area on the surface of the sphere is $dA = r^2 d\Omega$, where $d\Omega = \sin\theta d\theta d\phi$ is termed the solid angle. The differential element of volume associated with the point is $d^3 r = r^2 dr d\Omega = r^2 dr \sin\theta d\theta d\phi$.

The unit orthonormal vectors in spherical polar coordinates, $\hat{\mathbf{e}}_r$, $\hat{\mathbf{e}}_\theta$, and $\hat{\mathbf{e}}_\phi$ can be expressed in terms of the unit vectors, $\hat{\mathbf{e}}_x$, $\hat{\mathbf{e}}_y$, and $\hat{\mathbf{e}}_z$ in Cartesian coordinates according to the following relations:

$$\hat{\mathbf{e}}_r = \hat{\mathbf{e}}_x \sin\theta\cos\phi + \hat{\mathbf{e}}_y \sin\theta\sin\phi + \hat{\mathbf{e}}_z \cos\theta \tag{A.6.3d}$$

$$\hat{\mathbf{e}}_\theta = \hat{\mathbf{e}}_x \cos\theta\cos\phi + \hat{\mathbf{e}}_y \cos\theta\sin\phi - \hat{\mathbf{e}}_z \sin\theta \tag{A.6.3e}$$

$$\hat{\mathbf{e}}_\phi = -\hat{\mathbf{e}}_x \sin\phi + \hat{\mathbf{e}}_y \cos\phi \tag{A.6.3f}$$

Likewise, the unit vectors in rectangular coordinates can be expressed in terms of the unit vectors in spherical polar coordinates, according to the following relations,

$$\hat{\mathbf{e}}_x = \hat{\mathbf{e}}_r \sin\theta\cos\phi + \hat{\mathbf{e}}_\theta \cos\theta\cos\phi - \hat{\mathbf{e}}_\varphi \sin\phi \tag{A.6.3g}$$

$$\hat{\mathbf{e}}_y = \hat{\mathbf{e}}_r \sin\theta\sin\phi + \hat{\mathbf{e}}_\theta \cos\theta\sin\phi + \hat{\mathbf{e}}_\varphi \cos\phi \tag{A.6.3h}$$

$$\hat{\mathbf{e}}_z = \hat{\mathbf{e}}_r \cos\theta - \hat{\mathbf{e}}_\theta \sin\theta \tag{A.6.3i}$$

In spherical polar coordinates, the gradient operator is

$$\nabla = \hat{\mathbf{e}}_r \frac{\partial}{\partial r} + \hat{\mathbf{e}}_\theta \frac{\partial}{r\partial\theta} + \hat{\mathbf{e}}_\phi \frac{1}{r\sin\theta}\frac{\partial}{\partial\phi} \tag{A.6.3j}$$

The gradient of a scalar function, $F(r, \theta, \phi)$, is

$$\nabla F = \hat{\mathbf{e}}_r \frac{\partial F}{\partial r} + \hat{\mathbf{e}}_\theta \frac{\partial F}{r\partial\theta} + \hat{\mathbf{e}}_\phi \frac{1}{r\sin\theta}\frac{\partial F}{\partial\phi} \tag{A.6.3k}$$

The Laplacian in spherical polar coordinates is

$$\nabla^2 F = \left(\hat{\mathbf{e}}_r \frac{\partial}{\partial r} + \hat{\mathbf{e}}_\theta \frac{\partial}{r\partial\theta} + \hat{\mathbf{e}}_\phi \frac{1}{r\sin\theta}\frac{\partial}{\partial\phi}\right) \cdot \left(\hat{\mathbf{e}}_r \frac{\partial F}{\partial r} + \hat{\mathbf{e}}_\theta \frac{\partial F}{r\partial\theta} + \hat{\mathbf{e}}_\phi \frac{1}{r\sin\theta}\frac{\partial F}{\partial\phi}\right) \tag{A.6.3l}$$

Upon computing the indicated dot products in Eq. (A.6.3l), one obtains

$$\nabla^2 F = \left(\frac{\partial^2 F}{\partial r^2} + \frac{\partial^2 F}{r^2 \partial \theta^2} + \frac{1}{r^2 \sin^2\theta}\frac{\partial^2 F}{\partial \phi^2}\right)$$
$$+\left(\hat{\mathbf{e}}_r \cdot \frac{\partial F}{\partial r}\frac{\partial \hat{\mathbf{e}}_r}{\partial r} + \hat{\mathbf{e}}_r \cdot \frac{\partial F}{r\partial \theta}\frac{\partial \hat{\mathbf{e}}_\theta}{\partial r} + \hat{\mathbf{e}}_r \cdot \frac{1}{r\sin\theta}\frac{\partial F}{\partial \phi}\frac{\partial \hat{\mathbf{e}}_\phi}{\partial r}\right)$$
$$+\left(\hat{\mathbf{e}}_\theta \cdot \frac{\partial F}{\partial r}\frac{\partial \hat{\mathbf{e}}_r}{r\partial \theta} + \hat{\mathbf{e}}_\theta \cdot \frac{\partial F}{r\partial \theta}\frac{\partial \hat{\mathbf{e}}_\theta}{r\partial \theta} + \hat{\mathbf{e}}_\theta \frac{1}{r\sin\theta}\frac{\partial F}{\partial \phi}\frac{\partial \hat{\mathbf{e}}_\phi}{r\partial \theta}\right)$$
$$+\left(\hat{\mathbf{e}}_\phi \cdot \frac{\partial F}{\partial r}\frac{1}{r\sin\theta}\frac{\partial \hat{\mathbf{e}}_r}{\partial \phi} + \hat{\mathbf{e}}_\phi \cdot \frac{\partial F}{r\partial \theta}\frac{1}{r\sin\theta}\frac{\partial \hat{\mathbf{e}}_\theta}{\partial \phi} + \hat{\mathbf{e}}_\phi \cdot \frac{1}{r\sin\theta}\frac{\partial F}{\partial \phi}\frac{1}{r\sin\theta}\frac{\partial \hat{\mathbf{e}}_\phi}{\partial \phi}\right) \tag{A.6.3m}$$

The derivatives with respect to r of the unit vectors are identically zero:

$$\frac{\partial}{\partial r}\hat{\mathbf{e}}_r = \frac{\partial}{\partial r}\hat{\mathbf{e}}_\theta = \frac{\partial}{\partial r}\hat{\mathbf{e}}_\phi = 0 \tag{A.6.3n}$$

The derivatives of the unit vectors with respect to θ are, respectively,

$$\frac{\partial}{\partial \theta}\hat{\mathbf{e}}_r = \hat{\mathbf{e}}_x \cos\theta\cos\phi + \hat{\mathbf{e}}_y \cos\theta\sin\phi - \hat{\mathbf{e}}_z \sin\theta = \hat{\mathbf{e}}_\theta \tag{A.6.3o}$$

$$\frac{\partial}{\partial \theta}\hat{\mathbf{e}}_\theta = -\hat{\mathbf{e}}_x \sin\theta\cos\phi - \hat{\mathbf{e}}_y \sin\theta\sin\phi - \hat{\mathbf{e}}_z \cos\theta = -\hat{\mathbf{e}}_r \tag{A.6.3p}$$

$$\frac{\partial}{\partial \theta}\hat{\mathbf{e}}_\phi = -\frac{\partial}{\partial \theta}\hat{\mathbf{e}}_x \sin\phi + \frac{\partial}{\partial \theta}\hat{\mathbf{e}}_y \cos\phi = 0 \tag{A.6.3q}$$

The derivatives of the unit vectors, $\hat{\mathbf{e}}_r$ and $\hat{\mathbf{e}}_\theta$, with respect to ϕ are respectively,

$$\frac{\partial}{\partial \phi}\hat{\mathbf{e}}_r = -\hat{\mathbf{e}}_x \sin\theta\sin\phi + \hat{\mathbf{e}}_y \sin\theta\cos\phi = \hat{\mathbf{e}}_\phi \sin\theta \tag{A.6.3r}$$

$$\frac{\partial}{\partial \phi}\hat{\mathbf{e}}_\theta = -\hat{\mathbf{e}}_x \cos\theta\sin\phi + \hat{\mathbf{e}}_y \cos\theta\cos\phi = \hat{\mathbf{e}}_\phi \cos\theta \tag{A.6.3s}$$

The derivative of $\hat{\mathbf{e}}_\phi$ with respect to ϕ is more complicated. The calculation proceeds as follows:

$$\begin{aligned}
&\frac{\partial}{\partial\phi}\hat{\mathbf{e}}_\phi = -\hat{\mathbf{e}}_x\cos\phi - \hat{\mathbf{e}}_y\sin\phi \\
&= -\hat{\mathbf{e}}_r\sin\theta\cos^2\phi - \hat{\mathbf{e}}_\theta\cos\theta\cos^2\phi + \hat{\mathbf{e}}_\varphi\sin\phi\cos\phi \\
&-\hat{\mathbf{e}}_r\sin\theta\sin^2\phi - \hat{\mathbf{e}}_\theta\cos\theta\sin^2\phi - \hat{\mathbf{e}}_\varphi\sin\phi\cos\phi \\
&= -\hat{\mathbf{e}}_r\sin\theta\cos^2\phi - \hat{\mathbf{e}}_\theta\cos\theta\cos^2\phi - \hat{\mathbf{e}}_r\sin\theta\sin^2\phi - \hat{\mathbf{e}}_\theta\cos\theta\sin^2\phi \\
&= (-\hat{\mathbf{e}}_r\sin\theta - \hat{\mathbf{e}}_\theta\cos\theta)\cos^2\phi + (-\hat{\mathbf{e}}_r\sin\theta - \hat{\mathbf{e}}_\theta\cos\theta)\sin^2\phi \\
&\frac{\partial}{\partial\phi}\hat{\mathbf{e}}_\phi = (-\hat{\mathbf{e}}_r\sin\theta - \hat{\mathbf{e}}_\theta\cos\theta)
\end{aligned} \tag{A.6.3t}$$

where we have used Eqs. (A.6.3g) through (A.6.3i). When the derivatives computed in Eqs. (A.6.3n) through (A.6.3t) are substituted into Eq. (A.6.3m), the result is

$$\begin{aligned}
\nabla^2 F = &\left(\frac{\partial^2 F}{\partial r^2} + \frac{\partial^2 F}{r^2\partial\theta^2} + \frac{1}{r^2\sin^2\theta}\frac{\partial^2 F}{\partial\phi^2}\right) \\
&+ \left(\hat{\mathbf{e}}_\theta\cdot\frac{\partial F}{\partial r}\frac{\hat{\mathbf{e}}_\theta}{r} + \hat{\mathbf{e}}_\theta\cdot\frac{\partial F}{r\partial\theta}\left(\frac{-\hat{\mathbf{e}}_r}{r}\right)\right) \\
&+ \left(\hat{\mathbf{e}}_\phi\cdot\frac{\partial F}{\partial r}\frac{1}{r\sin\theta}\hat{\mathbf{e}}_\phi\sin\theta + \hat{\mathbf{e}}_\phi\cdot\frac{\partial F}{r\partial\theta}\frac{1}{r\sin\theta}\hat{\mathbf{e}}_\phi\cos\theta\right. \\
&\left.+\hat{\mathbf{e}}_\phi\cdot\frac{1}{r\sin\theta}\frac{\partial F}{\partial\phi}\frac{1}{r\sin\theta}(-\hat{\mathbf{e}}_r\sin\theta - \hat{\mathbf{e}}_\theta\cos\theta)\right)
\end{aligned} \tag{A.6.3u}$$

After computation of the indicated unit vector dot products, $\hat{\mathbf{e}}_\theta\cdot\hat{\mathbf{e}}_\theta = 1$, $\hat{\mathbf{e}}_\theta\cdot\hat{\mathbf{e}}_r = 0$, $\hat{\mathbf{e}}_\phi\cdot\hat{\mathbf{e}}_\phi = 1$, $\hat{\mathbf{e}}_\phi\cdot\hat{\mathbf{e}}_\phi = 1$, $\hat{\mathbf{e}}_\phi\cdot\hat{\mathbf{e}}_r = 0$, and $\hat{\mathbf{e}}_\phi\cdot\hat{\mathbf{e}}_\theta = 0$, Eq. (A.6.3s) reads,

$$\nabla^2 F = \left(\frac{\partial^2 F}{\partial r^2} + \frac{\partial^2 F}{r^2\partial\theta^2} + \frac{1}{r^2\sin^2\theta}\frac{\partial^2 F}{\partial\phi^2} + \frac{1}{r}\frac{\partial F}{\partial r} + \frac{1}{r}\frac{\partial F}{\partial r} + \frac{\cos\theta}{r^2\sin\theta}\frac{\partial F}{r\partial\theta}\right) \tag{A.6.3v}$$

The two $\frac{1}{r}\frac{\partial F}{\partial r}$ terms can be combined, and after factoring out $\frac{1}{\sin\theta}$, the various terms can be rearranged to read,

$$\nabla^2 F = \frac{1}{\sin\theta}\left(\sin\theta\frac{\partial^2 F}{\partial r^2} + \frac{2\sin\theta}{r}\frac{\partial F}{\partial r} + \frac{\sin\theta}{r^2}\frac{\partial^2 F}{\partial\theta^2} + \frac{\cos\theta}{r^2}\frac{\partial F}{r\partial\theta} + \frac{1}{r^2\sin\theta}\frac{\partial^2 F}{\partial\phi^2}\right) \tag{A.6.3w}$$

Finally, upon factoring out $\frac{1}{r^2}$, we obtain

$$\nabla^2 F = \frac{1}{r^2\sin\theta}\left(\sin\theta r^2\frac{\partial^2 F}{\partial r^2} + 2r\sin\theta\frac{\partial F}{\partial r} + \sin\theta\frac{\partial^2 F}{\partial\theta^2} + \cos\theta\frac{\partial F}{\partial\theta} + \frac{1}{\sin\theta}\frac{\partial^2 F}{\partial\phi^2}\right) \tag{A.6.3x}$$

which can be simplified to obtain the desired result,

$$\nabla^2 F = \frac{1}{r^2 \sin\theta}\left[\sin\theta \frac{\partial}{\partial r}\left(r^2 \frac{\partial F}{\partial r}\right) + \frac{\partial}{\partial\theta}\left(\sin\theta \frac{\partial F}{\partial\theta}\right) + \frac{1}{\sin\theta}\frac{\partial^2 F}{\partial\phi^2}\right] \quad \text{(A.6.3y)}$$

The Laplacian in spherical polar coordinates is thus

$$\nabla^2 = \frac{1}{r^2 \sin\theta}\left[\sin\theta \frac{\partial}{\partial r}\left(r^2 \frac{\partial}{\partial r}\right) + \frac{\partial}{\partial\theta}\left(\sin\theta \frac{\partial}{\partial\theta}\right) + \frac{1}{\sin\theta}\frac{\partial^2}{\partial\phi^2}\right] \quad \text{(A.6.3z)}$$

A.6.4 Direction Cosines

The unit vectors in rectangular coordinates are $\hat{\mathbf{e}}_x$, $\hat{\mathbf{e}}_y$, and $\hat{\mathbf{e}}_z$, respectively. The vector distance, $\mathbf{r}$, from the origin to a point (x, y, z) in the rectangular coordinate system is

$$\mathbf{r} = x\hat{\mathbf{e}}_x + y\hat{\mathbf{e}}_y + z\hat{\mathbf{e}}_z \quad \text{(A.6.4a)}$$

Define the direction cosines, $\cos\alpha$, $\cos\beta$, and $\cos\gamma$ by

$$x = r\cos\alpha \quad \text{(A.6.4b)}$$

$$y = r\cos\beta \quad \text{(A.6.4c)}$$

$$z = r\cos\gamma \quad \text{(A.6.4d)}$$

By virtue of the Pythagorean theorem,

$$r^2 = x^2 + y^2 + z^2 \quad \text{(A.6.4e)}$$

Upon substitution of Eqs. (A.6.4b) through (A.6.4d) into Eq. (A.6.4e), the result is

$$\cos^2\alpha + \cos^2\beta + \cos^2\gamma = 1 \quad \text{(A.6.4f)}$$

A.7 Simplification of the Radial Part of the Laplacian in Spherical Polar Coordinates

After cancellation of factors of $\sin\theta$ in Eq. (A.6.3z), the radial part ∇_r^2 of the Laplacian acting on an arbitrary function, $f(r)$, is

$$\nabla_r^2 f(r) = \frac{1}{r^2}\frac{d}{dr}\left[r^2\frac{df}{dr}\right] = \frac{1}{r^2}\left[2r\frac{df}{dr} + r^2\frac{d^2f}{dr^2}\right] = \frac{2}{r}\frac{df}{dr} + \frac{d^2f}{dr^2} \tag{A.7.1}$$

Now consider

$$\frac{1}{r}\frac{d^2(rf)}{dr^2} = \frac{1}{r}\frac{d}{dr}\left[f + r\frac{df}{dr}\right] = \frac{1}{r}\left[\frac{df}{dr} + \frac{df}{dr} + r\frac{d^2f}{dr^2}\right] = \frac{2}{r}\frac{df}{dr} + \frac{d^2f}{dr^2} \tag{A.7.2}$$

Hence, upon comparing Eqs. (A.7.1) and (A.7.2), we conclude that

$$\nabla_r^2 f(r) = \frac{1}{r}\frac{d^2(rf)}{dr^2} \tag{A.7.3}$$

A.8 Vector Identities

$$\mathbf{A}\cdot\mathbf{B} = \mathbf{B}\cdot\mathbf{A} = AB\cos\theta \tag{A.8.1}$$

$$\mathbf{A}\times\mathbf{B} = -\mathbf{B}\times\mathbf{A} \tag{A.8.2}$$

$$|\mathbf{A}\times\mathbf{B}| = AB\sin\theta \tag{A.8.3}$$

$$(\mathbf{A}\times\mathbf{B})\cdot\mathbf{C} = (\mathbf{B}\times\mathbf{C})\cdot\mathbf{A} = (\mathbf{C}\times\mathbf{A})\cdot\mathbf{B} \tag{A.8.4}$$

$$\mathbf{A}\times(\mathbf{B}\times\mathbf{C}) = (\mathbf{A}\cdot\mathbf{C})\mathbf{B} - (\mathbf{A}\cdot\mathbf{B})\mathbf{C} \tag{A.8.5}$$

$$(\mathbf{A}\times\mathbf{B})\cdot(\mathbf{C}\times\mathbf{D}) = \mathbf{A}\cdot[\mathbf{B}\times(\mathbf{C}\times\mathbf{D})] = (\mathbf{A}\cdot\mathbf{C})(\mathbf{B}\cdot\mathbf{D}) - (\mathbf{A}\cdot\mathbf{D})(\mathbf{B}\cdot\mathbf{C}) \tag{A.8.6}$$

$$(\mathbf{A}\times\mathbf{B})\times(\mathbf{C}\times\mathbf{D}) = [(\mathbf{A}\times\mathbf{B})\cdot\mathbf{D}]\mathbf{C} - [(\mathbf{A}\times\mathbf{B})\cdot\mathbf{C}]\mathbf{D} \tag{A.8.7}$$

A.9 Differential Operator Identities

$$\nabla(\phi\psi) = \varphi\nabla\psi + \psi\nabla\phi \tag{A.9.1}$$

$$\nabla \cdot (\psi\mathbf{A}) = \psi\nabla \cdot \mathbf{A} + \mathbf{A} \cdot \nabla\psi \tag{A.9.2}$$

$$\nabla \times (\psi\mathbf{A}) = \psi(\nabla \times \mathbf{A}) - \mathbf{A} \times \nabla\psi \tag{A.9.3}$$

$$\nabla \cdot (\mathbf{A} \times \mathbf{B}) = \mathbf{B} \cdot (\nabla \times \mathbf{A}) - \mathbf{A} \cdot (\nabla \times \mathbf{B}) \tag{A.9.4}$$

$$\nabla \times (\mathbf{A} \times \mathbf{B}) = \mathbf{A}(\nabla \cdot \mathbf{B}) - \mathbf{B}(\nabla \cdot \mathbf{A}) + (\mathbf{B} \cdot \nabla)\mathbf{A} - (\mathbf{A} \cdot \nabla)\mathbf{B} \tag{A.9.5}$$

$$\nabla(\mathbf{A} \cdot \mathbf{B}) = \mathbf{A} \times (\nabla \times \mathbf{B}) + \mathbf{B} \times (\nabla \times \mathbf{A}) + (\mathbf{B} \cdot \nabla)\mathbf{A} + (\mathbf{A} \cdot \nabla)\mathbf{B} \tag{A.9.6}$$

$$\nabla^2\psi = \nabla \cdot \nabla\psi \tag{A.9.7}$$

$$\nabla \times \nabla \times \mathbf{A} = \nabla(\nabla \cdot \mathbf{A}) - \nabla^2\mathbf{A} \tag{A.9.8}$$

$$\nabla \times \nabla\psi = 0 \tag{A.9.9}$$

$$\nabla \cdot (\nabla \times \mathbf{A}) = 0 \tag{A.9.10}$$

$$\nabla(\phi + \psi) = \nabla\phi + \nabla\psi \tag{A.9.11}$$

$$\nabla \cdot (\mathbf{A} + \mathbf{B}) = \nabla \cdot \mathbf{A} + \nabla \cdot \mathbf{B} \tag{A.9.12}$$

$$\nabla \times (\mathbf{A} + \mathbf{B}) = \nabla \times \mathbf{A} + \nabla \times \mathbf{B} \tag{A.9.13}$$

$$\mathbf{B} \times (\nabla \times \mathbf{A}) = \sum_{i=x,y,z} B_i\nabla A_i - (\mathbf{B} \cdot \nabla)\mathbf{A} \tag{A.9.14}$$

A.10 Special Vector Relations

$$\nabla \cdot \mathbf{r} = 3 \tag{A.10.1}$$

$$\nabla \times \mathbf{r} = 0 \tag{A.10.2}$$

$$\nabla r = \frac{\mathbf{r}}{r} \tag{A.10.3}$$

$$\nabla\left(\frac{1}{r}\right) = -\frac{\mathbf{r}}{r^3} \tag{A.10.4}$$

$$\nabla\frac{1}{|\mathbf{r} - \mathbf{r}'|} = -\frac{\mathbf{r} - \mathbf{r}'}{|\mathbf{r} - \mathbf{r}'|^3} \tag{A.10.5}$$

$$\nabla \cdot \left(\frac{\mathbf{r}}{r^3}\right) = -\nabla^2\left(\frac{1}{r}\right) = \delta(\mathbf{r}) \tag{A.10.6}$$

$$(\mathbf{A} \cdot \nabla)\mathbf{r} = \mathbf{A} \tag{A.10.7}$$

A.11 Theorems Involving Integrals

Let V be a volume bounded by a closed surface S where the vector differential element of the surface area, $d\mathbf{S}$, points in a direction outward from the enclosed volume.

Theorem I *Gauss–Green Divergence Theorem*

$$\int_S \mathbf{A} \cdot d\mathbf{S} = \int_V (\nabla \cdot \mathbf{A}) d^3 r \tag{A.11.1}$$

Proof In order to remain close to the physical applications of the divergence theorem, we shall prove Eq. (A.11.1) by considering a current density traversing a dielectric. We imagine the dielectric to be exposed to a steady source of ionizing radiation, which produces charge separation at the steady rate $\dot{Q}(\mathbf{r})$ Coulombs/m^3 s at position, $\mathbf{r}$. An external electric field, $\mathbf{E}$, applied to the material will give rise to a drift current density,

$$\mathbf{J}(\mathbf{r}) = J_x(\mathbf{r})\hat{\mathbf{e}} + J_y(\mathbf{r})\hat{\mathbf{e}}_y + J_z(\mathbf{r})\hat{\mathbf{e}}_z \tag{A.11.2a}$$

Imagine that the dielectric is held between two parallel planes, each with normal vector pointed in the x-direction. The planes are separated by a differential distance, dx. If A is the cross-sectional area of each of the planes, then the current flowing in the normal direction through a point (x, y, z) on one plane is $J_x(x, y, z)A$. The current flowing in the normal direction through a point $(x+dx, y, z)$ on the neighboring plane is $J_x(x+dx, y, z)A$. We can expand the latter current in a Taylor series. Keeping only the first two terms, the result is

$$J_x(x + dx, y, z)A = J_x(x, y, z)A + \frac{\partial J_x(x, y, z)}{\partial x} Adx \tag{A.11.2b}$$

The net current flowing into the volume between the two planes is the difference, i.e.

$$J_x(x, y, z)A - J_x(x + dx, y, z)A = -\frac{\partial J_x(x, y, z)}{\partial x} Adx \tag{A.11.2c}$$

The rate of production of charge within the differential volume, Adx, by the ionizing radiation is $\dot{Q}Adx$. Since charge is necessarily conserved, if the system is to be in a steady state, the net charge per second produced in the volume must be balanced by the net charge flowing out. Hence,

$$\dot{Q}Adx + J_x(x, y, z, t)A - J_x(x + dx, y, z)A = \dot{Q}Adx - \frac{\partial J_x(x, y, z)}{\partial x} Adx = 0 \tag{A.11.2d}$$

or

$$\dot{Q} = \frac{\partial J_x(x, y, z)}{\partial x} \tag{A.11.2e}$$

If in the y-direction, as well as the z-direction, there are parallel planes surrounding the differential volume, $dxdydz$ then to take these additional surfaces into account, Eq. (A.11.2e) is replaced by

$$\dot{Q} = \frac{\partial J_x(x, y, z)}{\partial x} + \frac{\partial J_y(x, y, z)}{\partial y} + \frac{\partial J_z(x, y, z)}{\partial z} = \nabla \cdot \mathbf{J}(\mathbf{r}) \tag{A.11.3a}$$

The total rate of charge production is the volume integral of Eq. (A.11.3a), which is

$$\int_V \dot{Q} d^3 r = \int_V (\nabla \cdot \mathbf{J}) d^3 r \tag{A.11.3b}$$

where V means that the domain of integration is over the entire volume. If, as we have assumed, the surface surrounding the volume is closed, then the current produced in the volume must ultimately flow out through the surface surrounding V. The total surface area is S. The differential of this area is $d\mathbf{S} = \hat{\mathbf{n}} dS$, where $\hat{\mathbf{n}}$ is the outward normal. By conservation of charge, the integral of $\mathbf{J}$ over the surface must satisfy

$$\int_V \dot{Q} d^3 r = \int_S (\mathbf{J} \cdot \hat{\mathbf{n}}) dS \tag{A.11.3c}$$

Upon elimination of the volume integral between Eqs. (A.11.3b) and (A.11.3c),

$$\int_V (\nabla \cdot \mathbf{J}) d^3r = \int_S \mathbf{J} \cdot d\mathbf{S} \tag{A.11.3d}$$

which upon identification of **J** with **A**, the theorem is proved. ■

Theorem II *Stokes theorem*

$$\oint_C \mathbf{A} \cdot d\mathbf{l} = \int_S (\nabla \times \mathbf{A}) \cdot d\mathbf{S} \tag{A.11.4}$$

Proof Let **A**be an arbitrary vector field defined within a right circular cylinder. The volume of the cylinder is V, and the total surface area including top, bottom, and lateral boundary is S. The symbol, C, denotes a closed contour in S. The outwardly drawn normal to any one of the surfaces making up the boundary of the cylinder is given the generic designation, $\hat{\mathbf{n}}$. If the outward drawn normal to the top of the cylinder is $\hat{\mathbf{n}}_{top} = \hat{\mathbf{n}}$, then the outward drawn normal to the bottom of the cylinder is $\hat{\mathbf{n}}_{bot} = -\hat{\mathbf{n}}$. These unit vectors are constants over the top and bottom of the cylinder, respectively. Consider the volume integral of an arbitrary vector $\mathbf{A} \times \hat{\mathbf{n}}$. Substitute $\mathbf{A} \times \hat{\mathbf{n}}$ for **J** in the Gauss – Green divergence theorem, which is Eq. (A.11.3d). The result of this substitution is

$$\int_V \nabla \cdot (\mathbf{A} \times \hat{\mathbf{n}}) d^3r = \int_S (\mathbf{A} \times \hat{\mathbf{n}}) \cdot \hat{\mathbf{n}} dS \tag{A.11.5a}$$

On the top/bottom, the integrand in the integral on the right is $\pm\hat{\mathbf{n}} \cdot (\mathbf{A} \times (\pm\hat{\mathbf{n}})) = \hat{\mathbf{n}} \cdot (\mathbf{A} \times \hat{\mathbf{n}})$. Since $\mathbf{A} \times \hat{\mathbf{n}}$ is perpendicular to $\hat{\mathbf{n}}$, the dot product, $\hat{\mathbf{n}} \cdot (\mathbf{A} \times \hat{\mathbf{n}}) = 0$. The contributions to the surface integral on the right due to the top and the bottom of the cylinder are thus zero. On the lateral surface of the cylinder, the normal is $\hat{\mathbf{n}}_{side} = \hat{\mathbf{u}} \times \hat{\mathbf{n}}$ where $\hat{\mathbf{u}}$ is a unit vector, which is tangent to the lateral surface and orbits it in a direction such that the surface is always on the left. On the lateral surface, the dot product in the integrand of the integral on the right in Eq. (A.11.5a) is

$$\hat{\mathbf{n}}_{side} \cdot (\mathbf{A} \times \hat{\mathbf{n}}) = (\hat{\mathbf{u}} \times \hat{\mathbf{n}}) \cdot (\mathbf{A} \times \hat{\mathbf{n}}) = (\hat{\mathbf{u}} \cdot \mathbf{A})(\hat{\mathbf{n}} \cdot \hat{\mathbf{n}}) - (\hat{\mathbf{u}} \cdot \hat{\mathbf{n}})\big(\hat{\mathbf{n}} \cdot \hat{\mathbf{A}}\big) \tag{A.11.5b}$$

where we have used Eq. (A.8.6) to expand $(\hat{\mathbf{u}} \times \hat{\mathbf{n}}) \cdot (\mathbf{A} \times \hat{\mathbf{n}})$. Because $\hat{\mathbf{u}}$ and $\hat{\mathbf{n}}$ are orthogonal, and $\hat{\mathbf{n}} \cdot \hat{\mathbf{n}} = 1$, Eq. (A.11.5b) simplifies to

$$\hat{\mathbf{n}}_{side} \cdot (\mathbf{A} \times \hat{\mathbf{n}}) = (\hat{\mathbf{u}} \cdot \mathbf{A}) \tag{A.11.5c}$$

We can expand the integrand of the integral on the left hand side by using Eq. (A.9.4). The result is

$$\nabla \cdot (\mathbf{A} \times \hat{\mathbf{n}}) = \hat{\mathbf{n}} \cdot (\nabla \times \mathbf{A}) - \mathbf{A} \cdot (\nabla \times \hat{\mathbf{n}}) \tag{A.11.5d}$$

Because $\hat{\mathbf{n}}$ is a constant, $\nabla \times \hat{\mathbf{n}} = 0$, and Eq. (A.11.5d) simplifies to

$$\nabla \cdot (\mathbf{A} \times \hat{\mathbf{n}}) = \hat{\mathbf{n}} \cdot (\nabla \times \mathbf{A}) \tag{A.11.5e}$$

Upon substitution of Eqs. (A.11.5c) and (A.11.5e) into Eq. (A.11.5a), the result is

$$\int\limits_V \hat{\mathbf{n}} \cdot (\nabla \times \mathbf{A}) d^3 r = \int\limits_S (\hat{\mathbf{u}} \cdot \mathbf{A}) dS_{side} \tag{A.11.5f}$$

where the surface integral on the right is confined to the lateral surface, S_{side}, of the cylinder.

If dS is a differential of the area of the bottom of the cylinder, h is the height of the cylinder, and $d\ell$ is a differential of the length of the circumference, then the differential volume is $d^3 r = dSdh$, while the differential of the lateral surface area is $dS_{side} = dhd\ell$. With these substitutions, Eq. (A.11.5f) becomes

$$\int\limits_0^h dh \int\limits_S dS\, \hat{\mathbf{n}} \cdot (\nabla \times \mathbf{A}) = \int\limits_0^h dh \oint\limits_C d\ell (\hat{\mathbf{u}} \cdot \mathbf{A}) \tag{A.11.5g}$$

where C indicates a circumferential path on the lateral surface. We can set the integrands of the integrals over dh equal to obtain

$$\int\limits_S \hat{\mathbf{n}} \cdot (\nabla \times \mathbf{A}) dS = \oint\limits_C (\mathbf{A} \cdot \hat{\mathbf{u}}) d\ell \tag{A.11.5h}$$

Because $d\mathbf{l} = \hat{\mathbf{u}} d\ell$, and $\mathrm{d}\mathbf{S} = \hat{\mathbf{n}}\mathrm{dS}$, Eq. (A.11.5h) is the same as Eq. (A.11.4), which is known as Stokes' theorem. ■

The following theorems are offered without proof:

Theorem III

$$\int\limits_S \psi d\mathbf{S} = \int\limits_V \nabla \psi d^3 r \tag{A.11.6}$$

Theorem IV

$$\int\limits_S d\mathbf{S} \times \mathbf{A} = \int\limits_V (\nabla \times \mathbf{A}) d^3 r \tag{A.11.7}$$

Theorem V *Let S is an open surface bounded by a closed contour C. If the differential* *line element of the contour is* $d\mathbf{l}$, then

$$\oint_C \psi d\mathbf{l} = \int_S d\mathbf{S} \times \nabla\psi \tag{A.11.8}$$

A.12 Matrix Addition and Matrix Multiplication

The law of matrix addition is

$$\begin{bmatrix} a_{11} & a_{12} \\ a_{21} & a_{22} \end{bmatrix} + \begin{bmatrix} b_{11} & b_{12} \\ b_{21} & b_{22} \end{bmatrix} = \begin{bmatrix} a_{11} + b_{11} & a_{12} + b_{12} \\ a_{21} + b_{21} & a_{22} + b_{22} \end{bmatrix} \tag{A.12.1}$$

The law of matrix multiplication is

$$\begin{bmatrix} a_{11} & a_{12} \\ a_{21} & a_{22} \end{bmatrix}\begin{bmatrix} b_{11} & b_{12} \\ b_{21} & b_{22} \end{bmatrix} = \begin{bmatrix} a_{11}b_{11} + a_{12}b_{21} & a_{11}b_{12} + a_{12}b_{22} \\ a_{21}b_{11} + a_{22}b_{21} & a_{21}b_{12} + a_{22}b_{22} \end{bmatrix} \tag{A.12.2}$$

A.13 Kronecker Delta and the Dirac Delta Function

The Kronecker delta symbol, δ_{ij}, appears in sums where the index values, i and j, take on discrete values. The definition of δ_{ij} is

$$\begin{aligned} \delta_{ij} &= 0, \text{ if } i \neq j \\ \delta_{ij} &= 1, \text{ if } i = j \end{aligned} \tag{A.13.1}$$

By virtue of Eq. (A.13.1), the sum of the Kronecker delta over one of its indices is

$$\sum_{i=1}^{N} \delta_{ij} = 1 \text{ for } 1 \leq j \leq N \tag{A.13.2}$$

The Kronecker delta symbol appears in summations,

$$\sum_{i=1}^{N} a_i \delta_{ij} = a_j \tag{A.13.3}$$

Due to its definition in Eq. (A.13.1), $\delta_{ij} = \delta_{ji}$, which allows us to conclude that the Kronecker delta is a symmetric function of its indices.

The Dirac delta function is a generalization of the Kronecker delta designed to treat the case where the integer indices are replaced by continuous variables whose values can extend over the domain, $-\infty$ to $+\infty$. The analog of Eq. (A.13.2) is

$$\int_{-\infty}^{+\infty} dx\delta(x - x_o) = 1 \tag{A.13.4}$$

The delta function is sharply peaked around $x = x_o$. The analog of Eq. (A.13.3) is

$$\int_{-\infty}^{+\infty} dxf(x)\delta(x - x_o) = f(x_o) \tag{A.13.5}$$

As noted above, the Kronecker delta has the property, $\delta_{ij} = \delta_{ji}$. By analogy, we can conclude that the Dirac delta is an even function of its argument, i.e. $\delta(x - x_o) = \delta(x_o - x)$. Whereas δ_{ij}, has little meaning outside a sum, the Dirac delta function makes little sense other than as a factor in an integrand.

A.14 Solution to an Inhomogeneous Differential Equation by the Green's Function Method

Consider the inhomogeneous differential equation

$$\frac{d^2\Psi(x)}{dx^2} - \alpha^2\Psi(x) = -F(x) \qquad \text{for } -\infty \le x \le +\infty \tag{A.14.1}$$

where α is a constant. The boundary conditions are

$$\Psi(-\infty) = 0 \tag{A.14.2a}$$

$$\Psi(+\infty) = 0 \tag{A.14.2b}$$

Equations (A.14.1), (A.14.2a), and (A.14.2b) are said to constitute a boundary value problem.

In order to solve the above boundary value problem, we consider the simpler differential equation where $F(x)$ is replaced by the Dirac delta function centered on $x = x_o$:

$$\frac{d^2G(x)}{dx^2} - \alpha^2G(x) = -\delta(x - x_o), \quad \text{for} \quad -\infty \le x \le +\infty \tag{A.14.3}$$

The associated boundary conditions are chosen to have the form,

$$G(-\infty) = 0 \tag{A.14.4a}$$

$$G(+\infty) = 0 \tag{A.14.4b}$$

The function, $G(x)$, that satisfies Eqs. (A.14.3), (A.14.4a), and (A.14.4b) is known as the Green's function. The solution for $G(x)$ can be used to construct an integral representation for $\Psi(x)$. The procedure is known as Green's Theorem:

Green's Theorem: The solution to the inhomogeneous boundary value problem for $\Psi(x)$ is the integral,

$$\Psi(x) = \int_{-\infty}^{+\infty} dx_o F(x_o) G(x, x_o) \tag{A.14.5}$$

Proof Multiply Eq. (A.14.1) by $G(x)$. The result is

$$G(x)\frac{d^2\Psi(x)}{dx^2} - \alpha^2 G(x)\Psi(x) = -G(x)F(x) \tag{A.14.6a}$$

Multiply Eq. (A.14.3) by $\Psi(x)$. The result is

$$\Psi(x)\frac{d^2G(x)}{dx^2} - \alpha^2\Psi(x)G(x) = -\Psi(x)\delta(x - x_o) \tag{A.14.6b}$$

Subtract Eq. (A.14.6b) from Eq. (A.14.6a). The result is

$$G(x)\frac{d^2\Psi(x)}{dx^2} - \Psi(x)\frac{d^2G(x)}{dx^2} = -G(x)F(x) + \Psi(x)\delta(x - x_o) \tag{A.14.7}$$

Integrate Eq. (A.14.7) over the domain of x. The integral is

$$\int_{-\infty}^{+\infty} dxG(x)\frac{d^2\Psi(x)}{dx^2} - \int_{-\infty}^{+\infty} dx\Psi(x)\frac{d^2G(x)}{dx^2} = -\int_{-\infty}^{+\infty} dxG(x)F(x) + \int_{-\infty}^{+\infty} dx\Psi(x)\delta(x - x_o) \tag{A.14.8}$$

Use the integral property of the Dirac delta function in Eq. (A.13.5) to evaluate the second integral on the right in Eq. (A.14.8) to obtain

$$\int_{-\infty}^{+\infty} dx\Psi(x)\delta(x - x_o) = \Psi(x_o) \tag{A.14.9}$$

The first integral on the left in Eq. (A.14.8) can be evaluated by parts. The result is

$$\int_{-\infty}^{+\infty} dx G(x) \frac{d^2\Psi(x)}{dx^2} = \left[G(x)\frac{d\Psi(x)}{dx}\right]_{-\infty}^{+\infty} - \int_{-\infty}^{+\infty} dx \frac{dG(x)}{dx}\frac{d\Psi(x)}{dx}$$

$$= -\int_{-\infty}^{+\infty} dx \frac{dG(x)}{dx}\frac{d\Psi(x)}{dx} \qquad \text{(A.14.10a)}$$

where the term in brackets is zero by virtue of Eqs. (A.14.4a) and (A.14.4b). The second integral on the left in Eq. (A.14.8) can also be evaluated by parts. The result is

$$\int_{-\infty}^{+\infty} dx \Psi(x) \frac{d^2G(x)}{dx^2} = \left[\Psi(x)\frac{dG(x)}{dx}\right]_{-\infty}^{+\infty} - \int_{-\infty}^{+\infty} dx \frac{d\Psi(x)}{dx}\frac{dG(x)}{dx} = -\int_{-\infty}^{+\infty} dx \frac{dG(x)}{dx}\frac{d\Psi(x)}{dx} \qquad \text{(A.14.10b)}$$

where the term in brackets on the right is zero by virtue of Eqs. (A.14.2a) and (A.14.2b). Upon substitution of Eqs. (A.14.9), (A.14.10a), and (A.14.10b) into Eq. (A.14.8), the result is

$$-\int_{-\infty}^{+\infty} dx \left[\frac{dG(x)}{dx}\frac{d\Psi(x)}{dx} - \frac{dG(x)}{dx}\frac{d\Psi(x)}{dx}\right] = -\int_{-\infty}^{+\infty} dx G(x.x_o)F(x) + \Psi(x_o) \qquad \text{(A.14.11)}$$

The integrand of the integral on the left is identically zero, so that we can conclude,

$$\Psi(x_o) = \int_{-\infty}^{+\infty} dx G(x_o, x) F(x) \qquad \text{(A.14.12)}$$

If we can assume that $G(x_o, x) = G(x, x_o)$, we can interchange x and x_o and get our final result which is Eq. (A.14.5). ■

Equation (A.14.5) is of little use unless we can solve Eq. (A.14.3) and show in addition that $G(x_o, x) = G(x, x_o)$. To prove this relation, we begin the solution to Eq. (A.14.3) by dividing the domain of x into two parts. Since we know that $\delta(x - x_o)$ is non-zero only when $x = x_o$, we define two functions, $G_1(x)$ and $G_2(x)$, which have the following properties:

$$\frac{d^2G_1(x)}{dx^2} - \alpha^2 G_1(x) = 0 \qquad -\infty \le x < x_o \qquad \text{(A.14.13a)}$$

$$G_1(-\infty) = 0 \tag{A.14.13b}$$

$$\frac{d^2G_2(x)}{dx^2} - \alpha^2 G_2(x) = 0 \quad x_o < x \le +\infty \tag{A.14.14a}$$

$$G_2(+\infty) = 0 \tag{A.14.14b}$$

In order to specify $G_1(x)$ and $G_2(x)$, we need two more boundary conditions. These two boundary conditions apply at the inner boundary where, $x = x_o$. The first is the continuity condition, which is

$$G_1(x_o) = G_2(x_o) \tag{A.14.15}$$

The second is obtained by integrating Eq. (A.14.3) over $x = x_o$. If we let ε be an infinitesimal, the integral takes the form,

$$\int_{x_o-\varepsilon}^{x_o+\varepsilon} dx \frac{d^2G(x)}{dx^2} - \alpha^2 \int_{x_o-\varepsilon}^{x_o+\varepsilon} dx G(x) = -\int_{x_o-\varepsilon}^{x_o+\varepsilon} dx \delta(x - x_o) \tag{A.14.16}$$

The first integral on the left in Eq. (A.14.16) is $\left[\frac{dG(x)}{dx}\right]_{x_o-\varepsilon}^{x_o+\varepsilon}$. Since ε is an infinitesimal and $G(x)$ is continuous, the second integral on the left is zero. By virtue of Eq. (A.13.4), the integral on the right is -1. Hence, Eq. (A.14.16) reads,

$$\left[\frac{dG(x)}{dx}\right]_{x_o-\varepsilon}^{x_o+\varepsilon} = \left(\frac{dG_2(x)}{dx}\right)_{x_o} - \left(\frac{dG_1(x)}{dx}\right)_{x_o} = -1 \tag{A.14.17}$$

where in Eq. (A.14.17), we have noted that when $x = x_o + \varepsilon$, $G(x) = G_2(x)$, and when $x = x_o - \varepsilon$, $G(x) = G_1(x)$.

When $x \neq x_o$, the right hand side of Eq. (A.14.3) is zero. It reads,

$$\frac{d^2G(x)}{dx^2} - \alpha^2 G(x) = 0 \tag{A.14.18}$$

The general solution to Eq. (A.14.18) is

$$G(x) = Ae^{\alpha x} + Be^{-\alpha x} \tag{A.14.19}$$

where A and B are constants of integration.

Equation (14.19) allows us to write the solution to Eq. (A.14.13a) as

$$G_1(x) = A_1 e^{\alpha x} + B_1 e^{-\alpha x} \tag{A.14.20a}$$

To satisfy Eq. (A.14.13b), we must have $B_1 = 0$. With this substitution, Eq. (A.14.14) simplifies to read,

$$G_1(x) = A_1 e^{\alpha x} \tag{A.14.20b}$$

Equation (A.14.19) allows us to write the solution to Eq. (A.14.14a) as

$$G_2(x) = A_2 e^{\alpha x} + B_2 e^{-\alpha x} \tag{A.14.21a}$$

To satisfy Eq. (A.14.14b), we must have $A_2 = 0$. Hence, Eq. (A.14.21a) can be rewritten as

$$G_2(x) = B_2 e^{-\alpha x} \tag{A.14.21b}$$

Equation (A.14.15) is satisfied by Eqs. (A.14.20b) and (A.14.21b) if

$$A_1 e^{\alpha x_o} = B_2 e^{-\alpha x_o} \tag{A.14.22a}$$

If we compute the derivatives of Eqs. (A.14.20b) and (A.14.21b) and substitute the results into Eq. (A.14.17), we obtain the algebraic equation,

$$-\alpha B_2 e^{-\alpha x_o} - \alpha A_1 e^{\alpha x_o} = -1 \tag{A.14.22b}$$

If we substitute Eq. (A.14.22a) into Eq. (A.14.22b), the result is

$$-\alpha A_1 e^{\alpha x_o} - \alpha A_1 e^{\alpha x_o} = -1 \tag{A.14.22c}$$

or

$$A_1 = \frac{1}{2\alpha} e^{-\alpha x_o} \tag{A.14.22d}$$

Substitute Eq. (A.14.22d) into Eq. (A.14.22a). The result is

$$\frac{1}{2\alpha} e^{-\alpha x_o} e^{\alpha x_o} = B_2 e^{-\alpha x_o} \tag{A.14.22e}$$

or

$$B_2 = \frac{1}{2\alpha} e^{\alpha x_o} \tag{A.14.22f}$$

With these results, we can write Eq. (A.14.20b) in the form

$$G_1(x) = \frac{1}{2\alpha} e^{-\alpha(x_o - x)} \tag{A.14.23a}$$

and Eq. (A.14.21b) in the form

$$G_2(x) = \frac{1}{2\alpha} e^{-\alpha(x-x_o)} \tag{A.14.23b}$$

Equations (A.14.23a) and (A.14.23b) can be summarized by writing,

$$G(x, x_o) = \frac{1}{2\alpha} e^{-\alpha|x-x_o|} \tag{A.14.24}$$

On the basis of Eq. (A.14.24), it is clear that $G(x_o, x) = G(x, x_o)$.

A.15 Complex Numbers and Complex Number Algebra

Complex numbers are based upon the real numbers plus the imaginary number i which satisfies $i^2 = -1$. A complex number, z, can be written in the form of the sum,

$$z = x + iy \tag{A.15.1}$$

where both x and y are real numbers. The complex number, z, is said to have a real part, $\mathrm{Re}(z)$, and imaginary part, $\mathrm{Im}(z)$, defined with respect to Eq. (A.15.1) by

$$x = \mathrm{Re}(z) \tag{A.15.2a}$$

and

$$y = \mathrm{Im}(z) \tag{A.15.2b}$$

If $z_1 = x_1 + iy_1$ and $z_2 = x_2 + iy_2$ are two complex numbers, their sum is

$$z_1 + z_2 = (x_1 + x_2) + i(y_1 + y_2) \tag{A.15.3}$$

The product of two complex numbers, z_1 and z_2, can be computed using ordinary binomial multiplication,

$$z_1 z_2 = (x_1 + iy_1)(x_2 + iy_2) = x_1x_2 + x_1iy_2 + iy_1x_2 + i^2y_1y_2 \tag{A.15.4a}$$

Upon use of $i^2 = -1$, the product becomes

$$z_1 z_2 = (x_1x_2 - y_1y_2) + i(x_1y_2 + y_1x_2) \tag{A.15.4b}$$

The complex conjugate, i^* of i, is defined by $i^* = -i$. This definition permits the complex conjugate of z, denoted by z^* to be written as

$$z^* = x - iy \tag{A.15.5}$$

The square root of the product of a complex number z and its complex conjugate z^* is said to be the "norm", the "modulus", or the "absolute value" of z. This product is ordinarily denoted by $|z|$ and is defined by

$$|z| = \sqrt{zz^*} = \sqrt{(x+iy)(x-iy)} = \sqrt{x^2+y^2} \tag{A.15.6}$$

Using rectangular coordinates, the number, z, can be represented as a point in the plane identified by the coordinates, (x, y). The distance, r, from the origin, $(0, 0)$ to the point, (x, y) is $r = |z|$. The angle θ which r makes with the x-axis satisfies

$$\tan\theta = \frac{y}{x} \tag{A.15.7a}$$

where

$$\cos\theta = \frac{x}{\sqrt{x^2+y^2}} = \frac{x}{r} \tag{A.15.7b}$$

$$\sin\theta = \frac{y}{\sqrt{x^2+y^2}} = \frac{y}{r} \tag{A.15.7c}$$

According to Eq. (A.15.6),

$$r = |z| = \sqrt{x^2+y^2} \tag{A.15.7d}$$

By use of Eq. (A.3.3), the power series representation of the exponential $e^{i\theta}$ is

$$\begin{aligned} e^{i\theta} &= 1 + i\theta + \frac{1}{2!}(i\theta)^2 + \frac{1}{3!}(i\theta)^3 + \frac{1}{4!}(i\theta)^4 + \frac{1}{5}(i\theta)^5 + \cdots \\ &= \left(1 - \frac{1}{2!}\theta^2 + \frac{1}{4!}\theta^4 - \right)\cdots + i\left(\theta - \frac{1}{3!}\theta^3 + \frac{1}{5!}\theta^5 - \right) \end{aligned} \tag{A.15.8a}$$

According to Eqs. (A.3.6) and (A.3.7), Eq. (A.15.8a) can be rewritten as

$$e^{i\theta} = \cos\theta + i\sin\theta \tag{A.15.8b}$$

By combining Eqs. (A.15.1), (A.15.7b) and (A.15.7c) with Eq. (A.15.8b), we can write z in polar form as

$$z = r(\cos\theta + i\sin\theta) = re^{i\theta} \tag{A.15.9}$$

One can represent the ratio of two complex numbers in rationalized form according to the following calculation:

$$\frac{z_1}{z_2} = \frac{(x_1 + iy_1)}{(x_2 + iy_2)} = \frac{(x_1 + iy_1)}{(x_2 + iy_2)} \cdot \frac{(x_2 - iy_2)}{(x_2 - iy_2)} = \frac{(x_1x_2 + y_1y_2) + i(x_2y_1 - x_1y_2)}{\left(x_2^2 + y_2^2\right)} \tag{A.15.10}$$

Using Eq. (A.15.9), we can write the same ratio more compactly in polar form as

$$\frac{z_1}{z_2} = \frac{r_1}{r_2} \exp[i(\theta_1 - \theta_2)] \tag{A.15.11}$$

A.16 Fourier Transforms

A.16.1 The Fourier Transform Pair

Let $f(t)$ be a function of the real variable, t. The Fourier transform, $\tilde{f}(\omega)$, of $f(t)$ is defined by the integral,

$$\tilde{f}(\omega) = \int_{-\infty}^{+\infty} f(t)e^{i\omega t}dt \tag{A.16.1a}$$

According to Eq. (A.16.1a), $\tilde{f}(\omega)$ is a complex valued function. Fourier proved a theorem, which showed that the inverse of Eq. (A.16.1a), is

$$f(t) = \frac{1}{2\pi} \int_{-\infty}^{+\infty} e^{-i\omega t}\tilde{f}(\omega)d\omega \tag{A.16.1b}$$

Equation (A.16.1b) is said to be the Fourier integral representation of $f(t)$. The functions, $f(t)$ and $\tilde{f}(\omega)$, taken together are said to constitute a Fourier transform pair. The Fourier transform pair satisfies some general theorems, two of which we prove below.

A.16.2 The Convolution Theorem

We let $E(t)$ be a forcing function and $G(t)$ be the function governing the dynamics of the response to $E(t)$. The function, $F(t)$, will represent the time dependence of the response of the system to the forcing function. If the system responds instantaneously to $E(t)$, then $F(t) = G(t)E(t)$. On the other hand, if there is a delay in the response, then the signal, $F(t)$ is the sum of dynamic effects that have evolved over a time $(t - \tau)$ to the stimulus provided by the forcing function, $E(\tau)$, that started acting at the time $\tau < t$. Given that the time, t, is a continuous variable, this sum takes the form of an integral.

Theorem: Consider the three functions, $E(t)$, $F(t)$, $G(t)$, which are related according to

$$F(t) = \int_{-\infty}^{+\infty} d\tau G(t-\tau)E(\tau) \tag{A.16.2a}$$

Equation (A.16.2a) defines a convolution integral. If the Fourier transform of $F(t)$ is $\tilde{F}(\omega)$, the transform of $G(t)$ is $\tilde{G}(\omega)$, and the transform of $E(t)$ is $\tilde{E}(\omega)$, it is then true that

$$\tilde{F}(\omega) = \tilde{G}(\omega)\tilde{E}(\omega) \tag{A.16.2b}$$

Proof According to Eq. (A.16.1b), the Fourier integral representation of $G(t)$ is

$$G(t) = \frac{1}{2\pi}\int_{-\infty}^{+\infty} e^{-i\omega t}\tilde{G}(\omega)d\omega \tag{A.16.3a}$$

If in Eq. (A.16.3a), we replace t by $t - \tau$, the result is

$$G(t-\tau) = \frac{1}{2\pi}\int_{-\infty}^{+\infty} e^{-i\omega(t-\tau)}\tilde{G}(\omega)d\omega \tag{A.16.3b}$$

Upon substitution of Eq. (A.16.3b) into Eq. (A.16.2a), the result is

$$F(t) = \int_{-\infty}^{+\infty} d\tau \left[\frac{1}{2\pi}\int_{-\infty}^{+\infty} d\omega\tilde{G}(\omega)e^{-i\omega(t-\tau)}\right]E(\tau) \tag{A.16.3c}$$

The integrals in Eq. (A.16.3c) can be rearranged to read

$$F(t) = \frac{1}{2\pi}\int_{-\infty}^{+\infty} d\omega\tilde{G}(\omega)e^{-i\omega t}\left[\int_{-\infty}^{+\infty} d\tau e^{i\omega\tau}E(\tau)\right] \tag{A.16.3d}$$

By reference to Eq. (A.16.1a), the Fourier transform of $E(t)$ is

$$\tilde{E}(\omega) = \int_{-\infty}^{+\infty} E(t)e^{i\omega t}dt \tag{A.16.3e}$$

Upon substitution of Eq. (A.16.3e) into Eq. (A.16.3d), the result is

$$F(t) = \frac{1}{2\pi} \int_{-\infty}^{+\infty} d\omega \tilde{G}(\omega) \tilde{E}(\omega) e^{-i\omega t} \tag{A.16.3f}$$

The Fourier integral representation of $F(t)$ is

$$F(t) = \frac{1}{2\pi} \int_{-\infty}^{+\infty} d\omega \tilde{F}(\omega) e^{-i\omega t} \tag{A.16.3g}$$

If we equate Eqs. (A.16.3f) and (A.16.3g), we conclude that $\tilde{F}(\omega) = \tilde{G}(\omega)\tilde{E}(\omega)$. ■

Although the range of integration in Eq. (A.16.2a) extends from $-\infty$ to $+\infty$, it must be the case that $G(t-\tau) = 0$, for $\tau > t$. Otherwise, the response, $F(t)$, of the system at time t would be determined by the force, $E(\tau)$, acting on the system at future times. The recognition of the impossibility of such an event is said to form the basis for the principal of causality.

We have now proved that the Fourier transform of a convolution integral is the product of the Fourier Transforms of the two functions involved in the convolution. Looked at in reverse, if one has a function $\tilde{F}(\omega)$ of ω that can be factored as in Eq. (A.16.2b), then it is possible to represent its image, $F(t)$, in the t variable space by the integral in Eq. (A.16.2a).

A.16.3 The Inverse Convolution Theorem

If a function $f(t)$ in the t-variable space can be factored in the form, $f(t) = g(t)h(t)$, it is possible to enquire into the nature of its Fourier transform, $\tilde{f}(\omega)$, assuming that the transforms $\tilde{g}(\omega)$ and $\tilde{h}(\omega)$, of $g(t)$ and $h(t)$ exist. We shall approach this problem in reverse by proving the following theorem:

Theorem: *Suppose that $\tilde{f}(\omega)$ is related to $\tilde{g}(\omega)$ and $\tilde{h}(\omega)$ by the convolution integral in ω frequency space,*

$$\tilde{f}(\omega) = \frac{1}{2\pi} \int_{-\infty}^{+\infty} d\omega' \tilde{g}(\omega - \omega') \tilde{h}(\omega') \tag{A.16.4a}$$

then

$$f(t) = g(t)h(t) \tag{A.16.4b}$$

Proof The Fourier transform of $g(t)$ as a function of $\omega - \omega'$ is $\tilde{g}(\omega - \omega')$, which according to Eq. (A.16.1a), is given by

$$\tilde{g}\left(\omega - \omega'\right) = \int_{-\infty}^{+\infty} g(t)e^{i\left(\omega-\omega'\right)t}dt \tag{A.16.5a}$$

Substitute Eq. (A.16.5a) into Eq. (A.16.4a). The result is

$$\tilde{f}(\omega) = \frac{1}{2\pi}\int_{-\infty}^{+\infty} d\omega' \int_{-\infty}^{+\infty} dt\, e^{i\left(\omega-\omega'\right)t} g(t)\tilde{h}\left(\omega'\right) \tag{A.16.5b}$$

Rearrange Eq. (A.16.5b) to read

$$\tilde{f}(\omega) = \int_{-\infty}^{+\infty} dt\, e^{i\omega t} g(t) \frac{1}{2\pi}\int_{-\infty}^{+\infty} d\omega'\, e^{-i\omega' t}\tilde{h}\left(\omega'\right) \tag{A.16.5c}$$

According to Eq. (A.16.1b),

$$h(t) = \frac{1}{2\pi}\int_{-\infty}^{+\infty} d\omega' e^{-i\omega' t}\tilde{h}\left(\omega'\right) \tag{A.16.5d}$$

Equation (A.16.5d) allows us to replace the integral involving ω' in Eq. (A.16.5c) with $h(t)$, so that Eq. (A.16.5c) reads,

$$\tilde{f}(\omega) = \int_{-\infty}^{+\infty} dte^{i\omega t} g(t)h(t) \tag{A.16.5e}$$

Replace $\tilde{f}(\omega)$ in Eq. (A.16.5e) by

$$\tilde{f}(\omega) = \int_{-\infty}^{+\infty} dte^{i\omega t} f(t)dt \tag{A.16.5f}$$

The result is

$$\int_{-\infty}^{+\infty} dt\, e^{i\omega t} f(t)dt = \int_{-\infty}^{+\infty} dt\, e^{i\omega t} g(t)h(t) \tag{A.16.5g}$$

If the integral on the left is to equal the integral on the right, the integrands must be equal, which implies $f(t) = g(t)h(t)$, which proves the theorem. ■

A.17 Fourier Transform of the "Sign" Function, *S*(*t*)

The "sign" function, $s(t)$, is defined by

$$\mathrm{s}(t) = \begin{cases} 1, & t > 0 \\ 0, & t = 0 \\ -1, & t < 0 \end{cases} \tag{A.17.1}$$

The Fourier transform, $\tilde{s}(\omega)$, of the sign function is

$$\tilde{s}(\omega) = \int_{-\infty}^{+\infty} dt\, e^{i\omega t} \mathrm{s}(t) \tag{A.17.2}$$

The definition of $s(t)$ in Eq. (A.17.1) can be exploited to break up the range of integration in the integral on the right hand side of Eq. (A.17.2) into the domains $-\infty \le t < 0$ and $0 < t \le +\infty$. Upon substitution of Eq. (A.17.1) into Eq. (A.17.2), we obtain

$$\tilde{s}(\omega) = \int_{-\infty}^{0} dt(-1)e^{i\omega t} + \int_{0}^{+\infty} dt(+1)e^{i\omega t} \tag{A.17.3}$$

In the first integral on the right in Eq. (A.17.3), we can replace t by $-t$. The result is

$$\int_{-\infty}^{0} dt(-1)\, e^{i\omega t} = \int_{+\infty}^{0} d(-t)(-1)\, e^{-i\omega t} = -\int_{0}^{+\infty} dt\, e^{-i\omega t} \tag{A.17.4}$$

Upon substitution of Eq. (A.17.4), Eq. (A.17.3) reads,

$$\tilde{s}(\omega) = -\int_{0}^{+\infty} dt\, e^{-i\omega t} + \int_{0}^{+\infty} dt\, e^{i\omega t} \tag{A.17.5}$$

Because

$$\lim_{\varepsilon \to 0} e^{(\pm i\omega \pm \varepsilon)t} = e^{\pm i\omega t} \tag{A.17.6}$$

it is possible to rewrite the integrals in Eq. (A.17.5) in the form,

$$\begin{aligned}
\tilde{s}(\omega) &= -\int_{0}^{+\infty} dt \lim_{\varepsilon\to 0}\left[e^{-(i\omega+\varepsilon)t}\right] + \int_{0}^{+\infty} dt \lim_{\varepsilon\to 0}\left[e^{(i\omega-\varepsilon)t}\right] \\
&= -\lim_{\varepsilon\to 0}\int_{0}^{+\infty} dt\, e^{-(i\omega+\varepsilon)t} + \lim_{\varepsilon\to 0}\int_{0}^{+\infty} dt\, e^{(i\omega-\varepsilon)t} \\
&= -\lim_{\varepsilon\to 0}\left[\frac{e^{-i\omega t}\cdot e^{-\varepsilon t}}{-(i\omega+\varepsilon)}\right]_0^{+\infty} + \lim_{\varepsilon\to 0}\left[\frac{e^{i\omega t}\cdot e^{-\varepsilon t}}{(i\omega-\varepsilon)}\right]_0^{+\infty} \\
&= -\frac{(-1)}{-i\omega} + \frac{(-1)}{i\omega} = \frac{i}{\omega} + \frac{i}{\omega} \\
\tilde{s}(\omega) &= \frac{2i}{\omega}
\end{aligned} \tag{A.17.7}$$

A.18 Kramers–Kronig Relation for the Dielectric Response Function

In Sect. 12.6 of Chap. 12, we derived integral relations linking the real and imaginary parts of the molecular polarizability, $\tilde{\alpha}(\omega)$. These equations are known as Kramers–Kronig relations. Here, we derive the equivalent relations linking the real and imaginary parts of the relative permittivity, $\tilde{\varepsilon}_r(\omega)$. To derive these relations, we note that the relative permittivity is related to the electric susceptibility, $\tilde{\chi}_e(\omega)$, according to

$$\tilde{\varepsilon}_r(\omega) = 1 + \tilde{\chi}_e(\omega) \tag{A.18.1}$$

Upon writing $\tilde{\varepsilon}_r(\omega)$ and $\tilde{\chi}_e(\omega)$ in terms of their real and imaginary parts, Eq. (A.18.1) becomes

$$\varepsilon_r'(\omega) + i\varepsilon_r''(\omega) = 1 + \chi_e'(\omega) + i\chi_e''(\omega) \tag{A.18.2}$$

On the basis of Eq. (A.18.2), we conclude that

$$\chi_e'(\omega) = \varepsilon_r'(\omega) - 1 \tag{A.18.3}$$

and

$$\chi_e''(\omega) = \varepsilon_r''(\omega) \tag{A.18.4}$$

We now turn to the results of Sect. 12.6 in Chap. 12. Following the method of Norman, Ruud, and Saue [1], we introduce

$$G(t) = G_{even}(t) + G_{odd}(t) \tag{A.18.5}$$

According to Eq. (12.6.14) of Chap. 12,

$$\tilde{G}_{even}(\omega) = \alpha'(\omega) \tag{A.18.6}$$

where $\tilde{\alpha}(\omega) = \alpha'(\omega) + i\alpha''(\omega)$ is the complex valued molecular polarizability. If in Eq. (A.18.6), we replace $\alpha'(\omega)$ by $\chi_e'(\omega) = \varepsilon_r'(\omega) - 1$, the result is

$$\tilde{G}_{even}(\omega) = \chi_e'(\omega) = \varepsilon_r'(\omega) - 1 \tag{A.18.7}$$

According to Eq. (12.16.16) of Chap. 12,

$$\tilde{G}_{odd}(\omega) = i\alpha''(\omega) \tag{A.18.8}$$

If we replace $\alpha''(\omega)$ by $\chi_e''(\omega) = \varepsilon_r''(\omega)$, Eq. (A18.8) reads

$$\tilde{G}_{odd}(\omega) = i\chi_e''(\omega) = i\varepsilon_r''(\omega) \tag{A.18.9}$$

According to Eq. (12.16.19a) of Chap. 12,

$$G_{even}(t) = s(t)\, G_{odd}(t) \tag{A.18.10}$$

where $s(t)$ is the sign function. On the basis of Eq. (12.6.22),

$$\tilde{G}_{even}(\omega) = \frac{1}{2\pi}\int_{-\infty}^{+\infty} d\omega_o \tilde{s}(\omega - \omega_o)\tilde{G}_{odd}(\omega_o) \tag{A.18.11}$$

where $\tilde{G}_{even}(\omega)$ is the transform of $G_{even}(t)$, $\tilde{s}(\omega)$ is the transform of $s(t)$, while $\tilde{G}_{odd}(\omega)$is the Fourier transform of $G_{odd}(t)$. Upon substitution of Eqs. (A.18.7) and (A.18.9), Eq. (A.18.11) reads

$$\varepsilon_r'(\omega) - 1 = \frac{1}{2\pi}\int_{-\infty}^{+\infty} d\omega_o \tilde{s}(\omega - \omega_o)\big(i\varepsilon_r''(\omega_o)\big) \tag{A.18.12}$$

Evaluated with $\omega - \omega_o$, Eq. (A.17.7) reads,

$$\tilde{s}(\omega - \omega_o) = \frac{2i}{(\omega - \omega_o)} \tag{A.18.13}$$

Upon substitution of Eq. (A.18.13) into Eq. (A.18.12), the result is

$$\varepsilon_r'(\omega) - 1 = \frac{1}{2\pi}\int_{-\infty}^{+\infty} d\omega_o\left(\frac{2i}{\omega - \omega_o}\right)\left(i\varepsilon_r''(\omega_o)\right) \tag{A.18.14}$$

After use of $i^2 = -1$, Eq. (A.18.14) reads

$$\varepsilon_r'(\omega) = 1 - \frac{1}{\pi}\int_{-\infty}^{+\infty} d\omega_o \frac{\varepsilon_r''(\omega_o)}{(\omega - \omega_o)} \tag{A.18.15}$$

Equation (12.6.20a) of Chap. 12 states that

$$G_{odd}(t) = s(t)G_{even}(t) \tag{A.18.16}$$

According to Eq. (A.18.16) and the Inverse Convolution Theorem

$$\tilde{G}_{odd}(\omega) = \frac{1}{2\pi}\int_{-\infty}^{+\infty} d\omega_o \tilde{s}(\omega - \omega_o)\tilde{G}_{even}(\omega_o) \tag{A.18.17}$$

Upon substitution of Eqs. (A.18.7) and (A.18.9), Eq. (A.18.17) reads,

$$i\varepsilon_r''(\omega) = \frac{1}{2\pi}\int_{-\infty}^{+\infty} d\omega_o \tilde{s}(\omega - \omega_o)\left(\varepsilon_r'(\omega_o) - 1\right) \tag{A.18.18}$$

Use of Eq. (A.18.13) permits Eq. (A.18.18) to be written as

$$i\varepsilon_r''(\omega) = \frac{1}{2\pi}\int_{-\infty}^{+\infty} d\omega_o\left(\frac{2i}{\omega - \omega_o}\right)\left(\varepsilon_r'(\omega_o) - 1\right) \tag{A.18.19}$$

After cancellation of factors of i, Eq. (A.18.19) reads

$$\varepsilon_r''(\omega) = \frac{1}{\pi}\int_{-\infty}^{+\infty} d\omega_o \frac{\left(\varepsilon_r'(\omega_o) - 1\right)}{(\omega - \omega_o)} \tag{A.18.20}$$

Equations (A.18.15) and (A.18.20) are the Kramers–Kronig relations linking the real and imaginary parts of the permittivity.

A.19 Volume of a Sphere Calculated Using an Exterior Origin

In Chap. 14, the calculation of the London force exerted by the material in a sphere on a molecule exterior to the sphere involves an integral over the volume of the sphere. The integration is made laborious by the fact that the origin of coordinates must be taken at the location of the exterior molecule. To demonstrate that the evaluation of the force is not an artifact of this off-center coordinate system, we use these off-center coordinates to calculate the volume of the sphere, which has radius, R_1. We compare the result with the expected value, $V = 4\pi R_1^3/3$.

The integral representing V in the off-center coordinate system is obtained by setting $-\lambda Q/r^6 = 1$ in Eq. (14.3.4b). The remaining integral is

$$V = \pi \int_{R-R_1}^{R+R_1} dr\, r\left[\frac{R_1^2 - (R-r)^2}{R}\right] = \frac{\pi}{R}\left[R_1^2 \int_{R-R_1}^{R+R_1} dr\, r - \int_{R-R_1}^{R+R_1} dr\, r(R-r)^2\right] \tag{A.19.1}$$

Define

$$V = \frac{\pi}{R}(I_1 - I_2) \tag{A.19.2}$$

where the first integral is

$$I_1 = R_1^2 \int_{R-R_1}^{R+R_1} drr = R_1^2\left[\frac{r^2}{2}\right]_{R-R}^{R+R_1} = \frac{R_1^2}{2}\left[(R+R_1)^2 - (R-R_1)^2\right]$$
$$= \frac{R_1^2}{2}\left[\left(R^2 + 2RR_1 + R_1^2\right) - \left(R^2 - 2RR_1 + R_1^2\right)\right] = 2RR_1^3 \tag{A.19.3a}$$

The second integral,

$$I_2 = \int_{R-R_1}^{R+R_1} dr\, r(R-r)^2 \tag{A.19.3b}$$

can be simplified by a variable substitution. In I_2 let $x = r - R$, $dx = dr$. When $r = R - R_1$, x has the value, $x = -R_1$. When $r = R + R_1$, x has the value, $x = +R_1$. With this variable substitution, I_2 transforms to

$$I_2 = \int_{-R_1}^{+R_1} dx\,(x+R)x^2 = \int_{-R_1}^{+R_1} dx\, x^3 + R\int_{-R_1}^{+R_1} dx\, x^2 = \left[\frac{x^4}{4}\right]_{-R_1}^{+R_1} + R\left[\frac{x^3}{3}\right]_{-R_1}^{+R_1}$$

$$= \left[\frac{R_1^4}{4} - \frac{R_1^4}{4}\right] + R\left[\frac{R_1^3}{3} - \frac{(-R_1)^3}{3}\right] = \frac{2RR_1^3}{3} \tag{A.19.3c}$$

Upon substitution of Eqs. (A.19.3a) and (A.19.3c) into Eq. (A.19.2), one obtains

$$V = \frac{\pi}{R}(I_1 - I_2) = \frac{\pi}{R}\left(2RR_1^3 - \frac{2}{3}RR_1^3\right) = \frac{4\pi R_1^3}{3} \tag{A.19.4}$$

which is the expected result.

Reference

1. P. Norman, K. Ruud, T. Saue, *Principles and Practices of Molecular Properties: Theory, Modeling, and Simulations* (Oxford, UK, 2018)

Recommended Additional Reading

1. A. Zangwill, *Modern Electrodynamics* (Cambridge University Press, UK, 2013), Chap. 1

Author Index

J. K. Baird, *Electromagnetism for Chemistry*,
https://doi.org/10.1007/978-3-031-88508-2

Subject Index

J. K. Baird, *Electromagnetism for Chemistry*,
https://doi.org/10.1007/978-3-031-88508-2

D